Current Research in Cement and Building Materials

Current Research in Cement and Building Materials

Editors

Peng Liu
Lingkun Chen
Ying Chen

MDPI

Basel • Beijing • Wuhan • Barcelona • Belgrade • Novi Sad • Cluj • Manchester

Editors

Peng Liu
Central South University
Changsha
China

Lingkun Chen
Yangzhou University
Yangzhou
China

Ying Chen
Central South University of
Forestry and Technology
Changsha
China

Editorial Office
MDPI
St. Alban-Anlage 66
4052 Basel, Switzerland

This is a reprint of articles from the Special Issue published online in the open access journal *Coatings* (ISSN 2079-6412) (available at: https://www.mdpi.com/journal/coatings/special_issues/cement_building).

For citation purposes, cite each article independently as indicated on the article page online and as indicated below:

Lastname, A.A.; Lastname, B.B. Article Title. *Journal Name* **Year**, *Volume Number*, Page Range.

ISBN 978-3-7258-0277-7 (Hbk)
ISBN 978-3-7258-0278-4 (PDF)
doi.org/10.3390/books978-3-7258-0278-4

Contents

About the Editors

Peng Liu

Peng Liu is a professor and doctoral supervisor of the school of civil engineering in Central South University. He is engaged in research into concrete structure durability, concrete material and damage identification, ballastless track mechanical properties, prefabricated construction, and other topics, and he is credited with a series of innovative achievements. He has published more than 200 academic papers, 62 of which are included in SCI source journals, and 25 of which are included in EI. The applicant is in charge of two National Natural Science Youth Funds, one natural science fund in Hunan Province, one research innovation project for postgraduate students in Hunan Province, and one open sharing fund for valuable instruments and equipment in Central South University. He has participated in eight projects of the National Natural Science Foundation and China Railway. He served as a Chief Editor and Editorial Board Member for both domestic and foreign journals.

Lingkun Chen

Lingkun Chen received a Ph.D. in structural engineering from Central South University's School of Civil Engineering in 201206, studying high-speed rail bridge dynamics. Dr. Chen is a Postdoctoral Fellow at Edinburgh University School of Engineering, a Visiting Associate Professor at UCLA's Department of Civil and Environmental Engineering, and a Visiting Scholar at SUNY Buffalo's MCEER Lab. He is also a visiting scholar at the MCEER Laboratory at SUNY Buffalo, and a Specially Appointed Associate Professor (Faculty) at Yangzhou University. His research includes computational mechanics for seismic and wind resistance and dynamic bridge and tunnel models. He has studied advanced train–bridge–tunnel engineering system vibration and intelligent transportation theory, numerical modeling, and experimental dynamic testing.

Ying Chen

Ying Chen is a professor of the school of civil engineering in Central South University of Forestry and Technology. She is engaged in research into concrete structures and chemical materials and has published more than 30 academic papers. She was supported by the National Natural Science Youth Fund.

coatings

MDPI

Article

Experimental Analysis of Bearing Capacity of Basalt Fiber Reinforced Concrete Short Columns under Axial Compression

Xinzhong Wang [1,2], Yiming Yang [1,*], Rihua Yang [1] and Peng Liu [3,*]

1 School of Civil Engineering, Hunan City University, Yiyang 413000, China; zhong811@126.com (X.W.); yrhzjh@163.com (R.Y.)
2 Hunan Engineering Research Center of Structural Safety and Disaster Prevention for Urban Underground Infrastructure, Hunan City University, Yiyang 413000, China
3 School of Civil Engineering, Central South University, Changsha 410013, China
* Correspondence: yangyiming@hncu.edu.cn (Y.Y.); 2015038@csu.edu.cn (P.L.); Tel.: +86-18173116706 (Y.Y.); +86-15116277646 (P.L.)

Abstract: Adding basalt fiber to concrete can improve the mechanical properties of concrete, and it is also one of the best ways to enhance the ultimate bearing capacity of concrete structure. In this paper, the construction performance and the compressive strength of basalt-fiber-reinforced concrete (BFRC) with five kinds of fiber lengths and eight kinds of fiber volume content subjected to an axial load are systematically investigated. The optimum fiber length and fiber volume content are obtained by comprehensively considering the construction performance and compressive strength. Moreover, the prediction model and finite element analysis method of the ultimate bearing capacity of basalt-fiber-reinforced concrete are developed. The results show that the optimum fiber length is about 12–24 mm and the fiber volume content is 0.15%. Adding an appropriate amount of basalt fiber can effectively improve the ultimate bearing capacity of concrete short columns, with maximum and average increases of 28% and 24%, respectively. In addition, the comparison with the experimental results shows that both the proposed prediction method and the finite element modeling method have good applicability, and they can be used to predict the ultimate bearing capacity of the BRFC short columns in practical engineering.

Keywords: basalt-fiber-reinforced concrete; concrete compressive strength; construction performance; ultimate bearing capacity; finite element analysis

Citation: Wang, X.; Yang, Y.; Yang, R.; Liu, P. Experimental Analysis of Bearing Capacity of Basalt Fiber Reinforced Concrete Short Columns under Axial Compression. *Coatings* **2022**, *12*, 654. https://doi.org/10.3390/coatings12050654

Academic Editor: Paolo Castaldo

Received: 15 April 2022
Accepted: 9 May 2022
Published: 11 May 2022

1. Introduction

Reinforced concrete (RC) short columns are some of the most basic components in structural engineering and are widely used in bridge piers, building frames, workshop columns, and other concrete structures. Their bearing capacity and durability are crucial to the safety, applicability, and economy of the entire structure. At present, they are effective at improving the mechanical properties of concrete by adding chopped fibers, such as steel fibers [1], glass fibers [2], synthetic fibers [3,4], basalt fibers [5], carbon fibers [6,7], etc. This is mainly because the appropriate fiber length and fiber volume contents can effectively combine with the weak matrix in concrete, so as to better control the development of internal cracks in concrete and finally improve the mechanical properties of concrete [8]. However, different types of fiber-reinforced concrete have different mechanical properties or application characteristics. For example, adding steel fiber to concrete can improve the toughness and tensile strength of concrete, but the processability and corrosion resistance of steel fiber are not good. Adding glass fiber into concrete can enhance the toughness of concrete, but its long-term strength will be reduced. Although carbon fiber has the characteristics of hardness and high strength, its use cost is high [9]. In view of the above problems, basalt fiber, which has the advantages of high tensile strength, high elastic modulus, corrosion resistance, good chemical stability, environmental protection, no pollution, and low cost, is gradually being studied [10–13]. Therefore, it is urgent to study the mechanical properties of the basalt-fiber-reinforced concret (BRFC) and its corresponding concrete members.

To date, researchers have mainly studied the reinforcement effect of basalt fiber and the mechanical properties of corresponding concrete. For example, both Ayub et al. [14] and Wang et al. [15] found that adding basalt fiber to concrete can improve its strength, and the maximum increase in compressive strength of basalt fiber concrete is 47.5%. According to previous research by Monaldo et al. [16], the tensile strength of concrete can be increased by 22.9% after 28 d by using 0.6% basalt fiber. A few scholars have determined the optimum content of basalt fiber based on the mechanical properties of concrete after adding basalt fiber. Based on the strength test of concrete with two kinds of fiber lengths and five kinds of fiber volume contents, Sun et al. [17] found that BFRC with 2% fiber volume content and 6 mm fiber length achieve the maximum strength. Tumadhir [18] believed that the optimum fiber volume content is about 0.3%, from the perspective of obtaining maximum compressive strength. However, most of the above studies obtain the optimal fiber length or fiber volume content based on the mechanical properties of specimens corresponding to a few fiber lengths and do not consider the construction performance of concrete. Therefore, it is urgent to carry out the mechanical property test of concrete under various fiber lengths and fiber volume contents and comprehensively determine the reasonable fiber parameters in combination with the construction performance.

Other than BFRC specimens, some scholars have studied the mechanical properties of basalt fiber concrete members, but most of them have focused on basalt fiber concrete beams. Based on the experimental research on BFRC beams, both Zhang [19] and Wang et al. [20] found the addition of basalt fiber can effectively prevent the development of cracks in reinforced concrete flexural members. Alnahhal and Aljidda [21] studied the flexural behavior and ultimate capacity of the BFRC beams experimentally and analytically based on the test results of 16 BFRC beams. At present, although a few scholars have conducted preliminary research on the performance of the BRFC short columns, such as Zhu [22], the research results on the ultimate bearing capacity of the BRFC short columns are highly deficient, and the relevant bearing capacity prediction methods have not been proposed. Therefore, it is necessary to further study the variation law of the bearing capacity of the BRFC short columns and propose corresponding prediction methods.

The objective of this study is to analyze the bearing capacity of the BRFC short columns under axial compression. First, the optimum fiber length and fiber volume content are obtained based on the construction performance and the concrete compressive strength. Then, the results of the axial compression test of the BRFC short columns are analyzed in depth. Finally, the theoretical and finite element calculation method of the ultimate bearing capacity of the BFRC short column are proposed, and their effectiveness is verified based on the test results. Among them, determining the optimum characteristic parameters of basalt fiber by comprehensively considering the construction and mechanical properties of concrete and proposing the assessment method of the ultimate bearing capacity of the BRFC short columns are the novelties of this paper.

2. Materials

The ordinary portland cement with the type of P.O 42.5 was selected. The crushed stone adopted two kinds of crushed stone of 5–10 mm and 10–25 mm, at a ratio of 2:3, to form a continuous secondary distribution. The corresponding crush value was 10.5, the sand fineness modulus was 2.85, and tap water was used. No water reducer was used during construction. The mix proportion data are shown in Table 1. The short cut basalt fiber produced by Zhejiang Hengdian Shijin Basalt Fiber Co., Ltd. Jinhua, China is adopted. Five lengths of basalt fibers are shown in Figure 1. The physical and mechanical properties of basalt fiber include the fiber diameter of 17 μm; the fiber density of 2650 kg/m^3; the tensile strength of 3000 MPa; the elastic modulus of 90 GPa; and fiber lengths of 12 mm, 18 mm, 24 mm, 30 mm, and 36 mm, respectively, as shown in Table 2.

Table 1. The mixed proportion of the concrete.

Target Intensity	Water Cement Ratio	Sand Ratio/%	Cement (kg/m^3)	Water (kg/m^3)	Sand (kg/m^3)	Crushed Stone (kg/m^3)
C30	0.55	34	355	195	703	1147

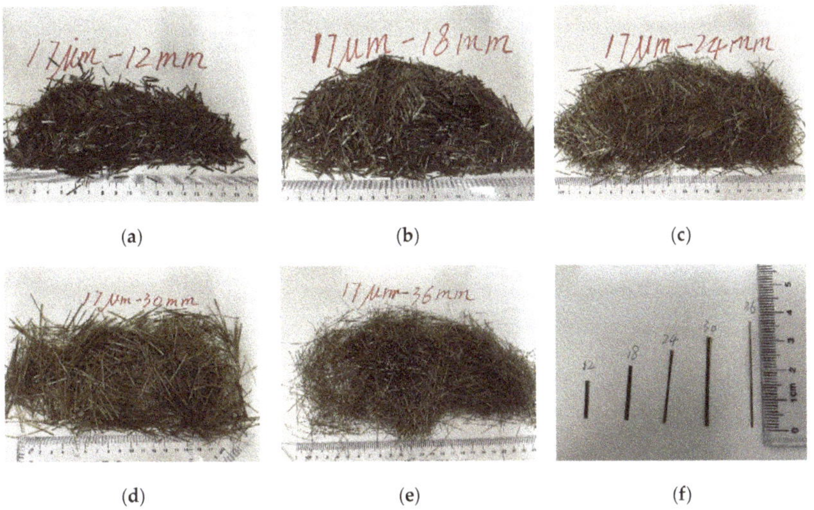

Figure 1. The five lengths of basalt fiber: (**a**) L = 12 mm; (**b**) L = 18 mm; (**c**) L = 24 mm; (**d**) L = 30 mm; and (**e**) L = 36 mm; and (**f**) a schematic diagram of the different fiber lengths.

Table 2. The physical and mechanical properties of the basalt fiber.

Index	Diameter (μm)	Length (mm)	Density (kg/m^3)	Tensile Strength (MPa)	Modulus of Elasticity (GPa)
Parameter	0.55	12/18/24/30/36	2650	3000	703

3. Construction Performance and Optimal Parameters of the BRFC

3.1. Construction Performance Test of the BRFC

The slump and expansion are selected here to analyze the construction performance of the BRFC, and the detailed test methods and procedures can be found in the Chinese specification of GB/T 50080-2016 [23]. A total of 13 groups of cube concrete specimens numbered ST1–ST13 are designed, and each group contains three parallel specimens. The STI, without adding fiber, is used as the reference specimen. Specimens ST2–ST6 are constructed using five kinds of basalt fibers with a fiber volume content of 0.15% and fiber lengths of 12 mm, 18 mm, 24 mm, 30 mm, and 36 mm. Five groups of the BRFC specimens with different fiber lengths are obtained to test the slump and expansion, and the average test results of each group of specimens are shown in Table 3. Specimens ST7–ST13 are constructed using seven kinds of basalt fibers with a fiber length of 12 mm and fiber volume contents of 0.075%, 0.10%, 0.15%, 0.20%, 0.25%, 0.30%, and 0.40%. The average test results of slump and expansion of ST7–ST13 are shown in Table 4. In order to better compare the difference between the slump and expansion of the BRFC and the ordinary concrete, the relative values of slump and relative expansion are calculated.

Table 3. The slump test results of the BRFC with different fiber lengths.

Group of Specimen	Fiber Length (mm)	Slump (mm)	Expansion (mm)	Relative Value of Slump (%)	Relative Value of Expansion (%)
ST1	0	30.4	61.5	100.0	100.0
ST2	12	27.8	53.4	91.4	86.8
ST3	18	25.7	48.6	84.5	79.0
ST4	24	23.2	43.2	76.3	70.2
ST5	30	24.7	47.1	81.2	76.6
ST6	36	26.8	49.4	85.1	80.3

Note: the relative value is the ratio of the measured value of the BRFC to that of the ordinary concrete.

Table 4. The slump test results of the BRFC with different fiber volume contents.

Group of Specimen	Fiber Volume Content (%)	Slump (mm)	Expansion (mm)	Relative Value of Slump (%)	Relative Value of Expansion (%)
ST1	0	30.4	61.5	100.0	100.0
ST7	0.075	28.2	56.3	92.8	91.5
ST8	0.10	27.8	54.2	91.4	88.1
ST9	0.15	25.4	52.1	83.6	84.7
ST10	0.20	23.1	43.4	76.0	70.6
ST11	0.25	20.2	38.3	66.4	62.3
ST12	0.30	18.3	32.4	60.2	52.7
ST13	0.40	15.8	30.1	52.0	48.9

As shown in Tables 3 and 4, the slump of the BRFC is lower than that of ordinary concrete. As the length of the added basalt fiber increases, the slump and expansion of the concrete first decreases and then increases. The slump and expansion of the BRFC is inversely correlated with the volume content of basalt fiber. This is due to the chaotic effect of basalt fibers, where a greater confinement effect on the concrete occurs with longer fiber and larger dosage and thus reduces the flow performance of the concrete. When the dosage is constant, the longer the fiber, the fewer the number of fibers, and thus the less restrictive the effect on the concrete.

3.2. Optimal Fiber Length and Fiber Volume Content

To further study the effect of fiber volume content on concrete compressive strength under different fiber lengths, we added BFRC specimens with a fiber length of 24 and fiber volume contents of 0, 0.075%, 0.10%, 0.15%, 0.20%, 0.25%, 0.30%, and 0.40% on the basis of ST1–ST13. Each case also contains three specimens. Then, based on the standard for the test methods of the mechanical properties of ordinary concrete (GB/T 50081-2002) [24], the 28 d compressive strength test is carried out on the cube concrete specimens with different fiber lengths and fiber volume contents, respectively. From the compressive failure modes of all specimens, the concrete specimens made an obvious cracking sound when they were damaged. The integrity of the BRFC specimen shows good resistance to damage regarding the rare cracking phenomenon. For the BFRC specimen that fails, the failure mode is brittle failure, although the characteristics of ductility are seen. The failure morphology of the BFRC specimens does not vary significantly with different fiber volume contents.

The compressive strength test results are presented in Figures 2 and 3. As indicated, with the increase of fiber length and fiber volume content, the concrete compressive strength basically conforms to the law of first increasing and then decreasing. This is mainly because under the action of short fiber length and small fiber volume content, the fiber distribution is more uniform, which can effectively prevent the generation of micro cracks in concrete, so as to improve the concrete compressive strength. For longer fiber length and larger fiber volume content, when they exceed a certain value, the fiber is easy to wind or agglomerate, which leads to the weakening of the connection effect between the fiber and the concrete matrix, thus leading to a failure to exert its reinforcing effect. It is worth noting that the average test strength of this group of concrete specimens with a fiber length of 36 mm does not comply with the above law, which may be caused by the good dispersion of fibers in the concrete. In addition, it can also be seen from Figures 2 and 3 that the concrete compressive strength is relatively high when the fiber length is about 12–24 mm and the fiber volume content is 0.15%. It is worth noting that although the construction performance cannot reach the optimal state under the effect of the above fiber length and content, it is sufficient to meet the normal construction needs. Thus, this paper focuses on the performance index of concrete compressive strength, and then obtains the optimal fiber length and fiber volume based on the analysis results.

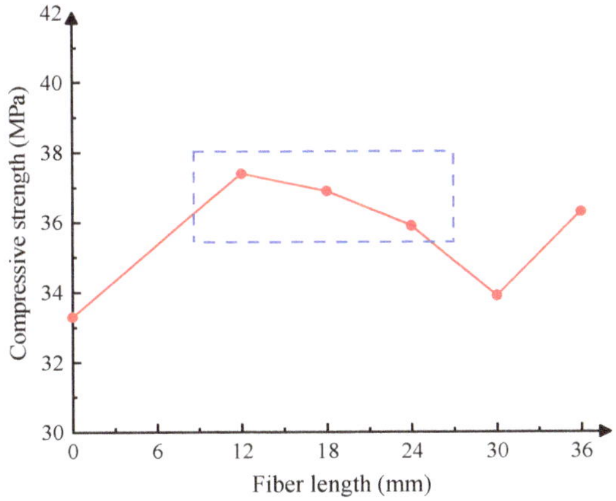

Figure 2. The 28 d compressive strength of the BRFC with different fiber lengths.

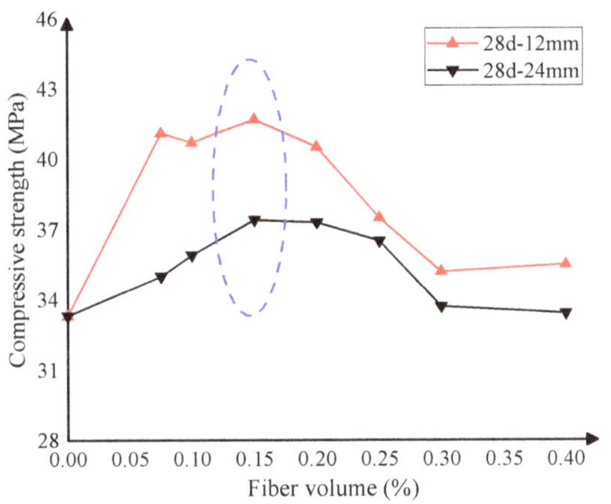

Figure 3. The compressive strength of the BRFC with different fiber volume contents.

4. Axial Compression Test of the BRFC Short Columns

4.1. Specimen Design

The analysis results of optimal design parameters in the previous section show that the optimal fiber length is 12–24 mm and the fiber volume content is 0.15%. As a result, in the subsequent experiments, only two fiber lengths of 12 mm and 18 mm, as well as fiber volume of 0.15%, are selected for the short column design. Here, a total of nine reinforced concrete short columns numbered S1–S9 are designed. Among them, the samples numbered S1–S3 are ordinary reinforced concrete short columns, and the samples numbered S4–S6 and S7–S9 are the BRFC short columns with fiber length of 12 mm and 18 mm, respectively. For all the BRFC short columns, the fiber volume content is selected as 0.15%, as previously described. The short column is featured with a cross-section size of width × height = 150 mm × 150 mm, and a slenderness ratio of 3.67 (length/width = 550/150 = 3.67), which is less than 8. The concrete cover is 10 mm. All short columns are arranged with four HRB400 threaded bars with a diameter of 12 mm as longitudinal reinforcement, and the stirrups are made of HRB335 plain round bars with a diameter of 6 mm and a spacing of 110 mm. The reinforcement configuration is provided in Figures 4 and 5. As required by GT/B 288.1-2010 [25], the yield and tensile strength of the two kinds of reinforcement are 486 MPa versus 360.1 MPa, and 590 MPa versus 570.4 MPa, respectively. Preloading is required before formal loading. Each short column is preloaded with 100 kN for 5 min to observe the reliability of the loading system and each measuring point. The formal loading is performed by the grading loading system. The loading of each stage is 50 kN, which is maintained for 2 min. When cracks or bulges appear, the loading is increased to 10 kN until the short column is damaged.

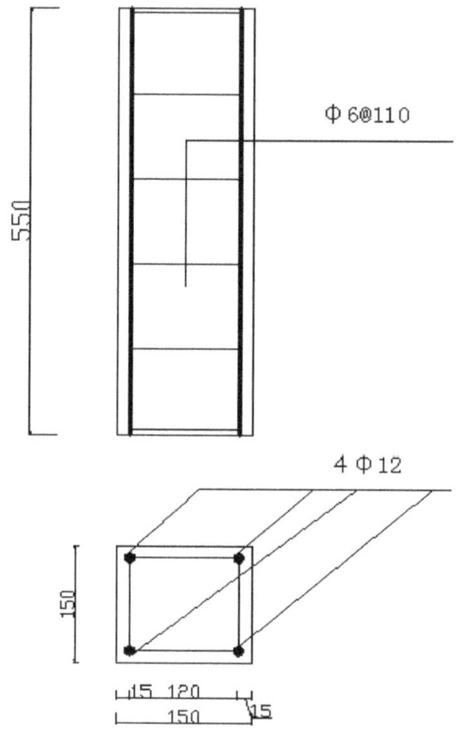

Figure 4. The structural dimension of the short column.

4.2. Experimental Phenomenon

During the loading process of all test short columns, no transverse cracks are found, while vertical cracks occur before failure. With the increase of load, the cracks develop gradually, the steel bars yield gradually, and the concrete is crushed and damaged. However, the stirrups remain intact when the short columns are damaged. The results reveal that the cracking of the BRFC short columns is significantly later than the ordinary RC short columns, which is also true for the failure load. The failure modes of the test specimens are shown in Figure 6.

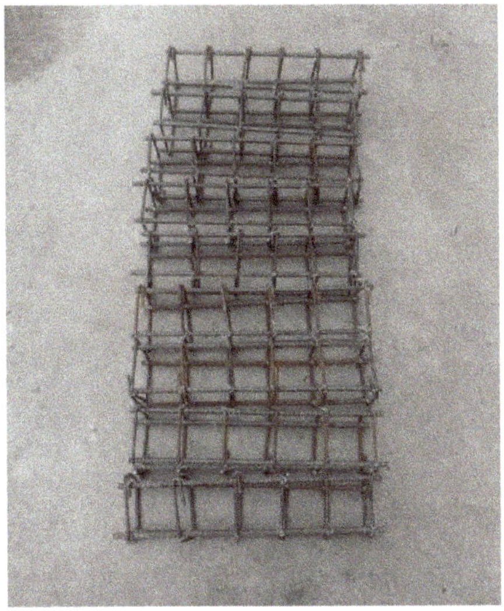

Figure 5. The reinforcement of the short column.

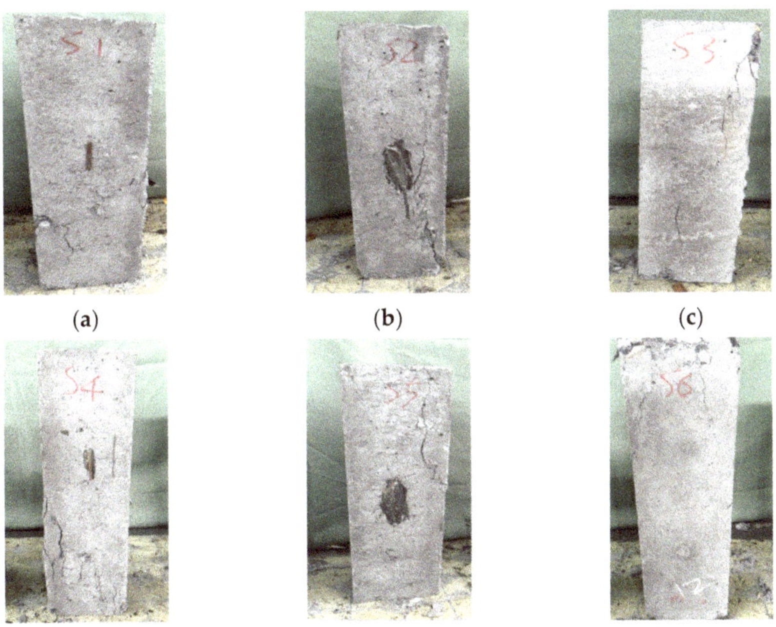

(**a**) (**b**) (**c**)

(**d**) (**e**) (**f**)

Figure 6. *Cont.*

(g) (h) (i)

Figure 6. The failure mode of the short columns: (**a**) S1; (**b**) S2; (**c**) S3; (**d**) S4; (**e**) S5; (**f**) S6; (**g**) S7; (**h**) S8; and (**i**) S9.

4.3. Results and Discussion

4.3.1. Ultimate Bearing Capacity

It can be seen in Table 5 that the ultimate bearing capacity (N_u) of the BRFCthe BRFC short columns is significantly higher than ordinary RC short columns. In particular, the maximum increase rate of axial compression ultimate bearing capacity is 28% for the BRFC short columns with fiber length of 12 mm and is 20% for the BRFC short columns with a fiber length of 18 mm. The average increase is 24%. The results highlight that basalt fiber is beneficial to the ultimate bearing capacity of reinforced concrete short columns. In addition, the average value of the ultimate bearing capacity of the BRFC short columns with a fiber length of 12 mm is 636.7 kN, which is 6.7% higher than the corresponding value (596.7 kN) with a fiber length of 18 mm. This conclusion matches the conclusion in Section 3.2 and other types of fibers reported in [26,27]. That is, when the fiber length exceeds a certain value, the reinforcement effect of the fibers will weaken.

Table 5. The axial compression test results of the BRFC short columns.

Specimen Number	Fiber Volume Content (%)	Fiber Length (mm)	N_u (kN)
S1	0	/	500
S2	0	/	500
S3	0	/	490
S4	0.15	12	640
S5	0.15	12	640
S6	0.15	12	630
S7	0.15	18	600
S8	0.15	18	590
S9	0.15	18	600

4.3.2. Load Strain Curve

The average value of the measured strain of the concrete and steel bar at the midpoint of each column is plotted in Figures 7 and 8. As indicated, the short columns are basically in the elastic stage at the beginning of loading and then enter the elastic–plastic stage with the increase of load. The concrete strain of the S1–S3 short columns reaches its peak around 0.0019, compared to a peak of around 0.0021 for the S4–S6 short columns. For the S7–S9 short columns, the peak value is about 0.0023. In addition to a slightly larger steel strain at peak value, the slope of the N-ε curve of the BRFC short columns in the elastic stage is also greater than that of the reinforced concrete short column.

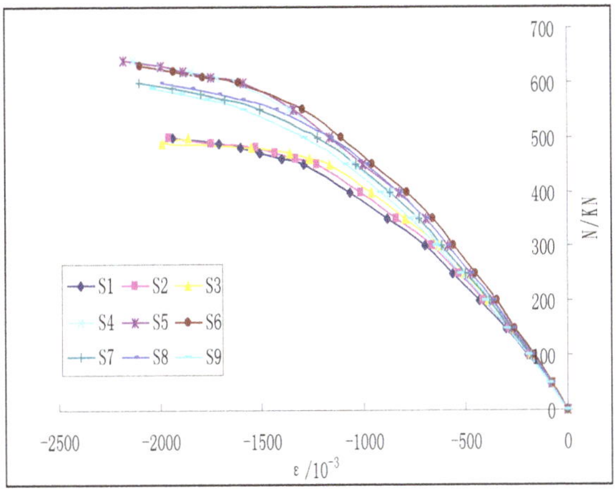

Figure 7. The average strain curve of the steel bars.

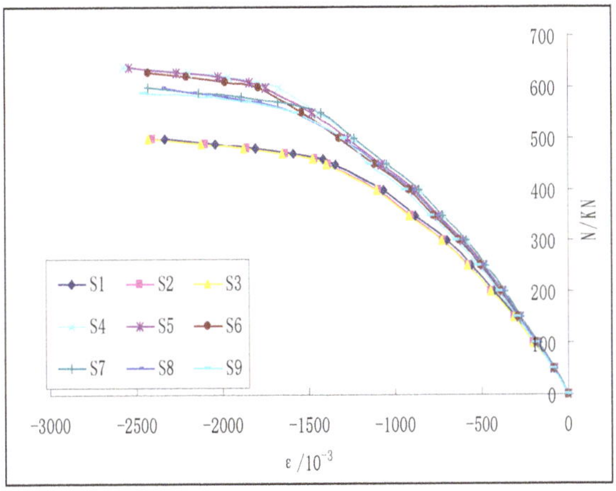

Figure 8. The average strain curve of the concrete.

Figure 9 shows the variation law of the vertical displacement at the top of each test short column. As indicated, the vertical displacement of the short column exhibits an obvious plastic stage with the increase of the load. The vertical displacement of the BRFC short columns is significantly greater than that of the ordinary RC short columns. More specifically, the vertical displacement of S4–S6 is the largest, followed by S7–S9, and is the smallest in S1–S3. Due to the restraint of fiber, the BFRC short column has greater displacement when the load reaches the peak value. From the elastic stage to elastic–plastic stage, the load displacement curve of the BRFC short columns is smoother and the plastic characteristics are obvious.

4.3.3. Influence Mechanism of Basalt Fiber

The chopped continuous basalt fiber is a hydrophilic material, which can be well combined with the cement-based material and form a spatial network structure between the concrete coarse aggregates. The structure encases the coarse aggregate and acts as a hindrance to aggregate movement during compression. Due to its high tensile strength, basalt fiber can provide circumferential restraint to resist transverse expansion when the specimen is compressed. As a result, its existence hinders the generation of a large number of microcracks or crosses countless microcracks. This makes the generation and development of cracks in the compression process require more energy. Thus, compared with the RC short columns, the ultimate bearing capacity of the BRFC short columns is significantly improved, and the initial crack time is also delayed. At the same time, when the basalt fiber concrete is equipped with an appropriate amount of stirrup and protective layer thickness, the anchorage ductility of the BRFC can be significantly increased. For the above reasons, the BFRC short column has higher ultimate bearing capacity and stronger deformation capacity.

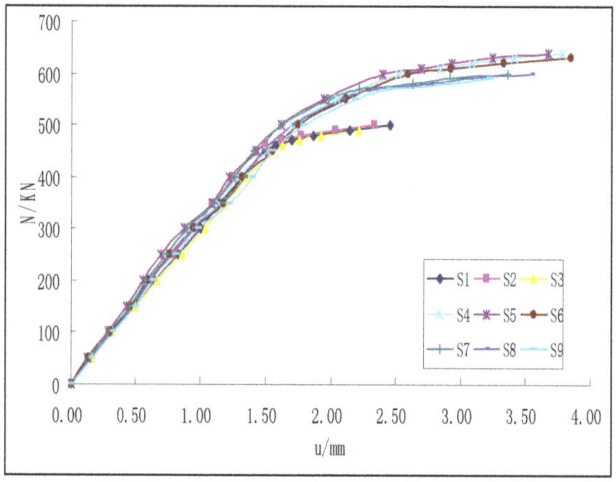

Figure 9. The load and the vertical displacement.

5. Calculation Method of Ultimate Bearing Capacity of BRFC Short Column

(1) Calculation method based on Standard approximate formula

In the specifications for the design of highway reinforced concrete and prestressed concrete bridges and culverts [28], the effect of the slenderness ratio is considered by solving the given calculation formula based on the sum of the maximum bearing capacity of concrete and steel bar. For the short columns with a rectangular cross-section, the calculation formula is as follows [28]:

$$N_u = 0.9\varphi(f_{cd}A + f'_{sd}A'_s) \tag{1}$$

where f_{cd} is the design value of the compressive strength of concrete, f'_{sd} is the design value of the yield strength of longitudinal reinforcement, A is the gross area of the cross section of a column, and A'_s is the area of longitudinal reinforcement. φ is the stability coefficient. The values or calculation methods of previous variables can be found in [28].

For the calculation of ultimate bearing capacity of the BRFC short column, the design value of concrete axial compressive strength in Equation (1) is replaced by the measured value of the compressive strength of the BRFC, and the other parameters remain unchanged. The calculation results are listed in Table 6.

(2) Finite element analysis method

The concrete constitutive model provided in [29] is adopted to calculate the ultimate bearing capacity of the BRFC short column, in which the peak strain, rising, and falling curves are modified by the measured strength of the BRFC. Moreover, considering the structural damage under the concrete stress, the elastic stiffness matrix is reduced and the correlation hardening is introduced into the constitutive model for the damage model of the ABAQUS software (version: 6.14.2), so as to better simulate the elastic–plastic behavior of the concrete in the loading. The C3D8R solid element and the T3D2 truss element are used for model concrete and reinforcement, respectively. The reference points are connected with the upper and lower surfaces by coupling. One end reference point is utilized to apply loads (with only one translational degree of freedom in the longitudinal direction of the column reserved), and another end reference point is used for boundary conditions (rigid junction). The reinforced concrete is connected by an embedded region and subject to load by a reference point according to the displacement. The point set is arranged at the loading point to facilitate reading the load in post-processing. A set of points is arranged at the core concrete to observe the relationship between the stress, the strain, and the load of the core concrete. The finite element model of the short column is illustrated in Figure 10, and the calculation results based on the finite element method are also shown in Table 6.

Table 6. The calculation value of the ultimate bearing capacity of the BRFC short column.

Specimen Number	Test Value (kN)	Calculation Value Based on Equation (1) (kN)	Calculation Value Based on Finite Element Analysis Method (kN)	η_1	η_2
S1	500	502	497	1.00	1.01
S2	500	502	497	1.00	1.01
S3	490	502	497	0.98	0.99
S4	640	630	637	1.02	1.00
S5	640	630	637	1.02	1.00
S6	630	630	637	1.00	0.99
S7	600	598	597	1.00	1.01
S8	590	598	597	0.99	0.99
S9	600	598	597	1.00	1.01

Note: η_1 = the ratio of the test value to the calculation value based on Equation (1); η_2 = the ratio of the test value to calculation value by the finite element analysis method.

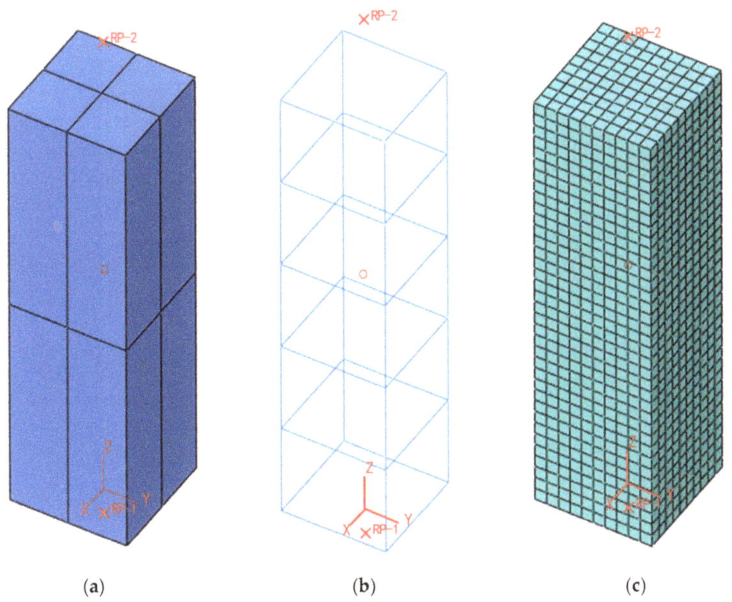

Figure 10. The finite element model of the short column: (**a**) the entity model; (**b**) the reinforcing element; and (**c**) the concrete unit.

As shown in Table 6, no matter which of the theoretical calculation methods based on the specification (i.e., Equation (1)) and the finite element simulation method is adopted, its calculation results are very consistent with the test results. More specifically, for the theoretical calculation method based on the specification, the calculated results of bearing capacity of the BRFC short columns are slightly lower than the measured values, while the corresponding results of the ordinary concrete short columns are slightly higher than the measured values. For the finite element simulation method, the calculation results of the BRFC and ordinary concrete short columns are mostly lower than the measured values. Although there is a certain deviation between the calculated results and the measured values, the maximum deviation is no more than 5%. This also verifies the feasibility of the bearing capacity prediction method of the BRFC short columns obtained by bringing the stress–strain relationship obtained from the test into the specification formula and the proposed finite element method. It is worth noting that due to the loading method of increasing 10 kN each time during the bearing capacity test of the concrete short columns (as described in Section 4.1), there is a certain deviation between the measured bearing capacity from the test and the actual bearing capacity of the concrete short columns. However, the deviation caused by this loading method is estimated to be between 1.5% and 2%. Therefore, it will not have a substantial impact on the effectiveness of the previous prediction methods.

6. Conclusions

In this paper, we obtained the optimum fiber length and fiber volume content based on the construction performance and the concrete compressive strength. The experimental phenomenon, the ultimate bearing capacity, the load strain curve, and the influence mechanism of basalt fiber are analyzed based on the results of the axial compression test of the BRFC short columns. In addition, the theoretical and finite element calculation method of the ultimate bearing capacity of the BRFC short column is proposed. The conclusions are summarized as follows:

(1) The optimum fiber length is about 12–24 mm, and the fiber volume content is 0.15%. In this case, the concrete has better slump and expansion properties and higher compressive strength.
(2) Adding appropriate basalt fiber can effectively improve the ultimate bearing capacity of the concrete short columns, and the maximum and average increases are 28% and 24%, respectively.
(3) No matter which of the theoretical calculation methods and the finite element simulation methods is adopted, its calculation results are very consistent with the test results. Even considering the deviation caused by the loading mode, the maximum deviation between the calculated results and the measured values is no more than 5%.

The limitation of this study is that only one diameter of basalt fiber is considered. In future research, the effect of the fiber diameter on the construction and mechanical properties of concrete needs to be further studied. Moreover, obtaining more measured data on the ultimate bearing capacity of the BRFC short columns to verify the effectiveness of the prediction method is also the focus of future research.

Author Contributions: Conceptualization, X.W.; methodology, X.W., R.Y. and P.L.; software, X.W.; validation, X.W. and Y.Y.; writing—original draft preparation, X.W. and Y.Y.; writing—review and editing, X.W., Y.Y., R.Y. and P.L.; supervision, R.Y. and P.L.; and funding acquisition, X.W., Y.Y. and R.Y. All authors have read and agreed to the published version of the manuscript.

Funding: This research was funded by the Hunan Provincial Natural Science Foundation of China (Grant Nos. 2021JJ50153, 2022JJ40024, and 2021JJ50156) and the Research Foundation of Education Bureau of Hunan Province (Grant Nos. 18A401, 21B0723, and 19C0343).

Institutional Review Board Statement: Not applicable.

Informed Consent Statement: Not applicable.

Data Availability Statement: Not applicable.

Acknowledgments: We would like to thank the anonymous reviewers and the editor for their valuable comments and remarks, which helped us to improve the original manuscript.

Conflicts of Interest: The authors declare that they have no known competing financial interests or personal relationships that could have influenced the work reported in this paper.

References

1. Iqbal, S.; Ali, A.; Holschemacher, K.; Bier, T.A. Mechanical properties of steel fiber reinforced high strength lightweight self-compacting concrete (SHLSCC). *Constr. Build. Mater.* **2015**, *98*, 325–333. [CrossRef]
2. Xin, H.; Liu, Y.; Mosallam, A.S.; He, J.; Du, A. Evaluation on material behaviors of pultruded glass fiber reinforced polymer (GFRP) laminates. *Compos. Struct.* **2017**, *182*, 283–300. [CrossRef]
3. Shaikh, F.U.A. Review of mechanical properties of short fibre reinforced geopolymer composites. *Constr. Build. Mater.* **2013**, *43*, 37–49. [CrossRef]
4. Hannawi, K.; Bian, H.; Prince-Agbodjan, W.; Raghavan, B. Effect of different types of fibers on the microstructure and the mechanical behavior of ultra-high performance fiber reinforced concretes. *Compos. Part B Eng.* **2016**, *86*, 214–220. [CrossRef]
5. Santarelli, M.L.; Sbardella, F.; Zuena, M.; Tirillo, J.; Sarasini, F. Basalt fiber reinforced natural hydraulic lime mortars: A potential bio-based material for restoration. *Mater. Des.* **2014**, *63*, 398–406. [CrossRef]
6. Bai, Y.; Nguyen, T.C.; Zhao, X.L.; Al-Mahaidi, R. Environment-assisted degradation of the bond between steel and carbon-fiber-reinforced polymer. *J. Mater. Civ. Eng.* **2014**, *26*, 04014054. [CrossRef]
7. Alabduljabbar, H.; Alyousef, R.; Mohammadhosseini, H.; Topper, T. Bond behavior of cleaned corroded lap spliced beams repaired with carbon fiber reinforced polymer sheets and partial depth repairs. *Crystals* **2020**, *10*, 1014. [CrossRef]
8. Meyyappan, P.L.; Carmichael, M.J. Studies on strength properties of basalt fiber reinforced concrete. *Mater. Today Proc.* **2021**, *43*, 2105–2108. [CrossRef]
9. Ramesh, B.; Eswari, S. Mechanical behaviour of basalt fibre reinforced concrete: An experimental study. *Mater. Today Proc.* **2021**, *43*, 2317–2322. [CrossRef]
10. Alaskar, A.; Albidah, A.; Alqarni, A.S.; Alyousef, R.; Mohammadhosseini, H. Performance evaluation of high-strength concrete reinforced with basalt fibers exposed to elevated temperatures. *J. Build. Eng.* **2021**, *35*, 102108. [CrossRef]
11. Branston, J.; Das, S.; Kenno, S.Y.; Taylor, C. Mechanical behaviour of basalt fibre reinforced concrete. *Constr. Build. Mater.* **2016**, *124*, 878–886. [CrossRef]
12. Wang, X.Z.; He, J.; Mosallam, A.S.; Li, C.; Xin, H. Effects of fiber length and volume on material properties and crack resistance of basalt fiber reinforced concrete (BFRC). *Adv. Mater. Sci. Eng.* **2019**, *4*, 7520549. [CrossRef]
13. Li, J.J.; Niu, J.G.; Wan, C.J.; Ling, X.Q.; Jin, Z.Y. Comparison of flexural property between high performance polypropylene fiber reinforced lightweight aggregate concrete and steel fiber reinforced lightweight aggregate concrete. *Constr. Build. Mater.* **2017**, *157*, 729–736. [CrossRef]
14. Ayub, T.; Shafifiq, N.; Khan, S.U. Compressive stress-strain behavior of HSFRC reinforced with basalt fibers. *J. Mater. Civ. Eng.* **2015**, *28*, 06015014. [CrossRef]
15. Wang, H.L.; Yuan, L.; Song, H. Experimental study on mechanical property of chopped basalt fiber reinforced concrete. *Build. Struct.* **2013**, *43*, 562–564.
16. Monaldo, E.; Nerilli, F.; Vairo, G. Basalt-based fiber-reinforced materials and structural applications in civil engineering. *Compos. Struct.* **2019**, *214*, 246–263. [CrossRef]
17. Sun, X.J.; Gao, Z.; Cao, P.; Zhou, C.J. Mechanical properties tests and multiscale numerical simulations for basalt fiber reinforced concrete. *Constr. Build. Mater.* **2019**, *202*, 58–72. [CrossRef]
18. Tumadhir, M. Thermal and mechanical properties of basalt fibre reinforced concrete. *Int. J. Civ. Environ. Eng.* **2013**, *7*, 334–337.
19. Zhang, Y. *Research on Basic Mechanical Properties of Chopped Basalt Fiber Reinforced Concrete*; Northeast Forestry University: Harbin, China, 2011.
20. Wang, J.; Luan, Y.; Ye, H.J. Experimental research on crack and deformation of basalt fiber reinforced concrete beams. *J. Archit. Civ. Eng.* **2016**, *33*, 76–81.
21. Alnahhal, W.; Aljidda, O. Flexural behavior of basalt fiber reinforced concrete beams with recycled concrete coarse aggregates. *Constr. Build. Mater.* **2018**, *169*, 165–178. [CrossRef]
22. Zhu, C.R. *Experimental Study on Fatigue of Basalt Fiber Reinforced Concrete Beams*; Inner Mongolia University of Technology: Hohhot, China, 2013.
23. *GB/T 50080-2016*; Standard for Test Method of Performance on Ordinary Fresh Concrete. MOHURD: Beijing, China, 2016.
24. *GB/T 50081-2002*; Standard for Test Method of Mechanical Properties on Ordinary Concrete. MOHURD: Beijing, China, 2002.
25. *GT/B288.1-2010*; Metallic Materials-Tensile Testing—Part 1, Method of Test at Room Temperature. National Technical Committee for Steel Standardization: Beijing, China, 2010.
26. Abirami, R.; Sangeetha, S.P. Study on fiber reinforced concrete beam-column connection—A review. *Mater. Today Proc.* **2020**, *33*, 415–419. [CrossRef]
27. Yoo, D.Y.; Banthia, N. Mechanical properties of ultra-high-performance fiber-reinforced concrete: A review. *Cem. Concr. Comp.* **2016**, *73*, 267–280. [CrossRef]
28. *JTG 3362-2018*; Specifications for Design of Highway Reinforced Concrete and Prestressed Concrete Bridges and Culverts. MOT: Beijing, China, 2018.
29. Shao, X.D.; Fan, W.; Huang, Z.Y. Application of ultra-high-performance concrete in engineering structures. *China Civ. Eng, J.* **2021**, *54*, 1–13.

coatings

MDPI

Article

Long-Term Performance of Recycled Asphalt Pavement with Recycled Engine Oil Bottom Based on Accelerated Loading Test

Jin Li [1], Li Zhu [1,*], Miaozhang Yu [1], Shen Zuo [1], Xinzhuang Cui [1,2] and Peng Liu [3]

1 School of Transportation Civil Engineering, Shandong Jiaotong University, Jinan 250357, China; sdzblijin@163.com (J.L.); 204153@sdjtu.edu.cn (M.Y.); zuoshen2006@163.com (S.Z.); cuixz@sdu.edu.cn (X.C.)
2 School of Civil Engineering, Shandong University, Jinan 250061, China
3 School of Civil Engineering, Central South University, Changsha 410083, China; liupeng868@csu.edu.cn
* Correspondence: 19011059@stu.sdjtu.edu.cn

Abstract: The practical performance of recycled asphalt obtained from recycled engine oil bottom (REOB) was evaluated by paving a test road of base asphalt and REOB-recycled asphalt mixture in a laboratory, where accelerated loading tests with 700,000 cycles were conducted. During accelerated loading, pavement temperature, layer bottom strain, pavement skid resistance, and rutting were monitored. The performance of pavement core material was tested after loading to analyze the index change of binder and the compaction degree of the mixture. The results show that the long-term anti-rutting ability of REOB-recycled asphalt pavement is approximately 10% higher than that of the base asphalt pavement, although the long-term anti-sliding force and anti-fatigue performance are poor. A developed model of rut with loading time can better predict the development trends of these parameters with loading cycles. The performance test of the pavement material after loading shows that 700,000 cycles can only degrade the performance of the test pavement, not damage it. The recycled asphalt pavement with 7% REOB has basically the same performance as the base asphalt under 700,000 cycles, indicating that REOB-recycled asphalt pavement can ensure basic road performance, while providing economic and environmental advantages. These results provide a reference for the application and form optimization of REOB-recycled asphalt pavement.

Keywords: recycled engine oil bottom; accelerated loading test; long-term performance; secondary aging

Citation: Li, J.; Zhu, L.; Yu, M.; Zuo, S.; Cui, X.; Liu, P. Long-Term Performance of Recycled Asphalt Pavement with Recycled Engine Oil Bottom Based on Accelerated Loading Test. *Coatings* **2022**, *12*, 522. https://doi.org/10.3390/coatings12040522

Academic Editor: Joaquim Carneiro

Received: 12 March 2022
Accepted: 8 April 2022
Published: 12 April 2022

1. Introduction

As a secondary by-product of engine oil, recycled engine oil bottom (REOB) presents a similar homology and compatibility with asphalt in production. The feasibility of using REOB as an asphalt modifier has previously been effectively verified [1–3]. Moreover, its product features, such as those related to economy and environmental protection, excellent performance, and waste treatment, have attracted global attention.

Golalipour et al. [4] studied ten types of oil-modified asphalt, including two kinds of recycled asphalt with REOB, and they found that both REOB blends that they tested were the least susceptible to aging among all of the oil blends they used. In addition, compared with the other eight kinds of oil, REOB most effectively improved the low-temperature performance of the binder. As early as 1998, Herrington et al. [5] paved a test road with a hot-recycled asphalt mixture produced by waste oil disintegration bottoms (WODB). After 57 months of testing, they found that all parts performed well, and the surface did not deteriorate, crack, or rut. Rostyslav et al. [6] used REOB-regenerated aging asphalt from different sources, before designing a mixture. They found that the permanent deformation and dynamic modulus of the REOB-modified asphalt mixture and the base asphalt mixture were similar when the performance grade (PG) of the mixture was consistent. Cooper et al. [7] designed five 12.5 mm asphalt mixtures for high-, medium-, and low-temperature performance tests using two base asphalts and three different REOB contents for recycled asphalt, and they established that the use of REOB can increase the temperature performance of the mixture.

REOB can effectively restore the performance of old asphalt, but during use, the low-temperature performance is often poor. Li et al. [8] tested the turning point temperature, freezing fracture temperature, and failure strength of asphalt mixtures mixed with 0%, 2.5%, 6.0%, and 15% REOB using the thermal stress restrained specimen test, and found that a high REOB content can have adverse effects on the low-temperature crack resistance of the asphalt mixture. Zaumanis et al. [9] studied aged REOB-recycled asphalt and discovered that its penetration index usually increases, which indicates that it is a more structural and brittle material and hence is easier to crack.

Related research has been conducted by means of basic indoor tests of materials. However, to date, no studies have been performed on the operation of REOB-recycled asphalt pavements owing to the long on-site test period required and high costs. In recent years, the simulated accelerated loading test method has received significant attention and is considered to be an important method for designing pavement based on service performance. This method is closest to simulating an actual traffic load in current research on road engineering.

In the history of pavement structural design, the AASHO full-scale pavement accelerated loading test is the most representative test method. By establishing the relationship between the service performance of a pavement structure and the cumulative number of axle loads in traffic, the AASHO test verified the mechanical design of pavement structure. This method also opened up new ways to study asphalt pavement. Sirin et al. [10] conducted tests on five test sections using a heavy vehicle simulator to predict pavement rutting in Florida. The five test sections adopted different running directions and different combinations of loading wheels. The results showed that when evaluating the rutting resistance of pavement under the traffic and climate conditions in Florida, driving the loading wheels in one direction is more effective than two-way driving. Choubane et al. [11] used accelerated loading test equipment to evaluate the long-term performance of high-performance asphalt pavement with either coarse or fine aggregate gradation. They found that fine aggregate asphalt pavement provides superior rutting resistance than coarse aggregate material. Theys et al. [12] proposed a semi-empirical–semi-mechanical subgrade design model using subgrade deflection as a road design standard instead of the generally accepted vertical stress by conducting a large number of heavy-duty simulator (HVS) tests in California, USA. Hugo [13], a researcher in South Africa, developed the fourth generation of the previous pavement accelerated loading simulation systems of MLS and MMLS and successfully applied it to evaluate pavement fatigue performance. Accelerated loading tests were carried out on cement- and lime-stabilized sandy soil base in Mozambique. During the test, indexes such as surface cracks, grooves, pits, dynamic deflection, pavement flatness, and rutting were examined. At the same time, the long-term performance of asphalt pavement was investigated by testing the mechanical properties of materials using indoor tests, and the fatigue performance laws of the asphalt mixtures were obtained. Clearly, accelerated loading tests can provide many useful conclusions to guide the design of pavement structures and materials.

In this study, pavement performance is compared and analyzed on the basis of the material and structural form by paving base asphalt and REOB-recycled asphalt roads. Accelerated loading equipment was used to quickly accumulate load on the test section to simulate the real service environment of pavement. Pavement performance in the loading process was monitored, which provides the basis for the research and application development of REOB-recycled asphalt pavement.

2. Materials and Methods

2.1. REOB Regenerant

The Recycled Engine Oil Bottom (REOB) selected in this test comes from a qualified recycled oil treatment plant in Zibo, Shandong Province. Its main treatment process is: filter waste oil—membrane distillation—uronic refining—clay process.

The United States of America state the requirements for REOB quality control in the "Standard specification for refined engine oil bottoms/vacuum tower asphalt extender (VTAE)" promulgated in 2017 [14]. Based on this specification, as well as the similarity between REOB and asphalt compatibility, the related physical indicators and components of REOB were tested, and the results are shown in Table 1.

Table 1. Basic technical properties of REOB.

Physical Properties	Ash Content/%	Relative Density	60 °C Viscosity/Cst	Flash Point/°C	Thin Film Oven Test	
					Viscosity Ratio	Quality Change/%
	3.7	0.911	484	247	1.69	−1.284
Four-component analysis		Asphaltene/%	Colloid/%	Aromatics/%	Saturation/%	
		0.2	17.7	81.2	0.9	

In this study, a representative REOB from factory machinery was selected. According to the results of testing the performance of this REOB shown in Table 1, it meets the following functions and characteristics of a regenerant:

(1) Security. The flash point index reflects its construction safety. If this value is too low, it is a fire hazard under high-temperature operation. After testing, the flash point value of REOB is 247 °C, and the flash point value as asphalt regeneration agent meets the requirements of construction safety standards shall not be less than 220 °C.

(2) Components. Aromatic components mainly play the role of dissolving and dispersing asphaltene in aged asphalt; thus, regenerants require a large amount of aromatic components. The four-component analysis revealed that the aromatic content of the REOB regenerant is high at 81.2%, which can supplement the aromatic content missing in the asphalt aging process, while the saturated content also meets the specification requirements of ≤30% regenerant.

(3) Anti-aging properties. The viscosity ratio and mass change reflect the increase in viscosity and the decrease in mass after the film is heated, respectively. The smaller these two values, the better the anti-aging properties of the material. Owing to the high content of light components in REOB, it easily volatilizes at high temperature. Certain amounts of mass loss and viscosity increase are inevitable, but they are within the specification and controllable ranges, namely, a viscosity ratio ≤3 and a mass loss from −4% to 4%.

2.2. Asphalt

This test design required two kinds of asphalt: base asphalt and recycled asphalt. The conventional base binder was A-grade 70# petroleum asphalt with a performance classification of PG64-28. Recycled asphalt was prepared by adding the REOB regenerant to the aged asphalt. The aged asphalt was prepared as follows: The base asphalt was weighed to 600 g and poured into a 40 cm × 30 cm × 4.8 cm iron plate. The iron plate with the base asphalt was transferred into a 163 °C oven and stirred every 2 h to ensure even heating and aging of the asphalt. The aging time was prolonged to 48 h. In this way, one oven could perform the aging test of at least 3.6 kg asphalt, thereby significantly improving the efficiency of aging asphalt preparation [15]. The results of our previous research showed that the ideal regeneration effect was obtained when the REOB content was approximately 7% [16]. The specific preparation method of recycled asphalt is as follows: 7% REOB was added to the aged base asphalt at 160 °C for mixing, followed by shearing at 3000 r/min at 165 ± 5 °C for 30 min. Thereafter, it was placed in an oven at 160 °C for 1 h. The performance of the different types of asphalt used in the test is shown in Table 2.

Table 2. Asphalt test parameters.

Test Index		Unit	Base Asphalt	Aged Asphalt	7% REOB-Recycled Asphalt
			Test Parameters		
Penetration (25 °C)		0.1 mm	72.7	29	71
Softening point		°C	49.7	68	51.9
Ductility (10 °C, 5 cm/min)		cm	56.4	0.6	50.0
Relative density (25 °C)		/	1.024	1.036	1.022
Viscosity (135 °C)		Cst	338.2	1651.5	391
RTFOT (163 °C, 5 h)	Mass loss	%	−0.17	−0.09	−0.18
	Residual Penetration ratio	%	79.1	67.0	80.2
	Residual ductility (10 °C)	cm	21.2	0	16.2

Taking the three index values of base asphalt as a reference, the penetration, softening point, and ductility of the aged asphalt could be restored to 97.7%, 104.4%, and 88.7%, respectively, after adding 7% REOB, which could achieve the complete regeneration of the aged asphalt.

2.3. Asphalt Mixture

The AC-13 and AC-20 asphalt mixtures were used for the upper and lower layers of the test road, respectively, and the use of the binder is shown in Figure 1. The new coarse and fine aggregates used in the mixture were all obtained from the Zhang Qiu limestone quarry in Jinan. Among the components, the recycled asphalt surface is made of REOB-recycled asphalt and new asphalt as the binder, which was then mixed with the new and old aggregate to yield a mixture. The Reclaimed Asphalt Pavement (RAP) in the recycled pavement was taken from the asphalt concrete surface course of an expressway in Shandong, in which the content of RAP was 40%. Calculations using the Marshall design method indicated that the oil–stone ratio was 5.0%, and the gradation designs are shown in Figures 1 and 2.

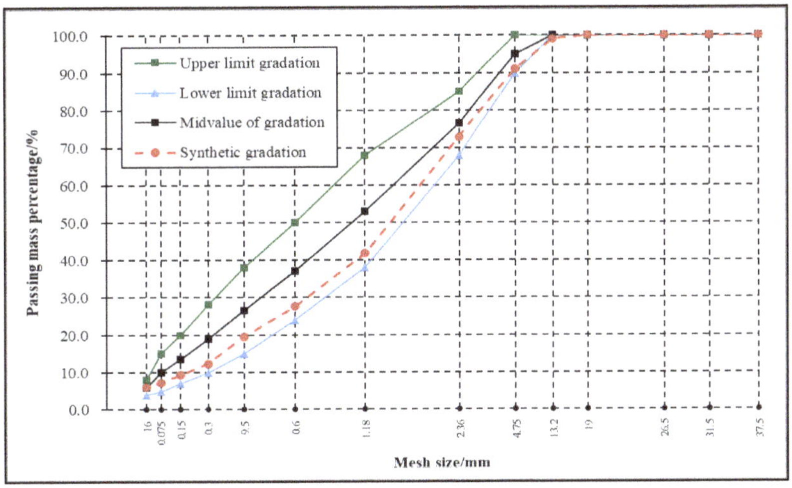

Figure 1. AC-13 aggregate gradation curve.

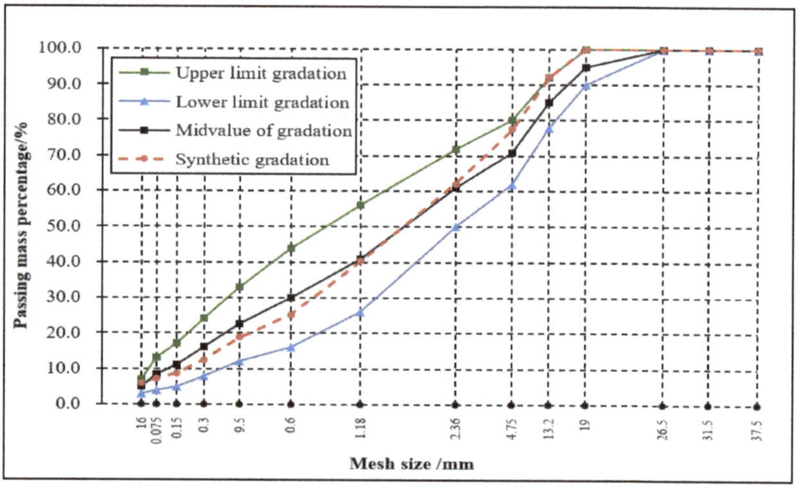

Figure 2. AC-20 aggregate gradation curve.

2.4. Test Plan

An accelerated loading test of the three pavement structures was conducted to compare the performance of different combinations of pavement. Five mixing pots of Shandong Provincial Key Laboratory of Asphalt Pavement Structure were used for mixing the mixture, and each mixing pot mixed 20 kg asphalt mixture each time. After research, it was decided that the temperature of the mixture that gets mixed from the completion to the paving stage be maintained by a large tray oven containing 1000 kg hot material in the Key laboratory of Asphalt Pavement structure in Shandong Province for hot material insulation process. The insulation temperature was uniformly set to 150 °C, which is conducive to the paving work of the test section.

This study was carried out according to the following flow chart, as shown in Figure 3. It involved two parts: (1) testing the pavement performance under loading and (2) extracting asphalt materials from the pavement after loading to examine changes in the properties of REOB-recycled asphalt.

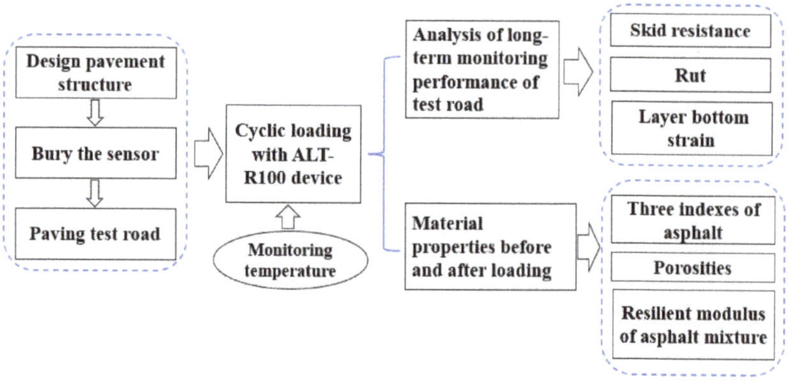

Figure 3. Flow chart of the accelerated loading test.

As shown in the paving plan in Figure 4, the upper and lower layers of the test pavement were made of dense-graded asphalt concrete. Fine-grained AC-13 gradation was selected for the upper layer, and medium-grained AC-20 gradation selected for the lower layer. The base layer cement stabilized the macadam base. The width and length of the pavement were designed to be 100 and 400 cm, respectively. To accelerate the destruction of the test road, the upper and lower layers were designed to be 4 and 6 cm thick, respectively. The lower layers of Pavements I and II were both an AC-20 base asphalt mixture However, the asphalt binders in their respective upper layers were different, namely, base asphalt and REOB-recycled asphalt, respectively. Using these materials in the upper layer enabled studying the performance difference between the REOB-recycled asphalt mixture and ordinary base asphalt concrete pavement. Meanwhile, Pavements II and III used base asphalt and REOB-recycled asphalt in the lower layer, respectively, which allowed investigating the long-term performance of pavement with REOB-recycled asphalt mixture in the underlying layer.

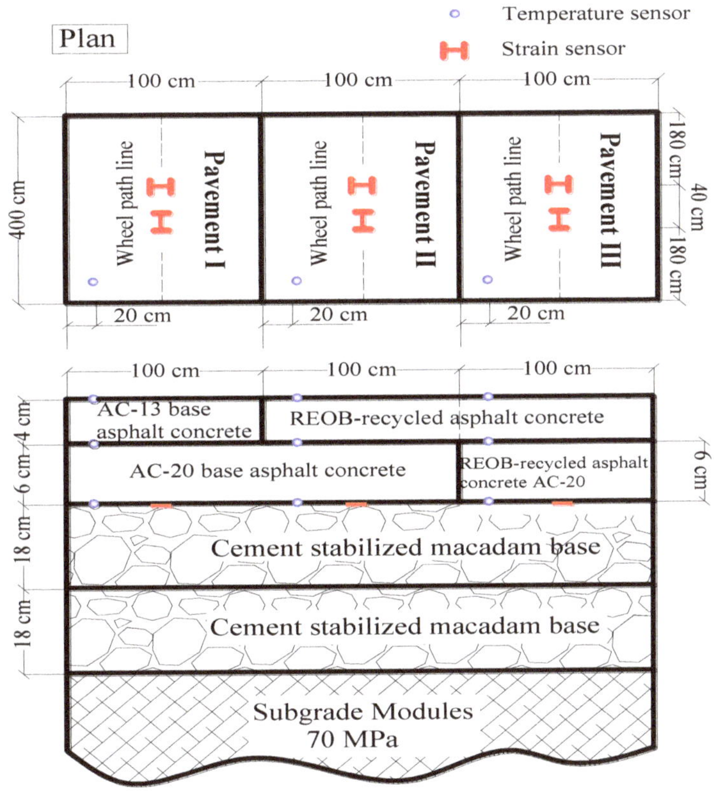

Figure 4. Road structure and sensor location map.

The equipment used in this model test was a small rotary road acceleration loading system (ALT-R100) independently developed by Shandong Jiaotong University. The equipment could be used in a three-wheel unidirectional cycle loading method, and the maximum loading speed reached 35 km/h (15,000 cycles/h). This approach was convenient to use and could quickly simulate the actual traffic load and environmental damage to the road surface in a short time. The reasonable structure and damage mechanism in the life cycle of a road could be effectively and accurately analyzed by monitoring the road conditions and test parameters during the loading process.

To accelerate the destruction speed of the test road, the upper and lower layers were designed to be 4 and 6 cm thick, respectively. The full-scale asphalt test pavement structure was paved with a 100 KN cyclic loading mode with a driving speed of 22 km/h, and the effective loading length of the equipment was 100 cm. To reduce the impact of environmental factors during the loading process, a reciprocating cyclic loading method was adopted: Pavement I was loaded for 50,000 cycles, translated to the loading area of Pavement II and loaded for 50,000 cycles, and then translated to the loading area of Pavement III and loaded for 50,000 cycles. This was the reciprocating cycle loading method. During the loading process, the pavement temperature, layer bottom strain, pavement friction coefficient, and rut depth were monitored regularly and at fixed points.

(1) Pavement temperature: Temperature monitoring was conducted using an embedded thermocouple temperature sensor with a measurement range able to reach −200–500 °C, while the accuracy reached 0.002 °C. A temperature sensor was embedded at the top of the upper layer, the bottom of the upper layer (the top of the lower layer), and the bottom of the lower layer of the three pavements of the test road in the form of pre-burying, which recorded temperature values every 1 h. The layout position is shown in Figure 4. Meanwhile, the indoor temperature of the corresponding time was recorded using a mercury thermometer.
(2) Layer bottom strain: After comparison, the layer bottom strain monitor was used to select the KM-100HAS "I"-shaped resistance strain gauge produced by Japan's TML company, which provided superior comprehensive performance. The three-dimensional position where the strain sensor was buried was located between the top of the cement-stabilized gravel layer and the bottom of the lower layer along the direction of the wheel track belt, and between the bottom of the upper layer and the top of the lower layer. The strain response data were collected every 10,000 cycles with a 100 Hz acquisition frequency, and the acquisition time was approximately 2 min/12 loading cycles. The strain data acquisition system included a DC-004P dynamic strain gauge that was made by Tokyo Measuring Instruments Lab in Japan, and visual log DC-7004P control software to collect and analyze the strain data.
(3) Anti-sliding performance: Each wheel of the loading equipment was rolled for 5000 cycles, and the friction coefficient data were collected once by a pendulum friction meter, which was the British Pendulum Number (BPN) condition value, representing the anti-sliding ability of the pavement. The friction coefficient of the asphalt pavement was monitored by setting three detection points along the wheel trace direction, as shown in Figure 5a.
(4) Anti-rutting performance: A laser car that was independently developed by Shandong Jiaotong University was used for the rutting test at points A and B in Figure 5. The test range of the laser car was 1 m^2. The rutting data were collected after every 5000 loading cycles, and the rutting of a cross-section was tested every 0.2 m, as shown in Figure 5b.

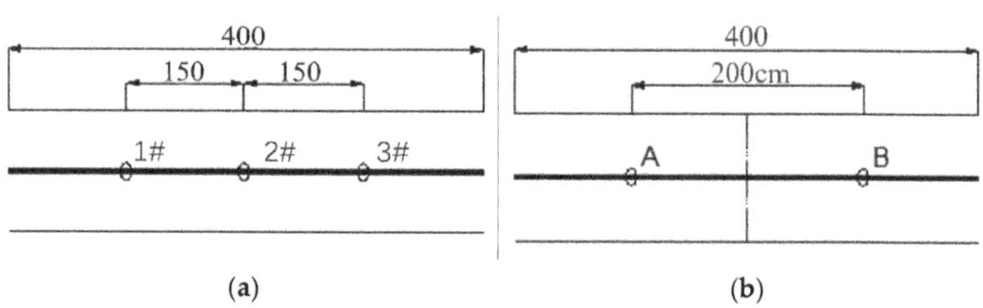

Figure 5. Layout of monitoring points. (**a**) Layout of friction coefficient monitoring points; (**b**) Layout of track monitoring points.

3. Results and Discussion

3.1. Analysis of Long-Term Monitoring Performance of Test Road

3.1.1. Temperature

The relationship curve between the loading cycles and temperature of the three types of pavements is shown in Figure 6, with 20,000 loading cycles as a statistical unit. Owing to the reciprocating loading mode of the three pavements, the difference in pavement temperature among the pavements was relatively small. However, owing to the large time span from summer to autumn, the overall temperature showed a downward trend with decreasing room temperature (15–35 °C), and the temperature dropped sharply after loading for 420,000 cycles. Therefore, the experiment was divided into two stages based on the temperature gradient.

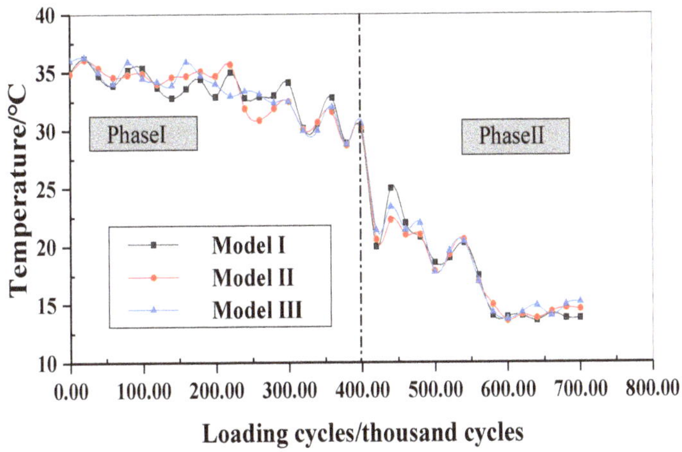

Figure 6. Temperature curve during loading.

Stage I: The average temperature of asphalt pavement was 33.3 °C, and the total number of loading cycles was 400,000.

Stage II: The average temperature of asphalt pavement was 17.4 °C, and the total loading number of loading cycles was 300,000.

3.1.2. Skid Resistance

The measured BPN values were all obtained for conventional dry road, and the thickness of the upper layer of the test road was 4 cm; thus, the measured values did not require temperature correction. The decay law of the friction coefficient of the three types of asphalt pavement structures under accelerated loading was obtained using a statistical method, as shown in Figure 7.

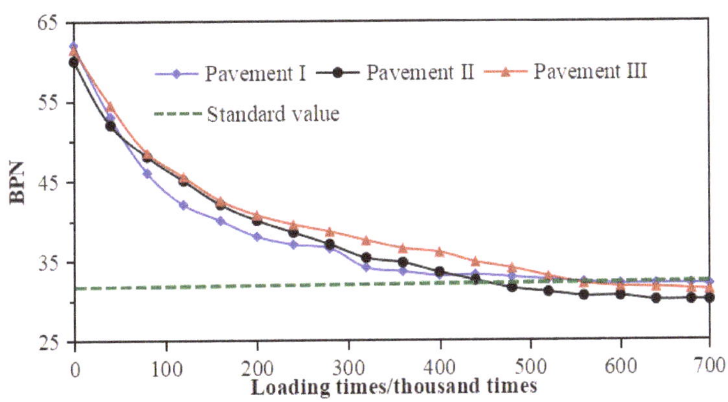

Figure 7. Variation law of pavement friction coefficient.

The initial friction coefficients of the three accelerated loading test sections were all above 60. The friction coefficient decreased gradually with increasing loading cycles and decreased rapidly in the early stage of loading. After loading for 400,000 cycles, the pendulum value dropped below 40, which is below 60% of the initial value, and Pavement I exhibited the fastest rate of decrease, falling to approximately 55% of the initial value. The difference in the skid-resistance performance of the three types of pavement was small, which indicates that REOB-recycled asphalt will not reduce the skid-resistance performance of asphalt pavement, which is mainly controlled by the characteristics of the aggregate and gradation. The pendulum value decreased slowly and tended to gradually stabilize after 400,000 loading cycles until the end of the test. The skid-resistance performance of Pavements II and III decayed rapidly. Their BPN values were lower than those of Pavement I at 440,000 and 560,000 loading cycles, respectively, and lower than the required value of 32 for class 2. Further, these were below highway maintenance specifications at 600,000 and 480,000 loading cycles, respectively.

Therefore, the skid-resistance performance of conventional asphalt mixture pavement is inferior to that of REOB-recycled asphalt mixture pavement in the short term because of the high viscosity of recycled asphalt, which will also cause a more serious degree of rutting in Pavement I compared to other pavements. With the increasing number of loading cycles, from the perspective of long-term pavement service, the anti-sliding performance of conventional asphalt mixture pavement is better than that of REOB-recycled asphalt mixture pavement.

3.1.3. Rut

The rut formed on the test road is typically unstable (flow type), owing to the reciprocating rolling action of the ALT-R100 system. Deformation curve of wheel rut section after 700,000 times of loading is shown in Figure 8a. Figure 8b shows the overall evolution of the road ruts of the three different pavement structures under the action of loading and rolling at room temperature.

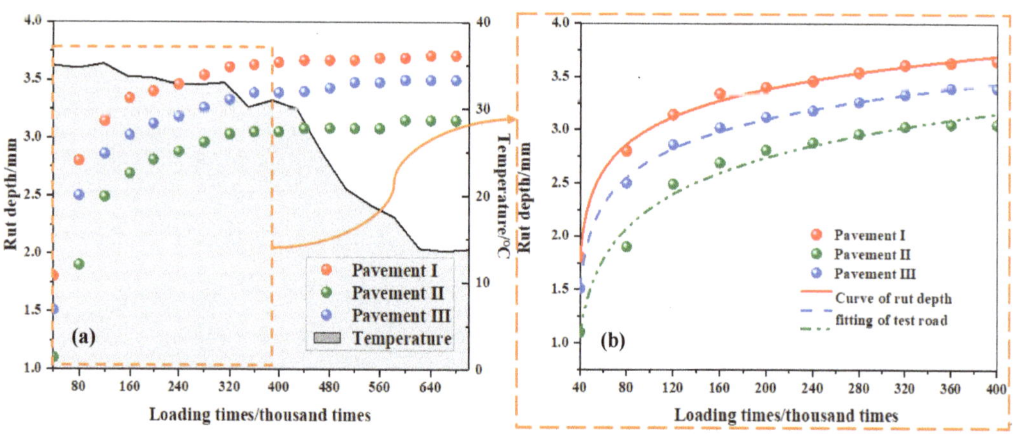

Figure 8. Variation law of pavement rutting. (**a**) Deformation curve of wheel rut section after 700,000 times of loading; (**b**) The overall evolution of the road ruts of the three different pavement structures under the action of loading and rolling at room temperature.

The tests reveal several results:

(1) The overall rutting shapes of the three pavement structures are similar after cycle loading, as shown in Figure 9. The rutting depths of the three types of pavement structures developed rapidly before 220,000 loading cycles with the acceleration loading system; at 160,000 loading cycles, the first inflection point appears in the

rutting development, and the second inflection point appears at 220,000 loading cycles. However, after 220,000 loading cycles, the rutting depth tended to increase at a slower rate and remained almost unchanged after 400,000 loading cycles. Because the compaction of the aggregate and the discharge of air were relatively fast, the pavement could be in the stage of aggregate compaction before 220,000 loading cycles.

(2) The gray area shown in Figure 8 is the change in room temperature; hence, higher room temperature and road temperature also played a certain role in promoting the generation of rutting, causing a greater change in the early rutting depth. At the end of 700,000 loading tests, the rutting depths of the first, second, and third structural pavements were 3.7, 3.2, and 3.5 mm, respectively.

(3) Compared with structural Pavements I and II, which had the same form of their lower layer, the anti-rutting ability of recycled Pavement II was approximately 13% better than that of conventional pavement "Pavement I," when the upper layer of asphalt mixture is comprised of 4 cm of base asphalt AC-13 and REOB-recycled asphalt AC-13, respectively. The upper and lower layers of Pavement III contained the recycled mixture, whereas Pavement II only contained the recycled mixture in the upper layer and showed better anti-rutting performance than Pavement III. Comprehensive analysis shows that the REOB-recycled asphalt mixture has good anti-rutting performance when used in the upper layer.

(4) At high temperature conditions, above 30 °C, damage is related to rutting but not fatigue. At intermediate temperature conditions, below 20 °C, damage is related to fatigue, and the ruts change slightly and gradually flatten out. According to the above test data results and relevant research data, the rutting changes with the loading cycles in a logarithmic form [17], and the regression curve above 30 °C (the first 400,000 loads) is shown in Figure 8b. The fitting model is shown in Table 3. The correlation coefficient R^2 of the fitted logarithmic curve is over 0.98, indicating that the fitted function is reliable.

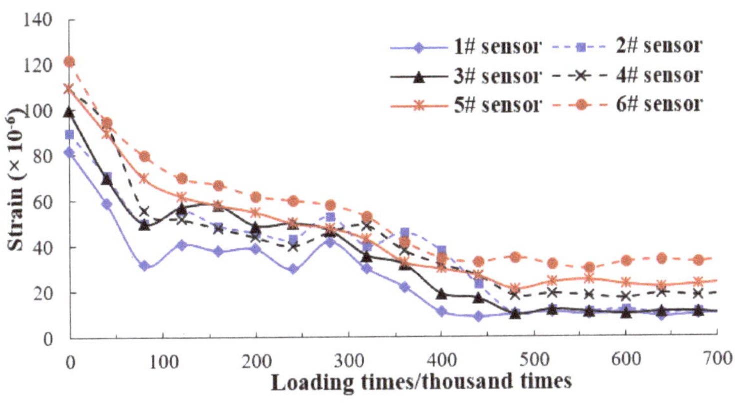

Figure 9. Relationship between layer bottom strain of each sensor and loading cycles.

Table 3. Rutting prediction Pavements of three pavement structures.

Pavement Structure Model	Rutting Prediction Model	R^2
Structural Pavement I	rut = 0.38959 ln(N − 3.71988) + 2.9481	0.99598
Structural Pavement II	rut = 0.41277 ln(N − 3.65412) + 1.94652	0.99743
Structural Pavement III	rut = 0.52695 ln(N − 3.27094) + 1.25467	0.98152

Where: rut is rutting depth, N is the number of loading cycles.

3.1.4. Layer Bottom Strain

An effective method to study the stress and fatigue of asphalt pavement is to analyze the layer bottom strain, as many problems with pavement are closely related to the layer bottom strain [18].

During the accelerated loading test, sensors No. 1, No. 3, and No. 5 were transverse strain sensors of Pavements I, II, and III, respectively, and sensors No. 2, No. 4, and No. 6 were longitudinal strain sensors. According to the monitoring results, the relationship between the layer bottom strain and loading time is shown in Figure 9.

(1) As seen in Figure 9, the change trends of the transverse and longitudinal sensors of each pavement structure are consistent, whereas the numerical values differ. The initial strain values of Pavements I, II, and III are above 80×10^{-6}, 100×10^{-6}, and 110×10^{-6}, respectively. After 80,000 loads, these values drop to approximately 40×10^{-6}, 50×10^{-6}, and 70×10^{-6}, respectively. At this time, the strain changes are related to the high temperature and compaction of the mixture. As there is only a limited number of load cycles for intermediate temperatures, fatigue does not occur yet, and thereafter, the strain is only related to the temperature conditions.

(2) The strain change during the entire loading process may be divided into three stages. In the initial stage of loading, the temperature is relatively high, in which the layer bottom strain is also high. In the middle stage of loading, the temperature drops sharply from 350,000 to 500,000 loading cycles. With the long-term effect of the load, the strain decreases gradually. Before the end of loading, from 500,000 to 700,000 loading cycles, the temperature drops, causing a halt in the change in strain value.

(3) In the later stage of loading (from 500,000 to 700,000 cycles), the strain values of Pavements I, II, and III are stable at 9–11 $\times 10^{-6}$, 10–19 $\times 10^{-6}$ and 21–33 $\times 10^{-6}$, respectively. The strain at the bottom of the third layer of the pavement is at a relatively high level compared to other pavements. This result shows that the pavement using REOB-recycled asphalt mixture on both the upper and lower layers has less sufficient mechanical properties than the pavement using this asphalt mixture only in the upper layer and the conventional asphalt pavement in the lower one. Excessive strain could directly cause pavement damage, and the anti-fatigue performance of the third layer of the pavement is poor.

Studies have been conducted on the fatigue performance of asphalt mixtures in different ways, and different fatigue pavements of asphalt mixtures have been proposed.

The British Transport and Road Research Institute has proposed fatigue design standard Formulas (1) and (2) [19,20] for asphalt pavement, with a reliability of 85% and an equivalent temperature of 20 °C.

$$N_f = 1.66 \times 10^{-10} \times (\varepsilon_t)^{-4.32}(\mathrm{AC}) \tag{1}$$

and

$$N_f = 4.169 \times 10^{-10} \times (\varepsilon_t)^{-4.16}\ (\mathrm{AM}). \tag{2}$$

Elliot and Thomson established a relationship between the tensile strain and pavement fatigue life. For a typical Illinois dense graded I asphalt mixture, the following fatigue pavement Formula (3) can be adopted [21]:

$$N_t = 5 \times 10^{-6} \times (\varepsilon_t)^{-3.0}. \tag{3}$$

The fatigue test results of different asphalt mixtures show that the resilient modulus plays an important role in the fatigue performance of materials. Monismith et al. established the following general Formula (4) [22]:

$$N_f = K(\varepsilon_t)^{-n}(S_{min})^{-m}. \tag{4}$$

Wang Xudong et al. conducted constant stress bending fatigue tests, combined with load transverse distribution, load intermittent time, and adverse season days. They modified the indoor fatigue model as follows [23]:

$$N_e = N_f \times 7 \times 40 \times \frac{365}{60} \times 2 = 13.7 \times A_a A_g / A_C \sigma^{-4.17} \tag{5}$$

where N_f is the laboratory fatigue life, $N_f = 4025 A_a A_g / A_C \sigma^{-4.17}$; ε_t is the tensile strain; A_a is the coefficient of asphalt type; A_b is the gradation coefficient of asphalt mixture; A_c is the coefficient of highway grade coefficient; and σ is stress.

In the actual service process of pavement, its stress and strain are constantly changing and uncertain under vehicle loads. Most of the above fatigue models were obtained by testing under the stress control and strain control modes, which differ from the actual situation of real pavement. For an unstable variable stress, damage accumulation hypothesis should be used to estimate the fatigue failure life of the road. Cyclic loading is adopted in this test, and the loading times are sufficient. A comparative analysis of the transverse and longitudinal tensile strains is performed by the Wohler equation, and load action cycles of Pavement III are used as an example, by which the relationship between tensile strain and load action cycles is established, as shown in Figure 10.

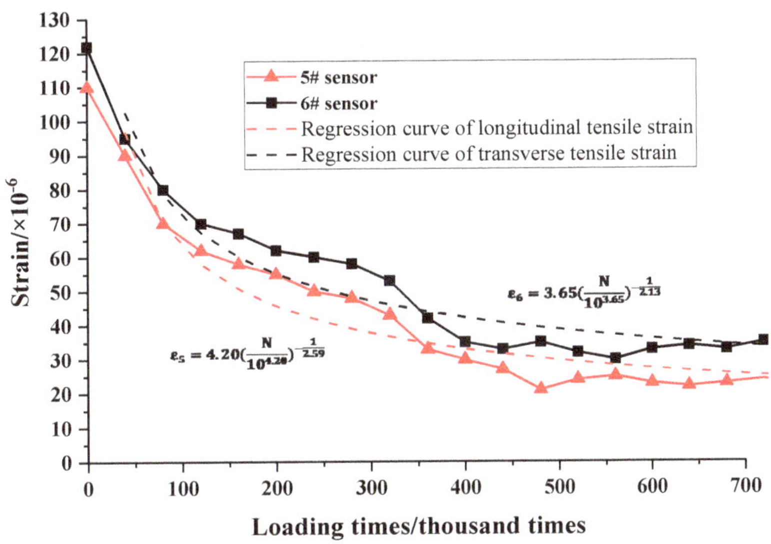

Figure 10. Relationship between strain and the number of loading cycles.

3.2. Material Properties before and after Loading

After 700,000 times of loading, cores were taken from the loading areas (including rutting pits and uplifts) and unloaded areas of the three pavements. Three index tests were performed on the extracted asphalt, and the mixture porosity test and resilience modulus test were performed on the core samples. Based on the original test data, the physical properties of the recovered binder and the porosity and modulus of the mixture of each group are shown in Figures 11 and 12 and Table 4.

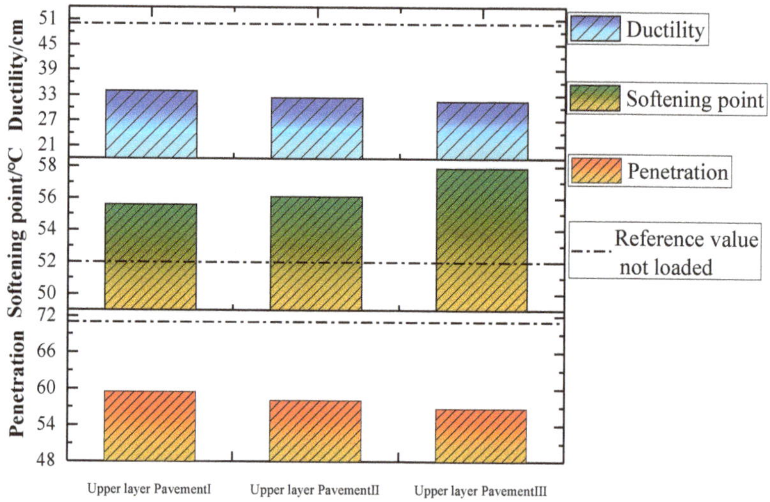

Figure 11. Three indexes of asphalt binder for each pavement.

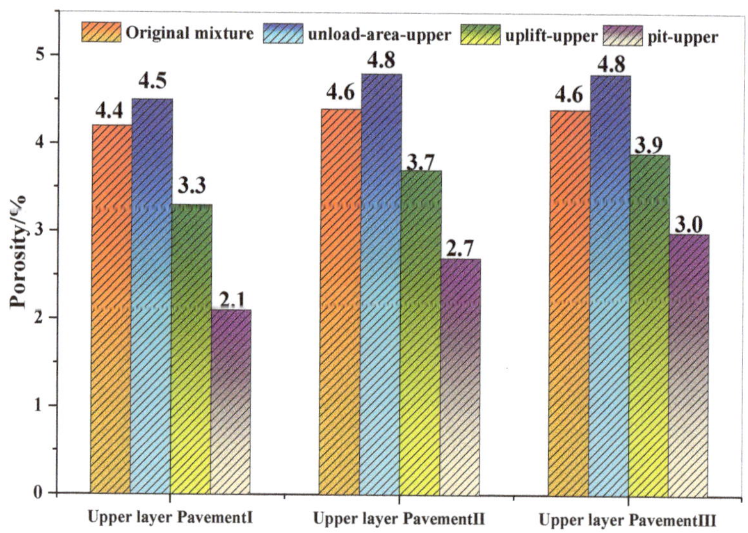

Figure 12. Porosities of cores taken from different pavements.

Table 4. Resilience modulus of cores taken from different pavements.

Coring Position	Resilience Modulus/MPa			
	Mixture	**Unload-Area-Upper**	**Uplift-Upper**	**Pit-Upper**
Upper layer Pavement I	673	666	785	826
Upper layer Pavement II	679	671	792	807
Upper layer Pavement III	679	670	788	801

The comparison of physical performance of the binder on the upper layer of each structural pavement after loading shows that among the three indexes, the degree of change in the softening point is the smallest, i.e., less than 6 °C, whereas the ductility changes to greatest degree, by more than 27 cm. Thus, the performance metrics of the asphalt binder decrease in varying degrees after mixing, paving, and wheel loading. The performance of the recycled asphalt of Pavement III decreases the most. The softening point of Pavement III is the most prominent; it is 6 °C and 1.2 °C higher than that of original asphalt and Pavement II with the same material, respectively. Pavement III has a penetration of 14.2 mm lower than original asphalt. In the rut fitting function $y = a - b \ln(x + c)$, b represents the slope of the curve. Model III has the largest b, and the rut develops the most rapidly, which verifies that the faster the asphalt performance declines, the faster the rutting resistance degrades.

Figure 12 shows the porosity comparison of materials at different positions in the upper layer of each pavement structure. Porosity index can reflect the compaction of pavement materials of each structure. The porosities of pavement II, III, and I are the highest, second highest, and lowest, respectively. The porosities of pit- and uplift-upper of pavement I are 53.3% and 26.7% lower than that of the unloaded area, respectively, whereas those corresponding to pavements II and III are 43.7% and 22.9% lower and 37.5% and 18.7% lower, respectively. The degree of wheel load compaction of pavement I is greater than that of other pavements. If the increase in porosity is attributed to the lateral flow of material, pavement I has more lateral material flow and is more affected by rutting. REOB-recycled asphalt is used as the mixture in the upper layer, which has obvious advantages over the base asphalt in terms of the rutting resistance of the whole asphalt layer. The porosity changes of Pavements II and III are basically equal, indicating that REOB-recycled asphalt can also provide satisfactory rutting resistance when used in the lower layer.

There is no significant difference in the modulus of compressive rebound between the new base asphalt Pavement I and the REOB-recycled asphalt Pavements II and III, as shown in Table 4. After 700,000 cycles, the modulus appreciations of uplift-upper of Pavements I, II, and III are 119 MPa, 121 MPa, and 118 MPa, respectively, and those of pit-upper are 160 MPa, 136 MPa, and 131 MPa, respectively. The moduli of Pavements II and III in the unload-area-upper and the uplift-upper remain large, and the possibility of damaging the pavement is greater under sufficient load. In addition, the moduli do not reach the attenuation stage, indicating that the pavement performance is only degraded under 700,000 cycles, but does not reach the damage stage. The performance of REOB-recycled asphalt pavement is basically equivalent to that of base asphalt surface under 700,000 loads.

To study the fatigue resistance of the material, loading times should be increased further.

4. Conclusions

In this study, REOB is used as an asphalt modifier in the field of aging asphalt recycling, which offers a certain amount of economic and environmental protection. To explore the practical long-term application performance of REOB-recycled asphalt pavement, the rotary accelerated loading test system (ALT-R100) independently developed by our university is used to carry out cyclic loading tests on three sections of REOB-recycled asphalt pavement, to simulate the long-term load behavior of service pavement in a short time. The following conclusions may be drawn:

(1) After using the ALT-R100 accelerated loading system to perform 700,000 loading cycles, the anti-skid performance and anti-rutting performance of concrete pavement with REOB-recycled asphalt and conventional asphalt are basically consistent, both meeting use requirements.

(2) The pavement with the REOB-recycled asphalt mixture in the upper and lower layers enters the compaction stage sooner; thus, it is prone to rutting damage earlier, and the asphalt binder is more prone to aging. This problem does not occur when REOB-recycled asphalt mixture is only used in the upper layer.

(3) A logarithmic model for predicting rutting and a Wohler-equation-based strain prediction model of the three-section test road under ALT-R100 accelerated loading were obtained, revealing the development laws for rutting and bottom strain within the applied loading times.
(4) The rutting depth, BPN value, and bottom strain of the pavement gradually stabilized after loading for 400,000 cycles. When ALT-R100 is used to evaluate the high-temperature rutting resistance and anti-skid performance of asphalt mixtures, the standard test can be performed for 400,000 to 500,000 loading cycles.

With sufficient time and funds, we can continue to conduct accelerated loading tests involving REOB-recycled asphalt pavement by constructing a field test road in the future to study the performance characteristics of base asphalts having different structures with REOB-recycled asphalt layers.

Author Contributions: Data curation, X.C.; Methodology, M.Y. and S.Z.; Project administration, P.L.; Writing—original draft, L.Z.; Writing—review & editing, J.L. All authors have read and agreed to the published version of the manuscript.

Funding: This research was funded by the project "Study on the Effect and Mechanism of Recycled Engine Oil Bottom on the Modification Performance of Asphalt" (ZR 2016EEM42) of Shandong Natural Science Foundation.

Institutional Review Board Statement: Not applicable.

Informed Consent Statement: Not applicable.

Data Availability Statement: The experimental data in this paper are from the pavement material laboratory of Shandong Jiaotong University, which is the provincial key laboratory.

Acknowledgments: The authors would like to acknowledge the assistance of staff at Shandong Natural Science Foundation and Shandong University Science and Technology Program of China, and thank Xiang Li, Jiabin Wang, Enzhou Di, and Zebin Guan for their help in the experiment.

Conflicts of Interest: The authors declare no conflict of interest.

References

1. Jia, X.; Huang, B.; Bowers, B.F.; Zhao, S. Infrared spectra and rheological properties of asphalt cement containing waste engine oil residues. *Constr. Build. Mater.* **2014**, *50*, 683–691. [CrossRef]
2. Hesp, S.A.; Shurvell, H.F. X-ray fluorescence detection of waste engine oil residue in asphalt and its effect on cracking in service. *Int. J. Pavement Eng.* **2010**, *11*, 541–553. [CrossRef]
3. Hesp, S.A.M.; Shurvell, H.F. Waste Engine oil Residue in Asphalt Cement. In Proceedings of the Seventh International Conference on Maintenance and Rehabilitation of Pavements and Technological Control, Thessaloniki, Greece, 12–17 June 2019.
4. Golalipour, A. Investigation of the Effect of Oil Modification on Critical Characteristics of Asphalt Binders. Ph.D Thesis, The University of Wisconsin—Madison, Madison, WI, USA, 2013.
5. Herrington, P.R.; Hamilton, P.G. *Recycling of Waste Oil Distillation Bottoms in Asphalt*; Transfund New Zealand Research Report; California Transit Association: Sacramento, CA, USA, 1998.
6. Bennert, T.; Ericson, C.; Pezeshki, D.; Haas, E.; Shamborovskyy, R.; Corun, R. Laboratory performance of refined engine oil bottoms (REOB) modified asphalt. *J. Assoc. Asph. Paving Technol.* **2016**, 163–207.
7. Cooper, S.B., Jr.; Mohammad, L.N.; Elseifi, M.A. Laboratory Performance of Asphalt Mixtures Containing Recycled Asphalt Shingles and Re-Refined Engine Oil Bottoms. *J. Mater. Civ. Eng.* **2017**, *29*, 04017106. [CrossRef]
8. Li, X.; Gibson, N.; Andriescu, A.; Arnold, T.S. Performance evaluation of REOB-modified asphalt binders and mixtures. *Road Mater. Pavement Des.* **2017**, *18*, 128–153. [CrossRef]
9. Zaumanis, M.; Mallick, R.B.; Frank, R. Evaluation of rejuvenator's effectiveness with conventional mix testing for 100% reclaimed asphalt pavement mixtures. *Transp. Res. Rec. J. Transp. Res. Board* **2013**, *2370*, 17–25. [CrossRef]
10. Sirin, O.; Tia, M.; Roque, R.; Choubane, B. Evaluation of performance characteristics of the heavy vehicle simulator in Florida. *Build. Environ.* **2007**, *42*, 1270–1277. [CrossRef]
11. Choubane, B.; Gokhale, S.; Sholar, G.; Moseley, H. Evaluation of coarse and fine graded superpave mixtures under accelerated pavement testing. *Transp. Res. Rec. J. Transp. Res. Board* **2006**, *1974*, 120–127. [CrossRef]
12. Theyse, H.L.; Hoover, T.P.; Harvey, J.T.; Monismith, C.L.; Coetzee, N.F. A mechanistic-empirical subgrade design model based on heavy vehicle simulator test results. *Pavement Mech. Perform.* **2006**, *154*, 195–202.

13. Hugo, F.; De Vos, E.R.; Tayob, H. Aspects of cement stabilized mozambique sand base material performance under MMLS3 and MLS10 APT trafficking. *SATC* **2007**. Available online: http://hdl.handle.net/10019.1/43003 (accessed on 6 April 2022).
14. *ASTM D8125-2017*; Standard Specification for re-Refined Engine Oil Bottoms (REOB)/Vacuum Tower Asphalt Extender (VTAE). American Society for Testing and Materials: West Conshohocken, PA, USA, 2017.
15. Yu, M.Z. Design and Road Performance Test of Recycled Asphalt Mixture with REOB. Ph.D. Thesis, Shandong Jiaotong University, Shandong, China, 2019. (In Chinese).
16. Li, J.; Yu, M.; Cui, X.; Wang, W. Properties and Components of Recycled Engine Oil Bottom Rejuvenated Asphalt and Its Grey Relationship Analysis. *Adv. Mater. Sci. Eng.* **2019**, *2019*, 1–11. [CrossRef]
17. Wang, P.C.; Hao, X.W.; Zhang, Y. Long term performance of warm mix asphalt mixture based on accelerated loading test. *J. Shenyang Jianzhu Univ.* **2020**, *36*, 860–869. (In Chinese)
18. Coutinho, R.P.; Babadopulos, L.F.A.L.; Freire, R.A.; Branco, V.T.F.C.; Soares, J.B. The use of stress sweep tests for asphalt mixtures nonlinear viscoelastic and fatigue damage responses identification. *Mater. Struct.* **2013**, *47*, 895–909. [CrossRef]
19. Luo, S.; Qian, Z.D.; Harvey, J. Fatigue decay characteristics test of epoxy asphalt mixture. *Chin. J. Highw.* **2013**, *26*, 20–25. (In Chinese)
20. Powell, W.D.; Potter, J.F.; Mayhew, H.C.; Nunn, M.E. The structural design of bituminous roads. *Transp. Road Res. Lab.* **1984**, *1132*, 1984.
21. Elliot, R.P.; Thompson, M.R. Mechanistic design concepts for conventional flexible pavement. *Transp. Eng. Ser. Urbana: Univ. Ill.* **1986**. Available online: https://www.researchgate.net/publication/269090247 (accessed on 6 April 2022).
22. Monismith, C.L.; Deacon, J.A. Fatigue of asphalt paving mixtures. *Transp. Eng. J.* **1969**, *95*, 122–131. [CrossRef]
23. Wang, X.D.; Sha, A.M.; Xu, Z.H. *Dynamic Characteristics and Dynamic Parameters of Asphalt Pavement Materials*; People's Communications Press: Beijing, China, 2002.

coatings

MDPI

Article

Shear Behavior of Stud-PBL Composite Shear Connector for Steel–Ceramsite Concrete Composite Structure

Hongbing Zhu [1], Zhenghao Fu [1], Peng Liu [2,3,*], Yongcan Li [1] and Benlu Zhao [1]

[1] School of Urban Construction, Wuhan University of Science and Technology, Wuhan 430065, China; zhuhongbing@wust.edu.cn (H.Z.); whfzh1998@163.com (Z.F.); wslyc2007@126.com (Y.L.); 13628659252@163.com (B.Z.)
[2] School of Civil Engineering, Central South University, Changsha 410075, China
[3] National Engineering Laboratory for High Speed Railway Construction, Central South University, Changsha 410075, China
* Correspondence: 2015038@csu.edu.cn

Abstract: For steel–concrete composite structure, a new type of stud–PBL composite shear connector can improve the shear resistance of steel–concrete interface, and polypropylene fiber ceramsite concrete can reduce the self-weight. Therefore, investigating the shear behavior of stud–PBL composite shear connectors for steel–ceramsite concrete composite structures bears significance. In this study, static testing and numerical simulation of the composite shear connector push-out specimen of polypropylene fiber ceramsite concrete were first conducted. The influencing factors of the shear bearing capacity were then analyzed. The formula for determining the shear bearing capacity of the steel–ceramsite concrete composite structure stud–PBL composite shear connectors was ultimately established. The results indicated that the new composite shear connector exhibited excellent shear resistance and good deformation ability. In addition, increasing concrete's strength, stud's diameter, and perforated plate's thickness could significantly improve the shear bearing capacity of the composite shear connector. The calculated value of the shear bearing capacity of the composite shear connector was well correlated the measured value of the test. Overall, the stud–PBL composite shear connector could effectively improve the interfacial shear bearing performance of the steel–ceramsite concrete composite structure. Moreover, the established formula demonstrated broad applicability.

Keywords: bridge engineering; shear bearing capacity; push-out test; stud–PBL composite shear connector; finite element analysis; ceramsite concrete; composite structure

Citation: Zhu, H.; Fu, Z.; Liu, P.; Li, Y.; Zhao, B. Shear Behavior of Stud-PBL Composite Shear Connector for Steel–Ceramsite Concrete Composite Structure. *Coatings* **2022**, *12*, 583. https://doi.org/10.3390/coatings12050583

Academic Editor: Andrea Nobili

Received: 29 March 2022
Accepted: 21 April 2022
Published: 24 April 2022

1. Introduction

Steel–concrete composite structures are characterized by reasonable stress and have been widely used in recent years. Shear connectors are the pivotal elements that bond the steel beam to the concrete slab. Stud and Perfobond Leiste (PBL) shear connectors are mostly used in engineering thus far [1]. Stud shear connectors can resist longitudinal shear force through the stud rod, and a large part resists the lifting force of the concrete slab. To resist the longitudinal shear force and the interface separation force, a PBL shear connector mainly consists of three parts: a perforated steel plate, the concrete tenon, and the penetrating steel bars [2].

Comprehensive studies have been conducted on the combined structural performance of studs or solely PBL shear connectors, and these reports have been verified in various engineering practices [1–6]. To meet engineering requirements such as large shear and pull-out resistance for a composite structure interface, flexible stud shear connectors have been combined with rigid PBL shear connectors; this integration is aimed at forming composite shear connectors that can be used in steel–concrete composite structures. However, the process of combining stud shear connectors and PBL shear connectors in space remains inconclusive. Zhang [7] added two rows of studs to both sides of the perforated steel plate to improve the shear bearing capacity; this connector, which was composed of the perforated plate and the studs, was used in the Nakano Viaduct Project in Japan. For this method, push-out tests and finite element simulation analysis were performed by Deng [8] and Zheng [9]. Their study verified the superior shear resistance of the perforated plate + stud shear connector. Meanwhile, to strengthen the vertical shear resistance of the transition between the longitudinal steel box girder section and the concrete girder section in the hybrid box girder bridge, Jin [10] arranged the studs and PBL shear connectors alternately vertically in the web of a hybrid box girder bridge and studied the ratio of the shear force borne by the PBL shear connectors to the studs. Chen [11] proposed a composite shear connector in which studs were transversely welded to the perforated plate. The formula for the shear bearing capacity of the composite shear connector was derived by finite element simulation analysis of push-out tests. The aforementioned studies demonstrated that the combination of studs and PBL to form a composite shear connector can indeed improve shear resistance; however, existing composite shear connectors (mainly used in special stress parts of the structure) are not universal. The PBL shear connectors arranged lengthwise can divide the concrete slab into strips longitudinally, reducing the integrity of the concrete slab. Meanwhile, the PBL shear connectors arranged lengthwise may increase the amount of steel. Widely distributed welding residual stresses also exist in PBL shear connectors and steel beams. Therefore, it is of good engineering value to design a more reasonable stud–PBL composite shear connector, which can optimize the welding residual stresses, enhance the connection between concrete and steel beams, and improve the force performance of the steel–concrete interface.

In this study, PBL and stud were alternately arranged in the same plane to form a new composite shear connector. The intermittently arranged PBL shear connectors can reduce the amount of steel used and increase the integrity of the concrete slab. The bearing effect of the end concrete can also enhance the shear resistance of the shear connectors [12,13]. Meanwhile, the combination of rigid PBL shear connectors and flexible studs can complement the advantages of the two shear connectors, endowing the interface between the concrete slab and the steel beam with sufficient stiffness, good ductility, and toughness.

Shear bearing capacity is an important performance index of shear connectors. Therefore, it is very important for its popularization and application to accurately evaluate the shear resistance of shear connectors and obtain a reliable calculation formula of shear bearing capacity. In a previous study, Xue [14] proposed the load–slip curve formula for the sheared stud and the calculation formula for the shear bearing capacity, considering the length to diameter ratio of the stud by testing the stud push-out specimen. Using push-out testing, Lee [15] and Hu [16] presented formulas for calculating the shear bearing capacity of high-strength concrete stud shear connectors and ultra-high-performance concrete stud shear connectors, respectively. Huo [17] studied the shear bearing capacity of stud connectors under impact loading. Based on integrating some group studs push-out tests and theoretical analysis results [18–21], "Code for Design of Steel and Concrete Composite Bridges" (GB 50917-2013) also provided the formula for calculating the shear bearing capacity of ordinary concrete composite beams when the studs were damaged, and highlighted that the diameter of the stud is an important influencing factor. The effects of diameter [12,13,22–30], quantity [22] and strength [25,26] of penetrating steel bars, diameter of openings [12,13,22–27,29,30], concrete strength [12,13,22,23,25,26,28,29], number of openings [12,27], thickness of perforated plates [13,21–23,25,27,30], spacing of perforated plates [26], and concrete stress state [12] on the shear bearing capacity of PBL shear connectors have been evaluated in other reports. Moreover, formulas for calculating the shear bearing capacity of PBL shear connectors have been proposed. The influencing factors and function forms considered in each calculation formula are essentially identical, but vary in coefficients. No mature formulas are currently recognized in the industry. Therefore, the formula for the shear bearing capacity of stud shear connectors is highly mature. By contrast, the formula for the shear bearing capacity of the PBL shear connector has yet to be verified by more tests and engineering examples. After the stud and the PBL shear connector are integrated into a composite shear connector, adding both in the shear bearing capacity is evidently inappropriate. Therefore, the determination of the mutual influence of the stud and the PBL shear connector still needs to be clarified by experimental and analytical research.

The ceramsite concrete with an appropriate number of polypropylene fibers presents several advantages, including lightness, high strength, and good seismic performance [31]. However, the physical and mechanical properties of ceramsite concrete still differ from those of ordinary concrete. Thus, the performance of composite structures made from it needs to be further defined using experiments and theoretical analysis. Applying the stud–PBL composite shear connector in a composite structure consisting of steel–ceramsite concrete can form a structural system characterized by lightweight, dead weight, and high bearing capacity. Accordingly, studying its mechanical properties is highly crucial.

This study explores the following three aspects. (i) First, push-out tests of one stud shear connector (LT-S), one PBL shear connector (LT-P), and three stud-PBL composite shear connectors (LT-C) were performed to research the interfacial shear resistance of steel–ceramsite concrete composite structures with different shear connectors. The finite element software Abaqus was then used to simulate push-out tests with 5 samples. The test phenomenon, load–slip curve, shear bearing capacity, stress illustration, and damage illustration (among others) were compared and then verified among one another. Finally, the shear damage mechanism and the failure form of shear connectors were revealed from multiple angles. (ii) Nine groups of orthogonal test simulations with four factors and three levels were conducted for 49 finite element push-out samples. Subsequently, the influences of parameters such as the stud diameter, PBL perforated plate thickness, opening diameter, penetrating steel bar diameter, and concrete strength on the shear resistance of the composite shear connector were analyzed. (iii) Multivariate linear regression analysis was performed on the analysis results of 49 finite element models. Subsequently, the formula for the shear bearing capacity of the stud–PBL composite shear connector in the steel–polypropylene fiber ceramsite concrete composite structure was derived. The formula was then verified by the test results. The objectives of the study are to clarify the shear resistance of composite shear connectors, to obtain their influencing factors, and to establish reasonable calculation formulas. The significance of this paper is to provide a new and scientific reference for the design of steel–concrete composite structure interfaces.

2. Push-Out Experiment

2.1. Experimental Design

To study the shear resistance of the combined arrangement of studs and PBL, five push-out specimens were designed, with reference to Eurocode 4 (EN 1994-2: 2005, Eurocode 4) [32]. They were divided into three groups based on the type of the shear connector: the LT-S group (studs shear connector, one specimen), the LT-P group (PBL shear connector, one specimen), and the LT-C group (composite shear connector with the combined studs and PBL, three specimens).

The polypropylene fiber ceramsite concrete used in the test had a 28-d cube compressive strength of 52.3 MPa and a content of 0.9 kg/m^3 (the volume content was 0.1%). The model of the penetrating steel bar was HPB300 and its diameter was 12 mm. Its measured yield strength, ultimate tensile strength, and elastic modulus of the penetrating steel bar were 328, 480, and 209.2 GPa, respectively. Both H-beam and PBL perforated plates consisted of Q235 steel, and their thickness was 16 mm. They had the following measurements: yield strength, 330.9 MPa; ultimate tensile strength, 440 MPa; elastic modulus, 205.7 GPa. The material of the stud was ML15A1 and the diameter of the rod part of the stud was 16 mm. Its measurements were as follows: yield strength, 360 MPa; ultimate tensile strength, 400 MPa; elastic modulus, 205 GPa. The opening of the perforated plate was set to 40 mm.

The structural form of the LT-C (composite shear connector) specimen is presented in Figure 1. Each sample contained four PBL shear connectors and four stud shear connectors.

To ensure that the studs as flexible connectors could fully exert their shear resistance when the sample was under compression, the studs in the specimen were arranged in the lower part of the specimen. The PBL shear connectors on the upper layer of the LT-C specimen were replaced with studs, which comprised the LT-S specimen. Similarly, the studs on the lower layer of the LT-C specimen were replaced with PBL shear connectors, which comprised the LT-P specimen. (See Table 1).

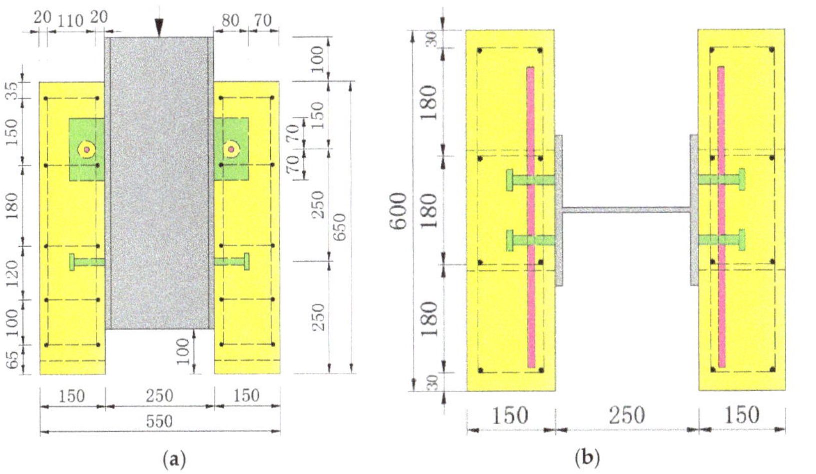

Figure 1. Details of push-out specimen (LT-C) (Unit: mm): (**a**) Side view of LT-C; (**b**) Top view of LT-C.

Table 1. Parameters of push-out test specimen.

Group	Number of Studs in a Specimen	Number of PBL Shear Connectors and Openings in a Specimen	Number of Penetrating Steel Bars in a Specimen
LT-S	8	-	-
LT-P	-	8	4
LT-C	4	4	2

As displayed in Figure 2, a 2000 kN electrohydraulic servo pressure testing machine is used for loading. Three preloads were conducted before the official loading (the loading step was 20 kN. During preloading, after loading to 60 kN, it was unloaded to 0 kN, and the loading rate was 0.5 kN/s). After preloading, monotonic hierarchical loading was employed for the formal loading. First, the loading level distance was set to 1/10 of the limit load, which was calculated using the finite element method. Moreover, the sample was loaded to 80% of the calculated limit load value, step by step (loading rate was 1 kN/s) in accordance with this loading level distance. The loading level distance was then set to 20 kN (0.2 kN/s), and the sample was loaded to the measured value of the limit load. Finally, loading continued after this and the test was terminated when the load was reduced to 85% of the actual ultimate load, or the sample was unloaded after the appearance exhibited obvious damage. The data of dial indicators were recorded, and the development of cracks was observed after each level of loading was stable for 3 min.

Figure 2. Sketch of loading.

2.2. Experimental Phenomena

The test demonstrated that the three types of shear connector specimens included the three working stages: elasticity, plasticity, and failure. At the initial stage of loading, small relative displacement was present between the H-beam and the concrete slab. The load increased substantially and varied linearly with the relative slip value. Diagonal cracks in concrete slabs were small in width and short in length and developed slowly. After the elastic phase, the relative slip value of the specimen and the load increment exhibited a nonlinear behavior. Meanwhile, the width and length of the diagonal cracks continued to increase. The loading progressed gradually, but the relative slip value increased rapidly, and both again exhibited a linear change. In addition, vertical cracks appeared, and each crack developed rapidly to penetrate the entire specimen in the height direction until a loud "bang" was heard from the specimen. Concrete spalling, concrete slab and H-beam detachment, and specimen and loading equipment shaking then occurred. The three types of shear connector specimens ultimately demonstrated different damage phenomena.

Figure 3 presents in detail the failure of the push-out specimen. The studs on one side of the LT-S specimen were sheared from the root, and those on the other side exhibited plastic deformation. Moreover, the studs on both sides indicated large plastic deformation at the root in the H-beam direction. The concrete under the studs was compressed, leading to plastic deformation, and fine cracks were observed at the bottom of the concrete slab. When the LT-P specimen was damaged, no apparent deformation of the perforated plate occurred, and the concrete tenon in the hole was crushed. In addition, the penetrating steel bars between the PBL shear connectors were deformed and yielded in the H-beam direction, and grooves appeared in the corresponding part of the PBL. The bottom of the concrete slab was also crushed. The studs of the LT-C group were broken at the root or had major plastic deformation. The concrete around the studs had plastic deformation, and the concrete tenon was crushed. Similarly, the penetrating steel bars exhibited plastic deformation, and the bottom of the concrete slab was crushed.

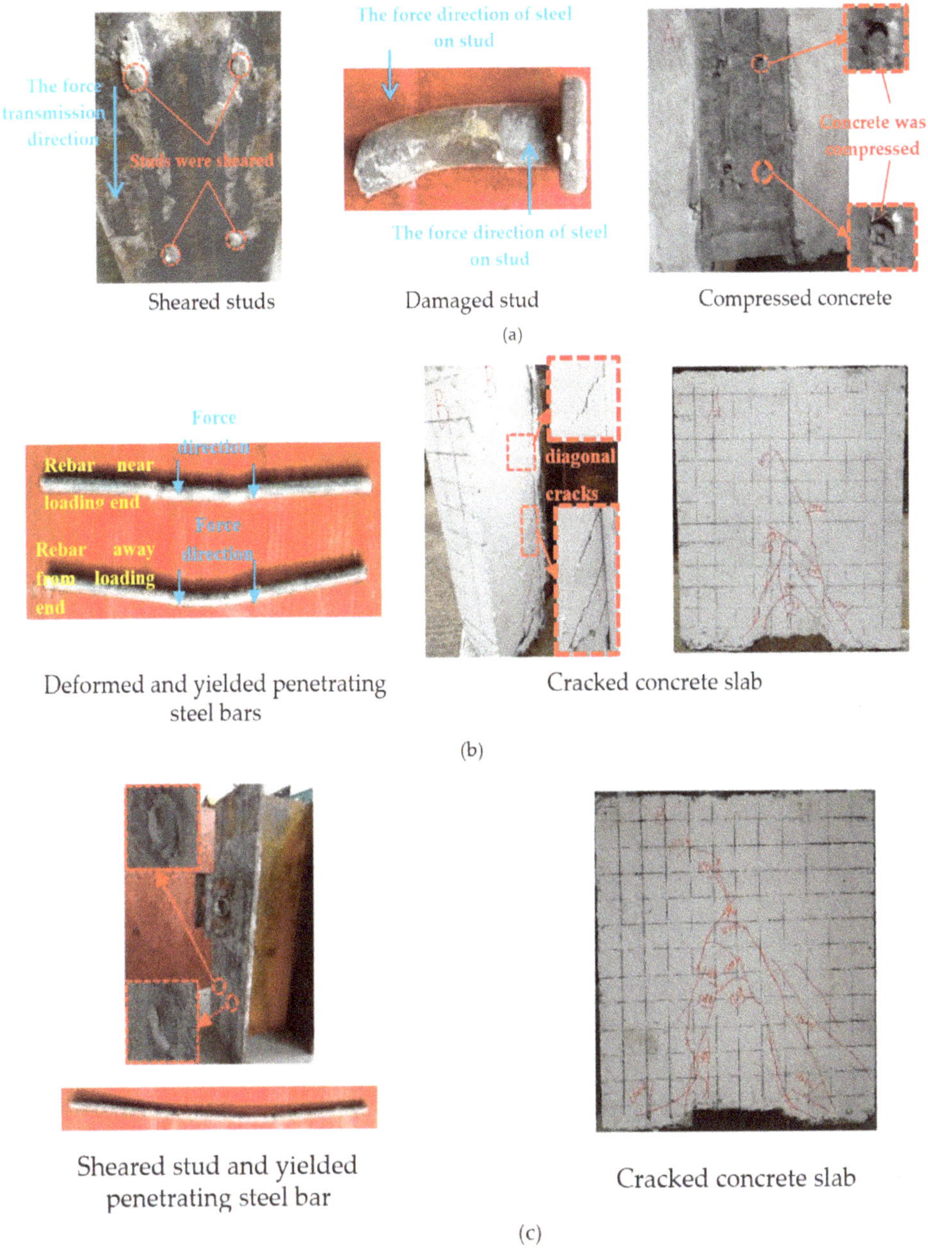

Figure 3. Typical failure model of push-out test specimens: (**a**) The failure details of LT-S specimen; (**b**) The failure details of LT-P specimen; (**c**) The failure details of LT-C specimen.

2.3. Experimental Results

Table 2 listed the shear bearing capacity of the specimens.

Table 2. Results of push-out test and finite element analysis (FEA).

Specimen Number	Test Value P_u/kN	Finite Element Calculation Value $P_{u,fea}$/kN	$P_{u,fea}/P_u$
LT-S	197.3	207.6	1.05
LT-P	325.0	334.0	1.03
LT-C-1	404.0		0.90
LT-C-2	375.1	363.9	0.97
LT-C-3	375.1		0.97
LT-C Average	384.7	363.9	0.95

Noted: P_u in the table was 1/4 of the shear bearing capacity of the test specimen. It meant the specimen LT-S contained 2 studs, the specimen LT-P contained 2 PBL, and the specimen LT-C contained 1 stud and 1 PBL.

As listed in Table 2, the measured shear bearing capacities of LT-S and LT-P are 197.3 and 325.0 kN, respectively. By superimposing the bearing capacity of a single stud and a single PBL, $(P_{u,\,LT\text{-}S} + P_{u,\,LT\text{-}P})/2 = 261.2$ kN (1 stud + 1 PBL)—that is less than the average shear capacity of LT-C (384.7 kN). This observation indicates the high shear resistance of the LT-C shear connector. The shear bearing capacity of LT-C was 18.4% higher than that of LT-P. The reason might be attributed to the high strength of the PBL shear connector. In addition, the concrete at the bottom of the perforated plate was in a state of concentrated stress, rendering it prone to cracking. Consequently, the effective bearing area of the concrete slab was reduced. Therefore, the shear bearing capacity of the specimen was decreased [9]. However, the flexible studs in LT-C could relieve the stress concentration of the rigid PBL to a certain extent to fully use the advantages of both the PBL and studs. This finding also suggests that the shear connector composed of studs and PBL could be better combined with the concrete slab and achieve the optimal performance of both the shear connector and the concrete slab.

Figure 4 illustrates the load–slip curve measured in the test. Only the load–slip curves of the samples in the elastic and plastic working stages were generated because of limitations to the equipment. The plastic stage of LT-S was the longest, followed by that of LT-C, and that of LT-P was the shortest. This reveals that under the condition of certain concrete strength, replacing the lower perforated plate with studs could improve the deformation performance of the push-out specimen. In the elastic stage, the stiffness of LT-S was significantly lower than that of LT-P, whereas the stiffness of the three LT-C specimens was similar to that of LT-P. It could be observed that the LT-C specimens were superior to the LT-P specimen in shear bearing capacity and ductility. However, they still had the same level of stiffness as that of LT-P.

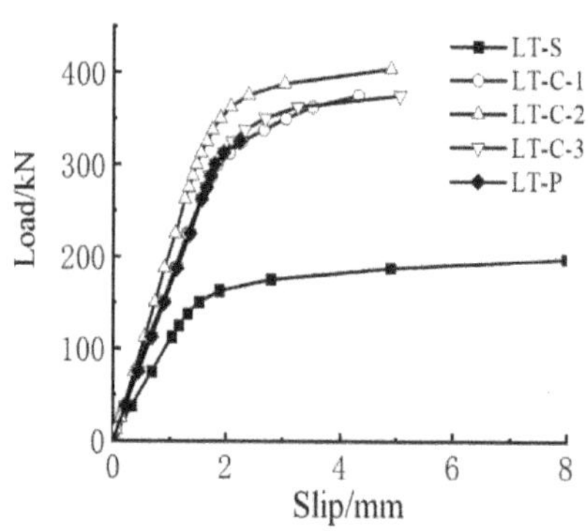

Figure 4. Load–slip curves of the push-out test.

After comprehensive analysis of the shear bearing capacity and deformation capacity of the three specimens, it was discovered that the composite shear connectors could improve the integrity of the concrete slab, and we also made full use of the flexibility of the stud and the rigidity of the PBL shear connector to improve the interface force performance. This made the LT-C specimen have the highest shear bearing capacity and stiffness while keeping an excellent deformation capacity.

3. Push-Out Experimental Finite Element Numerical Simulation

3.1. Overview of the Finite Element Model

As shown in Figure 5, a 1/4 model was established based on the actual size of the test sample, including H-beam, concrete slab, PBL shear connector, stud, penetrating steel bar, and rebar cages.

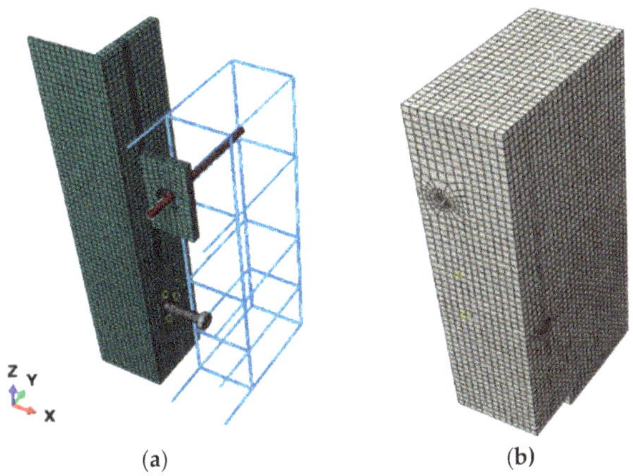

Figure 5. 1/4 Finite element model (FEM) of push-out test for the composite shear connector: (**a**) The model of Steel member and steel bars; (**b**) Concrete slab model.

(1) Material constitutive model

The plastic damage model was selected as the concrete material model in the Abaqus finite element software. Polypropylene fibers have previously been proven to strongly influence the control parameters of the descending section of the concrete stress–strain curve section [33,34]. Therefore, the stress–strain relationship of the concrete used in the test adopted the formula suggested by Ding [35], and the Poisson's ratio was taken as 0.2. Moreover, the damage factor was obtained by referring to the calculation method previously studied [36].

The stress–strain relationship of H-beam, the PBL shear connector, the penetrating steel bar, and the rebar cage adopted the tri-polyline model, and that of the studs used the double-line model. In addition, the elastic modulus used the measured value in Section 2.1, and Poisson's ratio was calculated as 0.3. All of the above used the von Mises yield surface to define isotropic yielding.

(2) Element type and mesh division

The H-beam, PBL shear connector, stud, concrete slab, and penetrating steel bar were simulated using the 3D eight-node reduced-integration element (C3D8R). The rebar cage was simulated using the 3D two-node truss element (T3D2). The size of the seeds was 10 mm. The meshes in the contact areas of different components were refined and divided.

(3) Interaction and boundary conditions

Bond contacts were applied between the penetrating steel bar and the surrounding concrete. The rebar cage was placed inside the concrete slab. The H-shaped steel was combined with stud shear connectors and PBL shear connectors to form a component. General contact was also used to simulate contact with concrete. The contact property was normal "hard" contact, and tangential friction was simulated using the penalty function.

Two symmetric surfaces of the model were positive symmetric constraints. After the top surface of the H-shaped steel was fixed (the actual test was the upward loading force on the bottom platform of the sample, and the simulation was consistent with the test), an upward displacement load was applied on the bottom surface of the concrete.

3.2. *Analysis of Finite Element Results*

Figure 6 indicates the failure mode of the finite element simulation of the samples, including the von Mises stress diagram of the steel member and the damagec and damaget diagrams of the concrete when the specimen reached the shear bearing capacity. The von Mises stress illustration indicates the following: (i) Among the three types of shear connectors, the stud root had the largest stress, which was consistent with the stud in Figure 3. (ii) The stress level of PBL steel plate was small, and practically no significant deformation occurs. (iii) The penetrating steel bar located within the PBL opening indicated the largest stress, which was consistent with the larger yield bending deformation of the penetrating steel bar in the test. The illustrations of damagec (LT-S specimen) and damaget (LT-P and LT-C specimens of the concrete) indicate that the concrete around the studs has different degrees of compressive damage. However, the concrete at the bottom of PBL was mostly damaged by tension. These were consistent with the failure modes such as plastic deformation of the concrete near the studs due to compression, and cracking of the concrete slab in the test. The above demonstrated that the failure modes of the samples simulated by the finite element were consistent with the test results.

Figure 7 presents the load–slip curves of the five specimens in Table 2, which were generated by finite element simulation. The rising stage of the simulated curve roughly resembled the experimental curve. Moreover, the curve variation conditions of the descending section were obtained by the element simulation. After the peak load was reached, the LT-P sample immediately entered the descending stage. In this stage, with the increase of the slip value, the load value displayed a tendency of first accelerating down and then decelerating down. However, the LT-C specimen did not enter the decreasing phase immediately after reaching the peak load; it entered the holding stage first. In this stage, the load changed very slowly with the slip value, and the decreasing trend was not obvious. The decreasing stage of LT-C specimen was similar to that of LT-P.

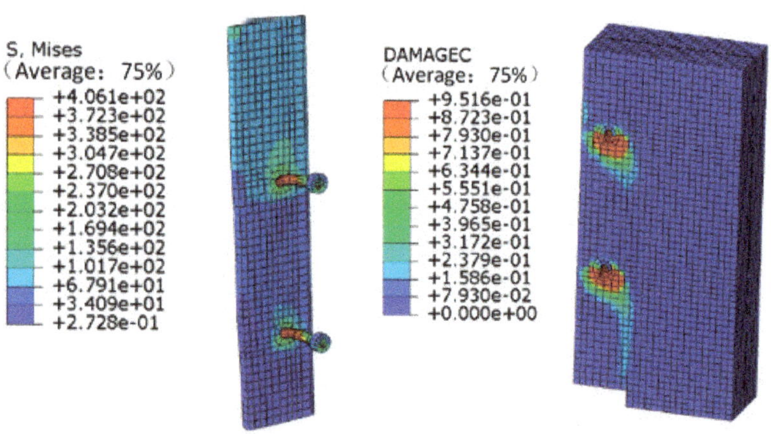

(a1) (a2)

Figure 6. *Cont.*

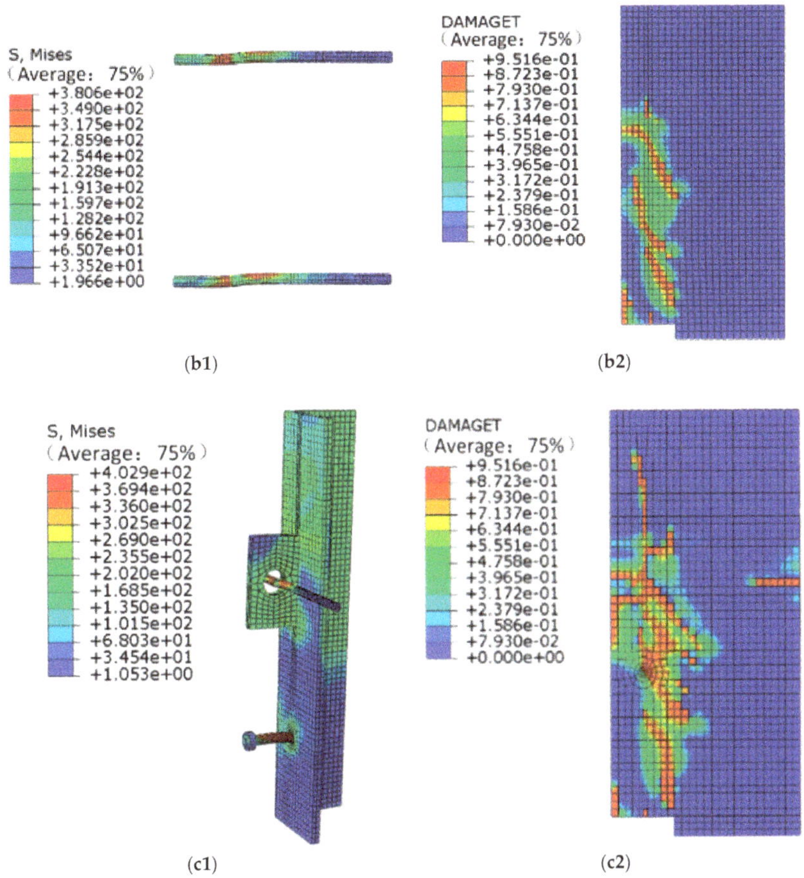

Figure 6. Von Mises stress and damage illustration of FEA models: (**a1**) The stress distribution of penetrating steel bars; (**a2**)The compressive damage to concrete; (**b1**) The stress distribution of penetrating steel bars; (**b2**) The Tensile damage to concrete slab; (**c1**) The Stress distribution of stud penetrating steel bar; (**c2**) The Tensile damage to concrete slab.

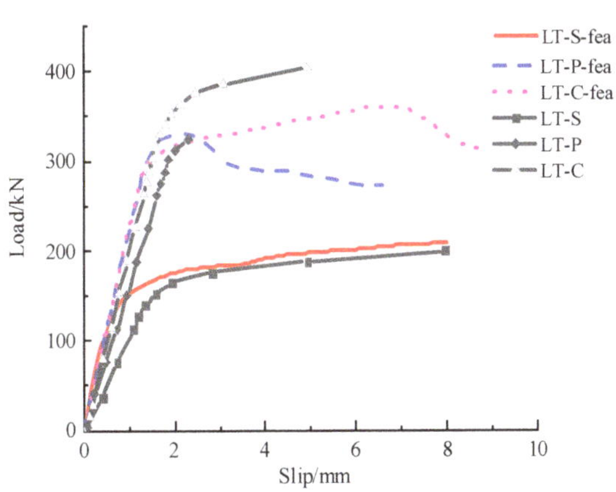

Figure 7. Load–slip curves of FEA models and push-out specimens.

The EC4 specification stipulates that the slip value corresponding to 0.9 Pu in the descending section of the load–slip curve of the shear connector is the slip characteristic value. The characteristic slip value of the composite shear connector exceeded 6 mm, as stated in the specification, which could be determined easily from Figure 7. Therefore, the composite shear connector had good deformation ability.

The shear bearing capacity results of the specimens obtained from the test and the finite element simulation (Table 2) illustrated that the composite shear connector possessed the highest shear bearing capacity, followed by the PBL shear connector and then by the stud shear connector. For the PBL shear connector specimen, the ratio of the finite element result to the test result was 1.03; for the stud shear connector specimen, the ratio was 1.05; and for the composite shear connector specimens, the ratio was 0.90–0.97. Therefore, the correlation between the simulated and measured values is excellent.

With the failure mode, load–slip curve, and shear bearing capacity of the specimen considered, the finite element simulation was consistent with the experimental phenomenon and results. Thus, finite element simulation was a feasible tool to reveal the force mechanism of the composite shear connector and conduct parametric analysis.

4. Finite Element Parameters Analysis

4.1. Parameters Design of Orthogonal Test Analysis Model

A total of 49 finite element models were established to evaluate the effects of different connector structures and different material parameters on the shear resistance of the specimens. According to existing research, the structural parameters set in this paper included the stud diameter, perforated plate opening diameter, perforated plate thickness, penetrating steel bar diameter, and concrete strength. Although the laws and mechanisms of the above parameters on the shear capacity of shear connectors have been revealed, they were for PBL shear connectors or stud shear connectors. For stud–PBL composite shear connectors, the above parameters were set in one member at the same time, and the effect on the shear resistance of the member would be variable and unclear, which was valuable to study. The first four factors were used to design nine groups of orthogonal experiments with four factors and three levels. In addition, the finite element analysis of changing a single parameter in the same finite element model (RP-C52.3) as the test sample construction was added. Table 3 lists the results of the finite element parametric analysis.

Table 3. FEA parametric analysis results of shear bearing capacity of composite shear connectors.

Group	Stud Diameter d_s/mm	Opening Diameter D/mm	Perforated plate Thickness t/mm	Penetrating Steel Bar Diameter d_r/mm	Concrete Cube Compressive Strength f_{cu}/MPa	FEA Shear Bearing Capacity Values $P_{u,fea}$/kN	Equation (6) Calculated Value of Shear Bearing Capacity $P_{u,c}$/kN	$P_{u,c}/P_{u,fea}$
1	13	30	16	12	40	287.4	229.31	0.80
					50	327.1	266.63	0.82
					60	355.0	303.37	0.85
2	13	40	20	16	40	326.2	289.21	0.89
					50	370.2	336.10	0.91
					60	388.0	382.27	0.99
3	13	50	24	20	40	345.1	356.92	1.03
					50	384.9	413.77	1.08
					60	401.0	469.73	1.17
4	16	30	20	20	40	349.9	324.60	0.93
					50	386.6	370.96	0.96
					60	408.3	416.58	1.02
5	16	40	24	12	40	339.1	341.59	1.01
					50	380.9	398.39	1.05
					60	417.2	454.24	1.09
6	16	50	16	16	40	325.2	307.75	0.95
					50	361.4	350.14	0.97
					60	391.7	391.41	1.00
7	19	30	24	16	40	349.4	380.04	1.09
					50	399.3	436.70	1.09
					60	437.2	492.34	1.13
8	19	40	16	20	40	335.7	345.87	1.03
					50	356.3	387.56	1.09
					60	378.6	428.11	1.13
9	19	50	20	12	40	339.2	366.93	1.08
					50	375.1	419.54	1.12
					60	401.9	470.75	1.17
RP	16	40	16	12	30	281.6	234.39	0.83
					40	327.9	276.38	0.84
					52.3	363.9	326.37	0.90
					60	388.7	356.43	0.92

Table 3. *Cont.*

Group	Stud Diameter d_s/mm	Opening Diameter D/mm	Perforated plate Thickness t/mm	Penetrating Steel Bar Diameter d_r/mm	Concrete Cube Compressive Strength f_{cu}/MPa	FEA Shear Bearing Capacity Values $P_{u,fea}$/kN	Equation (6) Calculated Value of Shear Bearing Capacity $P_{u,c}$/kN	$P_{u,c}/P_{u,fea}$
					40	300.9	243.55	0.81
LT-d_s13	13	40	16	12	52.3	355.9	291.61	0.82
					60	377.2	320.81	0.85
					40	335.2	316.01	0.94
LT-d_s19	19	40	16	12	52.3	372.9	368.31	0.99
					60	376.6	399.42	1.06
					40	312.4	262.14	0.84
LT-D30	16	30	16	12	52.3	360.8	310.08	0.86
					60	354.8	338.99	0.96
					40	300.3	294.69	0.98
LT-D50	16	50	16	12	52.3	351.8	347.30	0.99
					60	346.9	378.86	1.09
					40	333.7	308.99	0.93
LT-t20	16	40	20	12	52.3	375.7	369.00	0.98
					60	398.0	405.34	1.02
LT-t24	16	40	24	12	52.3	389.4	411.62	1.06
LT-d_r16	16	40	16	16	52.3	363.1	339.10	0.93
LT-d_r20	16	40	16	20	52.3	361.8	355.48	0.98

P_u in the table was 1/4 of the shear bearing capacity of the push-out specimen—that is, the specimen LT-C contained one stud and one PBL.

4.2. Analysis of the Orthogonal Test Simulation Results

With the C50 concrete grade as an example, the results in Table 3 were analyzed using the orthogonal test results of the experiment (4 factors, 3 levels).

As displayed in Table 4, K_{ij} is the sum of P_u corresponding to the level of factor i in column j, and $\kappa_{ij} = K_{ij}/3$. The optimal solution of each column of factors could be determined by analyzing the magnitude of κ_{ij}. Moreover, combining them yielded the optimal combination. Thus, the optimal structure of the composite shear connector was ds19-D50-t24-d16, which was obtained easily. Among them, the range R_j denotes the difference between the maximum value and the minimum value in the jth column factor, which can reflect the change in P_u when the level of the jth column factor changes. R_j in Table 4 indicates that the thickness of the perforated plate affects P_u the most, whereas the diameter of the opening influences P_u the least.

Table 4. Analysis of orthogonal test results.

Specimen Number	$j = 1$ Stud Diameter d_s/mm	$j = 2$ Opening Diameter D/mm	$j = 3$ Perforated Plate Thickness t/mm	$j = 4$ Penetrating Steel Bar Diameter d_r/mm	FEA Shear Bearing Capacity Values $P_{u,fea}$/kN
1-C50	13	30	16	12	327.1
2-C50	13	40	20	16	370.2
3-C50	13	50	24	20	384.9
4-C50	16	30	20	20	386.6
5-C50	16	40	24	12	380.9
6-C50	16	50	16	16	361.4
7-C50	19	30	24	16	399.3
8-C50	19	40	16	20	356.3
9-C50	19	50	20	12	375.1
K_{1j}	1082.2	1113	1044.8	1083.1	
K_{2j}	1128.9	1107.4	1131.9	1130.9	$\bar{P}_u = 371.3$
K_{3j}	1130.7	1121.4	1165.1	1127.8	
κ_{1j}	360.7	371	348.3	361.0	
κ_{2j}	376.3	369.1	377.3	377.0	
κ_{3j}	376.9	373.8	388.4	375.9	
R_j	16.2	4.7	40.1	15.9	
S_j	504.0	33.1	2573.4	476.9	$S_j = \sum_{i=1}^{3} (P_{ui} - \bar{P}_u)^2$
S_e	/	33.1	/	/	
F	15.2	/	77.7	14.4	$F = \frac{S_j}{f_j} / \frac{S_e}{f_e}$

P_u in the table is 1/4 of the shear bearing capacity of the push-out specimen—that is, the specimen LT-C contains 1 stud and 1 PBL.

The results of the variance analysis indicate that the test could be used to assess for significance. All columns of this orthogonal table contain factors; thus, the sum of the squared deviation S_j of the factors in the second column is regarded as the sum of the squared deviations of the errors S_e. f_j is the degree of freedom of the jth column factor, and f_j is equal to the number of levels of the j factor minus 1. The ratio F of the variance of each column factor could also be calculated. The factors could then be tested for significance $F \sim F_{\alpha}(f_j/f_e)$.

The results in Table 4 indicate that $F_3 > F_1 > F_4 > F_{0.10}$ (2,2) = 9.00, indicating that the thickness of the perforated plate, the diameter of the stud, and the diameter of the penetrating steel bar significantly affect shear capacity at $\alpha = 0.10$. Among the parameters, the thickness of the perforated plate has the greatest influence on the shear bearing capacity, followed by the diameter of the stud and then by the diameter of the penetrating steel bar.

4.3. Analysis of the Influencing Factors on the Shear Bearing Capacity of Composite Shear Connectors

By further parameter analysis of the model, the influencing factors of the shear bearing capacity of the composite shear connector were identified (Figure 8).

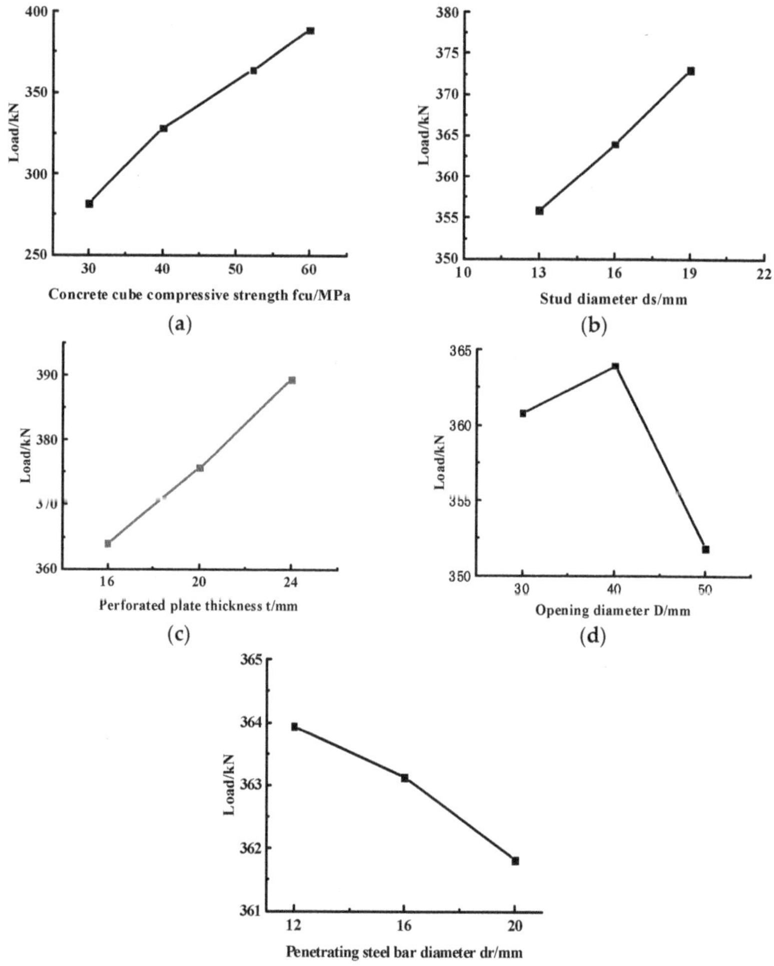

Figure 8. Results of FEM parametric analysis: (**a**) Influence of concrete cube compressive strength; (**b**) Influence of stud diameter; (**c**) Influence of perforated plate thickness; (**d**) Influence of opening diameter; (**e**) Influence of penetrating steel bar diameter.

(1) Influence of the strength of the concrete

The shear bearing capacities of the specimens RP-C30, RP-C40, RP-C52.3, and RP-C60 were 281.6, 327.93, 363.93, and 388.74 kN, respectively (Table 3 and Figure 8a). These results indicate that increasing the compressive strength of the concrete could improve the shear bearing capacity of the composite shear connector. When the concrete strength was increased from 30 to 52.3 MPa, the shear bearing capacity improved by about 29.2%.

(2) Influence of the diameter of the stud

The shear bearing capacities of the specimens LT-ds13-C52.3, RP-C52.3, and LT-ds19-C52.3 were 355.88, 363.93, and 372.94 kN, respectively (Table 3 and Figure 8b). These results indicate that increasing the diameter of the stud could improve the shear bearing capacity of the composite shear connector. When the diameter of the stud was increased from 13 mm to 19 mm, the shear bearing capacity improved by about 4.8%.

(3) Influence of the thickness of the perforated plate

The shear bearing capacities of the specimens RP-C52.3 and LT-t24-C52.3 were 363.93 and 389.41 kN, respectively (Table 3 and Figure 8c). These results reveal that increasing the thickness of the perforated plate could improve the shear bearing capacity of the composite shear connector. When the thickness of the perforated plate was increased from 16 mm to 24 mm, the shear bearing capacity improved by about 7%.

(4) Influence of the diameter of the opening

The bearing capacities of the specimens LT-D30-C52.3, RP-C52.3, and LT-D50-C52.3 were 360.78, 363.93, and 351.84 kN, respectively (Table 3 and Figure 8d). These results indicate that increasing the diameter of the opening could improve the shear bearing capacity, but an excessively large diameter could exert adverse effects. The reason was that increasing the diameter could enlarge the shear bearing area of the concrete tenon, thereby improving the shear bearing capacity. However, if the opening was too large, the strength of the perforated plate would decrease, reducing the shear bearing capacity.

(5) Influence of the diameter of the penetrating steel bar

The shear bearing capacities of the specimens RP-C52.3, LT-dr16, and LT-dr20 were 363.93, 363.12, and 361.82, respectively (Table 3 and Figure 8e). This demonstrated that increasing the diameter of the penetrating steel bar exerted no noticeable effect on the shear bearing capacity. The reason was that the shear bearing area of the concrete tenon in the hole would decrease with an increase in the diameter of the penetrating steel bar under the premise of a constant opening diameter; however, both the shear resistance of the concrete tenon and the flexural resistance of the penetrating steel bar could influence the shear bearing capacity of the specimen.

Based on the aforementioned analysis results, it can be observed that the concrete strength had the greatest influence on the shear bearing capacity, followed by the thickness of the perforated plate. When the concrete strength was constant in the actual project, it was the key to controlling the thickness of the perforated plate considering the structure of the composite shear connector.

5. Calculation Formula for Shear Resistance

The composite shear connector consists of studs and PBL shear connectors. The longitudinal shear resistance consists of the following parts: the dowel action of the concrete tenon in the hole of the perforated plate and the penetrating steel bar, the bearing effect of the bottom of the perforated plate, and the mechanical engagement of the stud. More research and theoretical analyses regarding these two shear connectors has been conducted by domestic and foreign scholars.

First, the "Code for Design of Steel-Concrete Composite Bridges" (GB 50917-2013) specified the formula for calculating the shear bearing capacity when the stud breaks:

$$P_u = 1.19A_{st}\left(\frac{E_c}{E_s}\right)^{0.2}\left(\frac{f_{cu}}{f_{st}}\right)^{0.1} \tag{1}$$

where P_u is the shear bearing capacity (N) when the stud breaks; A_{st}, f_{st}, E_s are the cross-sectional area (mm^2), ultimate tensile strength (MPa), and elastic modulus (GPa) of the stud, respectively; f_{cu} and E_c are the cubic compressive strength (MPa) and elastic modulus (GPa) of concrete, respectively.

Moreover, Xue [13] produced the calculation formula of shear bearing capacity under the failure mode of stud rupture considering the length to diameter ratio of studs by push-the experimental study as follows:

$$P_u = 3A_{st}\left(\frac{E_c}{E_s}\right)^{0.4}\left(\frac{f_{cu}}{f_{st}}\right)^{0.2}$$
$$\lambda = \begin{cases} 6 - \frac{H}{1.05d_s} & (H/d \leq 5) \\ 1 & (5 < H/d < 7) \\ \frac{H}{d_s} - 6 & (H/d \geq 7) \end{cases} \tag{2}$$

where *H/ds* is the aspect ratio of the stud, and the meanings of other symbols are the same as those of Equation (1).

Subsequently, Al-Darzi [37] performed a parametric analysis of PBL shear connectors by using the Ansys software. Based on the multiple linear regression analysis of the simulation results, the formula for calculating the shear bearing capacity of the PBL shear connectors with concrete end bearing effect is expressed as follows:

$$P_u = 255.31 + 7.62 \times 10^{-4}htf_c{}' - 7.59 \times 10^{-7}A_r f_{ry} + 2.53 \times 10^{-3}A_{sc}\sqrt{f_c{}'} \tag{3}$$

where P_u is the shear bearing capacity of PBL shear connector with concrete end bearing effect (kN); h and t are the height and thickness of the perforated plate (mm); A_{sc} and $f_c{}'$ are the area of the concrete tenon in the hole [2nπ(D2-dr2)/4, where n is the number of openings] (mm^2) and cylinder compressive strength of concrete (MPa); A_r and f_{ry} are the cross-sectional area (mm^2) and yield strength (MPa) of the penetrating steel bar, respectively.

In addition, Zhang [38] conducted a multiple linear regression analysis of the experimental and finite element simulation results and then proposed a formula for calculating the shear bearing capacity of the intermittently arranged PBL shear connectors as follows:

$$P_u = 3.5htf_{cu,k} + 32A_{sc}\sqrt{f_{cu,k}} + 5.2A_r f_{ry} \tag{4}$$

where P_u is the shear bearing capacity of the intermittently arranged PBL shear connectors; $f_{cu,k}$ is the standard value of cubic compressive strength of concrete (MPa); the meanings of other symbols are the same as those of Equation (3).

Therefore, based on the aforementioned research results and the failure form of the test specimens, the formula for calculating the shear bearing capacity of the composite shear connector is proposed as follows:

$$P_u = \beta_1 htf_{cu} + \beta_2 A_r f_{ry} + \beta_3 A_{sc} f_{cu}{}^{0.5} + \beta_4 A_{st}\left(\frac{E_c}{E_s}\right)^{0.2}\left(\frac{f_{cu}}{f_{st}}\right)^{0.1} \tag{5}$$

where P_u is the shear bearing capacity of the combined shear key (N); h and t are the height and thickness of the perforated plate (mm); A_r and A_{st} are the cross-sectional area of the penetrating steel bar and the stud rod, respectively (mm^2); A_{sc} is the area of the concrete tenon in the hole [2nπ(D2-dr2)/4, where n is the number of openings] (mm^2); f_{cu} is the cubic compressive strength of concrete (MPa); f_{ry} is the yield strength (MPa) of the penetrating steel bar; f_{st} is the ultimate tensile strength of the stud (MPa); E_c and E_s are the elastic modulus (GPa) of concrete and stud, respectively; β_1, β_2, β_3, β_4 are undetermined coefficients.

The results of the 49-model analysis in Table 3 were subjected to multivariate linear regression on Equation (5), resulting in the formula for calculating the shear bearing capacity of the composite shear connectors as Equation (6). The Pearson correlation coefficient of the regression analysis was 0.87 (see Figure 9), which demonstrates that the calculated value is consistent with the value obtained using finite element analysis.

$$P_u = 2.55htf_{cu} + 5.21 \times 10^{-1}A_r f_{ry} + 2.05A_{sc}f_{cu}{}^{0.5} + 2.34A_{sc}f_{st}\left(\frac{E_c}{E_s}\right)^{0.2}\left(\frac{f_{cu}}{f_{st}}\right)^{0.1} \quad (6)$$

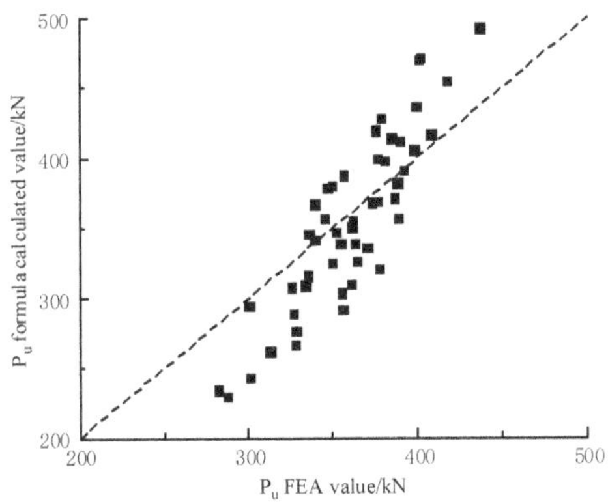

Figure 9. Results comparison of shear capacity Equation (6) and FEA.

Table 5 compares the measured shear bearing capacity of the specimens and the results calculated using Equation (6). The outcome indicates that the calculation results of Equation (6) are well correlated with the test results of the composite shear connector push-out specimens; however, the calculation results are conservative—about 0.85 times the measured value. With the safety factor considered, Equation (6) could be used to calculate the shear bearing capacity of the composite shear connector in practical engineering.

Table 5. Results comparison of Equation (6) and push-out test.

Specimen Number	The Test Measured Value of Shear Bearing Capacity P_u/kN	The Calculated Value of Equation (6) $P_{u,c}$/kN	$P_{u,c}/P_u$
LT-C-1	404.0	326.3	0.81
LT-C-2	375.1	326.3	0.87
LT-C-3	375.1	326.3	0.87
LT-C	384.7	326.3	0.85

P_u in the table is 1/4 of the shear bearing capacity of the push-out specimen—that is, the specimen LT-C contains one stud and one PBL.

6. Conclusions

Through the tests and numerical simulations of the shear bearing performance of the steel–polypropylene fiber ceramsite composite bonding interface, the following major conclusions are drawn:

(1) The push-out test on one stud shear connector (LT-S), one PBL shear connector (LT-P), and three stud–PBL composite shear connectors (LT-C) reveals the superior shear resistance of the composite shear connector. In addition, composite shear connectors have excellent plastic deformation properties.

(2) The finite element software Abaqus was used to simulate the push-out test with five specimens. The analysis results were in agreement with the experimental phenomenon, load–slip curve, and shear bearing capacity of the push-out test. In addition, the illustrations of stress and concrete damage can reveal the shear damage mechanism and failure form of the shear connector. It is feasible to use finite element simulation to reveal the force mechanism of the composite shear connector and perform parametric analysis.

(3) According to the results of the orthogonal test simulation analysis, increases in the stud diameter, perforated plate thickness, and concrete strength can improve the shear bearing capacity of the composite shear connector, among which the concrete strength has the most obvious effect. However, an increase in the diameter of the penetrating steel bar exerts no apparent effects on shear capacity. Regarding the opening diameter, the shear bearing capacity increases first and then decreases with its increase. When the diameter of the opening was nearly 40 mm, the shear bearing capacity was more favorable. The optimal construction combination for the composite shear connector is the stud with 19 mm diameter, the opening with 50 mm diameter, the perforated plate with 24 mm thickness, and the penetrating steel with 16 mm diameter. Moreover, when the concrete strength was constant in the actual project, it was the key to controlling the thickness of the perforated plate, considering the structure of the composite shear connector.

(4) By conducting multivariate linear regression analysis on the results of the finite element parameter analysis, the calculation formula of the shear bearing capacity of the composite shear connector was obtained. The calculated value is about 85% of the measured value, which displays a good correlation and verifies the feasibility of the calculation formula. The calculation results are conservative, and a formula can be used to calculate the shear bearing capacity of the composite shear connector in practical engineering with the safety factor considered.

(5) The experimental results, numerical simulation outcomes, and calculation formulas have a certain promoting effect on further reducing the self-weight of the composite structure and improving the shear resistance of the composite structure interface. However, the number of push-out test specimens is relatively small. In subsequent studies, experiments will be performed to further discuss the shear resistance of the stud–PBL composite shear connector and the method for calculating the shear bearing capacity.

Author Contributions: Conceptualization, H.Z. and Z.F.; methodology, H.Z.; software, P.L.; validation, Z.F. and B.Z.; formal analysis, Z.F.; investigation, Z.F.; resources, Y.L. and B.Z.; data curation, Z.F., Y.L. and B.Z.; writing—original draft preparation, Z.F. and Y.L.; writing—review and editing, H.Z. and Z.F.; visualization, Y.L.; supervision, H.Z.; project administration, H.Z.; funding acquisition, H.Z. All authors have read and agreed to the published version of the manuscript.

Funding: This study was funded by the National Natural Science Foundation of China (grant numbers 51778630, 52178182, and U1934217), China Railway Science and Technology Research and Development Plan Project (grant numbers 2020-Major project-02, 2021-Major project-02, 2021-Key projects-11).

Institutional Review Board Statement: Not applicable.

Informed Consent Statement: Not applicable.

Data Availability Statement: The data presented in this study are available on request from the corresponding author.

Conflicts of Interest: The authors declare no conflict of interest.

References

1. Liang, F.; Zhou, Z.X. New Shear Connectors Based on PBL Shear Connector for Composite Arch Members. *Struct. Eng. Int.* **2014**, *2*, 281–284.
2. Wu, F.; Liu, S.; Xue, C.; Yang, K.; Feng, Y.; Zhang, H. Experimental Study on the Mechanical Properties of Perfobond Rib Shear Connectors with Steel Fiber High Strength Concrete. *Materials* **2021**, *14*, 3345. [CrossRef] [PubMed]
3. Zhan, Y.L.; Yin, C.; Liu, F.; Song, R. Push out tests on headed studs and PBL shear connectors considering external pressure. *J. Bridge Eng.* **2020**, *25*, 04019125. [CrossRef]
4. Fang, Z.C.; Jiang, H.B.; Chen, G.F.; Dong, X.T.; Shao, T.F. Behavior of grouped stud shear connectors between precast high-strength concrete slabs and steel beams. *Steel Compos. Struct. Int. J.* **2020**, *34*, 837–851.
5. Kim, S.H.; Jung, C.Y.; Ann, J.H. Ultimate strength of composite structure with different degrees of shear connection. *Steel Compos. Struct. Int. J.* **2011**, *11*, 375–390. [CrossRef]
6. Ahn, J.H. Shear resistance of the perfobond-rib shear connector depending on concrete strength and rib arrangement. *J. Constr. Steel. Res.* **2010**, *66*, 1295–1307. [CrossRef]
7. Zhang, J.D.; Shoji, A.; Kobayashi, K. Design of a PC box girder bridge with corrugated steel webs in Hanshin expressway. In Proceedings of the First International Structural Engineering and Construction Conference, Honolulu, HI, USA, 24–27 January 2001; pp. 581–586.
8. Deng, W.Q.; Gu, J.C.; Liu, D.; Hu, J.; Zhang, J. Study of single perfobond rib with head stud shear connectors for a composite structure. *Mag. Concrete. Res.* **2019**, *71*, 920–934. [CrossRef]
9. Zheng, S.J.; Zhao, C.; Liu, Y.Q. Parametric push-out analysis on perfobond rib with headed stud mixed shear connector. *Adv. Civ. Eng.* **2019**, *2019*, 1–16. [CrossRef]
10. Jin, L.D. Analysis on spatial mechanical properties of steel-concrete segment with composite connectors. *Steel. Constr.* **2019**, *34*, 69–74.
11. Chen, H.; Guo, Z.X.; Liu, Y.; Guo, L.-T. Study on the shear resisting mechanism and strength for an innovative composite shear connector. *Eng. Mech.* **2019**, *36*, 159–168.
12. Gu, J.C.; Liu, D.; Deng, W.Q.; Zhang, J.D. Experimental study on the shear resistance of a comb-type perfobond rib shear connector. *J. Constr. Steel. Res.* **2019**, *158*, 279–289. [CrossRef]
13. Yang, Y.; Chen, Y. Experimental study on mechanical behavior of PBL shear connectors. *J. Bridge. Eng.* **2018**, *23*, 04018062. [CrossRef]
14. Xue, W.C.; Ding, M.; Wang, H.; Ziwen, L. Static behavior and theoretical model of stud shear connectors. *J. Bridge. Eng.* **2008**, *13*, 623–634. [CrossRef]
15. Young, H.L.; Min, S.K.; Heecheul, K.; Dae-Jin, K. Shear resistance of stud connectors in high strength concrete. *Struct. Eng. Mech.* **2014**, *52*, 647–661.
16. Hu, Y.; Yin, H.; Ding, X.; Li, S.; Wang, J.Q. Shear behavior of large stud shear connectors embedded in ultra-high-performance concrete. *Adv. Struct. Eng.* **2020**, *23*, 136943322093920. [CrossRef]
17. Huo, J.; Wang, H.; Zhu, Z.; Liu, Y. Experimental Study on Impact Behavior of Stud Shear Connectors between Concrete Slab and Steel Beam. *J. Struct. Eng.* **2018**, *144*, 04017203. [CrossRef]
18. Su, Q.T.; Yang, G.T.; Bradford, M.A. Static behaviour of multi-row stud shear connectors in high- strength concrete. *Steel Compos. Struct. Int. J.* **2014**, *17*, 967–980. [CrossRef]
19. Xu, C.; Sugiura, K.; Masuya, H.; Hashimoto, K. Experimental Study on the Biaxial Loading Effect on Group Stud Shear Connectors of Steel-Concrete Composite Bridges. *J. Bridge. Eng.* **2015**, *20*, 04014110. [CrossRef]
20. Xu, C.; Sugiura, K. FEM analysis on failure development of group studs shear connector under effects of concrete strength and stud dimension. *Eng. Fail. Anal.* **2013**, *35*, 343–354. [CrossRef]
21. Ding, J.; Zhu, J.; Kang, J.; Wang, X. Experimental study on grouped stud shear connectors in precast steel- UHPC composite bridge. *Eng. Struct.* **2021**, *242*, 112479. [CrossRef]
22. Zhang, Y.Z.; Li, Q.; Xia, S. Research on the Influence Factors to the Working Performance of PBL Shear Connectors. *Appl. Mech. Mater.* **2012**, *178–181*, 2192–2198. [CrossRef]
23. Xiao, L.; Qiang, S.Z.; Xu, X. Experimental Research on PBL Shear Connector for Steel-Concrete Composite Structures. *Adv. Mater. Res.* **2011**, *163–167*, 2137–2141. [CrossRef]
24. Zhang, Z.H.; Liu, P.; He, S.H.; Ji, T.G. Research on Influence Factors and Formulae for the Bearing Capacity of Twin-PBL Shear Connectors. *Appl. Mech. Mater.* **2013**, *253–256*, 3167–3171. [CrossRef]
25. Yu, B.C.; Zhang, Y.L. The Capacity Calculation and Finite Element Analysis to PBL Shear Key. *Appl. Mech. Mater.* **2014**, *501–504*, 722–726. [CrossRef]
26. Zhang, W.J.; Du, E.X.; Liu, B.Z.; Yang, S.C. Failure Mode Attribution Analysis of PBL Shear Connector. *Appl. Mech. Mater.* **2013**, *351–352*, 683–686. [CrossRef]
27. Zhao, C.; Li, Z.; Deng, K.; Wang, W. Experimental investigation on the bearing mechanism of Perfobond rib shear connectors. *Eng. Struct.* **2018**, *159*, 172–184. [CrossRef]
28. Li, Z.X.; Zhao, C.H.; Deng, K.L.; Wang, W. Load Sharing and Slip Distribution in Multiple Holes of a Perfobond Rib Shear Connector. *J. Struct. Eng.* **2018**, *144*, 264–276. [CrossRef]

29. Zhang, J.; Hu, X.; Kou, L.; Zhang, B.; Jiang, Y.; Yu, H. Experimental study of the short-term and long-term behavior of perfobond connectors. *J. Constr. Steel Res.* **2018**, *150*, 462–474. [CrossRef]
30. Wang, X.; Zhu, B.; Cui, S.; Lui, E.M. Experimental Research on PBL Connectors Considering the Effects of Concrete Stress State and Other Connection Parameters. *J. Bridge Eng.* **2018**, *23*, 04017125. [CrossRef]
31. Du, C.; Tian, W.L.; Wang, X.W.; Wang, D.J. Experimental Research on Ceramsite Concrete Beams. *Appl. Mech. Mater.* **2012**, *166–169*, 708–711. [CrossRef]
32. Eurocode 4: Design of Composite Steel and Concrete Structures – Part 2: General Rules and Rules for Bridges. European Standard: Brussels, Belgium, 2005; pp. 56–66.
33. Xu, L.H.; Li, B.; Ding, X.X.; Chi, Y.; Li, C.; Huang, B.; Shi, Y. Experimental Investigation on Damage Behavior of Polypropylene Fiber Reinforced Concrete under Compression. *Inter. J. Concr. Struct. Mater.* **2018**, *12*, 1–20. [CrossRef]
34. Cifuentes, H.; Garcia, F.; Maeso, O.; Medina, F. Influence of the properties of polypropylene fibres on the fracture behaviour of low-, normal- and high-strength FRC. *Constr. Build. Mater.* **2013**, *45*, 130–137. [CrossRef]
35. Ding, F.X.; Ying, X.Y.; Yu, Z.W. Unified calculation method of uniaxial mechanical properties of lightweight aggregate concrete. *J. Cent. South Univ. (Sci. Tec.).* **2010**, *41*, 1973–1979.
36. Qin, H.; Zhao, X.Z. Numerical analysis of composite columns with high steel ratio under cyclic loading. *Struct. Eng.* **2014**, *30*, 25–31.
37. Al-Darzi, S.Y.K.; Chen, A.R.; Liu, Y.Q. Finite element simulation and parametric studies of perfobond rib connector. *Am. J. Appl. Sci.* **2007**, *4*, 122–127. [CrossRef]
38. Zhang, J.D.; Gu, J.C.; Deng, W.Q.; Fu, H. Shear behavior of perfobond rib shear connectors for pre-fabricated composite bridges. *China J. Highw. Transp.* **2018**, *31*, 71–80.

coatings

MDPI

Article

Research on Bending Performance of Concrete Sandwich Laminated Floor Slabs with Integrated Thermal and Sound Insulation

Peng Liu [1,2,3], Sisi Xie [1,*], Lei Liu [1,*], Ao Luo [1], Ning Zhang [4], Sasa He [4], Yingye Wu [5], Wen Xu [6], Ying Chen [2,7] and Zhiwu Yu [1,2]

1 School of Civil Engineering, Central South University, 22 Shaoshan Road, Changsha 410075, China; liupeng868@csu.edu.cn (P.L.); ddt1205@163.com (A.L.); zhwyu0512@163.com (Z.Y.)
2 National Engineering Laboratory for High Speed Railway Construction, Central South University, Changsha 410075, China; 2015038@csu.edu.cn
3 Chenzhou Changxin Residence Technology Co., Ltd., 8 Huihuang Road, Chenzhou 423000, China
4 Hunan Zhongda Design Institue Co., Ltd., 68 Shaoshan Road, Changsha 410075, China; 115116277646@163.com (N.Z.); szwang863@163.com (S.H.)
5 Hunan Construction Project Investment Management Co., Ltd., 199 Xiangfu Road, Changsha 410007, China; llppp98@163.com
6 Railway Group 5 Mechanization of Engineering Co., Ltd., 32 HongtangCong Road, Hengyang 421002, China; 216056@csu.edu.cn
7 School of Civil Engineering, Central South University of Forestry and Technology, 498 Shaoshan Road, Changsha 410004, China
* Correspondence: xiesisicumtb@163.com (S.X.); llmysy@163.com (L.L.)

Abstract: In this study, a full-scale test on the bending performance of concrete sandwich laminated floor slabs with integrated thermal and sound insulation was carried out, and the effects of different reinforcement ratios on the bending performance of concrete sandwich laminated floor slabs were investigated as well as the variation law of the failure modes, characteristic loads, load-mid span deflection, load-rebar strain curves, and anti-slip performance. The results indicate that the concrete sandwich laminated floor slabs present typical bending failure characteristics. According to bending failure characteristics, the damage process can be divided into three stages, i.e., elasticity, cracking, and failure. The bearing capacity significantly increases with the increase in reinforcement ratio. The normal service, yield, and ultimate loads of bearing capacity of the floor slabs with a larger reinforcement ratio increase by 54.55%, 52.94%, and 46.46%, respectively. Moreover, the mid-span deflection decreases significantly with the increase in reinforcement ratio, and the cracking expansion is also delayed. Before cracking, the prefabricated layer and laminated layer can realize load bearing together, and the floor slab is in a state of complete interaction. When the floor slabs reach the ultimate state, the superimposed surface produces a sliding effect, and the floor slab is in a state of partial interaction. The finite element analysis software ABAQUS (with the version number of ABAQUS 2020, the chief creator of David Hibbitt, and the sourced location of the United States) was used to perform nonlinear numerical simulation. The test results accord well with the simulation results, which verifies the correctness of the finite element model. Based on finite element simulation, the influence of post-cast concrete strength on the ultimate load can be ignored.

Keywords: concrete sandwich laminated floor slab; bending properties; bond-slip; deflection

Citation: Liu, P.; Xie, S.; Liu, L.; Luo, A.; Zhang, N.; He, S.; Wu, Y.; Xu, W.; Chen, Y.; Yu, Z. Research on Bending Performance of Concrete Sandwich Laminated Floor Slabs with Integrated Thermal and Sound Insulation. *Coatings* **2022**, *12*, 1075. https://doi.org/10.3390/coatings12081075

Academic Editor: Paolo Castaldo

Received: 29 June 2022
Accepted: 28 July 2022
Published: 30 July 2022

1. Introduction

Laminated structures are widely used in architecture, aviation, train locomotives, and other fields. At present, extensive experimental research and theoretical analysis have been carried out [1–4]. Floor is an important inner envelope of the building to ensure safety, stability, and durability, which should be both heat insulation, sound insulation, waterproof, and moisture-proof at the same time. Laminated floor slab is a prefabricated concrete floor structure composed of prefabricated and laminated layers. It has good integrity and seismic resistance and has become an important horizontal bearing member of the prefabricated concrete structure [5,6]. However, its thermal and sound insulation effects fail to meet the code's requirements [7]. Therefore, it is necessary to study the structural system of thermal and acoustic bearing integrated floor slab that integrates energy saving, environmental protection, and green low-carbon [8,9]. The concrete sandwich laminated floor slab has the advantages of light weight, high strength, sound insulation, energy saving, fire resistance, good seismic performance, and convenient installation [10]. It is easy to realize product industrialization, standardization, and construction mechanization. Furthermore, it has become one of the most rapidly developing laminated floor structures.

Many studies regarding mechanical property tests and theoretical analysis of concrete sandwich laminated floor slabs have been conducted worldwide [11–14]. For example, Li et al. [15] derived a deflection calculation formula for the concrete sandwich plate in the ultimate limit state by experimental verification and numerical simulation technology. Chen et al. [16] investigated the factors that affect the bending performance of double reinforced concrete sandwich panels. Luo et al. [17] conducted bending tests of laminated slabs, and the results indicated that the prefabricated bottom plate construction form and core material significantly affect the bending performance of sandwich laminated floor slabs. Joseph et al. [18,19] analyzed the bending properties of prefabricated concrete sandwich panels under punching and bending loading conditions. Ahmad et al. [20] indicated that the cracks on the plate side were characterized by the combined effect of bending and shear stresses due to the sandwich layer. Tomlinson et al. [21] studied bending tests of laminated floor slabs with EPS as the core layer, and the results indicated that the floor slabs worked partially compositely when the shear connector failed. Currently, the expanded polystyrene (EPS) plate is mostly used as the core of concrete sandwich laminated floor slabs, which has advantages of light weight, low water absorption, good thermal insulation performance, etc. [22]. However, it has poor fire performance and is easy to burn, which can produce toxic smoke and has great safety hazards. Due to the less attention to the fire performance of insulation materials, fire accidents have been common in recent years [23]. Phenolic foam has low thermal conductivity, high fire safety coefficient, good thermal and sound insulation, etc. Therefore, when used in concrete sandwich laminated floor slabs, the thermal and sound insulation performance and the safety of the slab can be significantly improved.

The effective connection between the prefabricated and laminated layers is the basis that ensures the joint work of the laminated structure. According to research status worldwide, the primary method to solve this problem is the steel bar trusses in the laminated slabs. Wang et al. [24] investigated the bending performance of laminated slabs with trusses and stirrup reinforcement as the connections, respectively, and the results indicated that both structural forms could significantly improve the stiffness and shear resistance. Refs. [25–27] presented that truss webs can ensure the shear resistance of the laminated face. Huang et al. [28,29] investigated the bending performance of laminated slabs with different prefabricated bottom plates, and a design recommendation for structural measures was proposed. Thanoon et al. [30] regarded that truss reinforcement can provide good integrity to the laminated slab in all bending phases. Steel bar trusses can improve the shear capacity, limit the inter-story slip, and enhance the co-working property of the laminated floor slabs [31]. However, there are disadvantages such as numerous trusses, dense truss spacing, high cost, and complex vibration and compaction of post-cast layer concrete. Given the shortcomings of the steel bar truss connectors, the stirrup rebar is proposed to replace the steel bar trusses. Moreover, due to the advantages of low energy consumption, low cost, and saving the amount of steel reinforcement, the CRB600H high-strength steel bars are employed, which has remarkable economic and social benefits [32].

This paper proposed thermal and sound insulation integrated concrete sandwich laminated floor slab (Abbr. sandwich laminated floor slab). The sandwich laminated floor slab was combined with concrete layers on both sides and a phenolic core layer connected through the steel reinforcement grid. The phenolic foam board was the insulation layer, and the reinforcement grid consisted of stirrup rebars and upper and lower rebar mesh. The two-point bending tests of sandwich laminated floor slabs with different reinforcement ratios were carried out. The damage pattern, reinforcement strain, crack development distribution characteristics, and bending bearing capacity process were discussed. The influence law of reinforcement ratio on the characteristic load was also analyzed. Moreover, the slip resistance of sandwich laminated floor slabs in different stages was also studied. The force performance was numerically simulated by finite element analysis software ABAQUS. The results provided a scientific basis for the design and application of concrete sandwich laminated floor slabs with integrated thermal and sound insulation in practical projects.

2. Experimental Study

2.1. Design and Fabrication of Specimen

Two full-scale specimens were subjected to bending loads with the number of DBD01 and DBD02. Both the specimens were of the same size of 3600 mm × 2400 mm, and the thickness of the prefabricated and laminated layers was 90 mm and 70 mm, respectively. Figure 1 shows the schematic diagram of the interlayer structure.

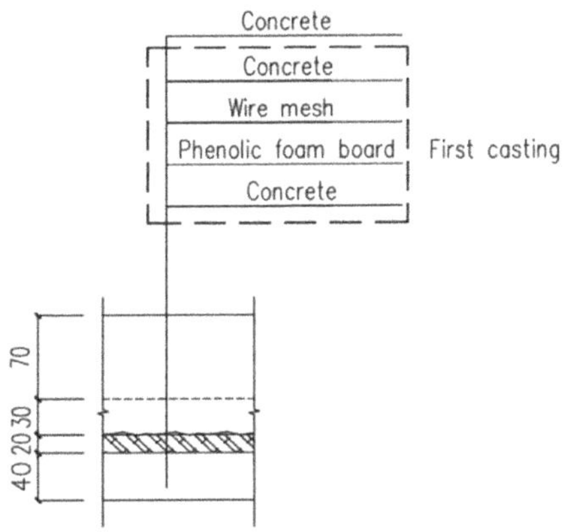

Figure 1. Schematic diagram of sandwich laminated floor slab interlayer structure.

Table 1 lists the principal design parameters of the specimens, Figure 2 shows the pouring process, and Figures 3–5 present the reinforcement diagram and the constructional detail.

The strength grade of concrete in the prefabricated and laminated layers of the specimen is C40. The type and diameter of reinforcement are the same, but their reinforcement ratios are different. The stirrup rebar is HPB300 with the design value of tensile strength f_y = 300 MPa. The longitudinal bearing and transverse distribution bars are CRB600H, with the design value of tensile strength f_y = 540 MPa. The diameters of longitudinal and distribution reinforcement are 8 mm and 6 mm, respectively. The surface reinforcement is HRB400 with a diameter of d = 8 mm, and the design value of tensile strength f_y = 400 MPa. Table 2 shows the mechanical property parameters of the steel bars.

Table 1. Detailed dimensions and reinforcement of sandwich laminated floor slabs.

Specimen Number	Size (mm)	Transverse Rebar	Longitudinal Rebar	Face Rebar	Stirrup Rebar	Rebar Ratio ρ/%
DBD01	3600 × 2400 × 160	$A^{RH}6@200$	$A^{RH}8@200$	C8@150	A6	0.26
DBD02			$A^{RH}8@100$			0.50

Figure 2. Pouring diagram of sandwich laminated floor slab.

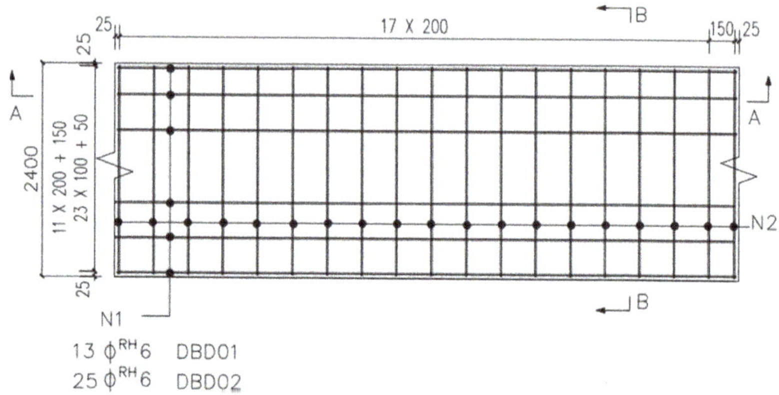

Figure 3. Reinforcement diagram of sandwich laminated floor slab (unit: mm).

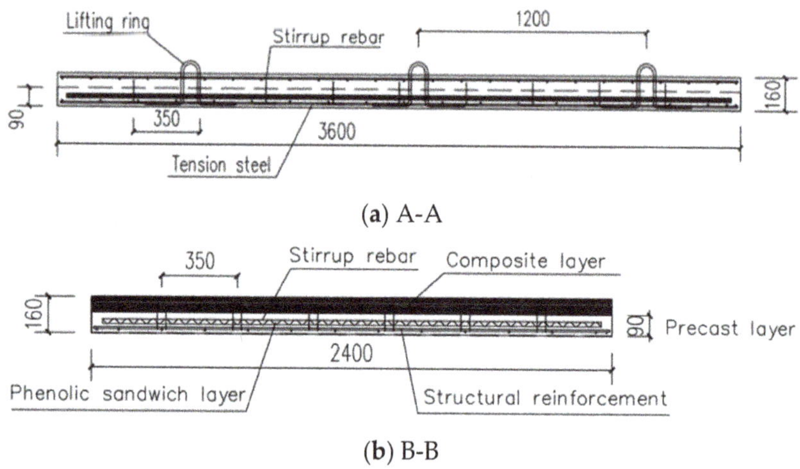

Figure 4. Section diagram of sandwich laminated floor slab. (**a**) Section A-A of floor slab; (**b**) Section A-A of floor slab (unit: mm).

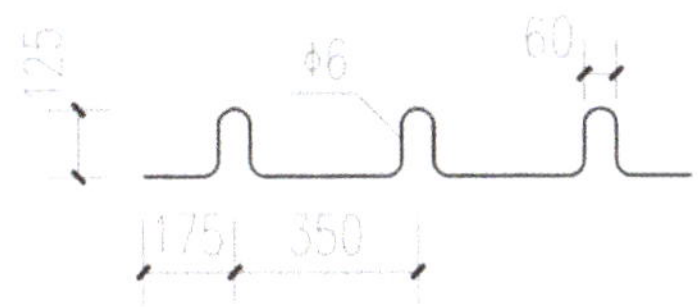

Figure 5. Schematic diagram of combined stirrup rebar (unit: mm).

Table 2. Mechanical properties parameters of steel bars.

Reinforcing Steel Type	f_{yk} (N/mm^2)	f_{stk} (N/mm^2)	E_s (N/mm^2)
CRB600H	540	600	190,000
HRB400	400	540	200,000
HPB300	300	420	210,000

The specimens were fabricated and cured according to GB/T50081-2019 [33] and were placed into a pool with the saturated solution of calcium hydroxide at (20 ± 2) °C for 28 d after demolding. There were two groups of pressure test blocks with the same batch of materials for the prefabricated and laminated layers. Moreover, three parallel specimens of each group were used to reduce the error caused by the randomness of the specimen. The loading equipment is a 100-ton electro-hydraulic servo material machine designed and produced by Shanghai New Sansi Measuring Manufacturing Co., Ltd. (Shanghai, China). Figure 6 shows the loading device and the failure mode under compression. Table 3 presents the mechanical performance parameters.

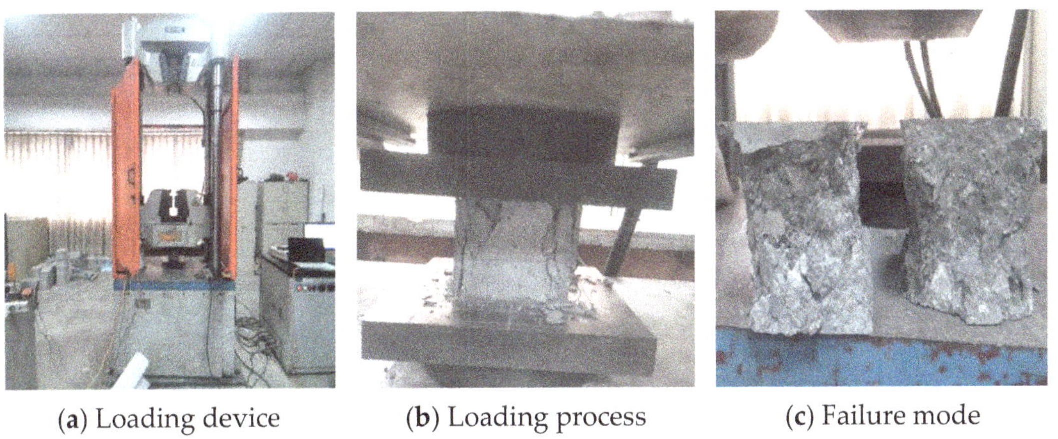

Figure 6. Loading device and failure mode under compression.

Table 3. Mechanical properties of concrete materials.

Type	Measured Value of Compressive Strength of Prefabricated Layer (MPa)	Measured Value of Compressive Strength of Laminated Layer (MPa)	f_c (MPa)	E_c (MPa)
DB01	49.2	48.7	26.8	32,500
DB02	48.3	48.8		

2.2. Test Methods

According to GB 50010-2010 [34], instead of the equalized load, two-point centralized loading was used to load with the loading point at the trisection of the span. Sand was pre-laid at the loading point for leveling to prevent stress concentration. Figure 7 shows the schematic diagram of the test device and loading.

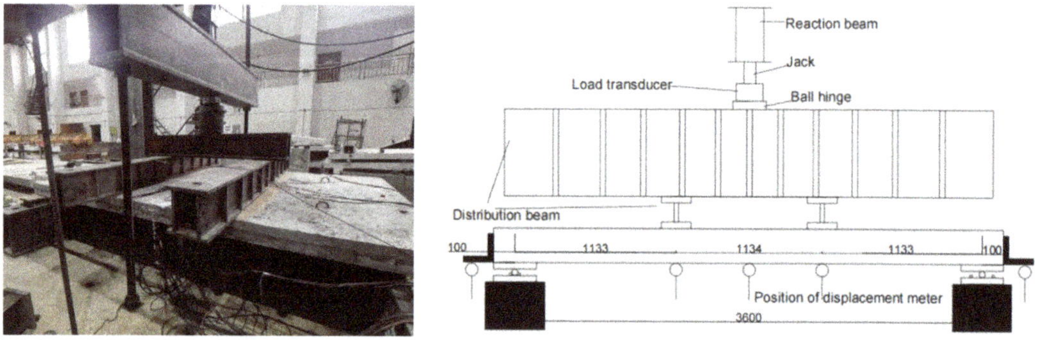

(**a**) Actual drawing of loading device (**b**) Schematic diagram of loading device

Figure 7. Schematic diagram of test device and loading (unit: mm).

The load-controlled loading system was used, including preloading and subsequent formal loading processes. The preloading process consisted of three levels, each taking 20% of the cracking load, then unloading after the preloading process and entering the formal loading stage. The formal loading process was divided into four stages. In the first stage, 20% of the cracking load was taken for each level and then entered the second stage after reaching 80% of the cracking load, and 5% of the cracking load value should be taken for each stage. Observing the strain change of the concrete at the bottom of the slab, the measured cracking load was determined when the strain deviated from the linear change significantly, and cracks appeared at the bottom of the slab. After the floor slab cracking, it entered the third stage, and 10% of the calculated value of the cracking load was taken for each level until the bearing capacity. Finally, it entered the fourth stage when reaching the ultimate state, and 5% of the limit load value was taken for each level until the specimen failed. The load holding time of each stage was 10 min. Figure 8 shows the loading system.

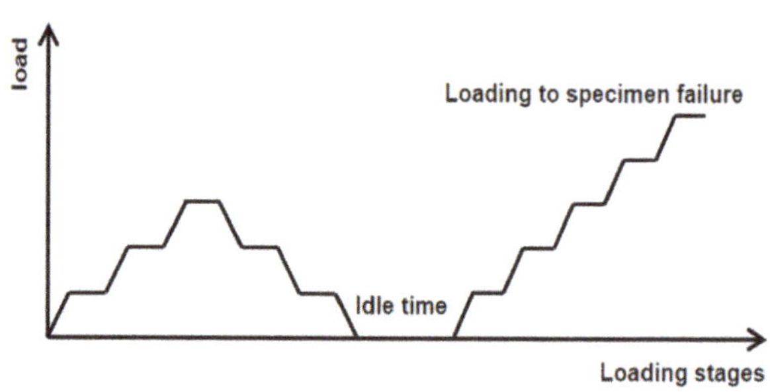

Figure 8. Loading mode diagram.

2.3. *Measuring Point Layout*

The reinforced strain gauges were arranged in the 1/2 and 1/3 of the bottom span and the middle of the top span, as shown in Figure 9. Concrete strain gauges were placed in the middle of both span bottom and side, as shown in Figure 10. The displacement gauges were placed at the 1/2 and 1/3 of the span and two-end supports to measure the corresponding deflection.

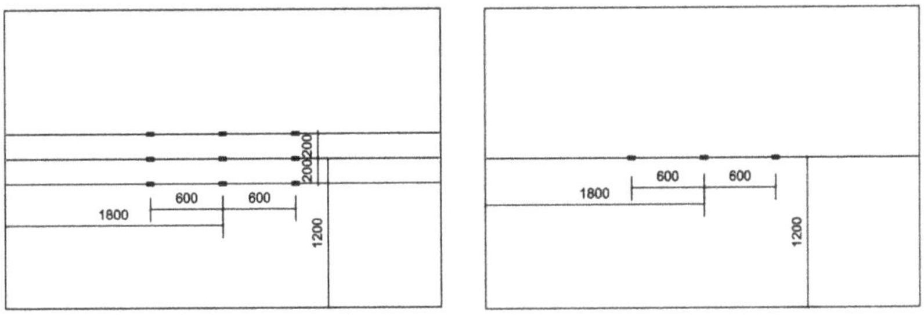

(**a**) Bottom of the floor slabs (**b**) Top of the floor slabs

Figure 9. Layout of measurement points for reinforcement strain gauges (unit: mm).

(**a**) Side of the floor slabs (**b**) Bottom of the floor slabs

Figure 10. Layout of measuring points of concrete strain gauge (unit: mm).

3. Results and Discussions

3.1. Failure Characteristics

The mid-span deflection, steel bar, and concrete strain increase slightly when the load is small. The first crack appears in the mid-span pure bending section with a load of 13 kN. At this time, the strain of steel bar, concrete, and the mid-span deflection increases slightly. Moreover, the cracks develop from the middle to both ends along the width with the load increases and mainly distributed symmetrically in the mid-span.

Firstly, the cracks on the side of the slab extend vertically upward in mid-span, and then the bending shear diagonal cracks successively appear within 1/3 of the span as the load increases. The interface between vertical and inclined cracks is a superposition surface, and the inclination angle of cracks decreases gradually. When the load is 66 kN, the deflection reaches 17.20 mm, corresponding to the width of 0.26 mm. According to GB 50010-2010, the deflection and crack width limit in the normal use stage is $L_0/200$ = 17 mm and 0.2 mm, respectively [34]. Based on the deflection and crack width control, the specimen reaches the normal use stage. When the load is 186 kN, the mid-span deflection instantly increases to 68.2 mm. According to GB 50010-2010, the deflection limit in the ultimate stage is $L_0/50$ = 68 mm [34], and the specimen reaches the ultimate state of the bearing capacity. The cracks at the bottom are mainly vertical to the axis of the short span and densely distributed within 1/3 of the span, as shown in Figure 11.

The first crack appears in the mid-span pure bending section when the load reaches 39 kN and then enters the limit stage of normal usage with the deflection of 17.10 mm when the load is 102 kN. The horizontal cracks appear at the superimposed surface when the load is 199 kN. According to GB50010-2010, the ultimate state of the bearing capacity is reached with a deflection of 17.20 mm and a width of more than 1.5 mm. Figure 12 shows the failure mode and crack distribution.

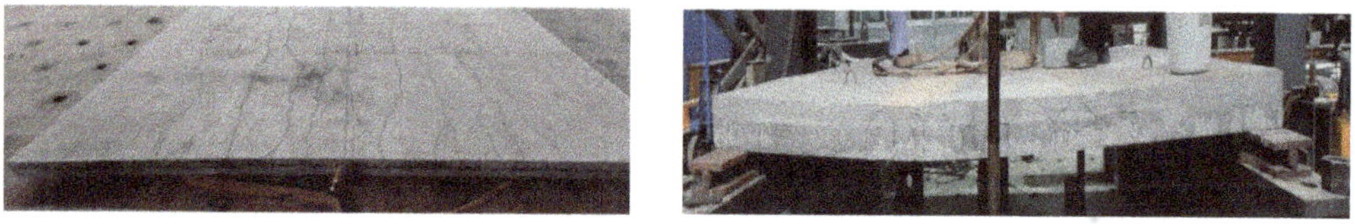

(**a**) Crack distribution at slab bottom (**b**) Crack distribution on slab side

Figure 11. Failure mode and crack distribution of DBD01.

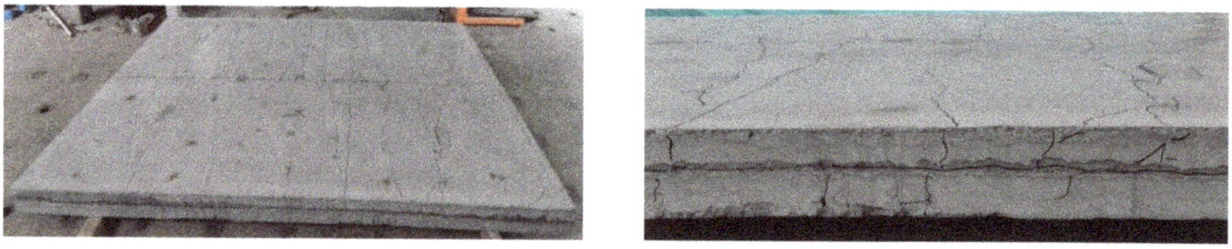

(**a**) Crack distribution at slab bottom (**b**) Crack distribution at slab bottom

Figure 12. Failure mode and crack distribution of DBD02.

The cracking load is significantly higher, and the crack number and growth rate of DBD02 are apparently lower than that of DBD01, which indicates that the mid-span deflection of the specimen is reduced and the cracking expansion is delayed with the increase in reinforcement ratio. Table 4 shows the characteristic load of the sandwich laminated floor slab. P_{cr} is the cracking load, $P_{1/200}$ is the normal use load, P_y is the yield load, and P_u is the ultimate load. Both sandwich laminated floor slabs present typical bending failure characteristics with similar early failure characteristics and good ductility. The cracking load of DBD02 is three times of DBD01, and the normal use load, yield load, and ultimate load of bearing capacity increase by 54.55%, 52.94%, and 46.46%, respectively. This indicates that the ultimate bearing capacity is significantly improved with the increase in the reinforcement ratio, but the bonding performance of the superimposed surface should be considered as well.

Table 4. Characteristic load of sandwich laminated floor slab.

Type	P_{cr} (kN)	$P_{1/200}$ (kN)	P_y (kN)	P_u (kN)	P_{cr}/P_u	$P_{1/200}/P_u$	P_y/P_u
DBD01	13	66	102	127	10.24%	51.97%	80.31%
DBD02	39	102	156	186	20.97%	54.84%	83.87%

Note: The characteristic loads of the sandwich laminated floor slab include the weight of the distribution beam.

3.2. *Bending Performance Analysis*

3.2.1. Verification of Plane-Section Assumption

Figure 13 shows the average strain distribution of concrete along the section height in the mid-span, where ε_c is the concrete strain and h is the section height. Since the strain gauges are damaged after cracking, only the strain data without the cracking of floor slab side are plotted. As seen in Figure 13, the concrete average strain distribution along the height of the mid-span section is linear, which conforms to the assumption of the flat section. Strain hysteresis is not apparent between the prefabricated and laminated layers, so the flat section assumption can be taken as the basic assumption in the bearing capacity calculation analysis.

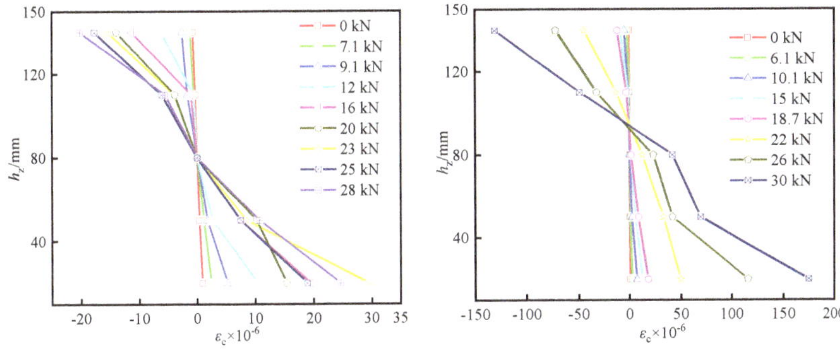

(**a**) DBD01 (**b**) DBD02

Figure 13. The concrete strain distribution along the section height in the mid-span.

3.2.2. Load-Mid Span Deflection Relationship

As shown in Figure 14, the load-mid span deflection curve is characterized by three straight lines, i.e., elasticity, cracking, and failure, indicating that the specimen has good ductility.

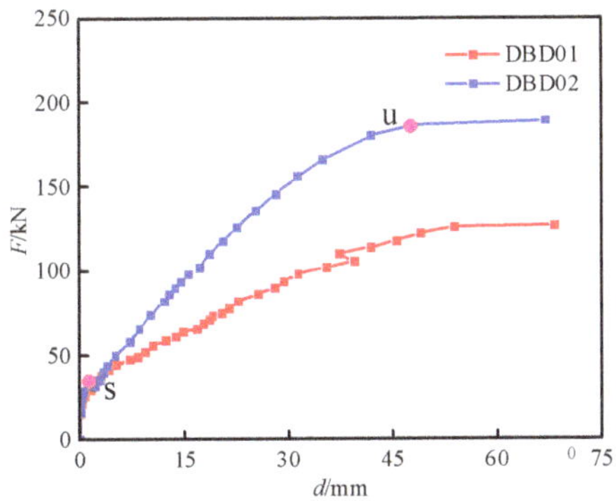

Figure 14. Load-mid span deflection curve.

a. Elastic stage (0~s). Before cracking, the specimen is in the elastic stage, and the load is proportional to the mid-span deflection. Due to the low tensile strength and high stiffness of concrete, the curve is short and steep. The cracking load of DBD02 is three times of DBD01 because the reinforcement ratio of DBD02 is twice of DBD01. The increase in reinforcement ratio can effectively delay the crack propagation.

b. Cracking stage (s~u). The slope of the curve and the bending stiffness decrease significantly after the crack appears. The load-mid span deflection curve nonlinearly increases as the floor slab enters the plastic stage. The cracking load, normal use load, and yield load of both specimens are 10%~20%, 50%~55%, and 80%~85% of the ultimate load, respectively. The curve slope of DBD02 is significantly larger than that of DBD01, and the deflection of DBD01 is lower than that of DBD01, which indicates that the increase in reinforcement ratio can effectively improve the late rigidity and reduce the mid-span deflection greatly.

c. Failure stage (u~). The curve slope sharply decreases, the load almost remains unchanged, and the deformation rapidly develops due to the low tensile strength of the phenolic foam board after the steel bars yield, which indicates that the sandwich layer significantly influences the latal rigidity and bearing capacity of specimens. The ultimate state of bearing capacity reaches when the concrete is crushed or deformed excessively, and the load-mid span deflection curve is almost horizontal.

3.3. Load-Reinforcement Strain Relationship

As shown in Figure 15, both sandwich laminated floor slabs have the same variation law. The curve development presents three-stage characteristics, i.e., elastic, elastoplastic, and failure. The curve develops linearly with a large curve slope when the load is small. When it reaches the cracking load, the reinforcement strain suddenly increases, and the slope of the curve decreases obviously. The variation trend is consistent with the load-mid span deflection curve. At this time, the concrete in the tensile zone fails, and the steel stress and strain instantly increase. The curve develops nonlinearly with the load increases, which indicates that the steel bar enters the plastic stage. Furthermore, the mid-span strain difference between the specimens increases with the increase in load, which demonstrates that the reinforcement stress is effectively retarded with the increase in reinforcement ratio.

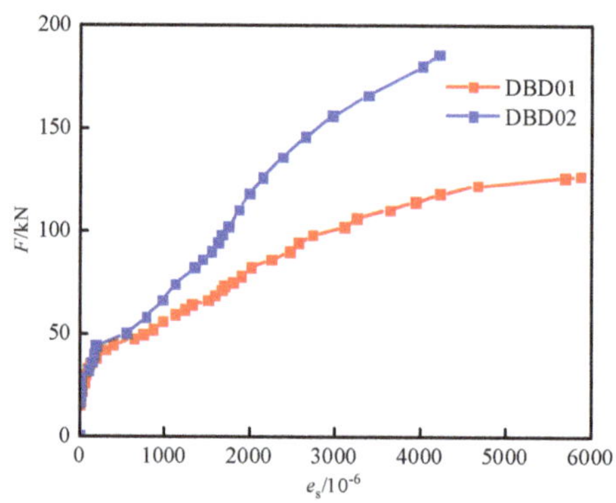

Figure 15. Load versus strain curve of reinforcement.

4. Analysis of the Slip Resistance of Sandwich Laminated Floor Slabs

4.1. Premise of Analysis

The concrete sandwich laminated floor slab is composed of reinforcement, concrete, and phenolic foam board with completely different physical and mechanical properties, which results in complex mechanical properties of the floor slab under load. The bonding performance of the laminated surface is the key to maintaining the integrity of the laminated floor slab, and the bending capacity will decrease rapidly when the laminated surface is damaged by the sliding [35], so the mechanical characteristics need to be studied before calculation.

The connection of the stirrup rebar is the key to the synergy of the upper and lower concrete slabs, whose bearing performance directly affects the overall mechanical properties of the structure. According to the stiffness of the connections, the forces of concrete sandwich laminated floor slabs can be divided into three types, i.e., non-composite, partially composite, and fully composite. For concrete sandwich laminated floor slabs of non-composite, there is no connection to transfer longitudinal shear force, and the upper and lower concrete layers are equivalent to two independent slabs with the same span and curvature. For concrete sandwich laminated floor slabs of partially composite, the magnitude of the transferred longitudinal shear force is related to the stiffness. The section strain effect also needs to be considered when performing equilibrium analysis. For concrete sandwich laminated floor slabs of fully composite, the stiffness of the connectors is large enough to avoid the slippage effect. The force pattern of the concrete sandwich plate is the same as that of the cast-in-place floor slab [36–38]. Figure 16 shows the distribution of shear and strain under bending moment.

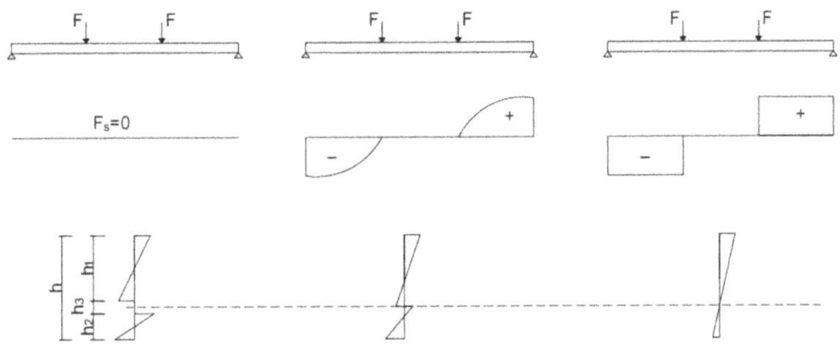

Figure 16. Distribution of shear force and strain.

4.2. Anti-Slip Performance before Cracking

The overall performance of the floor slab is good before cracking, which conforms to the plane-section assumption. Therefore, the cracking load M_{cr} can be calculated according to the fully composite action. Moreover, the bending performance of the phenolic foam board is ignored while calculating the elastic moment due to the small bending stiffness. According to GB 50010-2010 [34], the formulas for calculating the cracking load are as follows:

$$M_{cr} = \gamma f_{tk} W_0 \tag{1}$$

$$\gamma = (0.7 + \frac{120}{h})\gamma_m \tag{2}$$

$$W_0 = \frac{I_0}{h - y_0} \tag{3}$$

where M_{cr} is the cracking load of the positive section of the concrete sandwich laminated floor slab, γ is the coefficient of plasticity influence of the resisting moment, γ_m is the base value of the plastic influence coefficient of the resisting moment of the cross-section, and W_0 is the elastic resisting moment of the tensile edge of the cross-section.

The cracking moment can be determined according to Equations (1)–(3), and the calculated values are compared and analyzed with the measured values, as shown in Table 5.

Table 5. Composite working performance of sandwich laminated floor slabs.

Specimen Number	Before Cracking			At Limit State			
	M_{cs} (kN·m)	M_{cr} (kN·m)	$\lvert(M_{cs}-M_{cr})/M_{cs}\rvert$	M_{ut} (kN·m)	M_{un} (kN·m)	M_{uf} (kN·m)	M_{ut}/M_{uf}
DBD01	7.3	6.9	6.1%	71.8	12.5	128.7	55.7%
DBD02	22.0	22.5	2.1%	105.1	15.5	169.5	62.0%

Noted: M_{cs} is the measured cracking moment, and M_{cr} is the calculated cracking moment.

4.3. Anti-Slip Performance in Limit State

The ultimate moment of the normal section at the state of non-composite and fully composite states are calculated, respectively [17], and then the composite working performance of both specimens is judged. The formulas for calculating the ultimate load of non-composite state are as follows:

$$\alpha_1 f_{ct} b x_t = f_{yt} A_{st} \tag{4}$$

$$\alpha_1 f_{cb} b x_b = f_{yb} A_{sb} \tag{5}$$

$$M_{un} = \alpha_1 f_{ct} b x_t (h_{0t} - \frac{x_t}{2}) + \alpha_1 f_{cb} b x_b (h_{0b} - \frac{x_b}{2}) \tag{6}$$

where M_{un} is the normal section ultimate bending moment, x_t and x_b are the calculated compression zone of laminated and prefabricated layers, respectively, h_{0t} and h_{0b} are the effective height of laminated and prefabricated layers, respectively, f_{yt} and f_{yb} are the strength of reinforcement tensile reinforcement in the laminated and prefabricated layers, respectively, A_{st} and A_{sb} are the area of tensile reinforcement in the laminated and prefabricated layers respectively, and f_{cb} and f_{ct} refer to the axial compressive strength of laminated and prefabricated concrete layers, respectively, which is calculated and determined according to $f_c = 0.76f_{cu,k}$.

The formulas for calculating the ultimate load of a fully composite state are as follows:

$$\alpha_1 f_{ct} b x = f_{yb} A_{sb} + f_{yt} A_{st} \tag{7}$$

$$M_{uf} = \alpha_1 f_{ct} b x (h_0 - \frac{x}{2}) + f_{yt} A_{st} (h_0 - a_s) \tag{8}$$

where M_{uf} is the normal section ultimate bending moment, x is the height of the calculated compression zone, h_0 is the effective height of the section, and a_s is the distance from the resultant force point of the reinforcement in the compression area to the compression edge.

As shown in Table 5, the following conclusions can be drawn:

(1). Before cracking, the cracking moment error between the calculated and measured values is not larger than 7% under the fully composite state, and there is no relative slip, indicating good connection performance between the upper and lower concrete.
(2). When it reaches the ultimate state, the upper and lower concrete layers of the specimen are in a partially composite state. The reason is that the stirrup rebar is flexible and deforms under the transverse shear force, which results in the sliding effect and failure of co-working property.
(3). The slip resistance of DBD01 under the damage stage is less than that of DBD02, which indicates that the slip resistance at the later stage of loading can be enhanced with the increase in reinforcement rate.

5. Finite Element Analysis

The finite element calculation software ABAQUS is used to simulate the bending test. Except for the different reinforcement ratios, the other test conditions and loading methods of DBD01 and DBD02 are exactly the same. Therefore, only the finite element simulation analysis of DBD01 is carried out. The simulation is compared with the experimental results, and the finding can provide a theoretical basis for the structural mechanical analysis of the sandwich laminated floor slab.

The dimensions of the finite element model are the same as those of the actual specimen. The analysis includes the following steps: defining material properties and analysis types, defining loads and boundary conditions, defining interactions, meshing, and post-processing of results. The concrete damaged plasticity (CDP) model in the ABAQUS material library is selected to simulate concrete [39], and the constitutive relationship is calculated according to GB50010-2010 [34]. The bilinear model is used to simulate rebar, and the slope of the second section is 1/100 of the first section [40]. The three-dimensional solid element (C3D8R) with an eight-node reduced integral is used in concrete, and the two-node linear three-dimensional truss element (T3D2) is used in the steel bar. Establish four different material types, including a concrete material: C40, and three reinforcement types: CRB600H, HRB400 and HPB300. The constraint is applied at the bottom of the pad block to simulate the simply supported condition. Referring to the test value, the applied load value during finite element simulation is 140 kN. Assuming that the reinforcement and concrete are firmly bonded, they are coupled through embedded area constraint. The slip of the superimposed surface is ignored, and tie contact is used to simulate the bonding surfaces. The total number of elements is 91,163 and of nodes is 130,322. Figure 17 shows the constraints, loading, and mesh division of the model.

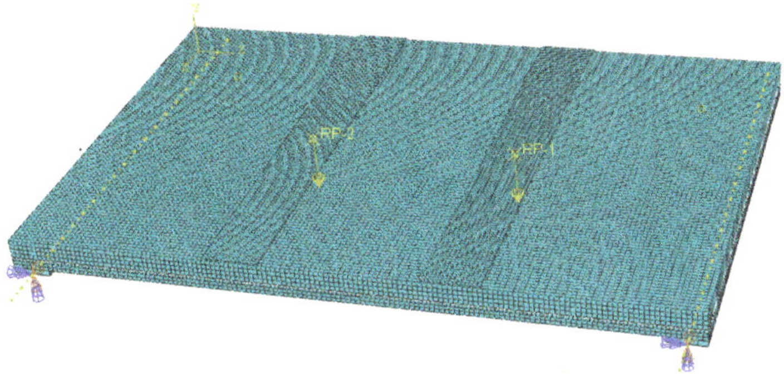

Figure 17. Finite element model of the specimen.

5.1. *Damage Analysis of Sandwich Laminated Floor Slab*

The damage and crack distribution of sandwich laminated floor slabs are closely related to the stress damage distribution, so it is necessary to carry out the stress damage analysis. As seen in Figure 18, the cracks at the bottom slab are mainly concentrated within the span of the loading point and are symmetrically distributed from the middle of the span to both ends. The crack direction is parallel to the short span of the sandwich laminated floor slabs. The cracks on the side of the slab gradually develop from the bottom to the top, with the inclination angle decreasing while passing through the sandwich layer. Moreover, the distribution of tensile damage at the bottom and side of the plate is roughly consistent with the crack location during the test. Therefore, the finite element simulation results can reflect the actual damage.

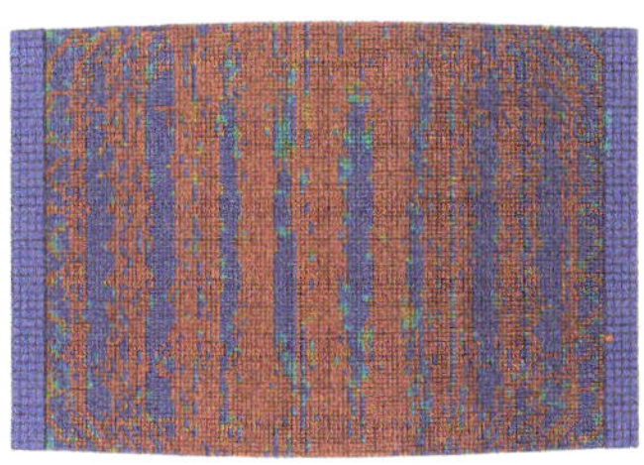

(**a**) Tensile damage nephogram of slab bottom

(**b**) Test failure mode of slab bottom

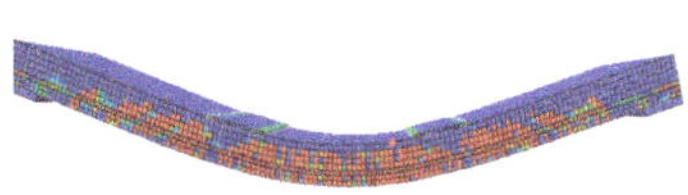

(**c**) Tensile damage nephogram of slab side

(**d**) Test failure mode of slab side

Figure 18. Tensile damage nephogram and failure mode.

5.2. *Comparative Analysis of Load-Deflection Curve, Load-Reinforcement Strain Curve, and Bearing Capacity*

Figure 19 shows the comparison between the finite element simulation and the test curve, which indicates that the simulated results accord well with the measured results, so the finite element method can be used for deeper analysis. The experimental values of cracking load are slightly smaller than the simulated values, which is mainly due to incomplete constraints of combined interface and bond-slip between reinforcement and concrete. However, the influences of material difference, manufacturing error, and curing conditions are ignored during simulation calculation, which is assumed under the ideal conditions, and accordingly results in the deviation between the simulated and the measured values.

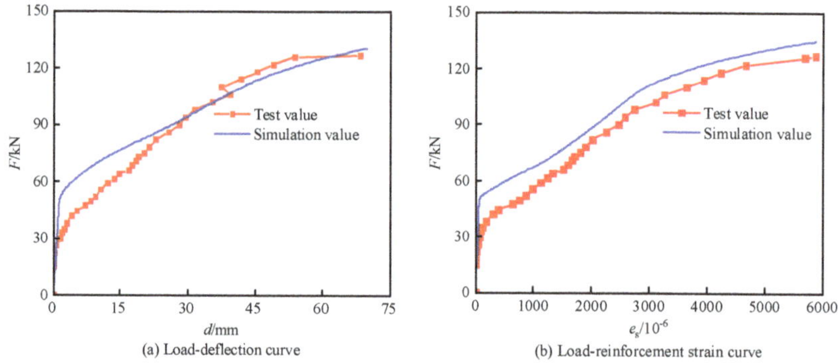

Figure 19. Load-deflection curve and load-reinforcement strain curve.

Table 6 shows the comparison of bearing capacity between the measured and the simulated value. The results show that the simulated values accord well with the measured values. The difference between the normal use load, yield load, and ultimate load of the specimen is 2%~16%. The root mean square error is 0.015, which indicates that the deviation between the simulated and the measured values between different characteristic loads is small. The deviation of cracking load is relatively large due to the discreteness of concrete properties and the deviation during manufacturing and loading. The simulation results reflect the test conditions well, especially the ultimate bearing capacity. Therefore, the finite element simulation method can be used to further analyze the mechanical performance of the sandwich composite floor.

Table 6. Comparison between simulated and measured values of specimen bearing capacity.

Specimen Number	Characteristic Load	Measured Value	Simulated Values	I Simulated Values-Measured Value I /Measured Value
DBD01	P_{cr} (kN)	13	24	0.46
	$P_{1/200}$ (kN)	66	79	0.16
	P_y (kN)	102	109	0.06
	P_u (kN)	127	130	0.02

5.3. *Effect of Post-Cast Layer Concrete Strength*

Concrete strength significantly influences the mechanical performance of the sandwich laminated floor slab. In order to further explore the influence of post-cast layer concrete strength on the normal use load and ultimate load of the specimen, the bearing performance under different post-cast concrete strength grades is compared and analyzed by numerical simulation. As shown in Figure 20, the load-deflection curves of the sandwich laminated floor slab with different concrete strength grades have the same variation trend, which shows an upward trend with the increase in concrete strength. The mid-span deflection of the same load decreases with the increase in concrete strength overall but is not significant. As the strength increases, the normal service load and ultimate load of the specimen gradually increase overall. When the concrete strength of the post-cast layer is C30, C35, and C40, the normal use load and ultimate load are the same at about 80 and 131 kN, respectively, and slightly increased to about 84 and 134 kN, respectively, when the concrete strength of the post-cast layer is C50 and C60. The tensile reinforcement firstly yields, and then the concrete is crushed, which is classified as an under-reinforced failure. The main reason is that during the period from the cracking to the ultimate bearing capacity, the post-cast layer concrete has not been damaged, and the load of the sandwich laminated floor slabs is borne by the tensile reinforcement at the bottom. The reinforcement ratio of the sandwich laminated floor slabs with different post-cast layer concrete strength is the same in the finite element simulation, so the difference in bearing capacity between the specimens is small. The effect of post-cast concrete strength can be ignored because of the small difference. Table 7 shows the effect of the concrete strength of the post-cast layer on the characteristic load.

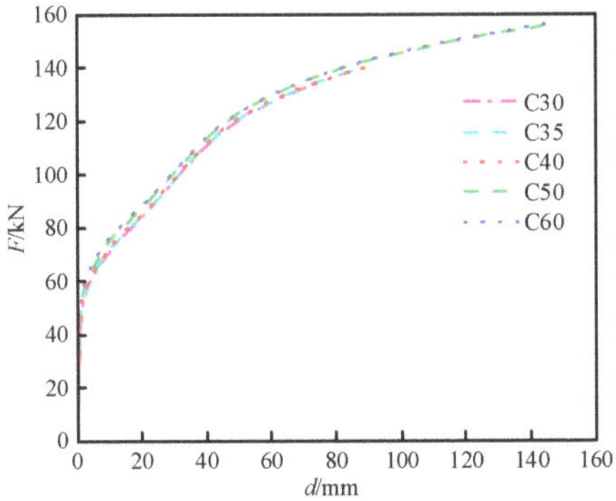

Figure 20. Load-deflection curve of specimens under different post-cast concrete strengths.

Table 7. Effect of concrete strength of post-cast layer on the characteristic load.

Concrete Strength of the Post-Cast Layer	C30	C35	C40	C50	C60
$P_{1/200}$ (kN)	80	80	80	83	85
P_u (kN)	131	131	132	134	134

6. Conclusions

The concrete sandwich laminated floor slabs with integrated thermal and sound insulation are proposed. The experimental research, theoretical analysis, and numerical simulation are carried out on two full-size specimens, and the major conclusions are as follows:

(1). The whole bending failure process of the sandwich laminated floor slab can be divided into three stages: elasticity, cracking, and failure. The demarcation points are concrete cracking and steel yielding, respectively. The bearing capacity of the sandwich laminated floor slab is significantly improved with the increase in reinforcement ratio. The normal use load, yield load, and the ultimate load with a larger reinforcement ratio are increased by 54.55%, 52.94%, and 46.46%, respectively. When the reinforcement ratio is increased, the late stiffness and the structural integrity are significantly improved. Moreover, the mid-span deflection decreases significantly with the increase in reinforcement ratio, and the cracking expansion is also delayed.

(2). The pure bending section of the sandwich laminated floor slab conforms to the plane-section assumption under the action of two-point symmetrical loads. The anti-sliding performance results show that the prefabricated and the laminated layer are fully composite before cracking, and the common bearing can be realized. The specimen is partially composite when reaching the ultimate state, with the slip effect on the superposition surface.

(3). The simulated and measured results accord well, which has the same variation of failure mode, load-deflection, and load-reinforcement strain. The difference between the simulated and the measured values of the ultimate load is less than 5%, which verifies the feasibility and effectiveness of the finite element analysis method. Further extending the analysis, the effect of the concrete strength of the post-cast layer on the normal use and ultimate load is not significant and can be neglected.

Author Contributions: Conceptualization, P.L. and S.X.; methodology, S.X. and L.L.; software, S.X. and A.L.; resources, N.Z.; data curation, S.H.; writing—original draft preparation, S.X.; writing—review and editing, P.L. and L.L.; visualization, Y.W.; supervision, W.X.; project administration, Y.C.; funding acquisition, Z.Y. All authors have read and agreed to the published version of the manuscript.

Funding: This study was funded by the National Natural Science Foundation of China (grant numbers 52178182, 52108262, and U1934217) and China Railway Science and technology research and development plan project (grant numbers 2020-Major project-02, 2021-Major project-02, 2021-Key projects-11). Authors also have received research grants from the Natural Science Foundation of Hunan Province of China (2020JJ5982), Science and Technology Innovation Program of Hunan Province (2020RC4049), Science and Technology Development Project of Chenzhou Municipal Science and Technology Bureau (ZDYF2020218), Hunan Innovative Province Construction Special Project (2020GK4058), and National Science Foundation for Distinguished Young Scholars of Hunan Province (Grant No. 2022JJ10075).

Institutional Review Board Statement: Not applicable.

Informed Consent Statement: Not applicable.

Data Availability Statement: Not applicable.

Conflicts of Interest: The authors declare no conflict of interest.

References

1. Huang, L.J.; An, Q.; Geng, L.; Wang, S.; Jiang, S.; Cui, X.P.; Zhang, R.; Sun, F.B.; Jiao, Y.; Chen, X.; et al. Titanium Matrix Composites: Multiscale Architecture and Superior High-Temperature Performance of Discontinuously Reinforced Titanium Matrix Composites. *Adv. Mater.* **2021**, *33*, 2170039. [CrossRef]
2. Kueh, A.B.H. Buckling of sandwich columns reinforced by triaxial weave fabric composite skin-sheets. *Int. J. Mech. Sci.* **2013**, *66*, 45–54. [CrossRef]
3. Soufeiani, L.; Ghadyani, G.; Kueh, A.B.H.; Nguyen, K.T.Q. The effect of laminate stacking sequence and fiber orientation on the dynamic response of FRP composite slabs. *J. Build. Eng.* **2017**, *13*, 41–52. [CrossRef]
4. Wang, S.; Huang, L.J.; An, Q.; Jiang, S.; Zhang, R.; Geng, L.; Qu, S.X.; Peng, H.X. Regulating crack propagation in laminated metal matrix composites through architectural control. *Compos. Part B Eng.* **2019**, *178*, 107503. [CrossRef]
5. Ding, K.W.; Zhang, Y. Study on Composite Slab Structure. *Appl. Mech. Mater.* **2012**, *214*, 311–314.
6. Wang, L.; Zhang, H.Y. Summary of Study on Composite Concrete Slabs. *Appl. Mech. Mater.* **2013**, *2544*, 695–698.
7. Wang, Q.J.; Huang, X. Configuration and construction technology of the heat preservation sound insulation mortar. *Concrete* **2012**, *5*, 108–110.
8. Joseph, J.D.R.; Prabakar, J.; Alagusundaramoorthy, P. Precast concrete sandwich one-way slabs under flexural loading. *Eng. Struct.* **2017**, *138*, 447–457. [CrossRef]
9. Sharkawi, N.; Baharun, A. Effect of Different Types of Paint to Residential Building. *J. Civ. Eng. Sci. Technol.* **2016**, *7*, 68–74. [CrossRef]
10. Bai, F.; Davidson, J.S. Analysis of partially composite foam insulated concrete sandwich structures. *Eng. Struct.* **2015**, *91*, 197–209. [CrossRef]
11. O'Hegarty, R.; Kinnane, O. Review of precast concrete sandwich panels and their innovations. *Constr. Build. Mater.* **2020**, *233*, 117145. [CrossRef]
12. Xu, D.F.; Chen, L.M.; Tao, S.N.; Cheng, S.C.; Xiao, F. The Study on Connector's Mechanical Performance of Energy-Saving Precast Concrete Sandwich Panel Structure. In Proceedings of the International Conference on Energy Development and Environmental Protection, Shenzhen, China, 17–18 September 2016; pp. 100–106.
13. Chen, Z.H.; Zhoum, T.; Lium, H.B.; Chen, Y.H.; Wang, X.D. Experimental and Theoretical Study on Structural Behavior of Fabric Reinforced Concrete Sandwich Panels. *Adv. Sci. Lett.* **2011**, *4*, 663–668. [CrossRef]
14. Zhu, M.C.; Liu, Z.R.; Chen, Z.H. Performance analysis of prestressed concrete sandwich composite slabs. *China Concr. Cem. Prod.* **2001**, *5*, 47–49.
15. Li, Y.B.; Du, Y.Y.; Men, D.H. Analysis on bending performance of concrete sandwich panels. *Build. Struct.* **2015**, *45*, 91–95.
16. Chen, H.Q.; Xie, J.; Zhu, L.M.; Dai, Z.Q. Experimental Research on Flexural Performance of Concrete Sandwich Two-Way Panels Reinforced by Bi-Steel. *J. Tianjin Univ.* **2009**, *42*, 35–40.
17. Luo, B.; Huang, W.; Ma, X.; Li, B.; Zhou, W.C.; Ren, S.S. Flexural behavior of sandwich composite panels with core material of expanded polystyrene thermal insulation motar. *J. Zhejiang Univ. Eng. Sci.* **2019**, *53*, 2185–2196.

18. Joseph, J.D.R.; Prabakar, J.; Alagusundaramoorthy, P. Flexural behavior of precast concrete sandwich panels under different loading conditions such as punching and bending. *Alex. Eng. J.* **2018**, *57*, 309–320. [CrossRef]
19. Joseph, J.D.R.; Prabakar, J.; Alagusundaramoorthy, P. Insulated precast concrete sandwich panels under punching and bending. *PCI J.* **2019**, *64*, 68–79. [CrossRef]
20. Ahmad, A.; Singh, Y. Flexural behavior of Expanded Polystyrene core Reinforced Concrete Sandwich Panels with different construction methods and end conditions. *Structures* **2021**, *34*, 2900–2911. [CrossRef]
21. Tomlinson, D.; Fam, A. Analytical approach to flexural response of partially composite insulated concrete sandwich walls used for cladding. *Eng. Struct.* **2016**, *122*, 251–266. [CrossRef]
22. Xu, M.G.; Wang, X.; Ding, Z.W. Development of Out-Wall External Thermal Insulation System Based on Rock-Wool. *Appl. Mech. Mater.* **2012**, *174–177*, 1589–1592.
23. Qi, B.; Zheng, C.; Lv, H.; Li, C.Z.; Le, B.L.; Zhang, F. Research and Application of Phenolic Resin in the External Thermal Insulation and Fire Resistance Composite System. *Eng. Plast. Appl.* **2012**, *40*, 28–30.
24. Wang, L.; Jin, L.Z.; Wang, D.W. Experimental study on flexural behavior of UHPC aminated plates with different superimposed surfaces. *J. Build. Struct.* **2018**, *39* (Suppl. S2), 43–51.
25. Zhang, X.Y. Experimental research on factors influencing the rigidity of concrete composite slab with steel bar truss. *J. Hefei Univ. Technol. Nat. Sci.* **2014**, *9*, 1122–1126.
26. De Seixas Leal, L.A.A.; de Miranda Batista, E. Composite floor system with CFS trussed beams, concrete slab and innovative shear connectors. *REM Int. Eng. J.* **2020**, *73*, 23–31. [CrossRef]
27. Cheng, J.Y.; Zhao, L.; Yang, J.J. Study on Short-Term Rigidity of Precast Composite Slab with Steel Truss and Concrete. *Adv. Mater. Res.* **2013**, *639–640*, 198–205.
28. Huang, W.; Luo, B.; Li, B.; Xu, X.K.; Su, Y.J. Experiment on Flexural Behavior of Green Concrete Composite Slab with Different Structural Forms. *J. Hunan Univ. Nat. Sci.* **2019**, *46*, 35–44.
29. Huang, H.L.; Wu, F.B.; Zhu, M.Q.; Zeng, C.J.; Lv, W.R. Influence of rib details on flexural behavior of concrete composite slab with precast prestressed ribbed panel. *J. Build. Struct.* **2015**, *36*, 66–72.
30. Thanoon, W.A.; Yardim, Y.; Jaafar, M.S.; Noorzaei, J. Development of interlocking mechanism for shear transfer in composite floor. *Constr. Build. Mater.* **2010**, *24*, 2604–2611. [CrossRef]
31. Chen, Z.H.; Wang, J.; Cui, X.T.; Jiang, J.Q.; Wang, Z.Y.; Qiu, H.F.; Huang, Z.Y. Field test study on bearing capacity of assembled concrete composite slab. *New Build. Mater.* **2020**, *47*, 49–52.
32. Hao, M.J.; Zheng, W.Z.; Chang, W. Evaluating Stress-Strain Properties of Reinforcing Steel for Reinforced Concrete. *Adv. Civ. Eng. Mater.* **2020**, *9*, 283–297. [CrossRef]
33. *GB/T 50081-2019*; Standard for Test Methods of Concrete Physical and Mechanical Properties. National Standard of the People's Republic of China: Beijing, China, 2019.
34. *GB 50010-2010*; Code for Design of Concrete Structures. National Standard of the People's Republic of China: Beijing, China, 2010.
35. Ma, X.; Huang, W.; Luo, B.; Zhang, J.R.; Xie, H.Q. Reviews of concrete composite slabs—Mechanical properties and design methods. *Concrete* **2020**, *364*, 129–133.
36. Huang, J.Q.; Xu, Y.Y.; Huang, H.; Dai, J.G. Structural behavior of FRP connector enabled precast geopolymer concrete sandwich panels subjected to one-side fire exposure. *Fire Saf. J.* **2022**, *128*, 103524. [CrossRef]
37. Lameiras, R.; Barros, J.A.O.; Valente, I.B.; Poletti, E.; Azevedo, M.; Azenha, M. Seismic behaviour of precast sandwich wall panels of steel fibre reinforced concrete layers and fibre reinforced polymer connectors. *Eng. Struct.* **2021**, *237*, 112149. [CrossRef]
38. Ma, S.; Hou, D.F.; Bao, P.; Wang, D.T. Influence of alkali-resistant glass fiber on seismic performance of precast ceramsite concrete sandwich wall panels. *Structures* **2022**, *38*, 94–107. [CrossRef]
39. Xu, G.L.; Xu, Q.; Wang, J.X.; Qi, H.; Bai, Y.S. Uniaxial elastic-plastic damage constitutive model and parameters of concrete. *J. Guangxi Univ. Nat. Sci.* **2016**, *41*, 332–338.
40. Dai, Y.; Nie, S.F.; Zhou, T.H. Finite element analysis of hysteretic behavior of square steel tube confined steel reinforced concrete column steel frame ring beam joint. *J. Jilin Univ. Eng. Technol. Ed.* **2018**, *48*, 1426–1435.

coatings

MDPI

Article

Research and Experimental Application of New Slurry Proportioning for Slag Improvement of EPB Shield Crossing Sand and Gravel Layer

Yongshuai Sun [1,*] and Dongjie Zhao [2]

[1] College of Water Resources & Civil Engineering, China Agricultural University, Beijing 100083, China
[2] Zhongbing Survey and Design Research Institute Co., Ltd., Beijing 100053, China
* Correspondence: causys666@163.com

Abstract: Based on the construction practice of the Beijing Metro Line 10 shield tunneling project, this paper describes research on soil improvement technology for Beijing stratum characteristics (sandy gravel stratum) and covers similar engineering conditions. It also describes the development of a new type of mud improver. Based on the laboratory tests with bentonite as the base mud and different additives, the effects of guar gum, CMC, xanthan gum, and clay medium particles on mud performance are analyzed. Then, two kinds of mud were used to conduct indoor simulated muck improvement tests (mixing test, slump test, sliding plate test, and adhesion resistance test), and the improvement effects of new mud and ordinary mud applied in pebble/round gravel and sandy soil layers were compared. The results show that xanthan gum is the best material to improve the performance of slurry, using the contrast test of bentonite-based slurry and different additives. The optimum slurry preparation scheme is 4% bentonite, 0.2% xanthan gum, 0.04% soda ash, and 1% clay particles. Using indoor simulated muck improvement tests (mixing test, slump test, slide plate test, and adhesion resistance test), the improvement effects of applying new mud and ordinary mud in pebble/round gravel and sandy soil layers are compared, and the advantages of the new mud in the application of the above two formations are verified. Among them, the new slurry has great advantages for improving the two parameters of the soil adhesion resistance coefficient and slump during shield tunneling. When the improved soil mass reaches the flowing plastic state, the proportion of new mud added to different soil mass is different. The proportion of new mud added to improved pebble/pebble soil is 28%, and that of sand and clay is 32%. It can be seen that new mud is more suitable for improving pebble/pebble soil.

Keywords: EPB shield; fluid-plastic state; muck improvement; new mud; xanthan gum; sandy gravel stratum

Citation: Sun, Y.; Zhao, D. Research and Experimental Application of New Slurry Proportioning for Slag Improvement of EPB Shield Crossing Sand and Gravel Layer. *Coatings* **2022**, *12*, 1961. https://doi.org/10.3390/coatings12121961

Academic Editor: Valeria Vignali

Received: 4 November 2022
Accepted: 10 December 2022
Published: 14 December 2022

Publisher's Note: MDPI stays neutral with regard to jurisdictional claims in published maps and institutional affiliations.

1. Introduction

With the rapid development of urban rail transit construction, subway shield construction has been widely used with its unique advantages [1]. The EPB shield tunneling in Beijing is often carried out in sandy gravel soil [2–4], and the excavated soil mass in the natural state cannot meet the requirements of plastic fluidization of soil mass [5–7]. Therefore, this can easily lead to the instability of the excavation surface [8–12], and cause the closure of the pressure chamber [13–15], the caking of the pressure chamber [16], gushing [17,18], and other technical problems often encountered in the construction of soil layers. To ensure the normal operation of shield tunneling, the soil must have good plasticity [19,20] and low permeability [21–23].

As one of the key technologies of EPB shield, muck improvement technology has been paid more and more attention in current shield research [13,24,25]. At present, the improvement additives mainly include foam agents, dispersant, clay minerals and flocculant. The clay minerals are predominantly bentonite and clay [26,27], which are mainly used to supplement the fine particles in order to reduce the internal friction angle of the soil mass in the pressure chamber [28]. Foam agent is composed of special foam agent and compressed air, which can be used to improve the fluidity [29,30] and impermeability [31,32] of the excavated soil mass. Flocculant is mainly applicable to water-rich consolidated clay, silt, sand and fine gravel formation, which can cause particles to agglomerate in the slag to improve the plastic flow behavior of the slag [33,34]. Dispersants can disperse substances in the water to form colloidal solutions, reducing the adhesion between particles [35,36]. At present, different scholars have conducted a great deal of research on the improvement effect of the selection and proportion of these soil improvers on the EPB shield construction, and these improvement evaluations are mostly seen in indoor tests.

Relying on a tunnel project in Fuzhou, Wang, S.M [37] carried out a study on the construction of an earth pressure shield machine in high permeability and high-pressure formation by using nano bentonite CMC as the muck improver, and they evaluated the effect of this soil improver through field tests. Zhu H. H and others [38–41] evaluated the stability of the earth pressure shield in the sandy gravel stratum by establishing the DEM model, and proposed measures to improve the fluidity and wear resistance. Zhao S. S. [36] discussed the effect of adding dispersant in mud on controlling clay cake, and verified the effectiveness of the dispersant in reducing viscosity through laboratory tests. Cheng Chihao and others [42,43] carried out laboratory soil improvement test research on the micro mechanism of soil improvement with different proportioning of foam and bentonite in the EPB shield construction, and evaluated the improvement effect of different schemes in combination with field tests. Jiang Houting and others [44–46] conducted relevant tests on foam improving round gravel formation during EPB shield construction and obtained the experimental law of foam's impact on plastic fluidity of round gravel formation and the optimal foam injection ratio suitable for EPB shield construction. Xiao Chao [47] studied the influence of modified bentonite grouting on the shield driving in the sand gravel layer by simulating the earth pressure balance shield system, and obtained the mixed proportion of modified bentonite slurry for the earth pressure shield driving in the sand gravel layer. Xu Q. W and others [48–51] injected different amounts of air bubbles into the soil with different water content to reflect the effect of air bubble mixed soil on torque reduction through the reduction rate of power consumption of the mixer. The results show that adding air bubbles to sandy pebble soil can reduce the power consumption of mixing by more than 50%.

In combination with the current research on the improvement of EPB shield tunneling, there is a lack of systematic theoretical evaluation standards and methods for the basic properties of the improver, the matching relationship between the improver and the improved soil, the improvement effect of the improver, and the relationship between the improver and the shield construction parameters, which leads to the blindness of the use of the improver in shield tunneling, and may cause problems such as large material consumption and environmental pollution. Therefore, based on the construction practice of Beijing Metro Line 10 shield tunneling project, this paper conducts research on soil improvement technology for Beijing stratum characteristics (sandy gravel stratum) and covers similar engineering conditions, as well as developing a new type of mud improver. Based on the laboratory tests with bentonite as the base mud and different additives, the effects of guar gum, CMC, xanthan gum, and clay medium particles on the mud performance are analyzed, and a new mud performance improvement material with the best performance is obtained. Then, two kinds of mud were used to conduct indoor simulated muck improvement tests (mixing test, slump test, sliding plate test, and adhesion resistance test), and the improvement effects of new mud and ordinary mud applied in pebble/round gravel and sandy soil layers were compared.

2. Study on Properties of New Mud Modifier

The soil state control in the pressure chamber is the key to EPB shield construction. The soil state in the pressure chamber is different in different strata. When most undisturbed soil enters the pressure chamber, it will cause construction difficulties, which are manifested in the increased torque of the cutterhead, the dumping efficiency of the screw conveyor, and the thrust of the shielding jack, which will lead to "blocking", "gushing", "caking" and other construction problems, and even lead to the instability of the excavation surface. To solve these problems, the ideal state of the soil mass in the pressure chamber, namely the "plastic flow state", is proposed. Its physical and mechanical indexes include the permeability coefficient, internal friction angle, slump, and compression coefficient. To effectively solve these problems in construction, it is necessary to add additives to the soil for soil improvement.

2.1. Site Mud Solution

The stability of the earth pressure balance shield excavation surface is achieved by the pressure of the sealed cabin. When the mud penetrates the soil, a layer of mud film with very low permeability is formed, and the mud pressure effectively acts on the excavation surface through the mud film, so it can prevent the deformation and collapse of the excavation surface and ensure the stability of the excavation surface. One of the key issues in earth pressure shield construction is to use of a muck modifier suitable for engineering geological conditions.

To compare the development of the mud, the mud configuration scheme adopted in the construction site was studied first. The mud used on the construction site is pure bentonite mud with a bentonite content of 9%, that is, 90~120 Kg of bentonite is added to one side of the water, and the performance parameters are shown in Table 1:

Table 1. Site mud performance parameters.

Mud Performance Index			
Density (g/cm^3)	1.04	pH	10
Marsh funnel viscosity (")	40	Static filtration (mL/30 min)	21
Apparent viscosity η_A(MPa·s)	13.5	Plastic viscosity η_p(MPa·s)	3
Dynamic shear τ_d(Pa)	10.7	Dynamic to plastic ratio τ_d/η_p	3.6
Initial static shear force (Pa)	9.2	Final shear force (Pa)	10.7
Liquidity Index	0.17	Consistency coefficient (Pa·sn)	4.2
Colloid rate	90%		

It can be seen from the above parameters that the mud of this formula has a high consistency and does not flow easily. After standing, it is even more difficult for the mud in the test tube to flow out. Therefore, the comprehensive performance of the mud should be improved by adding other additives. At the same time, the cost can be reduced by reducing the amount of bentonite.

2.2. New Mud Solution

In this paper, a new type of soil amendment that combines mineral improvement materials and water-soluble polymer improvement materials is developed. Bentonite is the main raw material of mineral modifiers, and the mineral material used in the test is the red sodium bentonite produced by the Shamaying Sodium Clay Factory in Weifang City, Shandong Province, which is used in the construction site. The solution is prepared as follows: add water to the container, stir with a mixer at medium speed, and at the same time add the sodium bentonite and other materials evenly into the container and stir. If the solution is thick, stop the mixer every few minutes. Stir the bentonite attached to the bottom layer with your hands or with sticks to speed up the dissolution, then stir at medium speed for about ten minutes.

On the one hand, if only bentonite is used to prepare slurry, the amount of bentonite is large, which leads to high slurry cost. On the other hand, the performance of the slurry is difficult to guarantee, which cannot meet the requirements for slag improvement in the EPB shield. It is therefore necessary to optimize the mud performance by adding appropriate additives. Common additives mainly include: 1. tackifier: guar gum, sodium carboxymethyl cellulose (CMC), xanthan gum, polyacrylamide; 2. lubricant; 3. PH value regulator, mainly sodium carbonate; 4. other additives such as clay particles.

Soda ash has greatly improved the performance of mud. Therefore, a proper amount of soda ash is added to the subsequent formula to improve the PH value of mud. According to the mud configuration plan of the construction site, the bentonite dosage is 9%, but the thickness is too high and the viscosity is not high. The topic of this lesson is to reduce the bentonite dosage to reduce the cost and improve the mud performance to adapt to the earth pressure balance shield. Therefore, the bentonite dosage is preliminarily determined to be 3%~5%.

In this paper, four additives will be added to evaluate the mud improvement effect, and the research scheme of the mud improvement test with different additives is shown in Table 2.

Table 2. Test Design Scheme of Different Additives.

Drugs / Scheme	Water (g)	Bentonite (g)	Sodium Carbonate (g)	Guar Gum (g)	Sodium Carboxymethylcellulose (g)	Xanthan Gum (g)	Polyacrylamide (g)
F1	1000	30 (4%)	0.5 (0.05%)	2 (0.2%)	/	/	/
F2	1000	40 (5%)	0.5 (0.05%)	2 (0.2%)	/	/	/
F3	1000	50 (3%)	0.5 (0.05%)	2 (0.2%)	/	/	/
F4	1000	30 (4%)	0.5 (0.05%)	/	2 (0.2%)	/	/
F5	1000	40 (3%)	0.5 (0.05%)	/	2 (0.2%)	/	/
F6	1000	50 (3%)	0.5 (0.05%)	/	2 (0.2%)	/	/
F7	1000	30 (3%)	0.5 (0.05%)	/	/	2 (0.2%)	/
F8	1000	40 (3%)	0.5 (0.05%)	/	/	2 (0.2%)	/
F9	1000	50 (3%)	0.5 (0.05%)	/	/	2 (0.2%)	/
F10	1000	30 (3%)	0.5 (0.05%)	/	/	/	2 (0.2%)
F11	1000	40 (3%)	0.5 (0.05%)	/	/	/	2 (0.2%)
F12	1000	50 (3%)	0.5 (0.05%)	/	/	/	2 (0.2%)

Where AV represents apparent viscosity, PV represents plastic viscosity, YP represents dynamic shear, Gelin represents initial shear, Gel10 represents final shear, FL represents filtration, K represents consistency index, n represents fluidity coefficient, and YP/PV represents a dynamic plastic ratio.

See Table 3 for data of mud properties with different additives.

Table 3. Mud Performance Test Data of Different Schemes.

Scheme	AV (Mpa·s)	PV (Mpa·s)	YP (Pa)	Gelin Pa	Gel10 (Pa)	FL (mL)	K (Pa.sn)	n	Marsh Funnel Viscosity S	Colloid Rate (%)	YP/PV
F0	13.5	3	10.7	9.2	10.7	21	4.2	0.17	40	90%	3.6
F1	9	7	2.04	1.53	2.04		1.48	0.21	35	80%	0.3
F2	15.5	10	5.6	2.6	3.6		0.33	0.56	38	80%	0.56
F3	18	13	8.2	3.1	4.1		0.71	0.47	40	82%	0.63
F4	9.5	8	1.53	1.53	2.04	19	0.04	0.79	36	99%	0.19
F5	14	11	3.07	2.04	2.56	18	0.10	0.72	41	99%	0.28
F6	18	13	5.11	3.57	4.09	17	0.20	0.65	54	99%	0.39
F7	14.5	7	7.67	4.09	6.13	21	0.93	0.40	47	99%	1.10
F8	21.5	10	11.8	7.67	9.20	19	1.58	0.38	61	99%	1.18
F9	26	9	17.4	8.69	9.20	17	2.82	0.33	94	100%	1.93

It can be seen from Table 3 that the overall viscosity of the slurry added with guar gum is low, and the colloid rate is about 80%, indicating that there is more water precipitation. After adding CMC, the mud performance is obviously improved, the viscosity is improved, and the filtration is within 20 mL/30 min, meeting the requirements of filtration. The colloid rate is also greatly improved. During the preparation of the three schemes F10-F12, a large number of insoluble matters were precipitated, most of which were white blocky colloids. During the API filtration test, the liquid continued to flow downwards, with a filtration loss of about 45 mL/30 min, which seriously exceeded the requirements. Therefore, the application of polyacrylamide in mud was extremely insoluble, and the viscosity increasing effect of the mud was not ideal, so it was abandoned in future tests.

2.3. Comparative Study on Mud Performance after Adding Different Additives

2.3.1. Viscosity Contrast

The plastic viscosity is the total reflection of the internal friction between solid particles, between solid particles and liquid phases, and between liquid molecules when the slurry flows. Apparent viscosity refers to the ratio of shear stress to the shear rate at a certain shear rate, which is the sum of plastic viscosity and structural viscosity.

The viscosity comparison of different mud schemes is shown in Figure 1.

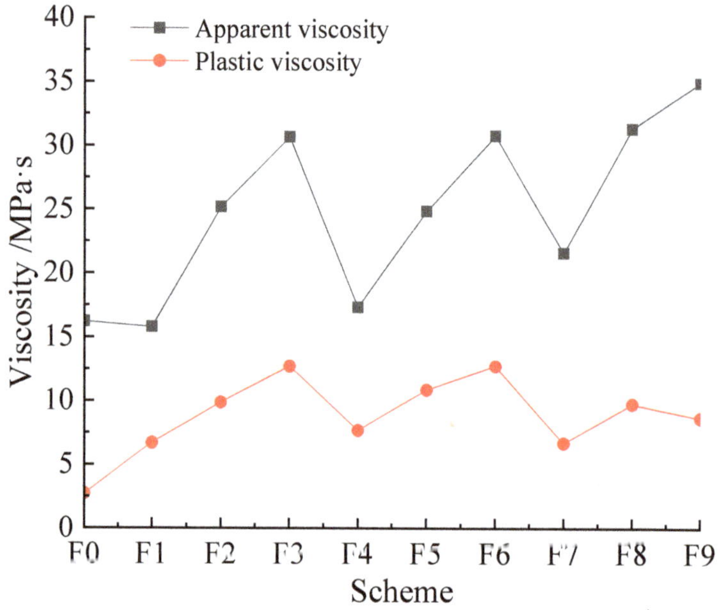

Figure 1. Viscosity Comparison of Different Mud Schemes.

It can be seen in Figure 1 above that the apparent viscosity of all schemes is higher than F0 except that the F1 and F4 schemes are lower than F0 of mud for a construction site. According to the requirements for the sand gravel layer of Beijing Metro Line 10, the slurry viscosity is high and favorable, so F1 and F4 schemes were abandoned. In the scheme of adding guar gum, CMC and xanthan gum, the formula with a bentonite dosage of 5% has the highest apparent viscosity value of mud, among which F9 is the highest, indicating that xanthan gum is superior to guar gum and CMC for improving the apparent viscosity of mud. In terms of plastic viscosity, the plastic viscosity of mud can be improved by adding a viscosity increaser, and the rule is roughly the same as the apparent viscosity. In comprehensive consideration, xanthan gum is higher than guar gum and CMC in improving mud viscosity.

2.3.2. Dynamic Shear Force Comparison

Microscopically, the dynamic shear force is the reaction of space grid structure force formed between clay particles when slurry flows, which can reflect the ability of slurry to carry sand and gravel to a certain extent. The dynamic shear force comparison of each scheme is shown in Figure 2.

As can be seen in Figure 2 above, among the schemes, only the F8 and F9 schemes have higher mud dynamic shear force values than the F0 scheme, indicating that only the addition of xanthan gum can effectively improve the mud dynamic shear force.

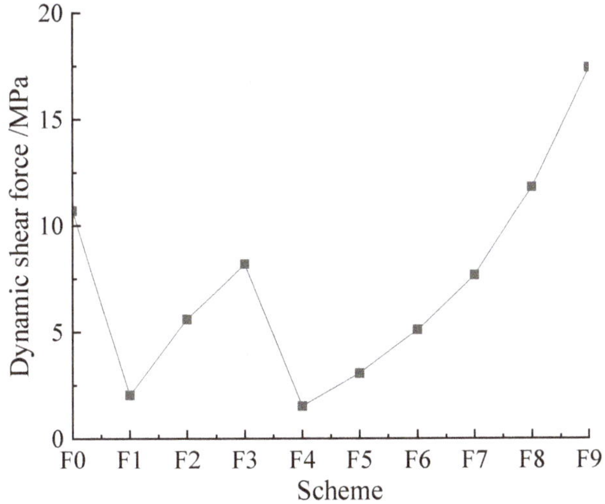

Figure 2. Dynamic shear force change diagram.

2.3.3. Static Shear Force Comparison

The static shear force comparison of each scheme is shown in Figure 3.

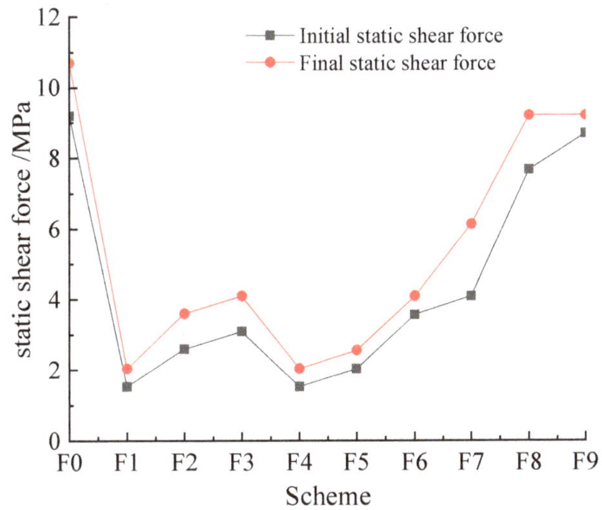

Figure 3. Variation diagram of static shear force of different muds.

The static shear force reflects the strength of the internal gel network structure when the drilling fluid is in a static state. A small initial shear force is conducive to reducing the starting power of the shield machine, so the initial shear force should be appropriately reduced; for larger sand and pebble formations, the large final shear force is conducive to the suspension and transportation of mud to the muck, so it should be appropriately increased. From the comparison of the above figures, it can be seen that except for schemes F8 and F9, the static shear forces of the other schemes are relatively small and unfavorable to the sandy cobble stratum. Only the plans F8 and F9 of adding xanthan gum are close to the construction site mud, and the initial shear force and final shear force of scheme F8 are more suitable.

2.3.4. Dynamic Plastic Ratio Comparison

The dynamic plastic ratio (τ_d/η_p) is an important indicator to measure the shear dilution characteristics of mud. Given the particularity of shield soil improvement materials, to prevent the excavated soil from settling and accumulating at the bottom of the earth ballast tank, the mud is required to have a greater rock-carrying capacity, and the dynamic-plastic ratio is recommended to be between 0.5–1.5 Pa/MPa·s in this test. The dynamic-plastic ratio comparison of each mud scheme is shown in Figure 4.

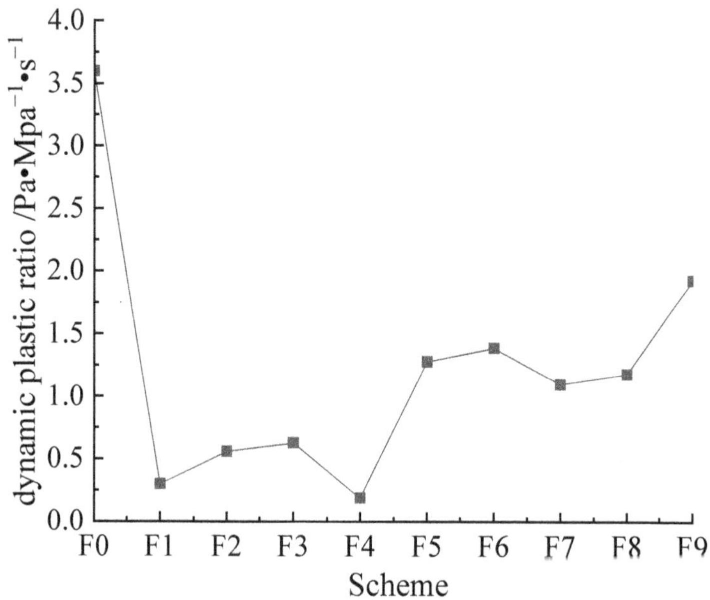

Figure 4. Changes in the dynamic-plastic ratio of different schemes.

An analysis of Figure 4 above shows that the F0 dynamic-plastic ratio of the mud scheme used on the construction site was too large, it is not conducive to the pumping of mud, and the dynamic-plastic ratio of the mud in the schemes F1~F6 was too small, which is not conducive to carrying cuttings. Scheme F8 is more suitable.

2.3.5. Flowability Index and Consistency Coefficient Comparison

The fluidity index n represents the non-Newtonian degree of the fluid within a certain flow velocity range, and the fluidity index is the representation of the fluidity of the mud. According to the knowledge of drilling fluids, the popularity index n is best around 0.5, and generally less than 1. The consistency coefficient k is the viscosity of the fluid under the 1 s^{-1} flow velocity gradient, and it is a reflection of the viscosity of the fluid; the larger the consistency coefficient, the more viscous the fluid. For the larger sand and pebble formation, the K value of the earth pressure balance shield should be 1.5~2. Figure 5 shows the fluidity index and consistency coefficient of the mud in each scheme.

An analysis of the Figure 5 above shows that the fluidity index of the mud F0 scheme used in the construction site is 0.17, which is too small; it will influence the effect of mud carrying gravel, and if the consistency coefficient is too large, the pumpability of the slurry will be poor. The consistency coefficients of schemes F2~F7 are all less than 1. It is more appropriate to comprehensively consider the fluidity index and consistency coefficient of scheme F8.

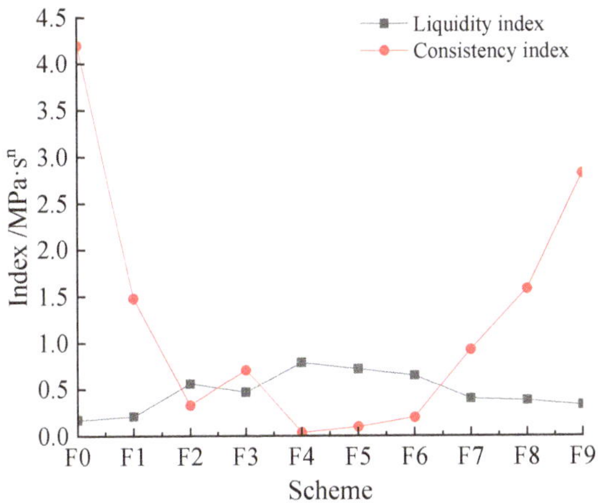

Figure 5. Changes of fluidity index and consistency coefficient for different schemes.

2.3.6. API Filtration Loss Comparison

Filtration loss is the degree of stability of the mud. The amount of water filtered by the mud is small, which can make the pore water pressure close to the excavation face rise less, ensuring that the effective mud-water pressure acting on the digging face remains unchanged after the mud film is formed, conducive to the stability of the excavation surface. To ensure mud performance, the smaller the water loss, the better. See Figure 6 for the comparison of mud filtration in each scheme.

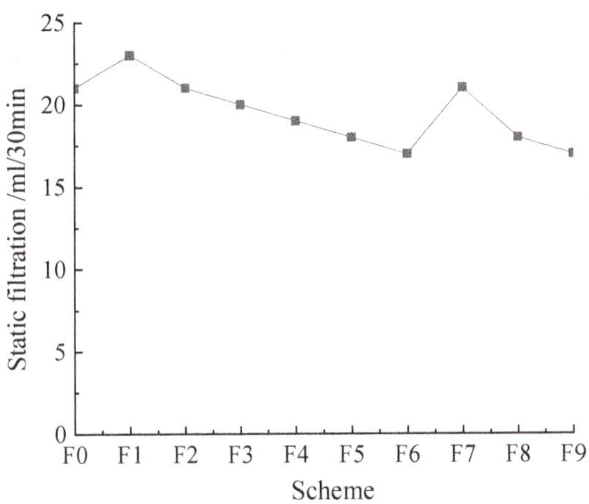

Figure 6. Variation diagram of filtration volume in different schemes.

For the earth pressure balance shield, the filtration loss is generally required to be less than 20 mL/30 min. Therefore, schemes F4, F5, F6, F8, and F9 can all meet the requirements.

2.4. Comprehensive Analysis of Mud Plan

Through the above comparative analysis of each index parameter of the mud of each scheme, it can be concluded that xanthan gum is superior to guar gum, CMC, and polyacrylamide in improving mud performance; in particular, the index parameters of scheme F8 are the most suitable, and follow-up programs were therefore carried out concerning the ratio of program F8.

2.5. Effect of Clay Particles on Mud Properties

When clay particles are added to the slurry, the particle surface dissolves, forming a layer of coating with viscous slurry. During shield construction, it will be filled in the space between sand and gravel, which plays a role in lubrication and drag reduction, increasing plasticity and reducing cutting power. To study the influence of clay particles on mud properties, the following experiments were conducted for scheme F8.

The additional amount of clay particles should not exceed 2%. The test plan is shown in Table 4.

Table 4. Test Scheme for Adding Clay Medium Particles.

Scheme \ Drugs	Water (g)	Bentonite (g)	Sodium Carbonate (g)	Xanthan Gum (g)	Clay Particles (g)
F13	1000	40 (4%)	0.5 (0.05%)	0.5 (0.05%)	5 (0.5%)
F14	1000	40 (4%)	0.5 (0.05%)	0.5 (0.05%)	10 (1%)
F15	1000	40 (4%)	0.5 (0.05%)	0.5 (0.05%)	15 (1.5%)

The data on mud properties are shown in Table 5.

Table 5. Test Data of Influence of Clay Particles on Mud Performance.

Scheme	AV (Mpa·s)	PV (Mpa·s)	YP (Pa)	Gelin (pa)	Gel10 (Pa)	FL (mL)	K (Pa.sn)	n	Marsh Funnel Viscosity S	Colloid Fraction(%)	YP/PV
F13	20.5	11	9.7	7.67	8.18	19	0.93	0.45	63	100%	0.88
F14	22.5	12	10.5	7.67	8.18	20	0.85	0.48	65	100%	0.88
F15	23	13	10	8.18	8.69	21	0.85	0.48	67	100%	0.77

See Figure 7 for a comparison of the parameters of each scheme.

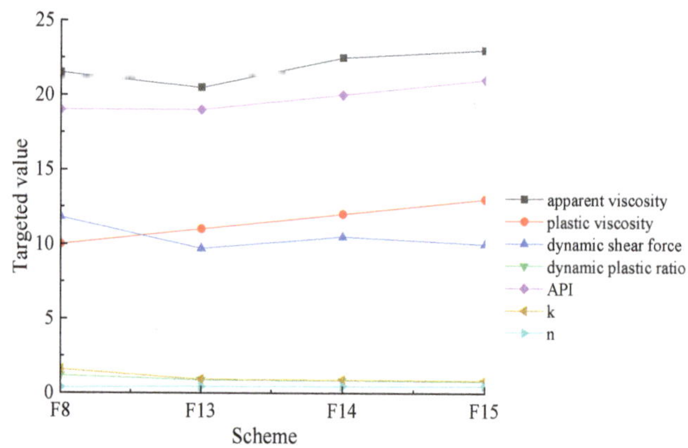

Figure 7. Comparison of clay particles' influence on mud.

It can be seen from the analysis of the above figure that the viscosity of medium particles has little impact on the overall performance of the mud. It is found in the test that the clay medium particles contain a small number of insoluble substances, and when the addition amount exceeds 1%, the insoluble substances in the mud will increase, which will have adverse effects on mud filtration and plastic viscosity. Therefore, the added amount will be controlled at 1% in the future.

3. Analysis of Indoor Muck Improvement Test

3.1. Soil Mixing Test Design

The soil mixing test mainly uses the mixing process of the mixer to simulate the soil mixing process in the pressure chamber of the earth pressure balance shield, to evaluate the improvement effect of the developed modifier on the simulated soil. The amount of soil amendment added can be controlled by changing the stirring power of the mixer. See Figure 8 for indoor mixer test. The steps of the soil stirring test are as follows:

(1) Use a plastic measuring cup to measure a certain volume of soil and pour it into the mixer, stir for 3 min, stop stirring after the reading of the power meter is stable, and record the stirring power when the reading is stable.
(2) Take out a certain amount of soil from the mixing tank and conduct the friction coefficient test, adhesion resistance test, and slump test in sequence.
(3) Pour the test soil back into the mixer, add a certain amount of soil amendment, stir the soil with a mixer for soil improvement, stir for 3 min and stop stirring after the reading of the power meter is stable. Record the stirring power when the reading is stable, and then remove the soil from the mixing drum for subsequent tests. The test cannot be terminated until the soil reaches the ideal plastic flow state. There were 3–5 groups in each experiment, and the results were averaged.

Figure 8. Indoor mixer test.

3.2. Friction Coefficient Test Design

The main purpose of the design of the friction coefficient test is to simulate the process of friction between the soil and the steel during the excavation process of the screw excavator. By measuring the external friction angle between the soil and the steel, the friction coefficient between the soil and the steel is obtained.

The specific test steps are as follows:

(1) Place the plastic ring on a stainless-steel plate, fill it with improved soil, and smooth the top of the plastic ring.
(2) Fix one end of the steel plate, and slowly lift the other end. When the ring with the improved soil starts to slide, stop and continue to lift.
(3) The angle at which the steel plate is tilted at this time is measured, which is the external friction angle φ between the improved soil and the steel plate.
(4) The friction coefficient between the improved soil and the steel plate can be obtained by the following formula. There were 3–5 groups in each experiment, and the results were averaged.

$$f = \tan \varphi$$

f—Improve the friction coefficient between soil and steel plate; *φ*—Improve the external friction angle between the soil and the steel plate.

3.3. Adhesion Resistance Test

The adhesion resistance test is to simulate the process of the soil adhesion to the steel surface during the excavation process of the screw excavator. By measuring the adhesion resistance of soil and steel, the adhesion resistance coefficient between soil and steel is obtained, which can be used to measure the fluidity of soil. The specific test steps are as follows:

(1) Place the iron cylinder on the iron box, fill it with the improved soil, and smooth the upper part of the improved soil.
(2) Lift the iron cylinder slowly, and stop lifting the iron cylinder when the iron cylinder is high enough.
(3) Measure the adhesion resistance F at this time, which is the adhesion resistance between the improved soil and the iron cylinder.
(4) After the adhesion resistance between the improved soil and the iron cylinder is measured, this does not provide the adhesion resistance between the reaction soil and the iron surface used, because this adhesion resistance only reflects the situation of this test. Therefore, this adhesion resistance should also be treated to a certain extent, obtaining the size of the adhesion resistance per unit area, to better reflect the change in the adhesion resistance between the soil and the iron cylinder. There were 3–5 groups in each experiment, and the results were averaged. The adhesion resistance coefficient k is obtained by the following formula. This adhesion resistance measuring device is shown in Figure 9.

Figure 9. Design of Adhesion Resistance Test Device.

Since the original adhesion resistance dimension measured in this test is a mass unit, to convert to weight units, first convert using the following formula:

$$F = M \times g$$

The surface area of the inner surface of the iron cylinder is:

$$S = \pi \times D \times h$$

The adhesion resistance coefficient between the improved soil and the iron cylinder is:

$$K = F/S = (M \times g)/(S = \pi \times D \times h)$$

M is the measured original lifting weight (kg); F is the adhesion resistance (N) between the improved soil and the iron cylinder; S is the surface area of the inner surface of the iron cylinder (m^2); D is the inner diameter of the iron cylinder (m); h is the height of the iron cylinder (m); K is the adhesion resistance coefficient between the improved soil and the iron cylinder (N/m^2).

3.4. Slump Test Design

The slump test uses standard test equipment, which is a standard slump tube. Its specifications are: the slump tube, the diameter of the upper mouth is 100 mm, the diameter of the lower mouth is 200 mm, and the height of the tube: 300 mm; Tamper, 16 mm diameter, 650 mm long, steel round bar with hemispherical ends. The slump test device is shown in Figure 10. Each experiment was conducted in 3-5 groups, and the average value of the data of each group was obtained.

Figure 10. Slump Test Device.

4. Results of Soil Improvement Test

The new mud formula is shown in Table 6.

Table 6. The new mud formula.

Raw Material	Bentonite	Xanthan Gum	Sodium Carbonate	Clay Particles
Increase the amount	4%	0.2%	0.04%	1%

4.1. Experimental Study on Improvement of Pebble/Gravel Soil

Through the qualitative analysis of the applicability of the new type of mud to pebble/gravel and sand, the improvement effect of the new mud soil amendment on the pebble/round gravel soil can be seen, and compared with the mud used in the construction site to determine the superiority of the new mud.

The test conditions are as follows: the amount of soil added is 7.5 L, after the initial test analysis, the amount of mud added was determined to be 18% of the soil volume (1350 mL), after each parameter measurement, the amount of mud added is 2% of the soil volume (150 mL) until the soil reaches the ideal plastic flow state.

4.1.1. Slump Test Design

The test soil is anhydrous round gravel soil, and the soil particle sieving is shown in Table 7.

Table 7. Particle size distribution of anhydrous round gravelly soil.

Particle Size(mm)	60~20	20~5	5~2	2~0.5	0.5~0.25	0.25~0.075	0.075~0.005
content(%)	30	20	15	15	10	5	5

The net power comparison curve is shown in Figure 11.

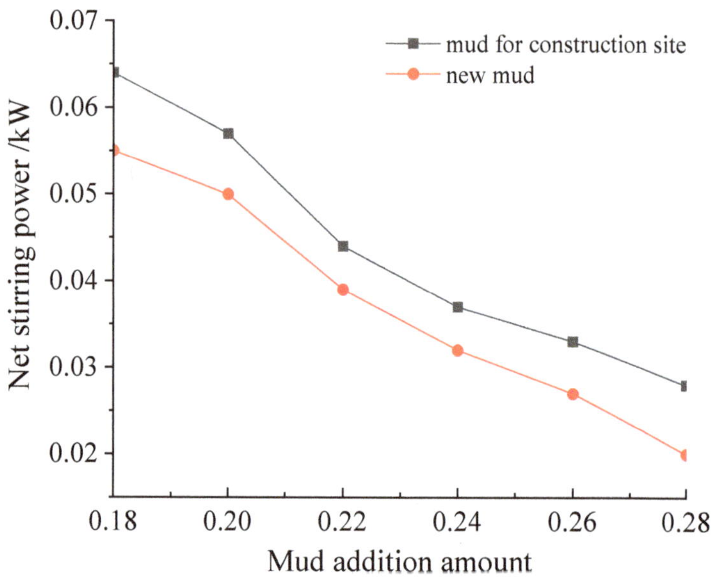

Figure 11. Comparison of net power of pebble/round gravel improved by two kinds of mud.

In Figure 11 above, it can be seen that the net stirring power of the soil is significantly reduced after adding the mud, and the influence of the two kinds of mud on the stirring power is roughly equal. It is found that when the net power is close to 0, the slag is in a better flow-plastic state. The net power of the new mud is smaller than that of the mud used on the construction site, therefore the effect of the new type of mud in reducing stirring power is better than that of the mud used in the construction site.

4.1.2. Friction Coefficient Test

The friction coefficient comparison curve is shown in Figure 12.

Through the comparison of friction coefficients, it can be seen that the friction coefficients tend to increase first and then decrease with the increase in mud addition; when the amount of mud used in the construction site is 26%, the friction coefficient has a maximum value of 0.6. The maximum value is 0.58 when the new mud addition is 22%. It was found in the test that when the friction angle was about 20°, the flow plasticity of the slag was better. When the amount of mud added was 28%, the friction coefficient of the new mud was smaller than that of the land-use mud, therefore, the performance of the new type of mud for improving the coefficient of friction of the muck is better than that of the mud used in the construction site.

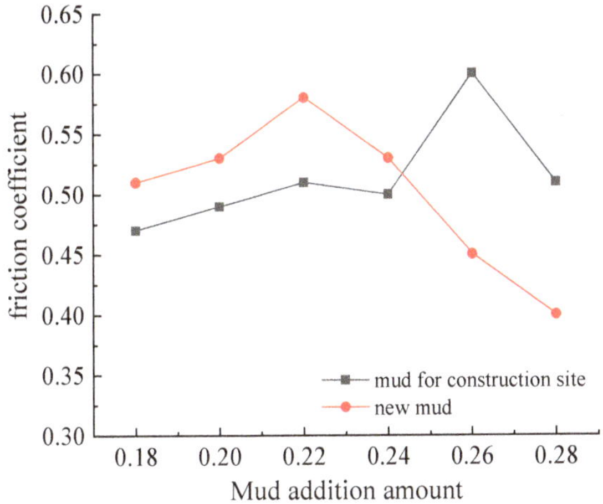

Figure 12. Comparison of friction coefficient of pebble/round gravel improved by two kinds of mud.

4.1.3. Adhesion Resistance Test

The adhesion resistance comparison curve is shown in Figure 13.

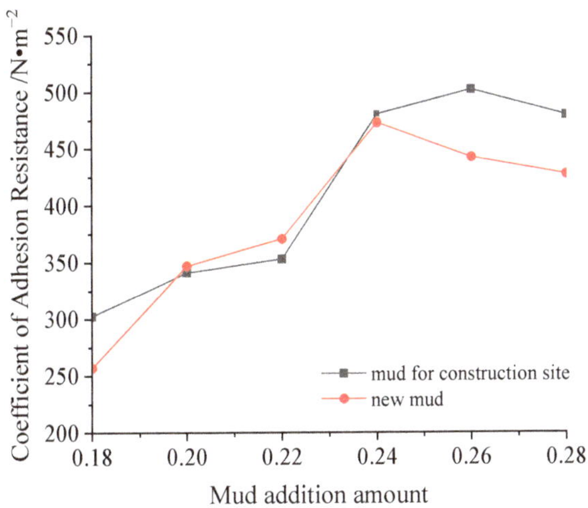

Figure 13. Comparison of adhesion resistance of pebble/round gravel improved by two kinds of mud.

It can be seen from the above curve that when the amount of mud added was less than 24%, the adhesion resistance of the slag increased with the increase in the amount of mud added. When the amount of mud added was about 28%, the adhesion resistance coefficient had a low, stable value, and the adhesion resistance coefficient of the new mud was smaller than that of the mud used on the construction site, therefore the new type of mud can better reduce the adhesion resistance coefficient of the muck, which is beneficial to the smooth discharge of the muck.

4.1.4. Slump Test

The slump comparison curve is shown in Figure 14.

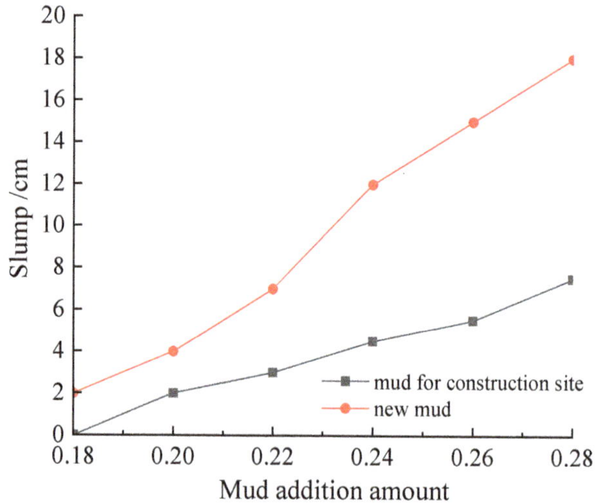

Figure 14. Comparison of the slump of pebble/round gravel improved by two kinds of mud.

It can be seen in Figure 14 above that with the increase of the amount of mud added, the slump of the muck increased significantly. When the amount of mud added was 28%, the slump of mud used on site was only 7.5 cm, and the soil slump requirement (12~20 cm) of the earth pressure balance shield was not met, while the slump of the new type of mud is 18 cm, which meets the requirement. Therefore, the new mud was better than the mud used on the construction site for improving the slump of the muck.

4.2. Experimental Study on Sandy Soil Improvement

Through qualitative analysis of the applicability of soil amendment and sandy soil, the applicability relationship between soil amendment and sandy soil was determined.

The soil volume used in the test analysis process was 7.5 L, and the amount of soil amendment added was determined according to the volume of the soil. The test soil was anhydrous sand, and the particle size distribution of the sand is shown in Table 8. After initial experimental analysis, the initial amount of mud added was determined to be 20% of the soil volume (1500 mL): After each parameter measurement, the amount of mud added was 4% of the soil volume (300 mL) until the soil reached the ideal plastic flow state. The following sub-soil improvement test categories were analyzed for the test results.

Table 8. Sand particle size distribution.

Particle Size (mm)	2~0.5	0.5~0.25	0.25~0.075	0.075~0.005
Content (%)	30	30	30	10

4.2.1. Sand Mixing Test

The net power comparison curve is shown in Figure 15.

When mud was added to the sand, the net stirring power was significantly reduced. When the mud addition amount was 32%, the net stirring power of the new mud was 0.002 KW, which means that the sandy soil is in a good fluid plastic state. The net stirring power of the mud used in the construction site was 0.009, significantly higher than the new type of mud, therefore the improved performance of the new mud on sandy soil was better than that of mud used on site.

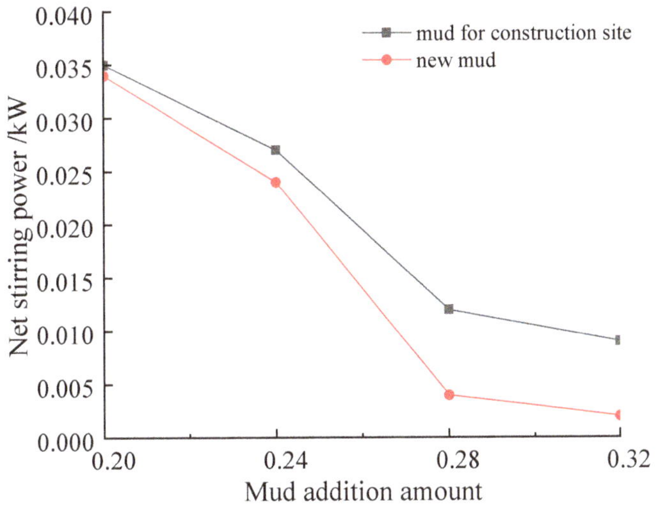

Figure 15. Comparison of net power of sand improved by two kinds of mud.

4.2.2. Sand Friction Coefficient Test

The friction coefficient comparison curve is shown in Figure 16.

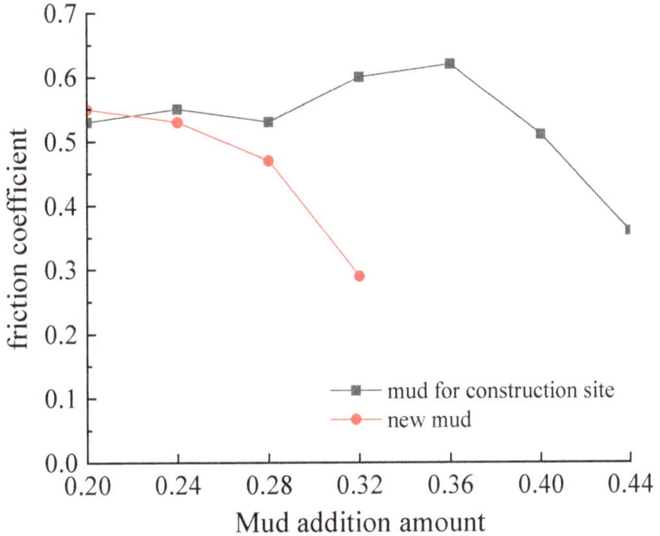

Figure 16. Comparison of friction coefficient of the sand improved by two kinds of mud.

As can be seen in Figure 16 above, when the mud addition amount was 32%, the friction coefficient of the new mud was 0.29, which meets the requirements of fluid plasticity. The construction site mud rose to a maximum value of 0.62, and then gradually decreased. When the amount of mud added to the construction site was 44%, the friction coefficient was reduced to 0.36, therefore the new type of mud was superior to the mud used on the construction site.

4.2.3. Sand Adhesion Resistance Test

The comparative analysis curve of adhesion resistance is shown in Figure 17.

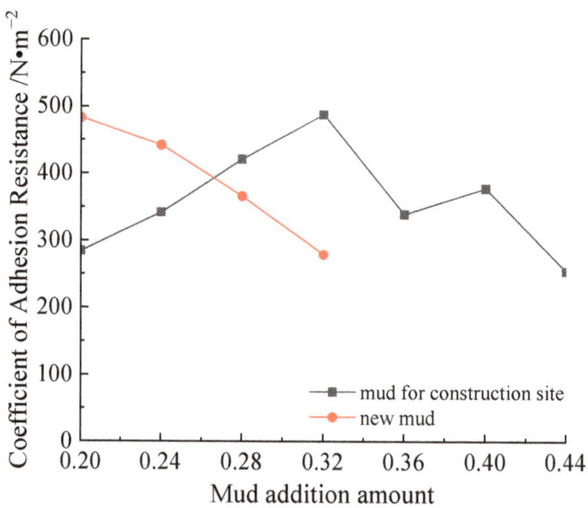

Figure 17. Comparison of adhesion resistance of sand improved by two kinds of mud.

It can be seen in the above figure that with the increase in the amount of mud added, the sandy soil increases, and the adhesion resistance coefficient first increases and then decreases. After adding the new mud to 20%, the adhesion resistance coefficient began to decrease. When the addition amount is 32%, the mud is in a fluid plastic state. The adhesion resistance reached the maximum value when the amount of mud used in the construction site was 32%; after that, it began to decrease. When the addition amount was 44%, the sand was in a fluid plastic state. Therefore, the improvement effect of the new mud was better than that of the mud used on the construction site.

4.2.4. Sand Slump Test

The slump comparison curve is shown in Figure 18.

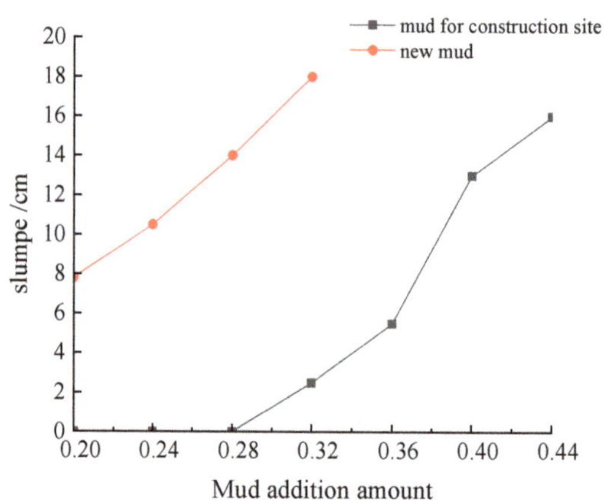

Figure 18. Comparison of the sand slump of pebble/round gravel improved by two kinds of mud.

From Figure 18 above, it can be seen that the sand slump gradually increased with the increase in the amount of mud. When the amount of new mud was 32%, the slump was 18 cm, which meets the requirements. The slump did not reach 16 cm until the amount of mud used on the construction site was 44%. Therefore, the new mud can effectively improve the sand slump.

5. Conclusions

(i) Through laboratory tests, the effects of guar gum, CMC, xanthan gum, and polyacrylamide, which are different additives, on the performance of the slurry are analyzed with bentonite as the base slurry. It was found that xanthan gum was the best material to improve the performance of the slurry. The best slurry preparation scheme is 4% bentonite, 0.2% xanthan gum, 0.04% soda ash, and 1% clay particles.

(ii) Through indoor simulated muck improvement tests (mixing test, slump test, slide plate test, and adhesion resistance test), the improvement effects of applying new mud and ordinary mud in pebble/round gravel and sandy soil layers were compared to verify the superiority of the new mud in the application of the above two kinds of formations.

(iii) In the soil improvement test, the effect of the new mud was similar to that of ordinary mud. However, it had greater advantages in improving the two parameters of the soil adhesion resistance coefficient and slumps during shield tunneling.

(iv) When used to improve pebble/round gravel soil, the test showed that when the net power was close to 0, the slag soil was in a good flow plastic state, and the net power of the new mud was less than that of the mud used in the construction site. When the friction angle was about 20°, the flow plasticity of slag was better. When the mud dosage was 28%, the friction coefficient of the new mud was smaller than that of the land mud. When the amount of mud was about 28%, the adhesion resistance coefficient had a low, stable value, and the adhesion resistance coefficient of the new mud was smaller than that of the mud used in the construction site. When the amount of mud added was 28%, the slump of the mud used at the construction site was only 7.5 cm, which cannot meet the requirements of the earth pressure balance shield on the slump of the muck, while the slump of the new mud was 18 cm, which meets the requirements.

(v) When used to improve the sand, the test found that when the amount of mud added was 32%, the net mixing power of the new mud was 0.002 KW, indicating that the sand was in a good flow plastic state, while the net mixing power of the mud used at the construction site is 0.009, which is significantly greater than the new mud. When the amount of mud added was 32%, the friction coefficient of the new mud was 0.29, which meets the requirements of flow plasticity. The site mud rose to the maximum value of 0.62, and then the friction coefficient decreased to 0.36 as the amount of mud added reached 44%. When the new mud dosage was 32%, the mud was in the flow plastic state, while when the mud dosage was 32%, the adhesion resistance reached the maximum value, and when the mud dosage reached 44%, the sand was in the flow plastic state. When the dosage of new mud was 32%, the slump was 18 cm, meeting the requirements. When the amount of mud used at the construction site was 44%, the slump reached 16 cm.

(vi) The new mud addition ratio was 28% for improved pebble/round gravel soil, and 32% for improved sand and clay. By comparison, it can be seen that the new mud is more suitable for improved pebble/round gravel soil and meets the requirements for the improvement of sandy pebble stratum.

Author Contributions: Data curation, D.Z.; Writing—original draft, Y.S.; Writing—review and editing, Y.S.; Supervision, D.Z.; Project administration, D.Z. All authors have read and agreed to the published version of the manuscript.

Funding: This research received no external funding.

Institutional Review Board Statement: Not applicable.

Informed Consent Statement: Not applicable.

Data Availability Statement: All data that support the findings of this study are available from the corresponding author upon reasonable request.

Conflicts of Interest: The authors declare no conflict of interest.

References

1. He, C.; Feng, K.; Fang, Y. Review and prospects on constructing technologies of metro tunnels using shield tunnelling method. *J. Southwest Jiaotong Univ.* **2015**, *50*, 97–109.
2. Qingke, N.; Guang, S.; Siyuan, G.; Hongtao, L.; Lichao, Z.; Jianpeng, H. Disturbance Process of Sandy Gravel Stratum Caused by Shield Tunneling and Ground Settlement Analysis. *Front. Earth Sci.* **2021**, *9*, 994. [CrossRef]
3. Hasheminezhad, A.; Farzalizadeh, R.; Rahimi, H.; Bahadori, H. Seismic performance assessment of wall-type gravel and rubber drains in liquefaction mitigation of sands. *Bull. Earthq. Eng.* **2022**, *20*, 3699–3714. [CrossRef]
4. Zhao, J.S.; Jiang, Q.; Lu, J.F.; Chen, B.R.; Pei, S.F.; Wang, Z.L. Rock fracturing observation based on microseismic monitoring and borehole imaging: In situ investigation in a large underground cavern under high geostress. *Tunn. Undergr. Space Technol.* **2022**, *126*, 104549. [CrossRef]
5. Chen, Z.T.; Bezuijen, A.; Fang, Y.; Wang, K.; Deng, R.Y. Experimental study and field validation on soil clogging of EPB shields in completely decomposed granite. *Tunn. Undergr. Space Technol.* **2022**, *120*, 104300. [CrossRef]
6. Wang, H.B.; Wang, S.Y.; Zhong, J.Z.; Qu, T.M.; Liu, Z.R.; Xu, T.; Liu, P.F. Undrained compressibility characteristics and pore pressure calculation model of foam-conditioned sand. *Tunn. Undergr. Space Technol.* **2021**, *118*, 104161. [CrossRef]
7. Farzalizadeh, R.; Hasheminezhad, A.; Bahadori, H. Shaking table tests on wall-type gravel and rubber drains as a liquefaction countermeasure in silty sand. *Geotext. Geomembr.* **2021**, *49*, 1483–1494. [CrossRef]
8. Dai, C.Q.; Sui, H.T.; Ma, C. Study on the Ultimate Supporting Force of Shield Excavation Face Based on Anisotropic Strength Theory. *Appl. Sci.* **2020**, *10*, 5222. [CrossRef]
9. Hasheminezhad, A.; Bahadori, H. Seismic response of shallow foundations over liquefiable soils improved by deep soil mixing columns. *Comput. Geotech.* **2019**, *110*, 251–273. [CrossRef]
10. Zhao, J.S.; Chen, B.R.; Jiang, Q.; Lu, J.F.; Hao, X.J.; Pei, S.F.; Wang, F. Microseismic Monitoring of Rock Mass Fracture Response to Blasting Excavation of Large Underground Caverns Under High Geostress. *Rock Mech. Rock Eng.* **2022**, *55*, 733–750. [CrossRef]
11. Liang, X.; Ye, F.; Ouyang, A.H.; Han, X.; Qin, X.Z. Theoretical Analyses of the Stability of Excavation Face of Shield Tunnel in Lanzhou Metro Crossing beneath the Yellow River. *Int. J. Geomech.* **2020**, *20*, 04020200. [CrossRef]
12. Sui, H.T.; Ma, C.; Dai, C.Q.; Yang, T.Z. Study on Stability of Shield Tunnel Excavation Face in Soil-Rock Composite Stratum. *Math. Probl. Eng.* **2021**, *2021*, 5579103. [CrossRef]
13. Huang, X.; Li, L.F.; Zhang, C.F.; Liu, B.; Li, K.J.; Shi, H.B.; Jing, B.Y. Multi-Step Combined Control Technology for Karst and Fissure Water Inrush Disaster During Shield Tunneling in Spring Areas. *Front. Earth Sci.* **2021**, *9*, 795457. [CrossRef]
14. Liu, P.F.; Wang, S.Y.; Ge, L.; Thewes, M.; Yang, J.S.; Xia, Y.M. Changes of Atterberg limits and electrochemical behaviors of clays with dispersants as conditioning agents for EPB shield tunnelling. *Tunn. Undergr. Space Technol.* **2018**, *73*, 244–251. [CrossRef]
15. Liu, W.; Shi, P.X.; Yu, M.; Jia, P.J. Analysis of working face stability during obstruction removal from a box tunnel machine. *Acta Geotech.* **2022**, *17*, 4627–4639. [CrossRef]
16. Zizka, Z.; Schoesser, B.; Thewes, M. Investigations on the transient support pressure transfer at the tunnel face during slurry shield drive Part 2: Case B-Deep slurry penetration exceeds tool cutting depth. *Tunn. Undergr. Space Technol.* **2021**, *118*, 104169. [CrossRef]
17. Newton, R.C. A Geo-Experimental Diagram for Garnet Amphibolite and Its Bearing on the Origin of Continents. *J. Geol.* **2018**, *126*, 531–539. [CrossRef]
18. Tomiczek, T.; Prasetyo, A.; Mori, N.; Yasuda, T.; Kennedy, A. Physical modelling of tsunami onshore propagation, peak pressures, and shielding effects in an urban building array. *Coast. Eng.* **2016**, *117*, 97–112. [CrossRef]
19. Zhu, C.H.; Wang, S.H.; Peng, S.; Song, Y.L. Surface settlement in saturated loess stratum during shield construction: Numerical modeling and sensitivity analysis. *Tunn. Undergr. Space Technol.* **2022**, *119*, 104205. [CrossRef]
20. Yu, H.J.; Mooney, M.; Bezuijen, A. A simplified excavation chamber pressure model for EPBM tunneling. *Tunn. Undergr. Space Technol.* **2020**, *103*, 103457. [CrossRef]
21. Cui, G.; Cui, J.; Fang, Y.; Chen, Z.T.; Wang, H.W. Scaled model tests on segmental linings of shield tunnels under earth and water pressures. *Int. J. Phys. Model. Geotech.* **2020**, *20*, 338–354. [CrossRef]
22. Wang, S.Y.; Huang, S.; Zhong, J.Z.; Zhang, S.; Hu, Q.X.; Qu, T.M.; Ye, X.Y. Permeability stability calculation model of foam-conditioned soil based on the permeability constant. *Int. J. Numer. Anal. Methods Geomech.* **2021**, *45*, 540–559. [CrossRef]
23. Zhang, P.; Chen, R.P.; Wu, H.N.; Liu, Y. Ground settlement induced by tunneling crossing interface of water-bearing mixed ground: A lesson from Changsha, China. *Tunn. Undergr. Space Technol.* **2020**, *96*, 103224. [CrossRef]
24. Lee, H.; Kim, D.Y.; Shin, D.; Oh, J.; Choi, H. Effect of foam conditioning on performance of EPB shield tunnelling through laboratory excavation test. *Transp. Geotech.* **2022**, *32*, 100692. [CrossRef]
25. Qu, T.M.; Wang, S.Y.; Hu, Q.X. Coupled Discrete Element-Finite Difference Method for Analysing Effects of Cohesionless Soil Conditioning on Tunneling Behaviour of EPB Shield. *KSCE J. Civ. Eng.* **2019**, *23*, 4538–4552. [CrossRef]

26. Kou, L.; Zhao, J.J.; Miao, R.H.; Lian, F.L. Experimental Study on Dynamic Mechanical Characteristics of Mud Slurry Penetrating into Excavation Surface of Large Diameter Slurry Shield. *Arab. J. Sci. Eng.* **2022**, *47*, 13139–13150. [CrossRef]
27. Ling, F.L.; Wang, S.Y.; Hu, Q.X.; Huang, S.; Feng, Z.Y. Effect of bentonite slurry on the function of foam for changing the permeability characteristics of sand under high hydraulic gradients. *Can. Geotech. J.* **2022**, *59*, 1061–1070. [CrossRef]
28. Huang, Z.Q.; Wang, C.; Dong, J.Y.; Zhou, J.J.; Yang, J.H.; Li, Y.W. Conditioning experiment on sand and cobble soil for shield tunneling. *Tunn. Undergr. Space Technol.* **2019**, *87*, 187–194. [CrossRef]
29. Liang, Y.; Huang, X.M.; Gao, S.J.; Yin, Y.H. Study on the Floating of Large Diameter Underwater Shield Tunnel Caused by Synchronous Grouting. *Geofluids* **2022**, *2022*, 2041924. [CrossRef]
30. Qian, Y.J.; Min, F.L.; Mo, Z.Z.; Fan, X.H. Experimental Study of the Influence of Excavation Surface Stability and Sand Flowability Caused by Dense Slurry-Earth Pressure Balance Shield Tunneling in Silty Sand Stratum. *Adv. Civ. Eng.* **2020**, *2020*, 8883190. [CrossRef]
31. Shevtsov, A.N.; Zhamaletdinov, A.A. Temperature and Rheological Parameters of the Baltic Shield Lithosphere from Electromagnetic Sounding Results. *Izv. Atmos. Ocean. Phys.* **2021**, *57*, 712–728. [CrossRef]
32. Yang, Z.Y.; Yang, X.; Ding, Y.J.; Jiang, Y.S.; Shao, X.K.; Qi, W.Q.; Liu, N. Effect of Soil Conditioning on Saturated Sand Layers in EPB Shields: A Microstructural Analysis Based on CT Scanning and SEM. *Arab. J. Sci. Eng.* **2022**, *47*, 12387–12397. [CrossRef]
33. Zhen, Z.; Ge, X.S.; Zhang, J. Soil Conditioning Tests on Sandy and Cobbly Soil for Shield Tunneling. *KSCE J. Civ. Eng.* **2021**, *25*, 1229–1238. [CrossRef]
34. Milligan, G. *Lubrication and Soil Conditioning in Tunnelling, Pipe Jacking and Microtunnelling: A State-of-the-Art Review*; Geotechnical Consulting Group: London, UK, 2000.
35. Wang, S.Y.; Liu, P.F.; Hu, Q.X.; Zhong, J.Z. Effect of dispersant on the tangential adhesion strength between clay and metal for EPB shield tunnelling. *Tunn. Undergr. Space Technol.* **2020**, *95*, 103144. [CrossRef]
36. Zhao, S.S.; Li, S.C.; Wan, Z.; Wang, X.W.; Wang, M.L.; Yuan, C. Effects of anti-clay agents on bubble size distribution and stability of aqueous foam under pressure for earth pressure balance shield tunneling. *Colloid Interface Sci. Commun.* **2021**, *42*, 100424. [CrossRef]
37. Wang, S.M.; Lu, X.X.; Wang, X.M.; He, C.; Xia, X.; Ruan, L.; Jian, Y.Q. Soil Improvement of EPBS Construction in High Water Pressure and High Permeability Sand Stratum. *Adv. Civ. Eng.* **2019**, *2019*, 4503219. [CrossRef]
38. Zhu, H.H.; Cheng, P.P.; Zhuang, X.Y.; Li, Y.H.; Li, P.N. Assessment and and structural improvement on the performance of soil chamber system of EPB shield assisted with DEM modeling. *Tunn. Undergr. Space Technol.* **2020**, *96*, 103092. [CrossRef]
39. Lee, H.; Choi, H.; Choi, S.W.; Chang, S.H.; Kang, T.H.; Lee, C. Numerical Simulation of EPB Shield Tunnelling with TBM Operational Condition Control Using Coupled DEM-FDM. *Appl. Sci.* **2021**, *11*, 2551. [CrossRef]
40. Liu, B.; Li, T.; Han, Y.H.; Li, D.Y.; He, L.L.; Fu, C.Q.; Zhang, G. DEM-continuum mechanics coupling simulation of cutting reinforced concrete pile by shield machine. *Comput. Geotech.* **2022**, *152*, 105036. [CrossRef]
41. Wu, L.; Qu, F.Z. Discrete element simulation of mechanical characteristic of conditioned sands in earth pressure balance shield tunneling. *J. Cent. South Univ. Technol.* **2009**, *16*, 1028–1033. [CrossRef]
42. Cheng, C.; Liao, S.; Chen, L.; Fan, M.; Huo, X.; University, T. Experiments of Soil Conditioning for Shield Tunneling in Watery Quartz Sand-Gravel Ground. *J. Shanghai Jiaotong Univ.* **2018**, *52*, 1492.
43. Shu-chao, Z.; Shao-hui, H.; Zi-peng, Z.; Cheng-hui, L. Research on soil conditioning for earth pressure balance shield tunneling in Lanzhou sandy pebble strata with rich water. *Rock Soil Mech.* **2017**, *38*, 279–286.
44. Jiang, H.; Gong, Q.; Du, X. Experimental study on soil conditioning in cobble layer by use of earth pressure balanced machine. *Chin. J. Geotech. Eng.* **2013**, *35*, 284–292.
45. Lee, H.Y.B.; Kwak, J.; Choi, J.; Hwang, B.; Choi, H. A lab-scale experimental approach to evaluate rheological properties of foam-conditioned soil for EPB shield tunnelling. *Tunn. Undergr. Space Technol.* **2022**, *128*, 104667. [CrossRef]
46. Zhao, S.S.; Li, S.C.; Wan, Z.; Wang, M.L. Dispersant for Reducing Mud Cakes of Slurry Shield Tunnel Boring Machine in Sticky Ground. *Adv. Mater. Sci. Eng.* **2021**, *2021*, 5524489. [CrossRef]
47. Xiao, C.; Tan, L.; Chen, R.; Zhong, Z.; Yang, J. Dynamic coupling simulation of shield construction mechanics considering characteristics of soils. *Chin. J. Geotech. Eng.* **2019**, *41*, 1108–1115.
48. Xu, Q.W.; Zhang, L.Y.; Zhu, H.H.; Gong, Z.Y.; Liu, J.G.; Zhu, Y.H. Laboratory tests on conditioning the sandy cobble soil for EPB shield tunnelling and its field application. *Tunn. Undergr. Space Technol.* **2020**, *105*, 103512. [CrossRef]
49. Li, L.P.; Sun, S.Q.; Wang, J.; Yang, W.M.; Song, S.G.; Fang, Z.D. Experimental study of the precursor information of the water inrush in shield tunnels due to the proximity of a water-filled cave. *Int. J. Rock Mech. Min. Sci.* **2020**, *130*, 104320. [CrossRef]
50. Wang, L.; Zhu, W.; Qian, Y.J.; Xu, C.; Hu, J.N.; Xing, H.T. Phenomenon and Critical Conditions of Chamber Soil Sliming during EPB Shield Tunneling in Water-Rich Weathered Diorite: Case Study of Jinan Metro, China. *Adv. Civ. Eng.* **2020**, *2020*, 6530832. [CrossRef]
51. Wei, Y.J.; Zheng, X.; Su, F.; Li, M.M.; Li, F.; Yang, Y.Y. Evaluation of Cutting Tool Wear of Earth Pressure Balance Shield in Granular Soil Based on Laboratory Test. *J. Test. Eval.* **2019**, *47*, 927–941. [CrossRef]

coatings

MDPI

Article

The Effect of Multi-Walled Carbon Nanotubes on the Compressive Strength of Cement Mortars

Nelli G. Muradyan [1], Harutyun Gyulasaryan [2], Avetik A. Arzumanyan [1], Maria M. Badalyan [1], Marine A. Kalantaryan [1], Yeghiazar V. Vardanyan [1], David Laroze [3], Aram Manukyan [2] and Manuk G. Barseghyan [1,*]

[1] The Educational, Scientific and Experimental Laboratory on Building Materials and Items, National University of Architecture and Construction of Armenia, 105 Teryan Street, Yerevan 0009, Armenia

[2] Laboratory for Solid State Physics, Institute for Physical Research, NAS RA, Ashtarak 0204, Armenia

[3] Instituto de Alta Investigación, CEDENNA, Universidad de Tarapacá, Casilla 7D, Arica 1000000, Chile

[*] Correspondence: manuk.barseghyan@nuaca.am or manuk.g.barseghyan@gmail.com

Abstract: In this work, multi-walled carbon nanotubes (MWCNTs) have been synthesized using a modified method of solid-phase pyrolysis. The MWCNTs are effectively dispersed using a simple and facile method such as ultrasonic energy without and with surfactant for two different sonication times (15 min and 40 min). In the present study, the effect of MWCNT concentration (0.001, 0.01, 0.05, 0.1 wt.%) on the compressive strengths of cement mortars has been investigated. Compressive tests were carried out on an automatic pressure machine (C089) with a loading rate of 0.5 kN/s at the age of 7 days and 28 days. It is shown that the optimal value of the nanotubes' concentration does not exist in the case of 15 min of sonication time, whereas the optimal value for 40 min of sonication time without and with surfactant is 0.01%. Moreover, in the absence of surfactants, the strength of the specimen over 7 days of hardening increased by 13%, and by 19.5% in the presence of surfactants. The compressive strength for a curing period of 28 days increased by 6.3% and 13.8%, respectively.

Keywords: carbon nanotubes; sonication time; surfactant; compressive strength; cement mortar

Citation: Muradyan, N.G.; Gyulasaryan, H.; Arzumanyan, A.A.; Badalyan, M.M.; Kalantaryan, M.A.; Vardanyan, Y.V.; Laroze, D.; Manukyan, A.; Barseghyan, M.G. The Effect of Multi-Walled Carbon Nanotubes on the Compressive Strength of Cement Mortars. *Coatings* **2022**, *12*, 1933. https://doi.org/10.3390/coatings12121933

Academic Editor: Peng Liu

Received: 17 October 2022
Accepted: 9 November 2022
Published: 8 December 2022

1. Introduction

Due to the increasing demand for high-performance cement composites, the properties of cement composites, such as architectural versatility, excellent mechanical properties, and durability, have been undergoing continuous improvements [1].

To this end, in the literature, various agricultural wastes [2,3], industrial wastes [4,5], natural minerals [6], and synthetic materials [7] have been successfully incorporated. These binders exhibited enhancement in the fresh, mechanical, durability, shrinkage, and microstructural properties of construction products. One such material sparingly used in the literature is carbon nanotubes (CNTs).

CNTs can be considered as seamless cylinders formed of one or more graphene roll sheets and can be classified based on the number of graphene layers [8]. Based on the number of graphene layers, there are two types: single-walled carbon nanotubes (SWCNTs) and multi-walled carbon nanotubes (MWCNTs) [9]. MWCNTs comprise manifold-wrapped graphene sheets arranged in concentric hollow tubes with outside diameters ranging from 2 nm to 100 nm [10]. High modulus of elasticity ($\geq$1 TPa), outstanding tensile strength (65–93 GPa), excellent thermal conductivity (two times more than that of a diamond), high aspect ratio (100–2500), and excellent electrical conductivity are the physical and mechanical characteristics of MWCNTs [11].

The use of CNTs in Portland cement matrices applied in civil engineering has a significant potential to enhance the mechanical properties of composites [12–14]. MWCNTs are used more frequently than SWCNTs because of their lower manufacturing costs and improved reinforcement. According to Lai and Basem [15], adding 0.25 percent MWCNTs to cement mass increased the flexural and tensile strengths of cement-based mortars by 25%. Using small amounts of MWCNTs at 0.08%, the authors reported enhanced flexural strength of 25% [16]. According to Jeevanagoudar et.al [17], MWCNT-reinforced mortars exhibit improved engineering properties compared to ordinary mortars. In particular, it has been discovered that 0.4 percent is the optimum concentration of MWCNT concentration in order to get the highest compressive strength.

CNTs' optimal dispersion is one of the important factors for the preparation of enhanced cement-based composite materials. There are two main methods for dispersing nanotubes: mechanical and chemical [18,19]. In this work, the dispersion of nanotubes by mechanical, in particular ultrasonication, method has been carried out. It is worth noting that the ultrasonication efficiency for MWCNT dispersion depends on many factors, including duration, sonicator type, energy, temperature, and the properties of MWCNTs [20–23]. Due to their high surface energy, MWCNTs have a tendency towards agglomeration, which may lead to the creation of week zones in the final product. By using surfactants, the efficiency of sonication increases; as a result, a more uniform dispersion of MWCNTs in cementitious matrices can be seen [24].

The effect of the concentration of MWCNTs synthesized by different methods, the duration of sonication, and the use of surfactants on the mechanical properties of cement-based materials, particularly cement mortar, still requires further investigation.

In the present work, the effects of concentration of MWCNTs, duration of sonication, and the use of surfactants on the compressive strength of cement-based mortar were investigated.

2. Experiment

2.1. Materials

Ordinary Portland cement 52.5 (GOST 31108-2020, which is available in Araratcement Factory, Yerevan, Armenia) has been used as a binder in the mortars within the framework of this study. The short MWCNTs have been synthesized (Figure 1) using a modified method of solid-phase pyrolysis of cobalt phthalocyanine. The pyrolysis was carried out in a closed quartz ampoule at a temperature of 900 °C and a duration of 30 min using pyrolysis analogs of Ni and Fe phthalocyanines [25,26]. The physical properties and the chemical composition of the used cement [27] (GOST EN 196-1-2002, 196-2-2002, 196-3-2002) are shown in Table 1, while the physical properties of the used sand are shown in Table 2.

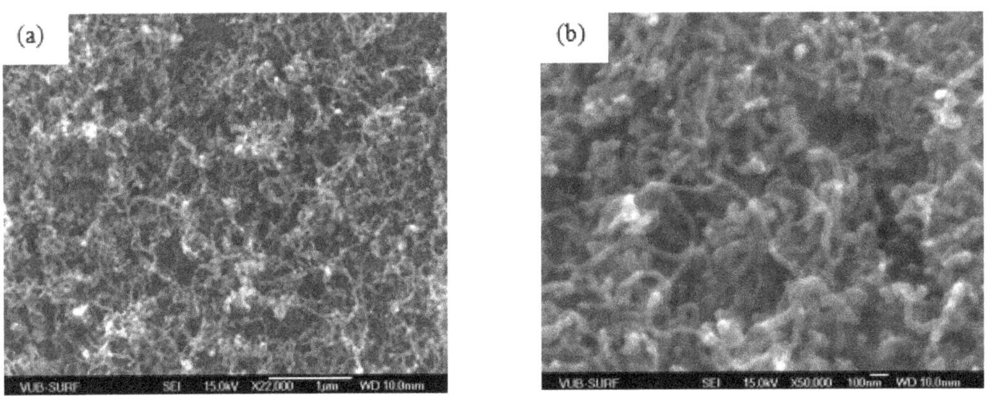

Figure 1. Scanning electron microscope (SEM) images (**a**) × 22,000 (**b**) × 50,000 of MWCNTs.

Table 1. Physical properties and chemical composition of cement.

Characteristics	Days	Results Obtained
Standard consistency (%)	-	31
Specific gravity (g/sm^3)	-	3.1
Blain's fineness (m^2/kg)	-	354.8
Compressive strength (MPa) (EN 196-1)	3 days	23
	7 days	38
	28 days	52
Setting time (min)	Initial	60
	Final	330

Chemical composition of cement (wt.%)

Al_2O_3	SiO_2	Fe_2O_3	CaO	MgO	SO_3	Loss of Ignition	Insol. Resid.	Free CaO
3.21	23.2	1.25	57.5	5.1	2.9	3.7	2.1	1.13

Table 2. Physical properties of sand and MWCNTs.

Sand	Fineness Modulus 2.43	Specific Gravity 2.17	Zone II	Bulk Density in Compact State (kg/m^3) 1829	Bulk Density in Loose State (kg/m^3) 1609
MWCNTS	Outer diameter 40–50 nm	Length < 1 μm		Purity > 90%	

2.2. Dispersion of MWCNTs

Many researchers have reported MWCNT dispersion using various techniques. In this work, MWCNTs, in the required amount, were mixed with water and stirred continuously to ensure proper mixing. Two different sonication times were considered (15 min and 40 min). The sonication process is conducted at room temperature using the ultrasonic device UP400S. In the present investigation, DISPERBYK 199 was also used in order to increase the efficiency of the MWCNT dispersion process in the water. A similar procedure was followed for solutions containing different wt.% of MWCNT content (0.001, 0.01, 0.05, 0.1).

2.3. Mixing and Sample Preparation

The w/c ratio used in the present work was 0.47, and the cement to sand proportion used was 1:4. First, cement and sand were mixed (E095 Mortar mixer, Matest, Treviolo, Italy) for 2.0 min, then the MWCNTS/water mixture was added and mixed for 5 min. The size of the molds were 40 mm × 40 mm × 160 mm. The mortar was compacted through a vibration machine (C278 Vibrating table, Matest, Treviolo, Italy) for 30 s. Similarly, a series of mortars and the one without different MWCNT contents (0.001%, 0.01%, 0.05%, and 0.1% by weight of cement) were cast. The specimens were also prepared by adding in the MWCNTS/water surfactant DISPERBYK-199 (produced by company BYK, Wesel, Germany). The weights of the surfactants were 4.4, 44, 220, and 440 mg, respectively. After 24 h, the specimens were de-molded, and the mortar sample was immersed in water at 20 ± 0.2 °C temperature (Figure 2).

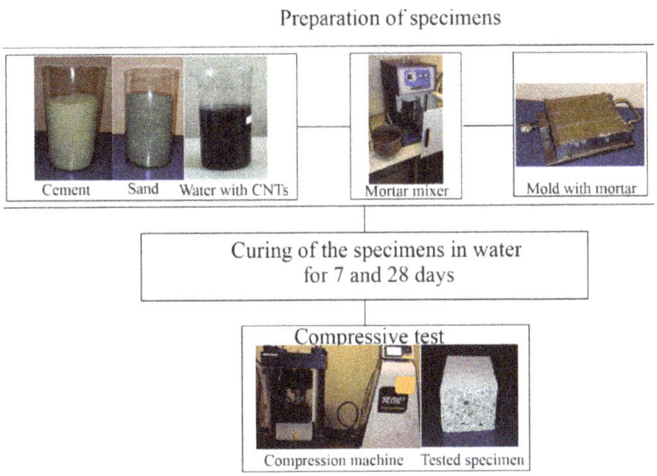

Figure 2. Diagram of the experimental procedure.

2.4. Compressive Strength Testing

Three cubes of each specimen were incidentally selected from each batch to test their average compressive strength according to the Concrete Compression Machine (Matest, Treviolo, Italy) 2000 kN automatic, Servo-Plus Progress (following the standard EN 196-1, and specimen sizes were 40 mm × 40 mm). Compressive tests were carried out on an automatic pressure machine (C089) (Matest, Treviolo, Italy) with a loading rate of 0.5 kN/s at the age of 7 days and 28 days.

3. Results and Discussions

First of all, the results obtained with 15 min sonication time and without surfactant are presented. Figure 3a,b show the compressive strength of the mortar with different wt.% of MWCNTs for 7 and 28 days, respectively. The results indicate that, for both cases of curing days of the cement mortar, the optimal value of MWCNT concentration does not exist. Furthermore, the compressive strength of the cement mortar in the presence of MWCNTs is lower than the compressive strength of the reference specimen. This can be explained due to the low-efficiency dispersion of the MWCNTs in the water, which will reduce the hydration degree. The results indicate that the compressive strength increases with increased curing period, which can be attributed to increased hydration with time.

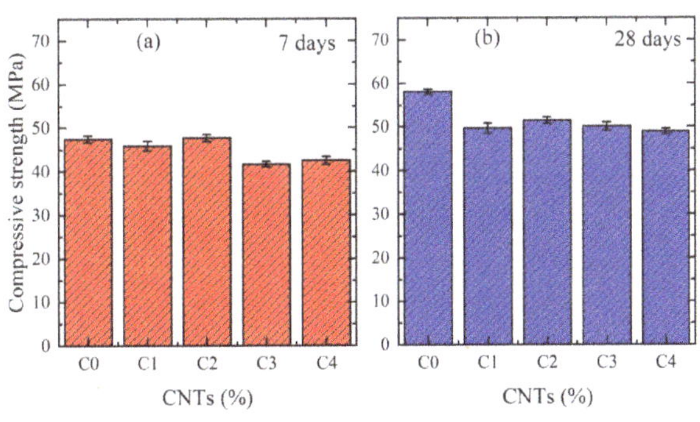

Figure 3. Compressive strength of cement mortars with different wt.% of MWCNTs. The results are for 15 min of sonication time without surfactant. (**a**) for 7 days, (**b**) for 28 days.

Figure 4a,b show the compressive strength of the mortar with different wt.% of MWCNTs for 7 and 28 days, respectively. C0, C1, C2, C3, and C4 correspond to the 0%, 0.001%, 0.01%, 0.05%, and 0.1% of MWCNTs, respectively. The results indicate that the compressive strength of each specimen increases as the curing period increases. This is associated with increased hydration over time. The addition of nanotubes with an appropriate concentration of MWCNTs leads to an increase in compressive strength. The compressive strength reaches the maximum value, then starts to decrease. In other words, the optimal value of the nanotube concentration at which the compressive strength for both cases of curing days reaches its maximum value was obtained. The indicated optimal value of nanotubes is 0.01 wt.%. This is due to the chosen composition of the mortar, as well as the physical and mechanical properties (structure and size) of nanotubes. It is known that in order to make the effect of nanotubes on the physical and mechanical properties of cement-based mortars or concretes effective, it is first necessary to ensure a homogeneous distribution of nanoparticles throughout the volume. For each case, with the increase of nanotube concentration, under the same conditions, the degree of homogeneous distribution decreases, as a result of which the compressive strength decreases. Elsewhere [17], the authors studied rather large values of nanotube concentration, and at 0.4 wt.% of MWCNTs the maximum compressive strength was found only for 28 days of curing (in this work for 7 curing days the optimal values of MWCNTs were not obtained). It can be seen from the figures that the strength of specimens with 7 curing days increased by 13%, and in the case of 28 curing days it increased by 6.3%.

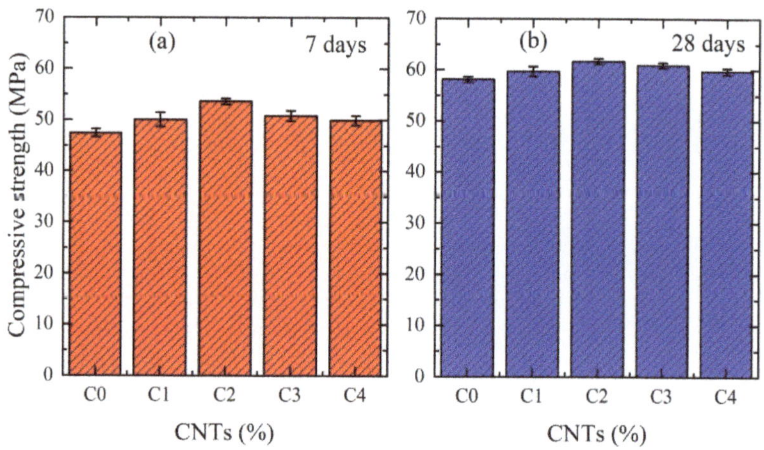

Figure 4. Compressive strength of cement mortars with different wt.% of MWCNTs. The results are for 40 min of ultrasonication time without surfactant. (**a**) for 7 days (**b**) for 28 days.

Figure 5a,b show the compressive strength of the mortar with different wt.% of MWCNTs for 7 and 28 days, respectively, when the surfactant in the MWCNT/water was added.

For both cases of curing days, the optimal values of the MWCNT concentration have been obtained at 0.01 wt.%. The improvement in the efficiency of the dispersion process of MWCNTs in the water brought an increase in the maximum value of the compressive strength. In particular, the strength of a sample with 7 curing days increased by 19.5%, and by 13.8% in the case of 28 curing days.

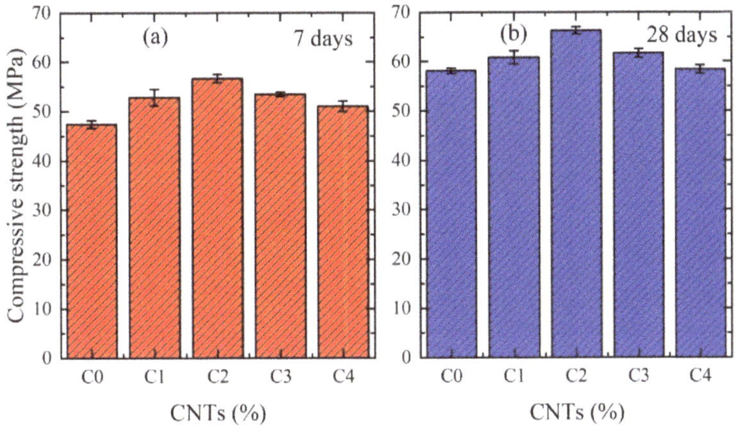

Figure 5. Compressive strength of cement mortars with different wt.% of MWCNTs. The results are for 40 min of ultrasonication time with surfactant. (**a**) for 7 days (**b**) for 28 days.

4. Conclusions

In the present research, the mechanical properties, such as compressive strength, of cement mortar with different concentrations of MWCNTs have been investigated. The sonication technique with and without surfactant, in particular DISPERBYK-199, was employed. For 15 min sonication time, the optimal values of MWCNT concentration to obtain maximum compressive strength do not exist, while in the case of 40 min of sonication it was found to be 0.01%, which is much less than the optimal value obtained in known works. In particular, in the absence of surfactants, the strength of the specimen over 7 days of hardening increased by 13%, and by 19.5% in the presence of surfactants. The compressive strength for a curing period of 28 days increased by 6.3% and 13.8%, respectively. Thus, the optimal mass of surfactants was found, and it was shown that when added to MWCNTs/water, the maximum value of compressive strength can increase by up to 7.5%.

Author Contributions: Conceptualization, M.M.B.; Data curation, N.G.M., H.G. and A.A.A.; Formal analysis, M.M.B., D.L., A.M. and M.G.B.; Funding acquisition, Y.V.V.; Investigation, N.G.M., H.G., A.A.A., M.M.B., M.A.K., A.M. and M.G.B.; Methodology, A.M.; Project administration, M.G.B.; Resources, M.A.K.; Supervision, Y.V.V.; Writing—original draft, A.A.A. and A.M.; Writing—review & editing, D.L. and M.G.B. All authors have read and agreed to the published version of the manuscript.

Funding: The authors would like to acknowledge the financial support by the Science Committee of the Republic of Armenia (Project no. 21AG-1C008). D.L. acknowledges partial financial support from the Centers of excellence with BASAL/ANID financing, Grant AFB180001, CEDENNA.

Institutional Review Board Statement: Not applicable.

Informed Consent Statement: Not applicable.

Data Availability Statement: Not applicable.

Conflicts of Interest: The authors declare no conflict of interest.

References

1. Xiao, J.; Han, N.; Li, Y.; Zhang, Z.H.; Shah, S.P. Review of recent developments in cement composites reinforced with fibers and nanomaterials. *Front. Struct. Civ. Eng.* **2021**, *15*, 1–19. [CrossRef]
2. Nasir, M.; Al-Kutti, W. Performance of Date Palm Ash as a Cementitious Material by Evaluating Strength, Durability, and Characterization. *Buildings* **2019**, *9*, 6. [CrossRef]
3. Nasir, M.; Al-Kutti, W.; Kayed, T.S.; Adesina, A.; Chernykh, T. Synthesis and SWOT analysis of date palm frond ash–Portland cement composites. *Environ. Sci. Pollut. Res.* **2021**, *28*, 45240–45252. [CrossRef]

4. Khan, M.U.; Nasir, M.; Baghabra Al-Amoudi, O.S.; Maslehuddin, M. Influence of in-situ casting temperature and curing regime on the properties of blended cement concretes under hot climatic conditions. *Constr. Build. Mater.* **2021**, *272*, 12186.
5. Nasir, M.; Johari, M.A.; Maslehuddin, M.; Yusuf, M. Sodium sulfate resistance of alkali/slag activated silicomanganese fume-based composites. *Struct. Concr.* **2020**, *79*, 1.
6. Ibrahim, M.; Salami, B.A.; Algaifi, H.A.; Rahman, M.K.; Nasir, M.; Ewebajo, A.O. Assessment of acid resistance of natural pozzolan-based alkali-activated concrete: Experimental and optimization modelling. *Constr. Build. Mater.* **2021**, *304*, 124657. [CrossRef]
7. Nasir, M.; Aziz, M.A.; Zubair, M.; Ashraf, N.; Hussein, T.N.; Allubli, M.K.; Manzar, M.S.; Al-Kutti, W.; Al-Harthi, M.A. Engineered cellulose nanocrystals-based cement mortar from office paper waste: Flow, strength, microstructure, and thermal properties. *J. Build. Eng.* **2022**, *51*, 104345.
8. Dresselhaus, M.S.; Dresselhaus, G.; Eklund, P.C.; Rao, A.M. Carbon nanotubes. In *The Physics of Fullerene-Based and Fullerene-Related Materials*; Andreoni, W., Ed.; Springer: Dordrecht, The Netherlands, 2000; Volume 23, pp. 331–374.
9. Tasis, D.; Tagmatarchis, N.; Bianco, A.; Prato, M. Chemistry of carbon nanotubes. *Chem. Rev.* **2006**, *106*, 1105–1136. [CrossRef] [PubMed]
10. US Research Nanomaterials, Inc. Layers of Multi Walled Carbon Nanotubes, MWNTs, MWCNTs, Short MWNTs, Short MWCNTs. Available online: https://www.us-nano.com/layers (accessed on 30 April 2020).
11. Makar, J.M.; Beaudoin, J.J. *Carbon Nanotubes and Their Application in the Construction Industry*; Royal Society of Chemistry: Cambridge, UK, 2004; pp. 331–341.
12. Silvestro, L.; Gleize, P.J.P. Effect of carbon nanotubes on compressive, flexural and tensile strengths of Portland cement-based materials: A systematic literature review. *Constr. Build. Mater.* **2020**, *264*, 120237. [CrossRef]
13. Han, B.; Yu, X.; Kwon, E.; Ou, J. Effects of CNT concentration level and water/cement ratio on the piezoresistivity of CNT/cement composites. *J. Compos. Mater.* **2012**, *46*, 19–25. [CrossRef]
14. Shi, T.; Li, Z.; Guo, J.; Gong, H.; Gu, C. Research progress on CNTs/CNFs-modified cement-based composites—A review. *Constr. Build. Mater.* **2019**, *202*, 290–307. [CrossRef]
15. Lai, Y.C.; Bassem, A. Finite element analysis of carbon nanotube/cement composite with degraded bond strength. *Comput. Mater. Sci.* **2010**, *47*, 994–1004.
16. Konsta-Gdoutos, M.S.; Metaxa, Z.S.; Shah, S.P. Highly dispersed carbon nanotube reinforced cement based materials. *Cem. Concr. Res.* **2010**, *40*, 1052–1059.
17. Jeevanagoudar, Y.V.; Krishna, R.H.; Gowda, R.; Preetham, R.; Prabhakara, R. Improved mechanical properties and piezoresistive sensitivity evaluation of MWCNTs reinforced cement mortars. *Constr. Build. Mater.* **2017**, *144*, 188–194. [CrossRef]
18. Assi, L.; Alsalman, A.; Bianco, D.; Ziehl, H.; El-Khatib, J.; Bayat, M.; Hussein, F.H. Multiwall Carbon Nanotubes (MWCNTs) Dispersion & Mechanical Effects in OPC Mortar & Paste: A review. *J. Build. Eng.* **2021**, *43*, 102512.
19. Hilding, J.; Grulke, E.A.; Zhang, Z.G.; Lockwood, F.J. Dispersion of Carbon Nanotubes in Liquids. *Dispers. Sci. Technol.* **2003**, *24*, 1–41.
20. Elkashef, M.; Wang, K.; Abou-Zeid, M.N. Acid-treated carbon nanotubes and their effects on mortar strength. *Front. Struct. Civ. Eng.* **2016**, *10*, 180–188. [CrossRef]
21. Alrekabi, S.; Cundy, A.; Lampropoulos, A.; Savina, I. Experimental investigation on the effect of ultrasonication on dispersion and mechanical performance of multi-wall carbon nanotube-cement mortar composites. *Int. J. Civil Environ. Struct. Constr. Archit. Eng.* **2016**, *111*, 268–274.
22. Metaxa, Z.S.; Boutsioukou, S.; Amenta, M.; Favvas, E.P.; Kourkoulis, S.K.; Alexopoulos, N.D. Dispersion of Multi-Walled Carbon Nanotubes into White Cement Mortars: The Effect of Concentration and Surfactants. *Nanomaterials* **2022**, *12*, 1031. [CrossRef]
23. Sobolkina, A.; Mechtcherine, V.; Khavrus, V.; Maier, D.; Mende, M.; Ritschel, M.; Leonhardt, A. Dispersion of carbon nanotubes and its influence on the mechanical properties of the cement matrix. *Cem. Concr. Compos.* **2012**, *34*, 1104–1113. [CrossRef]
24. Rashidi, Y.; Roudi, M.R.R.; Korayem, A.H.; Shamsaei, E. Investigation of ultrasonication energy effect on workability, mechanical properties and pore structure of halloysite nanotube reinforced cement mortars. *Constr. Build. Mater.* **2021**, *304*, 124610. [CrossRef]
25. Avakyan, L.; Manukyan, A.; Bogdan, A.; Gyulasaryan, H.; Coutinho, J.; Paramonova, E.; Sukharina, G.; Srabionyan, V.; Sharoyan, E.; Bugaev, L. Synthesis and structural characterization of iron-cementite nanoparticles encapsulated in carbon matrix. *J. Nanoparticle Res.* **2020**, *22*, 30. [CrossRef]
26. Manukyan, A.; Gyulasaryan, H.; Ginoyan, A.; Kaniukov, E.; Petrov, A.; Yakimchuk, D.; Shashov, S.; Nurijanyan, M.; Mirzakhanyan, A. Structural, morphological and magnetic properties of nickel-carbon nanocomposites prepared by solid-phase pyrolysis of Ni phthalocyanine. In *Fundamental and Applied Nano-Electromagnetics*; Maffucci, A., Maksimenko, S.A., Eds.; NATO Science for Peace and Security Series B: Physics and Biophysics; Springer: Dordrecht, The Netherlands, 2016; pp. 273–290.
27. Arzumanyan, A.A.; Tadevosyan, V.G.; Muradyan, N.G.; Navasardyan, H.V. Study of "Saralsk" Deposit for Practical Applications in Construction. *J. Arch. Eng. Res.* **2021**, *1*, 3–6. [CrossRef]

coatings

Article

The Influence of Extractant Composition on the Asphaltenes Extracted from Asphalt

Dachuan Sun [1,2,*,†], Yang Song [2,*,†] and Feiyong Chen [2]

1 School of Transportation Engineering, Shandong Jianzhu University, Jinan 250101, China
2 Huzhou Nanxun District Jianda Ecological Environment Innovation Center, Shandong Jianzhu University, Jinan 250101, China
* Correspondence: sundc@iccas.ac.cn (D.S.); songyang20@sdjzu.edu.cn (Y.S.); Tel.: +86-199-5312-0288 (D.S.)
† These authors contributed equally to this work.

Abstract: The compositions of extractants containing xylene (G) and n-heptane (P) recovered and reused in the asphaltene extraction process were determined by densimetry and the extracted asphaltenes were analyzed to study the influence of extractant composition on their elemental, spectral and structural properties. With increasing the number of extraction, the G/P ratio in extractant increases, which dissolves more asphaltenes of higher aromaticity and causes a decrease of asphaltene yield, the polarity and aromaticity in molecular structure of the extracted asphaltenes. Asphaltenes extracted at a higher G/P ratio were found to have less fractions of N and O elements as well as higher fractions of H and S elements, a lower C/H atom ratio and molecular unsaturation. Moreover, they have a smaller fraction of aromatic rings and aromatic carbon ratio, a higher substitution rate on aromatic rings, a higher fraction of alkyl chain and free ends in their average molecular structure. Compared with the base asphalt, asphaltenes' infrared absorptions generally move towards smaller wave numbers owing to more aromaticity in their molecules and subsequent stronger conjugative and inductive effects. The asphaltenes extracted at a higher G/P ratio have a denser packing of alkyl chains and a looser packing of aromatic rings, according to their spectra of X-ray diffraction.

Keywords: densimetry method; extractant composition; n-heptane and xylene; asphaltene extraction; molecular structure

Citation: Sun, D.; Song, Y.; Chen, F. The Influence of Extractant Composition on the Asphaltenes Extracted from Asphalt. *Coatings* **2022**, *12*, 1600. https://doi.org/10.3390/coatings12101600

Academic Editor: Ajay Vikram Singh

Received: 21 September 2022
Accepted: 20 October 2022
Published: 21 October 2022

1. Introduction

Asphalt is one kind of coating materials widely used for roof waterproofing, interface bonding and surface layers of high-grade highways [1,2]. Asphalt has a complex chemical structure and composition [3–5]. According to the different solubility, it can be divided into four components: saturate, aromatic, resin and asphaltene [1,5,6]. In our previous work [7], using the solubility discrepancy of asphaltene in different solvents [8], a simple scheme for extracting asphaltenes from asphalt at room temperature was proposed and verified experimentally. The scheme firstly dissolves the asphalt in a good solvent (denoted as G) and then uses a poor solvent (denoted as P) to precipitate the asphaltenes out. As the scheme is carried out at room temperature without heating, it avoids the problems of high energy cost, expensive devices, and safety risks of flammable steam in traditional methods (JTG E20-2011 or IP 143) of asphaltene extraction [6,9]. After extraction, the good solvent was mixed with the poor solvent. To reduce costs, these waste (used) solvents need to be recovered and reused as new extractants. However, the yield decreased significantly when the recovered solvent was used, probably caused by the variance of extractant composition. To better study the effect of extractant composition on asphaltene yield, a method is needed to analyze the extractant concentration or G/P ratio, which is also helpful to optimize the separation condition and improve the separation efficiency of the waste extractants.

Alkanes and aromatics are volatile components in coating materials and widely used as extractants or eluents [3,10]. When extracting asphaltenes from asphalt, the good solvents are generally aromatic hydrocarbons, such as xylene and toluene while the poor solvents are n-alkanes, such as n-heptane and n-hexane [3,5,9]. Xylene is a C_8 aromatic mixture containing isomers and a small amount of ethylbenzene [11,12]. Both xylene and heptane are volatile and miscible with each other in any ratio. Heptane is a poor solvent or precipitant for asphaltenes, owing to the difference of molecular structures (thus higher enthalpy penalty) between heptane and asphaltenes as well as high molecular weight of asphaltene (thus less mixing entropy) [3,6]. Asphaltene is precipitated only when the G/P ratio is below a critical precipitation point [9,13,14]. The analysis of G/P ratio in extractant is helpful to judge whether the critical precipitation point is exceeded, and helpful to guide the industrial extraction of asphaltenes. Compared with the analytical methods such as high performance liquid chromatography (HPLC), nuclear magnetic resonance (NMR) and gas chromatography-mass spectrometry (GC-MS) [15,16], the densimetry method may be the simplest and cheapest without requiring expensive equipment [17].

The physical properties [2,18] and molecular structures [4,5] of the extracted asphaltenes may vary with extraction number or G/P ratio. Asphaltenes have been analyzed via elemental analysis (EA), NMR, Fourier transform infrared (FTIR) and X-ray diffraction (XRD) methods [19,20] to get average molecular structure [3] to compare with other bitumen components [6] or with asphaltenes precipitated from different alkanes [10,21]. There are few studies for the effect of G/P ratio on the properties of extracted asphaltene. Such a study is helpful to evaluate the property discrepancy and to judge whether these asphaltenes can be mixed and used together. In this paper, the compositions of recovered extractants were analyzed via densimetry, while the properties of asphaltenes extracted at different G/P compositions were studied. Mechanisms were discussed for the variance of asphaltene properties and for the decrease of asphaltene yield with extraction number or G/P ratio.

2. Materials and Methods

2.1. Materials and Devices

The asphalt used is Qilu No. 70 base asphalt from Qilu Petrochemical Company (Zibo, China). Potassium bromide (KBr, crystals, of 99% purity) with chemical abstracts service number (CAS No.) 7758-02-3 is purchased from Merck & Co., Ltd. (Beijing, China). Deuterated chloroform ($CDCl_3$, CAS No. 865-49-6, 99.8% purity) containing 0.03% *v/v* tetramethylsilane (TMS) is purchased from Cambridge Isotope Laboratories, Inc. (Beijing, China). Xylene (xylene isomer + ethylbenzene, CAS No. 1330-20-7) and n-heptane (CAS No. 142-82-5) of analytical reagent (AR) grade are purchased from Macklin Biochemical Technology Company (Shanghai, China).

The glasswares were purchased from Chongqing Synthware Glass Company (Chongqing, China). The HJ-6A thermostatic magnetic stirrer is purchased from Jiangsu Zhongda Technology Instrument Company (Nanjing, China). The PX224ZH/E electronic balance and STX portable balance are purchased from Ohaus International Trading (Shanghai) Company (Shanghai, China). The sand-core filter device with a filter bottle of 1000 mL volume and a filter cup of 300 mL volume is purchased from Jiangsu Feida Glass Products Company (Nanjing, China) and used to get the precipitated asphaltenes by suction filtration. The RE-52AA rotary evaporator and SHZIII circulating water vacuum pump are purchased from Shanghai Yarong Biochemical Instrument Factory (Shanghai, China) and used to recover extractant via vacuum distillation. The medium-speed setting filter paper is purchased from Hangzhou Fuyang Northwood Pulp Company (Hangzhou, China). The 0.45 μm filter membrane is purchased from Shanghai Xinya Purification Device Factory (Shanghai, China). The polypropylene dropper and other consumables are supplied by Shandong Laboratory Experimental Instrument Company (Jinan, China).

2.2. Experimental Procedures

2.2.1. Preparation of Standard Samples of Extractants

The subscripts P and G represent poor solvent (n-heptane) and good solvent (xylene), respectively, in Table 1 and in the following. All operations were carried out in a fume hood. The room temperature was maintained at 20 °C by an air conditioner.

Table 1. The volume fraction v and mass fraction w of each component for standard samples of extractants containing xylene (G) and n-heptane (P) at room temperature.

No.	P (mL)	G (mL)	Sum (mL)	v_P	v_G	w_P	w_G
0	0	20	20	0%	100%	0.00	1.00
1	2	18	20	10%	90%	0.08	0.92
2	4	16	20	20%	80%	0.17	0.83
3	6	14	20	30%	70%	0.25	0.75
4	8	12	20	40%	60%	0.35	0.65
5	10	10	20	50%	50%	0.44	0.56
6	12	8	20	60%	40%	0.54	0.46
7	14	6	20	70%	30%	0.65	0.35
8	16	4	20	80%	20%	0.76	0.24
9	18	2	20	90%	10%	0.88	0.12
10	20	0	20	100%	0%	1.00	0.00

The procedures for preparing standard samples are as follows:

1. 30 mL cylindrical bottles are cleaned, dried and put on the desk. The bottles and their caps are marked with numbers from 0 to 10 by an oil-based pen;
2. The n-heptane is pipetted into graduated cylinders until a target volume is reached;
3. The n-heptane is poured into a glass bottle. The bottle number and corresponding volume of n-heptane for each sample are listed in Table 1;
4. The xylene is pipetted into another graduated cylinders until its target volume is reached. Then it is poured into the glass bottle with the designated number. The bottle number and volume of xylene are shown in Table 1;
5. The bottles are sealed and shaken for 1 min to mix sufficiently.

The sum of volume is 20 mL for each sample, as shown in Table 1. To simplify the subsequent analysis, the volume fraction v and mass fraction w of each component are calculated and listed in Table 1, according to the quantities [11,12] in Table 2. The graduated cylinders were calibrated by weighing 10 mL of ultrapure water on the PX224ZH/E balance and the density ρ of water at 20 °C is 0.9982 g/mL.

Table 2. Basic physical quantities of n-heptane (P) and xylene (G) at room temperature.

Physical Quantity	P	G	Physical Meaning
ρ	0.68 g/mL	0.86 g/mL	Density
M_W	100.20 g/mol	106.17 g/mol	Molecular weight
T_b	98 °C	140 °C	Boiling point

2.2.2. Procedures to Get Extracted Asphaltene and Real Samples of Extractants

A detailed method of recovering solvents and extracting asphaltenes from bitumen was documented in a previous report [7]. A total of 3 consecutive extractions were performed and the main steps are as follows:

1. In the first extraction, a total of 5 g asphalt is dissolved in 50 mL xylene of AR purity and then poured into 500 mL n-heptane of AR grade to precipitate the asphaltenes. The solution is then filtered to get the filtrate and filter cake, respectively. The filter cake is dried to obtain asphaltenes (denoted as A1). The weight ratio of A1 to asphalt is defined as asphaltene yield;

2. The above filtrate is recovered via vacuum distillation and the vacuum is provided by connecting SHZIII vacuum pump. The continuously supplied tap water (around 15 °C) is used for condensation and different distillates are obtained by controlling the temperature of the water bath of the rotary distillation equipment. The temperature firstly rises from room temperature to 45 °C and is then held for 1 h, during which the dripping of condensed liquid from the condenser is observable from 30 °C and usually completed within 30 min. The collected fraction is sealed and denoted as P1. Then, the temperature is raised from 45 °C to 80 °C and held for 1.5 h. The dripping of condensed liquid starts from 60 °C and is usually finished within 45 min. The collected fraction is sealed in another bottle and denoted as G1;
3. In the second extraction, asphalt is dissolved in G1, and then poured into P1 to precipitate the asphaltenes (A2). The ratio of asphalt/G1/P1 is maintained at 5 g/50 mL/500 mL for consistency. Subsequent operations are the same as steps (1) and (2), except that the collected condensates at 45 °C (heptane-rich fraction) and 80 °C (xylene-rich fraction) are denoted as P2 and G2, respectively;
4. In the third extraction, asphalt is dissolved in G2, and then poured into P2 to precipitate the asphaltenes (A3). The ratio of asphalt/G2/P2 is maintained at 5 g/50 mL/500 mL for comparison. Subsequent operations are the same as steps (1) and (2). The collected fractions at 45 °C and 80 °C are denoted as P3 and G3, respectively.

The concentration of real samples obtained above (P1, P2, P3 and G1, G2, G3) was measured by densimetry. The elements, spectra and structures are analyzed for the extracted asphaltenes (A1, A2, and A3).

2.2.3. Procedures of the Densimetry Method

A fitting curve or equation is made based on the standard samples and then used to analyze real samples. The main steps are as follows:

1. From the standard samples prepared according to Table 1, a 10 mL liquid is transferred to the graduated cylinder with a pipette for each sample;
2. The graduated cylinder is weighed on the PX224ZH/E electronic balance to calculate the mass and density of the sample. Each sample is independently pipetted and measured four times, and the resulting densities are averaged as the final density;
3. A standard curve of densimetry is obtained by plotting the density versus the sample concentrations. The curve is fitted to obtain the standard equation via ORIGIN software (version: OriginPro 2018C 64-bit SR1 for Windows);
4. For the real samples recovered by rotary evaporation, their densities are measured according to the same procedures as the standard samples. The concentration is obtained by substituting the measured density into the equation.

As the blank signals (Y_B) are chosen as the starting points of standard curves, the limits of detections (LOD) are three standard deviations of the blank (s_B). Seven replicates are analyzed for blank samples at $v_G = 0$ to calculate the standard deviations and the LOD of the densimetry method.

2.3. Characterization Methods

For extracted asphaltenes (A1, A2, and A3), their elements, spectra and structures were analyzed via EA, NMR, FTIR and XRD methods. The same characterization was performed on the base bitumen for comparison.

EA was performed via the Elemantar Vario EL cube Elemental Analyzer for the C, H, N and S elements (CHNS mode) at 25 °C. Firstly, a standard sample of 2 mg of sulfanilamide was measured for three independent times to obtain its mass percentages of CHNS elements and then compared with its standard values to obtain the correction factor. The correction factor was updated every 20 measurements, and used for EA calibration during the measurement of real samples. Each real sample was measured four times independently to calculate the mean and deviation. A total of 2 mg was used for each measurement of real sample. The oxidation furnace temperature was 1150 °C and the reduction furnace temperature was 850 °C. The flow rate of carrier gas (helium) was 200 mL/min and the flow rate of combustion oxygen was 200 mL/min. The combustion time of oxygen addition was 90 s.

NMR measurements were performed on a Bruker BioSpin GmbH instrument at 25 °C. $CDCl_3$ was used as the deuterated solvent. A 50 mg sample was sufficiently mixed with 0.5 mL of $CDCl_3$ and transferred into a 5 mm NMR tube for one-dimensional (1D) liquid ^{1}H-NMR characterization. Pulse sequence was zg30 and pulse width was 9.27. Number of scans was 16 and receiver gain was 101. Relaxation delay was 20.0 s and acquisition time was 4.0894 s. Spectrometer frequency was 400 MHz and flip angle was 30°. Acquired size was 32,768 and spectral size was 65,536. The spectral width was 8012.8. The NMR spectra were analyzed via MestReNova software (version 6.1.0-6224 for Windows). The chemical shift of TMS was set as 0 ppm for reference. The n-heptane of AR purity was measured firstly to get the relative ratio of residual solvent (chloroform) signal in deuterated solvent to the TMS signal. The ratio was found to be 0.48 ± 0.04. Based on this ratio and TMS signal, the area for chloroform signal was subtracted and the rest areas integrated for the selected ranges in NMR spectrum were normalized to sum 1 for ease of comparison. Each real sample was measured 3 times independently to calculate the mean and deviation of integrated areas.

For FTIR analysis, the samples were uniformly ground with KBr in an agate mortar and then pressed into round flakes under a pressure of 10 MPa for 60 s. The samples were then measured in transmission mode (TR) on Bruker Tensor II FTIR spectrometry at 25 °C, with a scanning range of 400–4000 cm^{-1} and a resolution of 0.5 cm^{-1}. A round flake of pure KBr was scanned for baseline, which was automatically subtracted in each spectrum. Each sample was scanned 16 times and the averaged spectrum was recorded. The infrared (IR) spectrum was processed via the Omnic software (version 8.2.0.387 for Windows).

The XRD analysis was performed with a D8 Advance X-ray diffractometer produced by Bruker (Germany) at 25 °C. The Cu target source (0.15406 nm) was selected. The measurement angle (2θ) ranged from 5° to 60°, which was twice the angle of X-ray incidence (θ). The scan rate was 1°/min and the step size was 0.02°. The spot diameter was 1 mm with LynxEye XE array detector. Asphaltene was ground to fine powder in an agate mortar prior to analysis.

3. Results and Discussion

3.1. Extractant Compositions Determined via Densimetry

The extractant used in this paper is a binary mixture of xylene (G) and n-heptane (P) in different G/P ratio. For standard samples of extractants with compositions listed in Table 1, their densities ρ were measured and plotted versus the volume fraction v_G. ρ shows a linear relation with v_G in Figure 1 and the coefficient of determination (R^2) is 0.99967 for the fitting equation

$$\rho = 0.67783 + 0.18164\, v_G, \quad (1)$$

The densities ρ of recovered extractants (P1, P2 and G1, G2) were measured by densimetry and substituted into Equation (1) to get v_G. Xylene and heptane of AR purity were used for the 1st extraction of asphaltenes. G1 and P1 were used for the 2nd extraction while G2 and P2 were used for the 3rd extraction. For heptane-rich fractions P1 and P2, their volume fraction of n-heptane v_P ($v_P = 1 - v_G$) is 0.907 ± 0.026 and 0.832 ± 0.011, respectively. The xylene-rich fractions G1 and G2 have higher purity, whose v_G is 0.992 ± 0.014 and 0.966 ± 0.014, respectively, as shown in Figure 1. The negative pressure (vacuum) generated by the pump during vacuum distillation draws away a large amount of solvent and causes a solvent loss and air pollution issue. The average yield for distillate P1 was only 75%. Owing to volatility and good compatibility between xylene and n-heptane, a shorter time of vacuum distillation is preferred to improve the distillate yield and to avoid the xylene vapor dissolving into the heptane-rich fraction. Solvent separation conditions need further optimization, for example, under appropriate negative pressure and distillation time. To reduce the solvent loss in the recovery process, it may be necessary to connect a cold trap to the vacuum pump or use other separation methods instead of vacuum distillation [22,23]. For blank samples at $v_G = 0$, the s_B of calculated volume concentration is 1.2% and the LOD of v_G is 3.6% for the densimetry method.

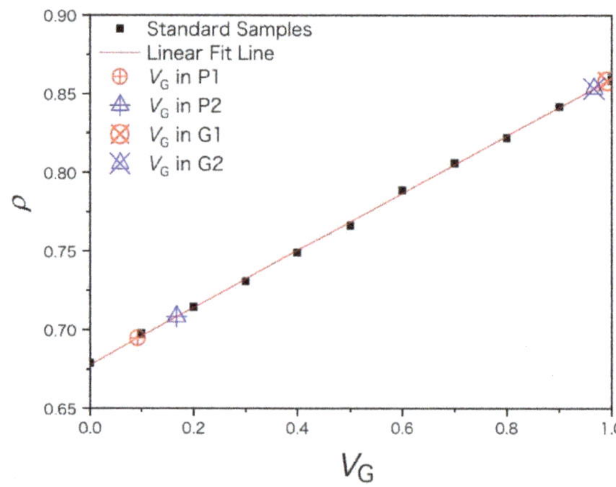

Figure 1. The red line is a linear fit of the volume fraction of xylene v_G and density ρ of standard samples (marked in black squares). The real samples of recovered extractants (P1, P2 and G1, G2) are marked by larger symbols along the fitted line. ρ is in unit of g/mL.

The asphaltene yield was 13.2 ± 0.4%, 10.8 ± 0.4% and 8.3 ± 0.4%, respectively, for the 1st, 2nd, and 3rd extractions. The asphaltene yield decreased with the extraction number but the mechanism was not known exactly [7]. To understand the scenarios of asphaltene precipitation more intuitively, the actual volumes of n-heptane and xylene in the solution for settling asphaltenes during the 1st, 2nd, and 3rd extractions were calculated according to their concentrations in Figure 1 and listed in Table 3, assuming 50 mL G and 500 mL P used in each extraction for ease of comparison. With increasing the extraction number, the heptane-rich fraction contains more and more xylenes, reducing the ability of n-heptane to sediment asphaltenes. Simultaneously, the concentration and total volume of xylene in the solution increase. Thus, more asphaltenes are dissolved while the sedimentation amount and final yield decrease. Even when the xylene-rich fraction is as low as $v_G = 96.6\%$, the asphalt can be completely dissolved, while the heptane-rich fraction is still capable to precipitate asphaltene as low as $v_P = 83.2\%$. Thus, the proposed extraction method [7] has good robustness and great prospects for industrialization, since asphaltenes can be efficiently extracted with solvents much lower than AR purity. Goual et al. found that the onset of asphaltene flocculation occurred near a toluene/heptane volume ratio of 70:30 [14]. The xylene/heptane volume ratio is 24:76 for the 3rd extraction in Table 3, well below the critical ratio of asphaltene precipitation. How the precipitation amount varies with the G/P ratio is an interesting question and may help address asphaltene deposits in oil pipelines [9,24].

Table 3. The actual volumes of n-heptane (P) and xylene (G) in extractant solution, calculated from the concentration of G and P. 50 mL G and 500 mL P were sequentially added for each extraction. Solvents G and P of AR purity were used for the 1st extraction. G1 and P1 were used for the 2nd extraction while G2 and P2 were used for the 3rd extraction.

Extraction No.	P (mL)	G (mL)	Total Volume (mL)	G/P Volume Ratio
1st	500	50	550	0.10
2nd	454	96	550	0.21
3rd	418	132	550	0.32

3.2. Properties of Asphaltenes Extracted at Room Temperature

Asphaltene is usually the most polar and cohesive component of asphalt, and forms the core of the asphalt colloid structure. In Table 4, with increasing the extraction number (from A1 to A3), the mass fractions of C, H and S elements increase while the fractions for element N and others (mainly oxygen) decrease. The fraction of N decreases sequentially from 1.50% in A1 to 1.38% in A3 then to 0.73% in asphalt. Because element N increases the molecular polarity [3], the polarity of asphaltenes from A1 to A3 may decrease with the extraction number and be larger than that of asphalt. C/H atomic ratio is defined as q, which decreases sequentially from 0.99 in A1 to 0.78 in asphalt. q reflects the degree of unsaturation of the molecule, usually in range of 0.5 to 1 for bitumen and its components [1,6,7]. q is larger than 1 for fused-ring molecules, and approaches to infinity for chars and graphites [25,26]. As asphaltenes and asphalt have apparently different q, their unsaturation discrepancy is much larger than the discrepancy between different asphaltenes [6]. Since the polarity and unsaturation increases from asphalt to A3 then to A1, the π-π conjugation and inter-molecular interactions increase, causing higher cohesion and hardness for asphaltenes than asphalt. Thus, asphaltenes extracted at larger G/P ratio have smaller unsaturation, polarity and inter-molecular interactions.

Table 4. Mass percentage of elements from EA (CHNS mode) and the calculated C/H atomic ratio (q) for extracted asphaltenes (A1, A2 and A3) and asphalt.

Fractions	N (%)	C (%)	H (%)	S (%)	Others (%)	q
A1	1.50 ± 0.03	82.46 ± 0.08	6.95 ± 0.04	5.97 ± 0.06	3.13 ± 0.15	0.99
A2	1.44 ± 0.02	82.62 ± 0.07	7.07 ± 0.06	6.09 ± 0.04	2.79 ± 0.10	0.97
A3	1.38 ± 0.03	83.31 ± 0.09	7.24 ± 0.06	6.20 ± 0.06	1.87 ± 0.18	0.96
Asphalt	0.73 ± 0.07	83.08 ± 0.12	8.84 ± 0.05	4.20 ± 0.06	3.15 ± 0.31	0.78

Asphaltenes with larger unsaturation are more easily dissolved in solvents with larger G/P ratio, allowing asphaltenes with smaller unsaturation to preferentially precipitate out. Thus, the averaged molecular structure of the extracted asphaltene may change with the extraction number. Owing to complex compositions and chemical structures of asphalt and asphaltenes [5], NMR is generally used to analyze their average molecular structure [1,27] rather than their content quantification [2,28]. As in Figure 2, the protons are divided into A, α, β, and γ categories [1,6,29], whose chemical shifts are, respectively, in ranges of 6.0–9.0 ppm, 2.0–4.0 ppm, 1.0–2.0 ppm and 0.5–1.0 ppm [1,6]. Their integrated areas are denoted as h_A, h_α, h_β, and h_γ, respectively. Protons directly attached to the single (S) and fused (F) rings are in the range of 6.0–7.4 ppm and 7.4–9.0 ppm, respectively [30], whose integral areas are denoted by h_S and h_F in Table 5. From A1 to A3 then to asphalt, both h_S and h_F decrease, which has a consistent trend with q and indicates a decrease of aromatic ring fractions. The aromatic carbon ratio f_A [1] shows a decrease trend while the hydrogen substitution rate around aromatic rings σ [6] shows an increase trend. Thus, with increasing the G/P ratio in extractant or with increasing the extraction number, the aromatic rings of extracted asphaltenes have more branched chains and occupy less proportion. From A1 to A3 then to asphalt, h_A decreases while h_β and h_γ increase. Therefore, the fraction of alkyl branches and (methyl) ends increases in extracted asphaltenes. In terms of averaged molecular structures, the extracted asphaltenes are significantly different from the base asphalt [29]. In terms of physical properties, the asphaltenes obtained from the 1st, 2nd and 3rd extractions are all brittle, hard and shiny, without apparent adhesion to glass, skin or plastics at room temperature, which are apparently different from the base asphalt. The remaining asphalt after asphaltene extraction is much softer and stickier than the original (base) asphalt, which can be used as a new coating material.

Table 5. Calculated values for the average molecular structure of the extracted asphaltenes (A1, A2 and A3) and the base asphalt according to their NMR spectra [1].

NMR Quantities	A1	A2	A3	Asphalt
h_F	0.08 ± 0.01	0.06 ± 0.01	0.04 ± 0.01	0.02 ± 0.01
h_S	0.11 ± 0.02	0.09 ± 0.01	0.08 ± 0.01	0.04 ± 0.01
f_A	0.59 ± 0.01	0.56 ± 0.01	0.54 ± 0.01	0.40 ± 0.02
σ	0.39 ± 0.02	0.43 ± 0.01	0.47 ± 0.02	0.52 ± 0.03
h_A	0.19 ± 0.02	0.15 ± 0.01	0.12 ± 0.01	0.06 ± 0.01
h_α	0.24 ± 0.03	0.23 ± 0.01	0.21 ± 0.01	0.12 ± 0.02
h_β	0.43 ± 0.03	0.47 ± 0.02	0.49 ± 0.02	0.62 ± 0.03
h_γ	0.13 ± 0.01	0.14 ± 0.01	0.17 ± 0.02	0.20 ± 0.01

[1] h_i is the normalized integral area (fraction) for protons of i category; f_A is the aromatic carbon ratio and calculated via $f_A = 1 - (h_\alpha + h_\beta + h_\gamma)/(2q)$; σ is substitution rate on aromatic rings and calculated via $\sigma = h_\alpha/(2h_A + h_\alpha)$.

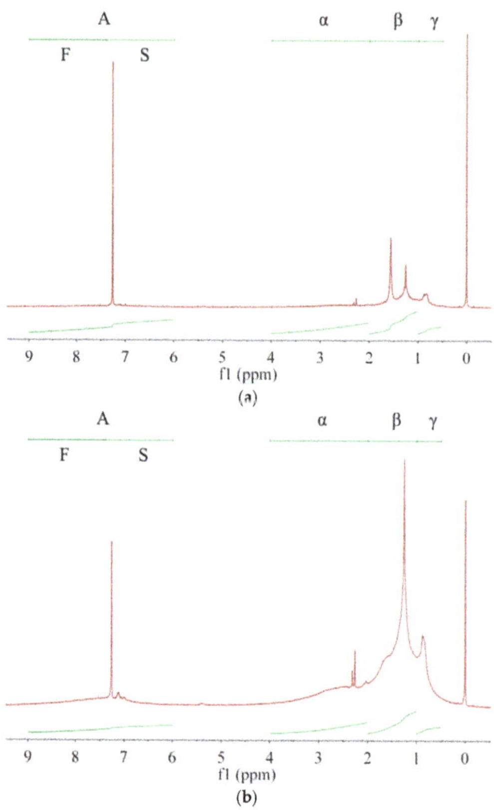

Figure 2. *Cont.*

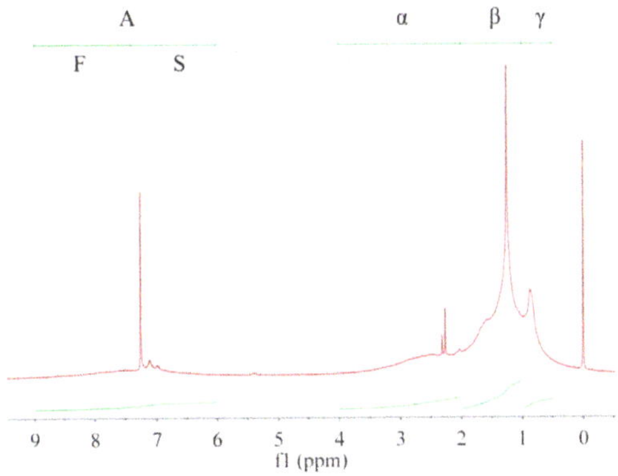

(c)

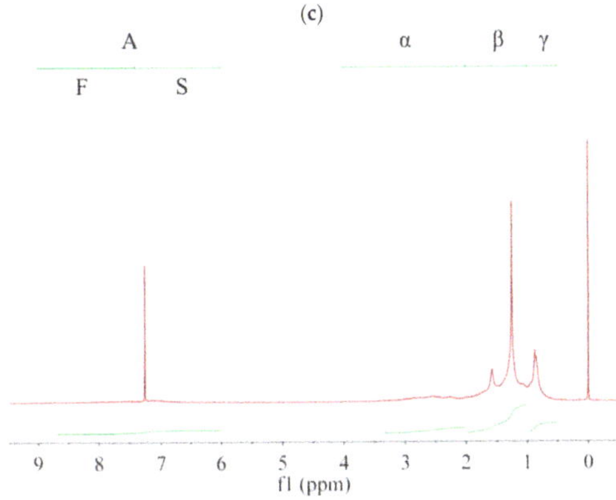

(d)

Figure 2. The ^{1}H-NMR spectra for (**a**) asphaltene A1, (**b**) asphaltene A2, (**c**) asphaltene A3 and (**d**) base asphalt. Protons are divided into A, α, β, and γ categories while the category A includes protons on single (S) and fused (F) aromatic rings.

In Figure 3, main absorption peaks are marked with purple letters on FTIR spectra and the wave numbers for these peaks are listed in Table 6 with peak assignments [6,31,32]. A general rule is that the peak of asphaltenes locates at smaller wave numbers than that of base asphalt. For example, the peak g arises from the skeleton stretching (C=C) of aromatic rings. It locates at 1592 cm^{-1} for A1 and locates at 1603 cm^{-1} for base asphalt. The shift of peak location towards higher wave number from A1 to asphalt arises from the decrease of molecular aromaticity. The aromatic ring makes the distribution of electron cloud density between single-bond and double-bond more uniform. This conjugative effect reduces the electron cloud density of double bond, and reduces its force constant and vibration frequency, so that the peak g moves towards smaller wave number in Figure 3 and Table 6. Moreover, aromatic rings have stronger electron absorption than saturated alkanes. This inductive effect reduces the electron cloud density and reduces the force constant and vibration frequency of the adjacent bond, thus its absorption peak moves towards lower wave number in Table 6. According to the previous EA and NMR results of Tables 4 and 5, from asphalt to A3 then to A1, the values of h_S, h_F and h_A increase while the unsaturation and aromaticity of the molecular structure increase. Such increase may improve the inductive and conjugation effects and shift the peak location towards lower wave number, which is consistent with results in Figure 3 and Table 6.

The aromatic rings can only influence a small proportion of chemical bonds adjacent to the rings while asphalt and asphaltene may have dramatically different proportion. In Table 6, the discrepancy between asphaltenes (A1, A2 and A3) is relatively smaller than that between asphaltenes and base asphalt. For example, the discrepancy of location for peak g is only 3 cm^{-1} between A1 and A3 while it is 11 cm^{-1} between A1 and asphalt, according to Table 6. The reason may be that the proportion of chemical bonds influenced by aromatic rings increases more dramatically (larger discrepancy) from asphalt to asphaltene A3 than it increases from asphaltenes A3 to A1. From asphalt to asphaltene A3, the asphaltene percentage increases from below 20% to nearly 100%. Such a shift of peak location was also observed for asphaltenes and even resins near 1603 and 1030 cm^{-1} in previous FTIR spectra [6]. Moreover, as asphalt was a complex mixture and its four components had significantly different spectra [6], the peak position and height may vary with the proportion of its components. Compared with asphaltenes, the saturates lacked peaks near 1030 cm^{-1} (peak d) and 1600 cm^{-1} (peak g) [6], owing to lacking S and O elements and double-bond in their molecules. The peak around 1377 cm^{-1} is used for the quantification of poly(styrene-*b*-butadiene-*b*-styrene) (AASHTO T302-15, JT/T 1329-2020) or other modifiers in asphalt [31,33,34]. According to Table 6, the position of this peak (e) moves towards a lower wave number from base asphalt to asphaltenes; thus, its specific position may vary with asphaltene proportion and should be checked carefully when calculating peak height or area integral for quantification.

Table 6. The wave number (in unit of cm^{-1}) for main peaks on IR spectra for the extracted asphaltenes (A1, A2 and A3) and the base asphalt [1].

IR Peak No.	A1	A2	A3	Asphalt
a	743	743	744	749
b	805	805	807	818
c	857	857	857	861
d	1027	1028	1028	1032
e	1372	1373	1374	1377
f	1452	1452	1452	1457
g	1592	1592	1595	1603
h	2849	2849	2849	2853
i	2919	2919	2919	2924
j	3053	3053	3054	3055

[1] Peaks a, b and c arise from the (C-H) aromatic ring substitution bands. The peak d may arise from the stretching vibration of sulfoxide (S=O) group. Peaks e and f arise from the bending vibration absorption of aromatic methyl ($-CH_3$) and methylene (CH_2) groups. The peak g is the skeleton stretching (C=C) of aromatic rings. The peak h arises from the symmetric stretching vibrations of aliphatic methyl ($-CH_3$) group. The peak i arises from the antisymmetric stretching vibrations of aliphatic methylene (CH_2) group. Peak j arises from the stretching vibration (=C-H) on the unsaturated bond or aromatic ring.

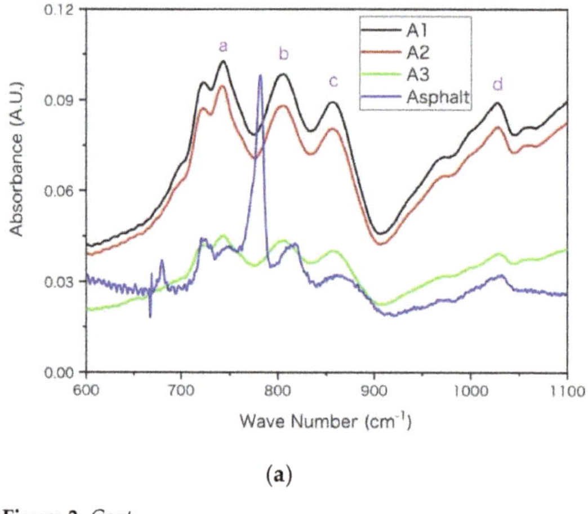

(**a**)

Figure 3. *Cont.*

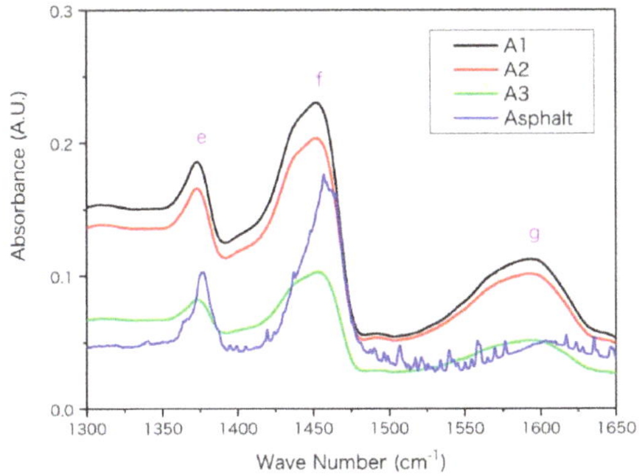

(**b**)

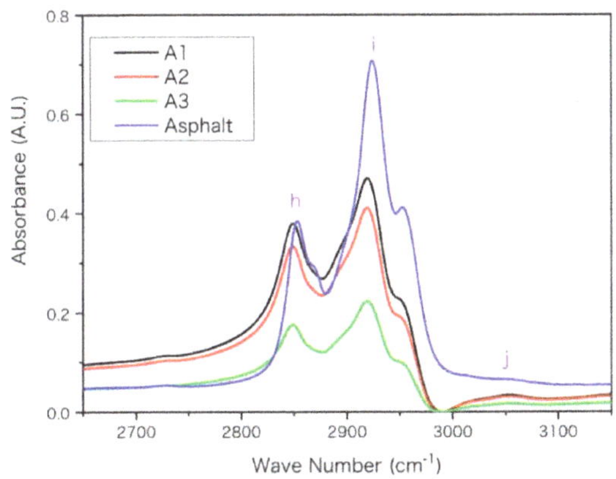

(**c**)

Figure 3. The FTIR spectra for the asphaltenes (A1, A2 and A3) and asphalt with wave number (in unit of cm^{-1}) in ranges of (**a**) 600–1100, (**b**) 1300–1650, (**c**) 2650–3150, respectively. The absorbance is in arbitrary unit (A.U.) and main peaks are marked in alphabetical order.

The XRD spectra of asphaltenes mainly have 3 characteristic peaks of γ, 002 and 10 band, which are labelled as γ, m and a, respectively, in Figure 4. The X-ray wavelength λ is 1.5406 angstrom (Å). According to their diffraction angles (2θ) listed in Table 7 and the Bragg equation $d = \lambda/(2\sin\theta)$ [35–37], the characteristic lengths d are calculated and listed in Table 7. The peaks γ and m reflect the packing structure of saturated alkyl chains and aromatic rings, respectively, while the peak a is related to the size of the aromatic sheet [35,37,38]. From A1 to A3 in Table 7, the decrease of average distance d_γ indicates a denser packing of alkyl chains, while the increase of layer spacing d_m implies a looser packing of aromatic rings. According to previous EA, IR and NMR results, the unsaturation, polarity and aromaticity of asphaltenes decrease while the number of substituted branches and fraction of methyl groups increase from A1 to A3. Thus, the packing of aromatic rings becomes looser owing to more substituted branches on aromatic rings and smaller inter-molecular interactions, while the packing of alkyl chains becomes denser because of the increased fraction of free chain (methyl) end. The decreased ring fraction and molecular unsaturation is also helpful to decrease the molecular rigidity and beneficial to the packing of alkyl chains. Therefore, the variation trend of molecular structure of asphaltenes reflected by XRD spectra is consistent with the previous experimental results.

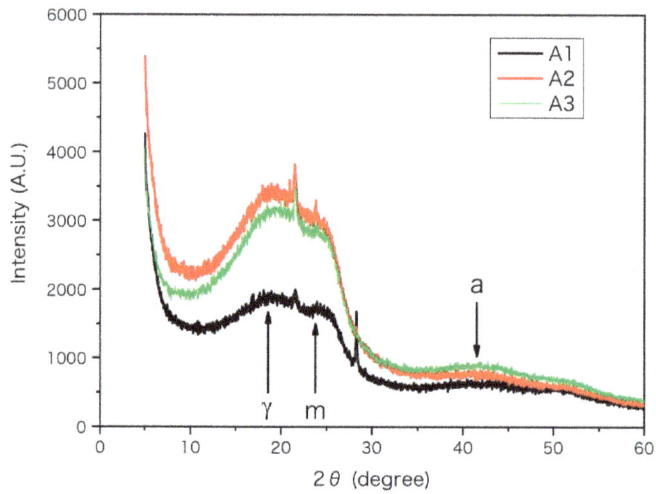

Figure 4. The XRD spectra for the asphaltenes (A1, A2 and A3). The diffraction intensity from position sensitive detector (PSD) is in arbitrary unit (A.U.) and plotted versus diffraction angle 2θ. The peaks of 002 and 10 band are labelled as m and a respectively.

Table 7. The packing length d (in unit of Å) calculated according to diffraction angles (2θ) from XRD spectra for the extracted asphaltenes (A1, A2 and A3). Subscripts m and a represent peaks of 002 and 10 band, respectively.

XRD Quantities	A1	A2	A3
$(2\theta)_Y$	19.07°	19.21°	19.49°
$(2\theta)_m$	23.90°	23.79°	23.76°
$(2\theta)_a$	41.89°	40.34°	41.61°
d_Y/Å	4.650	4.617	4.551
d_m/Å	3.720	3.737	3.742

4. Conclusions

The linear equation between the density and concentration for standard samples of extractant containing xylene (G) and n-heptane (P) was fitted with R^2 = 0.99967 and used to analyze the real samples of extractant recovered and reused in asphaltene extraction process. With the increase of extraction number, the G/P ratio in extractant increases. The proportion of asphaltenes dissolved in solution rises and the proportion of precipitation from solution falls, resulting in a decrease of the asphaltene yield with the number of extractions. The results of EA, IR, NMR and XRD reached a consistent conclusion. For asphaltenes (A1, A2 and A3) sequentially extracted at room temperature, the fractions of substituted alkyl chains on aromatic rings and chain ends increase while the fractions of N and O elements as well as the C/H atom ratio (unsaturation) decrease, with increasing G/P ratio of extractant or with the extraction number. With increasing the G/P ratio of extractant, the extracted asphaltenes have smaller fractions of C and H of aromatic rings (aromaticity). The decrease of unsaturation and aromaticity from A1 to asphalt leads to reduced conjugative and inductive effects, which shift IR peaks towards higher wave numbers. Since asphaltenes extracted at higher G/P ratio have a smaller fraction of aromatic rings, more substitution rate on aromatic rings and bigger fraction of alkyl and methyl groups in molecular structure; they have smaller molecular rigidity and weaker inter-molecular conjugation. Thus, asphaltenes extracted at a higher G/P ratio have a denser packing of alkyl chains and a looser packing of aromatic rings.

Author Contributions: Conceptualization, D.S.; data curation, Y.S.; methodology, D.S.; visualization, Y.S.; writing—original draft, D.S. and Y.S.; writing—review and editing, D.S. and F.C. All authors have read and agreed to the published version of the manuscript.

Funding: This research received funding from the Nanxun Collaborative Innovation Center Key Research Project (SYS01001), the Open-ended Fund of Key Laboratory of Urban Pollutant Conversion, Chinese Academy of Sciences (KLUPCKF-2020-4), the Special Research Funds in Shandong Jianzhu University (X20087Z0101) and Doctoral Fund in Shandong Jianzhu University (X20078Z0101).

Institutional Review Board Statement: Not applicable.

Informed Consent Statement: Not applicable.

Data Availability Statement: The data presented in this study and the source codes are available on request from the corresponding author Dachuan Sun.

Acknowledgments: The author Dachuan Sun acknowledged Xianglong Zhao and Rui Feng for helpful discussions.

Conflicts of Interest: The authors declare no conflict of interest.

References

1. Gong, Y.; Xu, J.; Yan, E. Intrinsic temperature and moisture sensitive adhesion characters of asphalt-aggregate interface based on molecular dynamics simulations. *Constr. Build. Mater.* **2021**, *292*, 123462. [CrossRef]
2. Que, Y.; Wang, Y.; Xu, S.; Fan, Y. Quantitative detection of polymer content in styrene-butadiene-styrene modified asphalt after aging. *Polym. Eng. Sci.* **2022**, *62*, 2197–2206. [CrossRef]
3. Hassanzadeh, M.; Abdouss, M. A comprehensive review on the significant tools of asphaltene investigation. Analysis and characterization techniques and computational methods. *J. Pet. Sci. Eng.* **2022**, *208*, 109611. [CrossRef]
4. Lemarchand, C.A.; Greenfield, M.L.; Dyre, J.C.; Hansen, J.S. ROSE bitumen: Mesoscopic model of bitumen and bituminous mixtures. *J. Chem. Phys.* **2018**, *149*, 214901. [CrossRef] [PubMed]
5. Strausz, O.P.; Mojelsky, T.W.; Lown, E.M. The molecular structure of asphaltene: An unfolding story. *Fuel* **1992**, *71*, 1355–1362. [CrossRef]
6. Guo, M.; Liang, M.; Fu, Y.; Sreeram, A.; Bhasin, A. Average molecular structure models of unaged asphalt binder fractions. *Mater. Struct.* **2021**, *54*, 173. [CrossRef]
7. Sun, D. A Simple Scheme for Extraction of Asphaltenes from Asphalt at Room Temperature. *Coatings* **2022**, *12*, 407. [CrossRef]
8. Gray, M.R.; Yarranton, H.W.; Chacón-Patiño, M.L.; Rodgers, R.P.; Bouyssiere, B.; Giusti, P. Distributed properties of asphaltene nanoaggregates in crude oils: A review. *Energy Fuels* **2021**, *35*, 18078–18103. [CrossRef]
9. Ali, S.I.; Awan, Z.; Lalji, S.M. Laboratory evaluation experimental techniques of asphaltene precipitation and deposition controlling chemical additives. *Fuel* **2022**, *310*, 122194. [CrossRef]
10. Shao, R.; Li, D.; Pei, L.; Yuan, Y.; Liu, X.; Li, W. Effect of Deasphalting Solvent on Structure of Coal Tar Asphaltene. *Acta Pet. Sin. (Pet. Process. Sect.)* **2017**, *33*, 1209–1217.
11. Yang, Y.; Bai, P.; Guo, X. Separation of Xylene Isomers: A Review of Recent Advances in Materials. *Ind. Eng. Chem. Res.* **2017**, *56*, 14725–14753. [CrossRef]
12. Cui, X.; Niu, Z.; Shan, C.; Yang, L.; Hu, J.; Wang, Q.; Lan, P.; Li, Y.; Wojtas, L.; Ma, S.; et al. Efficient separation of xylene isomers by a guestresponsive metal–organic framework with rotational anionic sites. *Nat. Commun.* **2020**, *11*, 5456. [CrossRef] [PubMed]
13. Mousavi-Dehghani, S.A.; Riazi, M.R.; Vafaie-Sefti, M.; Mansoori, G.A. An analysis of methods for determination of onsets of asphaltene phase separations. *J. Pet. Sci. Eng.* **2004**, *42*, 145–156. [CrossRef]
14. Goual, L.; Sedghi, M.; Wang, X.; Zhu, Z. Asphaltene Aggregation and Impact of Alkylphenols. *Langmuir* **2014**, *30*, 5394–5403. [CrossRef]
15. Roy, M.M. HPLC analysis of aldehydes in automobile exhaust gas: Comparison of exhaust odor and irritation in different types of gasoline and diesel engines. *Energ. Convers. Manag.* **2008**, *49*, 1111–1118. [CrossRef]
16. Vempatapu, B.P.; Kanaujia, P.K. Monitoring petroleum fuel adulteration: A review of analytical methods. *TrAC-Trend. Anal. Chem.* **2017**, *92*, 1–11. [CrossRef]
17. Hale, C.S.; Bhakta, H.C.; Jonak, C.R.; Yonan, J.M.; Binder, D.K.; Grover, W.H.; Rodgers, V.G.J. Differential densimetry: A method for determining ultra-low fluid flux and tissue permeability. *AIP Adv.* **2019**, *9*, 095063. [CrossRef]
18. Zhang, M.; Hao, P.; Dong, S.; Li, Y.; Yuan, G. Asphalt binder micro-characterization and testing approaches: A review. *Measurement* **2020**, *151*, 107255. [CrossRef]
19. Ismail, M.; Yang, Y.; Chaisoontornyotin, W.; Ovalles, C.; Rogel, E.; Moir, M.E.; Hoepfner, M.P. Effect of chemical inhibitors on asphaltene precipitation and morphology using ultra-small-angle X-ray scattering. *Energy Fuels* **2019**, *33*, 3681–3693. [CrossRef]
20. Nguyen, M.T.; Nguyen, D.L.T.; Xia, C.; Nguyen, T.B.; Shokouhimehr, M.; Sana, S.S.; Grace, A.N.; Aghbashlo, M.; Tabatabaei, M.; Sonne, C.; et al. Recent advances in asphaltene transformation in heavy oil hydroprocessing: Progress, challenges, and future perspectives. *Fuel Process. Technol.* **2021**, *213*, 106681. [CrossRef]

21. Fan, M.; Zhao, S.; Sun, X.; Xu, Z. Comparison of properties of C_5 and C_7 asphaltenes from different heavy oils. *China Pet. Process. Petrochem. Technol.* **2019**, *50*, 96–102.
22. Liu, S.; Zhou, G.; Cheng, G.; Wang, X.; Liu, G.; Jin, W. Emerging membranes for separation of organic solvent mixtures by pervaporation or vapor permeation. *Sep. Purif. Technol.* **2022**, *299*, 121729. [CrossRef]
23. AlTuwaim, M.S.; Alkhaldi, K.H.A.E.; Al-Jimaz, A.S.; Alanezi, K.M. Separation of Alkylbenzenes from n-Heptane Using Binary Mixtures of Ionic Solvents. *J. Chem. Eng. Data* **2019**, *64*, 1187–1194. [CrossRef]
24. Cimino, R.; Correra, S.; Sacomani, P.A.; Carniani, C. Thermodynamic modelling for prediction of asphaltene deposition in live oils. In Proceedings of the SPE International Symposium on Oilfield Chemistry, San Antonio, TX, USA, 14–17 February 1995.
25. Kreuger, T.; Bos, A.N.R.; Kersten, S.R.A. Predicting gasification rates of pyrolytic graphite deposited from methane. *Chem. Eng. J.* **2022**, *440*, 135487. [CrossRef]
26. Wang, C.; Zhang, X.; Wang, W.; Sun, J.; Mao, Y.; Zhao, X.; Song, Z. A stepwise microwave synergistic pyrolysis approach to produce sludge-based biochars: Optimizing and mechanism of heavy metals immobilization. *Fuel* **2022**, *314*, 122770. [CrossRef]
27. León, A.Y.; Alexander, G.M.; Picón, H.; Dionisio, L.C.; Daniel, R.M.V. Reactivity of Vacuum Residues by Thermogravimetric Analysis and Nuclear Magnetic Resonance Spectroscopy. *Energy Fuels* **2021**, *34*, 9231–9242. [CrossRef]
28. Zhao, H.; Chai, X.; Wang, Y.; Shi, H.; Zhang, R.; Dong, Z.; Han, L. Analysis of styrene-butadiene-styrene (SBS) content in modified asphalt with ^{1}H NMR Spectroscopy. *Chin. J. Magn. Reson.* **2017**, *34*, 323–328.
29. Huang, J. Characterization of asphalt fractions by NMR spectroscopy. *Pet. Sci. Technol.* **2010**, *28*, 618–624. [CrossRef]
30. Mondal, S.; Yadav, A.; Kumar, R.; Bansal, V.; Das, S.K.; Christopher, J.; Kapur, G.S. Molecular Level Structural Insight into Clarified Oil by NMR Spectroscopy: Estimation of Hydrocarbon Types and Average Structural Parameters. *Energy Fuels* **2017**, *31*, 7682–7692. [CrossRef]
31. Hou, X.; Lv, S.; Chen, Z.; Xiao, F. Applications of Fourier Transform Infrared Spectroscopy Technologies on Asphalt Materials. *Measurement* **2018**, *121*, 304–316. [CrossRef]
32. Wang, K.; Yuan, J.; Gao, N. Determination of Styrene-Butadiene-Styrene Block Copolymer in Modified Asphalt by Dissolution Coating Infrared Spectroscopy. *PCTA (Part B Chem. Anal.)* **2015**, *51*, 357–359.
33. Xu, S.; Fan, Y.; Feng, Z.; Ke, Y.; Zhang, C.; Huang, H. Comparison of quantitative determination for SBS content in SBS modified asphalt. *Constr. Build. Mater.* **2021**, *282*, 122733. [CrossRef]
34. Luo, S.; Tian, J.; Liu, Z.; Lu, Q.; Zhong, K.; Yang, X. Rapid determination of styrene-butadiene-styrene (SBS) content in modified asphalt based on Fourier transform infrared (FTIR) spectrometer and linear regression analysis. *Measurement* **2020**, *151*, 107204. [CrossRef]
35. Jennings, J.; Growney, D.J.; Brice, H.; Mykhaylyk, O.O.; Armes, S.P. Application of scattering and diffraction techniques for the morphological characterization of asphaltenes. *Fuel* **2022**, *327*, 125042. [CrossRef]
36. AlHumaidan, F.S.; Hauser, A.; Rana, M.S.; Lababidi, H.M.S.; Behbehani, M. Changes in asphaltene structure during thermal cracking of residual oils: XRD study. *Fuel* **2015**, *150*, 558–564. [CrossRef]
37. Caputo, P.; Ventruti, G.; Calandra, P.; Porto, M.; Teltayev, B.; Angelico, R.; Rossi, C.O. Searching effective indicators of microstructural changes in bitumens during aging: A multi-technique approach. *Colloid Surface A* **2022**, *641*, 128529. [CrossRef]
38. AlHumaidan, F.S.; Rana, M.S.; Vinoba, M.; Rajasekaran, N.; AlHenyyan, H.Y.; Ali, A.A. Synthesizing few-layer carbon materials from asphaltene by thermal treatment. *Diam. Relat. Mater.* **2022**, *129*, 109316. [CrossRef]

coatings MDPI

Article

On Strain Gradient Theory and Its Application in Bending of Beam

Anqing Li [1,2,*], Qing Wang [3], Ming Song [1,2], Jun Chen [1,2], Weiguang Su [1,2], Shasha Zhou [1,2] and Li Wang [1,2,*]

1 School of Mechanical Engineering, Qilu University of Technology (Shandong Academy of Sciences), Jinan 250353, China
2 Shandong Institute of Mechanical Design and Research, Jinan 250031, China
3 College of Mechanical and Electronic Engineering, Qingdao University, Qingdao 266071, China
* Correspondence: akin@qlu.edu.cn (A.L.); liwang@qlu.edu.cn (L.W.)

Abstract: The general strain gradient theory of Mindlin is re-visited on the basis of a new set of higher-order metrics, which includes dilatation gradient, deviatoric stretch gradient, symmetric rotation gradient and curvature. A strain gradient bending theory for plane-strain beams is proposed based on the present strain gradient theory. The stress resultants are re-defined and the corresponding equilibrium equations and boundary conditions are derived for beams. The semi-inverse solution for a pure bending beam is obtained and the influence of the Poisson's effect and strain gradient components on bending rigidity is investigated. As a contrast, the solution of the Bernoulli–Euler beam is also presented. The results demonstrate that when Poisson's effect is ignored, the result of the plane-strain beam is consistent with that of the Bernoulli–Euler beam in the couple stress theory. While for the strain gradient theory, the bending rigidity of a plane-strain beam ignoring the Poisson's effect is smaller than that of the Bernoulli–Euler beam due to the influence of the dilatation gradient and the deviatoric stretch gradient along the thickness direction of the beam. In addition, the influence of a strain gradient along the length direction on a bending rigidity is negligible.

Keywords: strain gradient theory; beam bending; size effect; plane-strain; couple stress theory

Citation: Li, A.; Wang, Q.; Song, M.; Chen, J.; Su, W.; Zhou, S.; Wang, L. On Strain Gradient Theory and Its Application in Bending of Beam. *Coatings* **2022**, *12*, 1304. https://doi.org/10.3390/coatings12091304

Academic Editor: Bohayra Mortazavi

Received: 12 August 2022
Accepted: 1 September 2022
Published: 5 September 2022

1. Introduction

A micro-beam is a typical component in micro-electro-mechanical systems (MEMSs). Accurate description of mechanical behavior of a beam is necessary for the design and performance prediction of MEMSs [1–3]. Although the traditional beam model has matured, the mechanical behavior of beams at the micro-scale is obviously different from that at the macro-scale, which cannot be captured by the traditional models [4–6]. Many experiments have reported that the mechanical properties of a micro-beam show obvious size effects [7–10]. McFarland and Colton observed that the bending stiffness of polypropylene cantilever beams with thicknesses of 15 and 30 μm is more than four times the value predicted by traditional theory [9]. Lam et al. found that the dimensionless bending stiffness of the epoxy polymeric beam with a thickness of 20 μm was approximately 2.3 times higher than that of 115 μm [7]. For the purpose of capturing the size-dependent mechanical properties, the strain gradient theory has been presented.

In strain gradient theory, the strain gradient is regarded as an additional (higher-order) deformation metric in addition to the traditional strain. Among the strain gradient theories, the couple stress theory is the earliest strain gradient theory that uses the rotation gradient as an additional deformation metric [11]. In fact, the stain gradient includes dilatation gradient, rotation gradient and deviatoric stretch gradient [12]. The rotation gradient is only the anti-symmetric part of the strain gradient. As with the development of strain gradient theory, the general strain gradient theory, including all strain gradient components, is proposed by Mindlin and Eshel by introducing five higher-order material constants [13]. Then, the general strain gradient theory of Mindlin has been re-formulated by Lam et al. on the basis of the equilibrium of moments of couples [7], and by Zhou et al. according to the independent strain gradient components [14], respectively. The two re-formulated strain gradient theories include three higher-order material constants. A simple strain gradient theory with only one higher-order material constant has been presented by Aifantis [15]. The general form of couple stress theory includes two higher-order material constants, which has been further simplified to include only one higher-order material constant by Yang et al. [16] and Hadjesfandiari and Dargush [17], respectively, by choosing only the symmetric or anti-symmetric parts as the higher-order deformation metrics. However, in our opinion, although these re-formulated strain gradient theories with three or fewer higher-order constants can also capture the size-effect phenomenon, these theories ignore the contribution of some strain gradient components and underestimate the stiffening effect of the strain gradient. In fact, Mindlin's strain gradient theory is a general strain gradient theory and should be used to develop micro-component models.

According to the strain gradient theory, many strain gradient beam models have been established to capture the size-dependent mechanical behaviors [18–23]. Based on Mindlin's strain gradient theory, a linear model of Bernoulli–Euler beams has been presented by Niiranen et al. [24], and a geometrically non-linear model was further developed by Tran and Niiranen [25] by adopting the von Kármán strain assumption. Based on the modified couple stress theory, Vo et al. presented the models of planar [26] and spatial [27] arbitrarily curved micro-beams and Li et al. investigated the bending and free vibration of bi-directional functionally graded graphene nano-platelets-reinforced composite micro-beams [28]. However, these existing models are contradictory [29]. The non-zero strain of Bernoulli–Euler beams is an axial strain, and its non-zero strain gradient includes the gradients of axial strain along the thickness direction and along the length direction. Currently, there are three main types of models available, depending on the stain gradient components used. The first category considers the strain gradient along the thickness direction and the length direction at the same time [30–33]. The second type considers only the strain gradient along the length direction and does not consider the strain gradient along the thickness direction [34]. The third type considers the strain gradient along the thickness direction rather than the length direction, since the strain gradient along the length direction can be neglected compared with that along the thickness direction for Euler beams [35]. The third type may be seen as an approximation of the first type. However, the second and third types are polar opposites, and one of them must be wrong. Thus, it is important to figure out which type is correct.

Lurie and Solyaev [29] checked the beam models of the first and second types and believed that only the second type (uniaxial stress state) could guarantee the satisfaction of boundary conditions on the top and bottom surfaces of the beams. However, in our opinion, the satisfaction of boundary conditions on the top and bottom surfaces of the beams depends on the definition of moment, which includes the higher-order stress. The first type can satisfy the boundary conditions on the top and bottom surfaces of the beams by defining a different moment from that of Lurie and Solyaev.

The purpose of this work is to establish a correct beam model and analyze the contribution of the strain gradient components to bending stiffness. This paper is organized as follows: Mindlin's general strain gradient elasticity theory is re-formulated by introducing a new set of strain gradient components so as to investigate the effects of each strain gradient component conveniently in Section 2. Then, the bending theory for plane-strain beams is presented in Section 3, and the semi-inverse solution for pure bending beams is developed in Section 4. Section 5 investigates the bending of Bernoulli–Euler beams and compares the results of plane-strain beams and Bernoulli–Euler beams. Finally, conclusions are summarized in Section 6.

2. Formulations of Strain Gradient Elasticity Theories

2.1. A Summary of Strain Gradient Theory

In strain gradient theory, not only the traditional strain ε_{ij} but also its gradients η_{ijk} are considered in the internal energy density U, written as:

$$U = U(\varepsilon_{ij}, \eta_{ijk}) \tag{1}$$

with

$$\varepsilon_{ij} = \frac{1}{2}(u_{i,j} + u_{j,i}),\ \eta_{ijk} = \varepsilon_{jk,i} \tag{2}$$

where u_i denotes the displacement vector and a comma represents the differentiation with respect to the coordinates. The work-conjugated Cauchy stress σ_{ij} and higher-order stress τ_{ijk} with respect to strain and strain gradient, respectively, are given by:

$$\sigma_{ij} = \frac{\partial U}{\partial \varepsilon_{ij}},\ \tau_{ijk} = \frac{\partial U}{\partial \eta_{ijk}} \tag{3}$$

Obviously, the stress tensor σ_{ij} is a symmetric tensor and the higher-order stress tensor τ_{ijk} is symmetric with respect to subscripts j and k.

For a solid of volume V with surface boundary S and sharp edge C, the virtual work principle of the strain gradient theory is given as [14]:

$$\int_V (\sigma_{ij}\delta\varepsilon_{ij} + \tau_{ijk}\delta\eta_{ijk})\mathrm{d}V = \int_V \overline{b}_k\delta u_k \mathrm{d}V + \int_S (\overline{t}_k\delta u_k + \overline{r}_k D\delta u_k)\mathrm{d}S + \oint_C \overline{f}_k\delta u_k \mathrm{d}C \tag{4}$$

Thus, the equilibrium equation can be obtained by applying the variation of internal energy, as shown in Appendix A in detail, given as:

$$\sigma_{ik,i} - \tau_{ijk,ij} + \overline{b}_k = 0 \tag{5}$$

and the boundary conditions on surface boundary S and sharp edge C are:

$$n_i(\sigma_{ik} - \tau_{jik,j}) + (D_p n_p)n_i n_j \tau_{ijk} - D_i(n_j\tau_{jik}) = \overline{t}_k \text{ or } u_k = \overline{u}_k \tag{6}$$

$$n_i n_j \tau_{ijk} = \overline{r}_k \text{ or } \overline{Du_k} = Du_k \tag{7}$$

and

$$[[k_j n_i \tau_{ijk}]] = \overline{f}_k \text{ or } u_k = \overline{u}_k \tag{8}$$

where $D = n_i\partial_i$ is the normal gradient operator with n_i representing a unit vector normal to the boundary surface S, $D_i = (\delta_{ik} - n_i n_k)\partial_k$ represents the surface gradient operator, the square brackets stands for the difference between the values of the enclosed quantity on the two sides of the edge and k_i denotes the outer co-normal vector.

For linear elastic isotropic materials, the internal energy density of Equation (1) is generally written as:

$$U = \frac{1}{2}k\varepsilon_{ii}\varepsilon_{jj} + \mu\varepsilon'_{ij}\varepsilon'_{ij} + g_1\eta_{iik}\eta_{kjj} + g_2\eta_{ijj}\eta_{ikk} + g_3\eta_{iik}\eta_{jjk} + g_4\eta_{ijk}\eta_{ijk} + g_5\eta_{ijk}\eta_{kji} \tag{9}$$

where $\varepsilon'_{ij} = \varepsilon_{ij} - \frac{1}{3}\delta_{ij}\varepsilon_{nn}$ denotes the deviatoric strain tensor. Then, the constitutive equations in terms of Equation (3) are given by:

$$\sigma_{ij} = k\delta_{ij}\varepsilon_{nn} + 2\mu\varepsilon'_{ij} \tag{10}$$

$$\begin{aligned}\tau_{ijk} = {} & \tfrac{1}{2}g_1(\delta_{ij}\eta_{knn} + \delta_{ik}\eta_{jnn} + 2\delta_{jk}\eta_{nni}) \\ & + 2g_2\delta_{jk}\eta_{inn} + g_3(\delta_{ij}\eta_{nnk} + \delta_{ik}\eta_{nnj}) + 2g_4\eta_{ijk} + g_5(\eta_{kji} + \eta_{jki})\end{aligned} \tag{11}$$

in which k and μ are the bulk and shear moduli, δ_{ij} denotes the Kronecker delta and g_i (i = 1, 2, 3, 4, 5) is the introduced higher-order constant.

2.2. Development of Strain Gradient Components, and Equilibrium and Boundary Conditions

Strain gradient has been decomposed into its orthogonal components to re-formulate the strain gradient theory [14]. In fact, strain gradient includes dilatation gradient γ_i, deviatoric stretch gradient $\eta_{ijk}^{(1)}$ and rotation gradient χ_{ij}. These strain gradient components are defined as:

$$\begin{aligned} &\gamma_i = \eta_{inn} \quad \chi_{ij} = e_{ilk}\eta_{ljk} \\ &\eta_{ijk}^{(1)} = \tfrac{1}{3}(\eta_{ijk} + \eta_{jki} + \eta_{kij}) \\ &\quad -\tfrac{1}{15}[\delta_{ij}(2\eta_{nnk} + \eta_{knn}) + \delta_{jk}(2\eta_{nni} + \eta_{inn}) + \delta_{ki}(2\eta_{nnj} + \eta_{jnn})] \end{aligned} \tag{12}$$

with e_{ilk} denoting the alternating tensor. The rotation gradient can be further decomposed into its symmetric and anti-symmetric parts, given as:

$$\chi_{ij}^s = \frac{1}{2}(\chi_{ij} + \chi_{ji}), \chi_{ij}^a = \frac{1}{2}(\chi_{ij} - \chi_{ji}) \tag{13}$$

Additionally, the anti-symmetric rotation gradient tensor χ_{ij}^a can be represented by its dual vector κ_i which is called as the curvature vector [17], written as:

$$\kappa_i = \frac{1}{2}e_{ijk}\chi_{kj}^a \tag{14}$$

Then, the strain gradient can be finally expressed as the sum of dilatation gradient γ_i, deviatoric stretch gradient $\eta_{ijk}^{(1)}$, symmetric rotation gradient χ_{ij}^s and curvature κ_i, given as [12]:

$$\eta_{ijk} = A_{ijks}\gamma_s + \eta_{ijk}^{(1)} + B_{ijkpq}\chi_{pq}^s + D_{ijks}\kappa_s \tag{15}$$

with the coefficients being:

$$\begin{aligned} A_{ijks} &= \tfrac{1}{5}(\delta_{ij}\delta_{ks} + \delta_{jk}\delta_{is} + \delta_{ik}\delta_{js}) \\ B_{ijkpq} &= \tfrac{1}{3}(\delta_{kq}e_{ijp} + \delta_{jq}e_{ikp}) \\ D_{ijks} &= \tfrac{1}{5}(2\delta_{jk}\delta_{is} - 3\delta_{ij}\delta_{ks} - 3\delta_{ik}\delta_{js}) \end{aligned} \tag{16}$$

Thus, the internal energy density becomes a function of strain and strain gradient components, written as $U = U(\varepsilon_{ij}, \gamma_i, \eta_{ijk}^{(1)}, \chi_{ij}^s, \kappa_i)$. The corresponding work-conjugated higher-order stress components p_i, $\tau_{ijk}^{(1)}$, m_{ij}^s, μ_i with respect to strain gradient components γ_i, $\eta_{ijk}^{(1)}$, χ_{ij}^s, κ_i, respectively, are given by:

$$p_i = \frac{\partial U}{\partial \gamma_i}, \ \tau_{ijk}^{(1)} = \frac{\partial U}{\partial \eta_{ijk}^{(1)}}, \ m_{ij}^s = \frac{\partial U}{\partial \chi_{ij}^s}, \ \mu_i = \frac{\partial U}{\partial \kappa_i} \tag{17}$$

The virtual work density can be rewritten in terms of the new strain metrics as:

$$\begin{aligned} \delta U &= \sigma_{ij}\delta\varepsilon_{ij} + p_i\delta\gamma_i + \tau_{ijk}^{(1)}\delta\eta_{ijk}^{(1)} + m_{ij}^s\delta\chi_{ij}^s + \mu_i\delta\kappa_i \\ &= \sigma_{ij}\delta\varepsilon_{ij} + (\delta_{jk}p_i + \tau_{ijk}^{(1)} + \tfrac{1}{2}m_{sj}^s e_{sik} + \tfrac{1}{2}m_{sk}^s e_{sij} + \tfrac{1}{2}\delta_{jk}\mu_i - \tfrac{1}{4}\delta_{ij}\mu_k - \tfrac{1}{4}\delta_{ik}\mu_j)\delta\eta_{ijk} \end{aligned} \tag{18}$$

From Equation (18), it can be concluded that the higher-order stress τ_{ijk} is associated with the higher-order stress components p_i, $\tau_{ijk}^{(1)}$, m_{ij}^s, μ_i by:

$$\tau_{ijk} = \delta_{jk}p_i + \tau_{ijk}^{(1)} + \frac{1}{2}m_{sj}^s e_{sik} + \frac{1}{2}m_{sk}^s e_{sij} + \frac{1}{2}\delta_{jk}\mu_i - \frac{1}{4}\delta_{ij}\mu_k - \frac{1}{4}\delta_{ik}\mu_j \tag{19}$$

Finally, applying the principle of virtual work yields the equilibrium equation in V as:

$$\sigma_{ik,i} - p_{i,ik} - \tau_{ijk,ij}^{(1)} - \frac{1}{2}e_{lik}m_{lj,ij}^s - \frac{1}{4}\mu_{i,ik} + \frac{1}{4}\mu_{k,ii} + \bar{b}_k = 0 \tag{20}$$

and the boundary conditions on S as:

$$\begin{array}{l} n_j(\sigma_{jk} - \delta_{jk}p_{i,i} - \tau_{ijk,i}^{(1)} - \frac{1}{2}e_{lik}m_{lj,i}^s - \frac{1}{2}e_{lij}m_{lk,i}^s - \frac{1}{2}\delta_{jk}\mu_{i,i} + \frac{1}{4}\mu_{k,j} + \frac{1}{4}\mu_{j,k}) \\ +(D_p n_p)(n_i n_k p_i + n_i n_j \tau_{ijk}^{(1)} + n_i n_j n_l n_k \tau_{ijl}^{(1)} + \frac{1}{2}n_i n_j e_{lik} m_{lj}^s + \frac{1}{4}n_i n_k \mu_i - \frac{1}{4}\mu_k) \\ -D_j(\delta_{jk}n_i p_i + n_i\tau_{ijk}^{(1)} + n_i n_l n_k \tau_{ilj}^{(1)} + \frac{1}{2}n_i e_{lik}m_{lj}^s + \frac{1}{2}n_i e_{lij}m_{lk}^s + \frac{1}{2}e_{lij}n_p n_i n_k m_{lp}^s \\ \quad +\frac{1}{2}\delta_{jk}n_i\mu_i - \frac{1}{4}n_j\mu_k - \frac{1}{2}n_k\mu_j + \frac{1}{4}n_j n_i n_k \mu_i) = \bar{t}_k \end{array} \quad \text{or } u_k = \bar{u}_k \tag{21}$$

$$n_j m_{kj}^s - n_j n_k n_i m_{ij}^s + 2n_j n_i n_l e_{kls}\tau_{ijs}^{(1)} + \frac{1}{2}n_l e_{slk}\mu_s = \bar{q}_k \text{ or } (\delta_{kl} - n_k n_l)\delta\theta_l = \bar{\theta}_k \tag{22}$$

$$p_i n_i + n_j n_i n_k \tau_{ijk}^{(1)} = \bar{r} \text{ or } \varepsilon_N = n_i n_j \varepsilon_{ij} = \bar{\varepsilon}_N \tag{23}$$

and along the sharp edge C as:

$$\begin{array}{c} [[k_k n_i p_i + k_j n_i \tau_{ijk}^{(1)} + k_l n_i n_j n_k \tau_{ijl}^{(1)} + \frac{1}{2}k_j e_{lik} n_i m_{lj}^s + \frac{1}{2}k_j e_{lij} n_i m_{lk}^s + \frac{1}{2}k_j e_{lij} n_p n_i n_k m_{lp}^s \\ +\frac{1}{2}k_k n_i \mu_i - \frac{1}{4}k_i n_i \mu_k - \frac{1}{2}k_j n_k \mu_j + \frac{1}{4}k_j n_j n_i n_k \mu_i]] = \bar{f}_k \\ \text{or } u_k = \bar{u}_k \end{array} \tag{24}$$

For linear elastic isotropic materials, the internal energy density in Equation (9) can be re-formulated by substituting Equation (15) into Equation (9), written as:

$$U = \frac{1}{2}k\varepsilon_{ii}\varepsilon_{jj} + \mu\varepsilon'_{ij}\varepsilon'_{ij} + a_1\gamma_i\gamma_i + a_2\eta_{ijk}^{(1)}\eta_{ijk}^{(1)} + a_3\chi_{pi}^s\chi_{pi}^s + a_4\kappa_i\kappa_i - a_5\kappa_i\gamma_i \tag{25}$$

where a_i (i = 1, 2, 3, 4, 5) is a higher-order material constant in another form which is related to g_i by:

$$\begin{array}{l} a_1 = g_1 + g_2 + g_3 + \frac{3}{5}g_4 + \frac{3}{5}g_5,\ a_2 = g_4 + g_5 \\ a_3 = \frac{2}{3}g_4 - \frac{1}{3}g_5,\ a_4 = 4g_3 + \frac{12}{5}g_4 + \frac{2}{5}g_5,\ a_5 = 2g_1 + 4g_3 + \frac{8}{5}g_4 + \frac{8}{5}g_5 \end{array} \tag{26}$$

By re-writing the higher-order material constants as:

$$a_1 = \mu l_0^2,\ a_2 = \mu l_1^2,\ a_3 = \mu l_2^2,\ a_4 = 2\mu l_3^2,\ a_5 = \mu l_4^2 \tag{27}$$

with l_i (i = 0, 1, 2, 3, 4) representing five material length scale parameters, the higher-order constitutive relations can be obtained from Equations (17) and (25) as:

$$p_i = 2\mu l_0^2\gamma_i - \mu l_4^2\kappa_i,\ \tau_{ijk}^{(1)} = 2\mu l_1^2\eta_{ijk}^{(1)},\ m_{ij}^s = 2\mu l_2^2\chi_{ij}^s,\ \mu_i = 4\mu l_3^2\kappa_i - \mu l_4^2\gamma_i \tag{28}$$

The present strain gradient theory is another expression of Mindlin's strain gradient theory, which is a general theory with five higher-order material constants. When the curvature vector κ_i is excluded (by letting $l_3 = l_4 = 0$), the present formulation of strain gradient theory will reduce to that of Lam et al.'s [7]. When the material length scale parameters are constrained as:

$$l_0 = \frac{3}{\sqrt{5}}\bar{l}_0,\ l_1 = \bar{l}_1,\ l_2 = \sqrt{2}\bar{l}_2,\ l_3 = 3\sqrt{\frac{2}{5}}\bar{l}_0,\ l_4 = 2\sqrt{\frac{6}{5}}\bar{l}_0 \tag{29}$$

the internal energy density of the present strain gradient theory will reduce to

$$\begin{aligned} U &= \tfrac{1}{2}k\varepsilon_{ii}\varepsilon_{jj} + \mu\varepsilon'_{ij}\varepsilon'_{ij} + \tfrac{9}{5}\mu\bar{l}_0^2\gamma_i\gamma_i + \mu\bar{l}_1^2\eta_{ijk}^{(1)}\eta_{ijk}^{(1)} + 2\mu\bar{l}_2^2\chi_{ij}^s\chi_{ij}^s + \tfrac{36}{5}\mu\bar{l}_0^2\kappa_i\kappa_i - \tfrac{24}{5}\mu\bar{l}_0^2\gamma_i\kappa_i \\ &= \tfrac{1}{2}k\varepsilon_{ii}\varepsilon_{jj} + \mu\varepsilon'_{ij}\varepsilon'_{ij} + \mu\bar{l}_0^2\gamma_i\gamma_i + \mu\bar{l}_1^2\eta_{ijk}^{(1)}\eta_{ijk}^{(1)} + \mu(\bar{l}_2^2 + \tfrac{9}{5}\bar{l}_0^2)\chi'_{ij}\chi'_{ij} + \mu(\bar{l}_2^2 - \tfrac{9}{5}\bar{l}_0^2)\chi'_{ij}\chi'_{ji} \end{aligned} \tag{30}$$

which is the internal energy density of Zhou et al. with $\chi'_{ij} = e_{ilk}\varepsilon'_{jk,l}$ denoting the deviatoric rotation gradient [14].

2.3. Couple Stress Theory

In couple stress theory, the rotation gradient rather than the strain gradient is included in the internal energy density. That is, the strain gradient theory will reduce to the couple stress theory when the dilatation gradient γ_i and deviatoric stretch gradient $\eta_{ijk}^{(1)}$ are excluded. Thus, when the higher-order stress components p_i and $\tau_{ijk}^{(1)}$ are excluded, the equilibrium equation and boundary conditions of the strain gradient theory in Equations (20)–(24) can reduce to those of the couple stress theory. The equilibrium equation is:

$$\sigma_{ik,i} - \frac{1}{2}e_{lik}m_{lj,ij}^s - \frac{1}{4}\mu_{i,ik} + \frac{1}{4}\mu_{k,ii} + \bar{b}_k = 0 \tag{31}$$

and the boundary conditions on S are:

$$\begin{aligned} &n_j(\sigma_{jk} - \tfrac{1}{2}e_{lik}m_{lj,i}^s - \tfrac{1}{2}e_{lij}m_{lk,i}^s - \tfrac{1}{2}\delta_{jk}\mu_{i,i} + \tfrac{1}{4}\mu_{k,j} + \tfrac{1}{4}\mu_{j,k}) \\ &\quad +(D_p n_p)(\tfrac{1}{2}n_i n_j e_{lik}m_{lj}^s + \tfrac{1}{4}n_i n_k \mu_i - \tfrac{1}{4}\mu_k) \\ &\quad -D_j(\tfrac{1}{2}n_i e_{lik}m_{lj}^s + \tfrac{1}{2}n_i e_{lij}m_{lk}^s + \tfrac{1}{2}e_{lij}n_p n_i n_k m_{lp}^s \\ &\quad +\tfrac{1}{2}\delta_{jk}n_i\mu_i - \tfrac{1}{4}n_j\mu_k - \tfrac{1}{2}n_k\mu_j + \tfrac{1}{4}n_j n_i n_k \mu_i) = \bar{t}_k \end{aligned} \quad \text{or } u_k = \bar{u}_k \tag{32}$$

$$n_j m_{kj}^s - n_j n_k n_i m_{ij}^s + \frac{1}{2}n_l e_{slk}\mu_s = \bar{q}_k \text{ or } (\delta_{kl} - n_k n_l)\delta\theta_l = \bar{\theta}_k \tag{33}$$

and along the sharp edge C is:

$$\begin{aligned} &[[\tfrac{1}{2}k_j e_{lik}n_i m_{lj}^s + \tfrac{1}{2}k_j e_{lij}n_i m_{lk}^s + \tfrac{1}{2}k_j e_{lij}n_p n_i n_k m_{lp}^s \\ &+\tfrac{1}{2}k_k n_i\mu_i - \tfrac{1}{4}k_i n_i\mu_k - \tfrac{1}{2}k_j n_k\mu_j + \tfrac{1}{4}k_j n_j n_i n_k\mu_i]] = \bar{f}_k \end{aligned} \quad \text{or} u_k = \bar{u}_k \tag{34}$$

In addition, the anti-symmetric couple stress component m_{ij}^a, which is work-conjugated with the anti-symmetric rotation gradient component χ_{ij}^a, is commonly used in many articles rather than the present higher-order stress component μ_i. In fact, according to the virtual work density of couple stress theory

$$\begin{aligned} \delta U &= \sigma_{ij}\delta\varepsilon_{ij} + m_{ij}^s\delta\chi_{ij}^s + \mu_i\delta\kappa_i = \sigma_{ij}\delta\varepsilon_{ij} + m_{ij}^s\delta\chi_{ij}^s + \tfrac{1}{2}e_{ijk}\mu_i\delta\chi_{kj}^a \\ &= \sigma_{ij}\delta\varepsilon_{ij} + m_{ij}^s\delta\chi_{ij}^s + m_{ij}^a\delta\chi_{ij}^a \end{aligned} \tag{35}$$

The anti-symmetric couple stress component m_{ij}^a is associated with the higher-order stress component μ_i by $m_{ij}^a = \frac{1}{2}e_{kji}\mu_k$. Additionally, in turn, the higher-order stress component μ_i is related to the anti-symmetric couple stress component m_{ij}^a through $\mu_i = e_{ijk}m_{kj}^a$. Then, from Equations (31)–(34), the equilibrium equation and boundary conditions of couple stress theory can be re-formulated in the form of couple stress components m_{ij}^s and m_{ij}^a,which are shown in Appendix B in detail.

For linear elastic isotropic materials, the internal energy density of the couple stress theory can be reduced from Equation (25), written as:

$$\begin{aligned} U &= \tfrac{1}{2}k\varepsilon_{ii}\varepsilon_{jj} + \mu\varepsilon'_{ij}\varepsilon'_{ij} + a_3\chi^s_{ij}\chi^s_{ij} + a_4\kappa_i\kappa_i \\ &= \tfrac{1}{2}k\varepsilon_{ii}\varepsilon_{jj} + \mu\varepsilon'_{ij}\varepsilon'_{ij} + a_3\chi^s_{ij}\chi^s_{ij} + \tfrac{1}{2}a_4\chi^a_{ij}\chi^a_{ij} \end{aligned} \tag{36}$$

Consider the material length scale parameters in Equation (27), the couple stress components can be given as:

$$m^s_{ij} = \frac{\partial U}{\partial \chi^s_{ij}} = 2\mu l_2^2\chi^s_{ij},\ \mu_i = \frac{\partial U}{\partial \kappa_i} = 4\mu l_3^2\kappa_i \text{ or } m^a_{ij} = \frac{\partial U}{\partial \chi^a_{ij}} = 2\mu l_3^2\chi^a_{ij} \tag{37}$$

3. Bending Theory for Plane-Strain Beams

Consider the same plane-strain beam problem as Lam et al. [7]. The length of a beam is L and the thickness is h. The thickness of a beam is much smaller than its length. The Cartesian coordinate system is adopted, and the X-axis coincides with the center line of the beam along the length direction, and the Z-axis coincides with the thickness direction. By ignoring the body forces and only considering the normal tractions $\overline{q}_{\pm h/2}$ in units of N/m^2 applied on the top or bottom surfaces of the beam, the governing equations along the length and thickness directions can be obtained from Equation (20), given by:

$$\left(\sigma_{11} - p_{1,1} - \tau^{(1)}_{111,1}\right)_{,1} + \left(\sigma_{13} - p_{3,1} - 2\tau^{(1)}_{113,1} - \tau^{(1)}_{133,3} - \frac{1}{2}m^s_{12,1} - \frac{1}{2}m^s_{23,3} - \frac{1}{4}\mu_{3,1} + \frac{1}{4}\mu_{1,3}\right)_{,3} = 0 \tag{38}$$

$$\left(\sigma_{33} - p_{1,1} - p_{3,3} - 2\tau^{(1)}_{133,1} - \tau^{(1)}_{333,3} + \frac{1}{2}m^s_{23,1} - \frac{1}{4}\mu_{1,1}\right)_{,3} + \left(\sigma_{13} - \tau^{(1)}_{113,1} + \frac{1}{2}m^s_{21,1} + \frac{1}{4}\mu_{3,1}\right)_{,1} = 0 \tag{39}$$

The boundary conditions on the top and bottom surfaces of the beam, $z = \pm h/2$, from Equations (21)–(23) are:

$$\sigma_{33} - p_{1,1} - p_{3,3} - 3\tau^{(1)}_{133,1} - \tau^{(1)}_{333,3} = \overline{q}_{\pm h/2} \tag{40}$$

$$\sigma_{31} - p_{3,1} - 2\tau^{(1)}_{113,1} - \tau^{(1)}_{133,3} - \frac{1}{2}m^s_{21,1} - \frac{1}{2}m^s_{23,3} - \frac{1}{4}\mu_{3,1} + \frac{1}{4}\mu_{1,3} = 0 \tag{41}$$

$$m^s_{23} + 2\tau^{(1)}_{133} - \frac{1}{2}\mu_1 = 0 \tag{42}$$

$$p_3 + \tau^{(1)}_{333} = 0 \tag{43}$$

On the surfaces normal to the X-axis, $x = 0$ or L, the boundary conditions obtained from Equations (21)–(23) are:

$$\overline{t}_1 = \sigma_{11} - \left(p_1 + \tau^{(1)}_{111}\right)_{,1} - \left(p_3 + 3\tau^{(1)}_{113}\right)_{,3} \tag{44}$$

$$\overline{t}_3 = \sigma_{13} + \frac{1}{2}\left(m^s_{21} + \frac{1}{2}\mu_3 - 2\tau^{(1)}_{113}\right)_{,1} + \frac{1}{2}\left(m^s_{23} - \frac{1}{2}\mu_1 + 2\tau^{(1)}_{133}\right)_{,3} - \left(p_1 + 3\tau^{(1)}_{133}\right)_{,3} \tag{45}$$

$$\overline{q}_2 = m^s_{21} - 2\tau^{(1)}_{113} + \frac{1}{2}\mu_3 \tag{46}$$

$$\overline{r} = p_1 + \tau^{(1)}_{111} \tag{47}$$

The boundary conditions at the edges of the top and bottom surfaces of the beam from Equation (24) are:

$$\overline{f}_1 = \pm\left[p_3 + 3\tau^{(1)}_{113}\right],\ \overline{f}_3 = \pm\left[p_1 + 3\tau^{(1)}_{133}\right],\ z = \pm\frac{h}{2} \tag{48}$$

The stress resultants can be re-defined due to the presence of higher-order stress components as:

$$
\begin{aligned}
&N = \int_{-h/2}^{h/2} \sigma_{11} \mathrm{d}z \;\; Q = \int_{-h/2}^{h/2} (\sigma_{13} - \tau_{113,1}^{(1)} + \tfrac{1}{2} m_{21,1}^{s} + \tfrac{1}{4}\mu_{3,1}) \mathrm{d}z \\
&M = \int_{-h/2}^{h/2} (z\sigma_{11} + \tau_{113}^{(1)} + p_3 + m_{21}^{s} + \tfrac{1}{2}\mu_3) \mathrm{d}z \\
&N^h = \int_{-h/2}^{h/2} (p_1 + \tau_{111}^{(1)}) \mathrm{d}z \;\; M^h = \int_{-h/2}^{h/2} z(p_1 + \tau_{111}^{(1)}) \mathrm{d}z
\end{aligned} \tag{49}
$$

where N, Q, M, N^h and M^h denote the axial stress resultant, shear stress resultant, moment, higher-order stress resultant and higher-order moment per unit width in the beam, respectively. According to the re-defined stress resultants in Equation (49) and the governing equations and boundary conditions in Equations (38)–(48), the same formulations from that of Lam et al. [7] can be obtained as:

$$\frac{dN}{dx} - \frac{d^2N^h}{dx^2} = 0 \tag{50}$$

where

$$\frac{dQ}{dx} + \overline{q} = 0 \tag{51}$$

$$\frac{dM}{dx} - \frac{d^2M^h}{dx^2} - Q = 0 \tag{52}$$

with $\overline{q} = \overline{q}_{h/2} + \overline{q}_{-h/2}$. Substituting Equation (52) into (51) yields:

$$\frac{d^2M}{dx^2} - \frac{d^3M^h}{dx^3} + \overline{q} = 0 \tag{53}$$

The stress resultants at the boundary are:

$$N - \frac{dN^h}{dx} = \overline{N} = \int_{-h/2}^{h/2} \overline{t}_1 dz + \overline{f}_1|_{z=h/2} + \overline{f}_1|_{z=-h/2} \tag{54}$$

$$\frac{dM}{dx} - \frac{d^2M^h}{dx^2} = Q = \overline{Q} = \int_{-h/2}^{h/2} \overline{t}_3 dz + \overline{f}_3|_{z=h/2} + \overline{f}_3|_{z=-h/2} \tag{55}$$

$$M - \frac{dM^h}{dx} = \overline{M} = \int_{-h/2}^{h/2} (z\overline{t}_1 + \overline{q}_2) dz + \frac{h}{2}\overline{f}_1|_{x=h/2} - \frac{h}{2}\overline{f}_1|_{x=-h/2} \tag{56}$$

$$N^h = \overline{N}^h = \int_{-h/2}^{h/2} \overline{r} dz \tag{57}$$

$$M^h = \overline{M}^h = \int_{-h/2}^{h/2} z\overline{r} dz \tag{58}$$

where $\overline{N}$, $\overline{Q}$, $\overline{M}$, $\overline{N}^h$ and $\overline{M}^h$ represent the prescribed axial force, shear force, moment, higher-order axial force, higher-order moment per unit width in the beam, respectively.

4. Semi-Inverse Solution for Pure Bending Beam

Consider a pure bending beam subjected to a moment M_0 acting on its two ends. The governing equations and boundary conditions are listed as Equations (38)–(48) with normal tractions $\overline{q}_{\pm h/2}$ being zero. The solution for the displacement field is suggested in the following form [29]:

$$
\begin{aligned}
u_1 &= C_1 xz - C_2 z + u_0 \\
u_3 &= -\tfrac{1}{2} C_1 x^2 + C_2 x + w(z) + v_0
\end{aligned} \tag{59}
$$

in which w is an unknown function with respect to coordinate z. C_1, C_2, u_0 and v_0 are four undetermined constants. According to the assumption of displacement, the non-zero strain components are:

$$\varepsilon_{11} = C_1 z,\ \varepsilon_{33} = w_{,3} \tag{60}$$

The non-vanishing components of strain gradient are calculated as:

$$\begin{array}{l} \gamma_3 = C_1 + w_{,33}\ \chi^s_{21} = \kappa_3 = \frac{1}{2}C_1 \\ \eta^{(1)}_{311} = \frac{4}{15}C_1 - \frac{1}{5}w_{,33}\ \eta^{(1)}_{333} = \frac{2}{5}w_{,33} - \frac{1}{5}C_1 \end{array} \tag{61}$$

According to the constitutive relations for plane-strain problems, the non-zero Cauchy stress is given by:

$$\begin{array}{l} \sigma_{11} = \frac{E}{(1+v)(1-2v)}[(1-v)C_1 z + v w_{,3}] \\ \sigma_{33} = \frac{E}{(1+v)(1-2v)}[(1-v)w_{,3} + vC_1 z] \end{array} \tag{62}$$

in which E is the Young's modulus and v is the Poisson's ratio. The work-conjugated higher-order stress components can be obtained from Equation (28), given as:

$$\begin{array}{l} p_3 = (2\mu l_0^2 - \frac{1}{2}\mu l_4^2)C_1 + 2\mu l_0^2 w_{,33}\ m^s_{21} = \mu l_2^2 C_1\ \mu_3 = (2\mu l_3^2 - \mu l_4^2)C_1 - \mu l_4^2 w_{,33} \\ \tau^{(1)}_{311} = 2\mu l_1^2(\frac{4}{15}C_1 - \frac{1}{5}w_{,33})\ \tau^{(1)}_{333} = \frac{2}{5}\mu l_1^2(2w_{,33} - C_1) \end{array} \tag{63}$$

According to the present bending theory of plane-strain beams, the governing equations and boundary conditions of the pure bending beam subjected to a moment M_0 acting on its two ends can be specified by substituting the stress components from Equation (62) and higher-order stress components from Equation (63) into Equations (38)–(48) along the length direction of Equation (38) and the boundary conditions of Equations (40), (42), (45) and (47) can be satisfied automatically. The current pure bending solution depends on the governing equations, which are Equation (39) and boundary conditions Equations (41), (43), (44), (46) and (48).

Substituting the stress components from Equation (62) and higher-order stress components from Equation (63) into Equation (69) yields the governing equation of the current pure bending beam, that is:

$$\frac{d^4 w}{dz^4} - \frac{E(1-v)}{d_1(1+v)(1-2v)}\frac{d^2 w}{dz^2} - \frac{Ev}{d_1(1+v)(1-2v)}C_1 = 0 \tag{64}$$

with the coefficient d_1 defined as $d_1 = 2\mu l_0^2 + \frac{4}{5}\mu l_1^2$. The solution of Equation (64) is given by:

$$w(z) = C_3\cosh(\frac{z}{\lambda}) - \frac{v}{2(1-v)}C_1 z^2 \tag{65}$$

where C_3 is an undetermined coefficient and the parameter λ is defined as:

$$\lambda = \sqrt{\frac{1-2v}{1-v}(l_0^2 + \frac{2}{5}l_1^2)} \tag{66}$$

It should be noted here that the constant term is not included in Equation (65) because it has been included in v_0 in the displacement solution of Equation (59).

Moreover, the assumption of displacement in Equation (65) satisfies the boundary condition in Equation (41) automatically. The coefficients C_1 and C_3 in Equation (65) can be determined from the boundary conditions Equations (43), (44), (46) and (48). Based on the boundary condition Equation (43), it can be obtained that

$$C_3 = (\frac{v}{1-v} - \frac{d_2}{d_1})\frac{\lambda^2}{\cosh(\frac{h}{2\lambda})}C_1 \tag{67}$$

with the coefficient d_2 representing $d_2 = 2\mu l_0^2 - \frac{2}{5}\mu l_1^2 - \frac{1}{2}\mu l_4^2$. According to the stress resultants at boundary Equation (56), actually, the applied moment M_0 at the two ends of beam equals

$$M_0 = \int_{-h/2}^{h/2} (z\bar{t}_1 + \bar{q}_2)dz + \frac{h}{2}\bar{f}_1|_{x=h/2} - \frac{h}{2}\bar{f}_1|_{x=-h/2} \tag{68}$$

Then, combining Equations (44), (46), (48) and (68), the coefficient C_1 can be solved as:

$$C_1 = \frac{M_0}{\frac{Eh^3}{12(1-v^2)} + \left(d_3 - \frac{v}{1-v}d_2\right)h + d_4\lambda} \tag{69}$$

with the coefficients d_3 and d_4 denoting

$$\begin{aligned} d_3 &= 2\mu l_0^2 + \tfrac{8}{15}\mu l_1^2 + \mu l_2^2 + \mu l_3^2 - \mu l_4^2 \\ d_4 &= \tfrac{Ehv\lambda}{(1+v)(1-2v)}\left(\tfrac{v}{1-v} - \tfrac{d_2}{d_1}\right) + 2\left(d_2 - \tfrac{Ev\lambda^2}{(1+v)(1-2v)}\right)\left(\tfrac{v}{1-v} - \tfrac{d_2}{d_1}\right)\tanh\left(\tfrac{h}{2\lambda}\right) \end{aligned} \tag{70}$$

Furthermore, the coefficients C_2, u_0 and v_0 can be obtained according to the displacement boundary conditions. For a simply supported beam, the displacement boundary conditions are:

$$u_1|_{\substack{x=0\\z=0}} = u_3|_{\substack{x=0\\z=0}} = 0,\ u_3|_{\substack{x=L\\z=0}} = 0 \tag{71}$$

Substituting Equations (59) and (65) into (71) yields

$$u_0 = 0,\ v_0 = -\left(\frac{v}{1-v} - \frac{d_2}{d_1}\right)\frac{\lambda^2}{\cosh(\frac{h}{2\lambda})}C_1,\ C_2 = \frac{1}{2}C_1 L \tag{72}$$

Thus, the displacement solution is finally given by:

$$\begin{aligned} u_1 &= C_1\left(x - \tfrac{L}{2}\right)z \\ u_3 &= C_1\left[-\tfrac{1}{2}x^2 + \tfrac{L}{2}x + \left(\tfrac{v}{1-v} - \tfrac{d_2}{d_1}\right)\tfrac{\lambda^2}{\cosh(\frac{h}{2\lambda})}\left(\cosh\left(\tfrac{z}{\lambda}\right) - 1\right) - \tfrac{v}{2(1-v)}z^2\right] \end{aligned} \tag{73}$$

in which the coefficient C_1 is given as Equation (69). Especially, for the axis of the beam ($z = 0$), its displacement solution is:

$$u_3|_{z=0} = C_1\left(-\frac{1}{2}x^2 + \frac{L}{2}x\right) \tag{74}$$

which is similar to that of traditional theory except for the bending rigidity. From Equation (69), it can be seen that the bending rigidity of the pure bending beam based on the strain gradient theory includes the classical bending rigidity $\frac{Eh^3}{12(1-v^2)}$ and the higher-order bending rigidity $(d_3 - \frac{v}{1-v}d_2)h + d_4\lambda$. By using the classical bending rigidity to normalize the total bending rigidity of the strain gradient theory (the denominator in Equation (69)), the dimensionless bending rigidity K can be obtained as:

$$K = 1 + 6(1-v)\left(\frac{l}{h}\right)^2\left[1 - \frac{v}{1-v}\frac{d_2}{d_3} + \left(\frac{v}{1-v} - \frac{d_2}{d_1}\right)\left[\frac{2v\lambda^2}{(1-2v)l^2} + \frac{2\lambda}{h}\left(\frac{d_2}{d_3} - \frac{2v\lambda^2}{(1-2v)l^2}\right)\tanh\left(\frac{h}{2\lambda}\right)\right]\right] \tag{75}$$

with the equivalent material length scale parameter l denoted as:

$$l = \sqrt{2l_0^2 + \frac{8}{15}l_1^2 + l_2^2 + l_3^2 - l_4^2} \tag{76}$$

When the Poisson's effect is ignored by v = 0, the dimensionless bending rigidity K reduces to

$$\widetilde{K} = 1 + 6\left(\frac{l}{h}\right)^2\left[1 - \frac{2\widetilde{\lambda}}{h}\frac{d_2^2}{d_1 d_3}\tanh\left(\frac{h}{2\widetilde{\lambda}}\right)\right] \tag{77}$$

in which the parameter $\widetilde{\lambda}$ is reduced from Equation (66) by letting v = 0.

$$\widetilde{\lambda} = \lambda_{v=0} = \sqrt{l_0^2 + \frac{2}{5}l_1^2} \tag{78}$$

Similarly, for a couple stress plane-strain beam subjected to a moment M_0 acting on its two ends, the displacement solution based on the couple stress theory can be obtained as:

$$\begin{aligned} u_1^c &= C_1^c\left(x - \tfrac{L}{2}\right)z \\ u_3^c &= C_1^c\left(-\tfrac{1}{2}x^2 + \tfrac{L}{2}x - \tfrac{v}{2(1-v)}z^2\right) \end{aligned} \tag{79}$$

where the coefficient is:

$$C_1^c = \frac{M_0}{\frac{Eh^3}{12(1-v^2)} + \mu(l_2^2 + l_3^2)h} \tag{80}$$

Then, the dimensionless bending rigidity for the couple stress theory is given as:

$$K^c = 1 + 6(1 - v)\left(\frac{l^c}{h}\right)^2 \tag{81}$$

with the equivalent couple stress length scale parameter l^c denoted as $l^c = \sqrt{l_2^2 + l_3^2}$. When the Poisson's effect is ignored, the dimensionless bending rigidity K^c reduces to

$$\widetilde{K}^c = 1 + 6\left(\frac{l^c}{h}\right)^2 \tag{82}$$

In order to illustrate the solution of the present plane-strain beam, a polyvinylidene difluoride (PVDF) beam with a unit thickness is considered here, in which the Young's modulus E is equal to 3.7 GPa, the material length scale parameters are assumed to be the same, the length of the beam is 20 times its thickness (L = 20 h), and the applied moment M_0 equals 10 μN·μm.

Based on Equations (75) and (81), the dimensionless bending rigidity is shown in Figure 1. It can be seen from Figure 1 that the dimensionless bending rigidity shows an obvious size effect. The dimensionless bending rigidity of beams decreases with an increase in thickness, and the bending rigidities based on the strain gradient theory, couple stress theory and classical theory tend to be consistent as the thickness increases. The effect of the strain gradient is a kind of stiffening effect. The bending rigidity based on the couple stress theory is larger than that of classical theory due to the consideration of a rotation gradient. Additionally, the bending rigidity based on the strain gradient theory is larger than that of the couple stress theory, since the dilatation gradient and deviatoric stretch gradient are also considered in addition to the rotation gradient. In addition, the dimensionless bending rigidity of a plane-strain beam based on higher-order theories increases with the decrease of the Poisson's ratio.

Figure 2 shows the displacement u_3 of the axis (Equation (74)) along the length of the beam. It can be found that the material length scale parameter l_0, which characterizes the dilatation gradient, exhibits a dominant stiffening effect, while the material length scale parameters l_1, l_2 and l_3 characterize the deviatoric stretch gradient, symmetric rotation gradient and curvature, respectively, exhibiting a secondary and decreasing stiffening effect. In particular, the material length scale parameter l_4 exhibits a softening effect. It should be noted here that the overall effect of the strain gradient is a stiffening effect, and the influence of l_4 is weak.

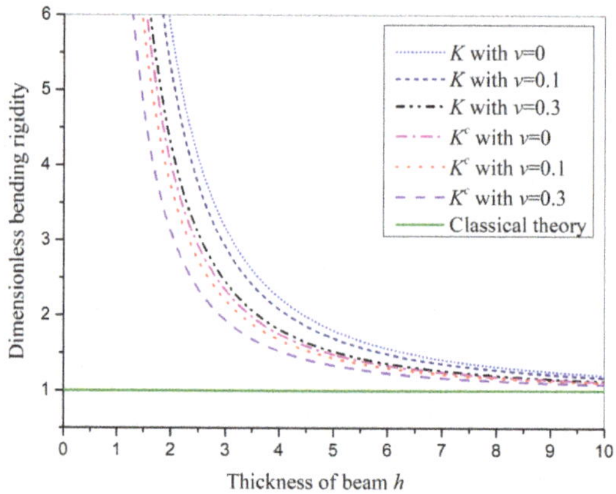

Figure 1. Size effect of plane-strain beams.

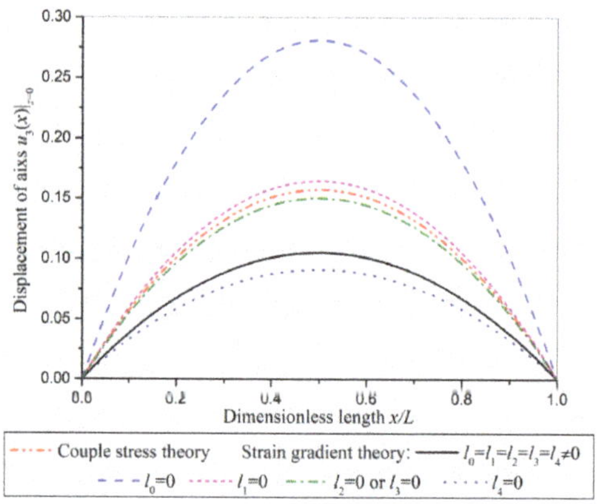

Figure 2. The displacement u_3 of the axis along the length direction of the beam.

Moreover, in order to reveal the deformation of a plane-strain beam along the thickness direction, the strain ε_{33} can be obtained by combining Equations (2) and (73) from the strain gradient theory, which is given by:

$$\varepsilon_{33} = C_1[(\frac{v}{1-v} - \frac{d_2}{d_1})\frac{\lambda^2}{\cosh(\frac{h}{2\lambda})}\frac{1}{l}\sinh(\frac{z}{l}) - \frac{v}{1-v}z] \tag{83}$$

While for couple stress theory, the strain ε_{33} is obtained by:

$$\varepsilon_{33}^c = -C_1^c\frac{v}{1-v}z \tag{84}$$

By defining the normalized strain ε_{33}/C_1 for the strain gradient theory and ε_{33}^c/C_1^c for the couple stress theory, the normalized strain of a plane-strain beam along its thickness direction is shown in Figure 3. Figure 3 reveals that the gradient of a strain ε_{33} along the thickness direction decreases as the Poisson's ratio decreases. When the Poisson's effect is neglected, the strain ε_{33} is equal to zero for the couple stress theory and non-zero for the strain gradient theory. Equation (83) demonstrates that the strain ε_{33} of a plane-strain beam based on the strain gradient theory includes the linear term and the non-linear term. The non-linear term is related to the parameter λ, which characterizes the dilatation gradient and the deviatoric stretch gradient along the thickness direction. When the dilatation gradient and the deviatoric stretch gradient along the thickness direction are neglected by letting $\lambda = 0$, the non-linear term will disappear.

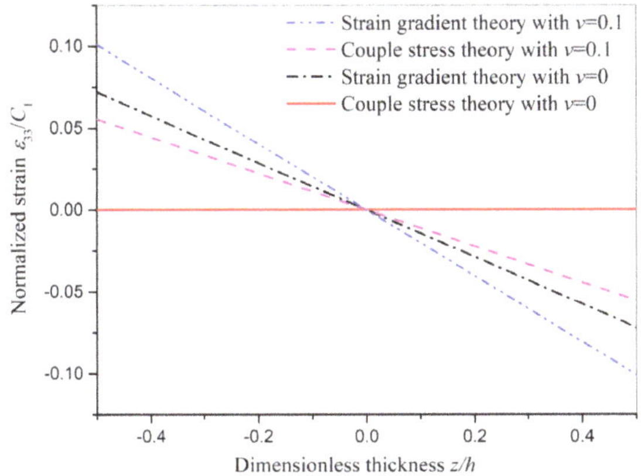

Figure 3. Normalized strain of a plane-strain beam along the thickness direction.

5. Strain Gradient Bernoulli–Euler Beam

Consider a Bernoulli–Euler beam with a unit thickness, the displacement components are set as:

$$u_1 = -z\frac{\partial \overline{w}}{\partial x} \quad u_2 = 0 \quad u_3 = \overline{w}(x) \tag{85}$$

According to the definitions of strain and strain gradient in Equation (2), the non-zero components are:

$$\varepsilon_{11} = -z\frac{\partial^2 \overline{w}}{\partial x^2},\ \eta_{111} = -z\frac{\partial^3 \overline{w}}{\partial x^3},\ \eta_{311} = -\frac{\partial^2 \overline{w}}{\partial x^2} \tag{86}$$

Hence, the non-vanishing strain gradient components are:

$$\begin{aligned}
&\gamma_1 = -z\tfrac{\partial^3 \overline{w}}{\partial x^3}\ \gamma_3 = -\tfrac{\partial^2 \overline{w}}{\partial x^2}\ \chi^s_{21} = \chi^s_{12} = -\tfrac{1}{2}\tfrac{\partial^2 \overline{w}}{\partial x^2}\ \kappa_3 = -\tfrac{1}{2}\tfrac{\partial^2 \overline{w}}{\partial x^2}\\
&\eta^{(1)}_{111} = -\tfrac{2}{5}z\tfrac{\partial^3 \overline{w}}{\partial x^3}\ \eta^{(1)}_{333} = \tfrac{1}{5}\tfrac{\partial^2 \overline{w}}{\partial x^2}\ \eta^{(1)}_{113} = \eta^{(1)}_{131} = \eta^{(1)}_{311} = -\tfrac{4}{15}\tfrac{\partial^2 \overline{w}}{\partial x^2}\\
&\eta^{(1)}_{122} = \eta^{(1)}_{212} = \eta^{(1)}_{221} = \eta^{(1)}_{313} = \eta^{(1)}_{331} = \eta^{(1)}_{133} = \tfrac{1}{5}z\tfrac{\partial^3 \overline{w}}{\partial x^3}\ \eta^{(1)}_{223} = \eta^{(1)}_{232} = \eta^{(1)}_{322} = \tfrac{1}{15}\tfrac{\partial^2 \overline{w}}{\partial x^2}
\end{aligned} \tag{87}$$

Substituting these non-zero components from Equation (87) into the constitutive equations, Equations (10) and (28), the corresponding Cauchy stress and higher-order stress components are:

$$\begin{aligned}
&\sigma_{11} = -Ez\tfrac{\partial^2 \overline{w}}{\partial x^2}\ p_1 = -2\mu l_0^2 z\tfrac{\partial^3 \overline{w}}{\partial x^3}\ p_3 = -2\mu l_0^2\tfrac{\partial^2 \overline{w}}{\partial x^2} + \tfrac{1}{2}\mu l_4^2\tfrac{\partial^2 \overline{w}}{\partial x^2}\\
&m^s_{21} = -\mu l_2^2\tfrac{\partial^2 \overline{w}}{\partial x^2}\ \mu_1 = \mu l_4^2 z\tfrac{\partial^3 \overline{w}}{\partial x^3}\ \mu_3 = -2\mu l_3^2\tfrac{\partial^2 \overline{w}}{\partial x^2} + \mu l_4^2\tfrac{\partial^2 \overline{w}}{\partial x^2}\\
&\tau^{(1)}_{111} = -\tfrac{4}{5}\mu l_1^2 z\tfrac{\partial^3 \overline{w}}{\partial x^3}\ \tau^{(1)}_{333} = \tfrac{2}{5}\mu l_1^2\tfrac{\partial^2 \overline{w}}{\partial x^2}\ \tau^{(1)}_{113} = \tau^{(1)}_{131} = \tau^{(1)}_{311} = -\tfrac{8}{15}\mu l_1^2\tfrac{\partial^2 \overline{w}}{\partial x^2}\\
&\tau^{(1)}_{122} = \tau^{(1)}_{212} = \tau^{(1)}_{221} = \tau^{(1)}_{313} = \tau^{(1)}_{331} = \tau^{(1)}_{133} = \tfrac{2}{5}\mu l_1^2 z\tfrac{\partial^3 \overline{w}}{\partial x^3}\ \tau^{(1)}_{223} = \tau^{(1)}_{232} = \tau^{(1)}_{322} = \tfrac{2}{15}\mu l_1^2\tfrac{\partial^2 \overline{w}}{\partial x^2}
\end{aligned} \tag{88}$$

From Equation (49), the defined bending moment and higher-order moment are given as

$$M = -(\frac{Eh^3}{12} + d_3 h)\frac{\partial^2 \overline{w}}{\partial x^2}\ M^h = -d_1 \frac{h^3}{12}\frac{\partial^3 \overline{w}}{\partial x^3} \tag{89}$$

where the coefficients d_1 and d_3 have been defined in Section 4. The bending governing equation of the beam is given as Equation (53) and the boundary conditions of stress resultants are listed as Equations (55), (56) and (58).

(1) Simply supported beam

For a simply supported beam subjected to a moment M_0 acting on the two ends of beam, the governing equation is:

$$\frac{\mathrm{d}^2 M}{\mathrm{d}x^2} - \frac{\mathrm{d}^3 M^h}{\mathrm{d}x^3} = 0 \tag{90}$$

and the boundary conditions on the two ends of beam are:

$$M - \frac{dM^h}{dx} = M_0,\ M^h = 0,\ w = 0 \text{ at } x = 0,\ L \tag{91}$$

Combining Equations (89)–(91), the deflection of pure bending beam is solved as:

$$\overline{w} = C_B(-\frac{1}{2}x^2 + \frac{L}{2}x) \tag{92}$$

where the coefficient is given by:

$$C_B = \frac{M_0}{\frac{Eh^3}{12} + d_3 h} \tag{93}$$

By using the classical bending rigidity $Eh^3/12$ to normalize the bending rigidity based on the strain gradient theory (the denominator in Equation (93)), the dimensionless bending rigidity K_B can be obtained as:

$$K_B = 1 + 6(\frac{l}{h})^2 \tag{94}$$

where the equivalent material length scale parameter l is defined as Equation (76). In addition, when the dilatation gradient γ_i and deviatoric stretch gradient $\eta_{ijk}^{(1)}$ are excluded, the strain gradient pure bending beam model will reduce to that of the couple stress theory. Additionally, the dimensionless bending rigidity will be the same as Equation (82).

The displacement distribution of the pure bending beam along its axis is shown in Figure 4. Figure 4 illustrates once again that the bending rigidity predicted by the strain gradient theory is greater than that predicted by the couple stress theory, and the bending rigidity decreases with the increase of the Poisson's ratio. Moreover, for the couple stress theory, the displacement of the Bernoulli–Euler beam is consistent with that of the plane-strain beam when the Poisson's effect is ignored. While for the strain gradient theory, the bending rigidity of the plane-strain beam without the Poisson's effect is smaller than that of the Bernoulli–Euler beam due to the influence of the dilatation gradient and the deviatoric stretch gradient along the thickness direction, which can also be explained by Equations (77) and (94).

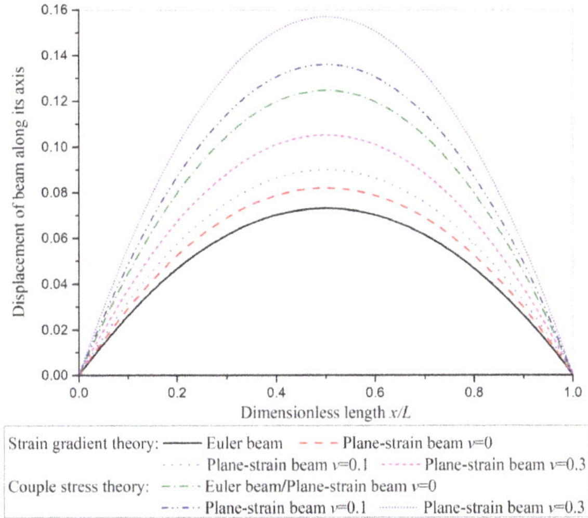

Figure 4. The displacement of beam along its axis.

(2) Cantilever beam

For a cantilever beam subjected to a concentrated force F_0 acting on its free end, the governing equation is the same as Equation (90) and the boundary conditions on the two ends of beam are given by:

$$\overline{w} = \overline{w}_{,1} = 0,\ M^h = 0 \text{ at } x = 0 \tag{95}$$

$$\frac{dM}{dx} - \frac{d^2M^h}{dx^2} = F_0,\ M - \frac{dM^h}{dx} = 0,\ M^h = 0 \text{ at } x = L \tag{96}$$

Combining Equations (90), (95) and (96), the deflection is solved as:

$$\overline{w} = B_1 + B_2x + B_3x^2 + B_4x^3 + B_5e^{\frac{\sqrt{K_B}}{\widetilde{\lambda}}x} + B_6e^{-\frac{\sqrt{K_B}}{\widetilde{\lambda}}x} \tag{97}$$

where $\widetilde{\lambda}$ and K_B are defined as Equations (78) and (94), respectively, and the coefficients $B_i(i = 1,2,3,4,5)$ are:

$$\begin{array}{l} B_1 = \frac{12F_0\widetilde{\lambda}^3(e^{\frac{\sqrt{K_B}}{\widetilde{\lambda}}L} + e^{-\frac{\sqrt{K_B}}{\widetilde{\lambda}}L} - 2)}{Eh^3K_B^{5/2}(e^{\frac{\sqrt{K_B}}{\widetilde{\lambda}}L} - e^{-\frac{\sqrt{K_B}}{\widetilde{\lambda}}L})}\ B_2 = -\frac{12F_0\widetilde{\lambda}^2}{Eh^3K_B^2}\ B_3 = \frac{6F_0L}{Eh^3K_B}\ B_4 = -\frac{2F_0}{Eh^3K_B} \\ B_5 = -\frac{12F_0\widetilde{\lambda}^3(e^{-\frac{\sqrt{K_B}}{\widetilde{\lambda}}L} - 1)}{Eh^3K_B^{5/2}(e^{\frac{\sqrt{K_B}}{\widetilde{\lambda}}L} - e^{-\frac{\sqrt{K_B}}{\widetilde{\lambda}}L})}\ B_6 = -\frac{12F_0\widetilde{\lambda}^3}{Eh^3K_B^{5/2}(1 + e^{-\frac{\sqrt{K_B}}{\widetilde{\lambda}}L})} \end{array} \tag{98}$$

In addition, many articles only consider the gradient of the axial strain along the thickness direction since the strain gradient along the length direction can be neglected compared with that along the thickness direction for the Euler beam. For a cantilever beam subjected to a concentrated force at its free end, when the strain gradient along the length direction is neglected, the deflection solution of Equation (97) will reduce to

$$\overline{w} = B_3x^2 + B_4x^3 \tag{99}$$

From Equations (97) and (99), the deflection of the cantilever beam is shown in Figure 5. It can be seen that the influence of the strain gradient along the length direction η_{111} on the bending rigidity is negligible. The main stiffening effect comes from the strain gradient along the thickness direction. In addition, compared with the couple stress model, the contribution of the dilatation gradient and deviatoric stretch gradient is non-negligible.

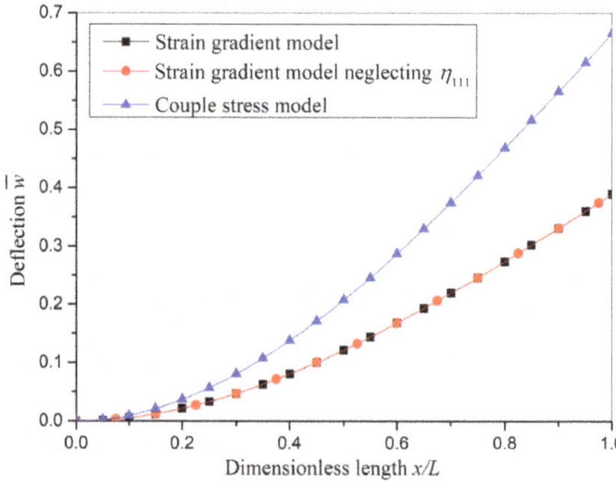

Figure 5. The deflection of cantilever beam.

6. Conclusions

In this paper, a new set of higher-order deformation metrics including dilatation gradient, deviatoric stretch gradient, symmetric rotation gradient and curvature is introduced to re-formulate the strain gradient theory for the sake of investigating the effect of each strain gradient components conveniently. All the constitutive equations, equilibrium equations and boundary conditions in the form of components are obtained. The present strain gradient theory with five higher-order constants is another form of Mindlin's strain gradient theory, which is a general theory. The present theory can be directly simplified to the couple stress theory by eliminating the dilatation gradient and deviatoric stretch gradient.

Based on the present strain gradient theory, the bending theory of plane-strain beam is proposed, in which the stress resultants are re-defined due to the presence of higher-order stress and the corresponding governing equations and boundary conditions of stress resultants are developed. The semi-inverse solution of a pure bending beam subjected to moments at its two ends is obtained. The results demonstrate that the material length scale parameter l_0, which characterizes the dilatation gradient, exhibits a dominant stiffening effect, while the material length scale parameters l_1, l_2 and l_3 characterizing the deviatoric stretch gradient, symmetric rotation gradient and curvature, respectively, exhibit a secondary and decreasing stiffening effect. The dimensionless bending rigidity increases with the decrease in Poisson's ratio.

The Bernoulli–Euler beam model is presented based on the present strain gradient theory and couple stress theory. For the couple stress theory, the displacement of the Bernoulli–Euler beam is consistent with that of a plane-strain beam without the Poisson's effect. While for the strain gradient theory, the bending rigidity of a plane-strain beam without considering the Poisson's effect is smaller than that of a Bernoulli–Euler beam, which is due to the influence of the dilatation gradient and the deviatoric stretch gradient along the thickness direction. In addition, it can be concluded that the influence of the strain gradient along the length direction on bending rigidity is negligible.

Author Contributions: A.L.: Methodology and Conceptualization. Q.W.: Writing—Original Draft. M.S.: Writing—Review and Editing. J.C.: Formal Analysis. W.S.: Editing. S.Z.: Data Curation. L.W.: Validation. All authors have read and agreed to the published version of the manuscript.

Funding: The work reported here is funded by the Pilot Program for the Integration of Science, Education and Industry of Qilu University of Technology (Shandong Academy of Sciences) (2022PX032), the

National Natural Science Foundation of China (NSFC) (52105554) and the Nature Science Foundation of Shandong Province, China (ZR2020QE187).

Institutional Review Board Statement: Not applicable.

Informed Consent Statement: Not applicable.

Data Availability Statement: Not applicable.

Conflicts of Interest: The authors declare no conflict of interest.

Appendix A

According to the definition of stain and strain gradient, the variation of internal energy is written as:

$$\int_V (\sigma_{ij}\delta\varepsilon_{ij} + \tau_{ijk}\delta\eta_{ijk})\mathrm{d}V = \int_V [-(\sigma_{ki,i} - \tau_{ijk,ij})\delta u_k + (\sigma_{ki}\delta u_k - \tau_{jik,j}\delta u_k)_{,i} + (\tau_{jik}\delta u_{k,i})_{,j}]\mathrm{d}V \tag{A1}$$

Applying the divergence theorem yields:

$$\int_V (\sigma_{ij}\delta\varepsilon_{ij} + \tau_{ijk}\delta\eta_{ijk})\mathrm{d}V = -\int_V (\sigma_{ki,i} - \tau_{ijk,ij})\delta u_k \mathrm{d}V + \int_S n_i(\sigma_{ki} - \tau_{jik,j})\delta u_k \mathrm{d}S + \int_S \tau_{jik} n_j \delta u_{k,i} \mathrm{d}S \tag{A2}$$

It should be noted that the variation $\delta u_{k,I}$ in Equation (A2) is not independent of δu_k on the surface boundary S, and only its normal component is independent [36]. Thus, the variation $\delta u_{k,i}$ on the surface S is separated into normal component and tangential component.

$$\delta u_{k,i} = D_i \delta u_k + n_i D \delta u_k \tag{A3}$$

where the normal gradient operator D and the surface gradient operator D_i are given as:

$$D = n_i \partial_i \; D_i = (\delta_{ik} - n_i n_k)\partial_k \tag{A4}$$

Substituting Equation (A3) into Equation (A2), the last integral can be rewritten as:

$$\int_S \tau_{jik} n_j \delta u_{k,i} \mathrm{d}S = \int_S n_j \tau_{jik} D_i \delta u_k \mathrm{d}S + \int_S n_i n_j \tau_{jik} D \delta u_k \mathrm{d}S \tag{A5}$$

Note that the integrand of the first integral in Equation (A5) can be further derived as [37]:

$$\begin{aligned} n_j \tau_{jik} D_i \delta u_k &= D_i(n_j \tau_{jik} \delta u_k) - D_i(n_j \tau_{jik})\delta u_k \\ &= (D_p n_p) n_i n_j \tau_{jik} \delta u_k + n_q e_{qpm} e_{mli} (n_l n_j \tau_{jik} \delta u_k)_{,p} - D_i(n_j \tau_{jik})\delta u_k \end{aligned} \tag{A6}$$

with e_{qpm} denoting the alternating tensor. According to the Stokes' theorem, we have

$$\int_S n_q e_{qpm} e_{mli} (n_l n_j \tau_{jik} \delta u_k)_{,p} \mathrm{d}S = \oint_C [[k_i n_j \tau_{jik} \delta u_k]] \mathrm{d}C \tag{A7}$$

in which $k_i = e_{ijk} s_j n_k$ is the outer co-normal vector with the unit vector s_j tangent to the edge C and the square brackets represent the difference between the values of the enclosed quantity on the two sides of the edge.

Finally, Substituting Equation (A5) into Equation (A2) and combining Equations (A6) and (A7), the variation of internal energy is given by:

$$\begin{aligned} \int_V (\sigma_{ij}\delta\varepsilon_{ij} + \tau_{ijk}\delta\eta_{ijk})\mathrm{d}V &= -\int_V (\sigma_{ki,i} - \tau_{ijk,ij})\delta u_k \mathrm{d}V \\ &+ \int_S [n_i(\sigma_{ki} - \tau_{jik,j}) + (D_p n_p) n_i n_j \tau_{jik} - D_i(n_j \tau_{jik})]\delta u_k \mathrm{d}S \\ &+ \int_S n_i n_j \tau_{jik} D \delta u_k \mathrm{d}S + \oint_C [[k_i n_j \tau_{jik} \delta u_k]] \mathrm{d}C \end{aligned} \tag{A8}$$

Appendix B

For the couple stress theory, the equilibrium equation is:

$$\sigma_{ik,i} - \frac{1}{2} e_{lik} \left(m_{lj}^{s} + m_{lj}^{a} \right)_{,ij} + \overline{b}_k = 0 \tag{A9}$$

The boundary conditions on S are:

$$\begin{gathered} n_j [\sigma_{jk} - \tfrac{1}{2} e_{lik} \left(m_{lj}^{s} + m_{lj}^{a} \right)_{,i} - \tfrac{1}{2} e_{lij} (m_{lk}^{s} + m_{lk}^{a})_{,i} + \tfrac{1}{2} (D_p n_p) n_i n_j e_{lik} (m_{lj}^{s} + m_{lj}^{a} \\ - \tfrac{1}{2} D_j [n_i e_{lik} (m_{lj}^{s} + m_{lj}^{a} + n_i e_{lij} (m_{lk}^{s} + m_{lk}^{a} + e_{lij} n_p n_i n_k (m_{lp}^{s} + m_{lp}^{a})] = \overline{t}_k \\ \text{or } u_k = \overline{u}_k \end{gathered} \tag{A10}$$

$$n_j (m_{kj}^{s} + m_{kj}^{a}) - n_j n_k n_i (m_{ij}^{s} + m_{ij}^{a}) = \overline{q}_k \text{ or } (\delta_{kl} - n_k n_l) \delta\theta_l = \overline{\theta}_k \tag{A11}$$

and along the sharp edge C is:

$$\frac{1}{2} [[k_j e_{lik} n_i (m_{lj}^{s} + m_{lj}^{a}) + k_j e_{lij} n_i (m_{lk}^{s} + m_{lk}^{a}) + k_j e_{lij} n_p n_i n_k (m_{lp}^{s} + m_{lp}^{a})]] = \overline{f}_k \text{ or } u_k = \overline{u}_k \tag{A12}$$

References

1. Roudbari, M.A.; Jorshari, T.D.; Lü, C.; Ansari, R.; Kouzani, A.Z.; Amabili, M. A review of size-dependent continuum mechanics models for micro- and nano-structures. *Thin-Walled Struct.* **2022**, *170*, 108562. [CrossRef]
2. Kong, S. A review on the size-dependent models of micro-beam and micro-plate based on the modified couple stress theory. *Arch. Comput. Methods Eng.* **2021**, *29*, 1–31. [CrossRef]
3. Thai, H.; Vo, T.P.; Nguyen, T.; Kim, S. A review of continuum mechanics models for size-dependent analysis of beams and plates. *Compos. Struct.* **2017**, *177*, 196–219. [CrossRef]
4. Fu, G.; Zhou, S.; Qi, L. On the strain gradient elasticity theory for isotropic materials. *Int. J. Eng. Sci.* **2020**, *154*, 103348. [CrossRef]
5. Rouhi, H.; Ebrahimi, F.; Ansari, R.; Torabi, J. Nonlinear free and forced vibration analysis of Timoshenko nanobeams based on Mindlin's second strain gradient theory. *Eur. J. Mech. A-Solid* **2019**, *73*, 268–281. [CrossRef]
6. Wang, B.; Zhao, J.; Zhou, S. A micro scale Timoshenko beam model based on strain gradient elasticity theory. *Eur. J. Mech. A-Solid* **2010**, *29*, 591–599. [CrossRef]
7. Lam, D.C.C.; Yang, F.; Chong, A.C.M.; Wang, J.; Tong, P. Experiments and theory in strain gradient elasticity. *J. Mech. Phys. Solids* **2003**, *51*, 1477–1508. [CrossRef]
8. Liu, D.; He, Y.; Tang, X.; Ding, H.; Hu, P.; Cao, P. Size effects in the torsion of microscale copper wires: Experiment and analysis. *Scripta Mater.* **2012**, *66*, 406–409. [CrossRef]
9. McFarland, A.W.; Colton, J.S. Role of material microstructure in plate stiffness with relevance to microcantilever sensors. *J. Micromech. Microeng.* **2005**, *15*, 1060–1067. [CrossRef]
10. Tang, C.; Alici, G. Evaluation of length-scale effects for mechanical behaviour of micro- and nanocantilevers: I. experimental determination of length-scale factors. *J. Phys. D Appl. Phys.* **2011**, *44*, 335501. [CrossRef]
11. Toupin, R.A. Theories of elasticity with couple-stress. *Arch. Ration. Mech. Anal.* **1964**, *17*, 85–112. [CrossRef]
12. Li, A.Q.; Wang, B.L.; Yang, S.Y. On some basic aspects of flexoelectricity in the mechanics of materials. *Int. J. Eng. Sci.* **2021**, *166*, 103499. [CrossRef]
13. Mindlin, R.D.; Eshel, N.N. On first strain-gradient theories in linear elasticity. *Int. J. Solids Struct.* **1968**, *4*, 109–124. [CrossRef]
14. Zhou, S.; Li, A.; Wang, B. A reformulation of constitutive relations in the strain gradient elasticity theory for isotropic materials. *Int. J. Solids Struct.* **2016**, *80*, 28–37. [CrossRef]
15. Aifantis, E.C. On the role of gradients in the localization of deformation and fracture. *Int. J. Eng. Sci.* **1992**, *30*, 1297–1299. [CrossRef]
16. Yang, F.; Chong, A.C.M.; Lam, D.C.C.; Tong, P. Couple stress based strain gradient theory for elasticity. *Int. J. Solids Struct.* **2002**, *39*, 2731–2743. [CrossRef]
17. Hadjesfandiari, A.R.; Dargush, G.F. Couple stress theory for solids. *Int. J. Solids Struct.* **2011**, *48*, 2496–2510. [CrossRef]
18. Tang, Y.; Qing, H. Elastic buckling and free vibration analysis of functionally graded Timoshenko beam with nonlocal strain gradient integral model. *Appl. Math. Model.* **2021**, *96*, 657–677. [CrossRef]
19. Zhang, B.; Li, H.; Kong, L.; Shen, H.; Zhang, X. Coupling effects of surface energy, strain gradient, and inertia gradient on the vibration behavior of small-scale beams. *Int. J. Mech. Sci.* **2020**, *184*, 105834. [CrossRef]
20. Barretta, R.; Faghidian, S.A.; de Sciarra, F.M.; Penna, R.; Pinnola, F.P. On torsion of nonlocal Lam strain gradient FG elastic beams. *Comp. Struct.* **2020**, *233*, 111550. [CrossRef]
21. Zhang, N.; Zhao, X.; Zheng, S.J.; Chen, D. Size-dependent static bending and free vibration analysis of porous functionally graded piezoelectric nanobeams. *Smart Mater. Struct.* **2020**, *29*, 045025.

22. Zhao, X.; Zheng, S.; Li, Z. Size-dependent nonlinear bending and vibration of flexoelectric nanobeam based on strain gradient theory. *Smart Mater. Struct.* **2019**, *28*, 075027. [CrossRef]
23. Barretta, R.; de Sciarra, F.M. Variational nonlocal gradient elasticity for nano-beams. *Int. J. Eng. Sci.* **2019**, *143*, 73–91. [CrossRef]
24. Niiranen, J.; Balobanov, V.; Kiendl, J.; Hosseini, S.B. Variational formulations, model comparisons and numerical methods for Euler-Bernoulli micro- and nano-beam models. *Math. Mech. Solids* **2019**, *24*, 312–335. [CrossRef]
25. Tran, L.V.; Niiranen, J. A geometrically nonlinear Euler-Bernoulli beam model within strain gradient elasticity with isogeometric analysis and lattice structure applications. *Math. Mech. Complex Syst.* **2020**, *8*, 345–371. [CrossRef]
26. Vo, D.; Suttakul, P.; Rungamornrat, J.; Nanakorn, P. Static analysis of planar arbitrarily curved microbeams with the modified couple stress theory and Euler-Bernoulli beam model. *Appl. Math. Model.* **2022**, *112*, 358–390. [CrossRef]
27. Vo, D.; Zhou, K.; Rungamornrat, J.; Bui, T.Q. Spatial arbitrarily curved microbeams with the modified couple stress theory: Formulation of equations of motion. *Eur. J. Mech. A-Solid* **2022**, *92*, 104475. [CrossRef]
28. Li, C.Y.; Zheng, S.J.; Chen, D.J. Size-dependent isogeometric analysis of bi-directional functionally graded microbeams reinforced by graphene nanoplatelets. *Mech. Based Des. Struct.* **2020**. [CrossRef]
29. Lurie, S.; Solyaev, Y. Revisiting bending theories of elastic gradient beams. *Int. J. Eng. Sci.* **2018**, *126*, 1–21. [CrossRef]
30. Li, A.; Ji, X.; Zhou, S.; Wang, L.; Chen, J.; Liu, P. Nonlinear axisymmetric bending analysis of strain gradient thin circular plate. *Appl. Math. Model.* **2021**, *89*, 363–380. [CrossRef]
31. Kong, S.L.; Zhou, S.J.; Nie, Z.F.; Wang, K. Static and dynamic analysis of micro beams based on strain gradient elasticity theory. *Int. J. Eng. Sci.* **2009**, *47*, 487–498. [CrossRef]
32. Lazopoulos, K.A. Post-buckling problems for long elastic beams. *Acta Mech.* **2003**, *164*, 189–198. [CrossRef]
33. Li, X.F.; Wang, B.L.; Lee, K.Y. Size effects of the bending stiffness of nanowires. *J. Appl. Phys.* **2009**, *105*, 074306. [CrossRef]
34. Polizzotto, C. A hierarchy of simplified constitutive models within isotropic strain gradient elasticity. *Eur. J. Mech. A-Solid* **2017**, *61*, 92–109. [CrossRef]
35. Yan, Z.; Jiang, L.Y. Flexoelectric effect on the electroelastic responses of bending piezoelectric nanobeams. *J. Appl. Phys.* **2013**, . [CrossRef]
36. Mindlin, R.D. Micro-structure in linear elasticity. *Arch. Ration. Mech. Anal.* **1964**, *16*, 51–78. [CrossRef]
37. Gao, X.L.; Park, S.K. Variational formulation of a simplified strain gradient elasticity theory and its application to a pressurized thick-walled cylinder problem. *Int. J. Solids Struct.* **2007**, *44*, 7486–7499. [CrossRef]

coatings

Article

Influence of Volume-to-Surface Area Ratio on the Creep Behavior of Steel Fiber Ceramsite Concrete Beams

Xiu Li [1], Hongbing Zhu [2,*], Zhenghao Fu [2], Peng Liu [3,4,*] and Changhuang Xia [2]

1 School of Transportation Engineering, Wuhan Technical College of Communications, Wuhan 430065, China; hubeilixiu@163.com
2 School of Urban Construction, Wuhan University of Science and Technology, Wuhan 430065, China; whfzh1998@163.com (Z.F.); 13628659252@163.com (C.X.)
3 School of Civil Engineering, Central South University, 22 Shaoshan Road, Changsha 410075, China
4 National Engineering Laboratory for High Speed Railway Construction, Central South University, Changsha 410075, China
* Correspondence: zhuhongbing@wust.edu.cn (H.Z.); lop868@163.com (P.L.)

Abstract: To explore the influence of the volume-to-surface area ratio (V/S) on the creep of ceramsite concrete beams, the creep performance of ceramsite concrete beams with different V/S was studied through a long-term deformation observation test, theoretical derivation analysis and finite element modeling. First, by observing the creep deflection of ceramsite concrete beams with five different V/Ss for 180 days, the relationship between creep deflection and loading time as well as the influence of V/S on creep deflection were obtained. Then, referring to the ACI209 and ACI435 creep coefficient calculation formula, the creep theory of ceramsite concrete beams involving V/S was established. Finally, the numerical model was built according to the test parameters. The results showed that the growth rate of the creep of ceramsite concrete beams increased rapidly in the early stage, but gradually slowed down with the passage of time and tended to be stable after 120 days of loading. The V/S had a significant impact on the creep of ceramsite concrete beams. In the first 7 days, the creep growth rate of each beam was approximately the same. Thereafter, the higher the V/S was, the lower the creep became. After 28 days, the creep of ceramsite concrete beams with varied V/Ss showed sharp differences. After the V/S exceeded 30, the increasing V/S could effectively reduce the creep value and the creep growth rate under a long-term load. The calculated results were in good agreement with the measured values, which fully reflected its creep variation. The finite element simulation further verified the influence of V/S on the creep of ceramsite concrete beams and the reliability of the creep calculation formula.

Keywords: ceramsite concrete beam; volume-to-surface area ratio (V/S); creep; experiment; calculation formula; numerical analysis

Citation: Li, X.; Zhu, H.; Fu, Z.; Liu, P.; Xia, C. Influence of Volume-to-Surface Area Ratio on the Creep Behavior of Steel Fiber Ceramsite Concrete Beams. *Coatings* **2022**, *12*, 977. https://doi.org/10.3390/coatings12070977

Academic Editor: Paolo Castaldo

Received: 13 June 2022
Accepted: 8 July 2022
Published: 10 July 2022

1. Introduction

Ceramsite concrete has the advantages of high strength, heat resistance, fire resistance, heat preservation, moisture retention and earthquake resistance. It not only can decrease the self-weight of the structure to improve the structural load-bearing capacity or increase the span of the structure, but also reduce the problems of arable land loss and forest destruction caused by natural aggregate mining [1,2]. Moreover, the strength and toughness of ceramsite concrete can be improved by adding steel fibers [3]. Recently, ceramsite concrete, which has been used in high-rise buildings, bridges and other structural projects, has good application prospects [4,5].

Creep is a long-term deformation property of concrete [6]. Concrete creep may cause creep deformation or stress relaxation phenomena in structural members, which leads to changes in the structural stress state, deformation patterns and mechanical properties [7]. Therefore, concrete creep patterns are of great theoretical and practical importance to structural analysis and design [8]. Moreover, the existing studies conclude that the creep characteristics of ceramsite concrete should be given sufficient attention [9] because the creep value of ceramsite concrete is larger than that of ordinary concrete and brings about a smaller creep coefficient.

Factors that affect the creep of concrete are usually the loading age [6,10,11] and body surface ratio [12,13] of structural members, the slump [14] and sand content [15] of concrete, the relative humidity [16] and air content [17] of the working environment where concrete members reside, etc. Existing studies showed: (i) The tensile creep of high-absorbent ceramsite concrete was lower than that of low-absorbent ceramsite concrete and ordinary ceramsite concrete [18,19]. (ii) High temperatures increased the creep of concrete specimens. The creep of specimens in dry an environment was higher than that of normal temperature conditions, for which temperature and humidity coefficients were added [20,21]. (iii) The creep of concrete changed significantly with the variation of reinforcement rate in the early stage, which is similar to the creep of plain concrete in the later stage [22]. Jiang [23] found that the early shrinkage creep of lightweight aggregate concrete was low. Its creep coefficient was half that of ordinary concrete in the same period, while the creep strain was 1.3 times that of early-age concrete. Wang [24] studied the effect of polyvinyl alcohol (PVA) fiber content on the creep of high-performance concrete and concluded that the maximum and minimum dry shrinkage creep could be obtained at 0.75% and 0.25% of PVA fiber content, respectively.

The size of the V/S (mm) affects the speed of temperature change inside the concrete and the rate and volume of internal moisture loss. Moreover, existing studies [25] demonstrated specimens with a small V/S, which feature fast-changing internal temperature, rapid water loss and high-water flow. It not only affected the strength and other properties of the concrete after forming, but also influenced the size of concrete. At present, the effect of the V/S of ceramsite concrete beams with a small amount of steel fibers on their creep properties has been not clear. This needs to be studied thoroughly by means of experimental and theoretical analysis.

This study explored the following three aspects. (i) Five steel fiber ceramsite concrete beams with different V/S were poured. Then, after the natural maintenance of sprinkling for 28 days and simple support for 212 days, the loading of 30% flexural ultimate bearing capacity was carried out. Next, the creep performance parameters were tested after 180 days under loading conditions, and the creep variation law influenced by the V/S was summarized. (ii) Based on the creep coefficient of concrete in the ACI209 model and the formula of shrinkage deformation in the ACI435 model, the formula of creep deformation of steel fiber ceramsite concrete beams was established by considering the effect of shrinkage deformation on the creep of steel fiber ceramsite concrete beams. Furthermore, the theoretical calculated values were compared with the experimental test values to verify the applicability of the modified creep calculation equation and the numerical simulation. (iii) ABAQUS finite element software was used to establish the long-term deformation model of steel fiber ceramsite concrete beams subjected to loads. On this basis, the calculation results were compared with the experimental results to analyze the applicability of the simulated model. The purpose of this study is to reveal the influence of the V/S on the creep characteristics of steel fiber ceramsite concrete beams. The research results can provide a reference for the design and analysis of steel fiber ceramsite concrete beam projects.

2. Mixed Proportion and Mechanical Properties of Steel Fiber Ceramsite Concrete

2.1. Materials

(1) Coarse aggregate crushed stone shale ceramsite of 900 grades (from Yichang Guangda, Yichang, China) was used. Its physical properties are shown as Table 1. It was pre-wetted for 4 h and dried for 10 h; then, the concrete was prepared in the end.

Table 1. Physical properties of shale ceramsite.

Particle Size /mm	Volume Density /(kg·m^{-3})	Apparent Density /(kg·m^{-3})	Compressive Strength of Concrete Cylinder/MPa	Water Absorption in 1 h/%
5~20	814	1517	6.8	2.42

(2) The fine aggregate was ordinary river sand (through 4.75 mm square hole sieve), and its physical properties are shown as Table 2.

Table 2. Physical properties of sand.

Apparent Density/(kg·m^{-3})	Volume Density/(kg·m^{-3})	Mud Ration/%	Fineness Number
2650	1570	≤2	2.7

(3) The cement was P.O.42.5 ordinary silicate cement with a density of 3150 kg/m^3 that was produced by Hubei Huaxin Co (from Wuhan, China). Its basic physical index is shown as Table 3.

Table 3. Basic physical index of cement.

Density/(g·cm^{-3})	Mineral Composition of Clinker				Fineness (Sieve Allowance by 80 μm Square Hole)/%
	C_3S	C_2S	C_3A	C_3AF	
3.15	45	25	12	8	6.5

(4) The steel fibers were SHWITCOM (from Wuhan, China) end-hooked steel fibers, and its physical properties are shown as Table 4.

Table 4. Physical properties of steel fiber.

Length/mm	Ratio of Length-Diameter	Density/(g·cm^{-3})	Tensile Strength/MPa	Material
30	60	7.8	≥1100	low carbon steel

(5) The water-reducing agent was high-performance polycarboxylic acid water-reducing agent produced by Qingdao Hongxia (from Qingdao, China). Its technical index is shown as Table 5.

Table 5. Technical index of water-reducing agent.

Color	PH	Relative Density	Solid Content/%	Water-Reducing Rate/%
Pale yellow	6~8	1.08 ± 0.02	40	25–35

(6) The water was Wuhan tap water.

2.2. *Mix Proportion Design*

Referring to The Technical Specification for Light Aggregate Concrete Structures (JGT 12–2006) in China, C40 was used as the target of the trial formulation. To increase the toughness of ceramsite concrete, steel fibers with a volume rate of 0.5% were incorporated with reference to The Technical Specification for the Application of Fiber Reinforced Concrete (JGJ/T 221-2010) in China. The test mix proportion is shown in Table 6. The measured concrete slump corresponded to 75.0 mm.

Table 6. Mixed proportion of ceramsite concrete beam specimens (Unit: kg/m^3).

Cement	Ceramsite	Sand	Water-Reducing Agent	Water	Ratio of Fiber Content
540	554	730	7.02	152	39 (0.5%)

2.3. Mechanical Property of Ceramsite Concrete

The test block was made according to the mix proportion listed in Table 6. Moreover, a 28 d cubic compressive strength test and flexural test were conducted. The results are shown in Table 7. It should be emphasized that the test blocks were poured in the same batch as the creep test beams and maintained under the same environment with water sprinkling.

Table 7. The mechanical parameters of ceramsite concrete at 28 d.

Mechanical Parameters	Strength			
	Specimen No. 1	Specimen No. 2	Specimen No. 3	Test Results
Compressive strength/MPa	43.82	42.57	43.12	43.17
Flexural strength/MPa	7.71	7.63	7.84	7.73
Elasticity modulus/GPa	2.46	2.40	2.42	2.43

3. Creep Test of Ceramsite Concrete Beams

3.1. Beam Specimens of Ceramsite Concrete

Creep observation tests of ceramsite concrete beams with five different body surface ratios were carried out under long-term loading conditions.

The ceramsite concrete beam specimens were 1500 mm long with protective layer of 20 mm thick. The main parameters are shown in Table 8. There were two HRB400 rebars of 8 mm diameter in both the upper and the bottom of the test beams. The stirrups were HPB235 steel bars of 6 mm diameter, as shown in Figure 1.

Table 8. Main parameters of test beam.

Specimens	Length/mm	Width/mm	Height/mm	V/S /mm	Loading Force/kN
TBB1	1500	100	120	26.316	3.25067
TBB2	1500	100	150	28.846	4.45055
TBB3	1500	100	180	30.823	5.65043
TBB4	1500	120	180	34.351	5.70859
TBB5	1500	150	180	38.793	5.76675

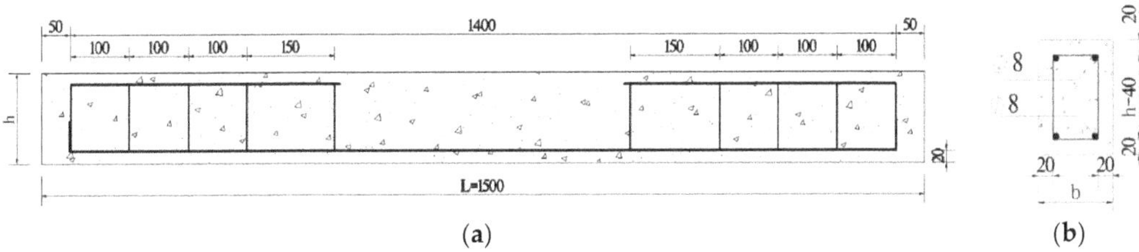

Figure 1. Structure of ceramsite concrete beam (Unit: mm). (**a**) Elevation. (**b**) Cross section.

Creep observation tests were conducted on five ceramsite concrete beams with different V/Ss under long-term loading conditions.

3.2. Loading and Testing Methods

The creep performance on beams TBB1~TBB5 were tested in a sealed environment of the interior. The temperature and the relative humidity were 20 °C and 60%, respectively. The measured temperature and humidity fluctuated slightly around the control values. The specimen beams were loaded after being maintained for 240 days. In general, concrete beams need to be kept in the stockyard for some time after being pouring. Moreover, before being officially used, they need to be kept for a period after being installed. Therefore, in practice, concrete beams are subjected to a long maintenance period before being subjected to load. To make the test close to the actual engineering situation, the test piece beams were maintained as follows. First, after being poured, the beams were sprinkled and maintained for 28 days. Then, in the case of simple support (the beam was subject to self-weight), the beams were naturally maintained for 212 days.

During the test, heavy loads were stacked by using the four-point loading method (See Figures 2 and 3). To ensure the safety of the test, two test beams were arranged side by side and parallel to each other. However, as can be seen from Table 8, the loading forces required at each specimen beam loading point were varied. In this case, combining the loading method in Figure 3 with the loading forces listed in Table 8, the following approach was taken. (i) First, the individual stacked test blocks were weighed and pre-stacked. Jacks were placed at the bottom of the loading points of the distribution beams. Then, adjusting the position of each loaded test block and distribution beam, the position of the loading point of the test beams making the reaction force provided by each jack was exactly the same as the loading force listed in Table 8. (ii) Second, position of each loaded concrete block, the distribution beam and the jacking action were marked in detail. (iii) Third, the supports and test beams were placed in order according to the positions that have been marked. Then, the concrete blocks were stacked on top of the specimen beams at the locations pre-marked in (ii) to achieve accurate loading of each specimen beam. For the loading force, referring to existing studies [10,14,16], its size was about 30% of the predicted ultimate load capacity of the ceramsite concrete beam. Dial indicators were used to measure the deflection of the specimen beams.

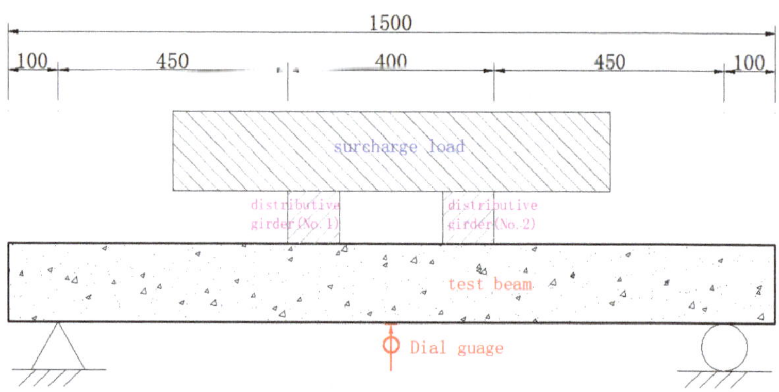

Figure 2. Loading diagram (Unit: mm).

Before loading, the loading points on the ceramsite concrete beam specimen were marked. After that, the upper surface of each contact surface was treated with sanding to ensure that the load can be transferred uniformly downward through the contact surface. To ensure the level of the loading surface, level measurement was performed, following the placement of the test beams. Then, the dial gauges were placed (a thin steel sheet was attached to the bottom contact area of the beam) and read. By doing all the procedures above, the weights of the heavy loads and their locations were measured by the weighting calculation to decrease the loading errors. Finally, the values of the dial indicators were read again, and the initial deflection was calculated.

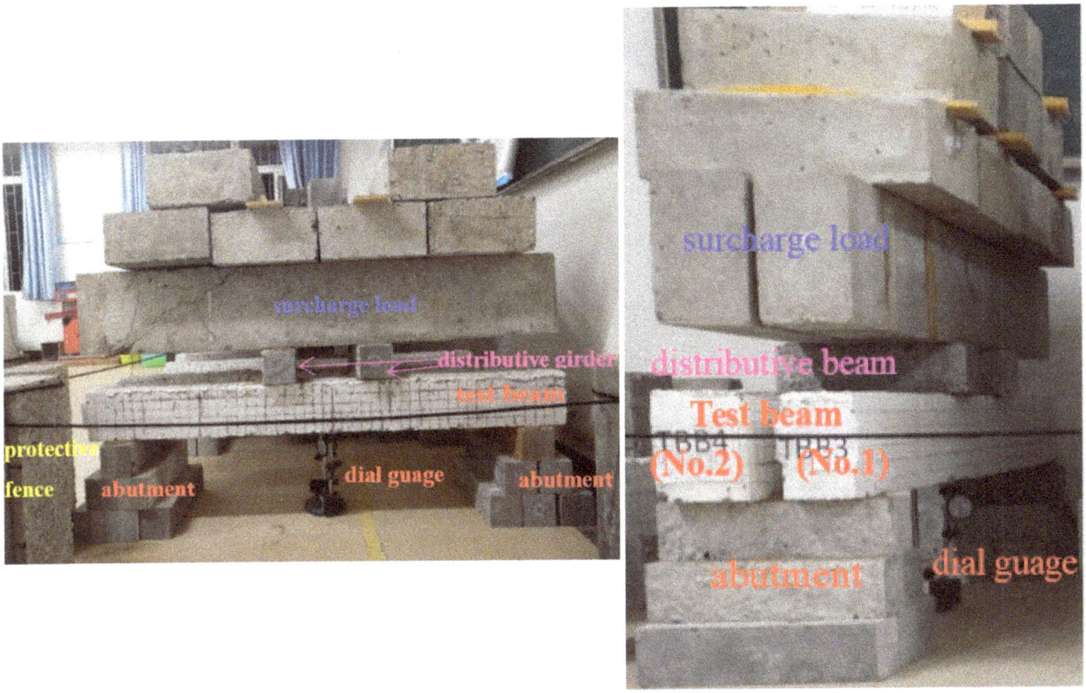

(a) (b)

Figure 3. Snapshots of loading. (**a**) Front view. (**b**) Side view.

3.3. Analysis of Test Results

The initial deflection (the deformation difference before and after initial loading), with a total deflection of 180 days, creep deflection and residual deflection after the unloading of each specimen beam are shown in Table 9. Figure 4 shows the development process of creep deflection with time growth. In addition, Figure 5 shows the influence of V/S on creep deflection at different loading times.

Table 9. Initial deflection, 180 d total deflection, creep deflection and residual deflection of the beams.

Specimens	V/S /mm	Initial Deflection f_0/mm	180 d Total Deflection f_{180}/mm	Creep Deflection f/mm	Residual Deflection f_r/mm
TBB1	26.316	1.3503	1.8426	0.5513	1.0233
TBB2	28.846	1.2064	1.6845	0.5331	0.9562
TBB3	30.823	1.1031	1.5673	0.5202	0.8765
TBB4	34.351	0.9452	1.3238	0.4106	0.6585
TBB5	38.793	0.8761	1.1645	0.2994	0.5544

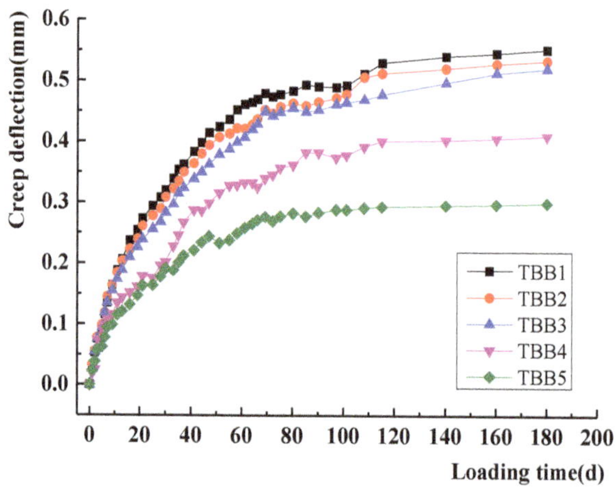

Figure 4. Curves of creep deflection–loading time of test beams with different V/Ss.

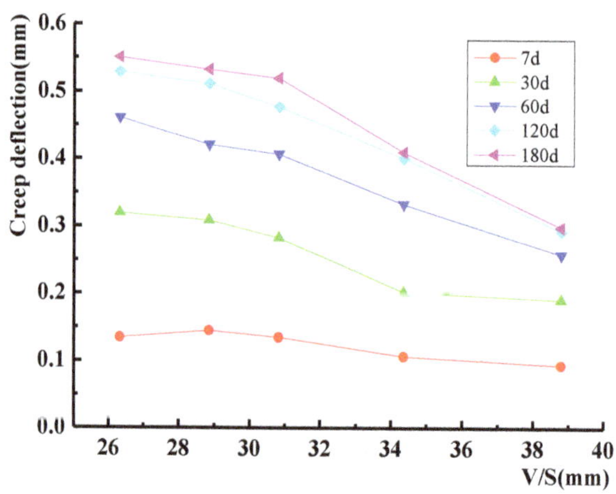

Figure 5. Influence of V/S on creep deflection.

It can be seen from Table 9 that there were differences in the initial deflection, total deflection within 180 days, creep deflection and residual deflection of each beam. In particular, the creep deflection decreased from 0.5513 to 0.2994 mm with an increasing V/S, which decreased by approximately 46%. This demonstrated the significant effect of V/S on the creep deflection of ceramsite concrete beams and proved the value of this paper's research.

Figure 4 shows that the growth rate of creep deflection of each beam was roughly the same in the first 7 days, but it became lower and lower with the rise of V/S after then. By day 28, the creep of ceramsite concrete beams differed significantly from each V/S. On the overall trend, the creep deflections of the TBB1, TBB2 and TBB3 beams had a small difference, but there was a great difference between the TBB4 and TBB5 beam. These indicated that when the V/S exceeded 30 mm, the increasing V/S could effectively decrease the creep under long-term loads.

Figure 5 shows that the creep deflection of the test beams decreased slightly with the increase of V/S at the 7th day. If the V/S was less than 35, the creep deflection of the test beams would decrease with the increase of the V/S at the 30th day. The V/S had little influence on the change of creep when it was lower than 35. However, from the 60th to the

180th day, the V/S had a great influence on the creep deflection, and the creep deflection decreased rapidly with the increase of the V/S. Therefore, the V/S had little influence on the creep of ceramsite concrete in the early stage of loading, but it had great influence on the creep of ceramsite concrete when the loading time exceeded 60 days.

4. Calculation of Creep of Steel Fiber Ceramsite Concrete Beams

The ACI209 model and ACI435 method are widely used in the study of concrete creep. They have been adopted in many countries' regulations. This model considered factors including the loading time, surrounding environment, length-to-height ratio of components, proportion and composition of coarse and fine aggregate, as well as the influence of concrete slump [20,21].

4.1. Calculation of Creep Coefficient

In this paper, the creep coefficient was calculated by the formula of the ACI209 model. The calculation results can be verified with experimental data and the conclusion of the model simulation to verify the applicability of the ACI209 model formula for ceramsite concrete.

The creep coefficient of concrete recommended by ACI 209 is expressed as follows:

$$\varphi(t,\tau) = \frac{(t-\tau)^{0.6}}{10+(t-\tau)^{0.6}}\varphi_u \tag{1}$$

where τ is loading age (d); t is calculated age (d); φ_u is the ultimate creep coefficient, $\varphi_u = 2.35\gamma_{cr}$; γ_{cr} is the product of various influence coefficients, $\gamma_{cr} = \gamma_\tau \cdot \gamma_{RH} \cdot \gamma_{VS} \cdot \gamma_s \cdot \gamma_\varphi \cdot \gamma_\alpha$; γ_τ is the influence coefficient of loading age, $\gamma_\tau = 1.25\tau^{-0.118}$ (when using wet maintenance.); γ_{RH} is the influence coefficient of relative humidity, and RH is the relative humidity of environment, $\gamma_{RH} = 1.27 - 0.67 \cdot RH$ (when RH > 40%); γ_{VS} is the influence coefficient of volume-to-surface area ratio, and V/S is the volume-to-surface area ratio (mm), $\gamma_{VS} = \frac{2}{3}[1 + 1.13e^{-0.0213V/S}]$; γ_s is the slump influence coefficient, and s is concrete slump (mm), $\gamma_s = 0.82 + 0.00264 \cdot s$; γ_ϕ is the influence coefficient of sand content, and ϕ is the sand rate (%), $\gamma_\phi = 0.88 + 0.0024 \cdot \phi$; and γ_α is the influence coefficient of air content, and α is the air content (%), $\gamma_\alpha = 0.46 + 0.09\alpha > 1$.

Based on the pouring of this test specimen beams and their actual condition of maintenance and loading, the values of each influence coefficient were calculated. Accordingly, the creep coefficient values of each test beams at different loading times can be further calculated, as shown in Table 10.

Table 10. Value of influence coefficient of creep coefficient.

Specimens	Loading Age (240 d) Influence Factor γ_τ	Relative Humidity Influence Factor γ_{RH}	V/S Influence Factor γ_{VS}	Slump Influence Factor γ_s	Influence Factor of Sand Content γ_ϕ	Influence Coefficient of Air Content γ_α	Ultimate Creep Coefficient φ_u
TBB1			1.0967			1.009	1.3287
TBB2			1.0742			1.045	1.3478
TBB3	0.6547	0.868	1.0574	1.0180	0.8832	1.090	1.3839
TBB4			1.0291			1.144	1.4136
TBB5			0.9964			1.171	1.4010

4.2. Calculation of Shrinkage Deformation

Assuming that concrete did not produce a downward-bending deformation under symmetric reinforcement, for simply supported beams, the calculation formula for shrinkage deformation in the beam span can be deduced as follows [21]:

$$\delta_{sh} = 0.125\varphi_{sh}l^2 = 0.125(\alpha_{sh}\frac{\varepsilon_{sh}}{h})l^2 \tag{2}$$

where l is beam span; ε_{sh} is the shrinkage strain of concrete, $\varepsilon_{sh} = 0.0004$; and α_{sh} is the comprehensive influence coefficient considering the reinforcement constraint, calculated according to Formula (3). In addition, Formula (3) is as follows:

$$\alpha_{sh} = \begin{cases} 0.7(100\rho - 100\rho')^{1/3}\left(\frac{\rho-\rho'}{\rho}\right)^{1/2} & (\rho - \rho' \leq 3.0\%) \\ 0.7(100\rho)^{1/3} & (\rho' = 0) \\ 1 & (\rho - \rho' > 3.0\%) \end{cases} \tag{3}$$

where ρ is the reinforcement ratio (%) of tensile reinforcement, and ρ' is the reinforcement ratio (%) of compressed reinforcement.

4.3. Calculation of Creep Deformation

The ACI435 method divides the long-term deformation of concrete into two parts: shrinkage deformation and creep deformation. The deflection increment caused by creep is calculated by plane assumption. Considering the influence of shrinkage deformation on the creep of ceramsite concrete beam, the increase coefficient of creep deformation is [11,21]:

$$\lambda_c = \frac{0.77\xi\varphi(t,\tau)}{1 + 6.3n\rho\prime} \tag{4}$$

Then, the additional deflection increase coefficient is:

$$\lambda = \lambda_c + \frac{\delta_{sh}}{\delta_d} \tag{5}$$

According to "the Standard for Test Methods of Physical and Mechanical Properties of Concrete (GB/T 50081-2019)" in China, the calculation formula of creep deformation is as follows:

$$\delta_{cr} = \lambda\delta_d = \lambda_c\delta_d + \delta_{sh} \tag{6}$$

where δ_{cr} is the creep deflection of component (mm) and δ_d is the instantaneous deflection of the component (mm).

Taking the elastic modulus of HRB400 reinforcement as 2.0×10^5 MPa, the creep deflection of the test beam can be calculated (Figures 6–10).

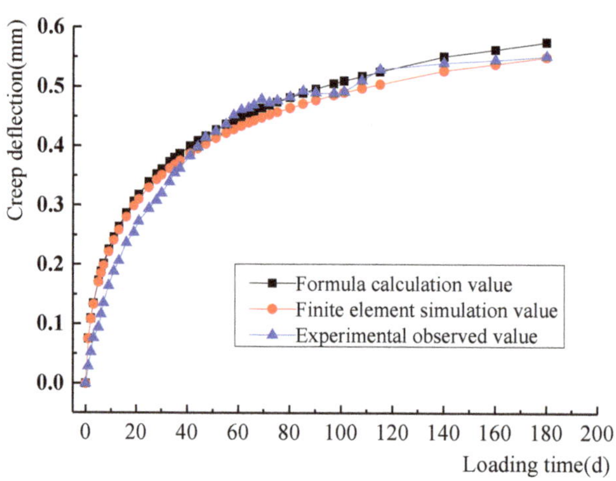

Figure 6. Creep-time curve of specimen TBB1.

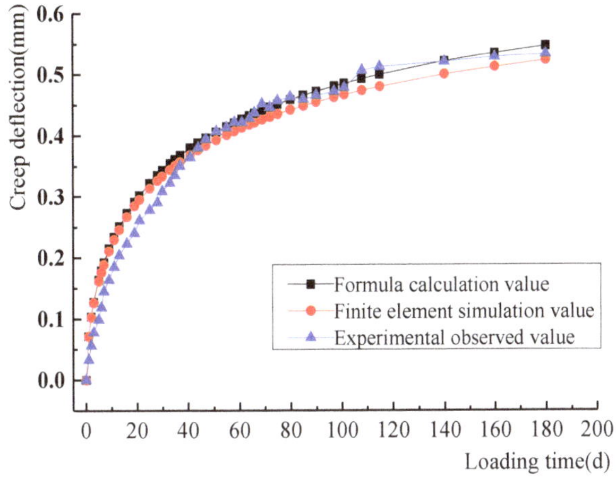

Figure 7. Creep-time curve of specimen TBB2.

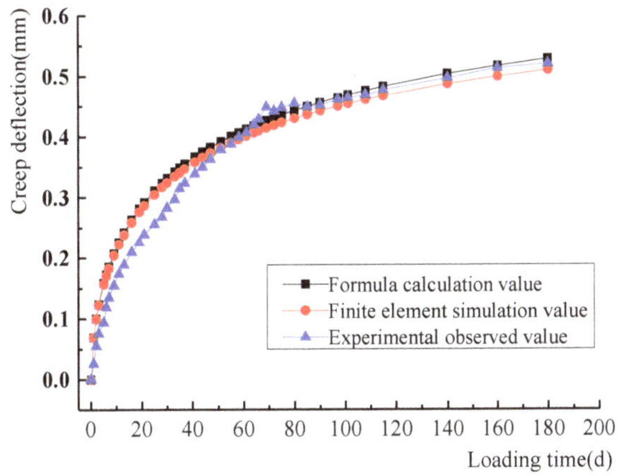

Figure 8. Creep-time curve of specimen TBB3.

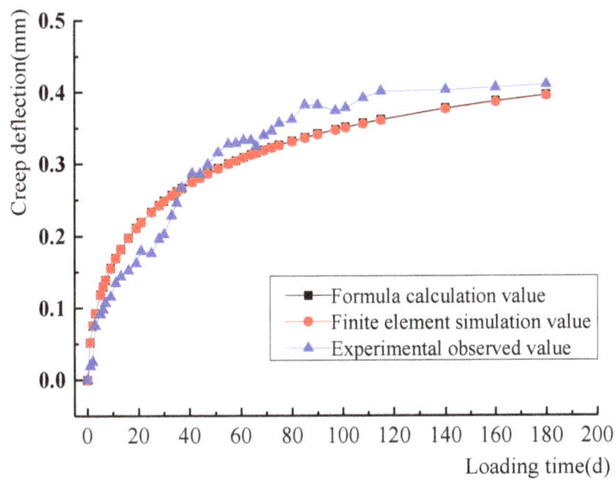

Figure 9. Creep-time curve of specimen TBB4.

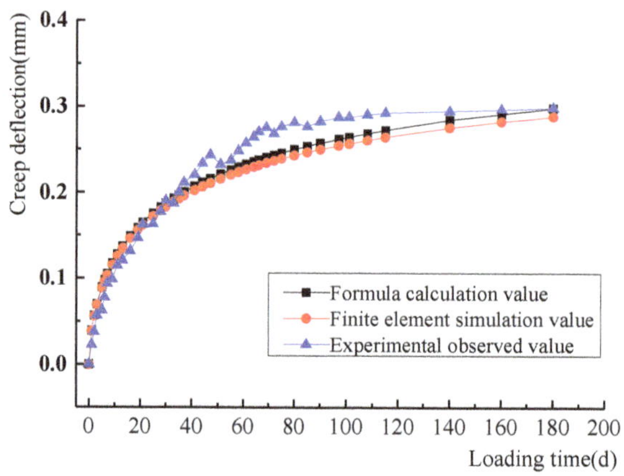

Figure 10. Creep-time curve of specimen TBB5.

From Figures 6–10, it demonstrated that the creep calculated based on the calculation formulas of ACI209 and ACI435 were in good agreement with the experimental and numerical simulated data. This suggested that it could better represent the creep variation rules of steel fiber ceramsite concrete beams. Therefore, it can be used as a formula for calculating the creep of ceramsite concrete beams in practical engineering.

5. Numerical Analysis of Long-Term Deformation of Ceramsite Concrete Beam

5.1. Constitutive Relation of Ceramsite Concrete

ABAQUS software was used to model and analyze the long-term deformation of ceramsite concrete beams under load. First, the plastic damage model provided by ABAQUS was used because it can accurately simulate the creep of ceramsite concrete beams [26]. Then, the stress–strain curve calculation model of concrete under a unidirectional load was used to describe the compression stress–strain curve of the plastic damage model, referring to the "Code for Design of Concrete Structures (GB50010-2019)" in China. The formulas are as follows:

$$\sigma = f(\varepsilon) = \begin{cases} nxf_{c,r}/(n-1+x^n) & (0 \le x \le 1) \\ xf_{c,r}/[\alpha_c(x-1)^b + x] & (x>1) \end{cases} \tag{7}$$

where $n = E_c\varepsilon_{c,r}/(E_c\varepsilon_{c,T} - f_{c,r})$ $x = \varepsilon/\varepsilon_{c,r}$; σ and ε are the stress and strain corresponding to a random point on the curve; $f_{c,r}$ and $\varepsilon_{c,r}$ are the stress and strain at the vertex of the curve; E_c is the elastic modulus of concrete; and α_c and b are the coefficients representing the shape of the curve in the descending part of the curve, which are 3.14 and 1.86, respectively, as referred to in reference [27].

5.2. Definition of Concrete Damage

Under the effect of the unidirectional compressive load, the compressive damage will occur when the deformation of concrete exceeds the elastic deformation. The calculation formula of the compressive inelastic strain ($\tilde{\varepsilon}_c^{in}$) was defined as this: the total compressive strain minus the elastic compressive strain of the material. After the compression damage data was input into ABAQUS, it can be converted into the plastic strain value ($\tilde{\varepsilon}_c^{pl}$) according to Equation (8). d_c is calculated by Equation (9).

$$\tilde{\varepsilon}_c^{pl} = \tilde{\varepsilon}_c^{in} - \frac{d_c}{(1-d_c)}\frac{\sigma}{E_0} \tag{8}$$

$$d_c = \left[\frac{1}{2}E_0\varepsilon^2 - \int f(\varepsilon)d\varepsilon\right] / \left(\frac{1}{2}E_0\varepsilon^2\right) \tag{9}$$

where E_0 is the initial tangent modulus of concrete, which corresponds to $f(\varepsilon)$.

If the plastic strain value $\tilde{\varepsilon}_c^{pl}$ output by the program is less than 0 or the inelastic strain $\tilde{\varepsilon}_c^{in}$ decreases, ABAQUS will display an error and stop the operation. When $\tilde{\varepsilon}_c^{pl} = \tilde{\varepsilon}_c^{in}$, compression damage will not occur.

The treatment of tensile damage to concrete is similar to that of compressive damage.

5.3. Finite Element Model

The finite element models for each of the five beams were built according to the parameters in Table 8 and Figure 1. The concrete was simulated using the 3D eight-node reduced-integration element (C3D8R). The rebar cage was simulated using the 3D two-node truss element (T3D2). The rebar cage was placed inside the concrete slab. Moreover, EMBED bond contacts were applied between the rebar and the surrounding concrete. The constitutive relation and plastic damage of concrete were expressed according to Equations (7)–(9). Poisson's ratio was taken as 0.2. The concrete creep coefficient was calculated using Equation (1) and imported into the ABAQUS model. The modeling and meshing are shown in Figure 11.

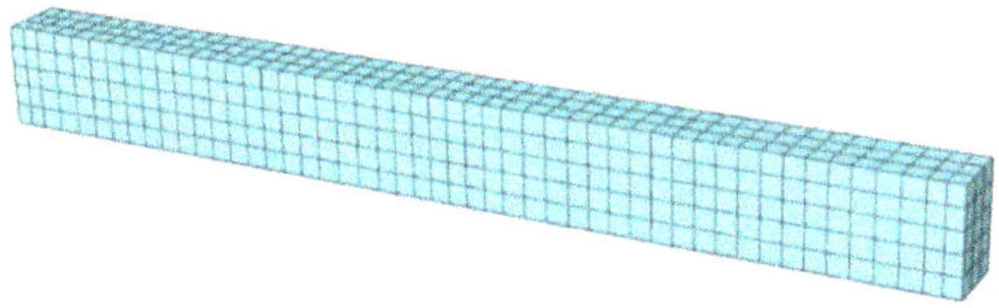

Figure 11. Finite element model and meshing of ceramsite concrete beam.

5.4. Long-Term Loading

Based on the program that comes with ABAQUS, the loading application program for ceramsite concrete beams was simulated using Python. User-defined field variables, state variables and custom expansion options were added to the concrete intrinsic structure definition. Additionally, the fitted creep procedure was written into the modeling file for the subroutine. After each output displacement of the ceramsite concrete beam, the loading state was maintained, the size was unchanged and the next cycle was continued. The variation of the elasticity modulus of ceramsite concrete with time was calculated in real time by using the field and state variables of the material in ABAQUS.

5.5. Finite Element Analysis Results

The parameters of the specimen beams were input into the model. The variation rule of the creep deflection of each model beam with the increase of time was obtained after loading.

To analyze and compare conveniently, the calculated creep, test results and fitting data of ceramsite concrete specimen beams are included in Figures 6–10.

5.6. Comparative Analysis of Simulatd, Experimental and Theoretical Value

The creep coefficient was calculated according to the proposed formula of ACI209. Then, the creep value of the beam was calculated by ACI435. From Figures 6–10, it shows that these values were in good agreement with the experimental and numerical results. In addition, it greatly reflected the creep changes in ceramsite concrete beam.

(1) Comparative analysis of creep at 180 d:

The test results, calculated values and simulated data of each specimen beam at 180 d are shown in Table 11.

Table 11. Comparison of creep value at 180 d.

Number of Test Beam	Test Values	Calculated Values	Error between Calculated and Test Values/%	Finite Element Simulation Values	Error between Simulated and Test Values/%
TBB1	0.5513	0.5749	4.36	0.5497	0.02
TBB2	0.5331	0.5467	2.63	0.5227	1.88
TBB3	0.5202	0.5285	1.54	0.5099	1.92
TBB4	0.4106	0.3960	3.41	0.3946	3.90
TBB5	0.2994	0.2982	0.33	0.2890	3.34

Table 11 and Figures 6–10 indicated that the theoretical and simulated values of creep of ceramsite concrete beams in these 180 days were close to the experimental results (the errors were within 5%). This showed that the finite element simulation can effectively analyze the creep variation patterns of ceramsite concrete beams. Meanwhile, Formulas (1)–(6) based on the method of ACI209 and ACI435 can be used as a predictive model for the creep of vitrified concrete beams. Moreover, they can fully predict the variations of creep values of ceramsite concrete beams over time.

(2) Comparative analysis of creep curve of deflection time:

It can be seen from Figures 6–10 that the creep variation rule of ceramsite concrete beam simulated by the finite element model was in good agreement with the results calculated referring to the ACI209 and ACI435 models. The finite element simulation value was slightly lower than the calculated value. The experimental value was in good agreement with the simulated and the calculated value. It was further shown that Formulas (1)–(6) based on the calculation method of ACI209 and ACI435 can well reflect the creep change in the ceramsite concrete beam, as well as the finite element model.

6. Conclusions

Long-term deformation tests, theoretical analysis and finite element modeling were carried out on five ceramsite concrete beams with different V/Ss. The following conclusions were obtained:

(1) The creep of ceramsite concrete specimens developed rapidly in the early stage of loading, but it gradually slowed down over time and tended to be stable after loading for 120 days.

(2) The V/S had an obvious influence on the creep of ceramsite concrete beams. In the first 7 days, the growth rate of each beam was roughly the same. The higher the V/S was, the lower the creep performance was. After 28 days, the creep of each beam was significantly different. When the volume–surface ratio exceeded 30, the increasing V/S can effectively decrease the creep and creep growth rate under long-term loading. In practical engineering, the V/S of the beam can be increased by controlling the beam length and increasing the beam width or height as appropriate to reduce the creep of the beam.

(3) The creep calculation Formulas (1)–(6) of ceramsite concrete beam were established according to the method of ACI209 and ACI435. The calculated results were in good agreement with the measured values, which can well reflect the creep variation rule of ceramsite concrete beams. This formula can be used for the calculation of creep in the design of ceramsite concrete beams.

(4) After adding 0.5% steel fiber, the creep law of ceramsite concrete beams can still be expressed by the modified formulas of ACI209 and ACI435.

(5) The finite element simulation can fully verify the calculation formula of creep and the influence of V/S on the creep of ceramsite concrete beams.

(6) The experiments, theory and numerical simulation in this paper initially revealed the findings that the creep of steel fiber ceramsite concrete beam was influenced by the V/S of the specimen. However, the quantities of specimens were small. As a result, the further experimental studies were planned to be carried out subsequently.

In the future, it will be planned to pour more steel fiber ceramsite concrete beams to expand the study scope of V/S. A combination of experiments and numerical simulations will be used to investigate the effect of V/S on the 180 d creep value of the beams. Furthermore, it can establish equations and reasonably predict the optimum V/S. The results are hoped to provide a scientific reference for engineering design. Moreover, the amount of steel fibers and the relative humidity of the environment may affect the creep of steel fiber ceramsite concrete beams, which will be viewed as the subsequent research directions.

Author Contributions: Conceptualization, X.L. and H.Z.; Methodology, X.L. and H.Z.; Software, P.L.; Validation, Z.F. and C.X.; Formal Analysis, X.L.; Investigation, C.X.; Resources, C.X. and X.L.; Data Curation, C.X., X.L. and H.Z.; Writing—Original Draft Preparation, X.L.; Writing—Review and Editing, X.L. and H.Z.; Visualization, C.X.; Supervision, X.L.; Project Administration, H.Z.; Funding Acquisition, H.Z. All authors have read and agreed to the published version of the manuscript.

Funding: This study was funded by the National Natural Science Foundation of China (grant numbers 51778630, 52178182 and U1934217), China Railway Science and technology research and development plan project (grant numbers 2020-Major project-02, 2021-Major project-02, 2021-Key projects-11) and Hubei Provincial Education Science Planning Project (2019GB201).

Institutional Review Board Statement: Not applicable.

Informed Consent Statement: Not applicable.

Data Availability Statement: The data presented in this study are available on request from the corresponding author.

Conflicts of Interest: The authors declare no conflict of interest.

References

1. Bu, C.; Zhu, D.; Liu, L.; Lu, X.; Sun, Y.; Yan, Z.; Yu, L.; Wei, Q. A study on the mechanical properties and microcosmic mechanism of basalt fiber modified rubber ceramsite concrete. *Buildings* **2022**, *12*, 103. [CrossRef]
2. Liu, X.; Meng, K.; Zhang, A.; Zhu, T.; Yu, C. Bearing capacity of H-section beam wrapped with ceramsite concrete. *Steel Compos. Struct.* **2021**, *40*, 679–696. [CrossRef]
3. Xie, Y.; Long, G.; Zhou, Q.; Chaktrimongkol, P.; Shi, Y.; Umar, H. Experimental investigation on mechanical property and microstructure of ultra-high-performance concrete with ceramsite sand. *Struct. Concr.* **2021**, *11*, 156. [CrossRef]
4. Bu, C.; Yang, H.; Liu, L.; Zhu, D.; Sun, Y.; Yu, L.; Ouyang, Y.; Cao, X.; Wei, Q. Quantification of ceramsite granules in lightweight concrete panels through an image analysis technique. *Materials* **2022**, *15*, 1063. [CrossRef]
5. Zhu, H.; Fu, Z.; Liu, P.; Li, Y.; Zhao, B. Shear behavior of stud-PBL composite shear connector for steel–ceramsite concrete composite structure. *Coatings* **2022**, *12*, 583. [CrossRef]
6. Mohebbi, A.; Graybeal, B.; Haber, Z. Time-dependent properties of ultrahigh-performance concrete: Compressive creep and shrinkage. *J. Mater. Civ. Eng.* **2022**, *34*, 04022096. [CrossRef]
7. Mei, S.Q.; Wang, Y.F.; Zhang, J.C. Creep of concrete-filled steel tube considering creep-recovery of the concrete core. *Adv. Struct. Eng.* **2020**, *23*, 997–1009. [CrossRef]
8. Liang, S.M.; Wei, Y. Methodology of obtaining intrinsic creep property of concrete by flexural deflection test. *Cem. Concr. Compos.* **2019**, *97*, 288–299. [CrossRef]
9. Wang, S.R.; Wu, X.G.; Yang, J.H.; Zhu, S. Acoustic emission characteristics and dynamic damage constitutive relation of shale-ceramsite concrete subjected to loading Tests. *J. Mater. Civ. Eng.* **2020**, *32*, 76. [CrossRef]
10. Hwang, E.; Kim, G.; Koo, K.; Moon, H.; Choe, G.; Suh, D.; Nam, J. Compressive creep and shrinkage of high-strength concrete based on limestone coarse aggregate applied to high-rise buildings. *Materials* **2021**, *14*, 5026. [CrossRef]
11. Ye, L.; Sun, H.L.; Lu, X.Z. *High Strength Lightweight Aggregate Concrete Structures: Performance, Analysis and Calculation*; Science Press: Beijing, China, 2009.
12. Chen, P.; Zheng, W.Z.; Zhou, X.Y. Creep of reinforced high-strength concrete-containing industrial by-products silica fume and slag. *Struct. Concr.* **2020**, *2*, 201900435. [CrossRef]
13. Zou, D.J.; Liu, T.J.; Teng, J.; Du, C.C.; Li, B. Influence of creep and drying shrinkage of reinforced concrete shear walls on the axial shortening of high-rise buildings. *Constr. Build. Mater.* **2014**, *55*, 46–56. [CrossRef]
14. Li, K.F.; Yang, C.Q.; Huang, W.; Zhao, Y.B.; Wang, Y.; Pan, Y.; Xu, F. Effects of hybrid fibers on workability, mechanical, and time-dependent properties of high strength fiber-reinforced self-consolidating concrete. *Constr. Build. Mater.* **2021**, *277*, 122325. [CrossRef]
15. Cai, X.; Jiang, M.M.; Guo, X.W.; Chen, J.J.; Zhao, Q. Experimental study on the creep behaviour of cemented sand and gravel (CSG) and temperature stress prediction of CSG dam under seasonal temperature change. *Adv. Civ. Eng.* **2020**, *1*, 8289520. [CrossRef]

16. Zheng, Z.H.; Hu, D.; Liu, P.; Sha, F.; Liu, L.; Yu, Z.W. Considering the effect of the randomness of concrete strength and relative humidity on concrete creep. *Struct. Concr.* **2020**, *10*, 201900435. [CrossRef]
17. Zhang, R.; Ma, L.; Wang, Q.; Li, J.; Wang, Y.; Chen, H.; Samosvat, V. Experimental studies on the effect of properties and micro-structure on the creep of concrete-filled steel tubes. *Materials* **2019**, *12*, 1046. [CrossRef]
18. Zhuang, Y.Z.; Chen, C.Y.; Ji, T. Effect of shale ceramsite type on the tensile creep of lightweight aggregate concrete. *Constr. Build. Mater.* **2013**, *46*, 13–18. [CrossRef]
19. Ji, T.; Chen, C.Y.; Chen, Y.B. Effect of ceramsite prewetting degree on tensile creep characteristics of lightweight aggregate concrete. *J. Build. Mater.* **2012**, *15*, 690–696.
20. James, A. *Prediction of Creep, Shrinkage and Temperature Effects in Concrete Structures*; ACI Committee 209: Farmington Hills, MI, USA, 1992.
21. Grouni, H.N. *Control of Deflection in Concrete Structures*; ACI Committee 435: Detroit, MI, USA, 1995.
22. Sun, G.J.; Xue, S.D. Experimental investigation of creep and shrinkage of reinforced concrete with influence of reinforcement ratio. *Adv. Concr. Constr.* **2019**, *7*, 211–218.
23. Jiang, W.; Yang, Z.H. Compressive creep of lightweight aggregate concrete at early age. *J. Build. Mater.* **2016**, *19*, 40–44.
24. Wang, W.L.; Li, J.T.; Peng, W.J. Effects of water-to-cement ratio, curing method and fiber on the autogenous shrinkage of early-age concrete. *J. Ceram. Process. Res.* **2019**, *20*, 77–85. [CrossRef]
25. Gedam, B.A.; Bhandari, N.M.; Upadhyay, A. An APT material model for drying shrinkage and specific creep of HPC using artificial neural network. *Struct. Eng. Mech.* **2014**, *1*, 97–113. [CrossRef]
26. Nie, J.G.; Wang, Y.H. Comparison study of constitutive model of concrete in ABAQUS for static analysis structures. *Eng. Mech.* **2013**, *30*, 59–67.
27. Zhao, S.B.; Zhao, M.S.; Zhang, X.Y. Study on complete stress curves of steel fiber reinforced lightweight aggregate concrete under uniaxial compression. *J. Build. Struct.* **2019**, *40*, 181–190.

coatings

Article

Dynamic Strain Response of Hot-Recycled Asphalt Pavement under Dual-Axle Accelerated Loading Conditions

Jin Li [1,*], Yingyong Li [2], Chongsheng Xin [3], Haoyu Zuo [1], Ping An [4], Shen Zuo [1,*] and Peng Liu [5]

1 School of Transportation Civil Engineering, Shandong Jiaotong University, Jinan 250357, China; 21107008@stu.sdjtu.edu.cn
2 Highway Service Department of Shandong Provincial Transportation Service Center, Jinan 250002, China; 21107006@stu.sdjtu.edu.cn
3 Jinan Kingyue Highway Engineering Company Ltd., Jinan 250409, China; 19011059@stu.sdjtu.edu.cn
4 Rizhao Transportation Development Group Co., Ltd., Rizhao 276800, China; 20107044@stu.sdjtu.edu.cn
5 School of Civil Engineering, Central South University, Changsha 410083, China; liupeng868@csu.edu.cn
* Correspondence: 204026@sdjtu.edu.cn (J.L.); 204096@sdjtu.edu.cn (S.Z.); Tel.: +86-5-(31)-8068-7989 or +86-13-67-882-4225 (J.L.)

Abstract: Accelerated pavement testing (APT) is an effective method to study the long-term performance of pavement. Therefore, the dynamic strain behavior analysis of asphalt pavement has important guiding significance in the study of pavement failure modes. To explore the dynamic response of a high-content plant-mixed hot-reclaimed asphalt mixture under a dynamic load of vehicles, a full-scale test road was paved, and ALT biaxial accelerated loading test equipment was used to simulate the dynamic loads of vehicles. Based on parameters such as axle load, temperature, speed, and loading times, the development law for the bottom strain of the three pavement structures was analyzed. The test results show that the most unfavorable position of the asphalt pavement load is located just below the centerline of the wheel track on one side, and the damage effect of a single double-axle wheel load is far greater than that of two single-axle wheel loads. Then, the longitudinal tensile strain of the pavement bottom always maintains the alternating state of compression-tension and compression. The longitudinal tensile strain of the pavement bottom is larger than the transverse tensile strain, and transverse fatigue cracks appear first. Under normal temperature conditions, the bottom tensile strains of the three composite pavement structures under different axial loads are close, and the pavement performance of the hot-recycled asphalt pavement of structure A and structure B can meet the specification requirements. The relationship between the bottom strain and axle load is nonlinear and is directly related to the tire ground pressure, and the difference in the tensile and compressive strain values of the bottom of the three composite pavement structures is small. Under high temperature conditions, the bottom layer temperature of structure A and structure B is lower than that of structure C, and the thermal heat transfer efficiency of hot-recycled asphalt pavement is lower than that of ordinary asphalt pavement. Additionally, the longitudinal tensile strain is about 1–1.5 times that of the transverse tensile strain. Based on the Boltzmann function, the accumulative tensile strain prediction model was established to reflect the relationship between the cumulative strain at the bottom and the number of loads.

Keywords: road engineering; hot-recycled asphalt pavement; dual-axle accelerated loading; dynamic response of strain

Citation: Li, J.; Li, Y.; Xin, C.; Zuo, H.; An, P.; Zuo, S.; Liu, P. Dynamic Strain Response of Hot-Recycled Asphalt Pavement under Dual-Axle Accelerated Loading Conditions. *Coatings* **2022**, *12*, 843. https://doi.org/10.3390/coatings12060843

Academic Editor: Andrea Simone

Received: 8 May 2022
Accepted: 10 June 2022
Published: 16 June 2022

1. Introduction

The vehicle dynamic load is one of the key factors that cause road damage and affect road life and service capacity [1,2]. This factor is closely related to the structure, load, speed, and other factors of the vehicle [3,4]. Dynamic strain is a direct reflection of vehicle's dynamic load on the pavement structure and also an important mechanical index to characterize the service performance of asphalt pavement. The value of strain in

China's highway asphalt pavement design code is still based on the theory of a static elastic layered system. Although the fatigue cracking test used for the asphalt mixture layer adopts dynamic compression modulus correction at 20 °C, this test cannot truly reflect the dynamic characteristics of the pavement structure under loads and environmental conditions. Thus, it is necessary to systematically study the dynamic strain behavior characteristics of the asphalt pavement structural layer under the action of vehicle dynamic loads [5–7].

Accelerated pavement testing (APT) can quickly accumulate loads on test pavements to evaluate long-term pavement performance in a short period of time [8] and analyze the pavement's failure mechanism. Through sensitivity analysis [9], an APT project can promote the use of innovative building materials and methods, improve pavement design and analysis procedures, promote aging-pavement survey work, and accelerate loading tests, which are all considered to be effective means of studying dynamic strain characteristics caused by the dynamic loads and service performance of asphalt pavements [10–13]. Dong Zhonghong [14] used ALF (accelerated loading facility) to analyze the effects of lateral distribution, axle weight, and temperature on the dynamic response of asphalt pavement structures under the action of a single-shaft and double wheels. Guan Zhiguang [15] introduced the two factors of vehicle driving speed and tire pressure into a study on the dynamic responses of road surface structures and established a speed–axle weight regression model. Ye Yali [16] preliminarily studied the dynamic strain evolution law of flexible base asphalt pavement structures through an APT test road. Chen Jingyun [17] studied the mechanical response law of a typical asphalt pavement structural layer under a positive load and partial load of traffic based on MLS66. Wu Jinting [18] used a seismic wave modulus measuring instrument to analyze the instantaneous dynamic response of the bottom of the pavement structure layer under the experimental conditions of overweight and high frequency and analyzed the relationship between the number of loads and the cumulative strain of the asphalt surface layer.

Ozer [19] obtained the pavement structural layer response under two different loading states of pure shear and compression shear through indoor testing and analyzed the influence of vehicle load (load and tire inflation pressure) and maneuvering (braking, acceleration, and cornering) on the strain behavior of the pavement structural layer. Saleh [20] established a model between roughness and load repetition, axial load, and asphalt layer thickness, establishing a theoretical basis for studying the influence of vehicle power size on the dynamic characteristics of flexible pavement. Liu Dawei [21] established a three-dimensional finite element analysis model of semi-rigid pavement, which analyzed the effect of changes in vehicle driving speed on the dynamic damage of semi-rigid asphalt pavement. Navarrina [22] studied the changes to power-axle weight applied by heavy vehicles on the road surface over time by extending the quarter-car model and established a comprehensive model of flexible road fatigue analysis considering the impact of vehicle dynamic loads. Lu Zheng [23] and Chen Jingyun [24] established a dynamic analysis model of vehicle-uneven pavement-subgrade structure coupling, which was used to analyze the influence of multiple parameter coupling on road dynamic responses such as the driving speed, road surface unevenness, road thickness, tire stiffness, and base modulus. Huang Zhiyi [24] used the 3D-MOVE Analysis finite-layer software to analyze the dynamic strain characteristics of the structural layer of regenerated asphalt pavement under the action of moving a non-uniform distribution load. M.S.H. Al-Furjan has derived and solved the governing equations of structures using the differential quadrature method (DQM). Afterward, a parametric study is conducted to present the effects of SMA fiber. Due to the difference between the parameter material constitutive relationship and the content gradation and the actual asphalt mixture, it is difficult to accurately describe the fatigue performance by the finite element method and the differential quadrature method, and there are large errors. Since accelerated loading is the most advanced technical means to study the fatigue properties of asphalt, it can provide a more reasonable explanation for the complexity of the material properties of asphalt mixtures. Although domestic and foreign scholars have conducted a number of studies on the dynamic strain behavior of asphalt pavement under the action of driving loads, foreign research has largely focused on flexible pavement. Consequently, the large differences in the pavement structure system decrease the applicability of foreign experimental research results. Moreover, due to the limitations of domestic accelerated loading equipment, funds, and technology, few scholars have conducted a complete study on the evolution of pavement strain behavior.

Plant-mixed hot recycling is one of the main means of producing reclaimed asphalt pavement, but RAP content, pavement structural stability, and durability have always restricted this type of pavement's wide application in China. Most scholars have only studied the strain behavior, dynamic characteristics, and long-term service performance of recycled asphalt pavement through theoretical analysis, material property testing, and numerical calculation methods, but no scholars have conducted a field-accelerated loading test on the omnidirectional strain response of the semi-rigid-base plant-mixed hot-recycled asphalt pavement structural system.

Based on these factors, we carried out an ALT acceleration loading test on an APT test road of the S222 West Lake Section of Rizhao City, Shandong Province, relying on the Applied Basic Research Project of the Ministry of Transport and the Transportation Science and Technology Project of Shandong Province. The bottom strain of the pavement structure in three different combinations was used as the experimental research object. We analyzed the evolution of the bottom-strain behavior of the plant-mixed hot-recycled asphalt pavement under the action of a high axle load, which provided a reference for studying the disease evolution mechanisms and long-term road performance of plant-mixed hot-recycled asphalt pavement.

2. Accelerated Loading Test Based on ALT

2.1. Test Equipment

The mobile road acceleration loading system, ALT, is shown in Figure 1. The equipment used was 26.342 m long, 4.220 m wide, and 7934 cm high. The hydraulic loading stepless adjustment system can apply single-sided single-axle or single-sided double-sided coupling loads. The double-sided wheelbase is 1.4 m, the effective load travel length is 10 m, the one-sided axle load range is 100–200 kN, and the one-way set wheel load can be applied 400–500 times per hour. The lateral offset load can be set, and the main and driven wheel double-axle double-wheel pack loading method is consistent with the actual road vehicle movement load state.

Figure 1. Mobile accelerated pavement loading equipment.

2.2. Road Surface Structure

A full-scale test road was paved at the S222 West Lake section of Rizhao City, Shandong Province, with a total length of 1200 m and a width of 10.5 m. This road is equipped with 3 kinds of pavement combination structures, where the surface layer is a typical 4 + 6 pavement structure, the base layer is cement-stabilized macadam, the thickness is 36 cm, and the bottom base layer is 18 cm low-dose cement-stabilized macadam. The rap dosage in the SMA-13 plant-mixed hot-reclaimed asphalt mixture is 20%, and the RAP dosage in the AC-20 plant-mixed hot-reclaimed asphalt mixture is 40%. To conduct a comparative test, three of the combined forms of pavement (where the width of the test road is 3, 3, 3, and 3 m, the total length is 9 m, and the length is 3.75 m) was selected to accelerate loading under the same test environment. The specific road structure form is shown in Table 1.

Table 1. Types of road surface structures.

Structure A	Structure B	Structure C
4 cm AC-20 Hot-recycled asphalt concrete	4 cm SMA-13 Plain asphalt concrete	4 cm SMA-13 Plain asphalt concrete
6 cm AC-20 Hot-recycled asphalt concrete	6 cm AC-20 Hot-recycled asphalt concrete	6 cm AC-20 Plain asphalt concrete
18 cm Cement-Stabilized Macadam	18 cm Cement-Stabilized Macadam	18 cm Cement-Stabilized Macadam
18 cm Cement-Stabilized Macadam	18 cm Cement-Stabilized Macadam	18 cm Cement-Stabilized Macadam
18 cm Cement-Stabilized Macadam	18 cm Cement-Stabilized Macadam	18 cm Cement-Stabilized Macadam

2.3. Sensor Placement

To analyze the transverse and longitudinal strain behavior at the bottoms of the three different combinations of pavement layers during the accelerated loading process, a sensor was embedded at the bottom of the lower layer. The km-100 HAS-type buried strain sensor (TML Corporation, Tokyo, Japan) is equivalent to the modulus of the asphalt mixture using the intermediate axial rod and the asphalt mixture modulus of Japan. To prevent interference, the horizontal spacing of the sensor was set to 0.400 m, and the vertical spacing was 0.500 m. To truly record the vertical change law for the temperature field of the asphalt pavement structure, the vertical position of the test pavement was 0 cm. Thermocouple temperature sensors with an accuracy of up to 0.002 °C were buried at 4 and 10 cm. The specific temperature and strain transducer layout is shown in Figure 2.

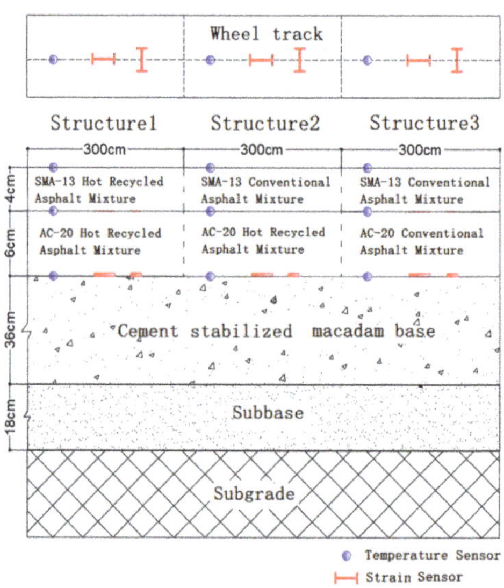

Figure 2. Schematic diagram of the test road sensor burial.

2.4. Implementation Program

This experimental study was carried out using a master–slave dual-axis accelerated loading system. The specific implementation setup for the accelerated loading test was as follows:

(1) Considering the difference in strain behavior under different axle weights, three axle weight classes were set up for comparison tests: 100, 130, and 150 kN.
(2) Considering the difference in strain behavior under different travel speeds, four speed classes were set: 12, 18, 22, and 24 km·h^{-1}.
(3) To test the long-term road performance of the plant-mixed hot-recycled asphalt pavement, the long-term loading test was set to 1.5 times the standard axle load, i.e., 150 kN, without considering the lateral offset moving load. The test environment was an open-air natural environment.
(4) Loading was performed 20 h/day for 60 days for a total of 400,000 times. The asphalt surface bottom layer tensile strain and asphalt surface layer permanent deformation were used as the test index and converted to a cumulative 5.5 million standard axle load.
(5) The base strain and pavement temperature gradient were continuously collected in real time, where the layer substrate strain acquisition frequency was 100 Hz, and the temperature gradient acquisition frequency was 10 Hz.

3. Analysis of Test Results

3.1. Determination of the Most Unfavorable Location for the Load

The dynamic load generated by driving a vehicle will act on the road surface, thereby affecting the road. It is beneficial to determine the most unfavorable position of the load to better study the pavement failure mode and improve road surface design. Therefore, we carried out a strain test under the center and side tracks of the gear train on a previous indoor full-scale test road. The sensor data from the two cases were then compared and analyzed, as shown in Table 2.

Table 2. The longitudinal strain at different position of the wheel bottom.

Sensor Number	Location	Longitudinal Tensile Strain /× 10^{-6}	Longitudinal Compressive Strain/× 10^{-6}	Total Amplitude /× 10^{-6}
11# sensor (Early foundation)	Under the center of the wheel gap	47.8	−3.4	51.2
	Under unilateral wheel tracks	96.3	−12.2	108.5
31# sensor (Early foundation)	Under the center of the wheel gap	44.2	−22.4	66.6
	Under unilateral wheel tracks	47.4	−49.0	96.4

As can be seen from Table 2, the amplitude of longitudinal tensile strain, compressive strain, and total strain under a single-wheel track in the case of two axles and two wheels are all greater than the values under the center of the wheel gap, so the degree of failure under a single-wheel track is much greater than that under the center of the wheel train. Therefore, the loading position of the embedded strain sensor is directly below the center of the single-wheel track.

3.2. Influence of Number of Axles

The acceleration loading equipment used at home and abroad is mainly of the single-axle double-wheel type. We adopted double-axle double-wheel-type acceleration loading equipment and analyzed the transverse and longitudinal bottom-strain test data of structure 3 at 13:00 on 21 August, as shown in Figure 3.

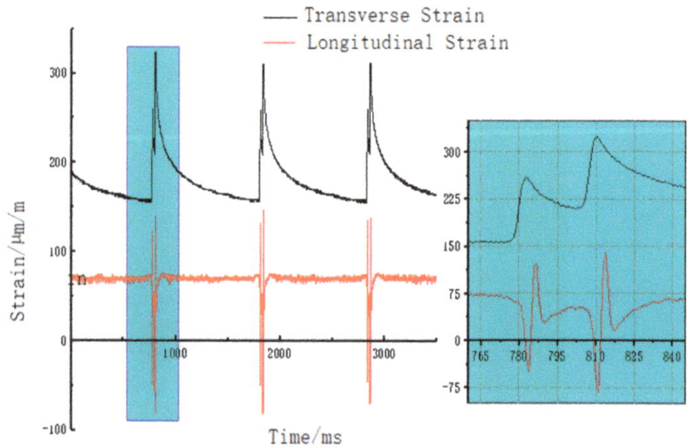

Figure 3. Transverse longitudinal strain at the base of the layer under biaxial action.

Figure 3 demonstrates the following:

(1) The transverse tensile strain value of the pavement is always positive, and the longitudinal tensile strain is alternately positive and negative. This phenomenon occurs mainly because the transverse strain gauge always experiences tension under the vertical action of the wheel load, the longitudinal strain gauge is subjected to horizontal and vertical combined loads, and the strain value alternates between positive and negative. When the tire is close to the strain timing, the horizontal and lateral components of the wheel load act on the longitudinal strain gauge, and the strain is negative. When the wheel load acts directly above the longitudinal strain gauge, the strain gauge is completely subjected to a vertical load, and the strain is positive.

(2) There are two peaks and troughs in the transverse and longitudinal strain data. Taking the transverse strain as the calculation object, the difference between the second peak and the first peak is 60~70 μm/m. Taking the longitudinal strain as the calculation object, the difference between the first wave crest and the second wave crest is 15~20 μm/m, and the difference between the second wave trough and the first wave trough is 40~45 μm/m. Asphalt mixture is a viscoelastic body, which will show elastic properties under the action of instantaneous load. Due to the deformation hysteresis effect caused by its viscoelastic characteristics, the strain generated by the first axle cannot be fully released, and the second axle produces strain superposition. Therefore, the dynamic effect of a biaxial load on pavement is much larger than that of a uniaxial load. Existing design codes still use the static load of a single-axle double-wheel set for the load, which is quite different from the actual multi-axle dynamic load of existing vehicles, which needs further study.

(3) The longitudinal strain reflects the complex stress state of the road surface under the action of the wheel load. The longitudinal compressive strain of the bottom layer is not significantly different from the lateral tensile strain, and the longitudinal tensile strain of the bottom layer is greater than the lateral tensile strain. The actual longitudinal stress on the bottom of the pavement is a variable strain state of alternating compression–tension–compression. Therefore, when designing asphalt pavement, the tensile and compressive strain and strain amplitude of the bottom layer should be considered.

3.3. *Influence of Axle Weight*

To analyze the effect of axle weight on the dynamic response of the pavement structure, the test data of longitudinal tensile strain and transverse tensile strain at the bottom of the three pavement layers at 08:00 on 21 August were selected for analysis, as shown in Table 3.

Table 3. Dynamic responses of layer substrates under different loads.

Pavement Structure	Longitudinal Tensile Strain/$\times 10^{-6}$			Longitudinal Compressive Strain/$\times 10^{-6}$			Transverse Tensile Strain/$\times 10^{-6}$		
	100 kN	130 kN	150 kN	100 kN	130 kN	150 kN	100 kN	130 kN	150 kN
Structure 1	76.9	82.6	87.2	−90.5	−99.0	−105.4	132.1	142.3	150.0
Structure 2	76.1	82.0	86.8	−87.5	−97.0	−103.3	132.4	141.6	149.3
Structure 3	78.3	84.3	89.7	−90.1	−100.3	−107.8	133.5	143.2	151.8

The following can be seen in Table 3:

(1) When the axle load increases by 30% and 50%, the longitudinal tensile strain of structure 1 increases by 7.41% and 13.39%, the longitudinal compressive strain increases by 9.39% and 16.46%, and the lateral tensile strain increases by 7.72% and 13.55%, respectively. The longitudinal tensile strain of structure 2 increases by 7.75% and 14.06%, the longitudinal compressive strain increases by 10.86% and 18.06%, and the transverse tensile strain increases by 6.95% and 12.76%, respectively. Finally, the longitudinal tensile strain of structure 3 increases by 7.66% and 14.56%, the longitudinal compressive strain increases by 11.32% and 19.64%, and the lateral tensile strain increases by 7.27% and 13.71%, respectively. The growth of underlayer strain is much less than that of the axial load, and the relationship between the underlayer strain and axial load is nonlinear.

(2) Under the same environment and different axle loads, the three kinds of pavement structures have small differences in their bottom strain, indicating that the pavement performance of the high-volume recycled asphalt pavement studied in this paper can basically achieve the new asphalt pavement standards.

During the test, to deeply study the nonlinear relationship between the axle load and bottom strain, the tire contact area under different axle loads (100, 130, and 150 kN) was tested on the spot using blue paint and white coordinate grid paper. We also measured and calculated the ground pressure (see Table 4 and Figure 4 for details).

Table 4. Wheel ground area under different ground areas.

Axial Load/kN	Single Wheel Ground Area/cm^2	Grounding Pressure/Mpa
100	304.87	0.82
130	318.63	1.02
150	353.77	1.06

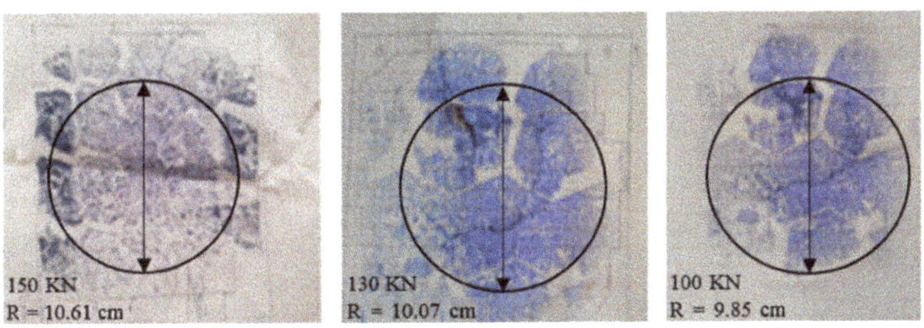

Figure 4. Diagram of wheel ground area.

It can be seen from Table 4 and Figure 4 that with an increase in the axle load of the vehicle, the contact area of the tire increases nonlinearly, and the contact pressure also increases nonlinearly. Therefore, the ground contact pressure can be used as an evaluation index to analyze the influence of overloaded vehicles on the road surface.

3.4. Influence of Temperature

To investigate the effect of temperature on the bottom strain of recycled asphalt pavement, accelerated loading tests were carried out at multiple temperature levels, and the laws of transverse maximum tensile strain and longitudinal maximum tensile strain for the bottom layer under different temperature conditions were analyzed. The driving speed of the vehicle was 24 km·h^{-1}, the axle load was set to 150 kN, and the tire pressure was 1.02 Mpa.

3.4.1. Analysis of Temperature Heat Transfer in Asphalt Pavements

In this paper, test data of Rizhao area in Shandong Province, taken during the high-temperature period from July to November, were used as an example to conduct the study. To intuitively observe the road temperature gradient and environmental temperature distribution, the temperatures during a representative high-temperature period (from 1 September to 22 September) were selected.

Figure 5 shows the temperature and ambient temperature distributions at different depths of asphalt pavement in structure 3. The variation rule of road surface temperature and ambient temperature is basically the same. With an increase in measuring-point depth, the road surface temperature experiences a certain lag with ambient temperature. The deeper the depth is, the more obvious the lag becomes, indicating that the road surface temperature and ambient temperature have obvious nonlinear characteristics.

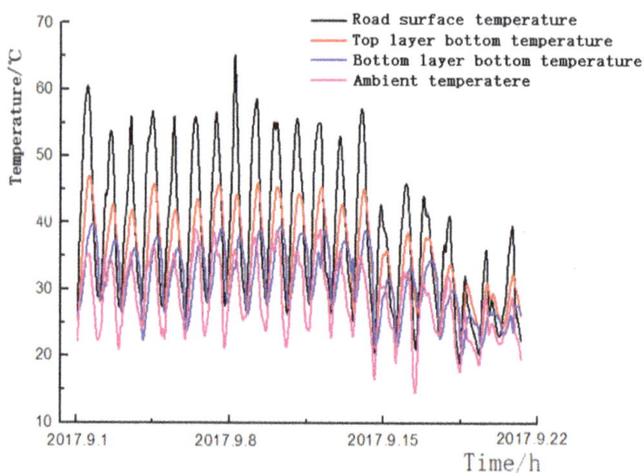

Figure 5. Measured ambient and pavement temperature data.

In this study, we used the temperature gradient calculation formula:

$$G = \frac{T_1 - T_2}{|y_1 - y_2|} \times 1000 \quad (1)$$

where G is the temperature gradient, °C/m; T_i is the temperature in the vertical direction y_i, °C; and y_i is the vertical distance of the i-th test point, mm.

Using to Formula (1), we carried out temperature gradient calculations on the temperature monitoring data from the three kinds of pavement structures. The maximum positive temperature gradient refers to the maximum positive temperature difference formed by the road surface temperature and the bottom layer temperature. The maximum positive temperature gradients of the three pavement structures were 359, 325, and 314 °C/m. The maximum positive temperature gradient occurred at 13:00 on 9 September when the weather was clear and the cloudiness was low. The ambient temperature was 36.3 °C; the surface temperature of the road surface was as high as 69.3 ± 1 °C under the action of strong infrared rays; the temperature of the upper bottom layer was 41.4, 41.5, and 41.8 °C; the bottom layer temperature was 34.4, 34.5, and 35.0 °C. Structure 2 and structure 3 had the same top layer material, and the temperature of the top bottom layer was basically the same. Structure 1 was 0.6 °C lower than the bottom layer temperature of structure 3. The test results show that the pavement used in structure 1 has poor heat transfer. This phenomenon is due to the presence of significant amounts of colloid, asphaltene, and other materials in the recycled asphalt mixture, which together reduce the heat transfer efficiency of the pavement.

3.4.2. Effect of Temperature on the Strain Behavior of the Layer Substrate

On September 18, the weather in Rizhao City was sunny and windy. The bottom temperatures and road surface temperatures of the three kinds of pavement structures, collected from 7:00 to 21:00, were compared and analyzed with the horizontal and vertical maximum strain data of the bottom layer. Specific data are shown in Figures 6–8.

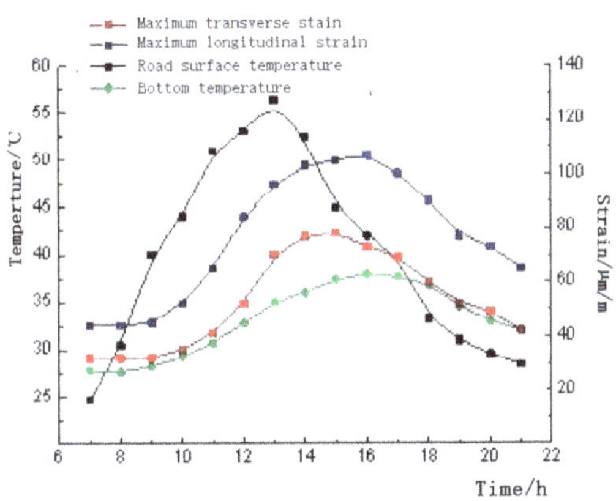

Figure 6. Structure A temperature versus maximum tensile strain in the transverse and longitudinal direction at the base of the layer.

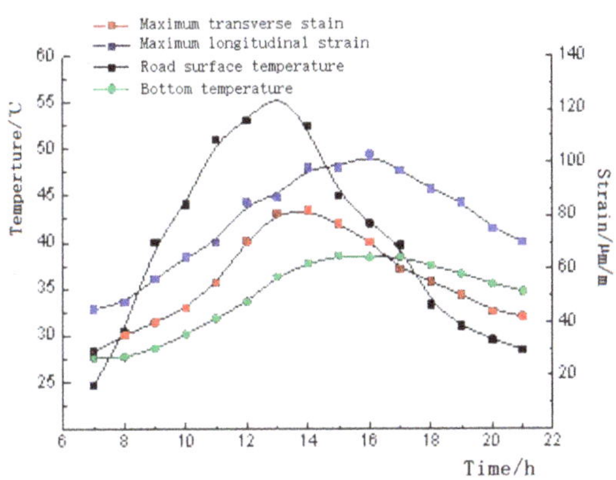

Figure 7. Structure B temperature versus maximum transverse and longitudinal tensile strain at the base of the layer.

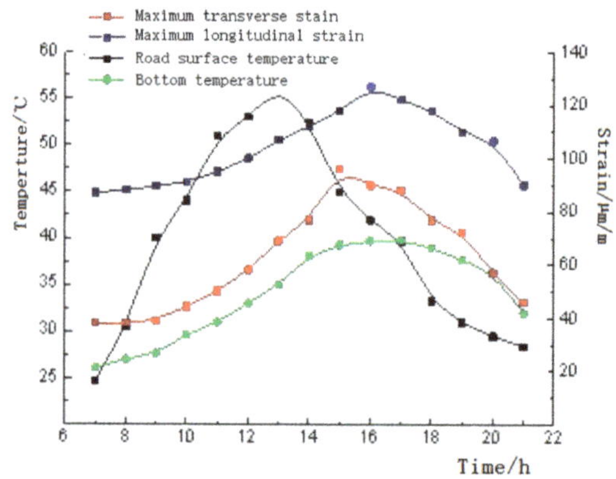

Figure 8. Structure C temperature versus transverse and longitudinal tensile strain at the base of the layer.

The following can be seen in Figures 6–8:

(1) The longitudinal and transverse tensile strains of the three pavement structures are basically consistent with the variation of the temperature at the bottom of the pavement. When the temperature at the bottom increases, the strain increases, and when the temperature decreases, the strain decreases. In Figure 6, we see that the layer bottom temperature, the maximum transverse strain, and the maximum longitudinal should reach a maximum value for structure A about an hour later than the road surface temperature; in Figure 7, in the time range 12–14, the road surface temperature is the highest, and the maximum transverse strain is more sensitive to the change in temperature, while the changes in the bottom temperature and the maximum longitudinal strain with temperature are slightly delayed, and the strain reaches the maximum value around 16 o'clock; the change in Figure 8 is closer to that in Figure 6.

(2) Under the same temperature conditions, the longitudinal tensile strain at the bottom of the three pavement structures is greater than the transverse tensile strain; under high temperature conditions, the longitudinal tensile strain is about 1.0–1.5 times the transverse tensile strain, so transverse fatigue cracking of the pavement bottom layer is the first to appear.

(3) At high temperatures, the bottom strain of structure C is obviously larger than that of structure A and B, with little difference between the bottom strain of structure A and B. Structure C belongs to new asphalt pavement, which has good heat transfer, good viscoelasticity of pavement materials, and a large bottom response. The reclaimed pavement has better high-temperature performance than the new pavement. At room temperature, the bottom temperatures of the three pavement structures are close to each other, but there is little difference in the bottom strain values.

(4) Under the action of repeated axial loads on the road surface, the cracks at the bottom of the layer cannot be directly reflected on the road surface. With the help of the highly consistent relationship between the temperature of the bottom of the layer and the longitudinal tensile strain, the cracks at the bottom of the layer can be analyzed. This analysis is used for road performance inspections and disease control. Assessment provides a new approach.

3.5. *Effect of Vehicle Speed*

First, we analyzed the relationship between the vehicle speed and the bottom strain of the asphalt pavement at 14:00 on 21 August, when the driving equipment was accelerated and loaded at 12, 18, 22, and 24 km·h^{-1}. Three kinds of lateral and longitudinal strains were collected from the pavement bottom layer during driving for comparative analysis. In this paper, the lateral and longitudinal tensile strains of the pavement bottom layer of structure A under four driving speeds were taken as an example, as shown in Figures 9 and 10.

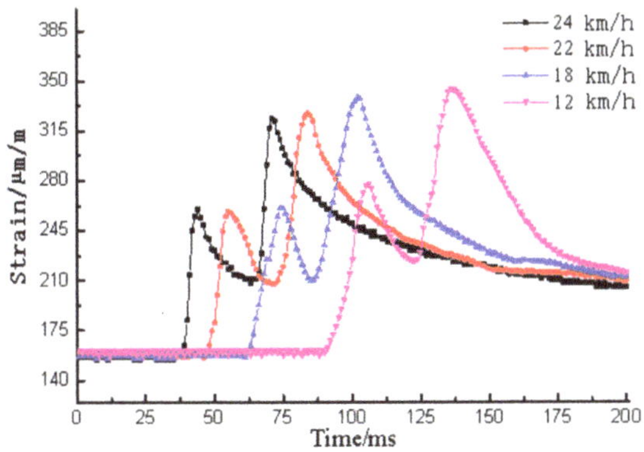

Figure 9. Relationship between vehicle speed and maximum transverse tensile strain at the base of the layer.

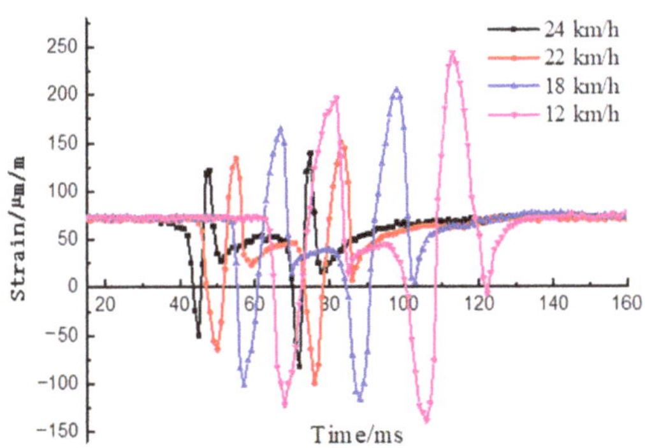

Figure 10. Relationship between vehicle speed and maximum longitudinal tensile strain at the base of the layer.

It can be seen from Figures 9 and 10 that the driving speed of the driving equipment directly affects the wheel load action time and the strain pulse time. As the viscoelastic material of asphalt mixture increases, so too does the vehicle travel speed (relative to the load action frequency). Moreover, the greater the modulus, the lower the strain value at the bottom of the layer.

According to the test data of structures B and C, there is little difference in the bottom strain value between the recycled asphalt pavement and the new asphalt pavement under the same driving speed conditions. The bottom strain values of the three pavement structures all decrease with an increase in the driving speed of the driving equipment, and this trend is consistent. It can be demonstrated that the initial material modulus of the recycled asphalt pavement is close to the modulus of the new asphalt pavement. This conclusion was also confirmed in the previous laboratory tests.

With an increase in vehicle driving degree, the maximum lateral and longitudinal strains of the bottom layer gradually decrease, and the trend of strain reduction does not decelerate. If the driving speed of the driving equipment continues to increase, the bottom layer strain value will reach a relatively stable value. According to the test results, a calculation model for the maximum strain value of structure A (12 km/h $\leq$ v $\leq$ 24 km/h) can be established:

$$\varepsilon = 368.5e^{-0.05v} \tag{2}$$

According to the measured data of the Jinan South Ring Expressway field test road (same road structure) paved by the research team, when the heavy-duty vehicle travelled at a speed of 80~120 km/h, the maximum strain value of the bottom layer was 239.3×10^{-6}~206.2×10^{-6}, and the error between the measured value and the calculated value was within 5.0%. The calculation models for different pavement structures and axle loads were established through accelerated loading test data corresponding to the test parameters.

3.6. Influence of the Number of Load Actions

The relationship between the number of loads and the cumulative bottom strain was also analyzed. As an example, we compared and analyzed the lateral and longitudinal cumulative tensile strains of pavement structure A and the number of loads. The relationship between the cumulative tensile strain and the number of loads is shown in Figure 11.

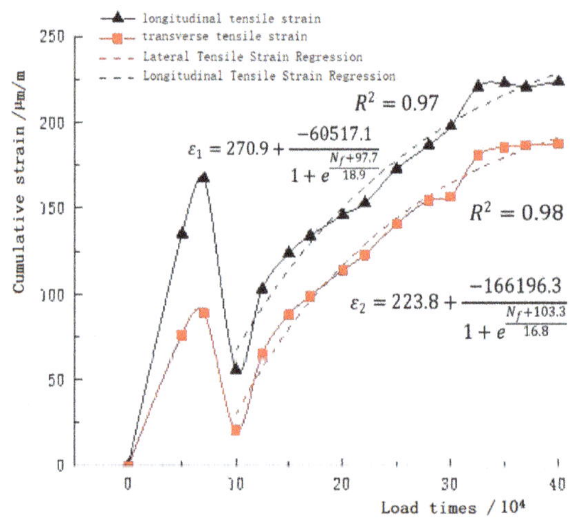

Figure 11. Cumulative strain versus number of load actions.

It can be seen from Figure 11 that in the initial stage of loading, the lateral and longitudinal cumulative strains at the bottom of each layer rise sharply, and the initial cumulative strain is about 75.0% of the total cumulative strain. With an increase in the number of loadings, the increasing trend of the strain decelerates. The cumulative longitudinal and transverse tensile strains tended to be stable when loaded 310,000 times, and the thermally recycled asphalt mixture entered the compaction stage. The average rut at this time was 1.30 cm, and the average rut at the end of the test was 1.55 cm.

According to the comparative analysis of Figures 9–11, the instantaneous strain is sometimes greater than the cumulative strain, indicating that hot-recycled asphalt pavement had self-healing abilities during the test process and in the intermittent stage. Thus, the number of loads here is not directly related to the instantaneous strain but directly related to the cumulative strain. The cumulative strain prediction model based on the Boltzmann function can better reflect the relationship between the cumulative strain at the bottom of the layer and the number of load actions, as shown in Equations (3) and (4):

$$\varepsilon_1 = 270.9 + \frac{-60517.1}{1+e^{\frac{N_f+97.7}{18.9}}} \tag{3}$$

$$\varepsilon_2 = 223.8 + \frac{-166196.3}{1 + e^{\frac{N_f + 103.3}{16.8}}} \quad (4)$$

The reason for the sharp decline stage shown in Figure 11 is that the bottom deformation was recovered after the test was interrupted for 6 days due to weather reasons and equipment failure. The test data indicate that the longitudinal accumulated strain recovered by 66.5%, and the lateral accumulated strain recovered by 76.4%, indicating that the thermal regenerative asphalt pavement had a strong ability to recover initial deformation.

4. Concluding Remarks

(1) In the case of dual axles and two wheels, the strain amplitude of the asphalt pavement under the center of the wheel track on one side was larger than that under the center of the wheel gap, and the load on the asphalt pavement under the center of the wheel track on one side was higher than that under the center of the wheel gap.

(2) The damage effect of a single dual-axle wheel load was far greater than that of the two single-axle wheel loads. The longitudinal strain at the bottom of the pavement layer always maintained an alternating state of compression–tension–compression, and the longitudinal tensile strain at the bottom of the pavement layer was always greater than the lateral tensile strain. Moreover, transverse fatigue cracks appeared first. Under high temperature conditions, the longitudinal tensile strain was about 1–1.5 times the transverse tensile strain.

(3) When the axle load increased by 30% and 50%, the longitudinal tensile strain increased by 7.41–14.56%, respectively; the longitudinal compressive strain increased by 9.39–19.64%; and the transverse tensile strain increased by 7.72–13.71%. The bottom strain of asphalt pavement had a nonlinear relationship with the axle load and a linear relationship with the ground pressure.

(4) The maximum positive temperature gradient of structure A was 12.50% higher than that of structure B and 9.50% higher than that of structure C. The heat transfer efficiency of the thermally recycled pavement was lower than that of the new asphalt pavement, and the temperature had a greater impact on the bottom strain of the asphalt pavement.

(5) The driving speed directly affects the load action time and pulse time of the response. The higher the vehicle speed was, the smaller the bottom layer strain value became. Based on the accelerated loading test data, a prediction model for the maximum bottom layer strain was established. The error rate in the calculations of the maximum strain value of the pavement bottom layer was controlled within 5%.

(6) No direct relationship was observed between the loading times and instantaneous strain, and the relationship between the loading times and accumulative bottom could be represented using the Boltzmann function. After 310,000 loads, the hot-recycled asphalt pavement entered the dense stage, and plastic deformation was basically completed. Under a load of 400,000 times, there was no obvious difference in rut performance between the recycled asphalt pavement and the new asphalt pavement.

(7) Certain limitations remain in evaluating the performance of hot-recycled asphalt pavement based on the bottom-strain behavior of asphalt pavement. The mechanical response of high-volume thermally recycled pavement can basically fulfill the new pavement standards, but the uniformity and migration law of thermally recycled materials cannot be determined in the strain behavior clearly and must be further verified in combination with the properties of thermally recycled materials.

(8) Based on the above conclusions, this paper accelerates road damage through the fast loading method of controllable axle load, and provides a real data model for analyzing the influence of various factors such as road load on the strain. Considering the actual operating conditions on the road, the poor accuracy, authenticity, practicability and real-time performance of the experimental data, and the resulting errors in the analysis results, the accelerated loading test is a cost-effective method. In addition, the dynamic strain behavior analysis of asphalt pavement also plays a very important

role in the study of pavement failure modes. Therefore, the popularization and application of the pavement accelerated loading test system is of great practical significance to study the long-term performance of the pavement. In this paper, the ALT biaxial accelerated loading test equipment is used to simulate the dynamic load of the vehicle, and the parameters such as axle load, temperature, speed, and loading times are synthesized, and the development law of the bottom strain of the three kinds of pavement structures is analyzed, which will also provide information for the advancement of related research.

Author Contributions: Conceptualization, J.L.; methodology, J.L. and Y.L.; software, J.L. and Y.L.; validation, H.Z.; formal analysis, H.Z.; investigation, H.Z.; resources, J.L. and S.Z.; data curation, J.L. and H.Z.; writing—original draft preparation, J.L. and H.Z.; writing—review and editing, J.L. and H.Z.; supervision, J.L. and P.L.; project administration, J.L., C.X. and P.A.; funding acquisition, J.L. and C.X. All authors have read and agreed to the published version of the manuscript.

Funding: Shandong Transportation Science and Technology Project: Preparation and road performance of REOB/SBS composite modified asphalt (2021B11).

Institutional Review Board Statement: Not applicable.

Informed Consent Statement: Not applicable.

Data Availability Statement: The experimental data in this paper are from the pavement material laboratory of Shandong Jiaotong University, which is the provincial key laboratory.

Conflicts of Interest: The authors declare that they have no conflict of interest regarding the publication of this paper.

References

1. Pan, T.C.; Li, J. Dynamic vehicle element method for transient response of coupled vehicle-structure systems. *J. Struct. Eng.* **2002**, *128*, 214–223. [CrossRef]
2. Watts, G.R.; Krylov, V.V. Ground-borne vibration generated by vehicles crossing road humps and speed control cushions. *Appl. Acoust.* **2000**, *59*, 221–236. [CrossRef]
3. Bilodeau, J.P.; Gagnon, L.; Doré, G. Assessment of the relationship between the international roughness index and dynamic loading of heavy vehicles. *Int. J. Pavement Eng.* **2017**, *18*, 693–701. [CrossRef]
4. Navarrina, F.; Ramírez, L.; París, J.; Nogueira, X.; Colominas, I.; Casteleiro, M.; Fernández-de-Mesa, J.R. Comprehensive model for fatigue analysis of flexible pavements considering effects of dynamic axle loads. *Transp. Res. Rec.* **2015**, *2524*, 110–118. [CrossRef]
5. Zhuang, C.Y.; Ye, Y.; Zhang, N. Analysis of dynamic response of semi-rigid base asphalt pavement based on APT test. *J. Shandong Transp. Inst.* **2014**, *22*, 55–61.
6. Lv, S.; Chen, J. Study on the decay law of asphalt mixture stiffness based on accelerated loading test. *J. Highw. Transp. Res. Dev.* **2016**, *33*, 1–6.
7. Hu, P.; Liu, Z.J. Analysis of fatigue damage of asphalt pavement under accelerated loading conditions. *J. Shandong Traffic Coll.* **2010**, *18*, 74–77.
8. Embacher, R.A.; Snyder, M.B.; Odden, T.D. *Using the Minnesota Accelerated Loading Facility to Test Retrofit Dowel Load Transfer Systems*; Transportation Research Record No. 1769; National Academy Press: Washington, DC, USA, 2001; Volume 1769, pp. 134–141.
9. Plessis, D.; Nokes, L.; Mahdavi, W.A.; Burmas, N.; Holland, J.; Lee, E.B. *Economic Benefits Assessment of Accelerated Pavement Testing Research in California: Case Study*; Transportation Research Record No. 2225; Research Board of the National Academies: Washington, DC, USA, 2011; Volume 2225, pp. 137–146.
10. Xu, Q. Accelerated loading test study of semi-rigid base and flexible base asphalt pavement. *Highw. Traffic Technol.* **2011**, *7*, 115–119.
11. Meng, S. Accelerated loading test study on the performance of semi-rigid base asphalt pavement. *Highw. Traffic Technol.* **1997**, 61–66.
12. Jiang, H.; Bian, X.; Chen, Y. Full-scale accelerated loading test for high-speed railway track-roadbed train moving load simulation. *J. Civ. Eng.* **2015**, *48*, 85–95.
13. Ran, W.; Ling, J.; Zhao, H. Accelerated loading dynamic response of epoxy asphalt pavement at high temperature and footage. *J. Tongji Univ.* **2015**, *43*, 1823–1828.
14. Dong, Z.H.; Xu, Q.L.; Lv, P.M. Dynamic response of semi-rigid base asphalt pavement based on accelerated loading test. *Chin. J. Highw.* **2011**, *24*, 1–5.
15. Guan, Z.-G.; Zhuang, C.-Y.; Lin, X.-X. Accelerated loading dynamic response of footprint asphalt concrete pavement. *J. Transp. Eng.* **2012**, *12*, 24–31.
16. Ye, Y.; Zhuang, C.; Wang, L.; Song, X.J. Dynamic response of flexible base asphalt concrete pavement based on accelerated loading test. *Highways* **2013**, *58*, 1–7.

17. Chen, J.Y.; Liu, J.Y.; Liu, Y.Q. Dynamic response of asphalt pavement structures under accelerated loading conditions. *J. Harbin Eng. Univ.* **2014**, *35*, 771–776.
18. Wu, J.T.; Ye, F. Analysis of strain behavior of asphalt pavement under heavy-frequency loading of MLS66. *Chin. J. Highw.* **2014**, *27*, 9–16.
19. Ozer, H.; Al-Qadi, I.L.; Hasiba, K.I.; Wang, H. Pavement layer interface shear strength using a hyperbolic Mohr-Coulomb model and finite element analysis. Airfield and Highway Pavement 2013: Sustainable and Efficient Pavements. In Proceedings of the Airfield & Highway Pavement Conference, Los Angeles, CA, USA, 9–13 June 2013; American Society of Civil Engineers: Los Angeles, CA, USA, 2013; pp. 1445–1456.
20. Saleh, M.F.; Owusu-Antwi, M. Mechanistic roughness model based on vehicle-pavement interaction. *Transp. Res. Rec.* **2000**, *1699*, 114–120. [CrossRef]
21. Liu, D.A.W.; Chen, J.; Huo, W. Research on the damage of road surface by vehicle speed. *J. Agric. Mach.* **2003**, *1*, 11–13.
22. Lu, Z.; Yao, H.L.; Hu, Z. Dynamic response analysis of uneven pavement based on coupled vehicle-road structure vibration. *J. Geotech. Eng.* **2013**, *35*, 232–238.
23. Chen, J.Y.; Sun, Y.I.R. Dynamic response analysis of a coupled vehicle-road system based on contact nonlinearity. *Vib. Shock.* **2013**, *32*, 119–124.
24. Huang, Z.; Chen, Y.; Zhang, Q. Response analysis of hot recycled asphalt pavement under moving load. *J. Railw. Sci. Eng.* **2019**, *16*, 107–113.

coatings

MDPI

Article

Analytical Study on Reinforced Concrete Columns and Composite Columns under Lateral Impact

Xianhui Li [1,2], Yao Yin [1], Tieying Li [1], Xiang Zhu [3,*] and Rui Wang [1]

1 College of Civil Engineering, Taiyuan University of Technology, Taiyuan 030024, China
2 College of Architectural Engineering, Shanxi Vocational University of Engineering Science and Technology, Jinzhong 030600, China
3 School of Electric Power, Civil Engineering and Architecture, Shanxi University, Taiyuan 030013, China
* Correspondence: zhuxiang@sxu.edu.cn

Abstract: This study investigates the lateral impact responses of reinforced concrete (RC) and composite columns through dynamic nonlinear analysis using LS-DYNA. The simulation results were first validated against experimental results performed earlier on four different cross sections. The finite element analysis results show that the simulation results of LS-DYNA can predict the experimental results well and can be used for further parametric analysis. The overall impact resistance of the four new composite columns is significantly better than that of RC columns. Among the composite columns, the solid concrete-filled double steel tube (S-DS) column has the best impact resistance with higher impact plateau force and smaller mid-span deflection under the same test conditions. It was found that the impact response process of all types of composite columns was similar. Finally, parametric analysis of the composite columns is performed to study the influence of load, material and other related parameters on the impact response of the composite columns. The results provide new information on the impact response of composite columns and the influence of materials and load parameters. The study provides a basis for the design and analysis of composite columns under lateral impact loading.

Keywords: RC columns; composite columns; lateral impact; numerical simulation; whole-process analysis; parameter sensitivity analysis

Citation: Li, X.; Yin, Y.; Li, T.; Zhu, X.; Wang, R. Analytical Study on Reinforced Concrete Columns and Composite Columns under Lateral Impact. *Coatings* **2023**, *13*, 152. https://doi.org/10.3390/coatings13010152

Academic Editor: Edoardo Proverbio

Received: 10 November 2022
Revised: 31 December 2022
Accepted: 7 January 2023
Published: 11 January 2023

1. Introduction

The collision load effect on structural components has to be considered in structural design and sometimes considered in whole life cycle analysis. Relevant reports on collision accidents are increasing year by year [1–5]. It is well-known that the compressive strength of concrete is much higher than its tensile strength. Furthermore, the compressive strength is enhanced under bi-axial or tri-axial restraint. For structural steel, the tensile strength is high while the shape may buckle locally in compression. In concrete-filled steel tubular members, steel and concrete are used such that their natural and most prominent characteristics are taken advantage of. The confinement of concrete is provided by the steel tube, and the local buckling of the steel tube is improved due to the support of the concrete core. To ensure the safety of buildings and other structures under impact loading, concrete-filled steel tube (CFST) composite columns were used to improve the impact resistance of new structural columns. In the past few decades, CFST has been extended to other forms of composite columns [6–8], such as R-ST columns, S-ST columns, H-DS columns and S-DS columns. These composite columns are expected to have improved impact resistance to reduce the losses caused by accidental collisions.

The mechanical properties of composite columns under impact have been studied by a number of researchers. Aghdamy et al. [9,10] studied the impact resistance of hollow concrete-filled double steel tube columns using parameter sensitivity analysis. Results show that the deflection is significantly affected by axial load. The impact velocity, slenderness ratio and thickness-to-diameter ratio of steel tube are the key factors in determining the dynamic response of CFST members as these are the parameters that determine the impact energy of the loads and stiffness of the members. Zhu et al. [11,12] studied the impact resistance of steel-reinforced concrete (SRC) columns and steel-reinforced concrete-filled circular steel tube (SRCFST) columns. The results show that the impact resistance of SRCFST columns is better than that of SRC columns under the same mass. Wang et al. [13] studied the behaviors of CFST under lateral impact by experimentation and finite element methods. The results show that the critical failure energy increases with an increase in the constraint factor, and the axial force has a significant influence on the lateral displacement of specimens. Xu et al. [14] and Wang et al. [15] studied the impact behaviors of carbon fiber-reinforced polymer (CFRP)-reinforced RC columns by experimental and numerical methods. The results show that CFRP-reinforced RC columns can effectively change the lateral impact failure mechanism from shear to bending, which can effectively reduce impact damage. Han et al. [16–18] studied the impact resistance of a CFST column and a CFST composite column by experimental and finite element methods. The results showed that the CFST column had good impact resistance, while the concrete in the outer layer of the composite column was seriously damaged; inner steel tube only had limited flexural deformation. Wang et al. [19,20] carried out lateral impact tests on 31 hollow concrete-filled double steel tube columns using a drop hammer device. The results showed that CFDST members had good ductility under impact loading, and the sandwich steel tube had an obvious restraint effect on concrete. Goldston et al. [21] investigated the behavior of glass fiber-reinforced polymer (GFRP) bar-reinforced high-strength concrete and ultra-high-strength concrete under impact loading. Swesi et al. [22] studied CFRP strengthening on the response of RC columns under lateral impact loads.

Recently, a total of 30 columns (6 RC columns and 24 composite columns) were tested under low-velocity impact by Zhu et al. [23]. The test results show that the change in the axial compression ratio has an important effect on the impact resistance of specimens. In this paper, LS-DYNA software is used to describe the nonlinear finite element analysis (FEA) of RC and composite columns under transverse impact. The FEA model is validated with experimental results. Then, the whole process of impact damage on the composite column is analyzed. The validated model is then used to conduct an extensive parametric study to investigate the effects of impact height, impact mass, unconfined compressive strength of concrete, yield strength of outer and inner steel tubes, diameter–thickness ratio of outer and inner steel tubes, influence of H-shaped steel on impact force and mid-span residual deflection.

2. Finite Element Model

2.1. Composite Column Parameters

The validation of the models is based on an impact test of composite columns [23]. This numerical study simulates the impact resistance of composite columns with five different cross sections (shown in Figure 1A) under axial compression ratios of 0, 0.3 and 0.5, resulting in 15 combinations, which are calculated as N_0/N_u, where N_0 is the axial force applied in the test and N_u is the axial ultimate capacity of the specimen obtained from references [24–27]. The parameters of the columns are shown in Table 1 and the dimensions of the different sections are shown in Figure 1A [23]. To compare the impact resistance of members with different section types, the composite columns and reinforced concrete columns are designed to have the same outer diameters and use the same amount of steel. The outer diameter of the steel tube is 114 mm and the wall thickness is 2 mm. For solid concrete-filled double steel tube (S-DS) and hollow concrete-filled double steel tube (H-DS) sections, the inner circular steel tube diameter is 50 mm and the wall thickness is 2.5 mm. The size of the cross-sectional structural steel of S-ST columns (Figure 1(Af)) is 57 mm × 14 mm × 2 mm × 3 mm ($h \times b \times t_1 \times t_2$). The longitudinal reinforcement diameter (D) of steel tubular-encased reinforced concrete (R-ST) and reinforced concrete (RC) members is 6 mm, and the stirrup diameter (d) is 4 mm. There is a 300 mm long plastic hinge zone at both ends of the specimen, where the stirrup spacing is 100 mm. Stirrup spacing is 200 mm outside the plastic hinge zone. The member length is 1800 mm for all.

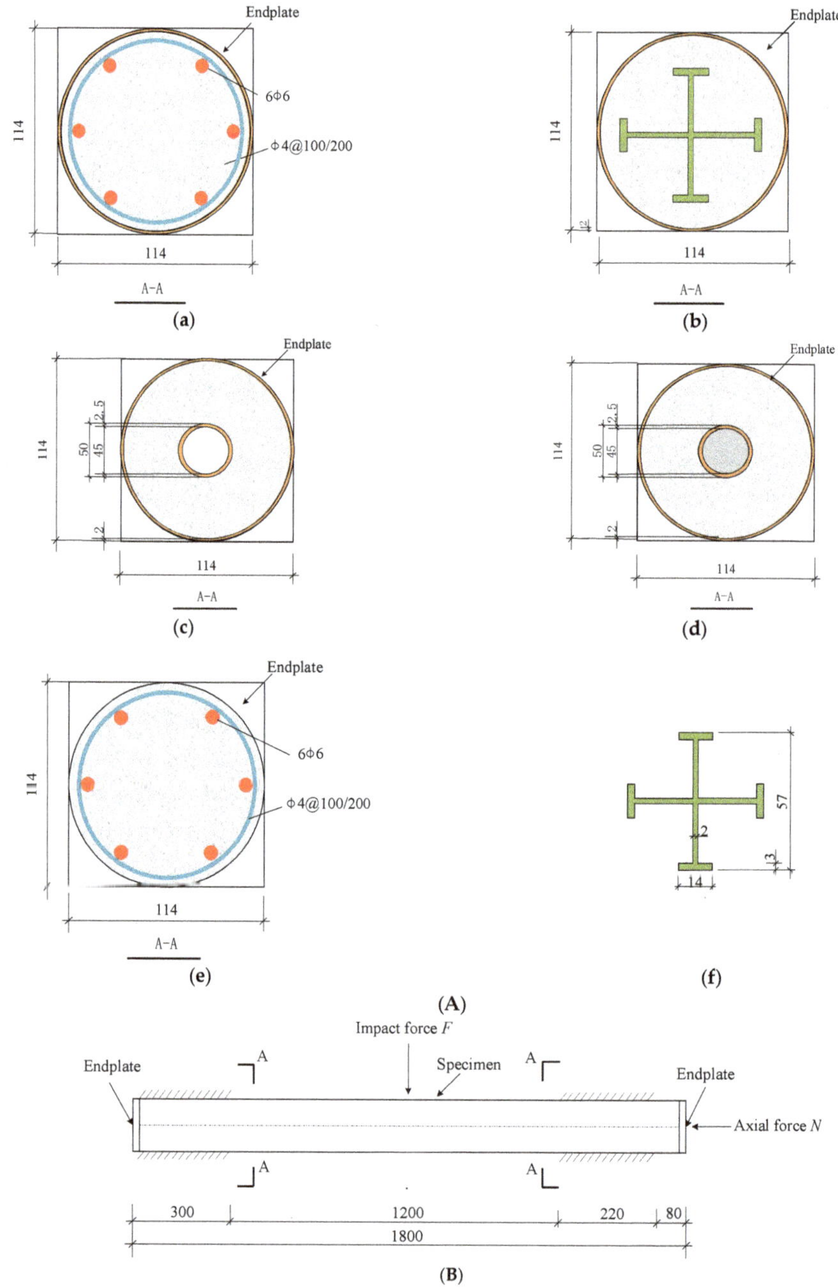

Figure 1. (**A**) Dimensions of the specimens (unit: mm). (**a**) Section of R-ST columns; (**b**) Section of S-ST columns; (**c**) Section of H-DS columns; (**d**) Section of S-DS columns; (**e**) Section of RC columns; (**f**) Section dimensions of the double H-shaped steel. (**B**) Plan view of the specimens.

Table 1. Specimen information table.

The Name of the Specimen	ID	N_0 (kN)	v_0 (m/s)	m_0 (kg)	F_{stab} (kN)		t_d (ms)		Δ_{res} (mm)	
					Test	FEA	Test	FEA	Test	FEA
Reinforced concrete (RC)	RC1	0	4.43	206.65	27.40	30.56	46.10	44.72	50.90	37.70
	RC2	59	4.43	206.65	22.45	22.09	45.45	31.52	80.88	63.93
	RC3	99	4.43	206.65	23.30	11.78	61.15	46.04	—	—
Steel tubular-encased reinforced concrete (R-ST)	R-ST1	0	9.39	206.65	135.24	129.33	23.60	21.19	56.72	60.82
	R-ST2	150	9.39	206.65	126.13	106.30	24.40	23.91	65.06	66.99
	R-ST3	250	9.39	206.65	104.70	97.66	27.55	24.33	73.87	77.53
Steel-reinforced concrete-filled steel tube (S-ST)	S-ST1	0	9.39	206.65	131.45	109.26	23.35	27.45	63.65	59.12
	S-ST2	190	9.39	206.65	118.69	88.68	24.90	29.90	75.51	69.22
	S-ST3	315	9.39	206.65	88.97	75.26	31.15	33.75	89.86	83.57
Hollow concrete-filled double steel tube (H-DS)	H-DS1	0	9.39	206.65	144.22	122.80	22.30	23.20	52.14	53.55
	H-DS2	170	9.39	206.65	129.50	108.11	23.40	24.25	70.05	61.67
	H-DS3	285	9.39	206.65	112.96	105.96	25.55	24.05	80.17	75.67
Solid concrete-filled double steel tube (S-DS)	S-DS1	0	9.39	206.65	125.34	112.95	21.50	24.25	57.82	51.14
	S-DS2	215	9.39	206.65	133.90	110.50	23.20	23.45	68.17	58.85
	S-DS3	355	9.39	206.65	113.27	112.80	24.65	22.85	76.37	69.54

The impact mass of the impact test is 206.65 kg, and the impact height of the hammer is 4.5 m, corresponding to an impact speed of 9.391 m/s. When the impact velocity is 9.391 m/s, the damage to the RC component is serious and has no analytical value, so the impact velocity of the RC component is set to 4.43 m/s. The test equipment is illustrated in Figure 2. During the test, the column is allowed to deform freely along the axial direction. The axial load was maintained at designed axial load ratios of 0, 0.3 and 0.5. The boundary conditions of the composite column in the test are fixed at both ends for rotations, and the effective length (L_0) of each specimen is 1200 mm (shown in Figure 1B).

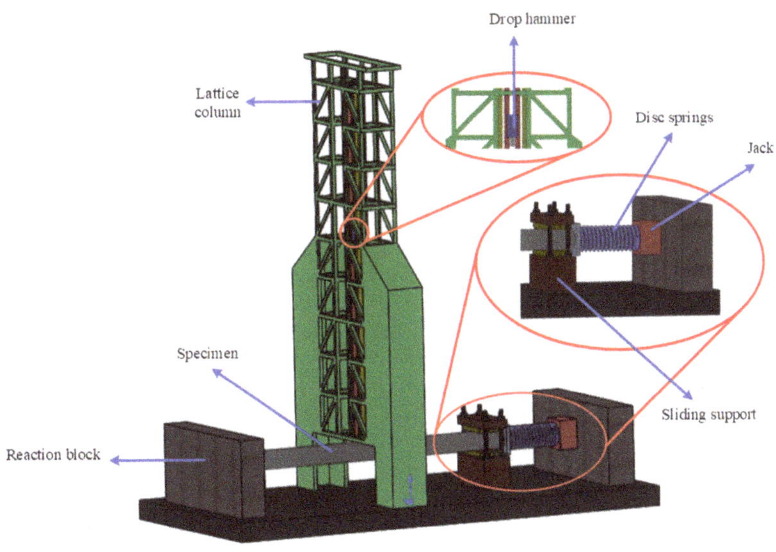

Figure 2. Test setup.

2.2. Finite Element Mesh and Contact Definition

In the finite element model, the outer and inner steel tubes and H-shaped beams are simulated with four-node SHELL163 elements, and the thickness of the steel tube has five integral points to capture the local bending of the steel tube. The filled concrete and drop hammer were modelled using eight-node SOLID164 elements. Through the study of mesh convergence, the appropriate density and mesh element are determined to ensure the simulation has sufficient accuracy and efficiency, and ensure that the hourglass energy is less than 5% of the total impact energy [28]. Figure 3 shows the typical finite element model with a minimum mesh size of 5 mm.

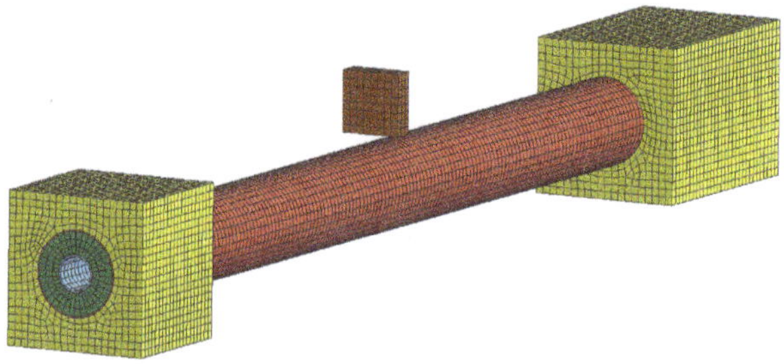

Figure 3. Schematic diagram of finite element model.

In the tests, it was observed that there is no slip between concrete and steel tubes, steel bars and H-shaped beams. Thus, the slip between steel tubes, steel bars and concrete is not considered in the simulation and the fluid–structure interaction (* CONSTRAINED LAGRANGE-IN-SOLID) method is adopted to bond the steel bars and concrete. All other contacts adopt automatic face contact (* CONTACT_AUTOMATIC _SURFACE_TO_SURFACE). The friction factor of the Coulomb friction model was taken as 0.15.

2.3. Materials Properties

Concrete is modelled by continuous cap material (* MAT-CSCM-CONCRETE), which is a cap model with a smooth intersection between the shear yield surface and the hardening cap. The initial damage surface coincides with the yield surface. Rate effects are modelled with viscoplasticity. The material model is widely used in numerical simulation due to its simplified input parameters and good simulation results. The unconfined compressive strength of a cylinder used in the simulation is 32.94 MPa, which is calculated as the average compressive strength of a concrete cube measured on the test day multiplied by 0.79 [29]; this coefficient is caused by the different calculation methods of the compressive strength of concrete in China [24] and the United States [30]. The size of the concrete aggregate is set to 5 mm.

Steel bars and steel plates are modelled by kinematic hardening plasticity with the option of including rate effects (* MAT_PLASTIC_KINEMATIC). This model is suited to model isotropic and kinematic hardening plasticity. The strain rate is accounted for using the Cowper and Symonds model [31] which scales the yield stress with a factor. It is a very cost-effective model and is available for beam (Hughes–Liu and Truss), shell and solid elements. The Young's modulus (E_s), yield strength (F_y), ultimate tensile strength of steel (F_u) and final failure strain (δ) of the material model parameters were obtained from the experimental test listed in Table 2, and the values of strain rate parameters D and p were set as 40.4 and 5, respectively [30].

Table 2. Material information sheet.

Steel Type	Diameter of Steel Bar or Thickness of Steel Plate (mm)	F_y (MPa)	F_u (MPa)	E_s (MPa)	δ
Stirrup	4	603.5	770.5	2.23	0.104
Longitudinal bar	6	529.2	681.1	2.17	0.206
Outer steel tube	2	330.1	396.7	2.21	0.192
Inner steel tube	2.5	330.4	398.4	1.75	0.235
Web of H-shaped steel	2	295.1	356.2	1.95	0.317
Flange of H-shaped steel	3	295.3	362.4	1.92	0.322

The drop hammer is modelled by rigid-body material (* MAT_RIGID). Mechanical properties of materials and degrees of freedom (excluding degrees of freedom in the direction of impact) are constrained. The impact velocity, impact mass and contact size of the drop hammer and specimen are consistent with the test.

3. Validation of the FEA Model

Figure 4 shows a comparison of predicted and observed failure modes for RC columns and composite columns. The residual deformations of the model include the overall flexural deformation, local buckling deformation of the outer steel tube and fractures. A more detailed comparison in terms of forces and displacements is presented in this section.

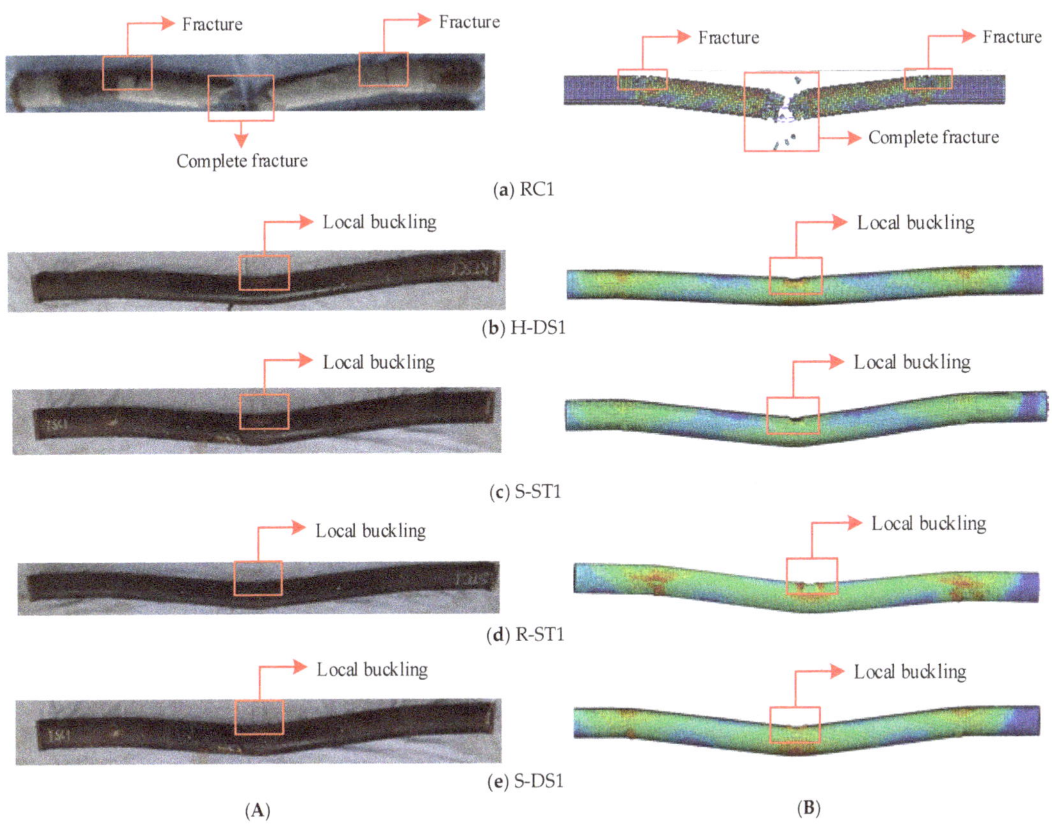

Figure 4. Comparison of observation and prediction of typical failure modes of RC columns and composite columns. (**A**) Observed in the test and (**B**) Predicted in FEA.

Figure 5 shows the comparison between the predicted and measured impact force–time histories of five composite column specimens. It can be seen that, except for the RC column, all composite sections fit well with the test results. This is because the RC column suffered severe impact damage and the CSCM concrete material model of LS-DYNA cannot accurately simulate concrete damage.

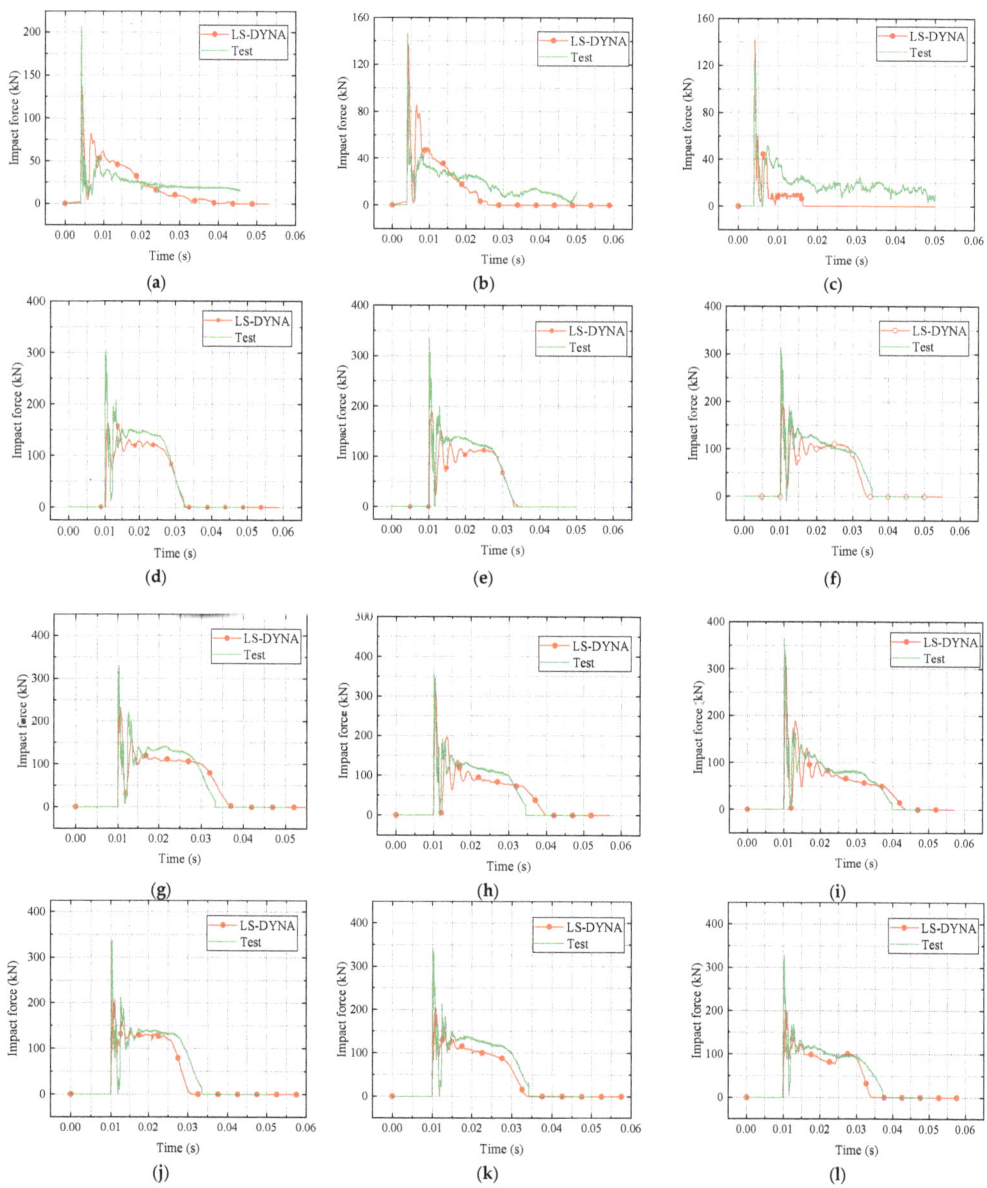

Figure 5. *Cont.*

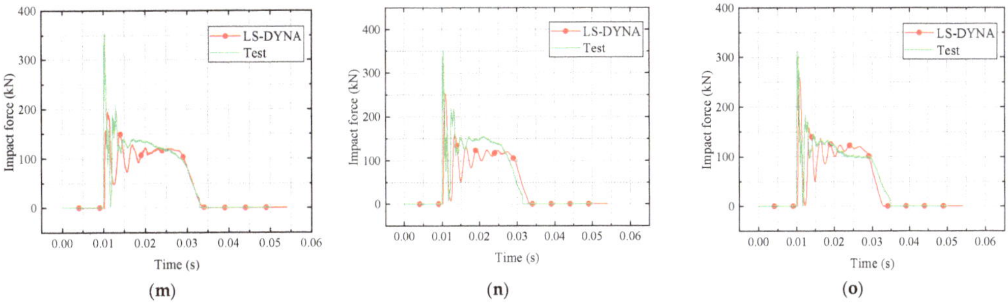

Figure 5. Impact force–time history curve. (**a**) RC; (**b**) RC0.3; (**c**) RC0.5; (**d**) H-DS; (**e**) H-DS0.3; (**f**) H-DS0.5; (**g**) S-ST; (**h**) S-ST0.3; (**i**) S-ST0.5; (**j**) R-ST; (**k**) R-ST0.3; (**l**) R-ST0.5; (**m**) S-DS; (**n**) S-DS0.3; (**o**) S-DS0.5.

Figure 6 shows the comparison of predicted and measured mid-span displacement–time histories of five column specimens with different cross sections. It can be seen that, consistent with the impact force–time history curves, the mid-span displacement of the simulated RC column deviates from the test result due to the limited capacity in simulating severe concrete damage.

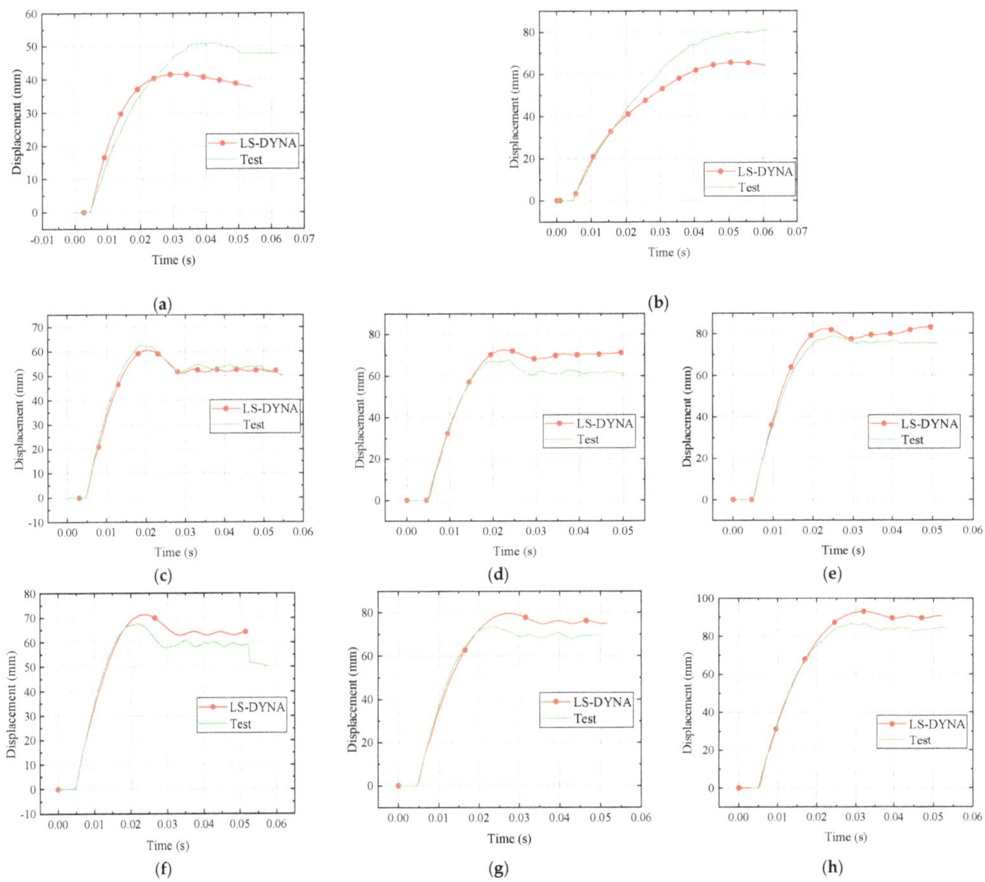

Figure 6. *Cont.*

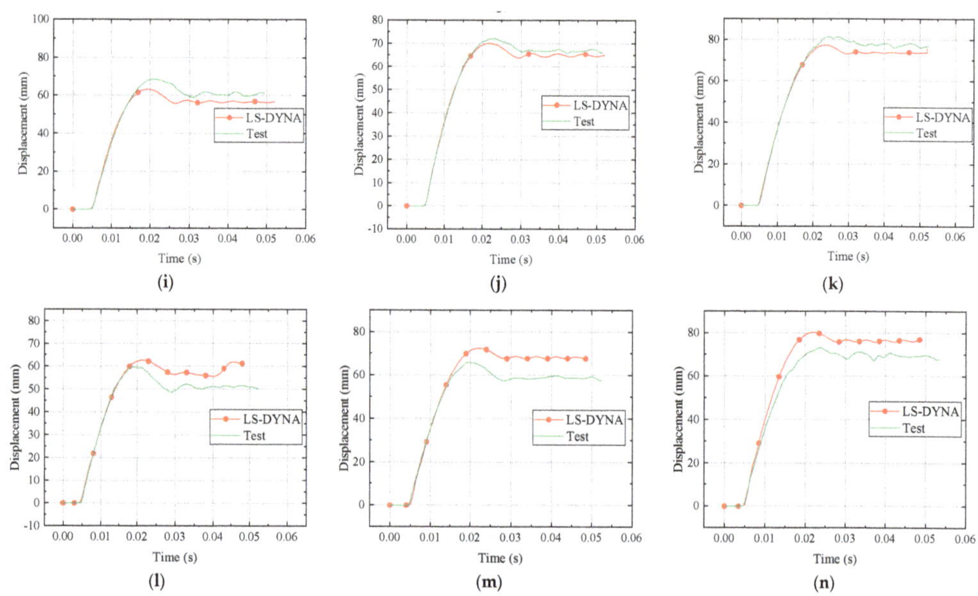

Figure 6. Mid-span deflection–time history curve. (**a**) RC; (**b**) RC0.3; (**c**) H-DS; (**d**) H-DS0.3; (**e**) H-DS0.5; (**f**) S-ST; (**g**) S-ST0.3; (**h**) S-ST0.5; (**i**) R-ST; (**j**) R-ST0.3; (**k**) R-ST0.5; (**l**) S-DS; (**m**) S-DS0.3; (**n**) S-DS0.5.

It can be seen from Figures 5 and 6 that in these composite columns, the existence of axial force reduces the impact resistance of specimens within the studied ranges. S-DS columns provide better impact-resistant performances; this is because the existence of the internal and external steel tubes enhances the effect on concrete constraints, thus enhancing the impact resistance of the component.

Figure 7 shows the difference between the numerical simulation and experimental results under various conditions. It can be seen that the difference between most numerical simulation results and experimental results is within 15%. Thus, a good agreement is generally achieved between the predictions and the test results. In Figure 7, F_{stab} is the average of the impact force in the platform stage, t is the time from impact initiation to complete unloading and δ is the global residual deformation at mid-span.

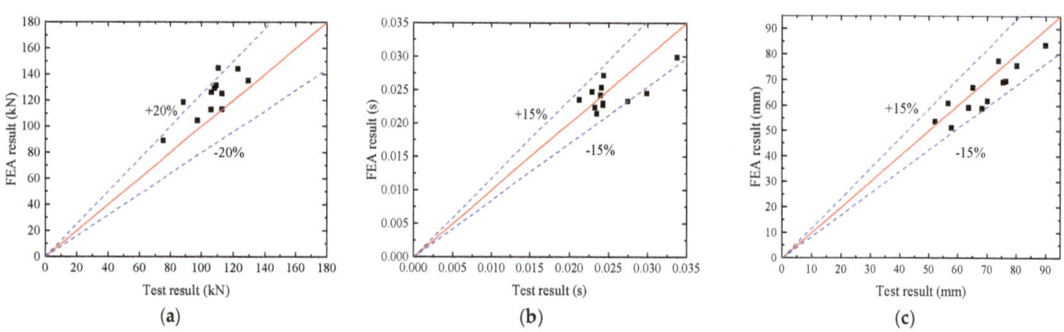

Figure 7. Comparisons between the plateau values (F_{stab}), load durations (t) of the impact forces and the mid-span deflections (Δ). (**a**) Plateau value (F_{stab}); (**b**) Load duration (t); (**c**) Mid-span deflection (Δ).

4. Impact Mechanism Analysis

4.1. The Impact Process

Figure 8 shows impact force (F)–, mid-span deflection (Δ)–, drop hammer impact velocity (V_0)– and component mid-span velocity (V_m)–time histories. In order to facilitate the study of the characteristics of the impact process, the curves are normalized to be dimensionless by F/F_{max}, Δ/Δ_{max}, V_0/V_{0max} and V_i/V_{imax}, respectively. The drop hammer velocity (V_0) and specimen velocity (V_i) are negative in the downward direction. Similarly, deflection (Δ) is defined as positive in the downward direction. Although the sections of the specimens were different and the axial forces applied were different, the impact force–time history curves of the specimens had similar stages throughout the impact, which are: peak stage, plateau stage and attenuation stage. Due to the limitation of article length, this paper only presents the normalized time history curves of each specimen without axial force. The stages are summarized as follows:

(1) Peak stage (OA): When the drop hammer touches the specimen, the impact force increases rapidly from zero to the peak value. The specimen demonstrates a downward velocity which rapidly increases to the peak value, which is equivalent to the impact velocity. At the same time, the velocity of the drop hammer decreases. During this stage, a small lateral displacement occurs in the specimen;

(2) Plateau stage (AB): After the impact force reaches ita peak, the specimen velocity decreases to a certain extent, and the velocity of the drop hammer fluctuates slightly. It is the difference in the two velocities that causes the sharp decrease in the impact force. Then, the drop hammer increases the specimen velocity again, and the corresponding impact force also increases after several fluctuations. After several fluctuations, the value remains stable. Subsequently, the drop hammer velocity stays consistent with the mid-span velocity of the specimen, and gradually decreases to zero;

(3) Attenuation stage (BC): At this stage, the drop hammer and the specimen begin to rebound, and the rebound velocity of the drop hammer is greater than that of the specimen, resulting in a decline in the impact force. When the specimen is separated from the drop hammer, the impact force drops to zero. After the specimen reaches the maximum mid-span displacement, the mid-span displacement of the specimen begins to decrease due to the rebound. With the recovery of elastic displacement, the mid-span displacement of the specimen remains stable and recovers to the residual deflection.

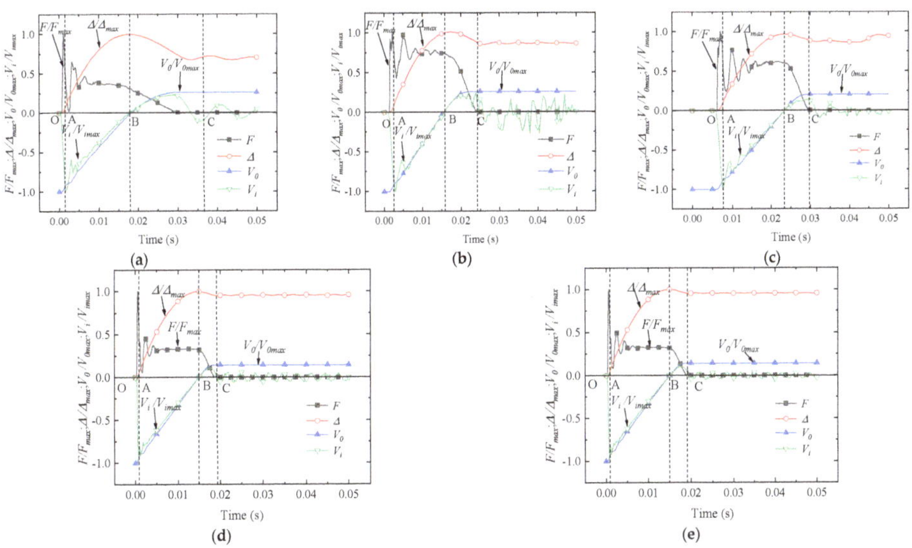

Figure 8. Normalized time history curve. (**a**) RC; (**b**) H-DS; (**c**) S-DS; (**d**) S-ST; (**e**) R-ST.

4.2. Deformation Mode

Figure 9 shows the equivalent plastic strain of RC, H-DS, S-DS, S-ST and R-ST components impacted by a drop hammer. The red area indicates high plastic strain. It can be seen from the figure that plastic deformation mainly occurs at the impact location and the support locations where there are maximum bending moments. The maximum deformation position of each component is similar, appearing in the middle and lower part of the span and the upper part of the support. The failure modes of internal and external steel tubes and concrete are similar.

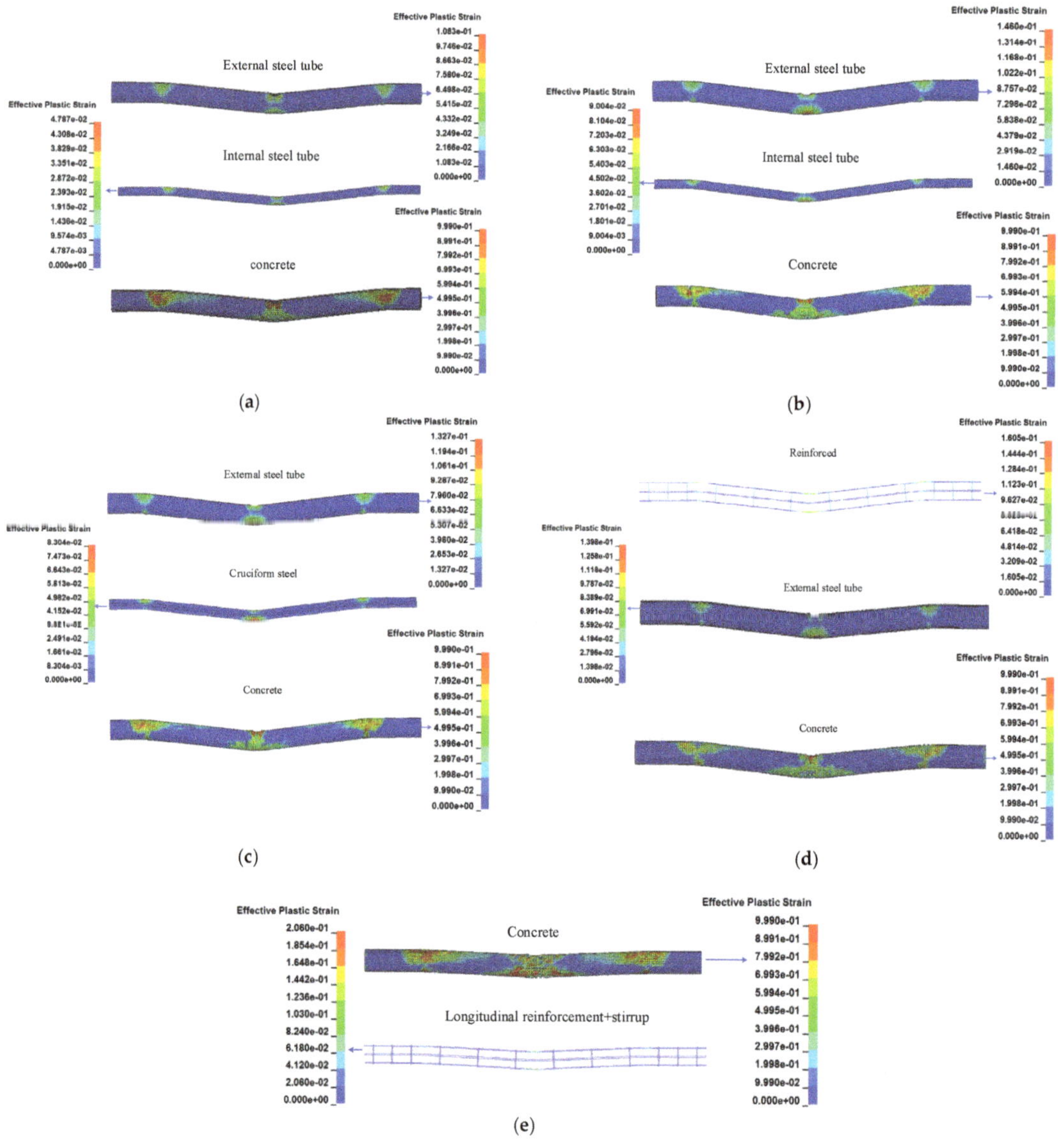

Figure 9. Plastic strain of members of different cross sections. (**a**) H-DS; (**b**) S-DS; (**c**) S-ST; (**d**) R-ST; (**e**) RC.

When comparing the different sections, the RC member has the most severe damage with excessive deformation which indicates that the impact resistance of RC columns is limited. For the H-DS member, it was found that the plastic strain of the inner steel tube is much smaller than that of the outer steel tube, which indicates that the energy absorbed by the inner steel tube is smaller than that of the outer steel tube. The maximum plastic strain of the outer steel tube is much smaller than that of the RC column, and the failure zone of concrete is also obviously smaller than that of the RC column, which indicates that the impact resistance of H-DS columns is better than that of RC columns. For S-DS columns, maximum plastic strains of outer and inner steel tubes are increased. For the S-ST member, the maximum plastic strain of the outer steel tube is greater than that of the double H-shaped steel, mainly because the double H-shaped steel is closer to the neutral axis, resulting in a slightly lower maximum plastic strain than that of the outer steel tube. For the R-ST member, the maximum plastic strain of steel bars is greatly improved compared with that of RC columns, which is mainly because the plastic deformation of the external steel tube is the main energy dissipation mechanism, making the deformation of steel bars and concrete smaller than that of RC columns.

4.3. Energy Dissipation

Figure 10 shows the internal energy–time history curves of RC, H-DS, S-DS, S-ST and R-ST components. As can be seen from Figure 10, for RC members, the energy absorbed by concrete is slightly greater than that absorbed by steel bars in the early stage of impact, mainly because at this time, concrete produces a large number of cracks due to impact and, thus, absorbs a large amount of energy. In the post-impact period, due to the concrete crack failure resulting in its main energy being absorbed by steel bars, at this time, the energy absorbed by steel bars is far greater than the energy absorbed by concrete. Due to the bending deformation of the specimen, the role of the stirrup is limited and it can hardly absorb energy. For H-DS and S-DS members, their deformation is nearly the same; the energy absorbed is similar. The main energy is absorbed by steel tubes, of which the outer steel tube absorbs most of the energy, about 4.7 kJ (52.24% of the impact energy), followed by concrete, about 2.1 kJ (22.95% of the impact energy), and the inner steel tube absorbs the least energy, about 1.0 kJ (10.93% of the impact energy). For S-ST members, most of the energy is absorbed by the external steel tube, about 5.43 kJ (59.34% of the impact energy), followed by concrete, about 2.23 kJ (24.37% of the impact energy), and the energy absorbed by double H-shaped steel is the lowest, about 1.11 kJ (12.13% of the impact energy). For R-ST members, the main energy dissipation is still absorbed by the external steel tube, about 5.43 kJ (59.34% of the impact energy), followed by concrete, about 2.19 kJ (23.99% of the impact energy), then the energy absorbed by longitudinal reinforcement, about 1.0 kJ (10.93% of the impact energy), and finally the energy absorbed by stirrups, about 2.75 J (0.003% of the impact energy).

Based on the above analysis, the amounts of energy absorbed by steel bars and concrete are roughly equal in the early stage of RC members, and when concrete cracks, steel bars become the main energy dissipation mechanism. For R-ST members and other composite columns, the plastic deformation of external steel tubes is the main energy dissipation mechanism, followed by concrete, and the energy absorption of internal steel tubes, steel bars and double H-beams is the lowest.

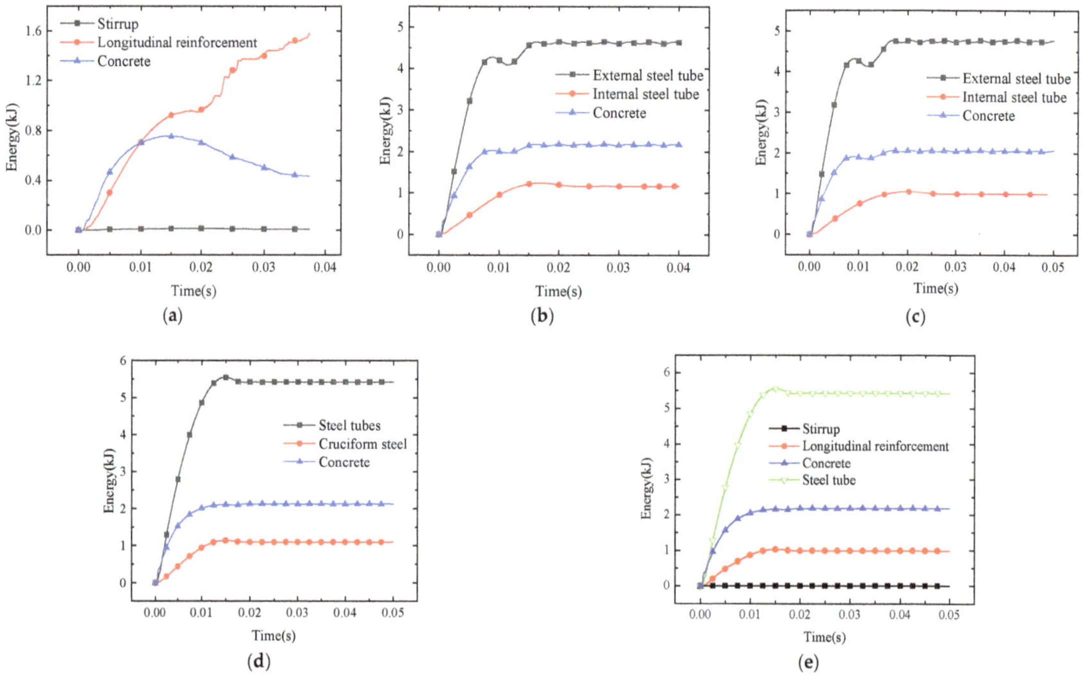

Figure 10. Energy–time history curve. (**a**) RC; (**b**) H-DS; (**c**) S-DS; (**d**) S-ST; (**e**) R-ST.

5. Parametric Sensitivity Analysis

Following the experimental test, a number of parameters are altered by a percentage to investigate the sensitivity. In the test, the mass is 206.65 kg, the velocity is 9.39 m/s, the compressive strength of concrete is 41.69 MPa, the yield strength of outer steel tubes is 330.1 MPa, the yield strength of inner steel tubes is 330.4 MPa, and the yield strength of steel bars is 529.2 MPa. The yield strength of double H-shaped steel is 295.2 MPa, the diameter–thickness ratio of outer steel tubes is 57, the diameter–thickness ratio of inner steel tubes is 20, the diameter of the longitudinal steel bar is 6 mm and the flange and web width–thickness ratios of double H-shaped steel are 4.67 and 12.25, respectively. These parameters are changed by −40%, −20%, +20% and +40%, and the corresponding changes in impact peak value, impact duration and deflection peak value are compared to conduct sensitivity analysis.

Figure 11 shows the influence of various parameters on the peak impact force, impact duration and deflection peak of H-DS members. The compressive strength of concrete, the diameter–thickness ratios of outer steel tubes and inner steel tubes and the impact velocity have a great influence on the peak impact force. With an increase in the compressive strength of concrete, the peak impact force increases. The diameter–thickness ratio of outer steel tubes and inner steel tubes is negatively correlated with the peak impact force; the greater the diameter-to-thickness ratio, the smaller the peak force. Impact velocity has the greatest influence in the decreasing region, but fluctuates in the increasing region. The impact mass and yield strength of outer and inner steel tubes have little effect on the impact peak force.

The diameter–thickness ratio, impact mass and impact velocity of the outer and inner steel tubes have a certain influence on the impact duration. With an increase in the steel tube diameter–thickness ratio, the wall thickness of steel tubes decreases, and bending stiffness and local stiffness are reduced, resulting in reduced impact force. According to the theorem of momentum, it is known that impact duration increases. The concrete strength and the yield strength of outer and inner steel tubes have the least influence on the impact duration. With an increase in the yield strength of steel tubes, the impact plateau value of H-DS columns increases, and the impact duration decreases.

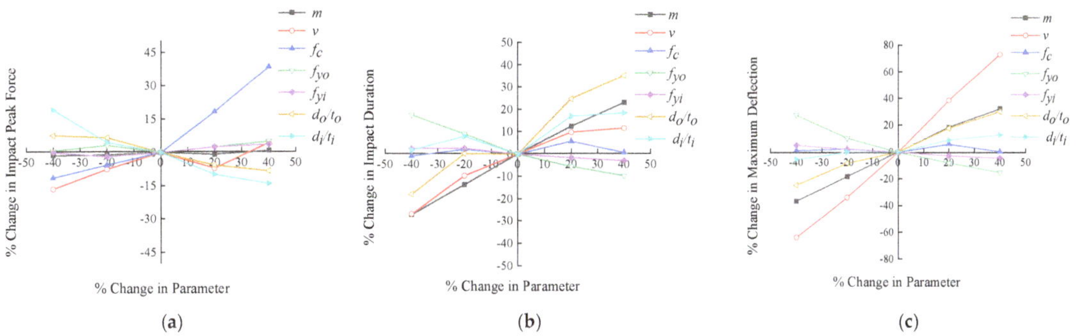

Figure 11. Sensitivity of H-DS corresponding to changes in structure-related variables. (**a**) Peak impact force; (**b**) Impact duration; (**c**) Maximum deflection.

Impact velocity has the greatest influence on the peak deflection, followed by impact mass and diameter–thickness ratio of the outer steel tube, diameter–thickness ratio of the inner steel tube, compressive strength of concrete and yield strength of outer and inner steel tubes. This is mainly because the impact energy is proportional to the quadratic power of velocity, so a larger displacement occurs. Changing the diameter–thickness ratio of the outer steel tube also has a great influence on the peak deflection, which is mainly because the bending stiffness of the structural column changes with an increase or decrease in outer steel tube wall thickness, while the inner steel tube has a small influence on the deflection due to the small moment of inertia. Change in other parameters do not effectively affect the bending stiffness of the structure column, so the influence on impact response is small.

From Figure 12, for S-DS columns, the compressive strength and impact velocity of concrete have the greatest influence on the peak impact force; the outer steel tube, the diameter–thickness ratio of the inner steel tube and the compressive strength of concrete have the greatest influence on the impact duration; and the impact velocity, impact mass and the diameter–thickness ratio of the outer steel tube have the greatest influence on the peak deflection.

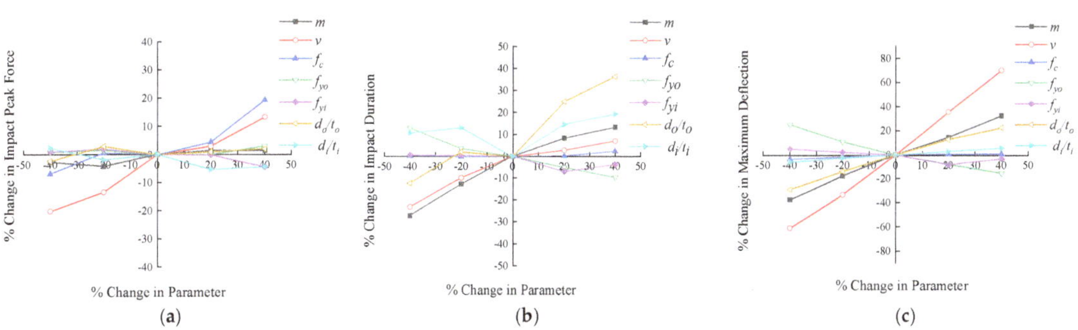

Figure 12. Sensitivity of S-DS corresponding to changes in structure-related variables. (**a**) Peak impact force; (**b**) Impact duration; (**c**) Maximum deflection.

Figure 13 shows the influence of various parameters of S-ST columns on impact performance. As can be seen from the figure, the impact velocity and diameter–thickness ratio of the outer steel tube have greater influences on peak impact value, concrete's compressive strength and double H-shaped steel have slightly greater influences, while steel's yield strength and impact mass have the least influence. This is because the peak impact force is mainly related to the impact velocity and local stiffness. The most effective way to increase the peak impact force is to increase the impact velocity and the stiffness of the external steel tube in direct contact with the drop hammer. Concrete's compressive strength also has a certain influence on the local stiffness. For impact duration, the impact mass and diameter–thickness ratio of the outer steel tube have a major influence, while the yield strength and width–thickness ratio of double H-shaped steel have the least influence; this is because bending stiffness is the most important factor affecting impact endurance. Because the H-shaped section's moment of inertia is small, it is difficult to change the bending stiffness of the section; therefore, it has little effect on impact duration.

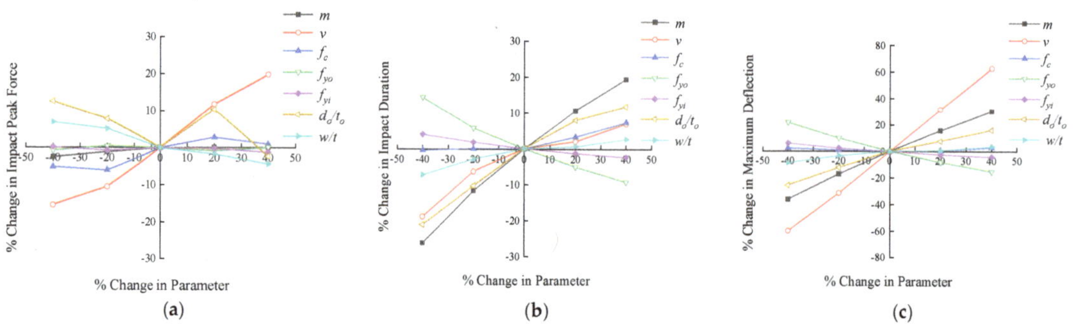

Figure 13. Sensitivity of S-ST corresponding to changes in structure-related variables. (**a**) Peak impact force; (**b**) Impact duration; (**c**) Maximum deflection.

Figure 14 shows the changes in energy dissipation of R-ST columns under various parameter changes. As can be seen from the figure, for the peak impact force, the factors that have a greater impact are the diameter–thickness ratio of the outer steel tube and the impact velocity, while other factors have a lesser impact, which is the same as the influence factors of the other three components. For impact duration, the most influential factors are the diameter–thickness ratio of the outer steel tube, impact velocity and impact mass, followed by the diameter of the steel bar, concrete's compressive strength and yield strength of steel bar. For the peak deflection value, the impact velocity and the diameter–thickness ratio of the external steel tube have the greatest influence, followed by the impact mass and the diameter of the steel bar, and the compressive strength of concrete and the yield strength of the steel bar have the least influence.

To summarize, the impact velocity and the diameter–thickness ratio of the outer steel tube have a greater influence on the energy dissipation of the structural column regardless of the cross section, while the influence of the inner steel tube and the double H-beam is the least. The impact velocity mainly increases the impact energy, resulting in greater impact force and impact displacement. The diameter–thickness ratio of the outer steel tube mainly changes the bending stiffness of the section of the structure. Because the outer steel tube is arranged at the outermost part of the section, it provides a larger moment of inertia, thus improving the bending performance of the specimen. Through the above analysis, it can be found that in terms of improving the flexural stiffness of the outer steel tube, changing the diameter–thickness ratio is obviously more effective than changing the yield strength of the component. When the compressive strength of concrete is increased, the interaction with the outer steel tube is strengthened and buckling of the outer steel tube can be prevented. Because the inner steel tube and the double H-shaped steel are closer to the neutral axis, the deformation is smaller than that of the outer steel tube, so the impact deformation cannot be completely deformed, and the impact on the bending stiffness of the section is less. Therefore, the impact energy dissipation parameters of the specimen are not sensitive to the parameters related to the inner steel tube being near the neutral axis and the double H-beam steel.

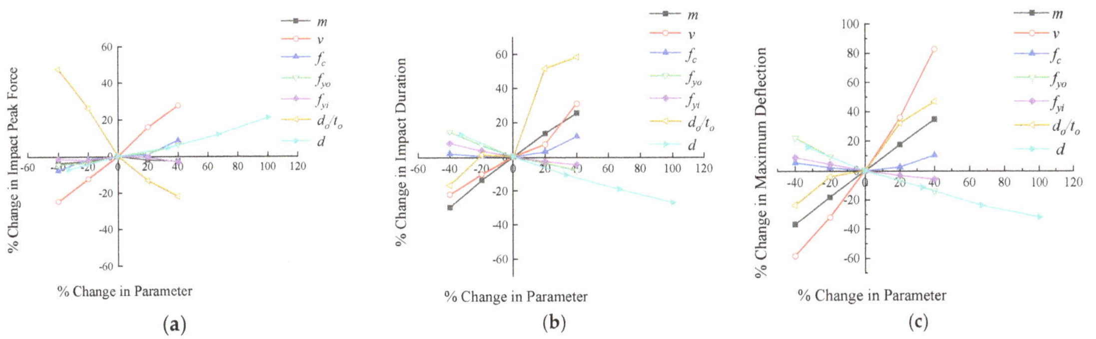

Figure 14. Sensitivity of R-ST corresponding changes to change in structure-related variables. (**a**) Peak impact force; (**b**) Impact duration; (**c**) Maximum deflection.

6. Conclusions

In this paper, the impact resistance of reinforced concrete columns and composite columns is studied on the basis of numerical simulation. Within the scope of parameter research in this paper, the following conclusions are drawn:

(1) The impact resistance of composite columns is better than that of RC columns. The axial force weakens the impact resistance of the composite column regardless of its cross-sectional form;

(2) In all composite columns, the impact process is similar, and the impact force–time curve shows three stages: peak stage, plateau stage and attenuation stage. Among them, S-DS columns have the best impact resistance, showing a higher impact platform value and a smaller peak deflection under an impact with the same energy;

(3) The simulation results of LS-DYNA fit well with the experimental results. The software is used to analyze the impact process of composite columns, and it is found that the outer steel tube is the main energy dissipation component of these components, which can absorb 50%–60% of impact energy, followed by concrete (about 24% of impact energy), the inner steel tube and the double H-shaped steel (about 12% of impact energy);

(4) LS-DYNA nonlinear finite element software was used to analyze the sensitivity of composite columns. Although the cross-sectional forms of different composite columns were different, the sensitivities of peak impact force, impact duration and peak deflection to parameter changes were similar. The impact velocity and diameter–thickness ratio of the outer steel tube have a great influence on the energy dissipation of the structural column, while the influence of the inner steel tube and the double H-shaped steel is the least, and the influence of concrete is in the middle;

(5) To increase the impact resistance, the best solution is to increase the thickness of the outer steel tube, followed by the yield strength of the outer steel tube and the compressive strength of concrete, and the least economic solution is to improve the impact resistance of components through the mechanical parameters of the inner steel tube.

Author Contributions: Conceptualization, X.L. and X.Z.; methodology, T.L.; software, Y.Y.; validation, T.L., X.Z. and R.W.; formal analysis, X.L.; investigation, X.Z.; resources, R.W.; data curation, X.L.; writing—original draft preparation, X.L.; writing—review and editing, X.Z.; visualization, Y.Y.; supervision, X.Z.; project administration, X.Z.; funding acquisition, X.Z. All authors have read and agreed to the published version of the manuscript.

Funding: This work was financially supported by the Fundamental Research Program of Shanxi Province (202103021223022), the Natural Science Young Foundation of Shanxi Province (201901D211178), the National Natural Science Foundation of China (51578274) and the Scientific and Technological Innovation Programs of Higher Education Institutions in Shanxi (2020L0052).

Institutional Review Board Statement: Not applicable.

Informed Consent Statement: Not applicable.

Data Availability Statement: Data is unavailable due to privacy.

Acknowledgments: The authors thank Qi Zhang from University of British Columbia for the assistance of proofreading.

Conflicts of Interest: The authors declare no conflict of interest.

References

1. Xu, L.; Lu, X.-Z.; Smith, S.; He, S.-T. Scaled model test for collision between over-height truck and bridge superstructure. *Int. J. Impact Eng.* **2012**, *49*, 31–42. [CrossRef]
2. Xu, L.; Lu, X.-Z.; Guan, H.; Zhang, Y.-Z. Finite-element and simplified models for collision simulation between over-height trucks and bridge superstructures. *J. Bridge Eng.* **2013**, *18*, 1140–1151. [CrossRef]
3. Zhu, X.; Lu, X.; Cheng, Q.; Li, Y. Simulation of the running attitude of a train after derailment. *Int. J. Crashworthiness* **2020**, *25*, 213–219. [CrossRef]
4. Lu, X.; Lin, K.; Cen, S.; Xu, Z.; Lin, L. Comparing different fidelity models for the impact analysis of large commercial aircrafts on a containment building. *Eng. Fail. Anal.* **2015**, *57*, 254–269. [CrossRef]
5. Sha, Y.; Amdahl, J.; Dørum, C. Dynamic responses of a floating bridge subjected to ship collision load on bridge girders. *Procedia Eng.* **2017**, *199*, 2506–2513. [CrossRef]
6. Zhao, X.-L.; Tong, L.-W.; Wang, X.-Y. CFDST stub columns subjected to large deformation axial loading. *Eng. Struct.* **2010**, *32*, 692–703. [CrossRef]
7. Zhu, M.; Liu, J.; Wang, Q.; Feng, X. Experimental research on square steel tubular columns filled with steel-reinforced self-consolidating high-strength concrete under axial load. *Eng. Struct.* **2010**, *32*, 2278–2286. [CrossRef]
8. Xian, W.; Wang, W.; Wang, R.; Chen, W.; Hao, H. Dynamic response of steel-reinforced concrete-filled circular steel tubular members under lateral impact losds. *Thin-Walled Struct.* **2020**, *151*, 106736. [CrossRef]
9. Aghdamy, S.; Thambiratnam, D.P.; Dhanasekar, M. Experimental Investigation on Lateral Impact Response of Concrete-Filled Double-Skin Tube Columns Using Horizontal-Impact-Testing System. *Exp. Mech.* **2016**, *56*, 1133–1153. [CrossRef]
10. Aghdamy, S.; Thambiratnam, D.; Dhanasekar, M.; Saiedi, S. Effects of load-related parameters on the response of concrete-filled double-skin steel tube columns subjected to lateral impact. *J. Constr. Steel Res.* **2017**, *138*, 642–662. [CrossRef]
11. Zhu, X.; Zhang, Q.; Zhang, D.; Du, Y.; Zhang, Q. Experimental and numerical study on the dynamic response of steel- reinforced concrete composite members under lateral impact. *Thin-Walled Struct.* **2021**, *169*, 108477. [CrossRef]
12. Zhu, X.; Kang, M.; Fei, Y.; Zhang, Q.; Wang, R. Impact behavior of concrete-filled steel tube with cruciform reinforcing steel under lateral impact load. *Eng. Struct.* **2021**, *247*, 113104. [CrossRef]
13. Wang, R.; Han, L.-H.; Hou, C.-C. Behavior of concrete filled steel tubular (CFST) members under lateral impact: Experiment and FEA model. *J. Constr. Steel Res.* **2013**, *80*, 188–201. [CrossRef]
14. Xu, J.; Demartino, C.; Shan, B.; Heo, Y.; Xiao, Y. Experimental investigation on performance of cantilever CFRP-wrapped circular RC columns under lateral low-velocity impact. *Compos. Struct.* **2020**, *242*, 112143. [CrossRef]
15. Wang, B.; Zhu, H.; Wu, X.; Zhang, N.; Yan, B. Numerical investigation on low-velocity impact response of CFRP wraps in presence of concrete substrate. *Compos. Struct.* **2019**, *231*, 111541. [CrossRef]
16. Han, L.-H.; Hou, C.-C.; Zhao, X.-L.; Rasmussen, K.J. Behaviour of high-strength concrete filled steel tubes under transverse impact loading. *J. Constr. Steel Res.* **2014**, *92*, 25–39. [CrossRef]
17. Hu, C.-M.; Han, L.-H.; Hou, C.-C. Concrete-encased CFST members with circular sections under laterally low velocity impact: Analytical behaviour. *J. Constr. Steel Res.* **2018**, *146*, 135–154. [CrossRef]
18. Hu, C.; Han, L. Experimental behaviour of circular concrete-encased concrete-filled steel tubes under lateral impact. *China Civ. Eng. J.* **2016**, *49*, 11–17. (In Chinese)
19. Wang, R.; Han, L.-H.; Zhao, X.-L.; Rasmussen, K.J. Experimental behavior of concrete filled double steel tubular (CFDST) members under low velocity drop weight impact. *Thin-Walled Struct.* **2015**, *97*, 279–295. [CrossRef]
20. Wang, R.; Han, L.-H.; Zhao, X.-L.; Rasmussen, K.J. Analytical behavior of concrete filled double steel tubular (CFDST) members under lateral impact. *Thin-Walled Struct.* **2016**, *101*, 129–140. [CrossRef]
21. Goldston, M.; Remennikov, A.; Saleh, Z.; Sheikh, M.N. Experimental investigations on the behavior of GFRP bar reinforced HSC and UHSC beams under static and impact loading. *Structures* **2019**, *22*, 109–123. [CrossRef]
22. Swesi, A.O.; Cotsovos, D.M.; Val, D.V. Effect of CFRP strengthening on response of RC columns to lateral static and impact loads. *Compos. Struct.* **2022**, *287*, 115356. [CrossRef]

23. Zhu, X.; Zhao, P.; Tian, Y.; Wang, R. Experimental study of RC columns and composite columns under low-velocity impact. *Thin-Walled Struct.* **2020**, *160*, 107374. [CrossRef]
24. *GB50010-2010;* Code for Design of Concrete Structures. China Architecture and Building Press: Beijing, China, 2015. (In Chinese)
25. *GB50936-2014;* Technical Code for Concrete Filled Steel Tubular Structures. China Architecture and Building Press: Beijing, China, 2014. (In Chinese)
26. Tao, Z.; Han, L.-H. Behaviour of concrete-filled double skin rectangular steel tubular beam–columns. *J. Constr. Steel Res.* **2005**, *62*, 631–646. [CrossRef]
27. Zhang, C.; Yin, Y.; Zhou, Y. Experiment study on the axial load-bearing capacity of high-strength concrete-filled double tube elements. *J. Guangzhou Univ.* **2004**, *3*, 61–65. (In Chinese)
28. Hu, R.-X.; Kang, S.-T. *Case Guidance for ANSYS 18.0/LS-DYNA Nonlinear Finite Element Analysis;* Machinery Industry Press: Beijing, China, 2018. (In Chinese)
29. Li, M.; Zong, Z.; Hao, H.; Zhang, X.; Lin, J.; Liao, Y. Post-blast performance and residual capacity of CFDST columns subjected to contact explosions. *J. Constr. Steel Res.* **2020**, *167*, 105960. [CrossRef]
30. *ACI 318-2011;* Building Code Requirements for Structural Concrete. American Concrete Institute: Farmington Hills, MI, USA, 2011.
31. Tanimura, S.; Tsuda, T.; Abe, A.; Hayashi, H.; Jones, N. Comparison of rate-dependent constitutive models with experimental data. *Int. J. Impact Eng.* **2014**, *69*, 104–113. [CrossRef]

coatings

MDPI

Article

Experimental Study on Shear Performance of Post-Tensioning Prestressed Concrete Beams with Locally Corroded Steel Strands

Rihua Yang [1,2], Yiming Yang [1,*], Peng Liu [3,*] and Xinzhong Wang [1]

1 School of Civil Engineering, Hunan City University, Yiyang 413000, China
2 Hunan Engineering Research Center of Structural Safety and Disaster Prevention for Urban Underground Infrastructure, Hunan City University, Yiyang 413000, China
3 School of Civil Engineering, Central South University, Changsha 410013, China
* Correspondence: yangyiming@hncu.edu.cn (Y.Y.); 2015038@csu.edu.cn (P.L.)

Abstract: To study the effect of the local corrosion of prestressed steel strands on the shear failure mode and shear bearing capacity of concrete beams, unilateral steel strands in four post-tensioning prestressed concrete (PC) beams are corroded, and the shear test of four PC beams are performed. Moreover, a simplified calculation method for the shear bearing capacity of concrete beams with diagonal steel bars is proposed considering the effect of cross-sectional reduction of steel bars, the degradation of mechanical properties, and the cross-sectional damage of concrete. Results shows that the crack propagation mode and failure mode are unrelated to the corrosion of prestressed steel bars when the shear span ratio of beam is the same. The shear capacity of the beam decreases with the increase of corrosion rate, but the decreasing rate is lower than the increasing rate of the corrosion rate. The growth rate of stirrup stress is much greater than that of load after concrete tension and compression loss cracking, and the yield of stirrup can be used as a sign of the ultimate bearing capacity of the beam. In addition, by comparing the experimental and numerical simulation results, the proposed simplified calculation method for the shear bearing capacity of concrete beams is of high accuracy.

Keywords: prestressed concrete beam; shear bearing capacity; corrosion; numerical simulation

Citation: Yang, R.; Yang, Y.; Liu, P.; Wang, X. Experimental Study on Shear Performance of Post-Tensioning Prestressed Concrete Beams with Locally Corroded Steel Strands. *Coatings* **2022**, *12*, 1356. https://doi.org/10.3390/coatings12091356

Academic Editor: Paolo Castaldo

Received: 1 August 2022
Accepted: 14 September 2022
Published: 17 September 2022

1. Introduction

Post-tensioning and pre-tensioning are the two most commonly used technologies for tensioning prestressed steel tendons in concrete structures [1]. Compared with pre-tensioning PC structures, post-tensioning PC structures have better structural performance and are more suitable for the curved arrangement of steel tendons in long-span structures [2]. As a result, post-tensioning PC structures has become the most widely used structural form in long-span bridge construction. However, with the increase of service time, various diseases such as corrosion of steel tendons, reduction of the effective area of steel tendons, and shedding of protective layers gradually appear [3–5]. This will not only lead to the degradation of structural safety performance but also reduce the normal service performance of the bridge [6,7]. Compared with bending failure, the shear failure of structure is mostly brittle, and there are almost no predictable symptoms. This means that the damage caused by the shear failure of corroded PC components is more serious [8,9]. Therefore, it is very important to study the degradation mechanism of shear mechanical properties of corroded PC structure and reasonably evaluate the residual shear capacity.

At present, some scholars have carried out research on the shear performance of corroded concrete structures, but they mainly focus on ordinary reinforced concrete (RC) specimens. For example, based on the bearing capacity test results of corroded reinforced concrete specimens, Ye [10] found that the structural shear strength decreased gradually with the increase of the corrosion degree of stirrups. Zhang [11] studied the corrosion-induced spalling of cover concrete and its effects on shear behavior of RC beams, and the results showed that both the concrete damage and bond degradation do not significantly affect the shear resistance of reinforced concrete beams at a small degree of corrosion. Based on the experimental and numerical analysis results, Huang [12] believed that the shear resistance of corroded RC beams in terms of initial stiffness, cracking load, ultimate bearing capacity, post-peak behavior and energy dissipation capacity is significantly degraded, and this degradation becomes more obvious with the increase of corrosion degree. Lu [13] studied the influence of different concrete strength grades and stirrup arrangement on the shear strength of RC structure, and found that the shear strength of structures is more sensitive to the change of stirrup yield strength. Compared with the RC structure, the stress characteristics of the PC structure and the elements that provide shear resistance inside the structure are significantly different, which leads to a more complex variation law of the shear resistance of the PC structure. Therefore, it is necessary to study the shear resistance of corroded PC structures.

To reasonably evaluate the shear performance of concrete structures, some scholars have proposed a calculation model of the shear capacity of corroded concrete members based on the experimental and theoretical analysis results. By refining the shear design model provided by American Concrete Institute (ACI), Soltani [14] proposed a model to estimate the reduced shear capacity for RC beams. Based on the modified compression field theory, Spinella [15] proposed a calculation model of the ultimate shear force of reinforced concrete beams with corroded stirrups. El-Sayed [16] presented a proposed design procedure for quantifying the residual shear strength of RC beams with corrosion-damaged stirrups. Da [17] developed a calculation model for the ultimate shear capacity of reinforced coral aggregate concrete beam by considering the influence of high-strength coral aggregate concrete and stirrup corrosion. Although the above models provide meaningful references for reasonably evaluating the degradation of shear capacity of corroded concrete structures, most of these models are proposed based on the assumption of uniform corrosion. In service concrete structures, the spatial variability of material properties, structural dimensions, and environmental parameters as well as grouting defects lead to uneven corrosion of steel strands, which makes the existing shear capacity calculation model no longer applicable. Therefore, it is necessary to further propose a shear capacity model that considers the effect of local corrosion of steel strands.

The main purpose of this paper is to study the change law of the shear capacity of PC structure considering the effect of local corrosion and to propose the corresponding theoretical calculation model. First, the crack development, failure mode, deflection change and ultimate bearing capacity development law are analyzed based on local corrosion test of PC member bearing capacity. Second, a simplified calculation model of shear capacity is proposed considering the effect of local corrosion of steel strands. Finally, the shear resistance of locally corroded PC structures is studied based on numerical simulations. Among them, studying the effect of local corrosion on the shear capacity of PC structures and proposing a calculation method of shear performance of PC structures considering the effect of local corrosion are the novelties of this paper.

2. Test Profile

A total of four post-tensioning PC beams, numbered as B1, B2, B3, and B4, are designed in this experiment. The test beams are 3000 mm in length and 160 mm × 450 mm in cross-sectional dimensions. The thickness of concrete cover is 30 mm. Prestressed tendons are made of seven wire twisted steel strands with a nominal diameter of 15.2 mm, and the controlled tensile stress is 1395 MPa. In addition, the 2Φ25 HRB400 reinforcing steel bars is used as ordinary tensile bars, 2Φ10 HRB400 steel bars are used as erection bars, and Φ6 HRB400 steel bars are selected as stirrups. The average measured cubic compressive strength of concrete is 42.2 MPa. The cross-sectional dimensions of test beams and the concrete layout of steel bars are shown in Figure 1, and the mechanical property parameters of ordinary and prestressed steel bars are presented in Table 1.

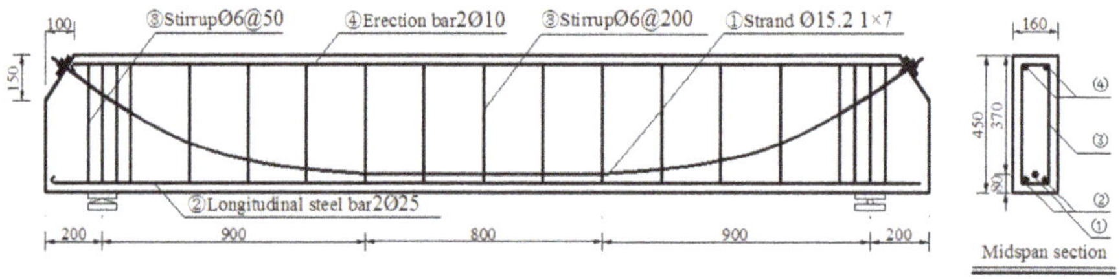

Figure 1. Dimensions and reinforcement of test beams (unit: mm).

Table 1. Mechanical parameters of steel strands and ordinary steel bars.

Type	Diameter/mm	Elastic Modulus/GPa GPAGGPa	Yield Strength/MPa	Ultimate Strength/MPa
Steel strand	15.2	195	1412.4	1876.3
Ribbed steel bar (HRB400)	6	200	412.5	561.2
Ribbed steel bar (HRB400)	10	200	403.9	551.6
Ribbed steel bar (HRB400)	25	200	424.1	574.8

Among these test beams, the B1 beam is not corroded, while the bending-shear segment at the side of B2 to B4 beams are corroded by electrochemical rapid corrosion method. In this experiment, the 5% NaCl solution is adopted. After completing the rapid corrosion, the static load tests of all beams are carried out, as shown in Figure 2. As a result, the spacing between center lines of supports for simply supported beams is 2.6 m, the pure bending segment is 1.0 m, and the shear span ratio is 2.1. Here, the multi-stage loading method is implemented through a jack. The loading level of the test beam before cracking is 5 kN, and the loading level after cracking is adjusted to 10 kN. The loading rate above two processes are 5 kN/min. When the actual loading value is close to the ultimate load, the loading level is adjusted to 5 kN, and the loading rate at this time is set to 2 kN/min. Here, the loading is stopped when the concrete is crushed or unstable deformation occurs, and then the structure is considered to be damaged at this time. Finally, after the static load test, the corroded steel bars inside the structure were taken out, and the corrosion rate of steel strands is determined by using the relatively accurate sectional profile method [18]. The corrosion time and corrosion rate of steel strands in different beams are presented in Table 2.

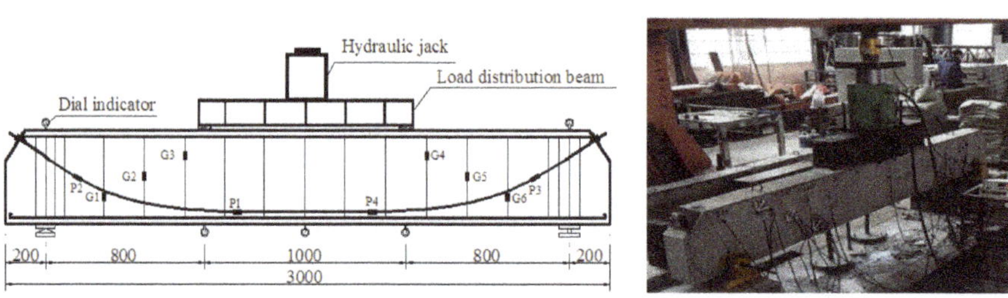

(a) (b)

Figure 2. Loading diagram and measuring point layout. (**a**) Loading and measuring point layout (unit: mm), (**b**) Loading photo.

Table 2. Corrosion time and corrosion rate η.

Beam No.	B1	B2	B3	B4
Corrosion time (d)	0	3	6	10
Area corrosion rate η (%)	0	7.9	19.4	31.7

3. Shear Performance Analysis

3.1. Crack Propagation

Although steel strands varied in corrosion rate, the crack propagation modes of test beams are basically consistent. Under the loading action, a vertical bending crack firstly appears in the pure bending segment between loading points. As the load continues to grow, multiple diagonal cracks pointing to loading points successively appear in the bending-shear segment at the corrosion treatment side, while diagonal cracks appear later at the side without corrosion treatment. Due to the difference between steel strands in corrosion rate, the cracking load of test beams is varied. The load value upon the appearance of the first bending crack and that upon the appearance of the first diagonal crack in B1–B4 are listed in Table 3. It could be seen that with the increase of corrosion rate, both the load upon the appearance of the first bending crack and that upon the appearance of the first diagonal crack in B1–B4 are gradually reduced, and both bending load and diagonal cracking load are inversely proportional to the change in corrosion rate. The more seriously the prestressed steel bars are corroded, the greater their effective cross-sectional area loss. The more serious the concrete bonding degradation, the greater the loss of the effective prestress, and the earlier the test beam cracked under the loading action. At the corrosion rate of 7.9%, 19.4%, and 31.7%, the loads leading to the initiation of bending crack and diagonal crack in test beams declines by 11.3%, 23.9%, and 40.8%, and 10%, 19% and 37.8%, respectively, in comparison with the uncorroded beam. This means that the bending crack is more sensitive to corrosion rate.

Table 3. Main test results.

No.	η/%	P_{cr1}/kN	P_{cr2}/kN	P_u/kN	θ/°	w/mm
B1	0	71	90	298	30	11.2
B2	7.9	63	81	281	36	11.7
B3	19.4	54	73	271	41	12.3
B4	31.7	42	56	252	35	12.9

Note: η% denotes the corrosion rate of steel strands in test beams, P_{cr1} stands for the load corresponding to the first bending crack, P_{cr2} represents the load corresponding to the first diagonal crack, P_u is the ultimate load, θ is the dip angle in the horizontal direction of the principal diagonal crack, and w is the midspan deflection under the ultimate load.

The crack distribution in B1–B4 under the ultimate load is given in Figure 3. For all test beams, the principal diagonal crack leading to the failure is a bending-shear crack, accompanied by avulsion phenomenon of longitudinal steel bars close to the beam bottom. The number of cracks in all test beams is within 11 to 12, the beam spacing is basically identical, the height distribution of each crack shows no obvious changes relative to the comparison beam (basically equivalent), and the crack distribution in the concrete beams is not evidently influenced by the corrosion of the prestressed steel bars.

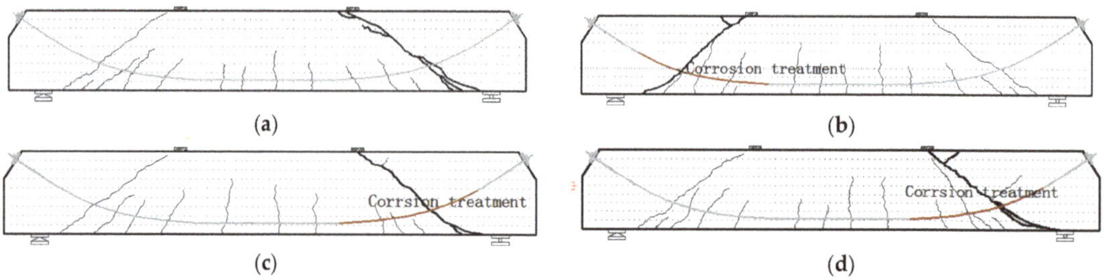

Figure 3. Crack distribution under the ultimate loading state. (**a**) B1 beam; (**b**) B2 beam; (**c**) B3 beam; (**d**) B4 beam.

3.2. Failure Mode

Despite different corrosion rates of steel strands, the failure modes of test beams are basically identical. In the initial loading stage, the pure bending segment is the first to start cracking (vertical bending tensile crack). With the increase in the load, the bending-shear segments at both sides are subjected to 3–5 diagonal cracks. As the load continued to increase, a web-shear diagonal crack pointing to the loading point firstly appears nearby the support of the bending-shear segment at the corrosion side, while the diagonal crack appears at the uncorroded side slightly later. Subsequently, the load is further increased, the vertical crack in the pure bending zone no longer developes towards the beam top, but the web-shear diagonal crack nearby the support at the corroded side developes simultaneously towards two directions and extended beneath the beam-top loading point, becoming a critical diagonal crack. The width of the principal diagonal crack increases with the load. When the ultimate load is reached and cannot be further increased, the test beam made a cracking sound and a transverse crack appears in the concrete nearby the lower part of load board at the top of the diagonal crack. As a result, the concrete is crushed and peels off. Moreover, the crack width in the zone between erection bars and steel strands at the lower end of the crack continued to grow, a horizontal avulsed crack appears on the upper surface of erection steel bars, and test beams are subjected to failures. The failure modes as shown in Figure 4. The failure modes of test beams do not go through any obvious change due to the corrosion of the steel strands, all of which experienced typical shear-compression failures.

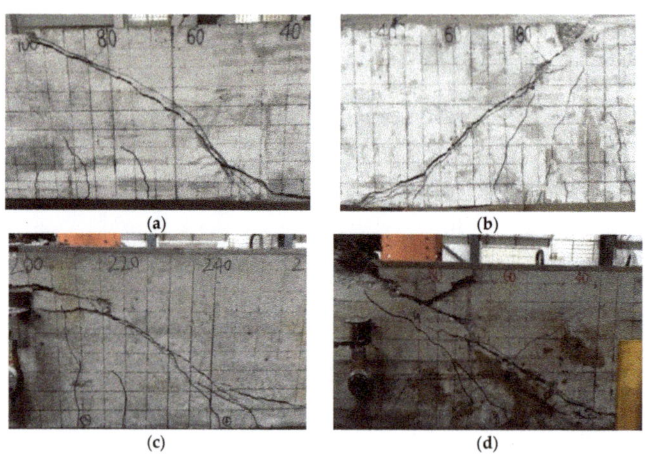

Figure 4. Failure sections of B1–B4. (**a**) B1 beam; (**b**) B2 beam; (**c**) B3 beam; (**d**) B4 beam.

3.3. Deflection Analysis

Deflection is the main index reflecting the overall stiffness of a structure. In this study, the deflections at midspan under the action of each stage of load are selected to appear in Figure 5. It appears that the curves presented indicate three-stage characteristics:

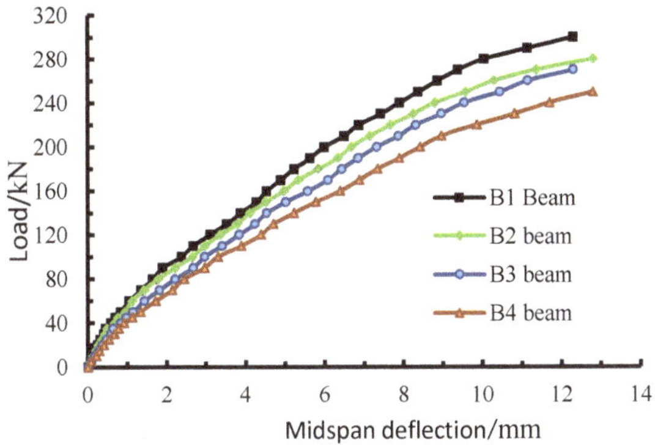

Figure 5. Load–midspan deflection curves.

In the initial loading stage, the midspan deflection of each test beam is basically linearly correlated with the load and showed elastic structural characteristics, and the deflection difference is very minor. This is because each test beam is in an elastic deformation stage, and the structural stiffness is mainly controlled by concrete. At this time, the structure is not crack yet, and the overall stiffness difference is not large. After the load reaches the cracking load, the slope of the load-deflection curve of each test beam declines. The growth rate of deflection is higher than that of the load, and the structure is inelastic, indicating that the structural stiffness declines due to cracking. However, the stiffness decrements of different test beams are inconsistent. The load–deflection curves of B1 and B2 are basically overlapped in this stage, but the slope of the load–deflection curves of B3 and B3 is apparently smaller than that of B1. This indicates that the decline degree of structural stiffness is associated with the corrosion rate of steel strands. The reason is that due to the corrosion of steel strands, the structure cracked in advance in comparison with B1. In this case, the concrete with a cracking cross section exit from the work, and thus the concrete stress is redistributed. With the appearance of cracks, the bonding performance between steel strands and concrete is damaged, between which slipping occurs with the degradation of the overall stiffness.

When the load approaches the ultimate load, the structure is evidently plastic; namely, the load growth is small but the structural deflection increases rapidly. Compared with the B1 test beam, the ultimate midspan deflections of the B2–B4 beams increases by 2.7%, 8.0%, and 11.6%, respectively.

3.4. Ultimate Load

The ultimate loads of the test beams (Table 4) are significantly influenced by the corrosion of steel strands, which could result in the degradation of structural shear bearing capacity. With the increase in the corrosion rate, moreover, the structural shear bearing capacity is degraded more significantly. In this test, the shear bearing capacity of the test beams decreases by 5.8%, 9.1%, and 15.5%, respectively, at the steel strands' area corrosion rate of 7.9%, 19.4%, and 31.7%. All test beams experienced a brittle shear failure. The relative ultimate shear strength ω is defined as the ratio of the ultimate load of corroded beam to that of uncorroded beam. The area corrosion rate η and relative ultimate shear strength ω of test beams are listed in Table 4.

Table 4. Corrosion rate η and relative shear strength ω of test beams.

Test Beam No.	B1	B2	B3	B4
Corrosion rate/η %	0	7.9	19.4	31.7
Ultimate load (KN)	298	281	271	252
Relative ultimate shear strength ω	1	0.9430	0.9094	0.8456

The data in Table 4 are linearly fitted to obtain η-ω relations, as shown in Figure 6. Due to the corrosion of the prestressed steel bars, the shear bearing capacity of component declines, and the area corrosion rate η shows a good linear relation with the relative ultimate shear strength ω.

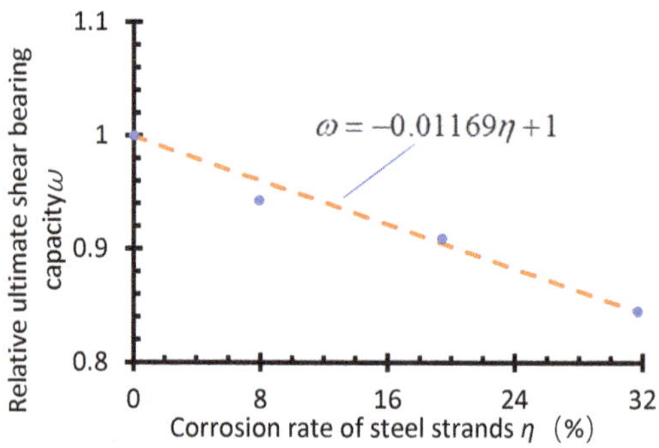

Figure 6. ω-η curves.

4. Calculation of Ultimate Shear Bearing Capacity

4.1. *Simplified Computational Analysis of Shear Bearing Capacity*

In accordance with Code for Design of Concrete Structures (GB 50010-2020) [19], the oblique cross-sectional shear bearing capacity of PC flexural members configured with stirrups and bent-up bars can be calculated through the following equations:

$$V_u = V_{cs} + V_p + 0.8 f_y A_{sb} \sin\alpha_s + 0.8 f_{py} A_{pb} \sin\alpha \quad (1)$$

$$V_{cs} = 1.75\, f_t b h_0/(\lambda + 1) + f_{yv} A_{sv} h_0/s \quad (2)$$

where V_u is the shear bearing capacity of component and V_{cs} represents the shear bearing capacity provided by stirrups and concrete. V_p denotes the shear bearing capacity provided by the prestress, which is 0, since the structure will reach the ultimate bearing capacity when cracking, and it is stipulated in the code that the contribution made by the prestress component to shear resistance should not be considered. A_{sb} and A_{pb} represent the cross-sectional areas of bent-up bars and prestressed steel bars, respectively. α_s and α_p are the included angle of the longitudinal axis of the component with bent-up bars and bent-up prestressed steel bars on the oblique section, respectively. λ is the shear span ratio of each test beam; f_t stands for the design axial tensile strength of concrete, which can be calculated as per $f_t = 0.26 f_{cu}{}^{2/3}$. f_{cu} is the cubic compressive strength of concrete, f_{yv} is the design tensile strength of stirrups, A_{sv} is the cross-sectional area of stirrups, s represents the spacing of stirrups, b and h_0 denote the rectangular cross-sectional with and without effective height, respectively.

According to the test beam design and the existing literature [20], the shear bearing capacity of test beams is expressed as

$$V_u = V_c + V_s + V'_p \quad (3)$$

$$V_c = \left(\frac{0.08}{\lambda - 0.3} + \frac{100\mu}{\lambda f_c}\right) \times f_c b h_0 \tag{4}$$

$$V_s = \frac{f_{yv} A_{sv} z}{s}(\cot\theta + \cot\alpha_s) \tag{5}$$

$$V_p' = f_{py} A_{pb} \sin\alpha_p \tag{6}$$

where V_c, V_s and V_p' represent the shear bearing capacity of concrete, stirrups and diagonal steel strands, respectively. μ denotes the longitudinal reinforcement ratio, and f_c is the compressive strength of concrete, taken as $f_c = 0.76 f_{cu}$. Z is the arm of force, taken as 0.875 h_0, θ stands for the included angle between principal diagonal crack and beam axis, α_s is the included angle between stirrups and beam axis, α_p represents the included angle between steel strands on the principal diagonal rupture plane and beam axis. It could be seen from the test results that stirrups already yielded upon a failure, so the actual yield value is taken during the stirrup calculation, and the standard value (1860 MPa) is selected for steel strands.

In Equations (3) and (4), the corrosion of steel strands is not yet considered. The design parameters of B1 are substituted into such equations to calculate the shear bearing capacity as 293.8 kN, whose error relative to the test value is −1.4%, indicating a high accuracy.

The corrosion of steel strands influences the structural shear bearing capacity mainly in two ways: On the one hand, the corrosion causes the cross-sectional damage of steel strands and reduces the strength of the steel strand. On the other hand, the corrosion led to the bonding degradation and uncoordinated strain between steel strands and concrete and stress hysteresis of prestressed steel bars. Consequently, steel strands fail to give full play to their strength, and they may not reach the standard strength upon the structural shear-compression failure. Hence, the effect of corrosion should be considered in order to correctly evaluate the shear bearing capacity of components with corroded steel strands. Given this, the concept of strength utilization rate γ_c [21] of the corroded prestressed steel bars for the PC beams is introduced to calculate the shear bearing capacity of the test beams with corroded steel strands. Through the strength utilization rate γ_c, the actual effective prestress of corroded steel strands under the ultimate load can be solved as:

$$T_{pc}' = \gamma_c(1 - 0.0981\eta) f_{pk} A_{pc} \tag{7}$$

$$A_{pc} = (1 - \eta) A_p \tag{8}$$

$$\gamma_c = \begin{cases} 1 & \xi_\eta \le 0.23 \\ 1 - k\left(2.24\xi_\eta - 1.61\xi_\eta^2 - 0.41\right) & \xi_\eta > 0.23 \end{cases} \tag{9}$$

$$k = 0.062\eta + 0.051 \tag{10}$$

$$\xi_\eta = \frac{(1 - \eta) T_e}{f_c b h_p} \tag{11}$$

where T_{pc}' is the actual prestress of corroded prestressed steel bars, f_{pk} is the standard strength of uncorroded prestressed steel bars, γ_c is the strength utilization rate; η is the area corrosion rate, A_{pc} is the actual effective area of prestressed steel bars, A_p is the cross-sectional area of uncorroded prestressed steel bars, k is the corrosion influencing factor, and h_p is the distance from the gravity center of prestressed steel bars to the concrete top.

Then, Equation (1) is corrected according to Equations (7)–(10), thereby acquiring the ultimate shear bearing capacity of test beams as follows

$$V_u' = V_c + V_s + T_{pc}' \sin\alpha_p \tag{12}$$

The calculation results for the ultimate bearing capacity of the test beams according to Equation (12) are presented in Table 5.

Table 5. Comparison between test results and theoretical calculation results of B1–B4.

No.	η (%)	θ	$\sin \alpha_p$	k	r_c	T'_{pc}/KN	V'_p/KN	Vc/KN	V_s/KN	V_u/kN	V_{u1}/kN	V_{u1}/V_u	V_{ue}/kN	V_{ue}/V_u
B1	0	30	0.271	0.051	0.9872	255.2	74.5	149.1	75.0	298	293.5	0.985	286.5	0.961
B2	7.9	36	0.221	0.056	0.9866	233.1	68.1	148.6	75.0	281	275.1	0.979	281.1	1.000
B3	19.4	41	0.210	0.063	0.9860	201.6	58.9	147.9	75.0	271	265.5	0.978	268.0	0.989
B4	31.7	35	0.223	0.071	0.9859	168.8	49.3	147.1	75.0	252	259.7	1.031	260.3	0.995

Note: η% is area corrosion rate; V_u is the test ultimate load; V_{u1} is ultimate load obtained through simplified calculation; V_{ue} is the ultimate load obtained through numerical analysis and calculation.

As seen in Table 5, the average error between test value and theoretical value of shear bearing capacity is 2.2%. The value obtained through simplified calculation as per the code coincides with the test value very well, thus verifying the reasonability of the simplified calculations.

4.2. Numerical Simulation Analysis of Shear Performance

A finite element calculation model of each damaged concrete beam is established via ABAQUS. The structure is constructed with a separated-type model. Therein, concrete is simulated using reduced integral hexahedral elements (C3D8R) and expressed by a constitutive concrete damage plasticity model. Load boards and supports are simulated using rigid elements (R3D4) to avoid stress concentration, and boundary conditions are set and loading is performed through the rigid body reference point (RP). Ordinary steel bars and prestressed steel bars are both simulated using truss elements (T3D2), and strong hardening constitutive relation models [22,23] such as double broken lines are chosen.

The bond-slip between steel strands and concrete is simulated using nonlinear spring elements (SPRING2), specifically by setting elements parallel and perpendicular to the steel bars at the co-node between concrete and steel strands. To simulate the rupture of steel strands, the elements are inactivated and activated by using the model change in the interaction module. In Stage 1, all structural elements are inactivated, and the elements in the corresponding stage are activated according to experimental steps, thus completing the simulation of the whole test process.

The calculated value and measured value of ultimate bending load of each test beam are listed in Table 5. It could be seen that the relative error between test value and the value obtained through numerical analysis and calculation was small, with the maximum error of only 3.9%, and all calculated values relatively agree with the test values.

The test value and the calculated value of midspan deformation of each beam under different load conditions are displayed in Figure 7. It could be observed from Figure 7 that the test value is relatively consistent with the calculated value, which further verifies the effectiveness of the calculation model.

To compare the concrete damage status and the stress state of stirrups under the same stage of loading action and theoretically analyze the influence of the corrosion of steel strands on structural shear bearing capacity, the load is taken as 252 kN by reference to the ultimate bearing capacity of each beam. The concrete damage and cracking conditions of beams under a tensile-compressive state as well as the stress state of stirrups are exhibited in Figures 8–10. It can be observed that in each test beam, tensile cracks are firstly generated at the bottom of the midspan pure bending segment. As the load continued to grow, and bending-shear diagonal cracks are formed in the bending-shear segment and pointed to the support along the loading point. It can be judged that the failure mode of the beam is not changed due to the increase in the corrosion rate of steel strands. That is, the shear-compression failure is a dominated factor. Relative to the comparison beam B1, the increasing corrosion rate of steel strands exerts a minor influence on the tensile damage and cracking of concrete in the pure bending segment but generates a great impact on the tensile-compression damage and cracking of concrete in the bending-shear segment. Meanwhile, as the load increases, the stress of stirrups develops at a high speed after the tensile-compressive damage and cracking of concrete. Under the action of 252 kN, the maximum stress of comparison beam B1, B2, B3, and B4 are 197 MPa, 200 MPa, 342 MPa, and 476 MPa, respectively. The greater the corrosion rate of steel strands, the earlier the stirrups would enter a yield state. The stress growth rate of stirrups is much greater than the load increasing rate, and meanwhile, the diagonal crack induced by the tensile-compressive damage of concrete already run through the whole beam body. By comparing the finite element simulation results (Figures 8 and 9) with the experimental results (Figure 3), we can find that the crack development laws of the two cases are consistent. This further proves the effectiveness of the calculation model proposed in this paper. Therefore, the yield of stirrups can serve as a sign for beams to reach their ultimate bearing capacity in practical applications.

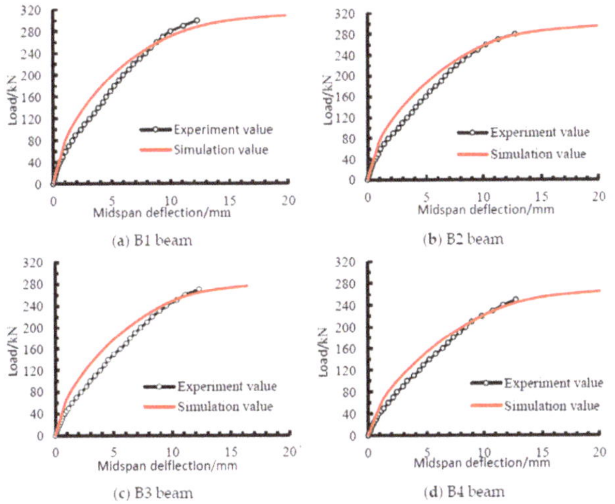

Figure 7. Comparison of midspan load–deflection curves between test results and calculation results.

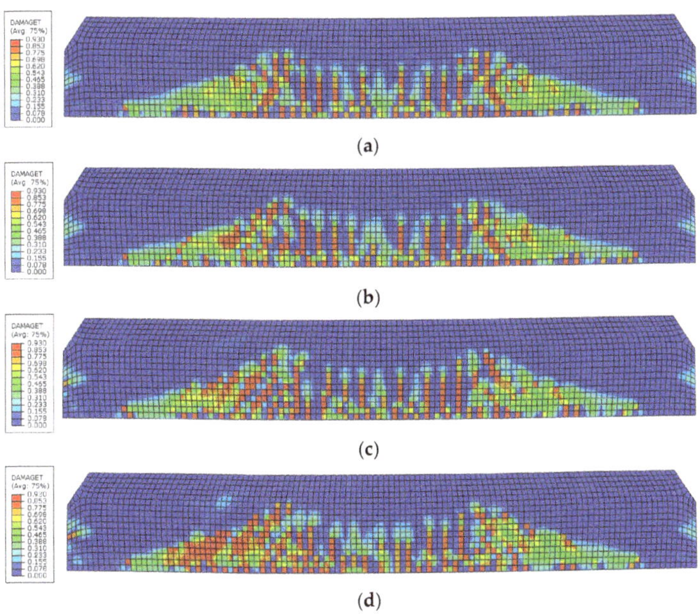

Figure 8. Tensile damage and cracking of concrete in test beams. (**a**) B1 beam; (**b**) B2 beam; (**c**) B3 beam; (**d**) B4 beam.

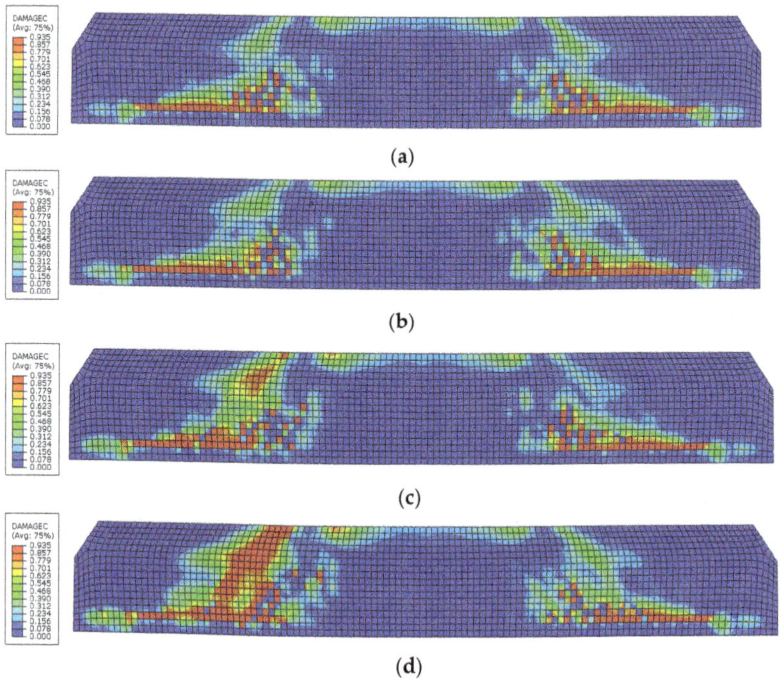

Figure 9. Compressive damage and cracking of concrete in test beams. (**a**) B1 beam; (**b**) B2 beam; (**c**) B3 beam; (**d**) B4 beam.

Figure 10. Stress diagram of stirrups. (**a**) B1 beam; (**b**) B2 beam; (**c**) B3 beam; (**d**) B4 beam.

5. Conclusions

In this paper, the crack development, failure mode, deflection change, and ultimate shear bearing capacity development law of the PC structure with locally corroded steel strands were analyzed, and the calculation model of ultimate shear bearing capacity considering the effect of local corrosion was also proposed. The conclusions are summarized as follows:

(1) The crack development mode and failure mode of PC structure are not affected by the corrosion of prestressed steel strands, but its shear bearing capacity gradually decreases with the increase of the corrosion rate of the steel strand section, and the corrosion rate of prestressed steel strand has a linear relationship with the degradation of the relative ultimate shear strength.

(2) The calculation method of shear capacity of PC structure considering the effect of local corrosion proposed in this paper has high prediction accuracy, and the average error of all prediction results in this experiment is only 2.6%.

(3) The larger the cross-section corrosion rate of the steel strand, the earlier the stirrup enters the yield state. In the service state evaluation of the corroded PC structure, the yield of stirrups can be used as a sign of the ultimate shear bearing capacity of the structure.

Author Contributions: Conceptualization, R.Y.; methodology, R.Y., X.W. and P.L.; software, R.Y.; validation, R.Y. and Y.Y.; writing—original draft preparation, R.Y. and Y.Y.; writing—review and editing, R.Y., Y.Y., X.W. and P.L.; supervision, X.W. and P.L.; funding acquisition, R.Y., Y.Y., P.L. and X.W. All authors have read and agreed to the published version of the manuscript.

Funding: This research was funded by the National Natural Science Foundation of China (grant no. 52208166), the Hunan Provincial Natural Science Foundation of China (grants no. 2021JJ50156, 2022JJ40024, and 2021JJ50153), the National Science Foundation for Distinguished Young Scholars of Hunan Province (grant no. 2022JJ10075), and the Research Foundation of Education Bureau of Hunan Province (grants no. 18A401, 21B0723, and 19C0343).

Institutional Review Board Statement: Not applicable.

Informed Consent Statement: Not applicable.

Data Availability Statement: Not applicable.

Acknowledgments: We would like to thank the anonymous reviewers and the editor for their valuable comments and remarks that helped us to improve the original manuscript.

Conflicts of Interest: The authors declare that they have no known competing financial interests or personal relationships that could have appeared to influence the work reported in this paper.

References

1. Yang, Y.M.; Peng, J.X.; Liu, X.H.; Cai, C.S.; Zhang, J.R. Probability analysis of web cracking of corroded prestressed concrete box-girder bridges considering aleatory and epistemic uncertainties. *Eng. Struct.* **2020**, *228*, 111486. [CrossRef]
2. Shen, X.W.; Wang, R.H. Mechanics Effect Analysis of Pre-tensioned and Post-tensioned Pre-stressed Concrete Simply-supported Box Beams on High-speed Railway. *China Railw. Sci.* **2004**, *25*, 100–104.
3. Yang, Y.M.; Tang, H.; Wang, X.Z. Failure probability analysis of corroded RC structures considering the effect of spatial variability. *Mag. Concrete Res.* **2022**. ahead of print. [CrossRef]
4. Yang, Y.M.; Peng, J.X.; Cai, C.S.; Zhang, J.R. Improved interval evidence theory–based fuzzy AHP approach for comprehensive condition assessment of long-span PSC continuous box-girder bridges. *J. Bridge Eng.* **2019**, *24*, 04019113. [CrossRef]
5. Peng, J.X.; Yang, Y.M.; Bian, H.B.; Zhang, J.R.; Wang, L. Optimisation of maintenance strategy of deteriorating bridges considering sustainability criteria. *Struct. Infrastruct. Eng.* **2022**, *18*, 395–411. [CrossRef]
6. Ma, Y.F.; Xu, F.Y.; Wang, L.; Zhang, J.R.; Zhang, X.H. Influence of corrosion-induced cracking on structural behavior of reinforced concrete arch ribs. *Eng. Struct.* **2016**, *117*, 184–194. [CrossRef]
7. Yang, Y.M.; Peng, J.X.; Cai, C.S.; Zhou, Y.D.; Wang, L.; Zhang, J.R. Time-dependent reliability assessment of aging structures considering stochastic resistance degradation process. *Reliab. Eng. Syst. Safe* **2022**, *217*, 108105. [CrossRef]
8. Juarez, C.A.; Guevara, B.; Fajardo, G.; Castro-Borges, P. Ultimate and nominal shear strength in reinforced concrete beams deteriorated by corrosion. *Eng. Struct.* **2011**, *33*, 3189–3196. [CrossRef]
9. Zhang, X.H.; Wang, L.; Zhang, J.R. Experimental RC beams research of shear performance of corroded arranged with diagonal reinforcement. *Bridge Constr.* **2017**, *47*, 77–82.

10. Ye, Z.; Zhang, W.P.; Gu, X.L. Modeling of Shear Behavior of Reinforced Concrete Beams with Corroded Stirrups Strengthened with FRP Sheets. *J. Compos. Constr.* **2018**, *22*, 04018035. [CrossRef]
11. Zhang, X.H.; Zhang, Y.; Liu, B.; Liu, B.W.; Wu, W.P.; Yang, C.Q. Corrosion-induced spalling of concrete cover and its effects on shear strength of RC beams. *Eng. Fail. Anal.* **2021**, *127*, 105538. [CrossRef]
12. Huang, L.; Ye, H.L.; Jin, X.Y.; Jin, N.G.; Xu, Z.N. Corrosion-induced shear performance degradation of reinforced concrete beams. *Constr. Build. Mater.* **2020**, *248*, 118668. [CrossRef]
13. Lu, Z.H.; Li, H.; Li, W.; Zhao, Y.G.; Tang, Z.; Sun, Z. Shear behavior degradation and failure pattern of reinforced concrete beam with chloride-induced stirrup corrosion. *Adv. Struct. Eng.* **2019**, *22*, 2998–3010. [CrossRef]
14. Soltani, M.; Abu-Abaileh, A.; Scott-Rowe, B. Statistical Approach to Modeling Reduced Shear Capacity of Corrosion-Damaged Reinforced Concrete Beams. *Pract. Period. Struct. Des. Constr.* **2021**, *26*, 04020073. [CrossRef]
15. Spinella, N.; Colajanni, P.; Recupero, A.; Tondolo, F. Ultimate Shear of RC Beams with Corroded Stirrups and Strengthened with FRP. *Buildings* **2019**, *9*, 34. [CrossRef]
16. El-Sayed, A.K. Shear capacity assessment of reinforced concrete beams with corroded stirrups. *Constr. Build. Mater.* **2017**, *134*, 176–184. [CrossRef]
17. Da, B.; Yu, H.F.; Ma, H.Y.; Yu, B.; Wu, Z.Y.; Guo, J.B. Study on shear behavior of reinforced coral aggregate concrete beam. *Adv. Struct. Eng.* **2020**, *23*, 2388–2398. [CrossRef]
18. Zhang, X.H. *Strain Compatibility between Corroded Prestressing Strand and Concrete & Calculation Theory of Beam Capacity*; Changsha University of Science & Technology: Changsha, China, 2016; pp. 22–23.
19. *GB 50010-2020*; Code for Design of Concrete Structures. China Architecture and Building Press: Beijing, China, 2019.
20. Guo, Z.H. *Reinforced Concrete Theory*, 3rd ed.; Tsinghua University Press: Beijing, China, 2012; pp. 246–259.
21. Yang, R.H.; Dai, L.Z.; Wang, L.; Zhang, J.R. Calculation of flexural capacity of PC beams considering strength utilization of corroded prestressing tendon. *J. Cent. South Univ. (Sci. Technol.)* **2018**, *49*, 2593–2602.
22. Fang, Z.H.; Zhou, H.J.; Lai, S.Y. Choose of ABAQUS concrete stress-strain curve. *Build. Struct.* **2013**, *43*, 559–561.
23. Alfarah, B.; López-Almansa, F.; Oller, S. New methodology for calculating damage variables evolution in plastic damage model for RC structures. *Eng. Struct.* **2017**, *132*, 70–86. [CrossRef]

coatings

MDPI

Article

Test and Mesoscopic Analysis of Chloride Ion Diffusion of High-Performance-Concrete with Fly Ash and Silica Fume

Huang Tang [1], Yiming Yang [1,*], Jianxin Peng [2,*], Peng Liu [3,*] and Jianren Zhang [2]

1 Hunan Engineering Research Center of Structural Safety and Disaster Prevention for Urban Underground Infrastucture, School of Civil Engineering, Hunan City University, Yiyang 413000, China; tanghuang@hncu.edu.cn

2 School of Civil Engineering, Changsha University of Science and Technology, Changsha 410114, China; jianrenz@hotmail.com

3 School of Civil Engineering, Central South University, Changsha 410114, China

* Correspondence: yangyiming@hncu.edu.cn (Y.Y.); jianxinpeng@csust.edu.cn (J.P.); 2015038@csu.edu.cn (P.L.); Tel.: +86-0737-4628297 (Y.Y.); +86-0731-85258698 (J.P.)

Abstract: High performance concrete (HPC) is a kind of concrete with mineral admixtures, which has better resistance ability to chloride ions diffusion than ordinary concrete. In the present study, the authors carried out a chloride ion diffusion experiment for the HPC with fly ash and silica fume. The influence of the water–binder ratio (W/B), binary (Portland cement–Fly ash (PC-FS) and Portland cement–Silica fume (PC-SF) and ternary (Portland cement–Fly ash–Silica fume (PC-FA-SF)) combinations on the concrete compressive strength and chloride ion diffusion was investigated. It was found that the compressive strength of normal concrete and HPC increase with the decrease in the W/B, the ratio of W/B deceasing value to strength increasing value for normal concrete is 0.74, and for the HPC is 0.20, so the influence of the W/B on the concrete strength for normal concrete was obviously more than the HPC. The influence of the contents of the SF or FA on developing the concrete strength was limited. The concrete compressive strength of ternary combination specimens decreases with the increase in FA or SF when the content of the other mineral admixture SF or FA remained unchanged. The ternary combination was more efficient in prohibiting chlorides ingress insider the specimens than the binary combination. The mesoscopic simulation and the tested value of the chloride ion under the same depth was close, the average ratio of simulation value to tested value was 0.91. The aggregate shape and distribution also had a negligible influence on chloride diffusivity in the HPC, but the chloride ion concentration increased with the increase in aggregate size.

Keywords: high performance concrete (HPC); chloride ion diffusion; mesoscopic simulation; fly ash; silica fume

Citation: Tang, H.; Yang, Y.; Peng, J.; Liu, P.; Zhang, J. Test and Mesoscopic Analysis of Chloride Ion Diffusion of High-Performance-Concrete with Fly Ash and Silica Fume. *Coatings* **2022**, *12*, 1095. https://doi.org/10.3390/coatings12081095

Academic Editor: Andrea Nobili

Received: 6 July 2022
Accepted: 28 July 2022
Published: 2 August 2022

1. Introduction

Many scholars have studied how adding other materials such as carbon nanotubes [1], basalt fiber [2] and mineral admixtures into normal concrete can improve the mechanical performance and durability. High performance concrete (HPC) is a kind of concrete with mineral admixtures, has high amounts of cementitious material and low water–cement ratio [3]. The addition of an appropriate amount of mineral admixture in the HPC can refine and improve the pore size distribution of the hardened cement paste, reduce the number of large pores, and increase the number of small and connected ones, which can improve the density of concrete and ultimately achieve the purpose of improving the performance of the concrete [4], so the HPC has been widely used in the erosion environment for its stronger resistance to all kinds of deleterious ions than ordinary concrete, in these ions, the aggressive of chloride ion concentration are the most obvious [5], so it is meaningful to explore the chloride ion diffusion mechanism in the HPC.

Many studies have confirmed that mineral admixtures in the HPC can improve its ability to resist chloride ion penetration compared with normal concrete, and there are two main reasons for the resistance of HPC to chloride ion penetration: firstly, the permeability of concrete mixtures, which is decreased because of the existence of the mineral admixtures. Secondly, the chloride ion binding capacity of the cementitious paste, which increases with calcium silicate hydrate gel formation and increased alumina levels inducing more chloride ion to be fixed as Friedel's salt [6]. As a common admixture, many authors focused on the durability of HPC with fly ash (FA). Sengul and Tasdemir [7] investigated the 50% replaced FA and 50% replaced finely ground granulated blast furnace slag on rapid chloride permeability. Chalee and Jaturapitakkul [8] used the original and classified FA used as a partial replacement of Portland cement at different weights of binder to investigate the effect of W/B and FA fineness on chloride diffusion coefficient (D_c) of concrete under a marine environment. The author concluded that the inclusion of pozzolans is more effective than decreasing the W/B to reduce the chloride permeability of concrete. Moffatt et al. [9] presented the durability performance of concrete incorporating high volumes of FA exposed to a harsh marine environment for 19 to 24 years. The depth of chloride penetration was found to be in excess of 100 mm for the concrete specimens without FA, whereas the presence of FA significantly decreased the depth of penetration to approximately 30 and 40 mm in specimens containing either normal density or lightweight aggregate, respectively. The above studies mainly focused on the influence of simply FA content on the durability performance of HPC, and a few studies deeply investigated the chloride permeability of HPC if other mineral admixtures as partial replacements of Portland cement.

At the same time, other authors explored the chloride ion penetration behavior in the HPC with other mineral admixtures. Cabrera and Nwaubani [10] measured the effective chloride ion diffusion coefficients of Portland cement (PC), PC-15% pulverized fuel ash (PFA), PC-15% metakaolin (MK) pastes using a chloride ion diffusion cell. Both the PC–MK and the PC–PFA pastes gave lower chloride ion diffusion coefficients than the PC paste, and the former gave particularly low values. Thomas et al. [11] and Hooton et al. [12] determined the chloride ion diffusion resistance of concrete replacement of PC with high-reactivity MK (HRM). The apparent diffusion coefficients were reduced on increasing the exposure time and on decreasing the W/B and showed marked decreases with increasing HRM content. Coleman and Page [13] also determined that the MK produces a significant increase in chloride ion binding capacity of the paste when used as a partial cement replacement. Seleem et al. [6] investigated the seawater resistance of concrete incorporating silica fume (SF), ground granulated blast furnace slag (GGBS), and metakaolin (MK) as an addition to cement in binary and ternary combinations. The authors confirmed that all kinds of pozzolanic materials are efficient in reducing the permeability of concrete far below the control one. Meanwhile, silica fume is the most efficient in this regard. The ternary blend (PC-MK-SF) is superior to all other mixtures in producing impermeable concrete. Geetha et al. [14] found that copper slag was marine application which could improve compressive strength and flexural strength and reduce sorptivity and chloride ion penetrability. Harilal et al's [15] work reported a novel high performance green concrete which has an better resistance ability of chloride ion than normal concrete. Jiang et al. [16] analyzed the influence of crack self-healing degree on chloride ion transport process, finding that chloride ions not only transport vertically along the crack, but also propagate in the horizontal direction of the crack. Bachtiar et al. [1] provided information about the mechanism and the influence of prepared high-performance concrete mixed with either freshwater or seawater and made to cure in freshwater, seawater, or air. Ding et al. [17] compared the different effects of FA and slag on anti-rebar corrosion ability of high strength concrete with chloride ion. FA and SF are the most common mineral admixtures used in the HPC, few above studies focused on the chloride ion diffusion behavior for HPC contained FA and SF. Though Khan [18] investigated the chloride ion penetration and diffusion for high performance concrete pulverized fuel ash and silica fume with various W/B, environmental factors such as temperature, humidity were not considered, and the influence of cementitious material combination (FA and SF) on the chloride ion diffusion needs to be further explored.

The corrosion of steel bar induced by chloride erosion is the main factor leading to the degeneration of structure mechanical performance. In recent years, many studies have focused on the mechanical property of structure after being corroded [19–25], at the same time, many studies have investigated chloride diffusion behavior in multiple phases of normal concrete by numerical simulations in mesoscopic level [26–30]. The HPC has the high amounts of cementitious material and low water–cement ratio [31], the chloride ion diffusion process in the HPC will be different with the normal concrete. However, the mesoscopic level research for the chloride ion diffusion process of HPC is few, scholars mainly focus on the material behavior of the HPC using mesoscopic model. Guo et al. [31] proposed a two-dimensional mesoscopic lattice model of high-performance concrete which accounts for fatigue damage. Huespe et al. [32] investigated the failure mode of high-performance fiber reinforced concrete composites (HPFRC), through proposing a finite element methodology. Forquin et al. [33] developed a mesoscopic numerical model for the ultra-high performance fiber reinforced concrete to investigate the dynamic tension behavior. Feng et al. [34] investigated the mechanical compression properties of UHPC through a 3D mesoscale model. However, the mesoscopic study of the chloride ion diffusion process of HPC containing FA and SF is limited.

In order to investigate the influence of different content combinations of FA and SF on the chloride permeability on HPC, in the present study, the authors carried out a chloride ion diffusion experiment for the HPC. The influence of W/B, binary (PC-FA and PC-SF) and ternary (PC-FA-SF) combinations on the concrete compressive strength and chloride ion diffusion was investigated. The chloride ion diffusion coefficient was modified and the chloride ion diffusion model of HPC was revised based on Fick's second law considering the W/B, temperature, humidity and time. The number simulation of chloride ion diffusion at the mesocopic level for HPC with FA and SF was carried out, and the influence of aggregate shape, distribution, size of HPC on the chloride ion diffusion is investigated.

2. Experimental Programs

2.1. Detail of Specimens

All specimens were divided into two groups in the experiment. Group I was mainly designed for investigating the influence of W/B, compressive strength on the chloride ion diffusion behavior in the HPC. Group II was mainly designed for analyzing the influence of different mixing methods of mineral admixtures on the chloride ion diffusion behavior.

Twenty-four standard cubic specimens with dimension of 150 mm × 150 mm × 150 mm are made in the group I as shown in Table 1. Eight mix proportions were designed, three specimens were made for every mix proportion. The mix proportions are shown in Table 1. The HPC80 and HPC100 adopted P52.5 ordinary Portland cement, others adopted P42.5 ordinary Portland cement. The coarse aggregate adopted macadam, and the max particle size was 20 mm continuous grading. The fine aggregate adopted fine sand, and the fineness modulus was from 2.5 to 3. The water-reducing rate of the water reducer was from 20% to 25%. The slump of concrete was controlled from 18 cm to 20 cm.

There are twelve specimens in the group II as shown in Table 2. The cement variety and the content of aggregate were the same for all specimens. The specimens A-60 were the HPC without any mineral admixtures. The specimens FA-10, FA-20 and FA-30 were mixed with 10%, 20% and 30% (in weight) FA, respectively. The specimens SF-5 and SF-10 were mixed with 5% and 10% silica fume, respectively. Other specimens were mixed with both FA and SF, the contents of mineral admixtures are shown in Table 2.

Table 1. Group I specimens mix proportion.

Specimen	Water (kg/m^3)	Cement (kg/m^3)	FA (kg/m^3)	SF (kg/m^3)	Fine Aggregate (kg/m^3)	Coarse Aggregate (kg/m^3)	Super Plasticizer (kg/m^3)	FA Ratio	SF Ratio	W/B
NC50 [1]	136.35	396	0	0	837.4	964.25	7.13	—	—	0.34
NC60	122.27	444	0	0	791.82	1008.91	8.44	—	—	0.28
NC70	108.145	488	0	0	747.3	1052.56	9.76	—	—	0.22
HPC50 [2]	123.135	396	64	35	795	1005.86	8.91	13%	7%	0.25
HPC60	119.63	409	66	36	783.34	1017.03	9.71	13%	7%	0.23
HPC70	115.215	421	69	37	769.56	1030.22	10.54	13%	7%	0.22
HPC80	144	501	81	44	647	1096	8.76	13%	7%	0.21
HPC100	129	516	84	45	607	1137	9.03	13%	7%	0.20

Note: [1] Normal concrete. [2] High performance concrete.

Table 2. Group II specimens mix proportion.

Specimen	Water (kg/m^3)	Cement (kg/m^3)	FA (kg/m^3)	SF (kg/m^3)	Fine Aggregate (kg/m^3)	Coarse Aggregate (kg/m^3)	Super Plasticizer (kg/m^3)	FA Ratio	SF Ratio	W/B
A-60	97.6	488	0	0	745	1044.5	10.5	—	—	0.2
FA-10	97.6	439.2	48.8	0	745	1044.5	10.5	10%	—	0.2
FA-20	97.6	390.5	97.6	0	745	1044.5	10.5	20%	—	0.2
FA-30	97.6	341.3	147	0	745	1044.5	10.5	30%	—	0.2
SF-5	97.6	463.2	0	24.6	745	1044.5	10.5	—	5%	0.2
SF-10	97.6	439.5	0	49.2	745	1044.5	10.5	—	10%	0.2
SF-15	97.6	414.2	0	73.8	745	1044.5	10.5	—	15%	0.2
F10-S5	97.6	415.0	48.8	24.4	745	1044.5	10.5	10%	5%	0.2
F20-S5	97.6	366.2	97.8	24.3	745	1044.5	10.5	20%	5%	0.2
F30-S5	97.6	316.8	146.8	24.6	745	1044.5	10.5	30%	5%	0.2
F10-S10	97.6	390.3	48.9	48.8	745	1044.5	10.5	10%	10%	0.2
F20-S10	97.6	341.3	98.2	48.5	745	1044.5	10.5	20%	10%	0.2
F30-S10	97.6	292.7	146.7	48.6	745	1044.5	10.5	30%	10%	0.2

2.2. Compressive Strength Test

The specimens were made using the standard maintenance method according to the Code for design of concrete structures (GT 50010-2010) [35]. All specimens were de-molded after 24 h and then cured in a standard curing chamber at 20 ± 2 °C and 97–100% relative humidity for 28 days. After curing, the standard cube compressive strength test was carried out to obtain the cube compressive strength.

2.3. Chloride Penetration Test

As shown in Figure 1, five surfaces (②–⑥ surfaces) of the specimens were painted using epoxy resin, and the last surface (① surface) was an erosion face. The process of painting epoxy resin is shown in Figure 1. All of the specimens were immersed in a sodium chloride solution with concentration of 3.5% (in weight) for two months, as shown in Figure 2a, the environmental temperature was 25. After that, the specimens were removed from the solution pool and dried naturally for seven days. An electric drill was used to obtain the concrete powders, as shown in Figure 2b. The drilling hole, 16 mm in diameter, was located in the center of specimen surface. The concrete powders were drilled with an interval from 5 mm to 30 mm away from the exposed surface (i.e., 5 mm, 10 mm, 15 mm, 20 mm, 25 mm and 30 mm depths from the surface). The chloride concentration of the concrete powders was determined by using the Rapid Chloride Test (RCT)-500 system produced by Germann Instruments company in Copenhagen, Denmark, as shown in Figure 2c. The RCT method was verified as having a good correlation with these methods recommended by ASTM C1202 [36] and AASHTOT260-97 [37].

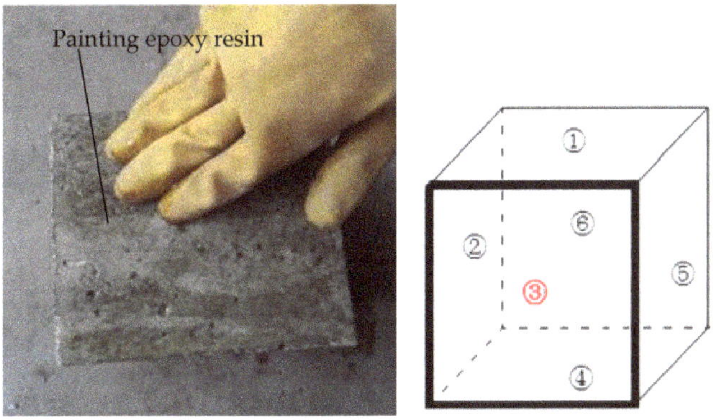

Figure 1. Process of painting epoxy resin.

(**a**) (**b**) (**c**)

Figure 2. Chloride penetration test set-ups: (**a**) specimens immersing; (**b**) drilling holes; (**c**) RCT test.

3. Experimental Result Analysis

The influence of the W/B and the mixing method of FA and SF on the compressive strength and chloride concentration of HPC is discussed through experimental results.

3.1. Influence of the W/B

3.1.1. Concrete Compressive Strength

Figure 3 shows the compressive strength of Group I specimens, the value in the bracket is the W/B. As seen in the Figure 3, no matter normal concrete or the HPC, the compressive strength of specimens increases with the decrease in the W/B. For the normal concrete, the W/B decreases 35.3% from 0.34 to 0.22, the concrete strength increases 47.6% from 47.59 MPa to 70.26 MPa. For the HPC, the W/B decreases 20% from 0.25 to 0.2, the concrete strength increases 95.9% from 54.08 MPa to 105.98 MPa. The ratio of W/B deceasing value and strength increasing value for normal concrete is 0.74, but for the HPC is 0.20, so the influence of the W/B on the concrete strength for normal concrete is obviously more than the HPC.

Sengul and Tasdemir [7], Chindaprasirt et al. [38] present that the compressive strength of concrete with mineral admixtures is smaller than the normal Portland concrete under the same W/B. This result also can be verified by the compressive strength of specimens

OC70 and HPC70. However, the compressive strength of specimens HPC 60 and HPC50 is 3.5% and 13.5% larger than specimens OC60 and OC50, respectively. This can be attributed to the smaller W/B.

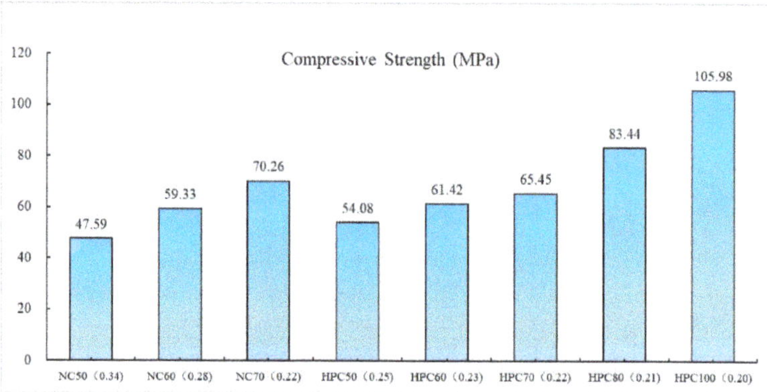

Figure 3. Compressive strength of the Group I specimens.

3.1.2. Chloride Ion Concentration (Permeability)

The W/B is an important factor impacting the chloride ion diffusion in the HPC. The chloride concentrations of part specimens were obtained as shown in Figure 4. For both the normal concrete and the HPC, from a depth of 7.5 mm to 17.5 mm, the influence of W/B on the chloride ion concentration is most significant; under the same diffusion depth, the chloride concentration decreases with the increase in W/B. The lower W/B reduces the amount of capillary pores and subsequently affects the resistance of concrete against chloride transport [6]. The concentration gradient of chloride ion is gradually decreased in the deeper regions of the concrete, and especially for depths beyond 22.5 mm, the influence of the W/B is very small.

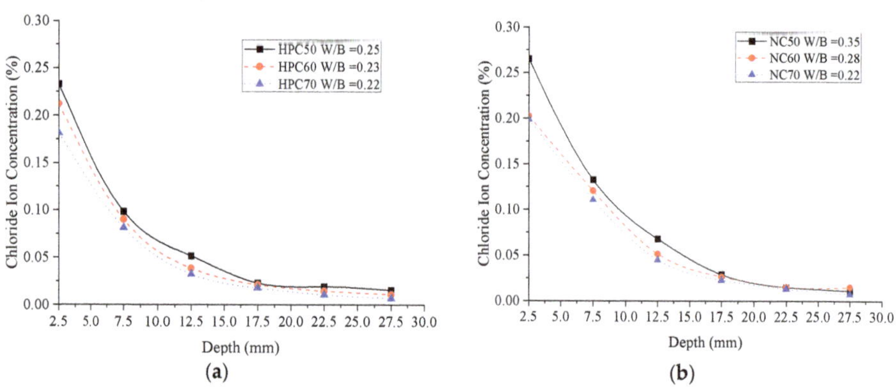

Figure 4. Chloride ion concentration of the specimens: (**a**) Normal concrete; (**b**) HPC.

Figure 5 shows the chloride ion concentration distribution of the normal concrete and HPC. From depth 0 mm to 7.5 mm, the chloride ion concentration of HPC decreases substantially compared to the normal concrete, the chloride ion concentration of specimen HPC 100 is about 16% of the specimen NC50 at depth 7.5 mm. The pozzolanic reaction is relatively slow compared to the hydration of Portland cement, which is the main reason of high early age permeability of the concretes containing pozzolanic materials. At later ages, however, the pozzolanic materials can be more effective in improving concrete properties due to the pozzolanic reaction continuing at a higher rate for a longer period, so the chloride ion concentration of the HPC decreases slower than the normal concrete from depths of 7.5 mm to 17.5 mm.

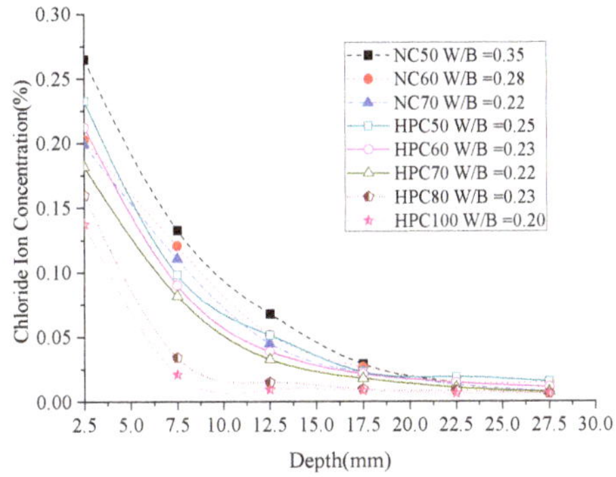

Figure 5. Chloride concentration distribution of Group I specimens.

3.2. *Influence of the Mixing Method of Mineral Admixtures*

3.2.1. Concrete Compressive Strength

The concrete compressive strengths of Group II specimens are presented in Figure 6. For binary combination specimens SF5, SF10 and SF15, the concrete compressive strength increases 11.93% when the content of the silica fume increases from 5% to 15%. However, the concrete compressive strength of specimen SF15 is 5.69% smaller than specimen SF10. The same situation appears for the compressive strength of specimens FA10, FA20 and FA30. The experimental results conclude that the contents of the SF or FA developing the strength is limited. Figure 7 shows the concrete compressive strength of specimens with different FA content obtained by Jerath and Hanson [39], which also concludes that the concrete compressive strength at the same age firstly increases, then decreases with the FA content increases.

Interestingly, it can be found from the strength of specimens F10-S5 to F30-S5 and F10-S10 to F30-S10, the compressive strength of specimens F10-S5 to F30-S5 decreases from 8.9% to 11.5% with the increase in FA when the content of the silica fume keeps unchanged. The compressive strength of specimens F10-S10 to F30-S10 decreases from 5.6% to 22.2% with the increase in SF when the content of the FA keeps unchanged.

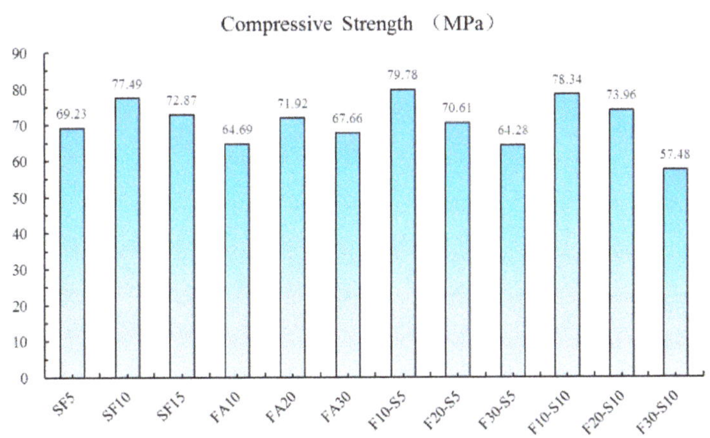

Figure 6. Compressive strength of the Group II specimens.

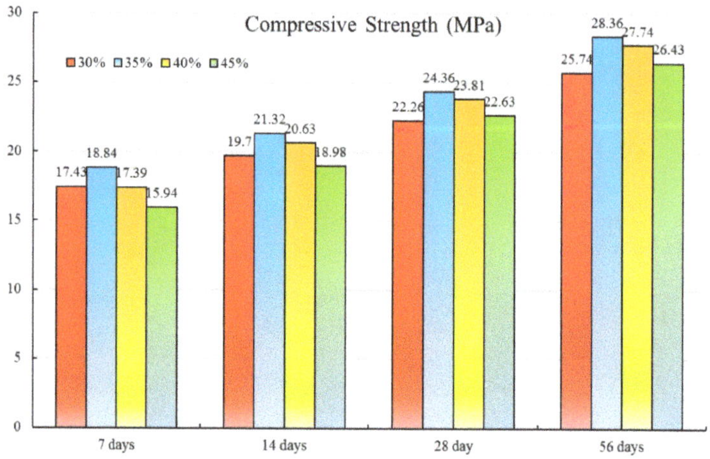

Figure 7. Compressive strength of the specimens in Jerath and Hanson's test.

3.2.2. Chloride Ion Concentration

It has been shown in laboratory and field studies that FA can mitigate corrosion by hindering chloride penetration as well as decreasing the content of free chlorides. The use of FA improves the distribution of pore size and pore shape of concrete, and the presence of C_3A in FA can absorb more chloride ions to form Friedel's salt ($C_3A{\cdot}CaCl_2{\cdot}10H_2O$) [40]. Moreover, fly ash reacts pozzolanically with calcium hydroxide to produce calcium-silicate hydrate (C-S-H), which absorbs more chloride ions and block the ingress path [9,40]. The SF is also more efficient in prohibiting chloride ingress inside the HPC as well as at the surface layer than round granulated blast furnace slag and metakaolin [6]. In this study, utilizing the specimens in Table 2, the influences of binary blends (Portland cement–fly ash (PC-FA) and Portland cement– silica fume (PC-SF)) and ternary blends (Portland cement–fly ash–silica fume (PC-FA-SF)) on the chloride permeability are investigated: the experimental results are shown in the Figures 8 and 9. In Figure 8, the chloride ion concentration of the specimens with FA or SF is smaller than normal concrete when the diffusion depth is smaller than 17.5 mm, but the chloride concentrations have no obvious difference beyond 17.5 mm. As shown in Figure 8a), the chloride ion concentration decreases with the FA increases from depth 2.5 mm to 12.5 mm. A similar situation also appears in Figure 8b: the chloride ion concentration decreases with the growth of the SF content from a depth of 2.5 mm to 7.5 mm.

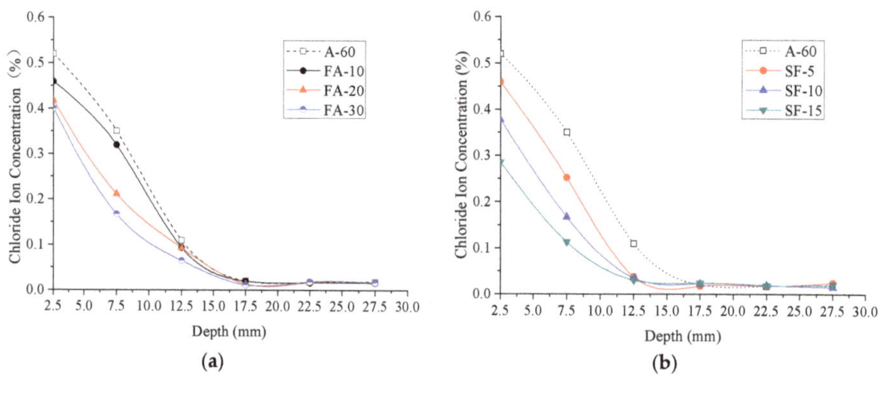

Figure 8. Chloride ion concentration of specimens with binary blends: (**a**) Portland cement–fly ash; (**b**) Portland cement–silica fume.

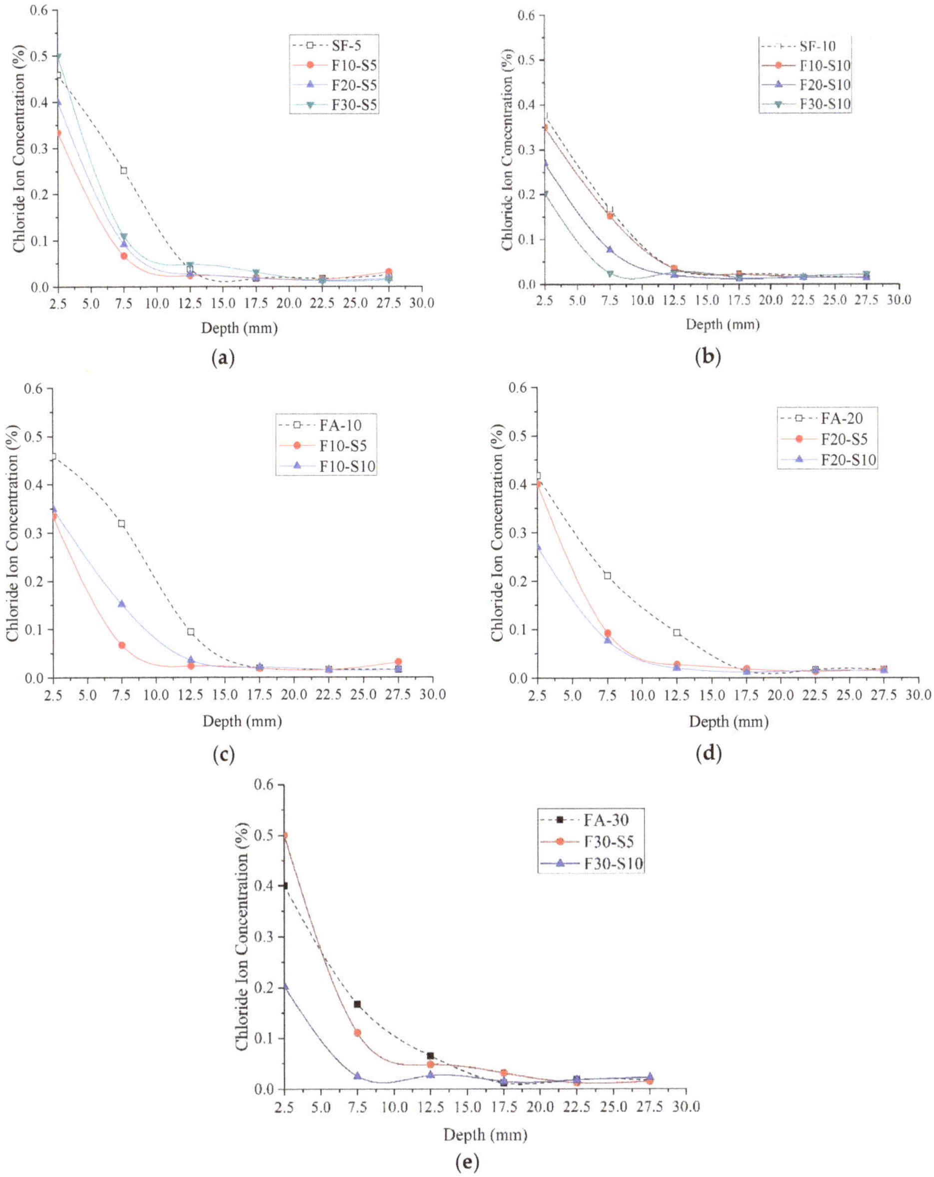

Figure 9. Chloride ion concentration of specimens with binary and ternary blends: (**a**,**b**) Portland cement–silica fume with Portland cement–fly ash–silica fume; (**c**–**e**) Portland cement–fly ash and Portland cement–fly ash–silica fume.

The profiles of chloride ion concentration of specimens with ternary blends are shown in Figure 9. When the diffusion depth is less than 12.5 mm (In Figure 9a,b,e) or less than or equal to 12.5 mm (In Figure 9c,d), the chloride ion concentration of specimens with ternary blends is smaller than specimens with binary blends under the same depth, which concludes that the ternary combination of PC-FA-SF has more efficient in prohibiting chloride's ingress inside the specimens than the binary combination of PC-FA or PC-SF.

Figure 9a presents the chloride ion concentration of specimens with different contents of FA and 5% SF. Keeping the silica fume content unchanged, the order of chloride ion concentration is F10-S5 < F20-S5 < F30-S5. When the depth is 12.5 mm, the chloride ion concentration of specimen F10-S5 is 50% less than specimen F30-S5. This situation is different from the chloride ion concentration distribution for the specimen containing only FA or SF. Interestingly, in Figure 9b, when the content of SF increases to 10%, the chloride ion concentration decreases with the FA increases under the same depth. In Figure 9c–e, the similar situation above is obviously more. When the content of FA is 10% (Figure 9c), the chloride ion concentration decreases with the SF increases. However, when the content of FA is 20% and 30% (Figure 9d,e), the chloride ion concentration decreases as the silica fume increases when the depth less than 17.5 mm. The above situation is attributed to the total content of mineral admixtures increasing, which leads to diffusion resistance improvement for improving the density of concrete.

4. Modification of the Chloride Ion Diffusion Model

4.1. Ideal Model-Fick's Second Law

Buenfeld et al. [41] considered that the progress of chloride transportation in the concrete is the unsteady diffusion process. For Fick's second law, in unit time and unit volume, the mass of material flowing out is different from that flowing into the internal structure of concrete, and the concentration of any point within it varies with time and space; this progress is also the unsteady process. Many scholars [42,43] concluded that the chloride ion diffusion process can be described by Fick's second law:

$$\frac{\partial C}{\partial t} = D\frac{\partial^2 C}{\partial x^2} \tag{1}$$

When the boundary condition is: $C(0, t) = C_s$, $C(\infty, t) = C_0$; the origin condition is $C(x,0) = C_0$, above equation can be solved as:

$$C(x,t) = C_0 + (C_s - C_0)\left[1 - erf\left(\frac{x}{2\sqrt{Dt}}\right)\right] \tag{2}$$

where $C(x, t)$ is the chloride ion concentration of the inner part concrete at time t and distance x, C_0 is the chloride ion origin concentration, C_s is the surface chloride ion concentration, D is the diffusion coefficient, $erf(z)$ is error function, $erf(z) = 2/\sqrt{\pi}\int_0^{\pi} e^{-z^2}dz$.

Fick's second law is an ideal model, which sets three suppositions: (1) the concrete is a semi-infinite homogeneous material; (2) the chloride ion does not react with concrete during the diffusion process; and (3) the diffusion coefficient is the constant. However, many studies have already concluded that: firstly, the concrete has micro-cracks and a void pore, and it is not a completely homogeneous material. Secondly, a chloride ion will bind with the cementitious paste and form Friedel's salt. At last, the chloride ion diffusion coefficient is not constant but time-dependent. The structure of HPC is more complicated than normal concrete because of the mineral admixture, the chloride diffusion process is affected by many factors. In this study, the diffusion coefficient was modified considering the influence of the water–bind ratio, time, temperature, humidity and deterioration effect of concrete, and a modified model is proposed.

4.2. Modified Diffusion Model

4.2.1. Relationship between the Diffusion Coefficient and W/B

It was reported that the diffusion coefficient was time-dependent, and it was affected by W/B, temperature, humidity, the combination affectation between concrete and chloride ion, etc. Due to having a different composition compared with normal concrete, the diffusion coefficient should be modified. As shown in Figure 10, the normalized chloride diffusion coefficients of the HPC were obtained in this experiment and research results from experimental results obtained by Chalee and Jaturapitakkul [8], Moffat et al. [9], Thomas and Matthews [44] and Shafq [45].

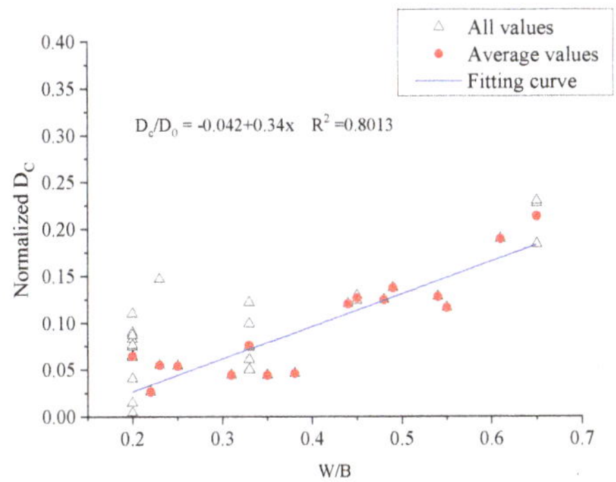

Figure 10. Chloride diffusion coefficient of concrete after different W/B.

Some scatters exist in the calculated chloride diffusion coefficients at different depths for each specimen. This could be attributed to the measurement errors of chloride concentration during the experiment. The corresponding average values, however, have relatively small scatters for the specimens with the same W/B. It can be shown in Figure 10 that the chloride diffusion coefficient shows the linear relationship with W/B, the diffusion coefficient of normal concrete is from 4.7 times to 12 times as compared with the HPC when the W/B is from 0.2 to 0.65. A coefficient is also developed to consider the W/B on the chloride diffusion coefficient based on the experimental data, which is expressed as:

$$R_C = \frac{D_C}{D_0} = -0.042 + 0.34\mathrm{W/B} \tag{3}$$

where R_C is the coefficient considering the effect of the W/B, which is the diffusion coefficient ratio of HPC to normal concrete. D_C and D_0 are the diffusion coefficient of HPC and normal concrete, respectively.

4.2.2. Relationship between the Diffusion Coefficient and Time, Temperature

The relationship between chloride ion diffusion coefficient $D(t, T)$ and time, temperature can be presented from the experimental results obtained by Mangat and Molloy [46]. The experimental results showed that the time and temperature effect should be considered to more accurate of chloride diffusion coefficient in concrete, so the chloride diffusion coefficient can be expressed as:

$$D(t,T) = D_0 f_t f_{T_b} \tag{4}$$

$$f_t = \left(\frac{t_{\exp}}{t - t_{build}}\right)^m \tag{5}$$

where $D(t, T)$ was the chloride diffusion coefficient for the real structures, f_{Tb} was the temperature correction coefficient. f_t was the time correction factor, which can be expressed by Equation (5). $t_{\exp}$ was the experimental total time, t was the calculated time, t_{build} was the construction completion time. m was the experimental parameter, which was obtained based on the different kind of cement. For normal Portland cement, m was in the range of 0.2–0.3, and for HPC with fly ash and slag, the higher values (0.5–0.7) could be used [41].

The chloride diffusion coefficient should also consider the temperature effect, which can be presented as [47]:

$$f_{T_b} = \exp\left[\frac{U}{R}\left(\frac{1}{T_{28}} - \frac{1}{T_b}\right)\right] \tag{6}$$

where U was the activated energy during the diffusion process of chloride ion in concrete and was taken as 35,000 J/mol, R was the gas constant and taken as 8.314 J/(mol.k), T_{28} was the absolute temperature corresponding to a curing of 28 days and was generally taken as 293 K, and T_b was the average absolute temperature related to calculation.

4.2.3. Relationship between the Diffusion Coefficient and Humidity

Bitaraf and Ammadis [48] pointed out that the transport of chloride ions in concrete must take pore water as continuous medium. The more pore water and the larger the section size, the smoother the diffusion of chloride ions would be. Therefore, the water content or saturation of concrete determined the number of chloride ion transport channels. At the same time, the relative humidity could be used to express the water content or saturation of concrete. When the water content inside concrete was relatively low, the chloride ion diffusion channel decreased, and its diffusion speed slowed down accordingly. Therefore, the humidity is also a factor that affects the diffusion of chloride ions: the relationship between chloride ion diffusion coefficient and humidity could be expressed as:

$$D_h = D_0\left[1+\frac{(1-h)^4}{(1-h_c)^4}\right]^{-1} \tag{7}$$

where D_h is the diffusion coefficient of chloride ion when the humidity is h, h_c is the critical relative humidity, and the value is 75% at the atmospheric environment, 90% at the splash zone and 95% at the tidal range zone.

In conclusion, the model of diffusion coefficient of chloride ion considering water–cement ratio, time, temperature, and humidity can be shown

$$D(t) = R_C D_0 \exp\left[\frac{U}{R}\left(\frac{1}{T_{28}}-\frac{1}{T}\right)\right]\left[1+\frac{(1-h)^4}{(1-h_c)^4}\right]^{-1}\left(\frac{t_{\exp}}{t-t_{build}}\right)^m \tag{8}$$

4.2.4. Initial Value Condition

The pouring and curing process of concrete is not eroded by chloride ion, so the chloride content inside the concrete is zero:

$$C_0(x,0) = 0 \tag{9}$$

4.2.5. Boundary Condition

The surface chloride ion concentration of concrete is not the fixed value, but changes with time. Liu et al. [49] proposed that the relationship between surface chloride ion concentration and time can be obtained using followed equation:

$$C_s(t) = C_{ref}\left(1-e^{-at}\right) \tag{10}$$

where C_{ref} was surface chloride concentration at the reference time t, and α was the regression parameter.

5. Random Aggregate Model of Concrete

5.1. Basic Theory

In aggregate gradation theory, the particle size and the gradation are the most important factors. In existing research, the definition of aggregate particle size is still difficult to unify and lack of practicability. Therefore, the sieving method is used to determine aggregate particle size, and the grain size can be calculated based on the upper and lower limits.

Aggregate gradation includes two aspects: particle size and the relative number of particles in each gradation. The sieving curve is used to determine aggregate gradation. Although this method can obtain the size and distribution of aggregate, it can be more vivid through numerical characterization. Commonly the modulus of fineness and specific surface area are the numerical indicators.

The existing research results show that the concrete gradation theory can be divided into two types: continuous gradation and discontinuous gradation. If the content of aggregates with different particle sizes is close to reasonable continuous gradation curve, the strength and maximum compactness of the concrete will reach the best effect according to this gradation. In this paper, the Walraven's model [50] is used to describe the aggregate gradation, which translate the spatial spherical aggregate to the flat round aggregate based on the Fuller curve. Based on the method of probability statistics, Walraven used the cumulative distribution function to transform the three-dimensional gradation curve into the probability that an aggregate diameter $D < D_0$ for any point on an internal section, where the probability can be shown as:

$$\begin{aligned} P_c(D < D_0) = P_k[1.065(D_0/D_{\max})^{0.5} - 0.053(D_0/D_{\max})^4 - \\ 0.012(D_0/D_{\max})^6 - 0.0045(D_0/D_{\max})^8 - 0.0025(D_0/D_{\max})^{10}] \end{aligned} \quad (11)$$

where D_0 was sieve diameter, $D_{\max}$ was maximum aggregate particle size, P_k was the percentage of aggregate volume in the total volume, the value was 0.75. The probability P_c of the different aggregate gradation of the two-dimensional section of the specimen could be obtained by replacing the boundary value of the three-dimensional aggregate content P_K with the diameter of the aggregate at all levels in formula (11), and then the two-dimensional gradation curve could be obtained.

5.1.1. Circular Aggregate Generation and Release Process

The release process of circular aggregate is introduced in this paper firstly. The radius of the circular aggregate is assumed to be R_i, and its corresponding circular coordinate is (x_i, y_i), then the calculation method of the geometric model of the aggregate is as follows:

(1) Defining the radius and the centroid coordinates of aggregate (x_i, y_i) The x_i and y_i need to satisfy the following requirements: $x_i = \frac{d}{2} + (a - d) \cdot \eta$, $y_i = \frac{d}{2} + (b - d) \cdot \eta$. Where, a and b are the side length of release area, respectively, $a = b = 150$ mm in this paper. η is a random number between 0 and 1.

(2) Overlapping judgment. This process is determining whether the generated i aggregate overlaps with the generate $i - 1$ aggregate. If un-overlapped, the area of the aggregate is calculated and the information of the aggregate is recorded; if overlapped, turning back to the step (1) and recoding the value. The basis of overlap judgment is: If the interval length d between two circular aggregate centroid is larger than the sum of their radius, namely $d > R1 + R2$, it can be concluded that the circular aggregate with radius R_1 does not overlap with the circular aggregate with radius R_2.

(3) Judging of the aggregate area. If the condition $\Sigma A_i \geq A_0 \cdot a \cdot b$ is not met, the researcher should be turn back to step (1) and continue to release. If the above condition is met, the aggregate is finished. Where A_0 is the area fraction of the calculated aggregate.

5.1.2. Polygonal Aggregate Generation and Release Process

In this paper, the generation process of polygonal random aggregate is realized as follows: Firstly, based on the simulation of circular aggregate, the diameter of random circle is assumed as the particle size of polygonal aggregate, and the location and size of the aggregate are determined by using the randomly generated circle. Secondly, a polygonal base aggregate is generated randomly in the random circle, and the area of the polygon is greater than or equal to 0.75~1.1 times of the area of the random circle. Finally, the longest edge is set as the reference edge, the polygonal random aggregate model is established by using the rule that the reference edge generates new points and determines whether the new points meet the outward extension condition.

5.1.3. Examples of Random Aggregate

The content of aggregate in each tested block was calculated based on the different mix proportion of test specimens in Table 1 and then the geometric model of circular aggregate was generated by MATLAB 2021a programming language according to the random number theory using the Monte Carlo method.

For a circular aggregate, the area S (volume fraction of aggregate) of a certain particle size in the two-dimensional model is calculated according to Equation (11) according to the size and grade of aggregate particle size, and then the amount of aggregate corresponding to each diameter is further determined. In the section of the specimen, the position of randomly generated aggregate centroid was firstly determined, and then circular aggregate is randomly generated according to the corresponding aggregate diameter at that position. Finally, the aggregate of all particle sizes is released, that is, the establishment of circular random aggregate model was completed

The mixing ratio of each specimen is shown in Table 1. According to the mixing ratio of each specimen in test, the volume percentage ratio of aggregate to concrete P_k can be obtained, as shown in Table 3.

Table 3. Proportion of aggregate volume to concrete for each strength grade.

Specimen	Mass of Aggregate	Volume of Concrete	Volume of Aggregate	Proportion of Aggregate Volume
NC50	6.08056875	0.003375	0.002252063	0.67
NC60	6.07746375	0.003375	0.002250913	0.67
NC70	6.074510625	0.003375	0.002249819	0.67
HPC50	6.077919375	0.003375	0.002251081	0.67
HPC60	6.07624875	0.003375	0.002250463	0.67
HPC70	6.074274375	0.003375	0.002249731	0.67
HPC80	5.882625	0.003375	0.00217875	0.65
HPC100	5.886	0.003375	0.00218	0.65

Specimens NC50 and HPC80 are used to build the randomly geometric of circular and polygon aggregates. Table 4 shows the distribution probability of aggregate particle size of specimens NC50 and HPC80. Combining the proportion of aggregate volume with the Equation (11), the probability P_c that aggregate diameter D less than D_0 in any cross section is calculated, then the aggregate number for every diameter is also confirmed. Figure 11 show the circle aggregate model of specimen NC50 and HPC80.

Table 4. Distribution probability of particle size of each aggregate in the specimens NC50 and HPC80.

Specimen	D_0	22	18	12	5
NC50	D_0/D_{max}	1	0.81	0.55	0.23
	P_c	0.65	0.61	0.51	0.33
HPC80	D_0/D_{max}	1	0.81	0.55	0.23
	P_c	0.63	0.59	0.49	0.32

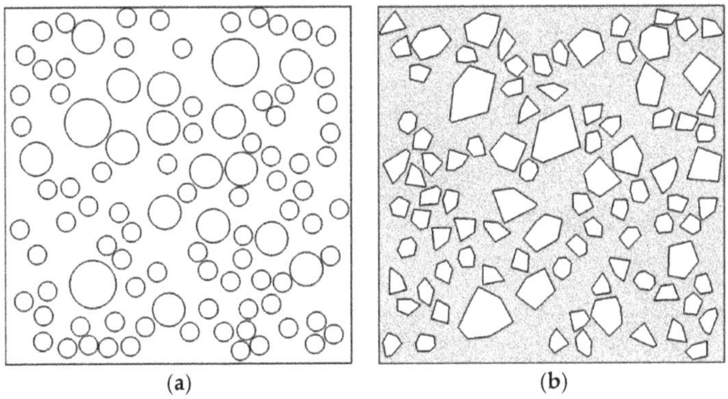

Figure 11. Circle aggregate model of specimens: (**a**) NC50; (**b**) HPC80.

6. Mesoscopic Simulation of Chloride Ion Diffusion

Simulation Method

Software COMSOL Multiphysics 5.1 provides modules for describing and analyzing various physical fields which including chemical transfer, fluid flow, and dynamic and solid mechanics. The Transport of Diluted Species Interface in this software is suitable to simulate the material transfer process under the consideration of convection–diffusion and diffusion processes, so the chloride ion diffusion process in high performance concrete in this paper is simulated using this module. The simulation steps are shown as follows:

i. Choosing the Transport of Diluted Species Interface in the COMSOL software and importing the two-dimensional concrete random aggregate geometric model proposed in Part 4 into the software and set the cell.
ii. Defining the parameter value in transfer equation and diffusion coefficient model based on the Equation (11).
iii. Defining the boundary condition C_s, diffusion coefficient and initial value condition C_0.
iv. Dividing the mesh.
v. Defining a two-dimensional transversal and extracting the chloride ion concentration on the transversal lines of 2.5 mm, 7.5 mm, 12.5 mm, 17.5 mm, 22.5 mm, and 27.5 mm, respectively.

Dividing the Mesh

Of the above steps, dividing the mesh is the most important. In this paper, concrete is regarded as a three-phase composite composed of aggregate, mortar, and the transition layer between them. The COMSOL software is used to carry out the grid-process for the two-dimensional random aggregate model generated in Part 4, and then the two-dimensional random aggregate finite element model is established. According to the research result from Basheer et al. [51], the thickness of transition layer is from 0 μm to 10 μm, in this paper, the thickness of the transition layer in the three-phase random aggregate model is 75 μm. The finite element model is shown in the Figure 12.

The process of mesh generation in COMSOL Multiphysics is defined by mesh sequence, which consists of operation characteristics and attribute characteristics. Among them, operation features are mainly for mesh type, copy mesh, refine mesh, or transform mesh, while attributed features are mainly for mesh type related features such as size, distribution, and proportion. Therefore, when creating a mesh in software, the operation characteristics of the mesh need to be defined first to generate or modify the corresponding geometry of the mesh. Then, the local attribute characteristics is defined. When adding local attribute characteristics in the software, the selected local attribute characteristics will appear on the corresponding sub-nodes under the operation characteristics node, especially when the selected target is the same. The global attribute node is automatically overwritten by the local attribute feature node.

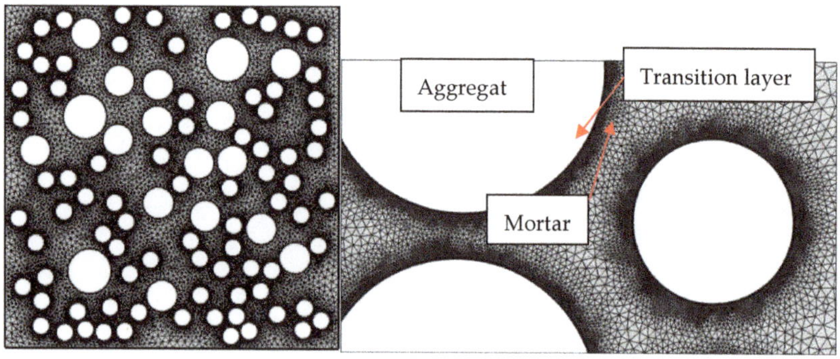

Figure 12. Random aggregate finite element model.

7. Microscopic Model Validation and Parameters Analysis

7.1. Model Validation

In order to verify the reliability and accuracy of the microscopic model in this paper, the measured concentration value in the experiment is compared with simulation value. In the specific test process, due to the presence of chloride crystals on the surface of the specimen and chloride salts in the micro-pores, it is impossible to accurately determine the chloride ion concentration on the concrete surface. Therefore, the chloride ion concentration at 2.5 mm is regarded as the chloride ion degree on the concrete surface, that is, chloride ion begins to diffuse into the concrete interior from 2.5 mm. Figure 13 shows the chloride ion concentration of test and microscopic simulation at different depth. Generally, it can be found from Figure 13 that the simulation and the tested value of the chloride ion concentration under the same depth is close, the average ratio of simulation value to tested value is 0.91. The simulation and tested chloride ion concentration of the specimens HPC60, HPC80, and HPC100 under the depths of 7.5 mm and 12.5 mm is slightly different, it may be the reason that the mortar content is not uniform.

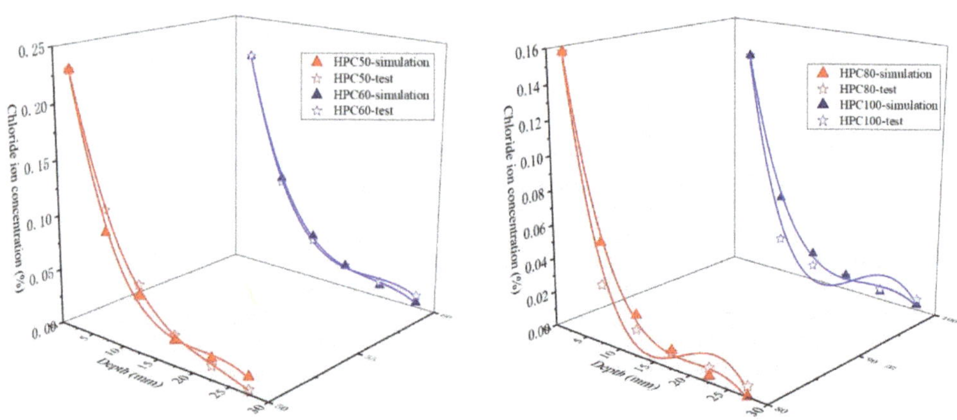

(a) (b)

Figure 13. Chloride ion concentration of test and microscopic simulation at different depth: (**a**) HPC50 and HPC60; (**b**) HPC80 and HPC100.

7.1.1. Effect of Aggregate Shape on Chloride Ion Permeability

In the mesoscopic model, the content, particle size, and position of aggregate are kept the same, and the shape of aggregate is changed to study its influence on the diffusion of chloride ion in concrete. The surface chloride ion concentration is 0.6, the thickness of transition layer is 75 μm, and the diffusion time is 88 days. The numerical simulation results of specimens HPC50 with circle and square aggregate are shown in Figure 14. Figure 14 shows the chloride ion concentration of specimen HPC50 with different aggregate shapes, the chloride ion concentration at the same depth in the Figure 15 is the average value of all chloride ion concentrations at a boundary.

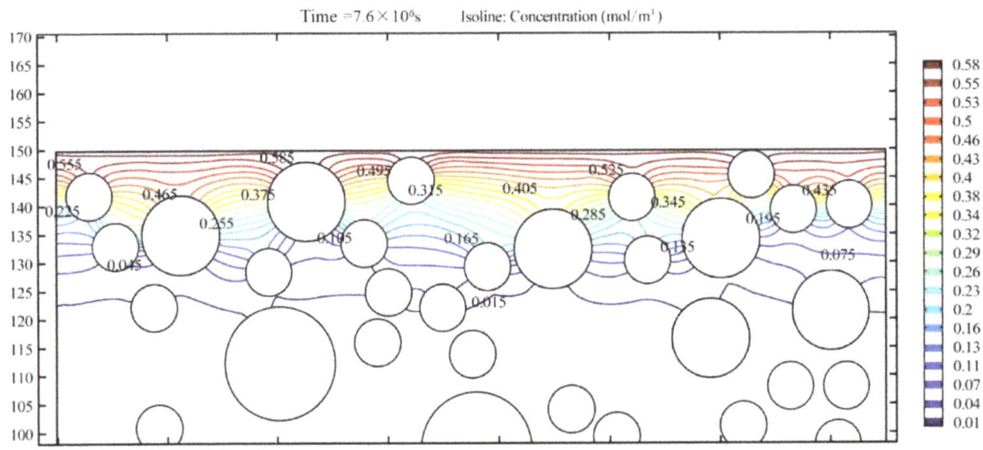

(**a**)

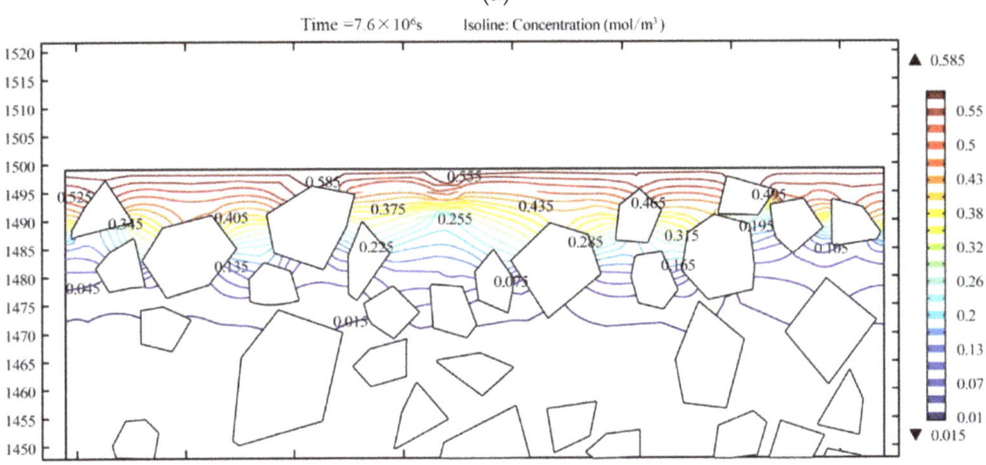

(**b**)

Figure 14. Numerical simulation results of specimens HPC 50 with circle and square aggregate: (**a**) Circle aggregate; (**b**) Square aggregate.

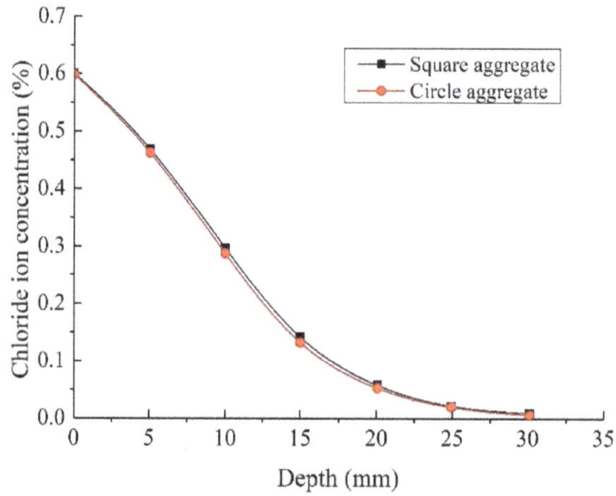

Figure 15. Chloride ion concentration of specimen HPC50 with different aggregate shape.

It can be found from Figure 15 that no matter the circle aggregate or square aggregate, the average chloride ion concentration under the depth from 0 mm to 30 mm is close. The average deviation ratio of chloride ion concentration between square aggregate and circle aggregate is 10%. Figure 14 shows that the aggregate shape effects the chloride ion concentration around the aggregate. From the above analysis it can be concluded that the aggregate shape has little influence on the chloride ion diffusion process in the concrete inside. This conclusion is same with normal concrete in Du's study [27]. Du et al. [27] found that aggregate shape has a negligible influence on chloride diffusivity in concrete.

7.1.2. Effect of Aggregate Distribution on Chloride Ion Permeability

Due to the influence of shape of aggregate on the chloride ion diffusion in concrete is not obvious, the circle aggregate is taken as an example in the microscopic model of HPC50. The content and particle size of aggregate are kept the same, and only the distribution of the aggregate is changed to study its influence on the chloride ion diffusion in concrete. The surface chloride ion concentration is still 0.6, the transition layer thickness is still 75 μm, and the erosion time is still 88 days. The numerical simulation results are shown in Figure 16. The average chloride ion concentration of specimen HPC50 with different aggregate distributions under the same depth is presented in Figure 17.

It can be seen from Figure 16 that aggregate only affects the transport path and director of chloride ions in concrete, and has little effect on the average chloride ion concentration at the same depth. Figure 17 also shows that the chloride ion concentrations between the different aggregate diffusion are close except under the 15 and 20 mm. The average deviation ratio of chloride ion concentration between Position 1 and Position 2 is 18.9%. Above finding is also similar with normal concrete in Du et al.'s study [27].

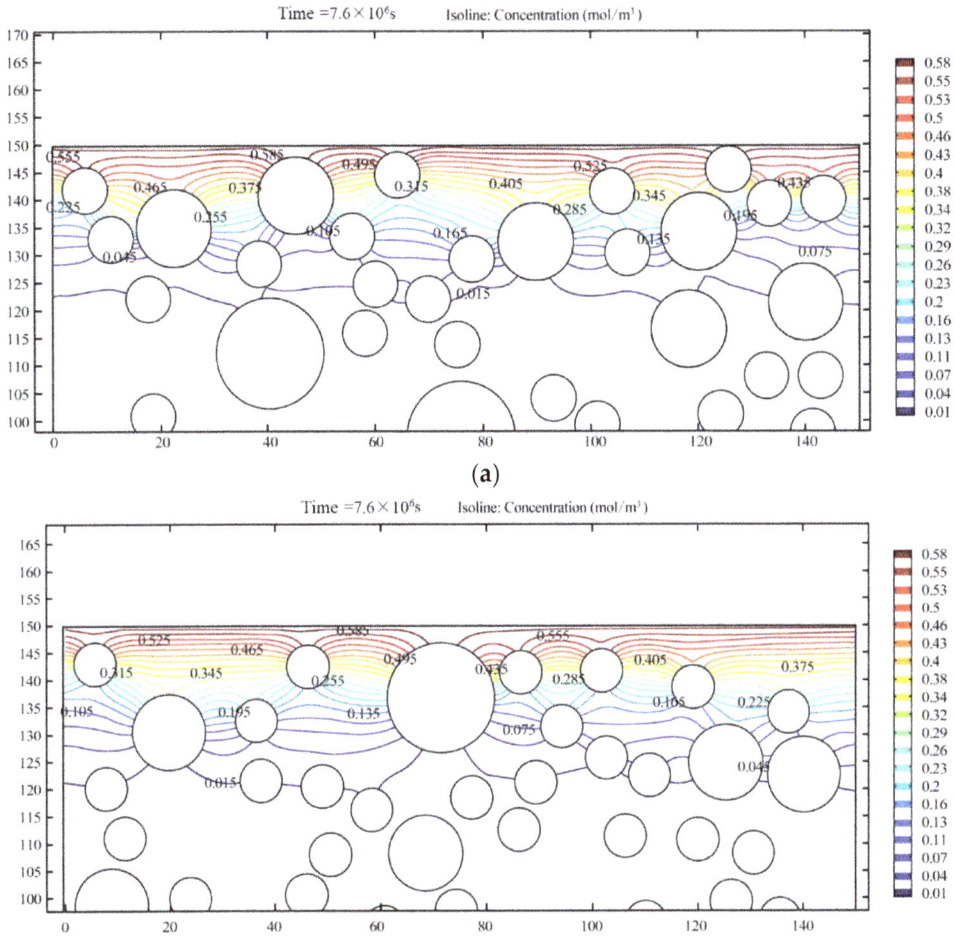

(a)

(b)

Figure 16. Numerical simulation results of specimens HPC 50 with different aggregate distribution: (**a**) Position 1; (**b**) Position 2.

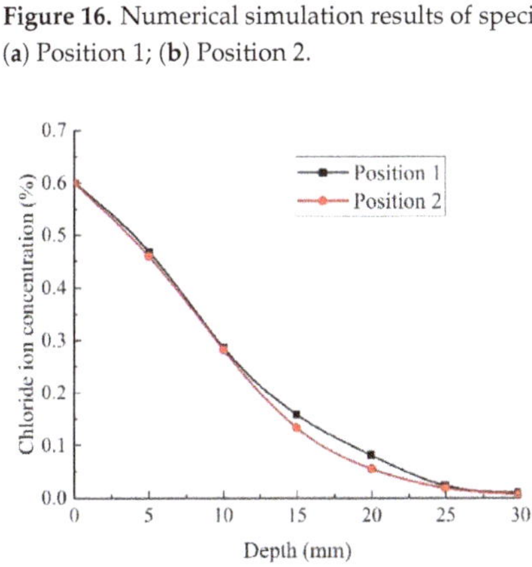

Figure 17. Chloride ion concentration of specimen HPC50 with different aggregate distribution.

7.1.3. Effect of Aggregate Size on Chloride Ion Permeability

Keeping the content of aggregate unchanged, the influence of aggregate size on the chloride ion concentration has been studied. The aggregate sizes in Figure 18a are 20 mm, 14 mm and 7.5 mm (Size 1), which are 30 mm, 15 mm, and 7.5 mm (Size 2) in Figure 18b. The surface chloride ion concentration is still 0.6, the transition layer thickness is still 75 μm, and the erosion time is still 88 d. The chloride ion distribution and average chloride ion concentration of specimen HPC50 with different aggregate size under the same depth are presented in Figure 19.

It can be seen from Figure 18 that when the aggregate size increases, and the proportion of aggregate remains unchanged, the aggregate number in concrete specimens will decrease, and the spacing between adjacent aggregates increases, which will be more conducive to the diffusion of chloride ions in the concrete.

Figure 19 also shows that when aggregate accounts is unchanged, only the particle size of aggregate is changed, the chloride ion concentration increases with the increase in aggregate size. With the increase in diffusion depth, the chloride ion concentration difference also increases gradually, the concentration difference is maximum when the diffusion depth is 15 mm. When the diffusion depth is greater than 15 mm, the concentration difference decreases, and the concentration tends to be equal due to at the depth 30 mm due to the limit of diffusion depth.

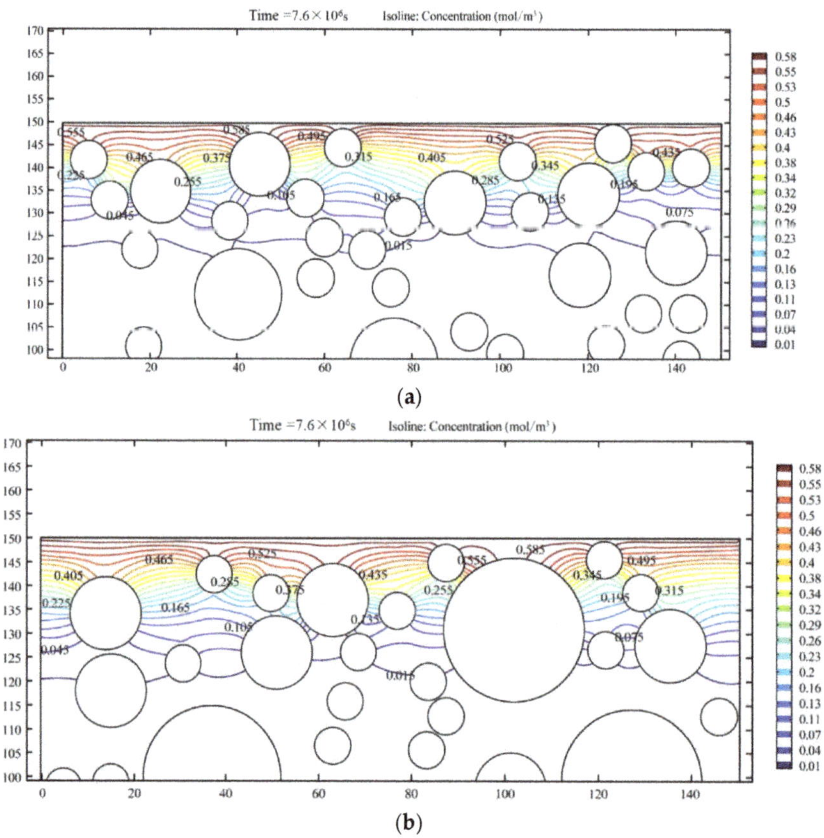

Figure 18. Chloride ion concentration distribution of specimen HPC50 with different sizes: (**a**) Size 1; (**b**) Size 2.

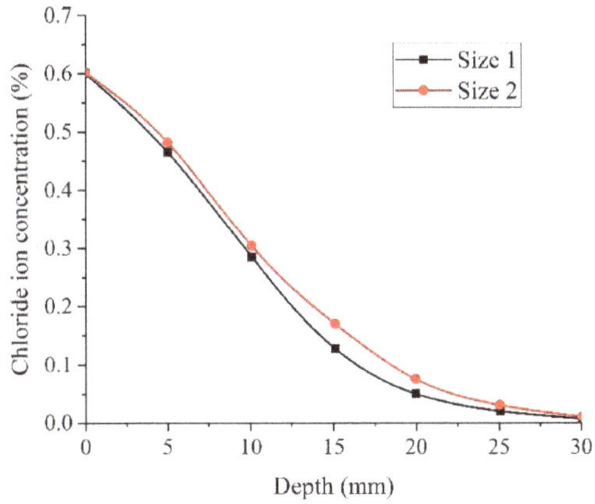

Figure 19. Chloride ion concentration of specimen HPC50 with different aggregate size.

8. Conclusions

In this paper, the chloride ion diffusion experiment for the HPC is carried out. The influence of W/B, binary (PC-FA and PC-SF), and binary (PC-FA-SF) combinations on the concrete compressive strength and chloride ion diffusion is investigated. The chloride ion diffusion coefficient is modified and the chloride ion diffusion model of the HPC is revised based on the Fick's second law considering the W/B, temperature, humidity, and time. A mesoscopic model is developed to investigate chloride diffusivity in concrete, the simulation results are compared with previous test data. The influence of aggregate shape, distribution, and size of HPC on the chloride ion diffusion is investigated. Some meaningful conclusions are obtained as follow:

(1) The influence of the W/B on the concrete strength for normal concrete is obviously more than the HPC. The contents of the SF or FA developing the compressive strength is limited. The concrete compressive strength of ternary combination specimens decreases with the increase in FA or SF when the content of the other mineral admixture SF or FA keeps unchanged.
(2) No matter normal concrete or the HPC, from depth 7.5 mm to 17.5 mm, the influence of W/B on the chloride ion concentration is most significant, under the same diffusion depth, the chloride concentration decreases with the increase in W/B. The ternary combination of PC-FA-SF is more efficient in prohibiting chlorides ingress inside the specimens than the binary combination of PC-FA or PC-SF.
(3) The mesoscopic simulation and the tested value of the chloride ion under the same depth is close, the average ratio of simulation value to tested value is 0.91. The aggregate shape and distribution have a negligible influence on chloride diffusivity in the HPC, but the chloride ion concentration increases with the increase in aggregate size.

Author Contributions: Conceptualization, H.T. and Y.Y.; methodology, H.T.; software, J.P. validation, H.T. and Y.Y.; formal analysis, H.T.; investigation, Y.Y.; resources, H.T.; data curation, H.T.; writing—original draft preparation, H.T.; writing—review and editing, Y.Y.; visualization, H.T.; supervision, J.P.; project administration, P.L.; funding acquisition, J.Z. All authors have read and agreed to the published version of the manuscript.

Funding: The work reported here was conducted with financial support from the National Natural Science Foundation of China (Grant Nos. 52108135 and 52078056), the National Key R&D Program of China (Grant No. 2021YFB2600900), National Science Foundation for Distinguished Young Scholars of Hunan Province (Grant No. 2022JJ10075), Natural Science Foundation of Hunan Province (Grant No. 2022JJ40024), the Science Fund for Creative Research Groups of Hunan Province (Grant No.

2020JJ1006), and the Scientific Research Outstanding Young Project of Education Department of Hunan Province (Grant No. 21B0721).

Institutional Review Board Statement: Not applicable.

Informed Consent Statement: Not applicable.

Data Availability Statement: All data, models, and code generated or used during the study appear in the submitted article.

Conflicts of Interest: The authors declare no conflict of interest.

References

1. Bachtiar, E.; Rachim, F.; Makbul, R.; Tata, A.; Irfan-Ul-Hassan, M.; Kırgız, M.S.; Syarif, M.; De Sousa Galdino, A.G.; Khitab, A.; Benjeddou, O.; et al. Monitoring of chloride and Friedel's salt, hydration components, and porosity in high-performance concrete. *Case Stud. Constr. Mat.* **2022**, *17*, e01208. [CrossRef]
2. Xie, H.; Yang, L.; Zhang, Q.; Huang, C.; Chen, M.; Zhao, K. Research on energy dissipation and damage evolution of dynamic splitting failure of basalt fiber reinforced concrete. *Constr. Build. Mater.* **2022**, *330*, 127292. [CrossRef]
3. Han, C.; Shen, W.; Ji, X.; Wang, Z.; Ding, Q.; Xu, G.; Lv, Z.; Tang, X. Behavior of high performance concrete pastes with different mineral admixtures in simulated seawater environment. *Constr. Build. Mater.* **2018**, *187*, 426–438. [CrossRef]
4. Felekoğlu, B.; Ramyar, K.; Tosun, K.; Musal, B. Sulfate resistances of different types of Turkish Portland cements by selecting the appropriate test methods. Constr. *Build. Mater.* **2006**, *20*, 819–823. [CrossRef]
5. Felekoğlu, B.; Sarıkahya, H. Effect of chemical structure of polycarboxylate-based superplasticizers on workability retention of self-compacting concrete. *Constr. Build. Mater.* **2008**, *22*, 1972–1980. [CrossRef]
6. Seleem, H.E.-D.H.; Rashad, A.M.; El-Sabbagh, B.A. Durability and strength evaluation of high-performance concrete in marine structures. *Constr. Build. Mater.* **2010**, *24*, 878–884. [CrossRef]
7. Sengul, O.; Tasdemir, M.A. Compressive strength and rapid chloride permeability of concretes with ground fly ash and slag. *J. Mater. Civil. Eng.* **2009**, *21*, 494–501. [CrossRef]
8. Chalee, W.; Jaturapitakkul, C. Effects of W/B ratios and fly ash finenesses on chloride diffusion coefficient of concrete in marine environment. *Mater. Struct.* **2009**, *42*, 505–514. [CrossRef]
9. Moffatt, E.G.; Thomas, M.D.A.; Fahim, A. Performance of high-volume fly ash concrete in marine environment. *Cem. Concr. Res.* **2017**, *102*, 127–135. [CrossRef]
10. Cabrera, J.G.; Nwaubani, S.O. The microstructure and chloride ion diffusion characteristics of cements containing metakaolin and fly ash. *ACI Symp. Public.* **1998**, *178*, 385–400.
11. Thomas, B.S.; Gupta, R.C.; Mehra, P.; Kumar, S. Performance of high strength rubberized concrete in aggressive environment. *Constr. Build. Mater.* **2015**, *83*, 320–326. [CrossRef]
12. Hooton, R.; Gruber, K.; Boddy, A. The chloride penetration resistance of concrete containing high reactivity metakaolin. In Proceedings of the PCI/FHWA International Symposium on High Performance Concrete, Orlando, FL, USA, 25–27 September 2001; pp. 172–183.
13. Coleman, N.J.; Page, C.L. Aspects of the pore solution chemistry of hydrated cement pastes containing metakaolin. *Cem. Concr. Res.* **1997**, *27*, 147–154. [CrossRef]
14. Geetha, S.; Madhavan, S. High Performance Concrete with Copper slag for Marine Environment. *Mater. Today Proc.* **2017**, *4*, 3525–3533. [CrossRef]
15. Harilal, M.; Rathish, V.R.; Anandkumar, B.; George, R.P.; Mohammed, M.S.H.S.; Philip, J.; Amarendra, G. High performance green concrete (HPGC) with improved strength and chloride ion penetration resistance by synergistic action of fly ash, nanoparticles and corrosion inhibitor. *Constr. Build. Mater.* **2019**, *198*, 299–312. [CrossRef]
16. Jiang, J.; Zheng, X.; Wu, S.; Liu, Z.; Zheng, Q. Nondestructive experimental characterization and numerical simulation on self-healing and chloride ion transport in cracked ultra-high performance concrete. *Constr. Build. Mater.* **2019**, *198*, 696–709. [CrossRef]
17. Ding, Q.; Geng, J.; Hu, S.; Sun, J.; Sun, B. Different effects of fly ash and slag on anti-rebar corrosion ability of concrete with chloride ion. Wuhan Univ. *J. Nat. Sci.* **2009**, *14*, 355–361. [CrossRef]
18. Khan, M.I. Comparison of chloride ion penetration and diffusion of high-performance concrete. *Ksce. J. Civ. Eng.* **2012**, *16*, 779–784. [CrossRef]
19. Wang, L.; Dai, L.; Bian, H.; Ma, Y.; Zhang, J. Concrete cracking prediction under combined prestress and strand corrosion. *Struct. Infrastruct. Eng.* **2019**, *15*, 285–295. [CrossRef]
20. Dai, L.; Bian, H.; Wang, L.; Potier-Ferry, M.; Zhang, J. Prestress Loss Diagnostics in Pre-tensioned Concrete Structures with Corrosive Cracking. *J. Struct. Eng.* **2019**, *146*, 04020013. [CrossRef]
21. Peng, J.; Tang, H.; Zhang, J. Structural Behavior of Corroded Reinforced Concrete Beams Strengthened with Steel Plate. *J. Perform. Constr. Facil.* **2017**, *31*, 04017013. [CrossRef]
22. Zhang, X.; Zhang, W.; Luo, Y.; Wang, L. Interface shear strength between self-compacting concrete and carbonated concrete. *J. Mater. Civil. Eng.* **2020**, *32*, 04020113. [CrossRef]

23. Tang, H.; Peng, J.; Zhang, J. Influence of further corrosion on structural behavior of corroded reinforced concrete beam strengthened with steel plate using different strengthening schemes. *J. Perform. Constr. Facil.* **2020**, *34*, 04019117. [CrossRef]
24. Xiao, L.; Peng, J.; Zhang, J.; Ma, Y.; Cai, C.S. Comparative assessment of mechanical properties of HPS between electrochemical corrosion and spray corrosion. *Constr. Build. Mater.* **2020**, *237*, 117735. [CrossRef]
25. Peng, J.; Xiao, L.; Jianren, Z.; Cai, C.S.; Wang, L. Flexural behavior of corroded HPS beams. *Eng. Struct.* **2019**, *195*, 274–287. [CrossRef]
26. Zheng, J.-J.; Zhou, X.-Z.; Wu, Y.-F.; Jin, X.-Y. A numerical method for the chloride diffusivity in concrete with aggregate shape effect. *Constr. Build. Mater.* **2012**, *31*, 151–156. [CrossRef]
27. Du, X.L.; Jin, L.; Ma, G.W. A meso-scale numerical method for the simulation of chloride diffusivity in concrete. *Finite Elem. Anal. Des.* **2014**, *85*, 87–100. [CrossRef]
28. Pan, Z.; Chen, A.; Ruan, X. Spatial variability of chloride and its influence on thickness of concrete cover: A two-dimensional mesoscopic numerical research. *Eng. Struct.* **2015**, *95*, 154–169. [CrossRef]
29. Hu, S.; Peng, J.; Zhang, J.; Cai, C.S. Influences of Time, Temperature, and Humidity on Chloride Diffusivity: Mesoscopic Numerical Research. *J. Mater. Civil. Eng.* **2017**, *29*, 04017223. [CrossRef]
30. Peng, J.; Hu, S.; Zhang, J.; Cai, C.S.; Li, L.-y. Influence of cracks on chloride diffusivity in concrete: A five-phase mesoscale model approach. *Constr. Build. Mater.* **2019**, *197*, 587–596. [CrossRef]
31. Guo, L.-P.; Carpinteri, A.; Roncella, R.; Spagnoli, A.; Sun, W.; Vantadori, S. Fatigue damage of high performance concrete through a 2D mesoscopic lattice model. *Comp. Mater. Sci.* **2009**, *44*, 1098–1106. [CrossRef]
32. Huespe, A.E.; Oliver, J.; Mora, D.F. Computational modeling of high performance steel fiber reinforced concrete using a micromorphic approach. *Comput. Mech.* **2013**, *52*, 1243–1264. [CrossRef]
33. Forquin, P.; Zinszner, J.L.; Lukic, B. Mesoscopic Modelling of Ultra-High Performance Fiber Reinforced Concrete Under Dynamic Loading. In *Dynamic Behavior of Materials*; Springer International Publishing: Cham, Switzerland, 2017; Volume 1, pp. 123–129.
34. Feng, T.; Jia, M.; Xu, W.; Wang, F.; Li, P.; Wang, X.; Tan, Y.; Jiang, J. Three-dimensional mesoscopic investigation of the compression mechanical properties of ultra-high performance concrete containing coarse aggregates. *Cem. Concr. Compos.* **2022**, *133*, 104678. [CrossRef]
35. MOHURD (Ministry of Housing and Urban-Rural Development). *GB50010-2010: Code for Design of Concrete Structures*; Ministry of Housing and Urban-Rural Development of the People's Republic of China (MOHURD): Beijing, China, 2010.
36. ASTM (American Society for Testing and Materials). *Standard Test Method for Electrical Indication of Concretes Ability to Resist Chloride Ion Penetration*; ASTM C1202; ASTM International: West Conshohocken, PA, USA, 2017.
37. AASHT (American Association of State Highway and Transportation Officials). *Standard Method of Test for Sampling and Testing for Chloride Ion in Concrete and Concrete Raw Material*; AASHTOT260-97; AASHT: Washington, DC, USA, 1997.
38. Chindaprasirt, P.; Jaturapitakkul, C.; Sinsiri, T. Effect of fly ash fineness on compressive strength and pore size of blended cement paste. *Cem. Concr. Comp* **2005**, *27*, 425–428. [CrossRef]
39. Jerath, S.; Hanson, N. Effect of Fly Ash Content and Aggregate Gradation on the Durability of Concrete Pavements. *J. Mater. Civil. Eng.* **2007**, *19*, 367–375. [CrossRef]
40. Leng, F.; Feng, N.; Lu, X. An experimental study on the properties of resistance to diffusion of chloride ions of fly ash and blast furnace slag concrete. *Cem. Concr. Res.* **2000**, *30*, 989–992. [CrossRef]
41. Bamforth, P. Spreadsheet model for reinforcement corrosion in structures exposed to chlorides. *Concr. Sev. Cond.* **1998**, *2*, 64–75.
42. Mangat, P.S.; Limbachiya, M.C. Effect of initial curing on chloride diffusion in concrete repair materials. *Cem. Concr. Res.* **1999**, *29*, 1475–1485. [CrossRef]
43. Thomas, M.; Bamforth, P.B. Modelling chloride diffusion in concrete: Effect of fly ash and slag. *Cem. Concr. Res.* **1999**, *29*, 487–495. [CrossRef]
44. Thomas, M.D.A.; Matthews, J.D. Performance of pfa concrete in a marine environment—10-year results. *Cem. Concr. Comp* **2004**, *26*, 5–20. [CrossRef]
45. Shafiq, N. Effects of fly ash on chloride migration in concrete and calculation of cover depth required against the corrosion of embedded steel reinforcement. *Struct. Concr.* **2004**, *5*, 5–9. [CrossRef]
46. Mangat, P.S.; Molloy, B.T. Prediction of long term chloride concentration in concrete. *Mater. Struct.* **1994**, *27*, 338–346. [CrossRef]
47. Peng, J.; Tang, H.; Xiao, L.; Chen, X.; Zhang, J. Chloride diffusion and time to corrosion initiation of reinforced concrete structures. *Mag. Concr. Res.* **2021**, *73*, 757–770. [CrossRef]
48. Bitaraf, M.; Mohammadi, S. Analysis of chloride diffusion in concrete structures for prediction of initiation time of corrosion using a new meshless approach. *Constr. Build. Mater.* **2008**, *22*, 546–556. [CrossRef]
49. Liu, T.; Weyers, R.W. Modeling the Dynamic Corrosion Process in Chloride Contaminated Concrete Structures. *Cem. Concr. Res.* **1998**, *28*, 365–379. [CrossRef]
50. Walraven, J.C.; Reinhardt, H.W. Concrete mechanics. Part A: Theory and experiments on the mechanical behavior of cracks in plain and reinforced concrete subjected to shear loading. *Nasa Sti/Recon Tech. Rep. N* **1981**, *82*, 25417.
51. Basheer, L.; Kropp, J.; Cleland, D.J. Assessment of the durability of concrete from its permeation properties: A review. *Constr. Build. Mater.* **2001**, *15*, 93–103. [CrossRef]

coatings

Article

Analysis of the Mechanical Properties of a Bidirectional Laminated Slab with Shear Keys

Ming Li [1], Zhijun Zhou [1,*], Qian Wu [1,*], Wei Yan [1] and Shuang Yuan [2]

[1] School of Civil Engineering, Shenyang Jianzhu University, Shenyang 110168, China
[2] School of Construction Engineering, Tianjin University, Tianjin 300072, China
* Correspondence: me572742890@163.com (Z.Z.); 18624028257@163.com (Q.W.); Tel.: +86-024-24691800 (Q.W.)

Abstract: The load-bearing capacity of a bidirectional laminated slab with shear keys arranged in a rectangular grid (BCSWSK) was investigated to determine its mechanical properties, using load tests on a fabricated laminated slab and finite element models. The shear force distribution across the shear keys was measured and analysed, and the effects of different parameters were identified. The interfaces between the cast bottom slab and the shear keys were strongly bonded, as was the interface between the precast bottom slab and the cast-in-place upper slab, despite it being a secondary concrete pouring. The shear force was distributed similarly along the X direction (columns) and the Y direction (rows) of the shear key arrangement. The shear forces along the Y direction were greater and reached a maximum value sooner, but the differences between columns were mostly less than 10%. A square cross-section is recommended for the shear keys. The number of rows, number of columns, and the cross-sectional area of the shear keys are the main factors influencing the mechanical properties of the composite slab, but as they individually increased, they reached a point at which further increase had little effect. Similarly, after a certain number of shear keys were used, the concrete grade cast in situ, the shear key concrete grade, and the friction coefficient between the precast bottom slab and the cast-in-place upper slab had little influence on the BCSWSK.

Keywords: laminated slab and plates; composite slab and plates; shear key; influencing factor; element modelling

Citation: Li, M.; Zhou, Z.; Wu, Q.; Yan, W.; Yuan, S. Analysis of the Mechanical Properties of a Bidirectional Laminated Slab with Shear Keys. *Coatings* **2022**, *12*, 1542. https://doi.org/10.3390/coatings12101542

Academic Editor: Andrea Nobili

Received: 20 September 2022
Accepted: 11 October 2022
Published: 13 October 2022

1. Introduction

A laminated concrete slab consists of a prefabricated bottom slab (PBS) and a cast-in-place concrete upper slab (CCUS). Its advantages over conventional cast-in-place slabs are ease of construction, green and low-carbon technology, and high-production efficiency. It is an important horizontal load-bearing member of assembled concrete structures and has been widely used in housing construction [1]. The most commonly used forms of laminated slabs are steel-truss-laminated slabs (STLSs) (Figure 1a) and ribbed laminated slabs (RLSs) (Figure 1b). Of which, the steel truss is made automatically by machine. An in-depth study shows that, in conjunction with the wider uses of STLSs, the increased use of steel reinforcement in them leads to greater slab thickness [2], resulting in higher construction costs. The ribs of RLSs extend for the full length of the slab, which can be problematic in crossing pipelines. Furthermore, the PBSs of them must be temporarily supported during construction due to their limited stiffness. To overcome the problems associated with the use of these types of laminated slabs, a concrete-laminated slab with shear keys (CSWSK) was proposed to connect the PBS to the CCUS (Figure 2a). The shear keys can also be connected to temporary I-beams to increase the stiffness of the PBSs to realise support-free construction (Figure 2b).

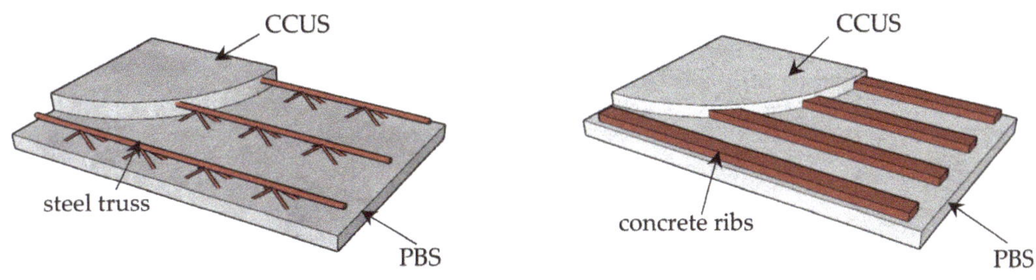

(a) (b)

Figure 1. The most commonly used forms of laminated slab: (a) STLS; (b) RLS.

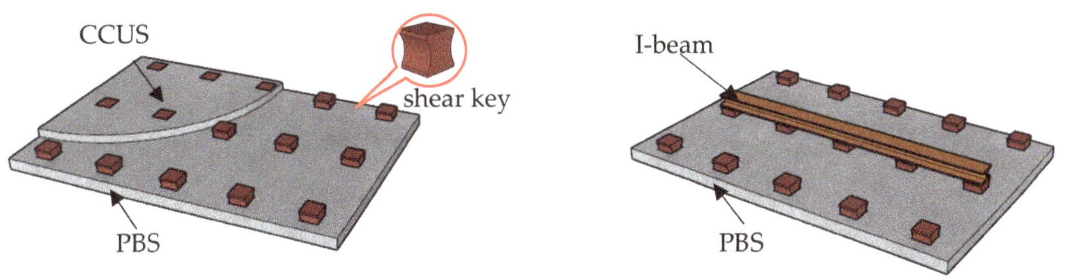

(a) (b)

Figure 2. CSWSK and precast bottom slab with I-beam bolted to it: (a) CSWSK; (b) PBS with I-beams bolted to it.

There are large bodies of systematic research on each [3–11]. Liu et al. [12] confirmed the practicality of using STLSs by empirical testing; Ma et al. [13] investigated its mechanical properties in construction and in service by testing and numerical modelling; Zhao et al. [14] analysed the influence of parameters, such as concrete strength and truss height and diameter, on the mechanical properties of STLSs using ANSYS software; and Cheng [15] developed a short-term stiffness model for STLSs from theoretical analysis. Wu et al. [16] confirmed the practicality of using RLSs by empirical testing; Qian et al. [17] experimentally investigated its mechanical properties, such as the flexural-bearing capacity and the shear resistance of the laminated surface, in construction and in service; Shi et al. [18] analysed the effects of parameters, such as rib height, and rib width and reinforcement rate, on the mechanical properties of RLSs using a finite element method and experimental testing; and Huang et al. [19] developed deflection and bending moment equations for RLSs in service, using load superposition and an orthotropic and anisotropic plate theory. However, for CSWSKs, only we have conducted a series of studies on its unidirectional mechanical properties [20,21], and this paper continues that series by investigating the mechanical properties of bidirectional laminated slabs with shear keys (BCSWSKs), including the flexural-bearing capacity, damage modes, and influencing factors, using experimental tests and finite element modelling. A flowchart of this study is shown in Figure 3.

Figure 3. Flowchart of this study.

2. Experimental Study

One BCSWSK was designed and fabricated, according to standards GB (50010-2010) [22] and JGJ (1-2014) [23], to be incorporated into an existing structure. The shear keys were designed according to finite element modelling, of which the generalisability was not verified. It was tested in the Civil Engineering Structure Test Centre (Shenyang, China) of Shenyang Jianzhu University.

2.1. Specimen Design

The dimensions and form of the shear keys used in the BCSWSK are shown in Figure 4. The vertical faces were concave arcs, and the top and bottom faces were flat. The concave arc faces were designed to prevent the PBS and the CCUS from separating in the direction perpendicular to them. The concrete grade of the PBS and the CCUS was C25, and the depth/thickness of each was 50 mm. The concrete grade of the shear keys was C30: this was greater than the strength of the two slabs to prevent the keys from failing before the slab failed. The PBS is shown in Figure 5. The reinforcements had a diameter = 8 mm and tensile yield strength = 335 MPa. Field curing was adopted for the specimen.

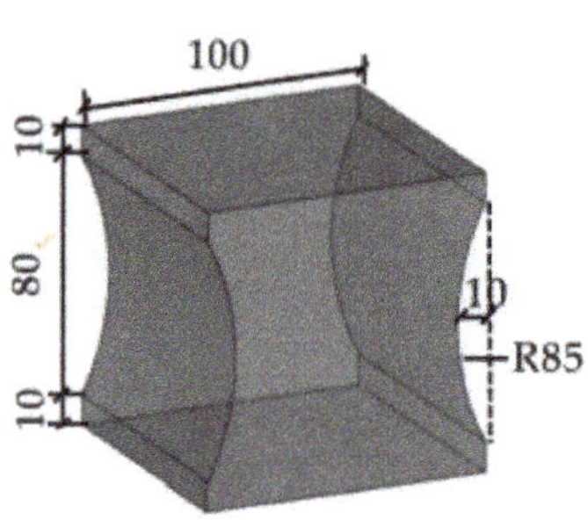

Figure 4. Dimensions and shape of shear key.

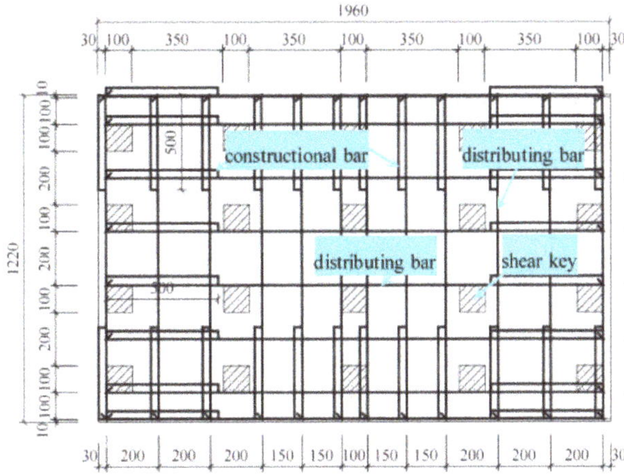

Figure 5. PBS of BCSWSK.

2.2. Fabrication Process

The fabrication process of the BCSWSK was as follows. The shear keys were made by cutting a long columnar concrete strip. The steel-reinforcing bars and shear keys were positioned in the mould, and concrete was poured to form the precast bottom slab (Figure 6a). After 28 days, the concrete of the CCUS was poured on top of the PBS to complete the lamination. The finished BCSWSK is shown in Figure 6b.

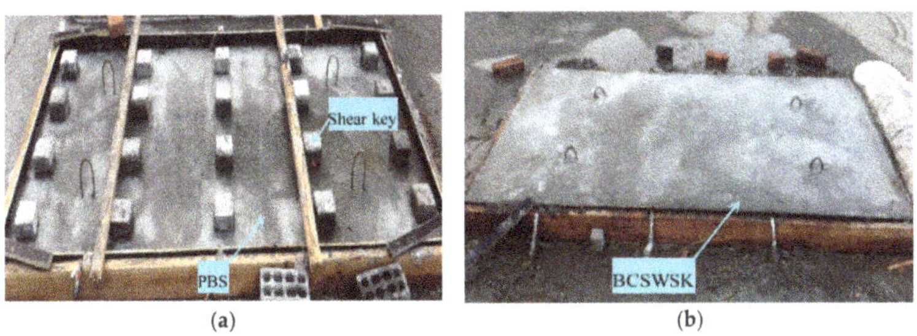

Figure 6. The production process for BCSWSK: (**a**) Precast bottom slab of BCSWSK; (**b**) BCSWSK.

2.3. Material Properties

The material properties of the steel bar and concrete were tested separately in the test rigs shown in Figure 7a,b. The compressive strength of the concrete (f_{ck}), the yield strength of the steel bar (f_y), the ultimate strength of the steel bar (f_u), and the corresponding Young's modulus ($E_c(E_s)$) are shown in Table 1.

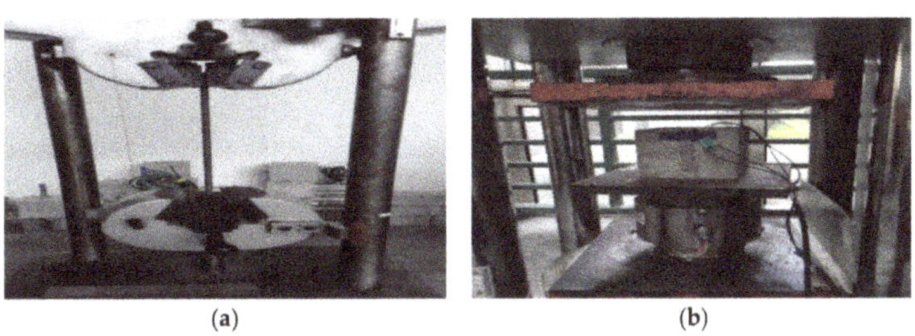

Figure 7. Steel bar and concrete materials tests: (**a**) Test rig for tensile strength of steel bar (HRB335); (**b**) Test rig for compressive strength of concrete.

Table 1. Measured mechanical property indexes of concrete and steel bar.

Batch	Material	f_{ck} (MPa)	f_y (MPa)	f_u (MPa)	$E_c(E_s)$ (MPa)
Precast bottom slab	Concrete	21.4	–	–	2.80×10^4
Post pouring concrete	Concrete	26.8	–	–	2.80×10^4
Shear keys	Concrete	27.6	–	–	3.00×10^4
HRB335	Steel bar	–	350	460	2.0×10^5

2.4. Experimental Equipment and Loading Scheme

The testing used a combination of sandbag loading and jack loading; the concentrated load exerted by the jack was transmitted to the slab by the sandbags. The loading device is shown in Figure 8. The slab was supported as follows. Two adjacent sides of the slab were supported by the immovable hinges of a triangular steel pipe, and the other two adjacent sides were supported by the rolling hinges of a cylindrical steel pipe. A steel slab was set between the bottom of the precast concrete slab and the hinge supports to avoid local compression failure. Seven layers of sandbags were stacked on the top slab surface, with eight bags per layer and each sandbag weighing 25 kg. A load-transfer device was placed on top of the sandbags. A jack was applied to the steel mat to apply the test load. For the loading system, the initial, maximum, and final settlement values were 0, 500, and 199 kN.

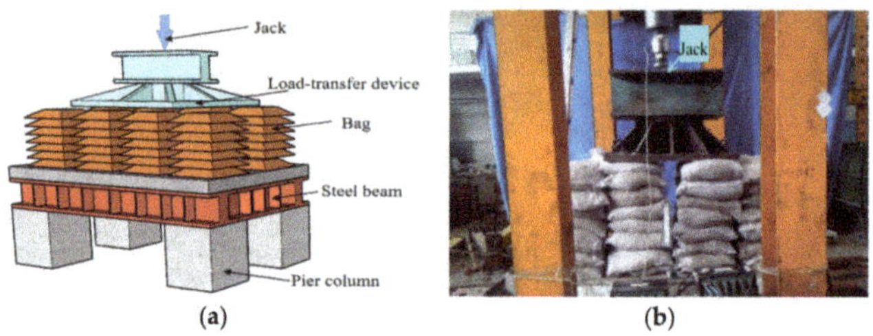

Figure 8. Loading system: (**a**) Schematic; (**b**) Photo.

2.5. Instrumentation Placement

Five displacement meters designated W1 to W5 were arranged symmetrically at the midpoints and the quarter points of the length and width of the slab bottom to measure the vertical displacement of the BCSWSK under vertical load. Strain gauges designated Y1 to Y12 were attached close to the midpoint and quarter points of the reinforcement to measure the strain of the tensile reinforcement at the bottom of the slab (Figure 9a). Displacement gauges designated W6 and W7 were arranged at the centre of the precast and laminated slabs to measure the relative displacement between the precast bottom slab and the cast-in-place concrete (Figure 9b).

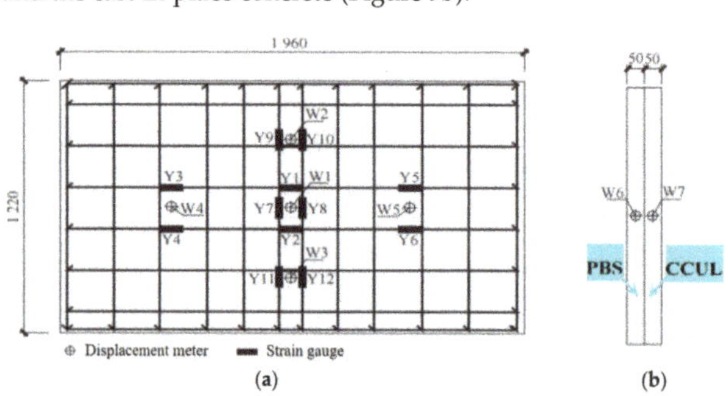

Figure 9. Placement of displacement meters and strain gauges for BCSWSK testing: (**a**) Placement of displacement meters and strain gauges at the bottom of the slab; (**b**) Placement of displacement meters at the centre of the slab.

3. Test Results

3.1. Failure Phenomena and Modes of Failure

The failure process of the BCSWSK was as follows. The first crack appeared on the side of the specimen when it was loaded to 80 kN; it was roughly vertical. When loaded to 160 kN, the cracks on the side of the slab extended to the slab bottom and developed to the mid-span. When loaded to the measured ultimate load (213 kN), destructive deformation occurred. The maximum crack appeared at the mid-span, where the crack width exceeded 1.5 mm, and the number of cracks also began to increase, indicating that the slab had failed (Figure 10).

(a)

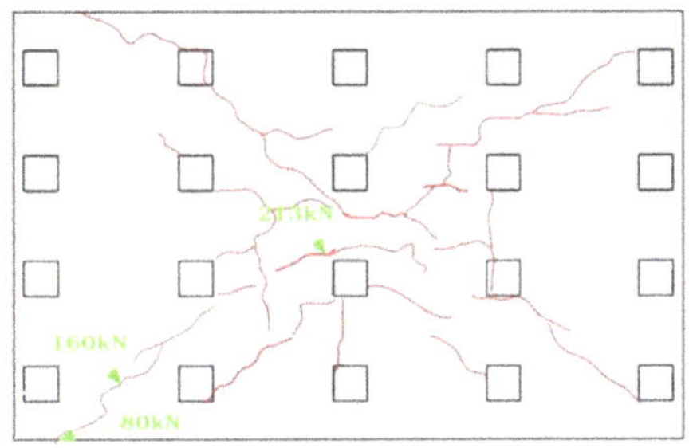

(b)

Figure 10. Crack patterns of BCSWSK after testing: (**a**) Photo; (**b**) Maps of failure patterns.

3.2. Analysis of the Curves from Testing

The load–strain curve of the tensile steel bar is shown in Figure 11, and the load–displacement curve of the concrete at the mid-span is shown in Figure 12.

It can be seen from the load–strain curve of the tensile-reinforcement steel bar of the BCSWSK (Figure 11) that, before the first crack occurred, the increase in load was approximately linearly proportional to the change in strain of the tensile steel bar, and the strain values were relatively small. As the load increased until cracks appeared in the specimen, the strain of the tensile reinforcement increased significantly, and there was a turning point in the curve. When the load increased to 160 kN, the tensile steel bar yielded, and the strain of the steel bar reached the yield value. It can be seen from Figure 12 that the load–displacement curve of the BCSWSK included an elastic section, an ascending inelastic section, and no descending section. The behaviour is similar to that of the cast-in-place slab, which indicates the integrity of the lamination [24].

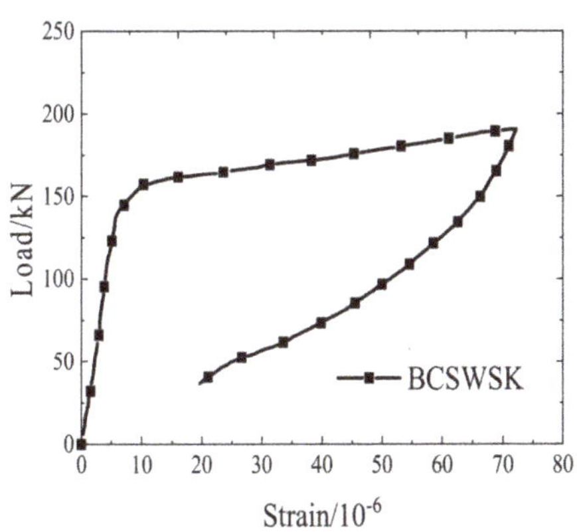

Figure 11. Load–strain curve of the tensile steel bar at mid-span.

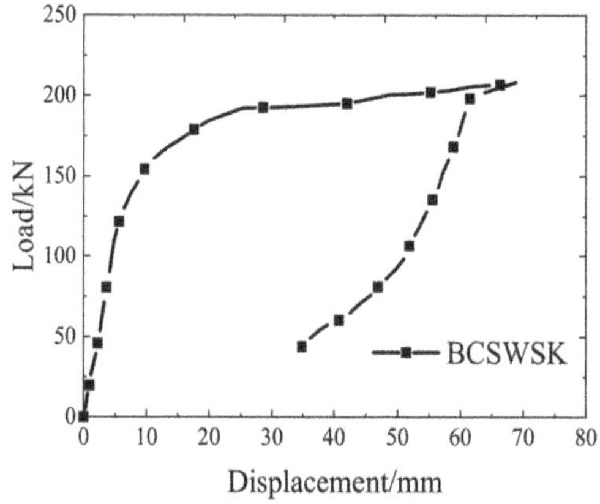

Figure 12. Load–displacement curve of the laminated concrete at mid-span.

3.3. Integrity of the Slab

In the fabrication of the BCSWSK, 28 d after the initial cast, there was a second concrete pour of the laminated slab, introducing concrete above the PBS and between the shear keys. Analysing the strength of the connection between the two was a key objective of the experiment. There was no disconnective cracking at the interfaces of the precast slab and the shear keys, which indicates that the precast slab and the shear keys were firmly connected and well-integrated (Figure 13). The horizontal displacement gauges (W6 and W7) showed that the relative displacement of the two slabs was almost zero, which also indicates the integrity of the PBS and the CCUL. It can also be seen that the crack pattern presents the failure characteristics of a two-way slab, which further shows its good integrity.

Figure 13. Failure of the precast bottom slab.

4. Finite Element Modelling and Analysis

Although the experimental results provide valuable information, they are limited to the material studied. Finite element analysis was used to generalise the results of the study. The main objectives of the numerical analysis were: (1) to determine the generalisability of the finite element model (FEM); (2) to determine the shear force distribution of the shear keys; and (3) to determine the effects of the main influencing parameters of the mechanical behaviour of the BCSWSK.

4.1. Generalisability of FEM

The load-bearing behaviour of the laminated concrete slab was modelled using ABAQUS. A concrete plastic-damage (CPD) model [24] was adopted for the constitutive relations of concrete. The equation of the compressive σ–ε curve used for it is:

$$\sigma_c = \begin{cases} \dfrac{E_c{}^2 f_{c.r}\varepsilon}{\dfrac{E_c{}^2\varepsilon_{c.r}}{\left(E_c{}^2\varepsilon_{c.r}-f_{c.r}\right)^2}-\left(\varepsilon_{c.r}E_c{}^3\varepsilon-E_c f_{c.r}\varepsilon\right)\left[1+\left(\frac{\varepsilon}{\varepsilon_{t.r}}\right)^{\frac{E_c\varepsilon_{c.r}}{E_c\varepsilon_{c.r}-f_{c.r}}}\right]} \\ \dfrac{f_{c.r}\varepsilon}{\varepsilon_{c.r}E_c\varepsilon\left[\left(0.157f_{c.r}^{0.785}-0.905\right)\left(\frac{\varepsilon}{\varepsilon_{t.r}}-1\right)^2+\frac{\varepsilon}{\varepsilon_{t.r}}\right]} \end{cases} \tag{1}$$

$$\varepsilon_{c.r} = \left(700+175\sqrt{f_c}\right)\times 10^{-6} \tag{2}$$

where $f_{c,r}$ is the representative value of concrete, $\varepsilon_{c,r}$ is the corresponding strain, E_c is the damage evolution parameters, and f_c is the standard compressive strength of the concrete.

A bifold model [25] was adopted for the constitutive relations of reinforcement. The equation of the σ–ε curve used for it is:

$$\sigma = \begin{cases} E\varepsilon & \varepsilon \le \varepsilon_y \\ f_y + 0.01E(\varepsilon-\varepsilon_y) & \varepsilon > \varepsilon_y \end{cases} \tag{3}$$

where f_y is the yield strength of steel bar and ε_y is the corresponding strain.

The PBS, shear keys, and CCUL were all C3D8R elements, and the steel bar was a T3D2 element incorporated into the concrete as an embedded region. The boundary conditions, loading modes, and meshing of the model are shown in Figure 14. The interface between the PBS and the CCUL was friction contact. The Newton iterative method was used to solve the FEM equations.

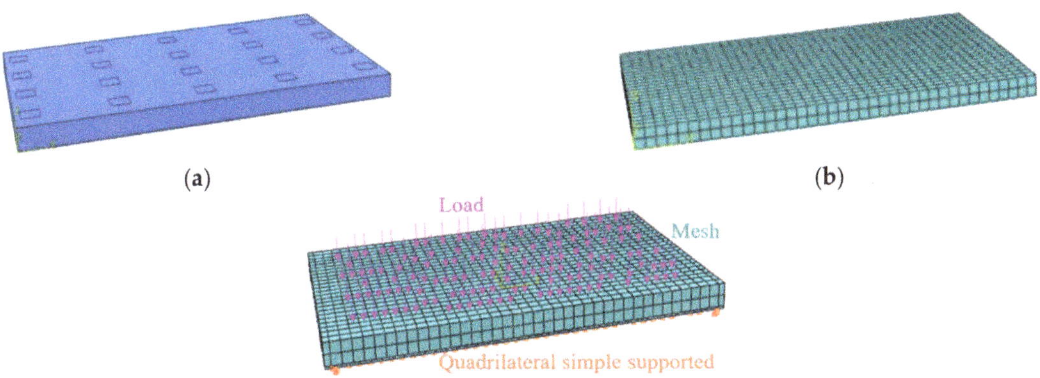

Figure 14. Boundary conditions, loading, and mesh generation of the model: (**a**) Unmeshed; (**b**) Load mesh preparation; (**c**) Load mesh.

The load–displacement curves and the load–strain curves of the reinforcement given by the FEM were compared with the test results (Figure 15). It can be seen that the simulated curve and the test curve were almost identical; the maximum difference was <1.5%, thus verifying the generalisability of the FEM.

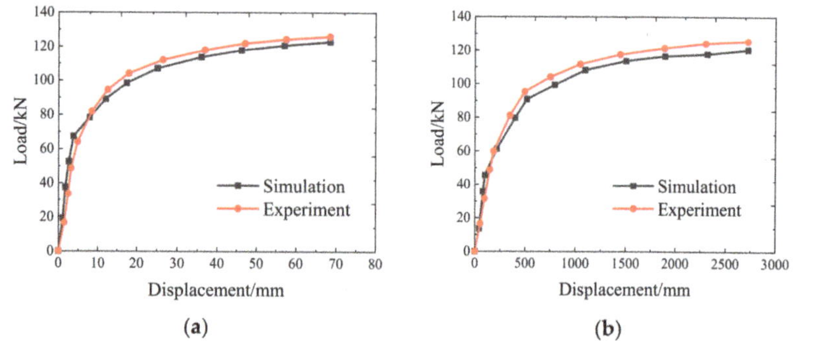

Figure 15. ABAQUS model and experimental results for comparison: (**a**) Load–displacement curves at mid-span; (**b**) Load–strain curves of steel bars.

4.2. Shear Force Distribution of the Shear Keys

Under vertical load, there was a relative misalignment between the PBS and the CCUL of the BCSWSK, which produced shear force in the shear keys. There were some differences between the keys in shear force along both the X and Y directions. A BCSWSK FEM, designated FJ1, was designed to enable us to analyse the shear distribution among the shear keys. The slab was grade C25 concrete, and the shear keys were grade C30. The remaining dimensions and physical parameters were identical to those of the test specimen. An identical loading process was simulated.

The shear force distribution within FJ1 under uniform vertical load was the same as for the original cast specimen because of the same symmetrical distribution of the shear keys along the X and Y directions. The shear keys in a quarter of the area of the slab were therefore selected for analysis, and they were numbered for convenience of description, as shown in Figure 16a.

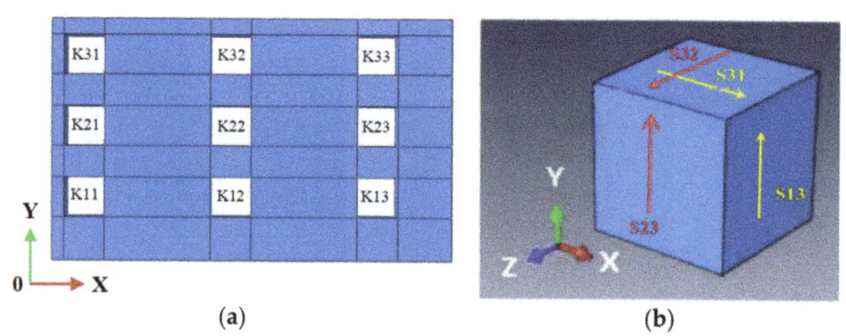

Figure 16. Number of shear keys and shear stress directions: (**a**) Number of shear keys in the quarter area of FJ1; (**b**) Shear stress directions on one shear key.

The shear keys were divided into eight elements. The shear stress directions of an element are shown in Figure 16b. X and Y are the length and width directions of FJ1, respectively. S13 is the shear stress along the Z direction in the YZ plane, and S23 is the shear stress along the Z direction in the XZ plane. According to the reciprocal theorem for shear stresses in the vertical plane, S31 = S13 and S32 = S23. The total shear force of a shear key at the contact surface of the PBS and the CCUL of FJ1 can be obtained: its value along the X direction is the product of S31 and its cross-sectional area, and its value along the Y direction is the product of S32 and its cross-sectional area. The relationship between the total load applied to FJ1, and the total shear force of each shear key in the X direction and the Y direction, can be plotted, as shown in Figure 17 (X) and Figure 18 (Y).

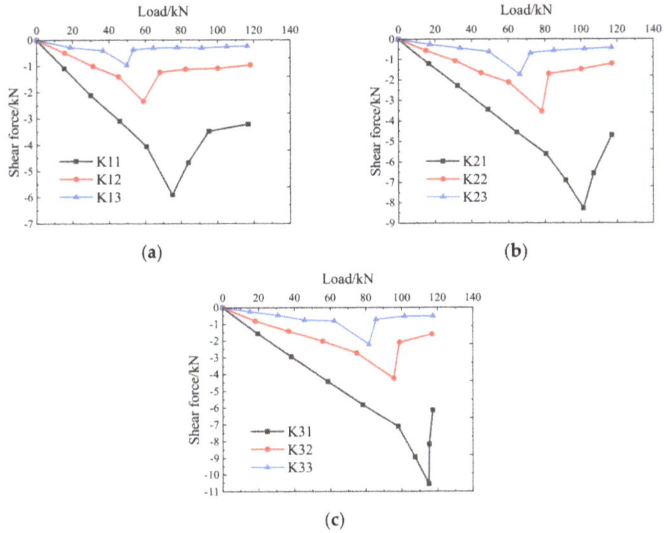

Figure 17. Load–shear force curves of the three groups of shear keys along the X direction: (**a**) K11–K13; (**b**) K21–K23; (**c**) K31–K33.

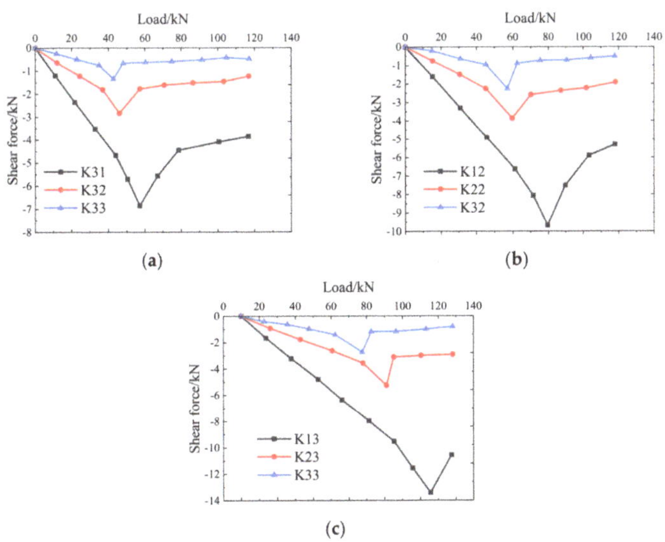

Figure 18. Load–shear force curves of the three groups of shear keys in the Y direction: (**a**) K11–K31; (**b**) K12–K32; (**c**) K13–K33.

4.2.1. Shear Distribution in the X Direction

Figure 17 shows the load–shear force curves of shear keys K11–K13, K21–K23, and K31–K33. It can be seen that for, all three groups of shear keys, the shear force increased linearly as the vertical load increased, and suddenly decreased when the shear force reached a maximum value. This behaviour indicates that the failure was brittle. The maximum shear force differed between the three groups: it was lowest near the mid-span and highest near the support. This is because the shear keys near the mid-span were primarily affected by bending moments. The vertical loads were different when the shear force reached its maximum value, so the shear keys did not all fail at the same time. As the vertical load increased, failure gradually developed from the shear keys in the region with a larger bending moment near the mid-span, to the supported region with a smaller bending moment.

A comparison of the maximum shear force between the three groups of shear keys shows that, although the load–shear curves of the three groups were very similar, the shear values of K31–K33 were the greatest, and the shear values of K21–K23 and K11–K13 were the least. This was mainly because, under a vertical uniform load, FJ1 deformed along the diagonal in the area nearest the support. Deformation increased towards the centre of the slab, as did the misalignment between the precast bottom slab and the cast-in-place slab. The group of shear keys K31–K33 was closest to the centre of the slab, so resistance to the shear force generated by the misalignment was greatest for this group.

4.2.2. Shear Force Distribution in the Y Direction

The load–shear force curves for the three groups of shear keys K11–K31, K12–K32, and K13–K33, in the Y direction of FJ1, are shown in Figure 18. It can be seen that, similar to the X direction, the shear force increased linearly as the applied vertical load increased and suddenly decreased when it reached a maximum value. In each of the three groups of shear keys, the shear force was at a maximum near the support and at a minimum near the mid-span of the slab. In each group, the load values differed when the shear force reached the maximum value. A comparison of the maximum shear values shows that the shear resistance of K13–K33 was greatest because they were close to the centre of the slab.

4.2.3. Comparison between the Shear Forces in the X and Y Directions

Figure 19 shows the load–shear force curves of K11, K22, and K33 in the X and Y directions. It can be seen that the maximum shear force was greater in the Y direction than in the X direction, and that the shear force in the Y direction reached the maximum value first. This is because the bending moment of FJ1 was greater in the short-span direction than in the long-span direction, and so the bending moment had a greater effect on the shear force of the keys in the Y direction. Thus, it would be better to give the shear keys a rectangular cross-section with the longer side parallel to the short span when designing a BCSWSK to make them more effective in resisting shear.

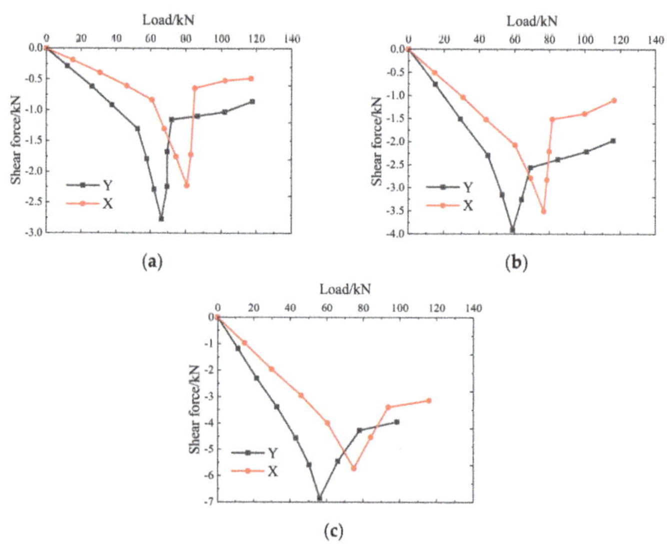

Figure 19. Load–shear force curves of different shear keys along the X and Y directions for comparison: (**a**) K11 along the X and Y directions; (**b**) K22 along the X and Y directions; (**c**) K33 along the X and Y directions.

4.2.4. Shear Key Cross-Section

The preceding analysis indicates that a rectangular shear key cross-section would be better than a square cross-section in terms of resisting shear. To clarify this issue, rectangular cross-section shear keys with different dimensions were modelled in the FEM. Figure 20 shows the different load–displacement curves obtained for the keys with respective dimensions in the X and Y directions: (L_{XY}) 50 × 200 mm, 100 × 100 mm, and 200 × 50 mm. Other dimensions and physical parameters were identical to those of FJ1. It can be seen that the bearing capacity was greatest for the 50 × 200 mm key until the displacement was about 53 mm, but after this point, the bearing capacity became greatest for the 200 × 50 mm key. However, the difference between the three curves was slight, mostly <10%, so the square cross-section is recommended for the shear keys.

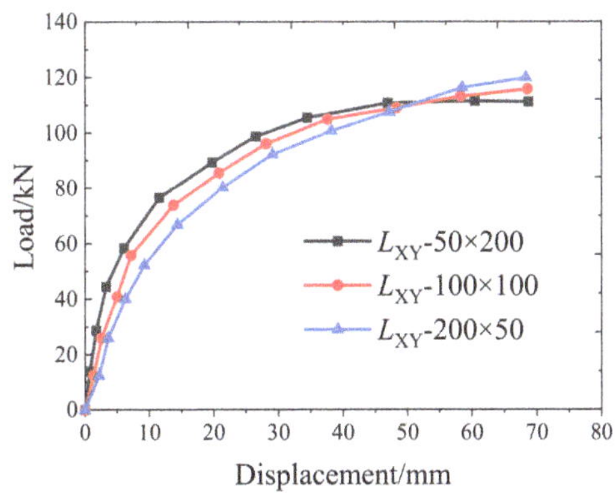

Figure 20. Load–displacement curves of BCSWSKs with different shear-key shapes for comparison.

4.3. Parameter Influence

Twenty-five laminated slabs were modelled to allow us to analyse the influence of different factors on the load–displacement curves of BCSWSKs, as shown in Table 2. All dimensions and physical parameters were identical to those of model FJ1, except for the parameters that were varied experimentally. The simulations show that changing the friction coefficient, the grade of the cast-in-place concrete, and the grade of the concrete used with shear keys had very little effect on the mechanical properties of BCSWSKs, mostly <3.2%, and their load–displacement curves almost coincided, but changing the other parameters had clearly observable effects, as shown in Figures 21–23.

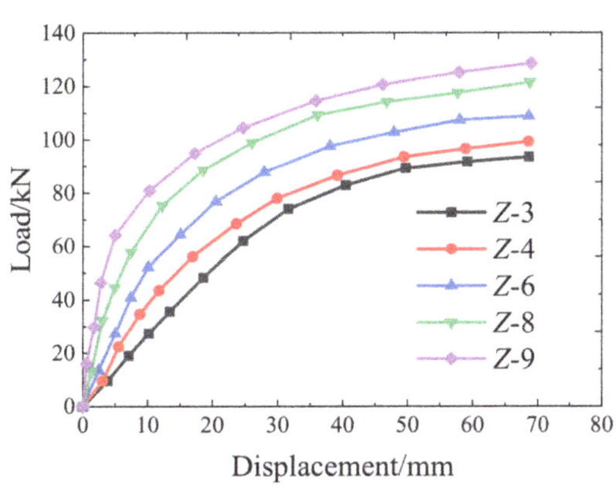

Figure 21. Load–displacement curves of BCSWSKs with different numbers of columns of shear keys for comparison.

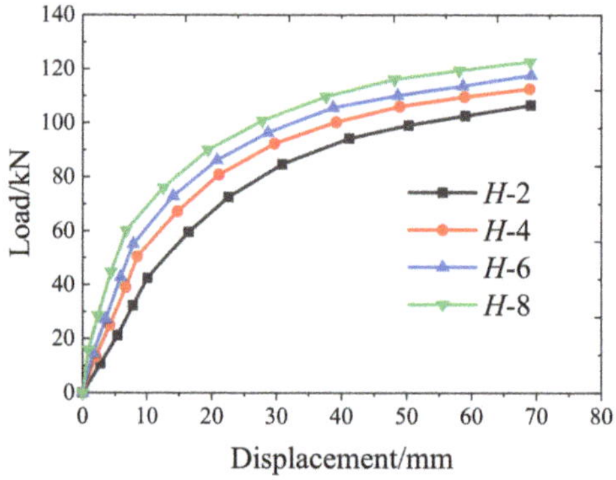

Figure 22. Load–displacement curves of BCSWSKs with different numbers of rows of shear keys for comparison.

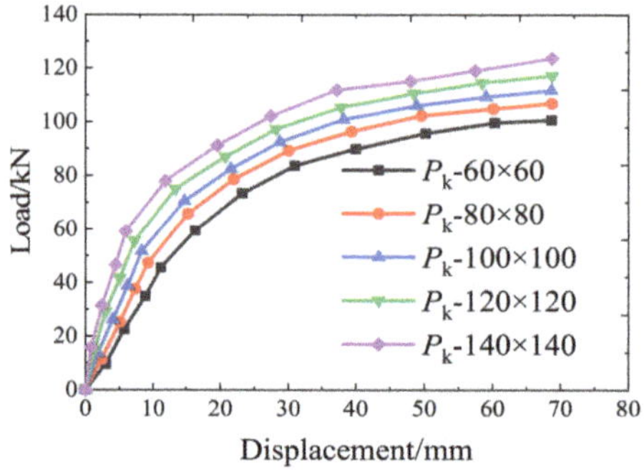

Figure 23. Load–displacement curves of BCSWSKs with different shear key dimensions for comparison.

Table 2. List of laminated slabs modelled.

Model Designation	Number of Shear Key Columns Z	Number of Shear Key Rows H	Shear Key Cross-Section Dimensions P_k (mm)	Concrete Grade		Friction Coefficient f	Yield Load (kN)	Yield Displacement (mm)
				Cast-in-Place Concrete S_i	Shear key S_k			
f-0	6	4	100 × 100	C25	C30	0	77.2	16.7
f-0.4	6	4	100 × 100	C25	C30	0.4	77.4	16.6
f-0.6	6	4	100 × 100	C25	C30	0.6	77.8	16.4
f-0.8	6	4	100 × 100	C25	C30	0.8	78.3	16.2
S_i-20	6	4	100 × 100	C20	C30	0.8	77.5	16.0
S_i-25	6	4	100 × 100	C25	C30	0.8	78.3	16.2
S_i-35	6	4	100 × 100	C35	C30	0.8	79.3	16.5
S_k-20	6	4	100 × 100	C25	C20	0.8	77.9	16.4
S_k-30	6	4	100 × 100	C25	C30	0.8	78.3	16.2
S_k-40	6	4	100 × 100	C25	C40	0.8	78.9	16.0
Z-3	3	4	100 × 100	C25	C30	0.8	70.4	23.5
Z-4	4	4	100 × 100	C25	C30	0.8	74.7	19.4
Z-6	6	4	100 × 100	C25	C30	0.8	78.3	16.2
Z-8	8	4	100 × 100	C25	C30	0.8	80.7	14.0
Z-9	9	4	100 × 100	C25	C30	0.8	81.7	12.5
H-2	6	2	100 × 100	C25	C30	0.8	72.4	22.4
H-4	6	4	100 × 100	C25	C30	0.8	75.7	18.2
H-6	6	6	100 × 100	C25	C30	0.8	78.3	16.2
H-8	6	8	100 × 100	C25	C30	0.8	79.5	15.0
P_k-60 × 60	6	4	60 × 60	C25	C30	0.8	70.6	23.2
P_k-80 × 80	6	4	80 × 80	C25	C30	0.8	75.4	18.8
P_k-100 × 100	6	4	100 × 100	C25	C30	0.8	78.3	16.2
P_k-120 × 120	6	4	120 × 120	C25	C30	0.8	79.6	15.3
P_k-140 × 140	6	4	140 × 140	C25	C30	0.8	81.0	13.0

4.3.1. Effect of Number of Shear Key Columns along the X Direction (Z)

Figure 21 shows the load–displacement curves of BCSWSKs for different numbers of shear key columns (Z = 3, 4, 6, 8, 9). Figure 21 and Table 2 show that the bearing capacity increased as the number of shear key columns increased. When the number of columns increased from three to nine, the yield load increased by 16.1% and the yield displacement decreased by 47%. When the number of columns increased from eight to nine, the yield load increased by 1.2% and the yield displacement decreased by 10.8%. These results show that increasing the number of columns increases the yield load and decreases the yield displacement of the BCSWSK, but this trend is not so pronounced after a certain point. This reduction in effectiveness occurs because, at some point, the number of shear keys in the slab is such that the relative misalignment between the PBS and the CCUL is very small, so increasing the number of shear keys has little effect.

4.3.2. Effect of Number of Shear Key Rows along the Y Direction (H)

Figure 22 shows the load–displacement curves of BCSWSKs for different numbers of rows of shear keys (H = 2, 4, 6, 8). Figure 22 and Table 2 show that the bearing capacity increased as the number of shear key rows increased. When the number of rows increased from two to eight, the yield load increased by 9.8% and the yield displacement decreased by 33.2%. When the number of rows increased from six to eight, the yield load increased by 1.5% and the yield displacement decreased by 7.5%. These results show that increasing the number of rows of shear keys increases the yield load and decreases the yield displacement of the BCSWSK, but this trend is not so pronounced after a certain point. The reason for this is the same as that for the increase in the number of columns. However, a change in the number of rows has a relatively small effect on the bearing capacity of BCSWSKs compared to a change in the number of columns.

4.3.3. Effect of Shear Key Cross-Section (P_k)

Figure 23 shows the load–displacement curves of the BCSWSK with different shear key cross-sections (P_k = 60 × 60, 80 × 80, 100 × 100, 120 × 120, and 140 × 140 mm). It can be seen from Figure 23 and Table 2 that bearing capacity increased as the shear key cross-section increased. When the cross-section increased from 60 × 60 mm to 140 × 140 mm, the yield load increased by 19.5% and the yield displacement decreased by 44%. When the cross-section increased from 120 × 120 mm to 140 × 140 mm, the yield load increased by 1.8% and the yield displacement decreased by 1.5%. These results show that increasing the cross-section areas of the shear keys increases yield load and decreases the yield displacement of BCSWSKs, but this trend is not so pronounced after a certain point. The reason for this is the same as that for the effect of the number of columns.

5. Conclusions

In this research, experimental testing and FEMs were used to analyse shear force distribution and the factors that influence the mechanical properties of BCSWSKs under the action of uniform vertical load. In view of the test results and analysis, the following conclusions can be drawn.

(1) The loading scheme of stacking and jacking to apply an equivalent uniform load is appropriate. The bonds at the interfaces between the PBS and the shear keys, and between the PBS and the CCUL, are strong despite the secondary concrete pouring for the upper slab.

(2) The shear force distribution among its shear keys is similar in the X and Y directions. The shear force increases linearly as the load increases and decreases suddenly when the load reaches a certain point, namely when the shear keys show brittle failure. The shear force is greater in the Y direction than in the X direction and reaches a maximum value earlier. The difference in shear force distribution among shear keys between the two directions is small, mostly <10%. A square cross-section is recommended for the shear keys.

(3) The number of rows, number of columns, and cross-sectional area of the shear keys are the main factors that influence the mechanical properties of BCSWSKs, of which the yield load and yield displacement varies by 9.8%~19.5% and 33.2%~47%, respectively. The effectiveness of all three factors decreases as the factors increase and, after a certain point, the effect is minimal, mostly <7.5%. After a certain number of shear keys have been placed, the cast-in-place concrete grade, the shear key concrete grade, and the friction coefficient between the precast bottom slab and the cast-in-place slab have little influence on the mechanical properties, mostly <3.2%.

6. Future Research

This study investigated the mechanical properties of BCSWSKs using load tests on a fabricated laminated slab and finite element models. The shear force distribution across the shear keys was measured and analysed, and the effects of different parameters were identified. These provide much reference for the design of BCSWSKs in the service stage. However, their mechanical properties in the construction stage was not studied, such as the bending properties of PBSs and PBSs with I-beams bolted to them. As such, more research should be conducted on them for their engineering practicality.

Author Contributions: Conceptualization, M.L. and Z.Z.; methodology, Z.Z.; software, W.Y.; validation and formal analysis, Z.Z.; data curation, S.Y.; writing—original draft preparation, M.L.; writing—review and editing, Q.W. All authors have read and agreed to the published version of the manuscript.

Funding: This research was funded by the National Natural Science Foundations of China (Grant No. 51978422), the General Project of Liaoning Provincial Department of Education (Grant No. LJKZ0561), and the Ministry of Housing and Urban-Rural Development Project (Grant No. 2019-K-080).

Institutional Review Board Statement: Not applicable.

Informed Consent Statement: Not applicable.

Data Availability Statement: The data used to support the findings of this study are available from the corresponding author upon request.

Conflicts of Interest: The authors declare no conflict of interest.

References

1. Xie, Q. Reflections on the development of assembled concrete structure buildings. *Eng. Constr. Des.* **2021**, *23*, 26–30. (In Chinese)
2. Nie, J.; Jiang, Y.; Nie, X.; Zhuang, L. Effect of truss reinforcement in laminated slabs on the force performance of precast slabs. *J. Build. Struct.* **2021**, *42*, 151–158. (In Chinese)
3. Liu, W.; Cui, S.; Liu, C.; Shi, L. Experimental and theoretical study on the flexural performance of prestressed concrete steel truss laminated slabs. *J. Build. Struct.* **2021**, *42*, 95–106. (In Chinese)
4. Yang, X.; Wang, Y.; Liu, Y.; Wei, Z. Experimental Study and Numerical Simulation on Mechanical Properties of the Bottom Plate in the Assembled Composite Slab with Additional Steel Trusses. *Adv. Civ. Eng.* **2021**, *2021*, 7240994. [CrossRef]
5. Kanchanadevi, A.K.; Ramanjaneyulu, K.; Srinivas, V. Behaviour of concrete composite slabs with truss type shear connectors of different orientation angle. *Adv. Struct. Eng.* **2021**, *24*, 3070–3084. [CrossRef]
6. Ran, H.; Yu, Q.; Qin, K.; Zhao, Q.; Zeng, X. Experimental study of truss reinforced concrete laminated slabs in service phase. *Struct. Eng.* **2019**, *35*, 164–171. (In Chinese)
7. Li, J.; Huang, H.; Zeng, C.; Zhang, M.; Huang, S. A simplified elastic calculation method for four-sided solidly supported precast ribbed bottom concrete two-way laminated slabs. *J. Hunan Univ. Sci. Technol. (Nat. Sci. Ed.)* **2020**, *35*, 33–43. (In Chinese)
8. Jiang, D.; Huang, H.; Zhu, M.; Zhang, M.; Huang, S. Experimental study on flexural fatigue performance of precast prestressed ribbed base concrete laminated slabs. *Ind. Constr.* **2020**, *50*, 76–83.
9. Huang, H.; Wu, F.; Zhu, M.; Zeng, C.; Lv, W. Study on the influence of plate rib form on the flexural performance of precast ribbed bottom concrete laminated slabs. *J. Build. Struct.* **2015**, *36*, 66–72. (In Chinese)
10. An, H.; Zheng, Z.; Cao, Y.; Li, X.; Wang, W. Experimental study on flexural performance of reinforced truss fiber cement composite slabs. *Build. Struct.* **2022**, *52*, 74–78+131. (In Chinese)
11. Huang, H.; Zhou, P.; Zhang, M.; Gao, Y. Simplified plastic design method of prefabricated composite rib concrete two-way superimposed slab cross beam floor. *Build. Struct.* **2022**, *52* (Suppl. S1), 1738–1743. (In Chinese)
12. Liu, Y. Study on the Performance of Self-Supporting Reinforced Truss Concrete Laminated Slabs. Master's Thesis, Zhejiang University, Hangzhou, China, 2006. (In Chinese)

13. Ma, L.; Chen, X.; Jiang, L.; Yu, Z.; Zhang, Z. Experimental and finite element analysis of reinforced truss concrete laminated floor slabs. *Build. Struct.* **2013**, *43*, 54–57+62. (In Chinese)
14. Zhao, L. Research on the Design and Calculation Method of Self-Supporting Reinforced Truss Concrete Laminated Slab. Master's Thesis, Central South University, Changsha, China, 2007. (In Chinese)
15. Cheng, J.Y.; Lei, Z.; Jian, J.Y. Study on short term rigidity of precast composite slab with steel truss and concrete. In *Advanced Materials Research*; Trans Tech Publications Ltd.: Wollerau, Switzerland, 2013; Volume 639.
16. Wu, F.; Huang, H.; Chen, W.; Zhou, X. Experimental study on the force performance of prefabricated ribbed thin concrete laminated slabs. *Civ. Constr. Environ. Eng.* **2011**, *33*, 7–12+19. (In Chinese)
17. Qian, H.; Zhou, K. Experimental study of PK prestressed concrete laminated slabs. *Chin. Foreign Archit.* **2006**, *3*, 96–97.
18. Shi, L.; Wang, H.; Wang, Q.; Li, X. Analysis of mechanical properties of prestressed ribbed concrete laminated slabs. *Build. Struct.* **2021**, *51* (Suppl. S1), 1135–1141. (In Chinese)
19. Huang, W.; Ma, X.; Luo, B.; Li, Z.; Sun, Y. Experimental study on flexural behaviour of lightweight multi-ribbed composite slabs. *Adv. Civ. Eng.* **2019**, *1*, 1–11. [CrossRef]
20. Li, M.; Wang, H.; Zhao, W. Testing the force performance of unidirectional laminated plates with shear keys. *J. Jilin Univ. (Eng. Ed.)* **2020**, *50*, 654–667. (In Chinese)
21. Li, M.; Wang, H.; Zhao, W. Mechanical properties and design methods of unidirectional laminated plates with shear bonds. *J. Build. Sci. Eng.* **2019**, *36*, 35–45. (In Chinese)
22. *JGJ1-2014*; Technical Specification for Precast Concrete Structures. Ministry of Housing and Urban-Rural Development of the People's Republic of China: Beijing, China; Architecture & Building Press: Beijing, China, 2014. (In Chinese)
23. Li, X.; Yin, J.; Xu, Q. Experimental research on mechanical properties of RC two-way cast-in-situ slab with existing cracks. *Earthq. Resist. Eng. Retrofit.* **2010**, *32*, 47–51. (In Chinese)
24. Du, X.; Dang, L.; Ren, Q. Plastic damage cracking study on the reinforced concrete slab. *Water Power* **2009**, *35*, 28–29+56. (In Chinese)
25. Ma, T. Study on Hybrid Testing of Continuous Beambridge Structure Based on Bilinear Model Renewal. Master's Thesis, Suzhou University of Science and Technology, Suzhou, China, 2019. (In Chinese)

coatings

Article

Experimental Study on Flexural Bearing Characteristics of Ceramsite Concrete Beams after Creep at Different Curing Ages

Hongbing Zhu [1,2,*], Jingyi Chen [1], Yuxin Wu [1], Jiapeng Li [1,3], Zhenghao Fu [1] and Peng Liu [4]

1 School of Urban Construction, Wuhan University of Science and Technology, Wuhan 430065, China
2 Institute of High-Performance Engineering Structure, Wuhan University of Science and Technology, Wuhan 430065, China
3 The National Prefabricated Construction Industry Base, Huasen Architecture and Engineering Design Consulting Company, Shenzhen 518000, China
4 School of Civil Engineering, Central South University, 22 Shaoshan Road, Changsha 410075, China
* Correspondence: zhuhongbing@wust.edu.cn

Abstract: Ceramsite concrete is increasingly used for structural applications due to its lightweight, high-strength, and high-temperature resistance advantages. However, research into the postcreep properties of ceramsite concrete has yet to keep pace with other areas. Therefore, this paper aims to investigate the flexural bearing properties of ceramsite concrete beams after creep and make recommendations for future research. To study the flexural load-bearing performance of ceramsite concrete beams after creep, a comparison test was carried out on four ceramsite concrete beams subjected to 200 d creep loading and four beams subjected to static load without creep loading for four curing ages. The results showed that, at first, the deformation capacity of the ceramsite concrete beams would increase after creep and the toughness and energy absorption capacity would also strengthen. Secondly, the creep could improve the ultimate bearing capacity of the ceramsite concrete beams. It was an approximate 10% increase after 200 d creep. Then, the stiffness, toughness, and ultimate bearing capacity of the ceramsite concrete beams would increase gradually with the increase of the curing age after the creep. In addition, the ultimate bearing capacity of the ceramsite concrete beams increased gradually with the increase of the curing age. The growth rate was faster in the early stage, slower after 120 d, and slower and more stable after 228 d. Finally, creep could cause the maximum crack width and depth of the ceramsite concrete beams decreased and the number of cracks would grow with the curing age. The conclusions obtained in the study provide a theoretical basis for the design of light aggregate structures for creep.

Keywords: ceramsite concrete beam; crack; creep; curing ages; ultimate bearing capacity

Citation: Zhu, H.; Chen, J.; Wu, Y.; Li, J.; Fu, Z.; Liu, P. Experimental Study on Flexural Bearing Characteristics of Ceramsite Concrete Beams after Creep at Different Curing Ages. *Coatings* **2023**, *13*, 158. https://doi.org/10.3390/coatings13010158

Academic Editor: Andrea Nobili

Received: 10 December 2022
Revised: 27 December 2022
Accepted: 9 January 2023
Published: 11 January 2023

1. Introduction

In recent years, as the spans of buildings and bridges have increased, the self-weight of structures is approaching the limits of what ordinary concrete can withstand. As a result, reducing the weight of such systems has attracted increasing research attention [1].

Ceramsite concrete can reduce the weight of ordinary concrete by 20% to 40% under the same load-bearing conditions [2,3]. Due to its lightweight, high strength, and high temperature resistance [4], it has been used in high-rise buildings and bridge projects and is of high research value. In practical engineering, most ceramsite concrete structures are subjected to long-term loading, causing creep in the structure. Jiang [5] found that ceramsite concrete has greater creep values than ordinary concrete, which can be up to 1.3 times greater. Under long-term loading conditions, this can lead to large creep deformations and relaxations. On the one hand, this may cause the structure to tilt and deform. On the other hand, excessive structural deformation may change the structure's mechanical properties and seriously affect its safety. Both aspects deserve attention.

For them, the former relates to the laws of concrete creep and the latter to the effects of concrete creep. Currently, many studies reflect concrete creep, and the creep values of ceramsite concrete are higher than those of ordinary concrete of the same age. The tensile creep of highly porous ceramsite concrete will be lower than that of low-absorbent ceramsite concrete, and plain ceramsite concrete [6,7]. In addition, the loading age of the structure [8–10], the curing age [11] and curing temperature [12], the concrete's collapse [13] and sand content [14], dosing of steel fibres [15,16], dosing of polymeric fibers [17], admixture [18], and the relative humidity of the working environment in which the concrete elements are placed [19,20] have all been shown to be influential factors in creep. As research continued, Zhu, Glanville [21,22] and others used experimental and theoretical hypotheses on creep to propose a creep model containing loading age and loading time parameters, respectively. Compared with the studies on the creep law of concrete, there are fewer studies involving the structure's mechanical properties after creep. Dong, Rong, and others [23–25] have suggested that long-term loading causes stress relaxation in concrete beams and reduces the stress at the tip of concrete cracks, resulting in an increase in the residual bearing capacity of the concrete beam and an improvement in crack development. However, Liu [26] obtained the opposite test results. He found that reclaimed concrete beams under long-term loading exhibited postductility and that secondary loading reduced their flexural bearing capacity. Furthermore, Saliba and Omar [27,28], among others, suggested that the development of microcracks in concrete during creep can reduce concrete residual strength. It is challenging to provide scientific references for practical engineering. Therefore, the effect of creep on the mechanical properties of ceramsite concrete structures needs to be urgently investigated. In addition, in existing studies, extending the loading age of plain concrete beams (i.e., increasing the curing age) reduces the effect of creep on the structure [8]. This gives good ideas for improving the mechanical properties of ceramsite concrete structures after creep. This is also a technical issue that deserves attention.

Analyzing the above, ceramsite concrete has good prospects for engineering applications because of its lightweight and high-strength properties. However, the effect of creep on its mechanical properties has yet to be unified, which does not provide a scientific reference for practical engineering. In addition, although extending the age of curing can reduce the effect of creep on plain concrete beams, it is still being determined whether it can improve the mechanical properties of ceramsite concrete beams after creep. Clarifying and improving the effect of creep on the mechanical properties of ceramsite concrete structures is a significant issue. Therefore, this study aims to clarify the effect of creep on the flexural properties of ceramsite concrete beams. On this basis, the influence of curing age on flexural bearing characteristics of ceramsite concrete beams after creep is found. In order to achieve these objectives, the following work was carried out in this paper. (i) Eight ceramsite concrete beams with four types of curing ages were produced. Then, the loading of 30% flexural ultimate bearing capacity was carried out on one group of them (four beams with different curing ages) for 200 days. Next, the two groups of the beam were tested under various loads by step loading, and the cracking load, failure load, and crack development of the beam were recorded. (ii) Based on the deformation law, ultimate bearing capacity, beam damage, and crack distribution, the influence of long-term creep on the flexural performance of ceramsite concrete beams was analyzed by comparing them with the experimental beams of the same age without creep loading. (iii) We drew the influence curve of ultimate bearing capacity and crack distribution curve of ceramsite concrete beams after creep with different curing ages during a static load test and fitted the data to analyze the influence of curing ages on mechanical properties of ceramsite concrete beams after creep.

2. Experimental Materials and Basic Mechanical Properties

2.1. Experimental Materials

So that related workers can repeat the experiment in the future, all materials used are listed as follows.

(1) The coarse aggregate is 900-grade crushed stone shale granules produced by Yichang Guangda Ceramic Granules Products Co. Prewetting ceramsite before concrete mixing can effectively improve the workability of concrete. Its physical properties are shown as Table 1. The ceramsite is prewetted for 4 h, and then dried for 12 h to remove the water on the ceramsite surface (shown in Figure 1).

Table 1. Physical properties of shale ceramsite.

Particle Size (mm)	Volume Density (kg/m^3)	Apparent Density (kg/m^3)	Compressive Strength of Concrete Cylinder (MPa)	Water Absorption in 1 h (%)
5~20	814	1517	6.8	2.42

(a) (b)

Figure 1. Shale ceramsite. (**a**) Before wetting process. (**b**) After wetting process.

(2) Fine aggregate was ordinary river sand, which was sieved before the test to avoid the influence of impurities in the sand. The filter sieve used for the test was a 4.75 mm side length square hole sieve. The physical properties are shown in Table 2.

Table 2. Physical properties of sand.

Volume Density (kg/m^3)	Apparent Density (kg/m^3)	Mud Ration (%)	Fineness
1570	2650	≤2	2.7

(3) The cement was P.O. 42.5 ordinary silicate cement produced by Hubei Huaxin Cement Company Limited, Wuhan, China and its physical indices all met the requirements of the test specification, as shown in Table 3.

Table 3. Basic physical index of cement.

Density (g/cm^3)	Mineral Composition of Clinker				Fineness(Sieve Allowance by 80 μm Square Hole)/%
	C3S	C2S	C3A	C3AF	
3.15	45	25	12	8	6.5

(4) The water-reducing agent is HsC polycarboxylat-type superplasticizers produced by Qingdao Hongxia. The technical specifications of the water-reducing agent are shown in Table 4.

Table 4. Technical index of the water-reducing agent.

Color	PH	Relative Density	Solid Content (%)	Water-Reducing Rate (%)
Pale yellow	6~8	1.08 ± 0.02	40	25–35

(5) The water was Wuhan tap water.

2.2. Mix Proportion

The strength of the concrete prepared for this test was based on the actual bridge project and the proposed strength class was LC40 ceramsite concrete. Referring to the technical standard for applying lightweight aggregate concrete (JGJ/T 12-019) in China [29], LC40 was used as the target of the trial formulation. Based on the research of the former research group [30], the mix proportion of ceramsite concrete with the strength grade of LC40 was determined after several trial mixes, as shown in Table 5. On the basis of this mix proportion, the density of ceramsite concrete was 1942 kg·m^{-3}.

Table 5. Mixed proportion of ceramsite concrete beam specimens (unit: kg/m^3).

Cement	Ceramsite	Sand	Water-Reducing Agent	Water	Sand Ratio
510	554	730	5.1	152	43%

2.3. Basic Mechanical Properties of Concrete

According to GB/T50081-2019 Standard for Physical and Mechanical Properties of Concrete [31], the compressive strength test used 100 mm × 100 mm × 100 mm cube test blocks. The axial compressive strength test used 100 mm × 100 mm × 300 mm prismatic test blocks. Flexural strength test use 100 mm × 100 mm × 400 mm prismatic test blocks. The test blocks are made from the mix proportions shown in Table 5. Tests were carried out after 28 d of curing to obtain their cubic compressive strength, axial compressive strength and flexural strength values. The specimens cured for 28 d were also tested for static compressive modulus of elasticity and the static compressive modulus of elasticity of ceramsite concrete was measured. The results are shown in Table 6.

Table 6. Basic mechanical indices of ceramsite concrete after 28-d maintenance.

Mechanical Parameters	Strength			
	Specimen No. 1	Specimen No. 2	Specimen No. 3	Test Results
Compressive strength/MPa	40.04	38.77	42.47	40.43
Flexural strength/MPa	6.92	7.05	6.43	6.80
Elasticity modulus/GPa	24.45	24.01	24.18	24.21

3. Design of the Test

This research aimed to study the change in the flexural capacity of ceramsite concrete under creep. Based on this goal, 12 specimens were divided into three groups.

The ceramsite concrete beams used in the test were designed depending on the principle of "strong shear and weak bending". Their cross-section size were 100 mm × 150 mm and span length *L* were 1500 mm. In addition, their net span L0 were 1400 mm. The longitudinal reinforcement of the beam was 24. The erection reinforcement was 48, type HRB400. the hoop reinforcement was φ6, type HPB300. Furthermore, the thickness of the concrete protective layer was 20 mm, as shown in Figure 2.

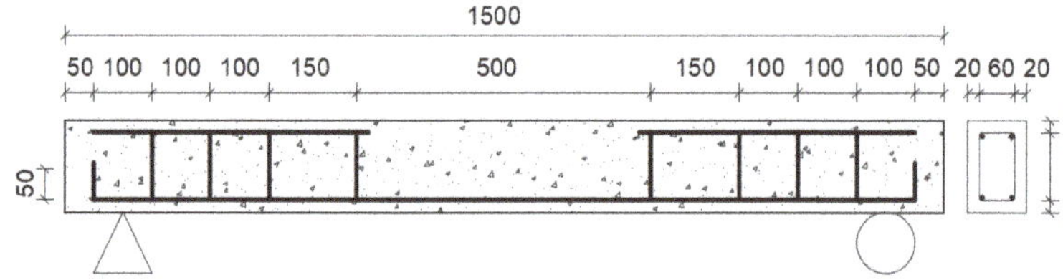

Figure 2. Structure of ceramsite concrete beam (unit: mm).

A total of 12 test beams were poured. To confirm the creep loading value of the test beam at each curing age, four beams were tested for the ultimate bending-bearing capacity under static load at 28 d, 60 d, 90 d, and 120 d of natural curing, respectively. As a reference group for the creep test beams, four beams were subjected to static load damage tests after 228 d, 260 d, 290 d, and 320 d of natural curing. The last four beams were subjected to 200 d of creep loading after 28 d, 60 d, 90 d, and 120 d of natural curing. Moreover, test setup for sustained load for reference Li [16], the four-point loading method was used for surcharge loads (shown as Figures 3 and 4).

(**a**) (**b**)

Figure 3. Snapshots of loading. (**a**) Front view. (**b**) Side view.

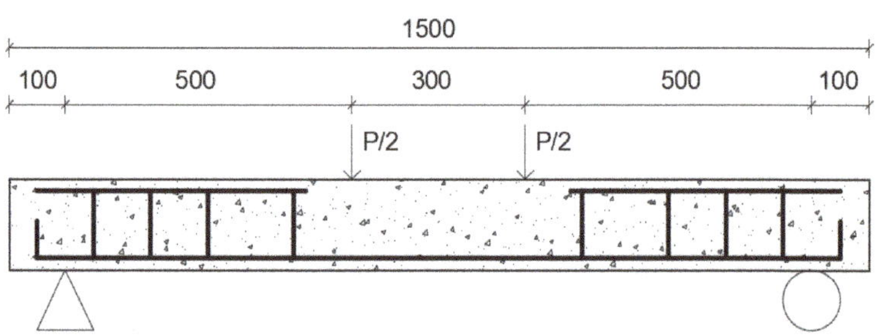

Figure 4. Diagram of loading (unit: mm).

This experiment takes into account laboratory specifics and safety issues, placing two test beams side by side. As the loading forces required for each beam are different, the following approach was taken for the experiment in order to combine the data from Table 7 with Figure 3. (i) First, we weigh and prestack each weight. We place a jack at the loading point of the distribution beam and then place the weight evenly on the distribution beam so that the force acting on the jacks are the same as the loading force shown in Table 7. (ii) Secondly, we mark in detail the distribution beam and the location of the jacks reaction points. (iii) Thirdly, the jacks are removed and the support, distribution beams and weight are placed in the marked positions. Referring to existing studies [14,32,33], the loading stress is taken to be 30% of the ultimate load of the specimen beams. the spanwise deflection of the specimen beams are measured with dial indicators during the test and the spanwise deflection values of the specimen beams are recorded periodically.

Table 7. Loading parameters of test beam.

Group No.	Curing Age/d	Loading Duration/d	Purpose	Ultimate Bearing Capacity/kN	Loading Force/kN	Ultimate Bearing Capacity after Creep/kN
L0D1	28	0	Used to confirm creep loading force.	48.0	–	–
L0D2	60			50.0	–	–
L0D3	90			52.5	–	–
L0D4	120			53.5	–	–
L1D1	228	0	The reference group of creep test beams.	55.9	–	–
L1D2	260			56.4	–	–
L1D3	290			56.8	–	–
L1D4	320			57.2	–	–
L2D1	228	200	Control group to study the influence of curing age.	–	14.40	61.9
L2D2	260			–	15.00	61.8
L2D3	290			–	15.75	62.8
L2D4	320			–	16.05	63.1

After the creep test, the flexural bearing capacity of ceramsite concrete beams was tested. The test was carried out by using four-point loading and the loading schematic is shown in Figure 5. Strain gauges were applied at the beam's midspan before the loading started. The strain values were measured during loading to verify the flat section assumption. First, we check whether there was a defect in the sensitive grid was checked and the multimeter was used to check whether there was a short-circuit phenomenon. Then, the surface of the concrete where the strain gauges were to be applied was polished with abrasive paper along the 45° direction of the strain gauges, cleaned with alcohol and scribed and positioned. Finally, strain gauges were applied. In addition, a layer of AB glue was evenly applied to the strain gauges to protect them. Before the loading test, they were allowed to dry naturally for one day.

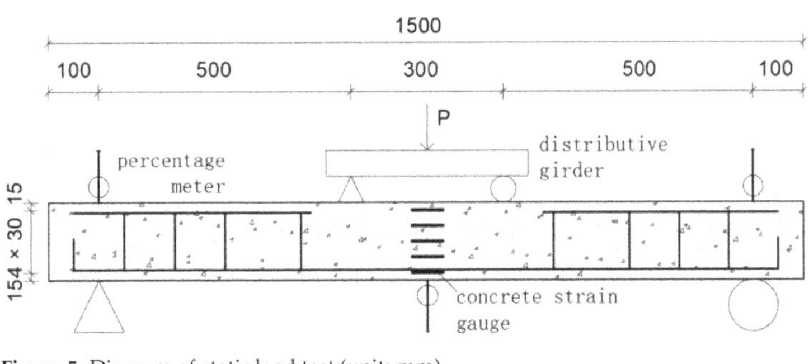

Figure 5. Diagram of static load test (unit: mm).

After the test preparation, a percentage meter was installed at the support and midspan of the ceramsite concrete beam to measure the midspan deflection. According to GB/T 50152-2012 Standard for Test Methods of Concrete Structures [34], preloading was carried out before formal loading. Moreover, the location of the specimens and instruments were checked to see if they were normal. The preload was carried out three times with continuous specimen position adjustment and instrumentation during the test. After passing the inspection, the official loading was performed.

The static load loading equipment of the test beam was a 200-t electrohydraulic servo pressure tester. The concrete strain value was collected by a static resistance strain gauge and the deflection was measured by a percentage meter. In addition, step-by-step loading was adopted. Each step's loading amount was 5 kN and the loading speed was 0.5 kN/s. When loading to 20 kN, the loading amount was adjusted to 2 kN per stage and the loading speed was set in 0.2 kN/s. When loading to 40 kN, the loading amount was adjusted to 1 kN per stage. The loading speed was controlled at 0.1 kN/s. Although the load value reached the ultimate bearing capacity of the ceramsite concrete beam, the pressure tester unloaded automatically. During the test, the loading time of each load level was controlled at 1–3 min. The midspan deflection and strain values imported into the computer were recorded and the crack development was observed. The field loading diagram is shown in Figure 6.

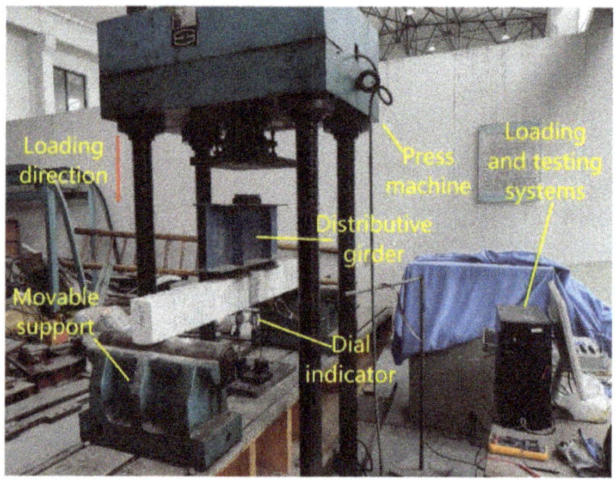

Figure 6. Snapshot of static load test.

Specimen beams L1D1, L1D2, L1D3 and L1D4 were subjected to direct static load damage tests when they were curing to their respective initial loading ages. Fore specimen beams L2D1, L2D2, L2D3, and L2D4, were first subjected to a 200-d period of creep loading at a value of 0.3 Pu when they were curing to their respective initial loading ages. Then, they were subjected to static load damage tests after the creep loading period.

4. Experimental Results and Analysis

4.1. Effect of Ceramsite Concrete Beam Performance after Creep

Figure 7 shows a typical damage pattern of ceramsite concrete beam. The cracks response of the beams were captured by the naked eye during the tests. It can be seen that all beams exhibit a similar cracking pattern whether they experience creep or not. That is, the first cracks in all volumes were produced in the bending region at the beginning of the loading. As the load is further increased, the cracks extend to the compressive side and additional cracks are produced in the shear zone. Compared with L1-type beams, L2-type beams displayed more cracks and less crack spacing. Moreover, the maximum crack width generated was also smaller during failure. In terms of the failure process, with the similar to ordinary concrete beams, both types of beams also experienced three stages. They were elastic working stage, elastic-plastic working stage and specimen failure. Figure 8 shows the load midspan deflection curves of beams L1D1 and L2D1 at static load loading. As can be seen from Figure 8, the direction and the changing trend of the load-mid span deflection curves of the ceramsite concrete beams were approximately the same, regardless of whether they had experienced creep loading. The load-deflection curves of the ceramsite concrete beams were essentially linear before the damage, with obvious yield points and roughly similar slopes of the curves. It exhibited that the preapplication of long-term loads did not result in an improved ability of the ceramsite concrete beams to limit their deformation. However, compared with L1D1 and L2D1 showed a higher strength of 11% but 40% lower stiffness. The midspan deflections in the yielding and failure states were raised by 85% and 82%, respectively. It was shown that the preapplication of long-term loads led to enhance flexural strength and reduced static stiffness of ceramsite beams. Furthermore, ceramsite concrete beams that had experienced creep had more significant midspan deflection than beams that had not experienced creep when they reached yielding or failure. as referred to in reference [35].

(a)

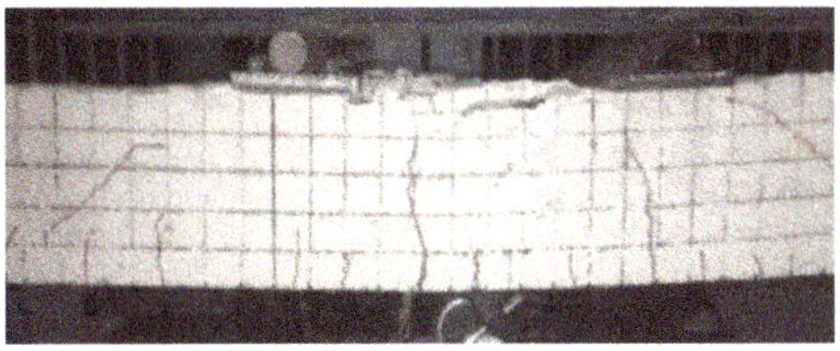

(b)

Figure 7. Ceramsite concrete beam damage pattern. (**a**) L1 beam damage pattern. (**b**) L2 beam damage pattern.

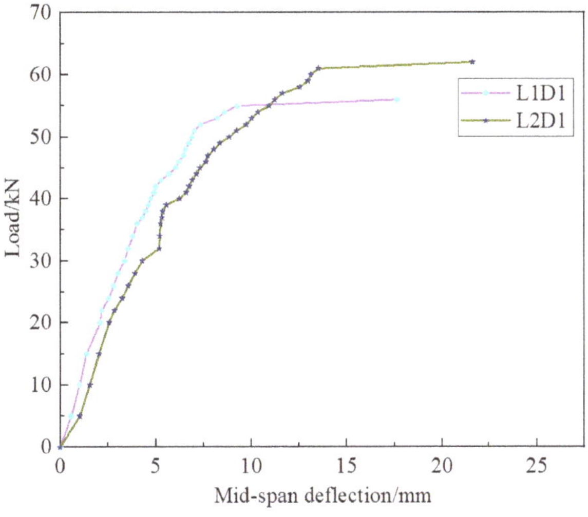

Figure 8. Load-deflection curve of L1D1, L2D1.

Figure 9 suggests the histogram of the ultimate bearing capacity of each beam. As shown in Figure 9, the ultimate bearing capacity of L2 beams enhanced by about 10% compared with L1 beams at the same curing age. It illustrated that long-term creep had an enhancing effect on the ultimate bearing capacity of ceramsite concrete beams. There were two plausible explanations for this experimental phenomenon [28,36]. On the one hand, it was caused by strengthening the concrete pressure zone during the long-term loading process. On the other hand, the stress relaxation reduced the stress at the tip of the concrete crack, resulting in the ultimate bearing capacity of the concrete beam being increased.

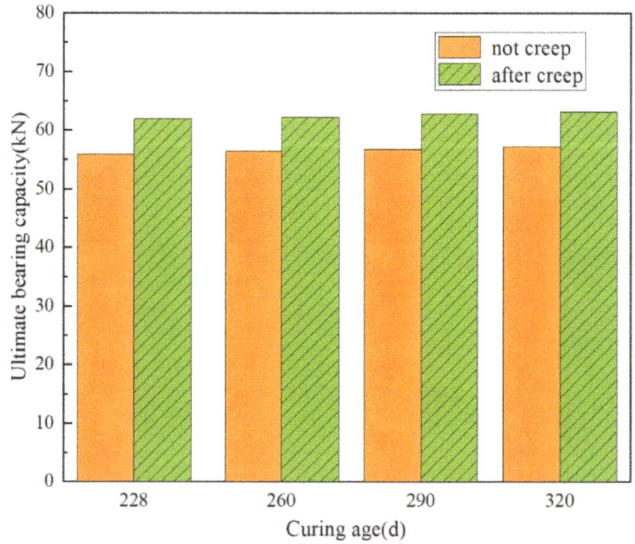

Figure 9. Ultimate bearing capacity of ceramsite concrete beams.

Figure 10 displays the distribution of cracks in ceramsite concrete beams. In the process of static loading, small vertical cracks appeared in the beam when it reached the cracking load. With continuously growing load values, the crack width, depth, and the number of cracks in the beams gradually increased. When beam reached the ultimate load, maximum cracking completely destroyed it.

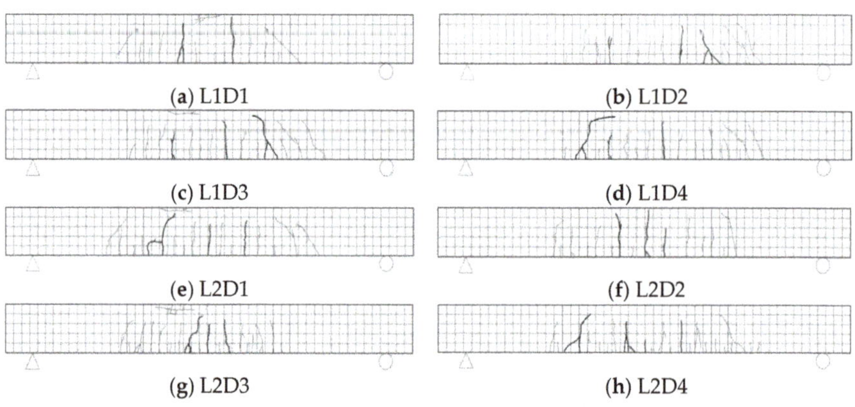

Figure 10. Distribution of cracks in ceramsite concrete beams.

Comparing the crack development of each test beam in Figure 10, it can be seen that the long-term creep effect caused the maximum crack width, crack depth, and average crack spacing to decrease when the ceramsite concrete beam (group L2) was damaged. Moreover, the number of cracks would increase. Because long-term loading causes stress relaxation in the concrete in the tension zone, stresses at the crack tip were reduced. In addition, cracks in the tension zone were difficult to develop when larger loads are applied. These could lead to that the maximum crack width and crack depth of ceramsite concrete beams were not significant when subjected to bending damage. When the ceramsite concrete beam reached the cracking load, although the crack width and the crack depth could not be further developed, the number of cracks would continue to increase. In addition, the average crack spacing remains the same [37].

Figure 11 exhibits the load-span deflection curves for the L1 and L2 series test beams when applying static load damage. As shown in Figure 11, the midspan deflection of the test beams in both the yielded and damaged states at the same age after casting with static load damage had a significant increase after creep.

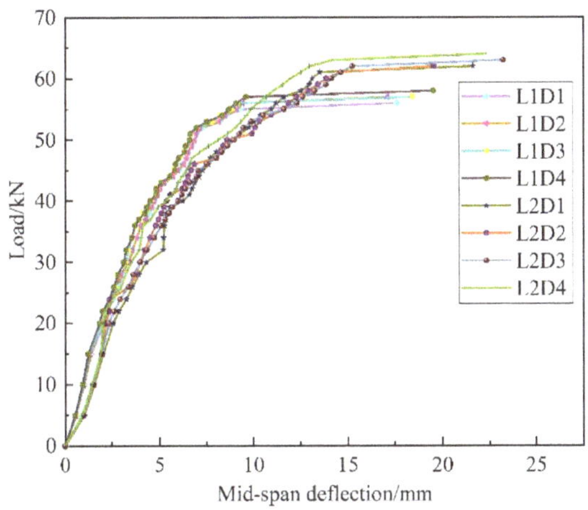

Figure 11. Load midspan deflection curve of ceramsite concrete beam in static load test.

It can also be seen from Figure 11 that the flexural stiffness of a beam undergoing creep enhanced as the age of caring increased. The deflection in the span corresponding to the transformation from elastic to plastic deformation became progressively larger and the deflection at ultimate load failure increased further. This indicated that the stiffness and ductility of the test beams gradually raised with increasing caring age. The deformation of the test beams at the same age of applied static load damage indicated that the change in spanwise deflection at static load yielding elevated slightly with increasing natural curing time for the same initial loading age. Moreover, the spanwise deflection at damage of the specimens raised significantly with increasing initial loading age [38]. All of the above studies demonstrated that the curing age affected the flexural performance of ceramsite concrete beams after creep. However, there was less research involving the effect of the curing age on the creep of ceramsite concrete beams, so it needed to be studied urgently.

4.2. *Effect of Age of Maintenance on Ceramsite Concrete Beams after Creeping*

As shown in Figure 12, the ceramsite concrete beams had a sizeable initial deflection deformation at loading. With the increase of the loading age, although the initial deflection value and the creep deflection value in the same holding time reduced gradually, the decreasing trend was less and less obvious. This was in accordance with Knaack [39] who proposed that the creep deflection decreases with the increase of the maintenance age.

As shown in Figure 13, as the age of loading increases, the reduction rate tended to zero. At this point, the total deflection reduction of the ceramsite concrete beam no longer raised significantly with the age of loading. This indicated a limit to the effect of the age of curing on the creep performance of the ceramsite concrete beams. As the load-holding time increased, the midspan deflection of the ceramsite concrete beams gradually increased. In the early stages of loading, the creep deflection rose at a faster rate and then slowly plateaued. The rate of change of the creep deflection with loading time varied for different loading ages of the ceramsite concrete beams. For the 28-d loading age, the deflection values enhanced fastest and the final total deflection values were the highest. As the loading age increased, the growth rate of the creep deflection of the ceramsite concrete beam at the same load-holding time slowed down and the total deflection tended to reduce. At 120-d loading age, the growth rate of the creep deflection and the final deflection value of the ceramsite concrete beam at the same load-holding time was also the smallest. From the above analysis, the ceramsite concrete's strength, shrinkage creep and long-term structural deflection improved with increasing loading age.

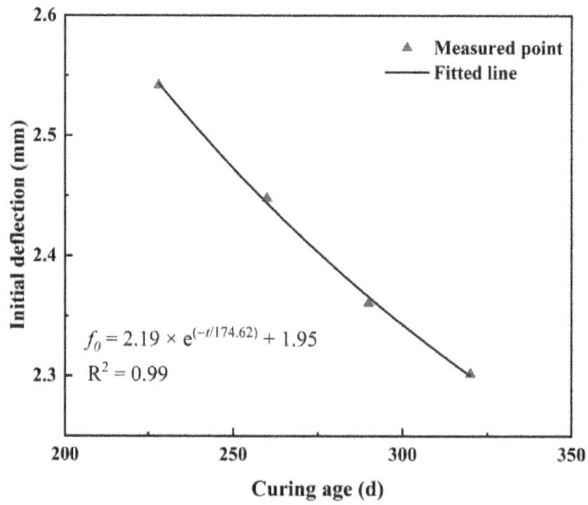

Figure 12. Effect of age of maintenance on the initial deflection.

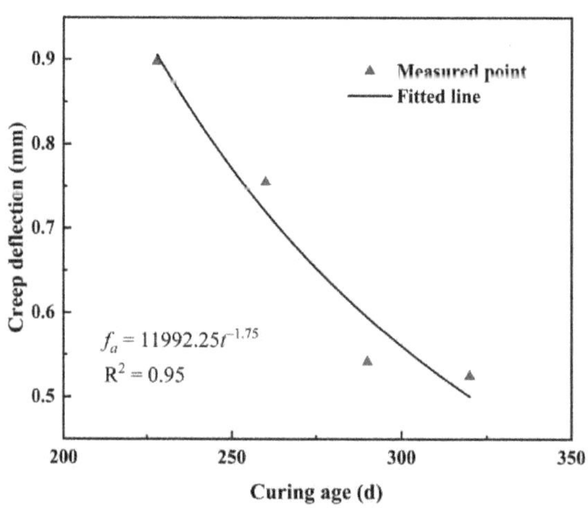

Figure 13. Effect of age of maintenance on creep deflection.

As displayed in Figure 14, The girders of group L2 were all loaded with 200 d of creep. The ultimate bearing capacity of the beams with a 320-d maintenance age increased by 1.81% compared with the beams with a 228 d maintenance age. Therefore, the age of curing also affected the ultimate bearing capacity of the ceramsite concrete beams subjected to creep. However, the extent of this effect was smaller at higher curing ages and the trend of increase in ultimate bearing capacity was not apparent.

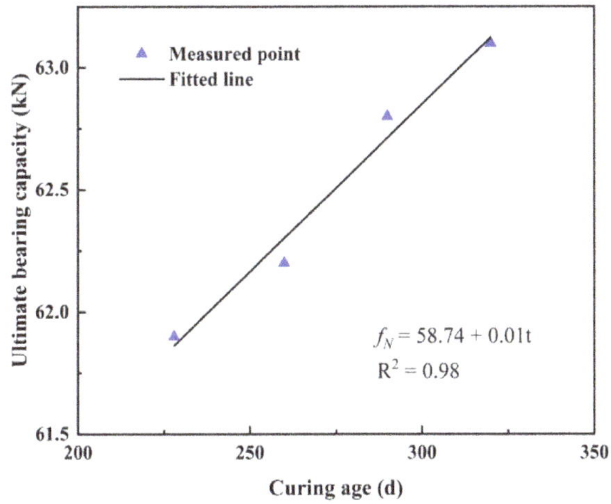

Figure 14. Ultimate bearing capacity curves at different maintenance ages.

Figures 15 and 16 and Table 8 suggest that as the age of curing increased, the number of cracks in the postcreep ceramsite concrete beams subjected to flexural compression damage gradually raised. In addition, it was with smaller maximum crack widths, crack depths, and average crack spacing. The ceramsite concrete beams with long curing periods had many dense cracks after bending and compression damage. Comparing with the test results of L2D1 and L2D4 beams showed the crack development pattern of the ceramsite concrete beams after creeping. This was because as the loading age of the ceramsite concrete beams increased, the internal and external moisture was continuously evaporated. In addition, the brittleness and strength class of the ceramsite concrete gradually enhanced. As a result, the number of cracks in the ceramsite concrete beams increased. The maximum crack width, depth, and average crack spacing decreased when the ultimate bearing capacity damage was reached. This indicated that as the age of maintenance increased, cracks in the beam were more likely to develop when subjected to external loads. The existing test results corroborated the crack-development pattern [40].

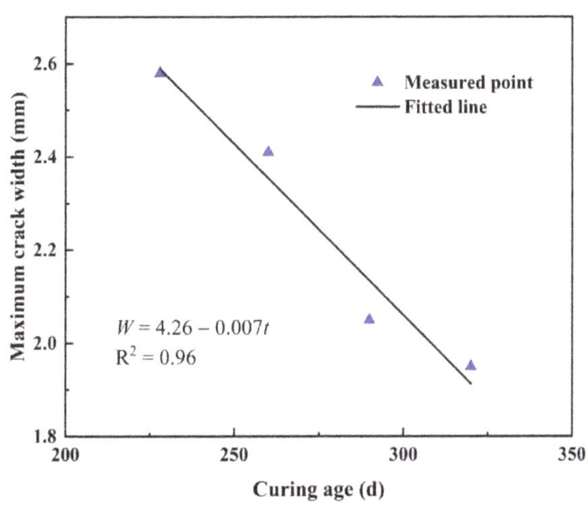

Figure 15. Maximum crack width curve after creep.

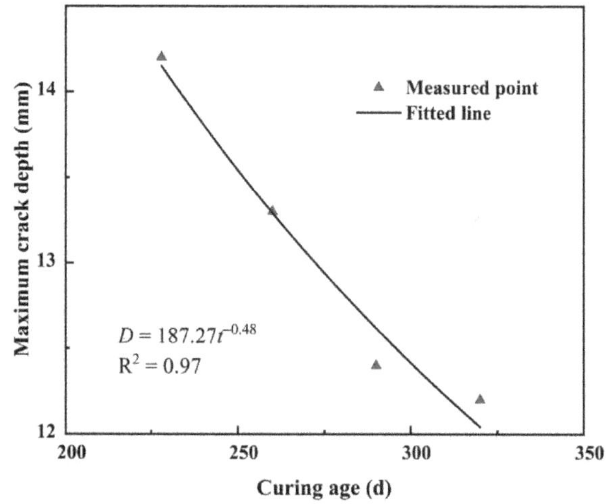

Figure 16. Maximum fracture depth curves after creep.

Table 8. Measured crack data.

Specimen Beam Grouping Number	Extreme Load Condition			Maximum Crack Depth at Damage (cm)	Average Crack Spacing at Damage (cm)
	Load (kN)	Maximum Crack Width (mm)	Number of Cracks		
L1D1	55.9	2.99	14	14.1	4.4
L1D2	56.4	2.73	16	13.5	4.3
L1D3	56.8	2.55	14	13.6	3.9
L1D4	57.2	2.41	17	13.2	3.6
L2D1	61.9	2.58	16	14.2	4.0
L2D2	61.8	2.41	17	13.3	3.6
L2D3	62.8	2.05	20	12.4	3.3
L2D4	63.1	1.95	22	12.2	3.2

5. Conclusions

Based on the results and discussion presented above, the following conclusions can be obtained.

(1) Creep will increase the deformation capacity of ceramsite concrete beams and enhance the toughness and energy-absorption capacity. L2D1 exhibits higher flexural strength than L1D1, with a reinforcement value of 11%.

(2) The law of load-bearing capacity of the ceramsite concrete beams was unaffected by creep loading. However, the ultimate load-bearing capacity of the ceramsite concrete beams was promoted by creep loading. The ultimate load-bearing capacity of the test beams increased by about 10% after 200 d of creep loading at the same curing age.

(3) Creep loading affected the development of cracks. The long-term creep effect reduced the maximum crack width and crack depth. Moreover, it increased the number and average crack spacing of cracks in ceramsite concrete beams.

(4) With the increase of loading age, the initial deflection value and the creep deflection value in the same holding time also gradually decreased. In addition, the decreasing trend was less and less noticeable. As the age of loading increases, the reduction rate tended to zero.

(5) The age of curing also affected the ultimate bearing capacity of ceramsite concrete beams influenced by creep. The growth rate of the ultimate bearing capacity is faster in the early stages of curing ages, slower after 120 d, and slower and more stable in

the 228–320 d range. The ultimate load capacity of L2D4 has increased by only 1.81% compared to L2D1.

(6) The age of loading affected the ability of the ceramsite concrete beams to develop cracks after creep. The shorter the age of loading, the fewer the cracks in the ceramsite concrete beams subjected to bending damage, the greater the maximum crack width, the depth of the cracks and the average crack spacing. The longer the loading age, the more dense the cracks in the concrete beams subjected to flexural damage.

In the future, it will be planned to pour more ceramsite concrete beams. A combination of experiments and numerical simulations will be used to study the effect of curing age on the load-bearing characteristics of creep-loaded ceramsite concrete beams. The results of the study will provide a scientific reference for engineering design. In addition, sulfate erosion affects the creep of ceramsite concrete beams, which will be viewed in the subsequent research directions.

Author Contributions: Conceptualization, H.Z. and J.L.; methodology, J.C. and H.Z.; software, J.C.; validation, Z.F. and Y.W.; formal analysis, J.C.; investigation, Y.W.; resources, H.Z. and J.L.; data curation, Z.F., Y.W. and H.Z.; writing–original draft preparation, J.C.; writing–review & editing, Y.W. and H.Z.; visualization, P.L. and J.C.; supervision, H.Z.; project administration, H.Z.; funding acquisition, H.Z. All authors have read and agreed to the published version of the manuscript.

Funding: This study was funded by the National Natural Science Foundation of China (grant numbers 52178182), Hubei Provincial Excellent Young and Middle aged Science and Technology Innovation Team Project of Colleges and Universities (grant numbers T2022002).

Institutional Review Board Statement: Not applicable.

Informed Consent Statement: Not applicable.

Data Availability Statement: The data presented in this study are available on request from the corresponding author.

Conflicts of Interest: The authors declare no conflict of interest.

References

1. Yin, Y.; Hu, S.; Wang, Y. Influence of self-weight on the fracture properties of three-point bending concrete beams. *Eng. Mech.* **2019**, *36*, 48–56.
2. Sofia Real, M.; Glória Gomes, A.; Moret Rodrigues, J.; Bogas, A. Contribution of structural lightweight aggregate concrete to the reduction of thermal bridging effect in buildings. *Constr. Build. Mater.* **2016**, *121*, 460–470. [CrossRef]
3. Zhu, H.; Fu, Z.; Liu, P.; Li, Y.; Zhao, B. Shear behavior of stud-PBL composite shear connector for steel–ceramsite concrete composite structure. *Coatings* **2022**, *12*, 583. [CrossRef]
4. Xie, J.; Zhao, J.; Wang, J.; Huang, P.; Liu, J. Investigation of the high-temperature resistance of sludge ceramsite concrete with recycled fine aggregates and GGBS and its application in hollow blocks. *J. Build. Eng.* **2021**, *34*, 101954. [CrossRef]
5. Jiang, W.; Yang, Z. Compressive creep of lightweight aggregate concrete at early age. *J. Build. Mater.* **2016**, *19*, 40–44.
6. Zhuang, Y.; Chen, C.; Ji, T. Effect of shale ceramsite type on the tensile creep of lightweight aggregate concrete. *Constr. Build. Mater.* **2013**, *46*, 13–18. [CrossRef]
7. Ji, T.; Chen, C.; Chen, Y. Effect of ceramsite prewetting degree on tensile creep characteristics of lightweight aggregate concrete. *J. Build. Mater.* **2012**, *15*, 690–696.
8. Zhang, Y.; Mao, J.; Jin, W.; Zhang, J. Creep model of high-performance concrete at different loading ages. *Constr. Build. Mater.* **2022**, *357*, 129379. [CrossRef]
9. Wang, S.; Wu, X.; Yang, J.; Zhu, S. Acoustic emission characteristics and dynamic damage constitutive relation of shale-ceramsite concrete subjected to loading Tests. *J. Mater. Civ. Eng.* **2020**, *32*, 76. [CrossRef]
10. Hwang, E.; Kim, G.; Koo, K.; Moon, H.; Choe, G.; Suh, D.; Nam, J. Compressive creep and shrinkage of high-strength concrete based on limestone coarse aggregate applied to high-rise buildings. *Materials* **2021**, *14*, 5026. [CrossRef]
11. Liang, S.; Wei, Y. Effects of water-to-cement ratio and curing age on microscopic creep and creep recovery of hardened cement pastes by microindentation. *Cement. Concr. Comp.* **2020**, *113*, 103619. [CrossRef]
12. Wang, Y.; Xu, Q.; Chen, S.; Zhou, X.; Liu, H. The influence of persistent curing-loading temperature on the basic creep of early-age concrete: Experimental investigation. *Constr. Build. Mater.* **2020**, *261*, 119950. [CrossRef]
13. Li, K.; Yang, C.; Huang, W.; Zhao, Y.; Wang, Y.; Pan, Y.; Xu, F. Effects of hybrid fibers on workability, mechanical, and time-dependent properties of high strength fiber-reinforced self-consolidating concrete. *Constr. Build. Mater.* **2021**, *277*, 122325. [CrossRef]

14. Cai, X.; Jiang, M.; Guo, X.; Chen, J.; Zhao, Q. Experimental study on the creep behaviour of cemented sand and gravel (CSG) and temperature stress prediction of CSG dam under seasonal temperature change. *Adv. Civ. Eng.* **2020**, *1*, 8289520. [CrossRef]
15. Chen, P.; Zheng, W.; Wang, Y.; Chang, W. Analysis and Modelling of Shrinkage and Creep of Reactive Powder Concrete. *Appl. Sci.* **2018**, *85*, 732. [CrossRef]
16. Li, Y.; Dong, J.; Tan, K.H. Reliability Analysis of SFRC Beams Under Ultimate Limit States of Long-Term Load. *J. Northeast. Univ. (Nat. Sci.)* **2017**, *38*, 1055–1059. (In Chinese)
17. Tošić, N.; Aidarov, S.; de la Fuente, A. Systematic review on the creep of fiber-reinforced concrete. *Materials* **2020**, *13*, 5098. [CrossRef] [PubMed]
18. Szydłowski, R.; Łabuzek, B. Experimental evaluation of shrinkage, creep and prestress losses in lightweight Aggregate concrete with sintered fly ash. *Materials* **2021**, *14*, 3895. [CrossRef]
19. Zheng, Z.; Hu, D.; Liu, P.; Sha, F.; Liu, L.; Yu, Z. Considering the effect of the randomness of concrete strength and relative humidity on concrete creep. *Struct. Concr.* **2020**, *10*, 201900435. [CrossRef]
20. Wei, Y.; Huang, J.; Liang, S. Measurement and modeling concrete creep considering relative humidity effect. *Mech. Time-Depend. Mat.* **2020**, *24*, 161–177. [CrossRef]
21. Yu, Q.; Bazant, Z.; Wendner, R. Improved algorithm for efficient and realistic creep analysis of large creep-sensitive concrete structures. *ACI Struct. J.* **2012**, *109*, 665.
22. Wang, W.; Gong, J. New relaxation function and age-adjusted effective modulus expressions for creep analysis of concrete structures. *Eng. Struct.* **2019**, *188*, 1–10. [CrossRef]
23. Dong, W.; Zhang, X.; Zhang, B.; Wu, Q. Influence of sustained loading on fracture properties of concrete. *Eng. Fract. Mech.* **2018**, *200*, 134–145. [CrossRef]
24. Dong, W.; Li, J.; Zhang, X.; Zhang, B. Evolutions of SIFs of concrete under sustained loading by considering the effects of stress relaxations. *J. Mater. Civil. Eng.* **2019**, *31*, 04019287. [CrossRef]
25. Rong, H.; Dong, W.; Zhang, X.; Zhang, B. Size effect on fracture properties of concrete after sustained loading. *Mater. Struct.* **2019**, *52*, 16. [CrossRef]
26. Liu, C.; Zhu, C.; Fan, J.; Bai, G.; Xiao, J. Experimental study on recycled concrete beam subjected to long-term loading and its flexural performance after unloading. *Chin. J. Build. Struct.* **2020**, *41*, 56–63. (In Chinese)
27. Saliba, J.; Loukili, A.; Grondin, F.; Regoin, J.-P. Experimental study of creep-damage coupling in concrete by acoustic emission technique. *Mater. Struct.* **2012**, *45*, 1389–1401. [CrossRef]
28. Omar, M.; Loukili, A.; Pijaudier-Cabot, G.; Le Pape, Y. Creep-damage coupled effects: Experimental investigation on bending beams with various sizes. *J. Mater. Civ. Eng.* **2009**, *21*, 65–72. [CrossRef]
29. *JGJ/T 12-2019*; Technical Standard for Application of Lightweight Aggregate Concrete. Ministry of Housing and Urban-Rural Development of the People's Republic of China: Beijing, China, July 2019; pp. 17–21. (In Chinese)
30. Chen, M.; Li, Z.; Wu, J.; Wang, J. Shear behaviour and diagonal crack checking of shale ceramsite lightweight aggregate concrete beams with high-strength steel bars. *Constr. Build. Mater.* **2020**, *249*, 118730. [CrossRef]
31. *GB/T 50081-2019*; Standard for Test Methods of Concrete Physical and Mechanical Properties. Ministry of Housing and Urban-Rural Development of the people's Republic of China: Beijing, China, June 2019; pp. 8–43. (In Chinese)
32. Sun, G.; Xue, S. Experimental investigation of creep and shrinkage of reinforced concrete with influence of reinforcement ratio. *Adv. Concr. Constr.* **2019**, *7*, 211–218.
33. Bažant, Z.P.; Baweja, S. Justification and refinements of model B3 for concrete creep and shrinkage 1. statistics and sensitivity. *Mater. Struct.* **1995**, *28*, 415–430. [CrossRef]
34. *GB/T 50152-2012*; Standard for Test Method of Concrete Structures. Ministry of Housing and Urban-Rural Development of the people's Republic of China: Beijing, China, January 2012; pp. 11–63. (In Chinese)
35. Chen, P.; Zhou, X.; Zheng, W.; Wang, Y.; Bao, B. Influence of high sustained loads and longitudinal reinforcement on long-term deformation of reinforced concrete beams. *J. Build. Eng.* **2020**, *30*, 101241. [CrossRef]
36. Altoubat, S.A.; Lange, D.A. Creep, shrinkage, and cracking of restrained concrete at early age. *ACI. Mater. J.* **2001**, *98*, 323–331.
37. Micelli, F.; Candido, L.; Vasanelli, E.; Aiello, M.; Plizzari, G. Effects of short fibers on the long-term behavior of RC/FRC beams aged under service loading. *Appl. Sci.* **2019**, *9*, 2540. [CrossRef]
38. Candido, L.; Micelli, F.; Vasanelli, E.; Aiello, M.; Plizzari, G. Cracking behaviour of FRC beams under long-term loading. *Concreep* **2015**, *10*, 1147–1156.
39. Knaack, A.; Kurama, Y. Sustained Service Load Behavior of Concrete Beams with Recycled Concrete Aggregates. *ACI Struct. J.* **2015**, *112*, 565–577. [CrossRef]
40. Hu, C.; SI, B. Experimental investigation on the cracks of existing concrete beams repaired with SHCC under sustained load. *Chinese Concrete* **2017**, *2*, 15–20. (In Chinese)

coatings

MDPI

Article

Effect of the Location of Broken Wires on Prestressed Concrete Cylinder Pipes under Working Pressure

Zhu Zhang [1], Tongchun Li [1,*], Lanhao Zhao [1] and Huijun Qi [2]

1 College of Water Conservancy and Hydropower Engineering, Hohai University, Nanjing 210098, China
2 College of Computer and Information, Hohai University, Nanjing 210098, China
* Correspondence: ltchhu@163.com

Abstract: The precise analysis of the overall mechanical performance of prestressed concrete cylinder pipes (PCCPs) with broken wires is a complicated problem. In this article, powerful finite element numerical means are applied to solve this problem. Firstly, the advantages and shortcomings of the current prestress simulation methods in the finite element analysis (FEA) literature are discussed, and the initial stress method based on a novel single-spring joint element method to model the pre-stress of wires is proposed. Then, different distributions of broken zones, including random broken distributions, are developed for two typical PCCP pipes with different wire-wrapped layers. The established nonlinear FEA model considers actual service processes such as manufacturing, installation and operation to investigate the mechanical behavior of two typical PCCP pipes with different breakage distribution regions and breakage ratios, and in particular, the overall mechanical behavior of a pipe with random breakage is determined first. To verify the accuracy of the proposed pre-stress simulation method and the established nonlinear finite element model, the overall mechanical response of a pipe before wires broke is compared with the results obtained via the specifications. The computed results under the corresponding breakage assumptions are consistent with the conclusions in the existing literature, providing important guidance for pipeline management and operation.

Keywords: finite element analysis; broken wires; random distribution; initial stress method

Citation: Zhang, Z.; Li, T.; Zhao, L.; Qi, H. Effect of the Location of Broken Wires on Prestressed Concrete Cylinder Pipes under Working Pressure. *Coatings* **2022**, *12*, 1361. https://doi.org/10.3390/coatings12091361

Academic Editor: Andrea Nobili

Received: 16 August 2022
Accepted: 15 September 2022
Published: 18 September 2022

1. Introduction

Prestressed concrete cylinder pipes (PCCPs) are broadly applied in urban water-diversion engineering projects. They have many advantages such as a good ability to bear high internal water pressure, high resistance to earthquakes and outstanding economic values. Their first application was in France, and they were then introduced to America in 1942. Since then, more than 22,000 miles of PCCPs have been in service in North America [1]. Recently, they have been applied to many water-diversion projects in China such as the Shenyang Dahuofang Water Conveyance Project, the East of Shenzhen Water Supply Project and the South-to-North Water Transfer Project. Four components, concrete, mortar coating, cylinders and pre-stressing wires, make up PCCP pipes. Concrete has high-strength properties and is the main component of the pipes, providing a smooth surface to maintain a low roughness coefficient and reduce hydraulic loss. The steel cylinder mainly plays an anti-seepage role and has a certain load-bearing capacity. The spiral-wound pre-stressing wire is the main load-bearing part, providing uniform pre-compressive stress to withstand the internal water pressure. The mortar coating serves as an essential material to protect the high-strength steel wires from corrosion by groundwater or soil around the pipeline for more than 50 years [2]. There are two major kinds of PCCPs: Lined-Cylinder Pipes (LCPs) and Embedded-Cylinder Pipes (ECPs) [3]. The wires of LCPs spirally wrapped on a steel cylinder. ECPs high-strength steel wires helically wrapping on the outer concrete of a tube core (see Figure 1), which are put to use in water transmission projects on a large scale at home and abroad with larger internal diameters. The overall mechanical performance of

ECPs is very important and has naturally been the focus of special attention from scholars and researchers.

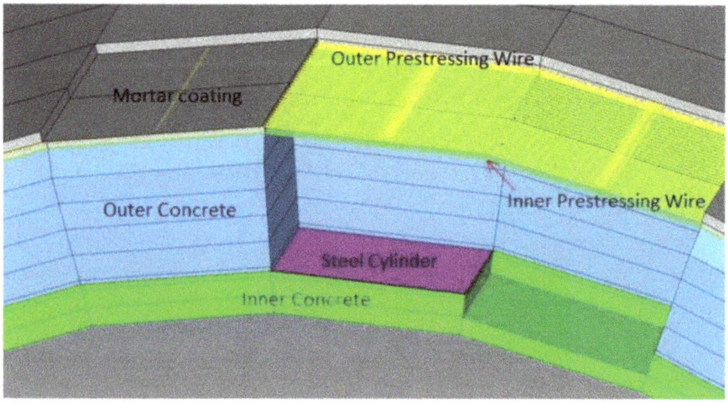

Figure 1. ECP components in ECP finite element model.

PCCP pipes are buried under the ground, so the surrounding groundwater and soil may corrode the pipes, resulting in the general deterioration of function and reductions in service life [4]. They are considered reliable and safe, with the low percentage of wire breakages, according to a recent study [5]. It showed that this percentage was only 3.7%. If they burst, however, this will cause catastrophic ruptures and leaks, resulting in serious losses of life and property. The well-known failure mode of PCCPs is bursting damage (see Figure 2), which is essentially caused by the corrosion and hydrogen embrittlement of wires [6–8]. Currently, wire break-off detection and leak detection methods are broadly used to evaluate the status of PCCPs regularly or in real time. However, it is difficult to determine the accurate number and location of broken wires through the mainstream electromagnetic eddy current inspection methods. In addition, common technologies such as acoustic emissions, which is the only technology applied in long pipelines, fail to detect leaks [9].

(**a**) (**b**)

Figure 2. Bursting pipe: (**a**) after the bursting event, (**b**) during the bursting event.

The substitution of an existing partially broken pipe with a whole undamaged pipe is not an economical approach [10], so it is very essential to assess the residual structure integrity of an ECP with broken wires. Numerical simulations, prototype tests and qualitative decision methods are generally used to assess the condition of pipes with broken wires.

Numerical simulations are widely used, which can provide further directions before an ECP fails or bursts. You [11] developed a simplified structural mechanics model based on the classical theory of cylindrical shells. Zarghamee et al. [12] calculated the circumferential thrust and bending moment of a tube wall due to broken wires and verified the results via finite element analysis (FEA). Though a simplified model and finite element method (FEM) model could be easily established, because it neglects the stiffness of wires, it does not reflect real life. Zarghamee et al. [13] considered two different geometries of a pre-stress loss zone. One was axisymmetric, in the form of a band, and the other was rectangular. Afterwards, he simulated the cracking process and the interlocking behavior of a cracked concrete core and investigated the strength of a cracked concrete outer core under the pressure of an expanding steel cylinder using the nonlinear FEA method [14]. The stiffness of wires, the earth load and pipe loads were calculated using the initial pipe design standard.

Gomez et al. [15] calculated the stress status of the middle zone of a pipe with different numbers of broken wires. Diab et al. [16] used both two-dimensional (2D) and three-dimensional (3D) models to present the mechanical behavior of the remaining structure via elastic FEA with three distribution zones. Ge et al. determined the effect of the adhesion quality of the mortar coating on the overall performance of PCCPs by only conducting elastic analysis [17]. The pipe performance as the water pressure and number of broken wires increased was studied by Alavinasab et al. [18,19]. Then, Hajali et al. [20,21] compared the effects of broken wire wraps in the middle of a pipe with those at a joint. In addition, Xiong et al. [22] proposed a new model called the wire-wrapping model to simulate pre-stress. Hu et al. [23] presented a novel bar-wrapping calculation model under pre-stress conditions for a new type of water conveyance pipe called a bar-wrapped cylinder concrete pressure pipe. Zhai et al. [24–26] considered two distributions (removing all of the broken wire and considering the partial loss of pre-stress through simplified contact interactions) in broken pipes and analyzed the strengthening effect via carbon fiber reinforced polymer (CFRP). Dong et al. [27] defined a cohesive contact interaction to simulate the delamination of the mortar coating in broken pipes.

Four methods are generally used to simulate the pre-stress of steel wires in FEA: the equivalent load method, the initial strain method, the equivalent cooling method, and the wire-wrapping model. Most researchers adopted the equivalent load method [11–20], which regards the pre-stress of steel wires as a form of radial load. The simulation is simple and direct, without consideration of the location of steel wires, and it is easy to converge. However, the stiffness of wires is neglected, causing distortion of the simulation results. In the initial strain method conducted by Lin [28], pre-stress was applied to steel wire by applying an initial strain to the wire. In this method, the pre-stress loss usually cannot be considered since it is represented by stress. Most researchers [24–26,29,30] used the equivalent cooling method. This method constructed wire elements to apply pre-stress and take the wire stiffness into account by increasing the temperature field and equalizing the initial pre-stress to the cooling temperature of the wire. However, cooling the wire keeps it in a compressed state, which is contrary to the initial tensile state seen in practice. In addition, if the nonlinear behavior of the material and contact is considered in the nonlinear FEA, adding the temperature field undoubtedly makes it a more complex nonlinear coupled problem, which increases the solution difficulty and reduces convergence. Xiong et al. [22] put forward the wire-wrapping model, which is very complex and inconvenient to apply.

A few scholars have carried out experimental analysis studies on pipes with broken wires. Hu [29] and Shen [31] conducted a three-edge bearing test under an external load test with a 4 m-diameter, single-layer, wire-wrapped pipe. Dou et al. [32] carried out an internal hydraulic test on a PCCP with an inner diameter of 2.6 m and compared the repairing effect with a PCCP reinforced with restoration materials.

Hu et al. [33] developed an internal pressure test for a full-scale, single-layer PCCP with an inner diameter of 2.8 m and a broken wire percentage of 0% (intact), 5% and 10%. Luo et al. [34] established the mathematical evaluation model of safety risk for PCCP pipelines by adopting the analytic hierarchy process (AHP) method.

However, due to the complexity of ECPs, it is difficult to further simulate the mechanical behavior of pipes with broken wires by considering construction process simulation, nonlinear constitutive models of materials and contact interaction. Due to the test costs and technical difficulty, it is difficult to carry out prototype field tests on large-diameter pipes with broken wires in actual operational status. Therefore, the most practical method is to evaluate the mechanical behavior of a broken pipe under actual operating conditions by using the numerical simulation method. In this paper, a location hypothesis of broken wires is presented, and a random broken case induced by congenital manufacturing defects, such as hydrogen embrittlement, is considered for the first time. Meanwhile, to avoid the shortcomings of modeling the pre-stress of wires, a pre-stress simulating method called the initial stress method, based on a novel single-spring joint element method, is proposed. By considering material nonlinearity, contact nonlinearity and the bond–slip relationship in reinforced concrete (RC) structures, and by choosing a proper pre-stress simulation method, we used the FEM method to model two large-diameter ECP pipes with different bar-wrapping layers. The results achieved from nonlinear FEA are analyzed and compared.

2. Location Hypothesis of Broken Wires

The deterioration process of PCCPs with clustered broken wires can be divided into four stages, as shown in Figure 3 [27]. If the number of broken wires continues to increase, the mortar coating will peel off, and the prestressed wires will pull away, revealing visible cracks on the concrete core. Finally, the steel cylinder yields and bulges out, and the outer concrete core falls off [27]. In fact, each wire layer of an ECP is a consistent pre-stressing wire helically wrapping around the tube core, which is meshed into bar elements with dependent circumferential distribution. If a wire breaks at any point, the whole circle of steel wire will release force based on the related test and will be removed in the model [29].

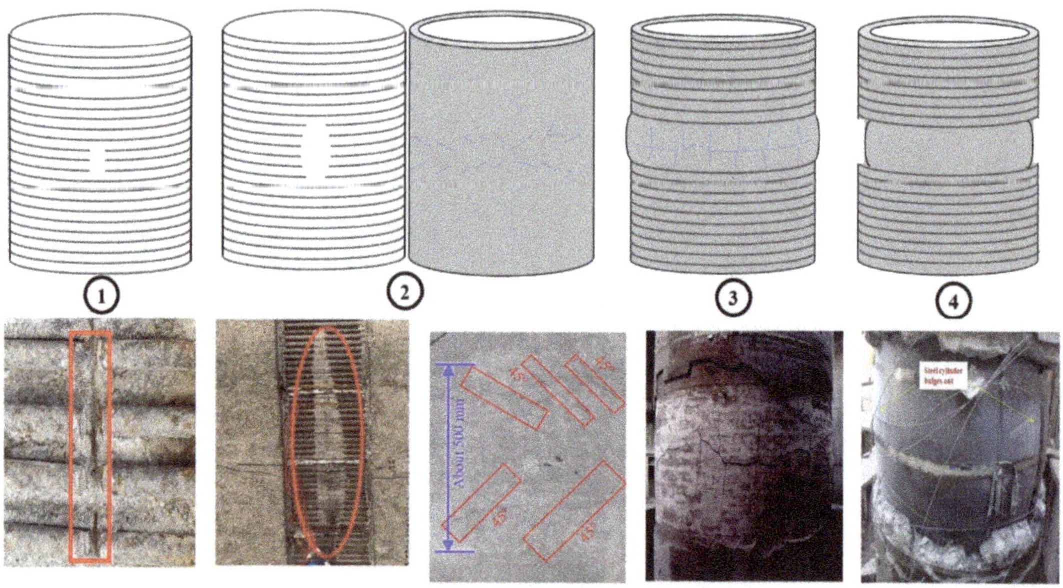

Figure 3. ECP components in ECP finite element model [27].

For pipes with two wire-wrapped layers, it is assumed that they break continuously in two layers at a joint (Figure 4a), the middle location (Figure 4b) and randomly in any location (Figure 4e). Because the existing breakage detection technology cannot detect whether a breakage occurs in the outer layer or the inner layer of a double-layer ECP, and accounting for the fact that stress in the outer layer is more than that in the inner layer, the scenarios of a continuous break in the outer layer at a joint (Figure 4c) and the middle location (Figure 4d) in a double-layer ECP are considered. For pipes with a single layer, it is presumed that they break consistently at the middle, joint and random locations (not drawn).

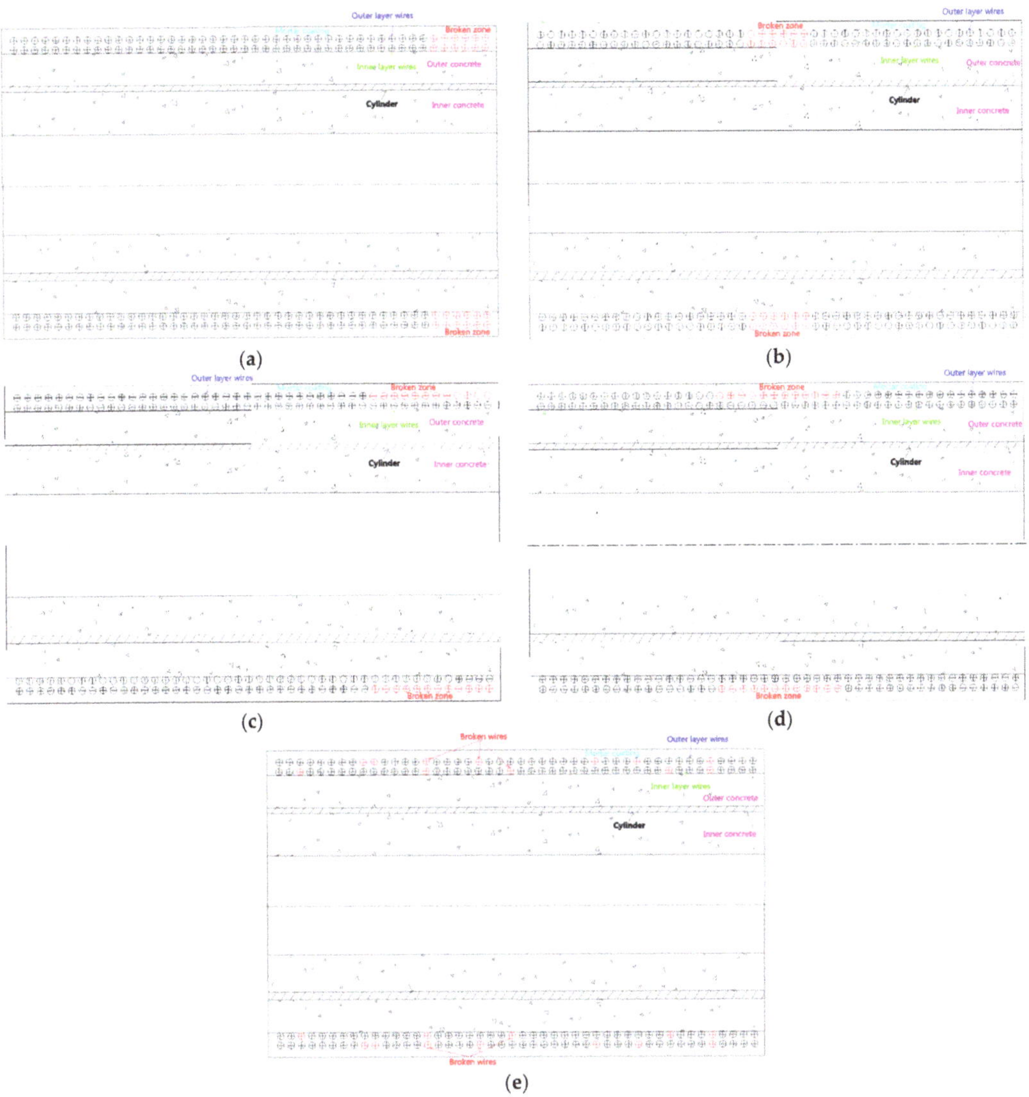

Figure 4. Location of broken wires: (**a**) two layers at joint, (**b**) two layers at middle, (**c**) outer layer at joint, (**d**) outer layer at middle, (**e**) random distribution.

3. Pre-Stress Simulating Methods

To avoid shortcomings when modeling the pre-stress of wires, a new pre-stress simulating method called the initial stress method, based on a novel single-spring joint element method, is proposed. In the novel single-spring joint element method, the tangential action between bars and concrete is modeled by the tangential spring, and the normal direction degree is compelled to be equal (see Figure 5). The related principles and derivation arrangements can be seen in the literature [35].

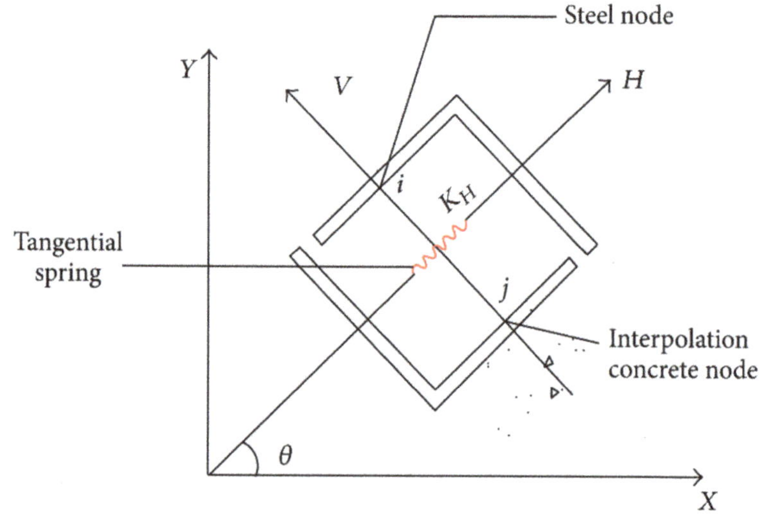

Figure 5. Single-spring joint element model for 2D problem [35].

The incremental equilibrium equation for the novel single-spring joint element method is as follows:

$$\begin{bmatrix} K + r^{\mathrm{T}}K_s r & -r^{\mathrm{T}}K_s \\ -K_s r & K^* + K_s \end{bmatrix} \begin{Bmatrix} \Delta u \\ \Delta u^* \end{Bmatrix} = \begin{Bmatrix} \Delta F \\ \Delta F^* \end{Bmatrix} \tag{1}$$

where K, u, F, K^*, u^*, F^* are the stiffness matrix and the displacement and load vector in the global and the nodal local coordinate system, respectively. K_s is the tangential stiffness matrix, where $K_s = \begin{bmatrix} k_s & 0 & 0 \\ 0 & 0 & 0 \\ 0 & 0 & 0 \end{bmatrix}$ and $k_s = \frac{\partial \tau}{\partial s}\pi Dl$, and D and l are the diameter and length of the reinforcement, respectively. r represents the coordinate conversion matrix between the global and the local coordinate system of the reinforcement node, and $\tau - s$ represents the bond–slip curve obtained from experiments.

When solving the initial stress field of a PCCP from the initial stress of prestressed wire, Equation (1) can be rewritten as follows:

$$\begin{bmatrix} K + r^{T}K_s & -{}^{\mathrm{T}}K_s \\ -K_s r & K^* + K_s \end{bmatrix} \begin{Bmatrix} u \\ u^* \end{Bmatrix} = \begin{Bmatrix} F_0 \\ F_0{}^* \end{Bmatrix} \tag{2}$$

where $F_0{}^* = -\int_l \mathbf{B}^T \sigma_0 \mathrm{d}s$, $\mathbf{B}$ is the strain matrix of wire elements, σ_0 is the initial stress vector of the wire elements and $F_0 = r^{\mathrm{T}} F_0{}^*$.

Because the wire-wrapping process is completed at once, Equation (2) is not solved by the increment method, but by the total iterative method. The simulating method is easy and direct with good convergence. Most importantly, the pre-stress loss can be readily considered.

4. Finite Element Model and Material Properties

4.1. Finite Element Model

Two classes of ECP grid models based on the China South–North Water Transfer Project with 4 m diameter were established. The initial design parameters of the two pipes are summarized in Table 1, and the established finite element models are shown in Figure 6.

For pipe No.1, the total number of elements was 123,488; for pipe No.2, 89,504 elements were used. We established five pipes along the axial direction and chose the middle pipe among the left pipeline pipes as a research target to minimize the boundary effect. Meanwhile, we simulated the pre-stressing wires as bar elements and modeled the cylinder with plate elements. The top surface of the FEA model was unconstrained, and normal fixed constraints were applied to other faces.

Table 1. The design parameters of pipes.

Pipe	Working Pressure (MPa)	Buried Depth (m)	Thickness of Tube Core (m)	Wrapping Area (mm^2)	Wire Diameter (mm)	Thickness of Cylinder (mm)
No.1	0.8	10.0	350	5497	7	2
No.2	0.4	3.0	320	2628	7	2

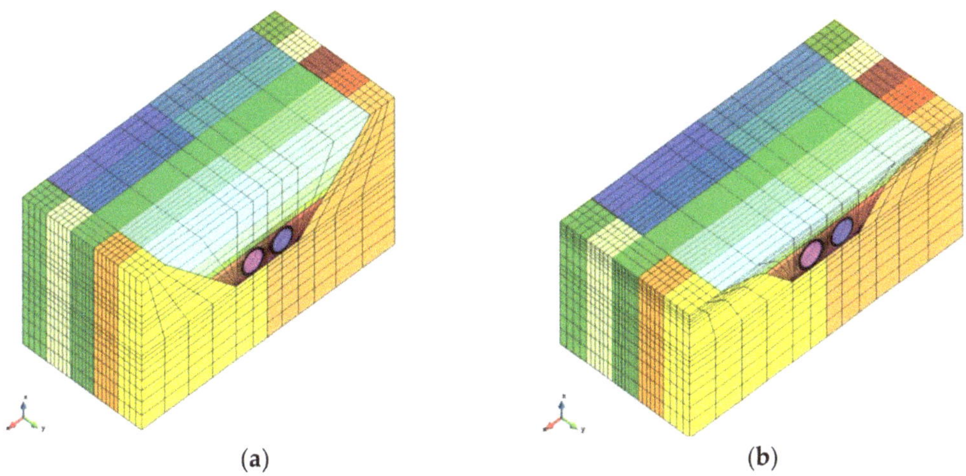

Figure 6. FEM model of pipe: (**a**) pipe No.1, (**b**) pipe No.2.

The contact interaction between pipes and soil and pipe segments were modeled. Meanwhile, the single-spring joint element method was chosen to model the interaction effect of bars and concrete. The ideal elastic–plastic model was adopted as the constitutive relationship of a cylinder based on the Mises yield criterion and supposing the cylinder has complete bond–slip ability with concrete due to its thin thickness. Three broken wires ratios were set to 10%, 20% and 30%.

4.2. Material Properties

The material properties are presented as Table 2.

Table 2. Material properties of pipes.

Material	Grade	Density ($kg{\cdot}m^{-3}$)	Young's Modulus (GPa)	Poisson's Ratio	Compressive Strength (MPa)	Yield Strength (MPa)	Tensile Strength (MPa)
Concrete	C60	2323	28.6	0.167	48.0	—	4.03
	C50	2323	27.0	0.167	40.0	—	3.67
Mortar	C60	2243	25.3	0.167	37.9	—	3.58
Wires	—	7833	191	0.29	—	1334.5	1570.0
Cylinder	—	7833	207	0.29	—	227.5	310.0

For pipe No.1, concrete is C60 grade, and the other one is C50.
The basic material properties of the soil can be seen in Table 3.

Table 3. The basic material properties of soil.

Kind	Density (kg·m^{-3})	Young's Modulus (MPa)	Poisson's Ratio
Foundation	2039	80.0	0.25
Bedding	2039	40.0	0.29
Backfill	2039	50.0	0.35

A four-parameter failure criterion [35,36] was adopted, which took the stress–strain curve reported in research document [3,22] as the damage constitutive curve of concrete and mortar to obtain the damage value. The Dun-Chang E-B model [37,38] was used as a constitutive model for soil. The parameters of the Dun-Chang E-B constitutive model for backfill soil are presented as Table 4. We employed the Goodman joint element [39] to model the contact effect.

Table 4. Dun-Chang E-B constitutive model coefficients of soil.

Kind	Nondimensional Parameter (K)	Nondimensional Index (n)	Failure Ratio (R_f)	Experimental Parameter (K_b)	Experimental Parameter (m)	Cohesion Force (Unit: Pa)	Internal Friction Angle (Unit: °)
Foundation	215.5	0.92	0.613	151.01	0.75	5.0 e3	30
Bedding	232.6	0.90	0.613	165.05	0.75	3.0 e3	30
Backfill	198.4	0.89	0.613	132.27	0.75	4.0 e3	30

We chose the nonlinear bond–slip constitutive curve developed by Di [40] due to the absence of bond–slip testing curves between high-strength concrete and plain wires, which were taken from the test of ordinary concrete and crescent wires, as presented in Equation (3):

$$\tau = 6.59 \times 10^2 s - 2.13 \times 10^4 s^2 + 0.22 \times 10^6 s^3 \tag{3}$$

where τ is the bond stress (MPa) and s is the slip displacement (mm). The bond stress was set only one-tenth of the value in Di's curve by taking the fact that plain wires were applied in this project into consideration. Meanwhile, the constitutive curve of wires can be seen in this paper [22].

5. FEA Results

In practice, a pipe is produced, then placed for a long time in an outdoor space and finally installed. After a series of water pressure tests, the pipe is applied in normal use; then, wires start to break. Combining the analyzing accuracy and calculation efficiency, based on the birth–death method, the FEA considering manufacturing, installation and operation was divided into eight steps:

(1) The tube core and inner layer wires are involved in the calculation, and the initial stress of inner layer wires is applied.

(2) The outer layer of wires is added to the calculation, and the initial stress is applied.

(3) The concrete shrinkage and creep factors are calculated according to the specification [3] and taken into account in FEA due to the long-time storage in outdoor space.

(4) Bedding soil and bottom backfill soil are added to the calculation, and meanwhile, the pipe weight is applied.

(5) Middle backfill soil is added to the calculation.

(6) The remaining backfill soil is added to the calculation.

(7) Apply internal water pressure.

(8) Broken wires are removed from the calculation based on different breakage hypotheses.

The backfill image of the FEA model for pipe No.1 is presented in Figure 7.

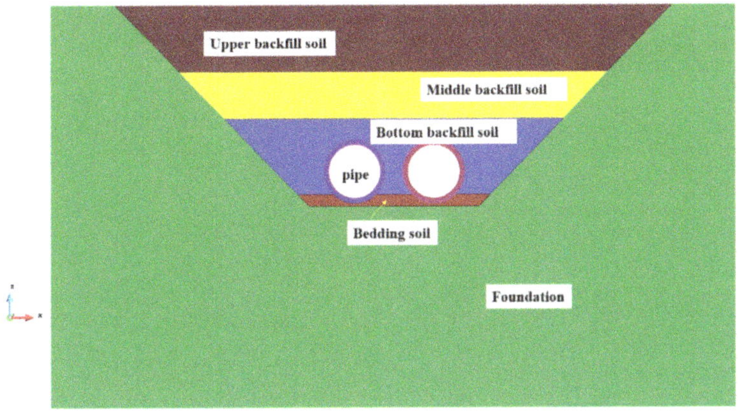

Figure 7. The backfill diagram of FEA model (pipe No.1).

For pipe No.2, step (2) and step (5) were neglected, since there was only a single layer and a low embedment depth.

The factors of creep, shrinkage and wire relaxation were determined by the specification [3] and are shown in Table 5.

Table 5. Wire-relaxation, creep and shrinkage factors.

Pipe	Wire-Relaxation Factor (Outer Wire Layer)	Wire-Relaxation Factor (Inner Wire Layer)	Shrinkage Factor	Creep Factor
No.1	0.127	0.108	6.6617×10^{-5}	1.177
No.2	—	0.082	9.7186×10^{-5}	1.182

The circumferential stress range of concrete for pipe No.1 in step (3) was 10.82 MPa to 12.63 MPa with even distribution, displayed in Figure 8. However, the specification [3] assumes that the cross-sectional stress of the entire tube core is 12.09 MPa. We present the results of pipe No.1 by the standard method and FEA in Table 6 to verify the accuracy of the finite element calculation before wires are broken. The results indicate that various indexes are basically consistent for both methods. Altogether, based on the proposed pre-stress simulation method, the 3D nonlinear FEA shows great consistency with the standard method in simulating ECPs before wires are broken, so the proposed FEA model is suitable for use to evaluate the overall performance of ECPs.

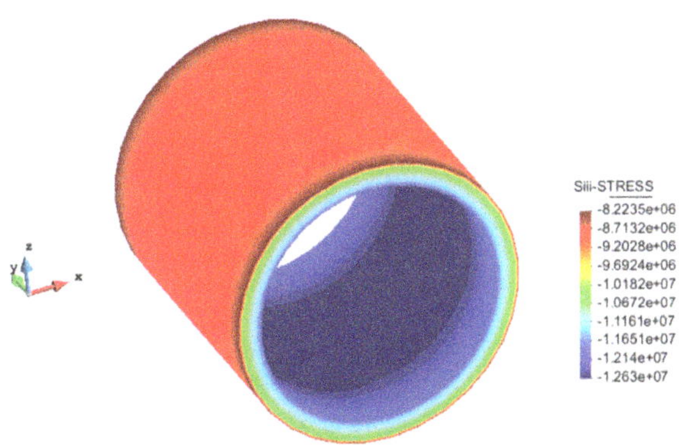

Figure 8. The circumferential stress of the tube core after wire−wrapping and outdoor storage (pipe No.1).

Table 6. Main analysis results of pipe No.1 before wires broke.

Method	Circumferential Stress in Concrete after Outdoor Storage (MPa)	Strain in Mortar Coating	Strain in Outer Concrete	3rd Principal Stress in Inner Concrete (MPa)	Stress in Wires (MPa)	Stress in Cylinder (MPa)
Standard method	−12.09	0.00056	0.00004	19.7	1024.1	−117.9
FEA	−10.82~−12.63	0.00035	0.000157	17.3	1039.7	−138.9

5.1. Finite Element Results of Pipe No.1 after Breaking Wires

Various scenarios combine the breakage ratios with distributional locations. Several results regarding the 10% broken wire ratio were selected to describe the state of pipes with broken wires.

5.1.1. Case 1: 10% Breakage Ratio and Breaking at the Outer Pre-stressing Layer in the Middle of Pipe

The resulting nephogram is shown in Figure 9.

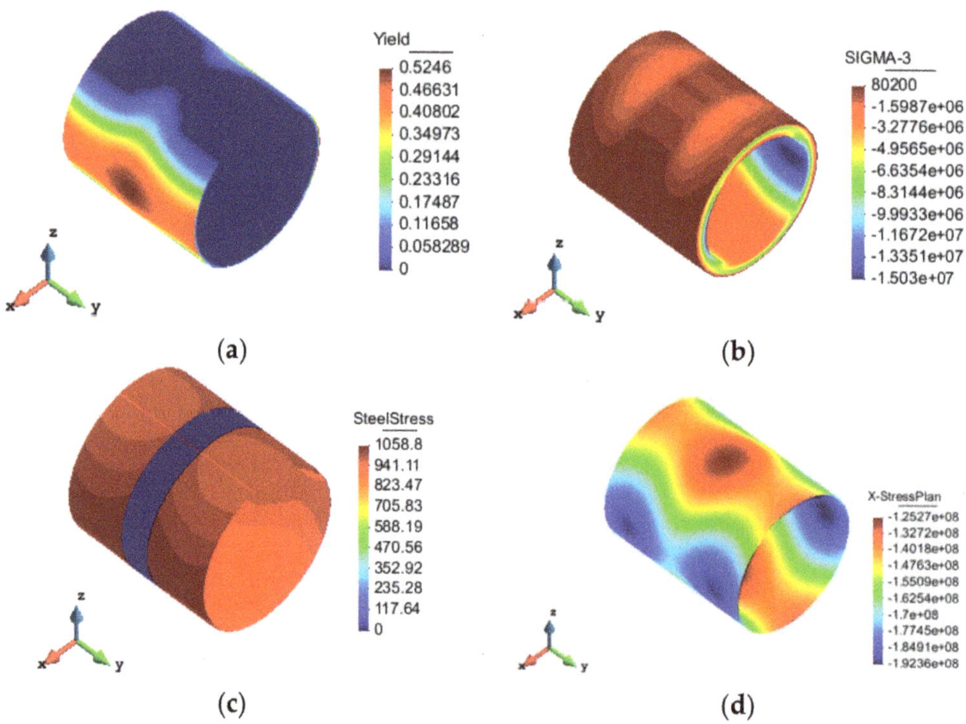

Figure 9. Nephogram of case 1: (**a**) damage, (**b**) 3rd principal stress, (**c**) wire stress, (**d**) cylinder stress.

The maximum damage value of the mortar is 0.52, which increases by 0.16 compared with the pre-breaking stage. The damage to the outer concrete starts growing at the springline in the broken area, with the maximum value of 0.24. The maximum stress in the outer-layer wires increases to 1058.8 MPa, with the increment of 19 MPa compared with the pre-breaking stage. Meanwhile, the stress in the inner wires near the pre-stressing loss zone also increases, which increases to around 11.0 MPa at the springline. Once the outer-layer wire in the middle of pipe is broken, the stress in the unbroken and inner wire on the side of the broken area will increase, and the stress in the cylinder will rise. The damage on the outer concrete will expand but not penetrate along the whole circumferential direction, and the inner concrete will not be damaged temporarily. The internal water load borne by the broken wire will be redistributed to the above materials.

5.1.2. Case 2: 10% Breakage Ratio and Breaking at the Outer Pre-stressing Layer at the Joint of Pipe

The resulting nephogram is shown in Figure 10. It shows that the maximum damage value of mortar is 0.55, which increases by 0.19 compared to the pre-breaking stage. At the springline in the broken area, the damage to the outer concrete starts growing, showing the maximum value of 0.30. The maximum stress in the outer-layer wires reaches 1058.9 MPa with the increment of about 19 MPa. At the same time, the stress in the inner wires and the stress in the cylinder around the pre-stressing loss zone increase. The damage to the outer concrete will not only expand but also penetrate along the whole circumferential direction, and the inner concrete is also damaged.

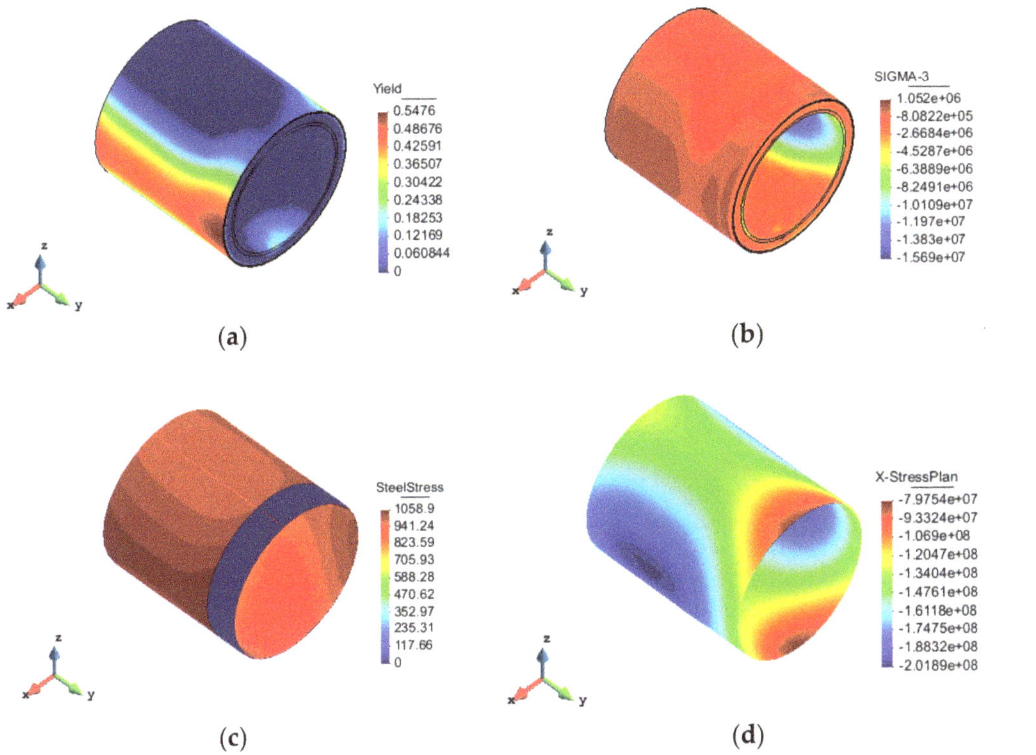

Figure 10. Nephogram of case 2: (**a**) damage, (**b**) 3rd principal stress, (**c**) wire stress, (**d**) cylinder stress.

5.1.3. Case 3: 10% Breakage Ratio and Breaking at Both Outer and Inner Layer in the Middle of Pipe

The obtained nephogram is shown in Figure 11. The maximum damage value of the mortar is 0.53. The damage to the outer concrete emerges at the springline in the broken area but does not penetrate along the whole circumferential direction, with the maximum value of 0.29. Damage also occurs at the invert and crown inside the pipe in the pre-stressing loss zone. The maximum stress in the outer wires approaches 1060.4 MPa. Meanwhile, the stress in the inner wires and the stress in the cylinder around the pre-stressing loss zone increase as well.

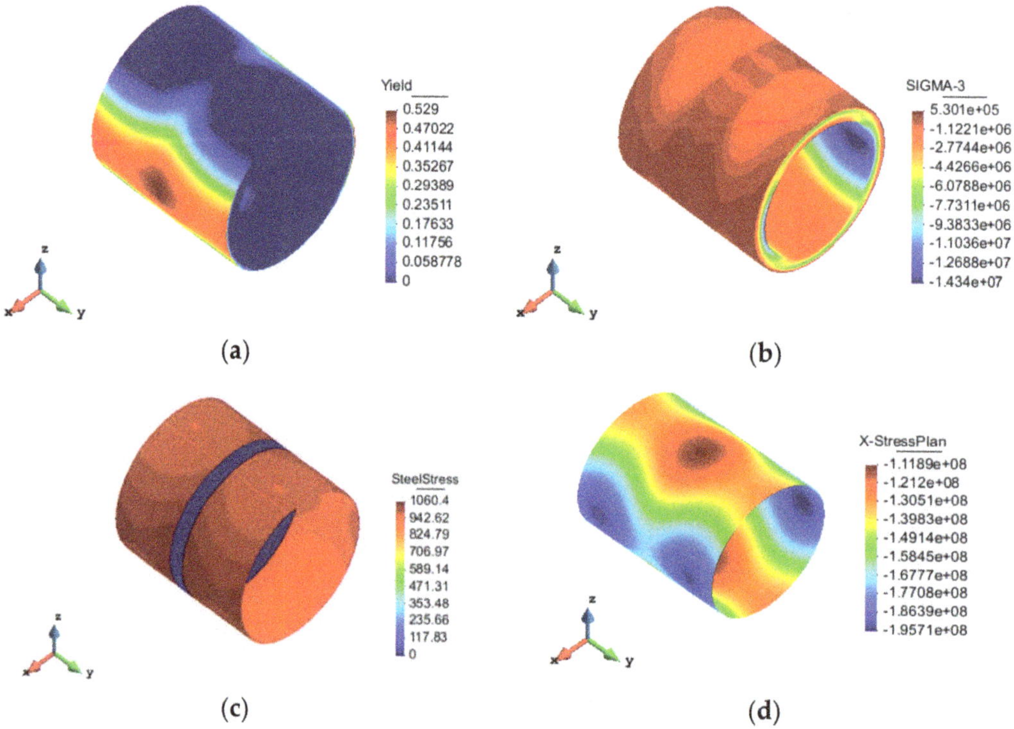

Figure 11. Nephogram of case 3: (**a**) damage, (**b**) 3rd principal stress, (**c**) wire stress, (**d**) cylinder stress.

5.1.4. Case 4: 10% Breakage Ratio and Breaking at Both Outer and Inner Layer at the Joint of Pipe

The resulting nephogram is presented in Figure 12. The maximum damage value of the mortar is 0.57. At the springline in the broken area, the damage to the outer concrete emerges, exhibiting the maximum value of 0.36. Damage occurs at the invert and crown inside the pipe in the pre-stressing loss zone. The maximum stress in the outer wires reaches 1076.4 MPa. In addition, the stress in the inner wires and the cylinder around the pre-stressing loss zone also increases.

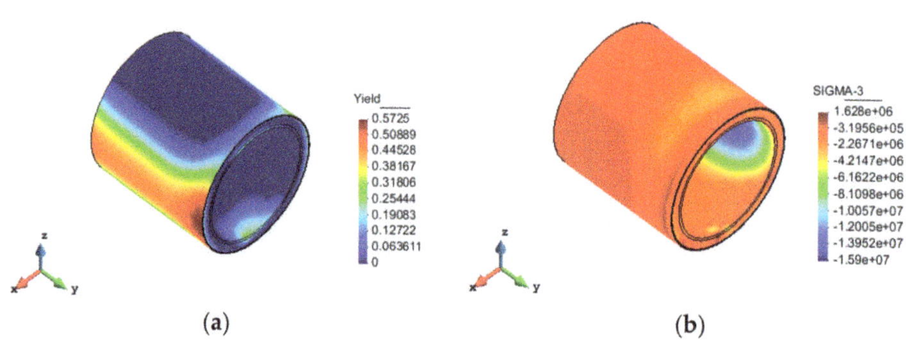

Figure 12. *Cont.*

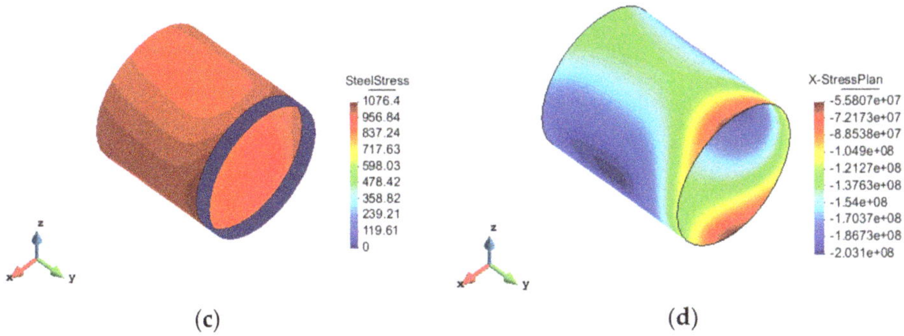

Figure 12. Nephogram of case 4: (**a**) damage, (**b**) 3rd principal stress, (**c**) wire stress, (**d**) cylinder stress.

5.1.5. Case 5: 10% Breakage Ratio and Breaking at Random Location

The obtained nephogram is illustrated in Figure 13. The mortar presents a maximum damage value of 0.39 with a small increment compared with the pre-breaking stage. The damage to the outer concrete emerges at the springline and penetrates along the whole circumferential direction, with the maximum value of 0.22. A small amount of damage to the inner concrete occurs at the invert and crown. The damage distribution in concrete and mortar is uniform along the axial direction. The maximum stress in the outer wires reaches 1057.1 MPa with the increment of 17.4 MPa. In addition, the stress in the cylinder also rises. Because the breakage ratio is small with a relatively scattered broken area, when the random broken situation occurs once, the stress in the wires next to the broken area increases, and the cylinder stress changes a little.

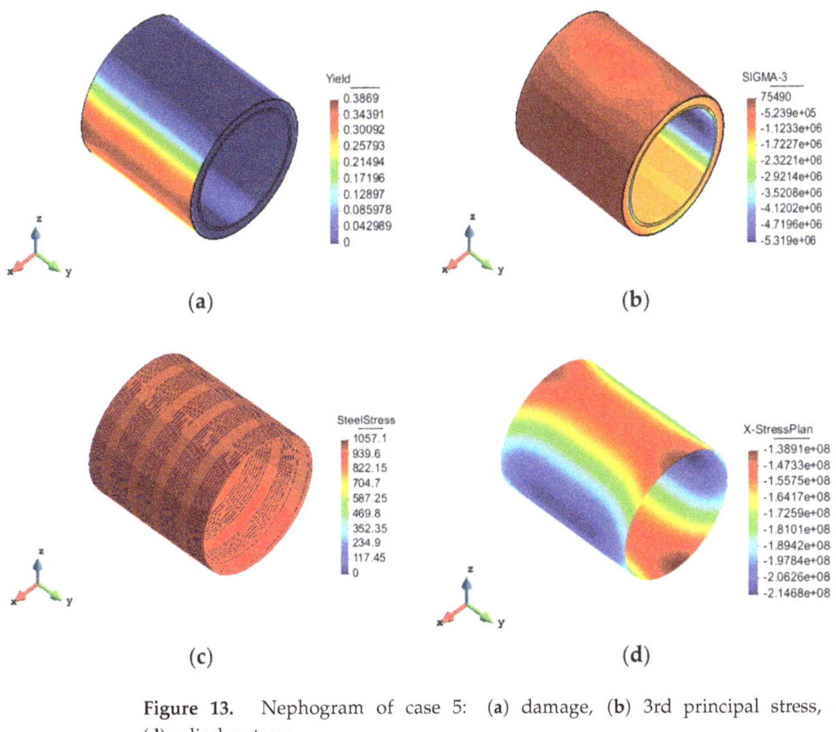

Figure 13. Nephogram of case 5: (**a**) damage, (**b**) 3rd principal stress, (**c**) wire stress, (**d**) cylinder stress.

5.2. Finite Element Results of Pipe No.2 after Breaking Wires

5.2.1. Case 6: 10% Breakage Ratio and Breaking in the Middle of Pipe

It can be seen from Figure 14 that the maximum damage value of mortar is 0.39, with an increment of 0.22 compared with the pre-breaking stage. At the springline in the broken area, the damage to the outer concrete emerges and penetrates along the whole circumferential direction, exhibiting the maximum value of 0.20. Damage occurs at the invert and crown inside the pipe around the pre-stressing loss zone with the value of 0.10. The maximum stress in the outer-layer wires grows to 1031.1 MPa. Meanwhile, the stress in the inner wires and the stress in the cylinder around the pre-stressing loss zone are also enhanced. The damage to the outer concrete will not only expand but also penetrate along the whole circumferential direction, and the inner concrete is also damaged.

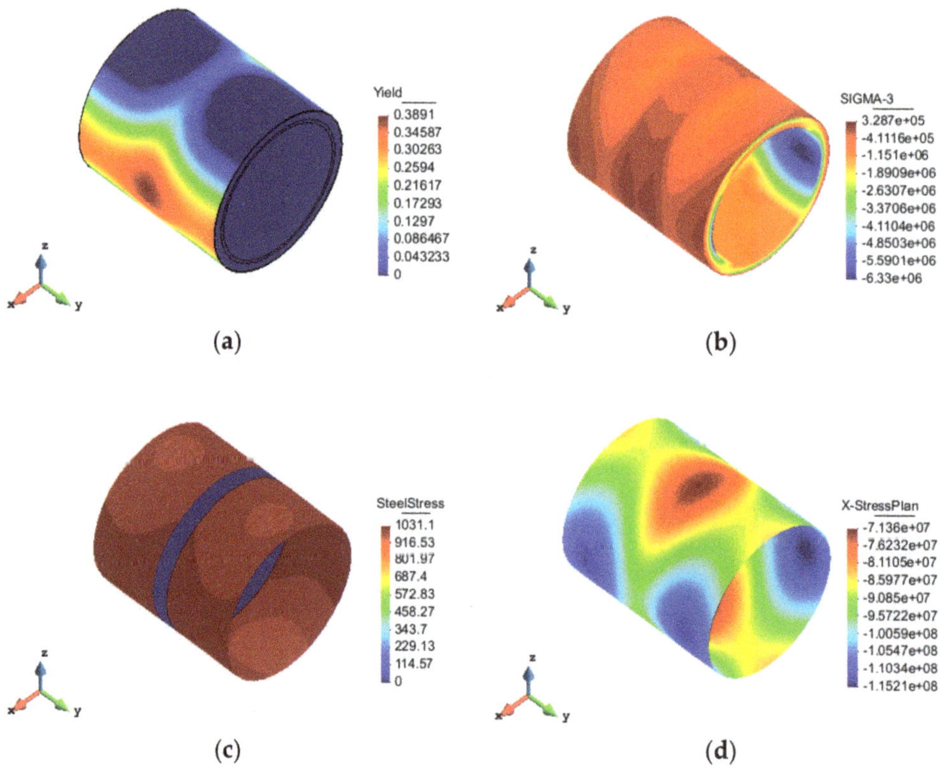

Figure 14. Nephogram of case 6: (**a**) damage, (**b**) 3rd principal stress, (**c**) wire stress, (**d**) cylinder stress.

5.2.2. Case 7: 10% Breakage Ratio and Breaking at the Joint of Pipe

The resulting nephogram is displayed as Figure 15. The maximum damage value of mortar is 0.42, with an increase in value of 0.25 compared with the pre-breaking stage. The damage to the outer concrete emerges at the springline in the broken area and penetrates along the whole circumferential direction, presenting the maximum value of 0.28. Damage can be observed in the invert and crown inside the pipe around the pre-stressing loss zone with the value of 0.12. The maximum stress in the outer-layer wires increases to 1037.4 MPa, along with the increased stress in the inner wires, the third principal stress in the inner concrete and the stress in the cylinder around the pre-stressing loss zone. The damage to the outer concrete will not only expand but also penetrate along the whole circumferential direction, and the inner concrete is also damaged.

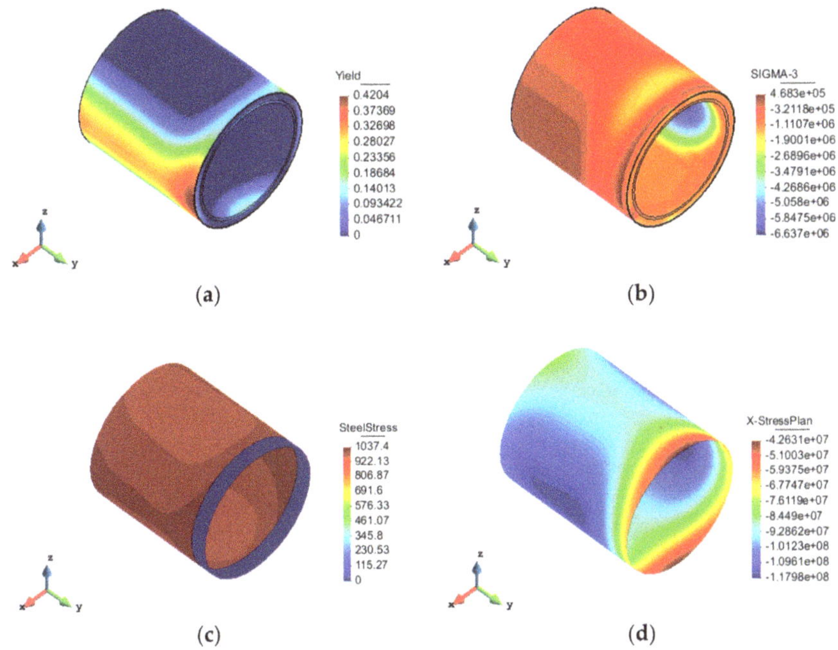

Figure 15. Nephogram of case 7: (**a**) damage, (**b**) 3rd principal stress, (**c**) wire stress, (**d**) cylinder stress.

5.2.3. Case 8: 10% Breakage Ratio and Breaking at Random Location

The obtained nephogram is shown in Figure 16.

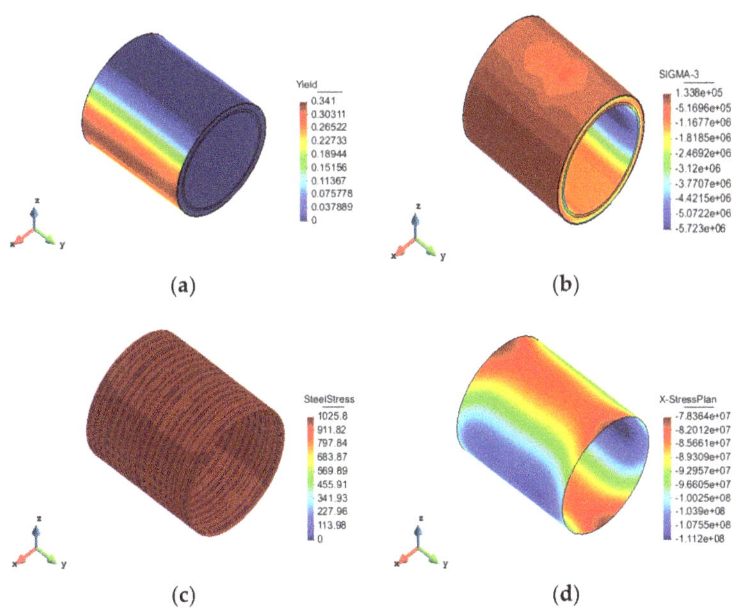

Figure 16. Nephogram of case 8: (**a**) damage, (**b**) 3rd principal stress, (**c**) wire stress, (**d**) cylinder stress.

As illustrated in Figure 16, the maximum damage value of mortar is 0.34, with an increment of 0.14 compared with the pre-breaking stage. At the springline, the damage to the outer concrete emerges and does not penetrate along the whole circumferential direction, exhibiting the maximum value of 0.16. A small amount of damage occurs at the invert and crown inside the pipe. The damage distribution in concrete and mortar is uniform along the axial direction. The maximum stress in the outer-layer wires approaches 1025.8 MPa with the increment of 11.0 MPa. At the same time, the third principal stress in the inner concrete and the stress in the cylinder increase. The damage to the outer concrete only expands, and the inner concrete is not damaged.

5.3. Summary of FEA Results

The FEA results of two pipes with 10% broken wire ratios are presented above. The other results are listed in the tables, with the results of pipe No.1 listed in Table 7, and those of pipe No.2 listed in Table 8. From the table, it can be seen that when the same number of wires are removed from the joint and middle of the pipe with the same breakage ratio, the area without pre-stressing wires at the joint is larger, so breaking at joint is more dangerous than that in the middle of the pipe [21]. For the double-layer broken pipe with a 10% breakage ratio, the outer concrete is circumferentially penetrated along the broken area in both assumptions of the outer layer or both layers at the joint, and the inner concrete also shows damage at the crown and invert, which does not occur in the middle breakage. As the ratio of wire breakage increases, the damage value and area of the outer concrete expands, penetration also occurs in the central wire breakage and damage occurs in the inner concrete. For the single-layer broken pipe with three wire break ratios, whether the wire is broken at the joint or in the middle, the outer concrete shows damage and circumferential penetration along the broken area, and the inner concrete also shows damage. Additionally, as the wire break rate increases, the concrete damage value and damage area further expand but still do not exceed their respective strength limits. Regardless of the single-layer or double-layer broken pipe, when breaks occur consistently in the middle or joint location, the wire stress and cylinder stress near the broken area further increase after the wire break, and the unevenness of the cylinder stress also worsens with the increase in the wire break rate, which may lead to the radial detachment of the cylinder and concrete, further leading to the deterioration of the overall service performance of the pipe.

Table 7. The main analyzing index value of pipe No.1.

Ratio	Maximum Stress of Cylinder (MPa)					Damage of Outer Concrete at Springline					Damage of Inner Concrete at Crown/Invert					Maximum Value of Wires (MPa)				
	Two Layers		Outer Layer		Random Location	Two Layers		Outer Layer		Random Location	Two Layers		Outer Layer		Random Location	Two Layers		Outer Layer		Random Location
	Middle	Joint	Middle	Joint		Middle	Joint	Middle	Joint		Middle	Joint	Middle	Joint		Middle	Joint	Middle	Joint	
10%	−118.9	−55.8	−125.3	−79.8	−138.9	0.29	0.36	0.24	0.30	0.22	0.07	0.30	0.03	0.20	0	1060.4	1076.4	1058.8	1058.9	1057.1
20%	−84.0	−18.8	−93.3	−78.5	−109.9	0.37	0.41	0.30	0.30	0.23	0.18	0.38	0.15	0.22	0	1081.4	1083.7	1062.0	1061.4	1060.8
30%	−59.0	−8.49	−84.0	−78.1	−99.8	0.42	0.50	0.31	0.43	0.23	0.24	0.39	0.17	0.23	0.10	1089.7	1089.8	1062.7	1061.5	1069.6

Table 8. The main analyzing index value of pipe No.2.

Ratio	Maximum Stress of Cylinder (MPa)			Damage of Outer Concrete at Springline			Damage of Inner Concrete at Crown/Invert			Maximum Value of Wires (MPa)		
	Middle	Joint	Random Location	Middle	Joint	Random Location	Middle	Joint	Random Location	Middle	Joint	Random Location
10%	−71.4	−42.6	−78.4	0.20	0.26	0.16	0.10	0.17	0	1031.1	1037.4	1025.8
20%	−53.7	−25.6	−73.5	0.28	0.31	0.19	0.12	0.21	0	1040.6	1041.2	1029.8
30%	−37.9	−21.7	−70.2	0.32	0.47	0.18	0.19	0.33	0.03	1044.9	1044.5	1032.5

Under the three random wire breakage rates, for the double-layer pipe, the damage value of the outer concrete expands but does not penetrate, and the inner concrete does not show damage, but its tensile stress increases. The stress in the wire and cylinder do not increase much, and the overall stress uniformity of the cylinder is excellent. For the single-layer pipe, the outer layer of concrete damage does not penetrate along the circumferential direction, and the inner concrete does not develop damage in the 10% and 20% of the broken wire ratio scenario, but it occurs in the 30% ratio scenario.

The results of continuous and random breakage calculations are compared under three breakage ratios. For one pipe, the material damage values and area distribution, wire stress and cylinder stress under the random breakage assumption are smaller than the statistical values of the corresponding parameters under continuous breakage with the same breakage ratio. In addition, under the low breakage ratio, the outer concrete damage under continuous breakage basically penetrates along the circumferential direction, and the concrete and cylinder stress are more uniformly distributed. Since the distribution of random breakage locations is more discrete, the consequences caused by wire breakage are lesser than those under the assumption of continuous breakage. The bearing resistance of the double-layer pipe in the random and outer breakage assumptions is better than that of the single-layer pipe, while in the case of both double-layer breakage it is worse than that of the single-layer pipe. Although the double-layer pipe suffers more inner design water pressure, its integrity is stronger, because the double-layer wire is similar to the double insurance action.

6. Conclusions

A nonlinear FEA was developed to investigate the mechanical behavior of two typical ECPs with broken wires. We considered the actual production, installation and operation stages by dividing the analysis process into six or eight steps. The results were obtained under the hypothesis of broken wires with three ratios and different distributions when suffering under working water pressure and other loads. The main conclusions are summarized as below:

(1) We summarized the advantages and shortcomings of various pre-stressing simulation methods and then proposed an initial stress method based on a novel single-spring joint element method for FEA. The structural response results before wire breakage based on the specification method and the FEA model are basically consistent.

(2) The damage to the outer concrete increases and penetrates along the entire circumferential direction as the percentage of wire breakage grows at continuous wire breakage. Damage to the inner concrete begins to occur, and indicators such as wire stress, cylinder stress and distribution also deteriorate. The risk of breakage is higher in the joint than that in the middle location, and the breakage that occurs with continuous centralized distribution is more dangerous than breakage with random distribution. The double-layer pipe has better overall destruction resistance than that of the single-layer pipe under random breakage and outer continuous breakage conditions.

(3) A cylinder can be used as both an anti-seepage body to prevent seepage and safety reserve under wire breakage. If the tearing cylinder is regarded as the pipe burst indicator, the pipe will not immediately burst in the drafted broken ratios and assumptions. However, micro-cracks in concrete will lead to corrosive groundwater or chemical substances to further corrode intact wires and cylinder. At the same time, the unevenness of the cylinder will potentially lead to radial detachment with concrete. The combined effect will further result in the deterioration of the overall service performance.

Existing prototype tests of broken pipes are only performed for individual internal water load or three-edge bearing tests, and it is still very difficult to conduct test simulations of broken pipes in service states due to technical and economic reasons. At the same time, this paper does not further discuss more breakage ratio sensitivity analysis and ultimate internal water pressure calculation but focuses on research into the mechanical effects of the breakage location of two different pipe types. If we consider the mechanical behavior of a broken pipe after concrete cracking, it will be related to the fine structure of concrete and crack path development problems, which will be further investigated in follow-up research.

Author Contributions: Z.Z.: analysis, programing, writing. T.L.: investigation, statistical analysis, draft editing, review. L.Z.: programing, validation. H.Q.: typesetting, modelling. All authors have read and agreed to the published version of the manuscript.

Funding: This research has been greatly supported by the National Key Research and Development Plan (No. 2018YFC0407102) and the Project of the Research on Long Term Monitoring and Safety Evaluation of Concrete Dams Based on BIM (DJ-ZDXM-2018-02).

Institutional Review Board Statement: Not applicable.

Informed Consent Statement: Not applicable.

Data Availability Statement: The datasets generated during the current study are available from the corresponding author on reasonable request.

Conflicts of Interest: We declare that this article has no competing interest with others.

References

1. Romer, A.E.; Bell, G.E.C.; Ellison, R.D. Failure of Prestressed Concrete Cylinder Pipe. *Pipelines* **2007**, *2007*, 1–17. [CrossRef]
2. Hassi, S.; Ebn Touhami, M.; Boujad, A.; Benqlilou, H. Assessing the effect of mineral admixtures on the durability of Prestressed Concrete Cylinder Pipe (PCCP) by means of electrochemical impedance spectroscopy. *Constr. Build. Mater.* **2020**, *262*, 120925. [CrossRef]
3. *ANSI/AWWA C304-14*; Standard for Design of Prestressed Concrete Cylinder Pipe. AWWA: Singapore, 2014.
4. Shatnawi, A.; Al-Sadder, Z.S.; Abdel-Jaber, M. Symmetrical and Anti-Symmetrical Buckling of Long Corroded Cylindrical Shell Subjected to External Pressure. *Jordan J. Civ. Eng.* **2010**, *4*, 106–118.
5. Zarghamee, M.S.; Ojdrovic, R.P.; Nardini, P.D. Prestressed concrete cylinder pipe condition assessment-What Works? What doesn't? What's Next? *Pipelines* **2011**, *2011*, 182–194. [CrossRef]
6. Mergelas, B.J.; Atherton, D.L.; Kong, X. Risk Management: A Case Study in Operating PCCP. *Pipelines 2002* **2002**, *116*, 1–9. [CrossRef]
7. Ge, S.; Sinha, S.K. Analysis of a 60-In. PCCP that Failed without Warning from Acoustic Fiber Optic System. In *Pipelines 2014*; American Society of Civil Engineers: Reston, VA, USA, 2014; pp. 86–95. [CrossRef]
8. Hassi, S.; Touhami, M.E.; Ejbouh, A.; Berrami, K.; Boujad, A.; Ech-Chebab, A. Case Study of the Performance of Prestressed Concrete Cylinder Pipes in the Greater Agadir of Morocco. *J. Pipeline Syst. Eng. Pract.* **2021**, *12*, 0502100. [CrossRef]
9. Luo, J.; Gao, L. In-Situ test experimental research on leakage of large diameter pre-stressed concrete cylinder pipe (PCCP). *Civ. Eng. J.* **2016**, 25. [CrossRef]
10. Lee, Y.; Lee, E.-T. Retrofit Design of Damaged Prestressed Concrete Cylinder Pipes. *Int. J. Concr. Struct. Mater.* **2013**, *7*, 265–271. [CrossRef]
11. You, R. Longitudinal Cracking Analysis of PCCP with Broken Wires. *Appl. Mech. Mater.* **2012**, *157–158*, 976–981. [CrossRef]
12. Zarghamee, M.S.; Fok, K.L. Analysis of Prestressed Concrete Pipe under Combined Loads. *J. Struct. Eng.* **1990**, *116*, 2022–2039. [CrossRef]
13. Zarghamee, M.S.; Eggers, D.W.; Ojdrovic, R.P. Finite-Element Modeling of Failure of PCCP with Broken Wires Subjected to Combined Loads. *Proc. Pipeline Div. Spec. Conf.* **2002**, 1–17. [CrossRef]
14. Zarghamee, M.S.; Eggers, D.W.; Ojdrovic, R.; Rose, B. Risk Analysis of Prestressed Concrete Cylinder Pipe with Broken Wires. *Publ. Am. Soc. Civ. Eng.* **2003**, 599–609. [CrossRef]
15. Gomez, R.; Muñoz, D.; Vera, R.; Escobar, J.A. Structural Model for Stress Evaluation of Prestressed Concrete Pipes of The Cutzamala System. *Am. Soc. Civ. Eng.* **2004**, *146*, 1–9.
16. Diab, Y.G.; Bonierbale, T. A Numerical Modeling and a Proposal for Rehabilitation of PCCPs. *Pipelines* **2001**, *2001*, 1–8. [CrossRef]
17. Ge, S.; Sinha, S. Effect of Mortar Coating's Bond Quality on the Structural Integrity of Prestressed Concrete Cylinder Pipe with Broken Wires. *J. Mater. Sci. Res.* **2015**, *4*. [CrossRef]
18. Alavinasab, A.; Padewski, I.E.; Higgins, M. Effects of the Location of Broken Prestressing Wire Wraps in Structural Integrity of a Damaged PCCP. In *Pipelines 2013*; American Society of Civil Engineers: Reston, VA, USA, 2013; Volume 4, pp. 767–774. [CrossRef]
19. Alavinasab, A.; Hajali, M. Do Broken Prestressing Wire Wraps at the Joint Impact the Safe Operation of a Pipe? In *Pipelines 2014*; American Society of Civil Engineers: Reston, VA, USA, 2014. [CrossRef]
20. Hajali, M.; Alavinasab, A.; Shdid, C.A. Effect of the Location of Broken Wire Wraps on the Failure Pressure of Prestressed Concrete Cylinder Pipes. *Struct. Concr.* **2015**, *16*, 297–303. [CrossRef]
21. Hajali, M.; Alavinasab, A.; Shdid, C.A. Effect of the Number of Broken Wire Wraps on the Structural Performance of PCCP with Full Interaction at the Gasket Joint. *J. Pipeline Syst. Eng. Pract.* **2016**, *7*, 04015026. [CrossRef]

22. Xiong, H.; Li, P.; Li, Q. FE model for simulating wire-wrapping during prestressing of an embedded prestressed concrete cylinder pipe. *Simul. Model. Pract. Theory* **2010**, *18*, 624–636. [CrossRef]
23. Hu, S.W.; Sun, Y.Y.; Xu, X.; Huan, Y.Q. Calculation model for bar-wrapping during prestressing of an embedded bar-wrapped cylinder concrete pressure pipe. *Thin-Walled Struct.* **2019**, *139*, 39–45. [CrossRef]
24. Zhai, K.; Fang, H.; Guo, C.; Ni, P.; Fu, B.; Wang, F.; Zhang, C. Strengthening of PCCP with broken wires using prestressed CFRP. *Constr. Build. Mater.* **2020**, *267*, 120903. [CrossRef]
25. Zhai, K.; Guo, C.; Fang, H.; Li, B.; Ma, B.; Hu, Q.; Wang, F. Stress distribution and mechanical response of PCCP with broken wires. *Eng. Struct.* **2021**, *245*, 112858. [CrossRef]
26. Zhai, K.; Fang, H.; Guo, C.; Ni, P.; Wu, H.; Wang, F. Full-scale experiment and numerical simulation of prestressed concrete cylinder pipe with broken wires strengthened by prestressed CFRP. *Tunn. Undergr. Space Technol.* **2021**, *115*, 104021. [CrossRef]
27. Dong, X.; Dou, T.; Cheng, B.; Zhao, L. Failure analysis of a prestressed concrete cylinder pipe under clustered broken wires by FEM. *Structures* **2021**, *33*, 3284–3297. [CrossRef]
28. Lin, J.H. Structural Analysis and Evaluation of Pre-Stressed Concrete Cylinder Pipe. Master's Thesis, National Cheng Kung University, Tainan City, Taiwan, 2008.
29. Hu, S.W.; Shen, J. Study on impact from wire-breaking inside of super-large diameter PCCP on its bearing capacity. *Water Resour. Hydropower Eng.* **2011**, *42*, 41–44. (In Chinese)
30. Li, H.; Feng, X.; Chen, B.; Zhao, L. Damage evaluation of a PCCP joint with a localized soil void and broken wires. *Adv. Struct. Eng.* **2021**, *24*, 3133–3143. [CrossRef]
31. Shen, J. Experimental Study on Structure Safety Performance and Theoretical Analysis for Large Caliber PCCP. Ph.D. Thesis, Nanjing Hydraulic Research Institute, Nanjing City, China, 2010. (In Chinese).
32. Dou, T.; Chen, B.; Hu, H. Internal hydraulic pressure test of PCCP repaired by CFRP. *China Concr. Cem. Prod.* **2017**, *6*, 35–40. (In Chinese)
33. Hu, B.; Fang, H.; Wang, F.; Zhai, K. Full-scale test and numerical simulation study on load-carrying capacity of prestressed concrete cylinder pipe (PCCP) with broken wires under internal water pressure. *Eng. Fail. Anal.* **2019**, *104*, 513–530. [CrossRef]
34. Luo, J.; Yao, X. Large diameter PCCP pipeline broken wires safety risk management based on APH method. *J. Civ. Archit. Environ. Eng.* **2016**, *38*, 65–72. (In Chinese)
35. Zhao, L.; Zhang, W.; Bai, X.; Yan, T.; Li, T. Single Spring Joint Element Based on the Mixed Coordinate System. *Math. Probl. Eng.* **2015**, *2015*, 1–16. [CrossRef]
36. Wei, W. Four-parameter failure criterion for concrete in strain space. *Adv. Sci. Technol. Water Resour.* **2004**, *24*, 27–29. (In Chinese) [CrossRef]
37. Duncan, J.M.B.P.; Wong, K.S. Strength, Stress-Strain and Bulk Modulus Parameters for Finite Element Analysis of Stresses and Movements in Soil Masses. In *Report. UCB/CT/80-01*; University of Caliafornia: Berkeley, CA, USA, 1980.
38. Lee, Y.; Feng, M.Q.; Lee, E.T. Deflection of buried prestressed concrete cylinder pipe with soil-pipe interaction. *Ksce J. Civ. Eng.* **2014**, *18*, 2191–2195. [CrossRef]
39. Goodman, R.E.; Taylor, R.L.; Brekke, T.L.A. A model for the mechanics of jointed rock. *J. Soil Mech. Found. Div.* **1968**, *94*, 637–659. [CrossRef]
40. Yi-feng, F. Research Status of the Constitutive Relationship of Bond-slip between Steel Bar and Concrete. *J. Xiangfan Vocat. Tech. Coll.* **2011**, *10*, 22–24. (In Chinese)

coatings

Article

Experimental Research on Dynamic Response of Layered Medium under Impact Load

Mingqing Liu [1,2,*] and Qinyu Gan [3]

1 College of Architecture and Energy Engineering, Wenzhou University of Technology, Wenzhou 325035, China
2 Wenzhou Key Laboratory of Intelligent Lifeline Protection and Emergency Technology for Resilient City, Wenzhou University of Technology, Wenzhou 325035, China
3 College of Design and Art, Wenzhou University of Technology, Wenzhou 325035, China
* Correspondence: liumvsq@126.com

Abstract: Stress waves propagating through multiple parallel fractures are attenuated due to the multiple wave reflections and transmissions at the fractures. The dynamic response of a layered medium under an impact load is systematically studied by a model test in this paper. The effects of the impact energy, number of joints, thickness of the medium and joint layer, material properties of the cementation layer, and other factors of the dynamic stress propagation and attenuation of layered media are analyzed. Studies have shown that Sadovsky's empirical formula fits the attenuation law of vertical peak velocity well, and the obtained attenuation coefficient k and index b can be used for reflecting the dynamic response characteristics of layered media. The attenuation coefficient k is positively correlated with the impact height and negatively correlated with the plate thickness in a single-layer plate impact test. The thickness of the medium layer is the main factor affecting the vibration response of the layered medium in the impact test of the multi-layer slab. The medium thickness, the number of medium layers, and the number of joints have little influence. The different cementation materials and the change of cementation thickness will have a great influence on the dynamic response of layered media, when the cementation layer is filled with joint surfaces for the impact test of multi-layer plates.

Keywords: layered medium; cementation layer; dynamic stress attenuation; fitting coefficient; parametric analysis

Citation: Liu, M.; Gan, Q. Experimental Research on Dynamic Response of Layered Medium under Impact Load. *Coatings* **2022**, *12*, 1474. https://doi.org/10.3390/coatings12101474

Academic Editor: Jose Ygnacio Pastor

Received: 7 September 2022
Accepted: 27 September 2022
Published: 5 October 2022

1. Introduction

The research on the dynamic properties of a layered medium has always been an important topic in geotechnical engineering, such as for the safety protection of a military underground structure, mine blasting or excavation blasting, the excavation of a metro tunnel or an underground pipe rack, the design of a roadbed or a pavement structure, or the vibration prediction of the blasting demolition of a structure. The propagation characteristics of stress waves in layered media need to be deeply studied and understood in these fields.

Generally, three methods are available for studies on wave propagation of a layered medium. One is the theoretical study from which the mechanism and the process of wave propagation can be revealed for some special geological cases. The experimental study is the second method, which is sometimes limited by the existing test techniques. The third is numerical modeling, which provides a convenient and economical approach, especially for complicated geological cases.

The most classical theoretical model is the displacement discontinuity model [1,2], which assumes that when waves propagate across a fracture, the stress field is continuous, but the displacement field is discontinuous due to the fracture deformation. On the basis of the displacement discontinuity model and Snell's law, propagation of wave incidence across a planar linear slip interface was investigated by Schoenberg [3,4]. Many scholars have further extended the displacement discontinuous model. The effects of different joint quantities, joint geometric characteristics, stress wave incident conditions, and rock mass and joint media material deformation characteristics on the propagation characteristics of the stress wave in jointed media were considered. The wave propagation across parallel fractures and the permeation and reflection characteristics of joint surfaces were studied by Pyrak-Nolte [5,6] and Cook [7,8]. Chai S.B. [9] and Zhu [10] studied the wave transmission across jointed rock masses, where multiple intersecting joint sets exist. Goodman R.E. [11], Lu [12], Li [13], and Ju [14] considered the influence of the characteristics of the joint layer, such as joint stiffness, physical and mechanical properties of joint-filled media, joint surface boundary condition, and transfer condition. Pyrak-Nolte [5] and Gu [15,16] discussed the relationship between the transmittance and the reflection coefficient, the waveform conversion phenomenon, and the incident angle and derived the transfer equation of the incident angle. Cai [17] and Zhao [18–20] analyzed the effects of the linear and nonlinear deformation characteristics of the joint layer.

Although the wave propagation across different topography joints, such as single or groups of parallel joints and simple intersecting joints, have been considered in the existing theoretical research. However, it is difficult to accurately analyze and evaluate by theoretical methods in actual engineering, due to the complexity of joint distribution and the inaccuracy of the mechanics and boundary conditions. Therefore, many researchers have done a lot of work through the more intuitive research method of experiment. Generally, the Split Hopkinson Pressure Bar (SHPB) apparatus was adopted to investigate the wave propagation of a layered medium. Ju [14], Wang [21], Lu [22], and Li [23] studied the propagation characteristics of one-dimensional stress waves in layered media through the fractal theory and analyzed the transmittance and reflection properties at the joints and the effects of the fractal dimension and roughness of the joints. Similarly, based on SHPB test equipment, the influence of impact rate, strain rate, joint angle, joint thickness, and other parameters was studied by Li [23], Yang [24], Wei [25], and Challita G. [26]. In addition to the SHPB test system, Tian [27] and Sun [28] carried out some research with independently developed impact test equipment. Based on actual blasting engineering, Peng [29] and Wang [30] analyzed the wave attenuation of a natural layered foundation and improved the empirical formula of stress wave propagation.

For the numerical modeling, the same problem was studied with the many available tools, such as the finite element method (FEM), boundary element method (BEM), finite difference method (FDM), and discrete element method (DEM). In FEM and BEM, joints are modeled using "joint elements", which simulate the interface discontinuity [31,32]. In FDM, joints are simulated using slide-lines [33]. Compared with the other methods, DEM has been recognized as a better alternative. The layered medium is treated as an assemblage of discrete blocks by fictitious or real joints, of which some are the contact interfaces between the distinct bodies [34]. In recent years, the universal distinct element code (UDEC), which is a 2D DEM-based numerical program, has been widely adopted to study wave propagation across the joints of layered media [35]. Brady [36] performed UDEC modeling on the wave attenuation across a single joint. Zhao et al. [37,38] carried out numerical studies on wave propagation across multiple non-linearly deformable joints with UDEC. Deng [39] performed a study aimed to verify the capability of 3DEC to model wave propagation across joints in 3D space. Zhu [40] developed a newly numerical code, the distinct lattice spring model (DLSM), to study normally incident wave propagation across a joint set.

However, no experimental work has been conducted to systemically study the dynamic stress propagation and attenuation of a layered medium with parallel joints. In this paper, a self-developed concrete-slab-impact testing system is used to study the dynamic properties of the layered medium through the impact test of single-layer slab, multi-layer slab with parallel joints, and multi-layer slab with cementation. Firstly, the attenuation parameters are obtained by fitting the peak particle velocity with the empirical formula.

Subsequently, parametric studies on impact energy, the number of joints, the thickness of the medium and joint layer, and the properties of the cementation layer are conducted. The dependence of wave propagation and attenuation on these parameters are discussed in detail.

2. Model Test Design

2.1. Specimen Dimension Determination

In order to eliminate the influence of the boundary effect, the numerical software LS-DYNA (960, 2001, ANSYS, Pittsburgh, PA, USA) was used to simulate the attenuation of peak particle velocity by modeling the hammer drop test on a three-layer concrete slab.

2.1.1. Calculation Model Establishment

The stress cloud diagram of the concrete slab was recorded to analyze the influence of boundary effect by applying the impact load of a 2.5 kg drop hammer on the surface, as shown in Figure 1. The SOLID164 element was used to build the concrete slab model, and the length of each slab is 4 m, the width is 1 m, and the thickness is 10 cm. The simulation in this paper adopts the projection boundary condition, which can simulate the effect of the wireless range in the finite range model. By applying artificial dampers to the boundary, the wave will be projected when it propagates to the boundary and then spreads to the infinite distance, as if it propagates in the infinite medium. The bottom concrete slab is fixed and restrained at the bottom. Surface-to-surface contact is set between each layer of the slab, and point-to-surface contact is set between the drop hammer and the upper layer of the slab.

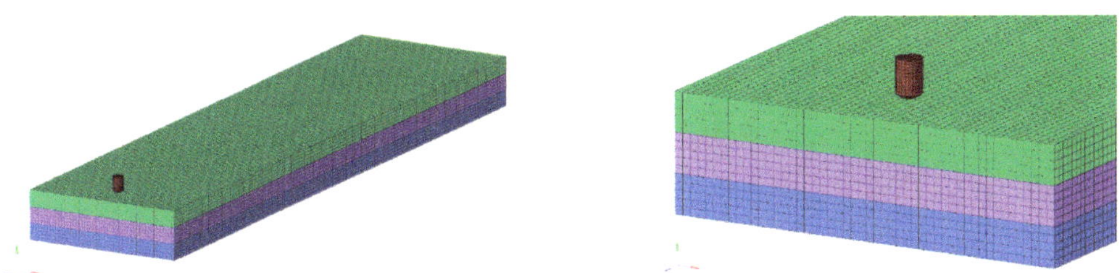

Figure 1. Calculation model of concrete slab impacted by drop hammer.

Surface-to-surface contact is adopted for the contact type between each layer of slabs, and this kind of contact is one of the most commonly used contact types. It is commonly used for contact problems between objects with an arbitrary shape and a large contact area, such as block sliding, cylinder sliding, etc. Point-to-surface contact is adopted for the contact type between the drop hammer and the uppermost concrete slab. A continuous cap model MAT_159 (MAT_CSCM_CONCRETE) was used for the concrete, and a follow-up hardening model MAT_003 (MAT_PLASTIC_KINEMATIC) was used for the drop hammer.

In the impact model of LS-DYNA, the material constitutive model of concrete is usually the continuous cap model MAT_159 (MAT_CSCM_CONCRETE). The model requires fewer parameters and can take into account the strain-rate effect, which is very suitable for media that are sensitive to compressibility. The main parameters required are a density of 2.35×10^{-9} t/mm^3, a compressive strength of 23.5 MPa, and an aggregate size of 10 mm. The material constitutive model of the drop hammer adopts the commonly used nonlinear plastic dynamic hardening model MAT_003 (MAT_PLASTIC_KINEMATIC). This model takes into account the strain-rate effect of the material, and the isotropy and dynamic hardening can be selected by adjusting the hardening parameters between 0–1. The strain rate is considered by the Cowper–Symonds model, and the factors related to the strain rate are used to express the yield stress, which is very suitable for metal materials under an explosive impact load.

The parameters of the concrete and drop hammer materials are shown in Table 1. The transmission boundary is adopted for the boundary condition, and the initial impact velocity of the drop hammer is defined as 4.849 m/s.

Table 1. The parameters of materials.

Concrete Slab			Drop Hammer		
Density (kg/m^3)	Compressive Strength (MPa)	Aggregate Size (m)	Density (kg/m^3)	Elasticity Modulus (GPa)	Yield Strength (MPa)
2.35×10^3	23.5	0.01	8.32×10^3	200	400

2.1.2. Numerical Results Analysis

The direction stress cloud diagram of the concrete slab at different moments in the impact process is shown in Figure 2. The dynamic stress of the concrete slab shows gradual attenuation with the increase in impact distance. The horizontal influence range of the impact load is 2.5 m, and the stress response is very small, outside the range of 2.5 m. In the vertical direction, the dynamic stress shows an obvious jump attenuation across the joints, and a certain stress concentration appears at the interface between the layers. It can be seen that the joints have a significant effect on the dynamic propagation characteristics of the concrete slabs.

Figure 3 shows the vertical and radial peak particle velocity (PPV) values of each measurement point of the concrete surface in the horizontal direction. It can be seen that the attenuation of the vertical PPV is obviously greater than that of the radial PPV. The peak particle velocity tends to be stable when the measure point distance is greater than 2.5 m. Therefore, the final length of the concrete slab is determined to be 2.5 m.

2.2. Model Making

C30 fine aggregate concrete is selected as the concrete for model pouring, and the concrete ratio is water:cement:sand:gravel = 0.38:1:1.11:2.72, with 42.5 ordinary Portland cement used as the cement. Due to the small size of the model structure, melon seed chips are used as the stones, and the particle size is less than 20 mm.

Three kinds of cementitious materials with different elastic moduli are selected in this test, which are 1 cm EPE pearl cotton, 1 cm XPS extruded board, and 1 cm sand, respectively. The XPS extruded board and EPE pearl cotton are cut and pasted according to the width of the concrete slab and fixed on the concrete slab with adhesive tape to form a multi-layer slab.

Figures 4 and 5 show the process of concrete-slab-mold making, steel binding, and specimen casting. Firstly, the model boxes were made according to the size of the specimen, as shown in Table 2. Then, the release agent was painted on the inner wall of the model boxes, and the steel was tied. The concrete was poured into the model in batches and vibrated fully with a vibrator. Meanwhile, the standard cube concrete test blocks were made and cured under the same condition during the casting processing of the specimens. The average uniaxial compressive strength of the five groups of specimens at 28 days was 23.64 MPa.

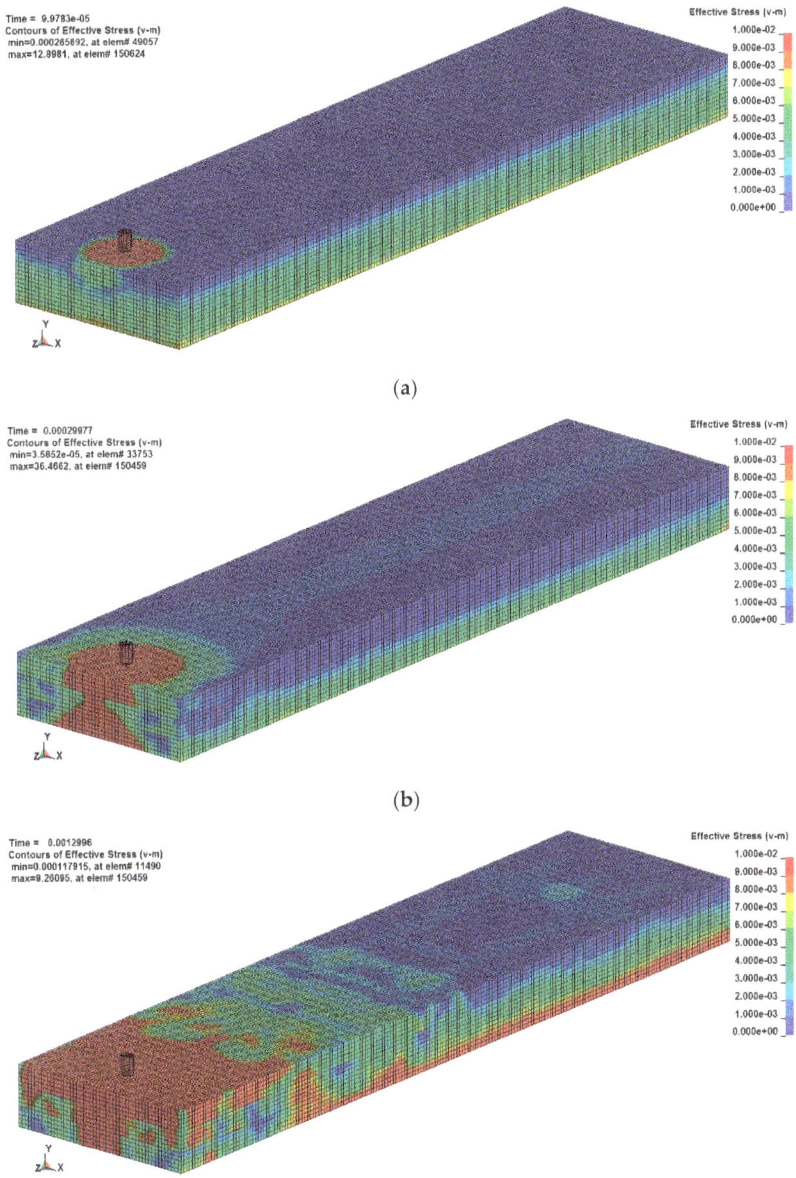

(**a**)

(**b**)

(**c**)

Figure 2. The stress cloud diagram at different moments: (**a**) t = 9.983 $\times$ 10^{-2} ms; (**b**) t = 2.9977 $\times$ 10^{-2} ms; (**c**) t = 1.2996 ms.

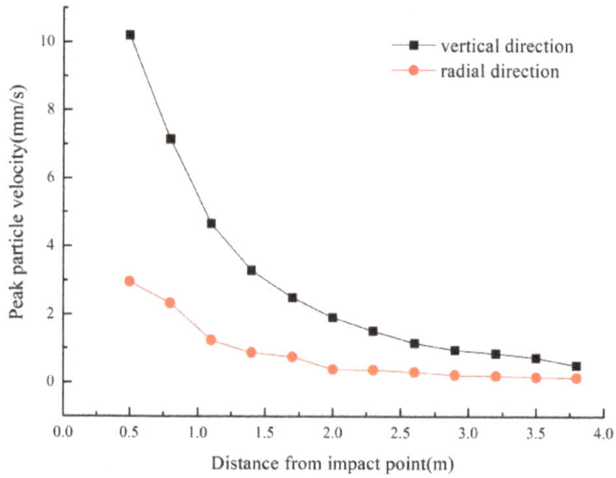

Figure 3. Attenuation of vertical and radial peak particle velocity with distance on concrete surface.

Figure 4. Mold making and steel binding.

Figure 5. Specimens casting.

Table 2. Type and size of concrete slab.

Type	Size (cm × cm × cm)	Number
B1	250 × 100 × 5	4
B2	250 × 100 × 10	4
B3	250 × 100 × 15	1
B4	250 × 100 × 20	1

2.3. Testing Process

This paper uses Canada's Minimate Pro4 vibration and overpressure monitor, which can be used to monitor the vibration velocity. The vibration sensors of the Minimate Pro4 vibration are arranged along the center line of the concrete surface, and the ranging is 0.3 m, 0.6 m, 0.9 m, 1.2 m, 1.5 m, 1.8 m, and 2.1 m, respectively, as shown in Figure 6. The concrete slab is placed on the ground freely without fixed constraint. In addition, the experimental acceleration of the fixed point was monitored. The 2.5 kg drop hammer was raised to the heights of 0.2 m, 0.4 m, 0.6 m, 0.8 m, 1.0 m, and 1.2 m, so that it can fall freely and impact the concrete slab. The JM5938 dynamic signal test and analysis system was used for signal acquisition, and its connection diagram is shown in Figure 7. Each group of tests was repeated three times. The larger discrete data will be eliminated, and the remaining data will be averaged to eliminate the error caused by the experimental contingency.

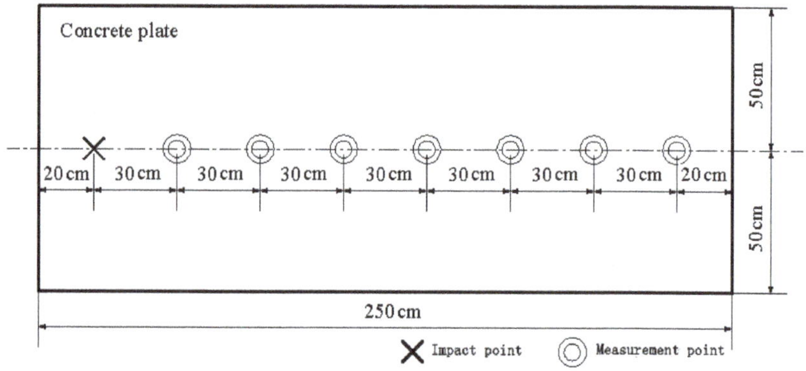

Figure 6. Schematic diagram of measurement point arrangement.

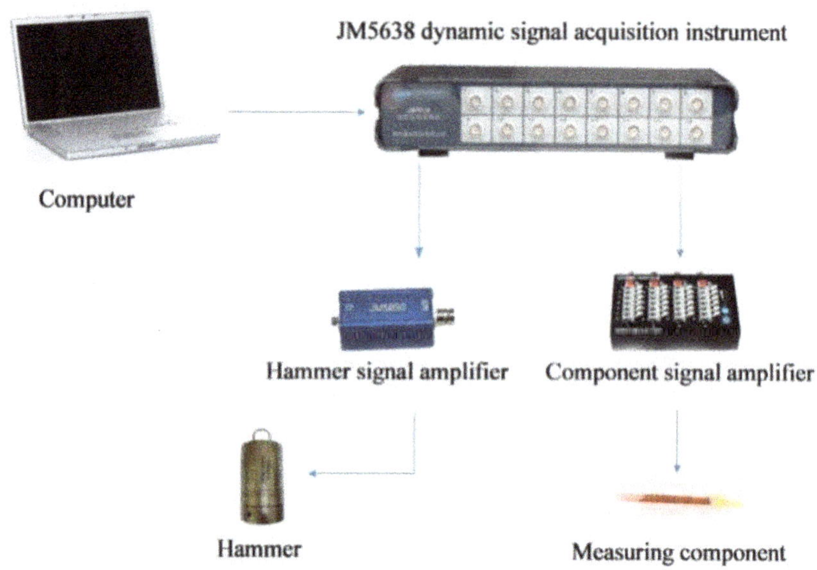

Figure 7. JM5938 dynamic signal test and analysis system.

3. Experimental Results Analysis

By changing the thickness, the number of layers, the types of the cementing layers, the thickness of the cementing layers, the stacking order of the slabs, and the material collocation of the cementing layers, the working conditions of the different layered medium models subjected to the impact load are simulated. The effects of the characteristics of the medium and the characteristics of the interlayer joint on the dynamic mechanical response of the layered medium under impact load were studied, by analyzing the vibration response and changing law of each measuring point under different working conditions.

There are many factors that affect the vibration strength of the medium, so it is difficult to obtain the specific mathematical function expression directly based on the theoretical analysis. Various kinds of empirical prediction formulas for the vibration strength from the experiments were obtained, combining dimensional analyses and empirical assumptions. It is generally recognized that the peak particle velocity shows the power decay with the increase in measuring distance. Sadovsky's empirical formula is widely used in vibration prediction, and its expression is, where V_{peak} is the peak particle velocity, k is the attenuation coefficient, b is the attenuation index, and Z is the ranging.

3.1. Dynamic Response Analysis of Single-Layer Plate

Figure 8 shows the attenuation law of the vertical peak particle velocity of different thicknesses of concrete slabs. It can be seen that the value of attenuation index b fluctuates less under different impact heights (H), when the thickness of concrete slab is the same, and the value of attenuation coefficient k increases monotonically with the increase in impact height in the same condition. However, the value of the attenuation coefficient decreases with the increase in the thickness of the concrete slab, when the drop height remains unchanged. In order to further analyze the influence of the impact height and the thickness of the slab on the dynamic response, a sensitivity analysis of the attenuation coefficient and its influencing factors is carried out through linear regression, as shown in Figure 9. It is obvious that the attenuation coefficient and the impact height show a significant positive linear correlation under the same slab thickness, while the thickness of the concrete slab has a great influence on its dynamic stress response and attenuation law under the same impact height, and the attenuation coefficient k of the vertical peak particle velocity shows a negative correlation with the thickness of the concrete slab.

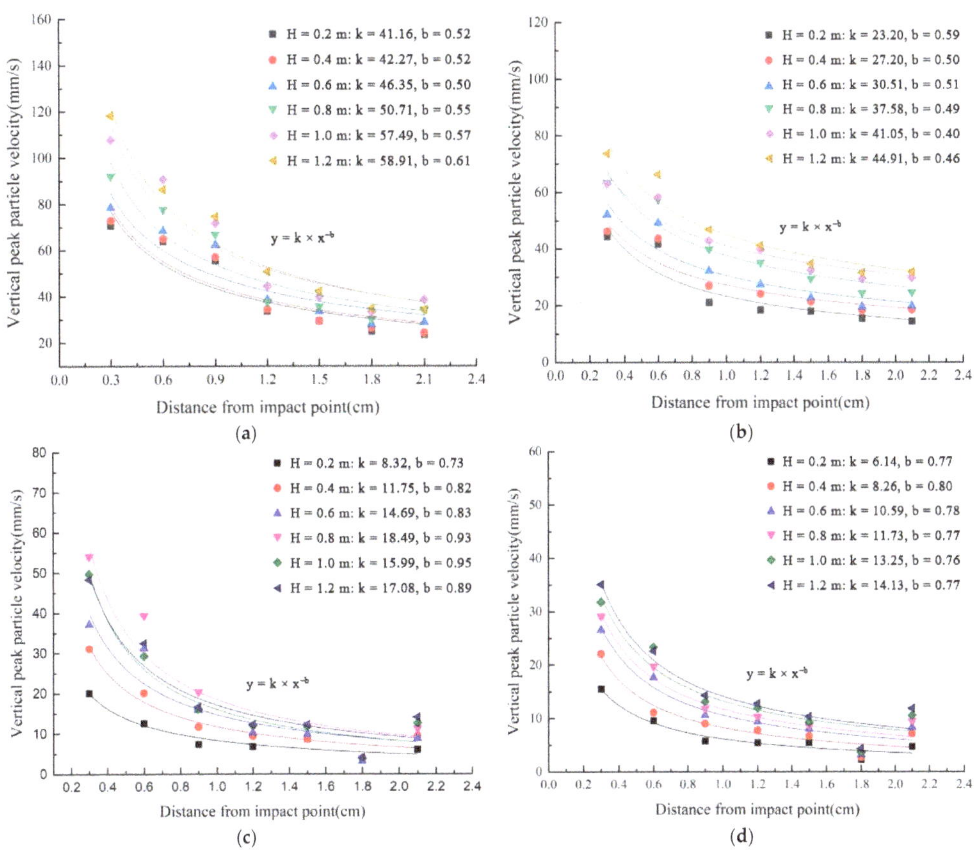

Figure 8. The attenuation law of the vertical peak particle velocity of different thicknesses of concrete slabs: (**a**) 5 cm; (**b**) 10 cm; (**c**) 15 cm; (**d**) 20 cm.

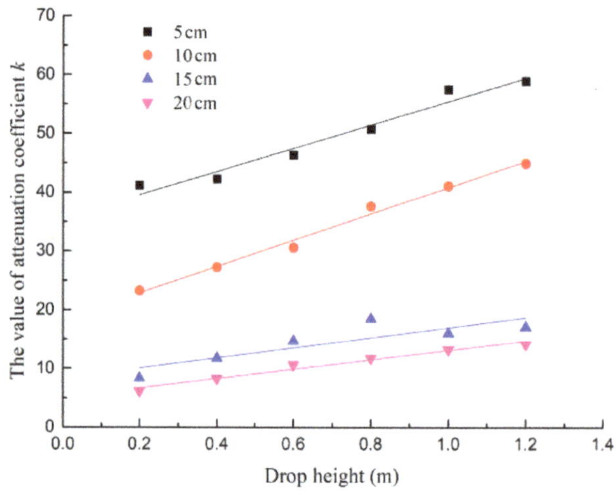

Figure 9. The change of the attenuation coefficient k of the vertical peak particle velocity with impact height under different thicknesses of concrete slabs.

3.2. *Dynamic Response Analysis of Multi-Layer Plate*

Compared with a single-layer plate, a multi-layer plate is stratified and discontinuous. The wave propagation becomes complicated due to inter-fracture wave reflection and transmission. Related research shows that the propagation of a stress wave in a layered medium is closely related to the properties of the joints, which are manifested in the coupling effect of the media and joint. The effects of the number and thickness of the layered medium (slab and joint) on the dynamic response of the layered medium are discussed under the same drop height.

Figure 10 shows the attenuation law of the vertical peak particle velocity under the same thickness of the layered medium and different numbers of slabs. Figure 11 shows the attenuation law of the vertical peak particle velocity under the same number but different thicknesses of slabs. Figure 12 shows the attenuation law of the vertical peak particle velocity under the same thickness but different numbers of slabs. Table 3 shows the fitting parameters based on Sadovsky's empirical formula. According to the test results, the numbers and thicknesses of the slabs have a significant effect on the dynamic response of a multi-layer plate with the same thickness. The value of the vertical peak particle velocity increases with a smaller slab thickness, and the attenuation law in the horizontal direction is more obvious. In other words, the attenuation parameters k and b increase with the decrease in the thickness of the slab. It is obvious that the experimental results show little difference from the change of the number of layers or the total thickness of the layered medium under the same slab thickness. In other words, the main influence factor of the dynamic response of the multi-layer plate is the layer thickness, so the influence of the number of layers and the total thickness can be neglected.

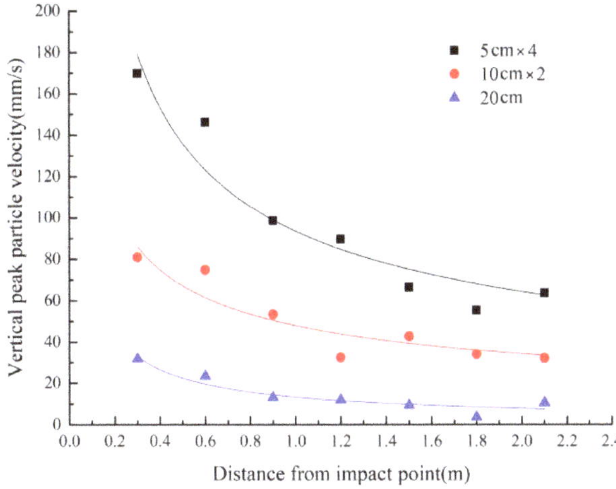

Figure 10. The impact test under the same thickness of layered medium and different number of slabs.

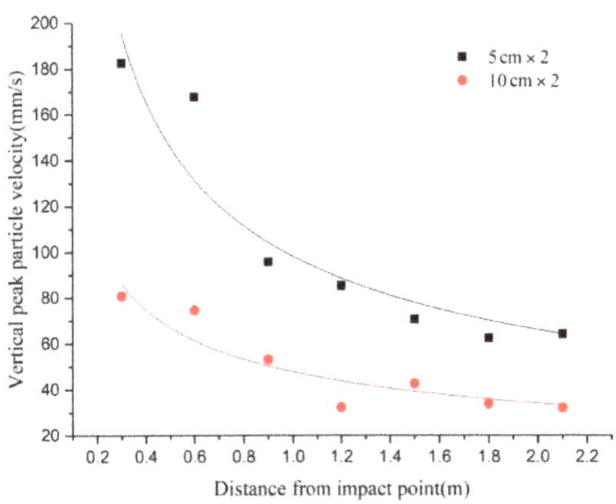

(**a**)

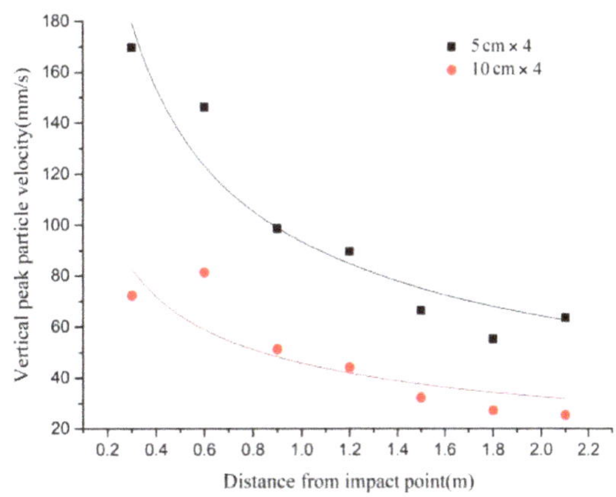

(**b**)

Figure 11. The impact test under the same number but different thicknesses of slabs: (**a**) 5 cm layer thickness double-layer plate and 10 cm layer thickness double-layer plate; (**b**) 5 cm layer thickness four-layer plate and 10 cm layer thickness four-layer plate.

3.3. *Dynamic Response Analysis of Multi-Layer Plate Considering Cementing Joints*

In the above test, the joint can be regarded as a two-dimensional plane without any thickness. However, the joint is generally presented as an interlayer, such as a clay thin interlayer, in the actual engineering. Not only the wave propagation across the joints but also the properties of the cementing joints, such as wave resistance, elastic modulus, interface roughness, contact area, cementing joint thickness, etc., need to be considered. So, the influence of the different filling materials and thicknesses of cementing joints on the dynamic response of the multi-layer plate are studied.

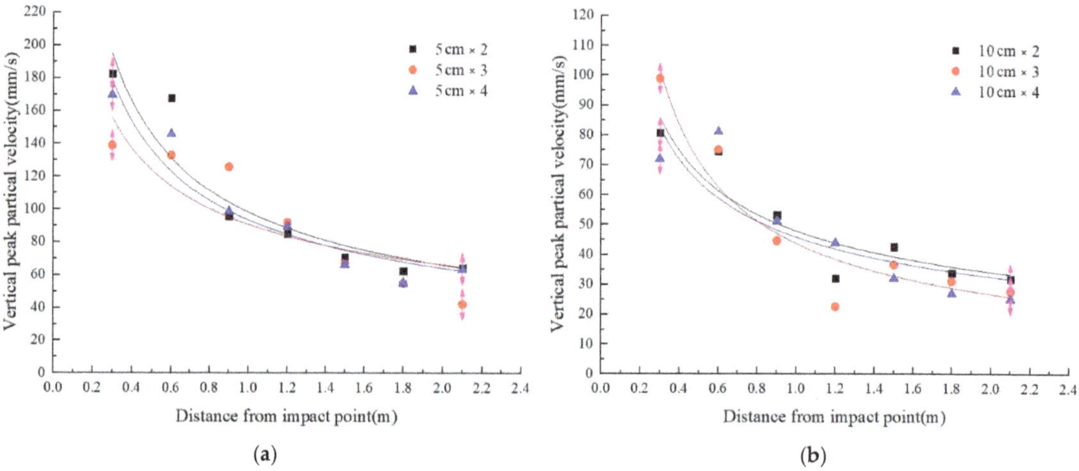

Figure 12. The impact test under the same thickness but different numbers of slabs: (**a**) 5 cm layer thickness multi-layer plate; (**b**) 10 cm layer thickness multi-layer plate.

Table 3. The fitting parameters of the vertical peak particle velocity attenuation curve under different test conditions.

Parameters	Single-Layer Plate			Double-Layer Plate		Triple-Layer Plate		Four-Layer Plate	
	10 cm	15 cm	20 cm	5 cm × 2	10 cm × 2	5 cm × 3	10 cm × 3	5 cm × 4	10 cm × 4
k	41.1	16.0	13.2	98.2	47.9	90.7	43.7	93.5	45.8
b	0.40	0.95	0.76	0.57	0.49	0.45	0.70	0.54	0.49

The filling materials of EPE, XPS, and sand are shown in Figure 13. Figure 14 gives the impact test results of the multi-layer plate, considering a cementing joint with different filling materials, as shown in Figure 13, their thicknesses are all 1 cm. Due to the diversity of the reflection and the transmission of the material, the value of the vertical peak particle velocity and its attenuation law are obviously different. When filled with sand and EPE, the attenuation law of the medium is close, but the value of the vertical peak particle velocity of the sand-filled multi-layer plate is larger. The postponed tendency is no longer apparent when XPS is selected as the filling material, and the value of the attenuation index is only 0.12, though the average value of the vertical peak particle velocity is the largest.

The effect of the cementing joint thickness on the dynamic response of the layered medium was investigated by the double-layer plate impact test. XPS was selected as the filling material of the cementing joint, and the thickness of XPS was set as 1 cm, 2 cm, and 3 cm. As shown in Figure 15, the dynamic response of the double-layer plate varies greatly with the change of the XPS slab's thickness. When the thickness of the XPS plate is 2 cm, the attenuation law of the vertical peak particle velocity in the horizontal direction is the most obvious, and the curve fitting degree is the highest. The fitting attenuation's indexes are only 0.12 and 0.29 when the thickness of the XPS plate is 1 cm and 3 cm, respectively. Apparently, the thickness of the cementation layer is the main factor affecting the dynamic response of the layered medium, but there is no regularity between the thickness and the attenuation parameters.

(**a**) (**b**) (**c**)

Figure 13. Different cementing joint filling materials: (**a**) EPE; (**b**) XPS; (**c**) sand.

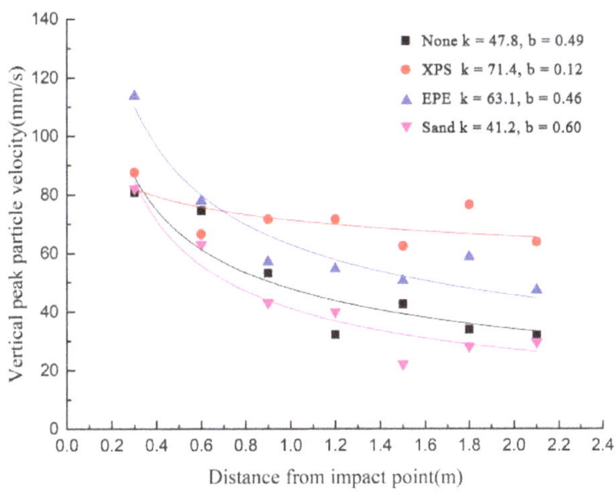

Figure 14. Dynamic response of double-layer plate with different filling materials.

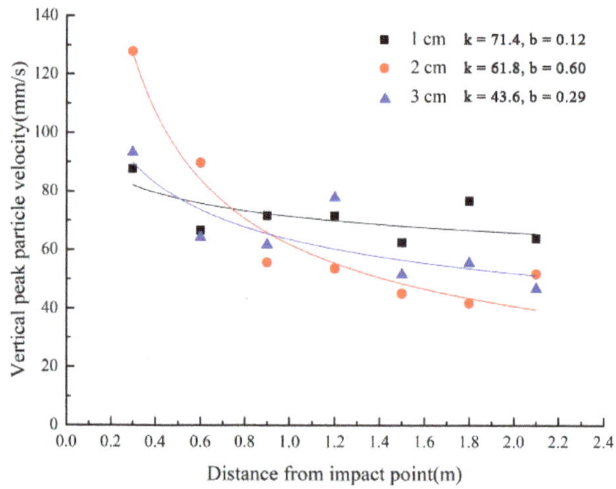

Figure 15. The dynamic response of double-layer plate with different cementing joint thicknesses.

4. Discussion and Conclusions

The influence factors of the dynamic response and attenuation law of the layered medium, such as impact energy, number and thickness of layers and joints, cementing joints materials, etc., are systematically investigated by impact tests of the single-layer plate and the multi-layer plate. The following conclusions are drawn:

(1) Sadovsky's empirical formula can fit the attenuation law of the vertical peak particle velocity well, with the ranging in the horizontal direction, and the fitting attenuation's coefficient k and index b can directly reflect the characteristics of the wave attenuation. The attenuation coefficient k has a linear positive correlation with the impact height under the same plate thickness, while it is negatively correlated with the thickness of the concrete plate under the same impact height.

(2) The thickness of the layer is the main factor affecting the dynamic response of the layered medium, so the number of plates and joints and the total thickness of the medium have little influence. The value of the vertical peak particle velocity is larger, and the attenuation law is more obvious, with an increase in the number of layers and a decrease in the thickness of layers under the same total thickness of the layered medium.

(3) The dynamic response of the multi-layer plate, when considering the cementing joints filled with different materials, varies significantly, and the thickness of the cementing joint has a great influence on the wave propagation of the layered medium. However, the effects of the material and thickness are only presented apparently, so the internal connection and the mechanism between the media and the joint need to be discussed deeply. The properties of the cementing joints and their effects need to be conducted systematically in subsequent studies.

Author Contributions: M.L. designed the study, performed the research, analyzed the data, and wrote the manuscript. Q.G. provided assistance for the data acquisition, data analysis, and statistical analysis. All authors have read and agreed to the published version of the manuscript.

Funding: This research received no external funding.

Institutional Review Board Statement: Not applicable.

Informed Consent Statement: Not applicable.

Data Availability Statement: The data used to support the findings of this study are available from the corresponding author upon request.

Acknowledgments: This work was supported by the Wenzhou Key Laboratory of Intelligent Lifeline Protection and Emergency Technology for Resilient City. In addition, the insightful comments and suggestions from the anonymous reviewers and the editor were sincerely appreciated.

Conflicts of Interest: The authors declare no conflict of interest.

References

1. Miller, R.K. An Approximate Method of Analysis of the Transmission of Elastic Waves Through a Frictional Boundary. *J. Appl. Mech.* **1977**, *44*, 652. [CrossRef]
2. Miller, R.K. The effects of boundary friction on the propagation of elastic waves. *Bull. Seismol. Soc. Am.* **1978**, *68*, 987–998. [CrossRef]
3. Schoenberg, M.; Sayers, C.M. Seismic anisotropy of fractured rock. *Geophysics* **1995**, *60*, 204–211. [CrossRef]
4. Schoenberg, M. Elastic wave behavior across linear slip interfaces. *J. Acoust. Soc. Am.* **1980**, *68*, 1516–1521. [CrossRef]
5. Pyrak-Nolte, L.J.; Myer, L.R.; Cook NG, W. Transmission of seismic waves across single natural fractures. *J. Geophys. Res. Solid Earth* **1990**, *95*, 8617–8638. [CrossRef]
6. Pyrak-Nolte, L.J. The seismic response of fractures and the interrelations among fracture properties. *Int. J. Rock Mech. Min. Sci. Geomech. Abstr.* **1996**, *33*, 787–802. [CrossRef]
7. Hopkins, D.L.; Myer, L.R.; Cook NG, W. Seismic wave attenuation across parallel fractures as a function of fracture stiffness and spacing. *Eos Trans. AGU* **1988**, *69*, 1427–1436.
8. Pyrak-Nolte, L.J.; Myer, L.R.; Cook NG, W. Anisotropy in seismic velocities and amplitudes from multiple parallel fractures. *J. Geophys. Res. Solid Earth* **1990**, *95*, 11345–11358. [CrossRef]
9. Chai, S.B.; Li, J.C.; Zhang, Q.B.; Li, H.B.; Li, N.N. Stress wave propagation across a rock mass with two non-parallel joints. *Rock Mech. Rock Eng.* **2016**, *49*, 4023–4032. [CrossRef]
10. Zhu, J.B.; Deng, X.F.; Zhao, X.; Zhao, J. A numerical study on wave transmission across multiple intersecting joint sets in rock masses with UDEC. *Rock Mech. Rock Eng.* **2013**, *46*, 1429–1442. [CrossRef]
11. Goodman, R.E.; Taylor, R.L.; Brekke, T.L. A model for the mechanics of jointed rock. *J. Soil Mech. Found. Div.* **1968**, *94*, 637–659. [CrossRef]
12. Lu, W. A Study on Interaction Between Stress Wave and Slipping Rock Interface. *Rock Soil Mech.* **1996**, *17*, 70–75.
13. Li, N.N.; Li, J.C.; Li, H.B.; Liu, T.T.; Chai, S.B. SHPB Experiment on Influence of Contact Area of Joints on Propagation of Stress Wave. *Chin. J. Rock Mech. Eng.* **2015**, *34*, 1994–2000.
14. Ju, Y.; Li, Y.; Xie, H.; Song, Z.; Tian, L. Stress Wave Propagation and Energy Dissipation in Jointed Rocks. *Chin. J. Rock Mech. Eng.* **2006**, *25*, 2426–2434.
15. Gu, B.; Suárez-Rivera, R.; Nihei, K.T.; Myer, L.R. Incidence of plane waves upon a fracture. *J. Geophys. Res. Solid Earth* **1996**, *101*, 25337–25346. [CrossRef]
16. Gu, B.; Nihei, K.T.; Myer, L.R.; Pyrak-Nolte, L.J. Fracture interface waves. *J. Geophys. Res. Solid Earth* **1996**, *101*, 827–835.
17. Cai, J.G.; Zhao, J. Effects of multiple parallel fractures on apparent attenuation of stress waves in rock masses. *Int. J. Rock Mech. Min. Sci.* **2000**, *37*, 661–682.
18. Zhao, J.; Zhao, X.B.; Cai, J.G. A further study of P-wave attenuation across parallel fractures with linear deformational behaviour. *Int. J. Rock Mech. Min. Sci.* **2006**, *43*, 776–788.
19. Zhao, J.; Cai, J.G.; Zhao, X.B.; Li, H.B. Experimental study of ultrasonic wave attenuation across parallel fractures. *Geomech. Geoengin. Int. J.* **2006**, *1*, 87–103. [CrossRef]
20. Zhao, J.; Cai, J.G. Transmission of elastic P-waves across single fractures with a nonlinear normal deformational behavior. *Rock Mech. Rock Eng.* **2001**, *34*, 3–22. [CrossRef]
21. Wang, Q.Z.; Zhang, S.; Xie, H.P. Rock dynamic fracture toughness tested with holed-cracked flattened Brazilian discs diametrically impacted by SHPB and its size effect. *Exp. Mech.* **2010**, *50*, 877–885. [CrossRef]
22. Lu, F.; Lin, Y.; Wang, X.; Lu, L.; Chen, R. A theoretical analysis about the influence of interfacial friction in SHPB tests. *Int. J. Impact Eng.* **2015**, *79*, 95–101. [CrossRef]
23. Li, J.C.; Li, N.N.; Li, H.B.; Zhao, J. An SHPB test study on wave propagation across rock masses with different contact area ratios of joint. *Int. J. Impact Eng.* **2017**, *105*, 109–116. [CrossRef]
24. Yang, Y.; Yang, R.S.; Wang, J.G. Simulation Material Experiment on Dynamic Mechanical Properties of Jointed Rock Affected by Joint Thickness. *J. China Univ. Min. Technol.* **2016**, *45*, 21–26.
25. Wang, W.; Li, K.; Yan, Z.; Tang, X. Study on the Closure Deformation Properties of Joint Fractal under SHPB Load. *Gold Sci. Technol.* **2017**, *25*, 75–83.
26. Challita, G.; Othman, R. Finite-element analysis of SHPB tests on double-lap adhesive joints. *Int. J. Adhes. Adhes.* **2010**, *30*, 236–244. [CrossRef]
27. Tian, Z.N.; Li, S.H.; Xiao, N. Experimental Studies and Numerical Simulation of Stress Wave Propagation in One-dimensional Rock Mass. *Chin. J. Rock Mech. Eng.* **2008**, *27*, 2687–2693.
28. Sun, B.; Guo, S.S.; Zeng, S.; Ma, A.Y.; Deng, X.S. Transmission and Reflection and the Attenuation Law of Stress Wave in Layered Joint Rock Mass. *J. Disaster Prev. Mitig. Eng.* **2015**, *35*, 828–832.

29. Peng, F.H.; Li, S.L.; Cheng, J.Y. Experimental study on characteristics of stress wave propagation in mesoscale and complex rock mass by microseismic monitoring. *Chin. J. Geotech. Eng.* **2014**, *36*, 312–319.
30. Wang, Z.J.; Li, X.L.; Ge, L.; Wang, P.Y. Free-field stress wave propagation induced by underground chemical explosion in granite. *Chin. J. Rock Mech. Eng.* **2003**, *22*, 1827–1831.
31. Ghaboussi, J.; Wilson, E.L.; Isenberg, J. Finite element for rock joints and interfaces: 11F, 8R. J. SOIL MECH. FOUND. DIV. V99, SM10, 1973, P833–848. *Int. J. Rock Mech. Min. Sci. Geomech. Abstr.* **1974**, *11*, 66.
32. Schwer, L.E.; Lindberg, H.E. A finite element slideline approach for calculating tunnel response in jointed rock. *Int. J. Numer. Anal. Methods Geomech.* **1992**, *16*, 529–540. [CrossRef]
33. Coates, R.T.; Schoenberg, M. Finite difference modelling of faults and fractures. *Geophysics* **1994**, *60*, 1514–1526. [CrossRef]
34. Fan, S.C.; Jiao, Y.Y.; Zhao, J. On modelling of incident boundary for wave propagation in jointed rock masses using discrete element method. *Comput. Geotech.* **2004**, *31*, 57–66. [CrossRef]
35. Cundall, P.A. A computer model for simulating progressive, large-scale movements in blocky rock systems. In Proceedings of the International Symposium on Rock Mechanics, Nancy, France, 4–6 October 1971; Volume 1, pp. 11–18.
36. Brady, B.H.; Hsiung, S.H.; Chowdhury, A.H.; Philip, J. Verification studies on the UDEC computational model of jointed rock. In Proceedings of the International Conference on Mechanics of Jointed and Faulted Rock, Vienna, Austria, 18–20 April 1990; pp. 551–558.
37. Zhao, X.; Zhao, J.; Cai, J.; Hefny, A.M. UDEC modelling on wave propagation across fractured rock masses. *Comput. Geotech.* **2008**, *35*, 97–104. [CrossRef]
38. Zhao, G.F.; Zhao, J. Microscopic Numerical Modelling of the Dynamic Strength of Brittle Rock. In Proceedings of the International Conference on Analysis of Discontinues Deformation: New Developments & Applications, Singapore, 25–27 November 2010.
39. Deng, X.F.; Zhu, J.B.; Chen, S.G.; Zhao, J. Some Fundamental Issues and Verification of 3DEC in Modeling Wave Propagation in Jointed Rock Masses. *Rock Mech. Rock Eng.* **2012**, *45*, 943–951. [CrossRef]
40. Zhu, J.B.; Zhao, G.; Zhao, X.; Zhao, J. Validation study of the distinct lattice spring model (DLSM) on P-wave propagation across multiple parallel joints. *Comput. Geotech.* **2011**, *38*, 298–304. [CrossRef]

coatings

Article

General Curve Model for Evaluating Mechanical Properties of Concrete at Different Ages

Qiuwei Yang [1,2], Xinhao Wang [1,2], Xi Peng [1,2] and Fengjiang Qin [3,4,*]

1 School of Civil and Transportation Engineering, Ningbo University of Technology, Ningbo 315211, China; yangqiuwei@nbut.edu.cn (Q.Y.); wangxinhao0709@163.com (X.W.); pengxi@nbut.edu.cn (X.P.)
2 Engineering Research Center of Industrial Construction in Civil Engineering of Zhejiang, Ningbo University of Technology, Ningbo 315211, China
3 Key Laboratory of New Technology for Construction of Cities in Mountain Area, Ministry of Education, Chongqing 400045, China
4 School of Civil Engineering, Chongqing University, Chongqing 400045, China
* Correspondence: qinfengjiang@cqu.edu.cn

Abstract: During the process of pouring and solidification of concrete, the compressive strength and elastic modulus of concrete exhibit dynamic growth patterns. The mechanical properties of concrete usually remain stable in the later stage (28 days after pouring). Studying appropriate curve models to accurately evaluate the changes in early mechanical properties of concrete has always been an important topic in the field of concrete materials. This work proposes a new dual parameter curve model for accurately evaluating the growth pattern of early compressive strength and elastic modulus of concrete. A comparative study was conducted between the proposed new curve model and existing curve models using 18 sets of experimental data from 10 literature sources. The research results indicate that the fitting average error and standard deviation of this new curve model are significantly smaller than the existing curve models. For some examples, the fitting error and standard deviation of the new model are only about 20%–30% of those of the existing models. The main advantages of this new curve model lie in two aspects. The first advantage is that this new curve model only contains two unknown parameters, so only a small amount of experimental data is required for data fitting and does not require complex mathematical operations. The second advantage is that this new curve model has a wide range of applications, which include compressive strength evaluation and elastic modulus evaluation and can also be extended to the evaluation of the mechanical properties of other materials similar to concrete.

Keywords: concrete; elastic modulus; compressive strength; two-parameter curve model; three-parameter curve model

Citation: Yang, Q.; Wang, X.; Peng, X.; Qin, F. General Curve Model for Evaluating Mechanical Properties of Concrete at Different Ages. *Coatings* **2023**, *13*, 2002. https://doi.org/10.3390/coatings13122002

Academic Editor: Paolo Castaldo

Received: 31 October 2023
Revised: 21 November 2023
Accepted: 23 November 2023
Published: 25 November 2023

1. Introduction

Concrete is a widely used building material for residential buildings, factories, and bridges. The mechanical properties of concrete, such as compressive strength and elastic modulus, exhibit dynamic growth patterns after pouring and remain stable until 28 days later. Evaluating the early mechanical properties of concrete can effectively guide the early construction and design of concrete engineering [1–4], such as prestressed tensioning or formwork removal. For example, in the production of large-volume concrete for bridge box girders, steel reinforcement prestressing should be applied as early as possible to prevent concrete cracks. Therefore, it is necessary to accurately evaluate the compressive strength and elastic modulus of concrete at an early stage to determine the time for steel reinforcement tensioning [5]. In high-rise buildings and other projects that use a large amount of concrete, the cost of formwork engineering is very high. If the formwork can be removed as early as possible, it can accelerate the turnover of formwork and thus save construction costs. Thus, accurately evaluating the strength of concrete at early ages is also very important for determining the time of formwork removal [6]. At present, there are two main methods for evaluating the early mechanical properties of concrete, namely machine learning-based evaluation methods and mathematical curve-based evaluation methods. Machine learning-based evaluation methods typically utilize large amounts of experimental data to train artificial neural network (ANN) models. The trained ANN models can then be used to evaluate the mechanical properties of concrete at various ages. Jaf et al. [7] used machine learning technology to evaluate the effects of silicon dioxide and calcium oxide in fly ash on the compressive strength of green concrete. Ahmed et al. [8] employed innovative machine learning techniques to predict the compressive strength of geopolymer concrete modified with nanoparticles. Trtnik et al. [9] analyzed the relationship between ultrasonic pulse velocity, static and dynamic Young's modulus, and shear modulus. Based on the experimental results, they demonstrated that the model constructed via artificial neural networks can estimate the compressive strength of concrete using only ultrasonic pulse velocity values and some mix ratio parameters of concrete. Yousif and Abdullah [10] found that water–cement ratio is the most important factor affecting the output of neural network models, and neural networks have strong potential as a tool for predicting concrete compressive strength. Duan et al. [11] constructed, trained, and tested a neural network model for predicting the compressive strength of recycled aggregate concrete using 146 sets of available data obtained from 16 different published literature sources. Nikoo et al. [12] used an artificial neural network model to calculate the compressive strength of concrete and optimized it using genetic algorithms. Khashman and Akpinar [13] designed an artificial neural network model for predicting the compressive strength of different concrete mixtures. Deng et al. [14] proposed a prediction model based on deep learning theory, which learns the deep features of concrete components via convolutional neural networks. The research results indicate that the prediction model based on deep learning has the advantages of high accuracy, high efficiency, and strong generalization ability and can be used as a new method for calculating the strength of recycled concrete. Feng et al. [15] proposed an intelligent prediction method for concrete compressive strength based on machine learning technology. Use adaptive boosting algorithm to predict the compressive strength of concrete. Muliauwan et al. [16] used three artificial intelligence methods: ANN, support vector machines (SVM), and linear regression (LR) to predict the compressive strength of concrete. Asteris and Mokos [17] studied the application of artificial neural networks in predicting the compressive strength of concrete in existing structures. Bka et al. [18] constructed three different neural network models using the Levenberg Marquardt (LM) algorithm to predict the compressive strength of recycled aggregate concrete. The results indicate that the prediction model for compressive strength of recycled aggregate concrete based on artificial neural networks has high accuracy and can be further used for mix ratio optimization design. Cui et al. [19] used seven machine learning algorithms to apply machine learning to predict the compressive strength of concrete. The experimental results indicate that the model has high prediction accuracy when the compressive strength is greater than 40 MPa. Kumar et al. [20] used machine learning algorithms such as Gaussian progressive regression (GPR), support vector machine regression (SVMR), ensemble learning (EL), and optimized GPR, SVMR, and EL to predict the compressive strength of lightweight concrete (LWC). Pakzad et al. [21] used machine learning (ML) and deep learning (DL) algorithms to predict the compressive strength (CS) of steel fiber-reinforced concrete (SFRC) containing hooked industrial steel fibers (ISF). Cünyt et al. [22] studied the performance of adaptive neural fuzzy inference system (ANFIS) in predicting the elastic modulus of ordinary concrete and high-strength concrete. The results indicate that their proposed ANFIS outperforms other models in terms of predictive ability. Demir [23] used collected data to establish, train, and test artificial neural network models and compared the predicted elastic modulus results with building codes and practical experience results. The results indicate that artificial neural networks have strong potential in predicting the elastic modulus of high-strength concrete and ordinary concrete. Ahmadi [24] further validated the method for predicting elastic modulus based on the

ANFIS model. The experimental comparison shows that the ANFIS model is superior to the other three nonlinear regression models. Han et al. [25] proposed a comprehensive machine learning (ML) model for predicting the elastic modulus of concrete prepared from recycled concrete aggregates. They compared the predictive performance of the integrated ML model with five commonly used ML models. The results indicate that compared to independent models, integrated ML models always produce more accurate predictions. Hasanzadeh et al. [26] utilized machine learning to predict the mechanical properties of basalt fiber-reinforced high-performance concrete by simulating the compressive strength and elastic modulus of concrete.

The second method is to evaluate the relationship between the mechanical properties of concrete and its age based on a certain function curve. The basic principle of this method is to obtain the functional equation between concrete mechanical parameters and age via data fitting based on early experimental data, and then calculate the concrete mechanical parameters at any time. The main advantage of this method is that it can be carried out with only a small amount of experimental data. At present, the commonly used concrete compressive strength–age curve models are the exponential model [27–29], the logarithmic model [30,31], the hyperbolic model [32,33], and the polynomial model [33,34]. Yang et al. [27] tested the compressive strength and elastic modulus of high-strength concrete within 28 days via experiments and proposed an exponential model for data fitting. Zhao et al. [28] measured the compressive strength and elastic modulus of C50 concrete at 1, 2, 3, 5, 7, 11, 14, 28, and 40 days and fitted the exponential curves of compressive strength and elastic modulus with these data. Yang et al. [29] combined regression fitting methods with exponential models to establish a prediction model for the relationship between the compressive strength and age of C60 concrete. Liu [30] analyzed the changes in compressive strength and elastic modulus of C50 concrete over time via experiments. Zhao [31] measured the compressive strength of concrete at 3, 7, 28, and 60 days via experiments and fitted the corresponding logarithmic function formula via data regression. Carino [32] and Viviani [33] fitted the changes in compressive strength of concrete over time using hyperbolic models, thereby establishing a prediction of compressive strength over age of concrete. Jin [34] used a polynomial model to fit the test data of early-age compressive strength and elastic modulus of high-strength concrete. Ling [35] identified the components that have an impact on the compressive strength of concrete by analyzing the composition of the concrete and built a prediction model for the compressive strength of mechanical sand concrete using a third-order nonlinear function. Jia et al. [36] obtained the compressive strength of ordinary Portland concrete at 1 to 28 days via concrete compressive strength experiments and fitted the obtained data using a cubic polynomial model to obtain a fitting curve of concrete compressive strength over time. Yeh [37] proposed a parametric trend regression and a four-parameter optimization method for predicting the compressive strength of concrete. Elaty [38] analyzed experimental data and revealed that a model can be constructed using two constants to predict the compressive strength of concrete at any age. Guo et al. [39] developed a time-dependent relationship model between concrete compressive strength and elastic modulus by referring to the existing curve models. Wang and Yue [40] analyzed the existing time-varying models of compressive strength, summarized the advantages and disadvantages of these models, and proposed further research directions.

Generally, although existing prediction models can predict the mechanical properties of concrete to a certain extent, they also have certain shortcomings in some aspects. The limitations of the ANN-based evaluation models lie in two aspects. Firstly, the accuracy of the ANN model is greatly affected by the number of neurons. However, there is currently no rigorous mathematical basis for determining the optimal number of neurons in the ANN model. Secondly, the ANN model can only be used for evaluation after being trained and matured. Usually, the training of the ANN model requires a lot of experimental data, which increases the cost of the experiment. The limitations of current curve regression models mainly lie in their low fitting accuracy and weak universality. For example, polynomial models are inaccurate in predicting the mechanical properties of concrete in the later stage. Although logarithmic and exponential models can converge to a certain extent in the later stage, they cannot predict the mechanical properties of concrete well in the later stage. These types of prediction models do not perform well with data with high dispersion, resulting in low fitting accuracy. The three-parameter model has one more unknown fitting parameter than the two-parameter model, so more experimental data is needed to perform curve fitting accordingly. When there is little experimental data, the fitting accuracy of the three-parameter model may not be as good as that of the two-parameter model. In addition, the three-parameter model is more complex to calculate than the two-parameter model. In order to improve evaluation accuracy, it is necessary to overcome the shortcomings of existing curves and integrate the advantages of existing curves to study a new mathematical curve model for predicting the mechanical properties of concrete. The goal of this work is to develop a new dual parameter curve model that can more accurately evaluate the compressive strength and elastic modulus changes of concrete at early age compared to the existing curve models. Based on the evaluation results of the proposed model, it can more reliably guide the construction or mechanical calculation of concrete components at early age. To this end, this work first analyzes the advantages and disadvantages of the existing logarithmic model, exponential model, hyperbolic model, and polynomial model. Subsequently, this work develops a new curve model that overcomes the shortcomings of existing curve models, which can be used for the evaluation of concrete compressive strength at early ages as well as for the evaluation of elastic modulus at early ages. The improvements provided by the new model mainly lie in three aspects. The first is that the mathematical formula of the new model is concise, including only two unknown parameters, which is very convenient for engineering applications. The second is that the curves used by the new model are more in line with the changes in mechanical properties of concrete at early age, so the fitting accuracy of the new model is higher than that of existing models. Thirdly, the new model has better universality, can be applied to more types of concrete, and may also be applicable to other cementitious materials. This work focuses on the performance of the proposed new model in evaluating the compressive strength and elastic modulus of concrete at early age. Using 18 sets of experimental data from existing literature, a comparative study was conducted on the computational accuracy of the new model and existing models to illustrate the superiority of the proposed new model.

2. Mathematical Models for Predicting the Mechanical Properties of Concrete

As mentioned earlier, the prediction method of concrete mechanical properties based on some assumed mathematical curve is a simple and feasible method in engineering practice, because it only needs a few early test data for concrete. Figure 1 shows the operational process of this type of method in practice.

Obviously, the suitable mathematical curve is an important factor in accurately predicting the mechanical properties of concrete. The commonly used curve models can be divided into two-parameter models and three-parameter models based on the number of unknown parameters that need to be fitted in the curve equations. In this section, the common two-parameter and three-parameter models are reviewed, and the advantages and disadvantages of each model are briefly analyzed. Finally, a new universal two-parameter model is proposed to predict the mechanical properties of concrete at different ages. Without losing generality, in the following mathematical equations of curve models, t represents the time (age) after concrete pouring, and $z(t)$ represents the mechanical parameters (compressive strength or elastic modulus) of the concrete at time t.

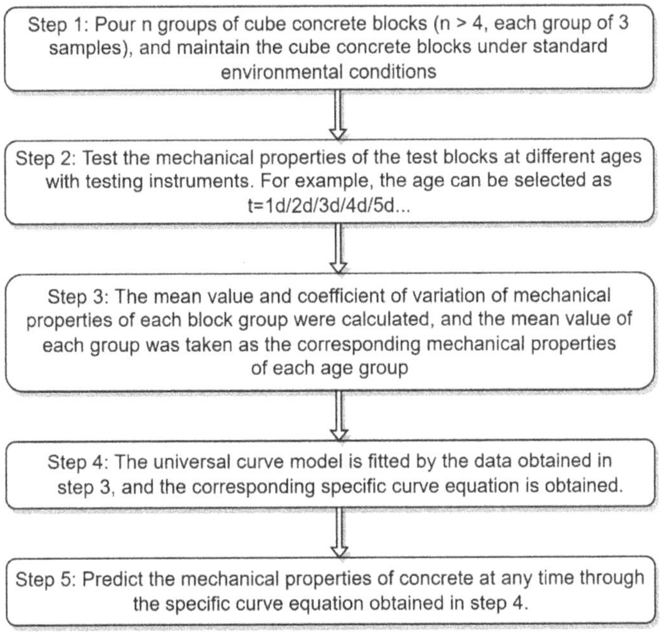

Figure 1. Flow chart for predicting mechanical properties of concrete.

2.1. *Two-Parameter Curve Model*

The commonly used two-parameter curve models include exponential model, logarithmic model, and hyperbolic model. The curve equation of exponential model is expressed by

$$z(t) = a \cdot e^{-b/t} \tag{1}$$

in which a and b represent two unknown non-negative parameters that need to be determined based on experimental data. As can be seen from Equation (1), the advantage of exponential model is that the exponential function increases monotonically, so with the increase of t, the value of $z(t)$ function will increase. This is consistent with the fact that the mechanical properties of concrete will increase with time in engineering practice. When $t \to \infty$, $z(t) \to \text{a}$, it is also consistent with the law that the mechanical properties of concrete will remain basically unchanged after a period of growth. However, the exponential model also has some shortcomings. The first drawback of this model is the lack of $t = 0$ point in the definition domain because t is the denominator of $\frac{b}{t}$ in Equation (1). In practical application, when $t = 0$, the mechanical property of concrete is 0. Therefore, it is important to note that the exponential model cannot include this point. In addition, because the equation of exponential model contains only two fitting parameters, the fitting accuracy may be low for experimental data with large dispersion, and it can easily produce large errors.

The curve equation of logarithmic model is expressed as:

$$z(t) = a + b \cdot \ln(t) \tag{2}$$

In Equation (2), a and b represent the two fitting parameters, which are obtained by combining partial experimental data with curve fitting. Comparing Equations (1) and (2), it can be seen that the logarithmic model can be seen as a modified equation obtained by taking the logarithms of both sides of the exponential model equation. Therefore, the advantages and disadvantages of the logarithmic model and the exponential model are similar. The advantage of the logarithmic model is that the function is a monotonically increasing function, which is consistent with the law of concrete mechanical properties increasing over time in practical engineering. The disadvantage is that its definition domain also lacks the point $t = 0$. Meanwhile, due to its monotonic increasing function, when $t \to \infty$, although the growth rate slows down, the results do not converge, which is inconsistent with the actual growth law of concrete mechanical properties. Since this equation only contains two fitting parameters, the fitting accuracy of this equation may be low for data with large dispersion.

The curve equation of hyperbolic model is expressed as:

$$z(t) = \frac{t}{a + b \cdot t} \tag{3}$$

In the above formula, a and b are two fitting parameters of the hyperbolic model. The advantage of hyperbolic model compared with other models is that the predicted mechanical properties of concrete will gradually approach a fixed number with the increase of time, which is consistent with the fact that the mechanical properties of concrete remain basically unchanged in the later stage. The disadvantage of hyperbolic model is that the fitting accuracy of the model is not very good when dealing with some very discrete data.

2.2. Three-Parameter Curve Model

The common three-parameter models include polynomial model and hybrid curve model. The curve equation of polynomial model is expressed as

$$z(t) = a + b \cdot t + c \cdot t^2 \tag{4}$$

Equation (4) is a quadratic polynomial model, where a, b and c are the three fitting parameters of this model. Compared to the previous three prediction models, the advantage of the polynomial model is that its definition domain includes point $t = 0$, and the fitted function is also a monotonically increasing function, which is in line with the development laws of concrete mechanical properties in practical engineering. Meanwhile, due to the fact that the prediction model has three fitting parameters, its fitting accuracy is correspondingly more accurate than the logarithmic model, the exponential model, and the hyperbolic model with only two fitting parameters. The disadvantage of the polynomial model is that at $t \to \infty$, the growth rate of $z(t)$ will be too fast, which is different from the law in engineering practice that the mechanical properties of concrete will increase slowly or even remain basically unchanged in the later stage. The reason for the faster growth rate in the later stage of the polynomial model is the square term of time (i.e., $c \cdot t^2$) in the model. The longer the time, the faster the square term of time increases. Therefore, this prediction model is not suitable for predicting the later mechanical properties of concrete.

The curve equation of the hybrid curve model is expressed as

$$z(t) = at\ln(1+\frac{1}{b+ct}) \tag{5}$$

Equation (5) is a mixed curve model, which includes a logarithmic model and a hyperbolic model. a, b and c are the three fitting parameters of this equation. The advantages of this model are that its definition domain includes point $t = 0$, the fitting function is also a monotonically increasing function, and the prediction model will eventually converge to a value, which is more in line with the development law of concrete mechanical properties in practical engineering. Meanwhile, due to the fact that the prediction model has three fitting parameters, its fitting accuracy is correspondingly higher than those of the logarithmic model and the hyperbolic model. However, although the three-parameter model has higher fitting accuracy than the two-parameter model, it is less convenient to use in practical engineering applications due to the need for an additional parameter for data fitting.

2.3. A Universal Two-Parameter Model

Based on the above analysis, a new mathematical model is proposed to accurately describe the changes in mechanical properties of concrete over time in practical engineering. The mathematical expression for the new dual parameter curve model is given as:

$$z(t) = \frac{a \cdot t}{t+1} + b \cdot \arctan(\frac{t}{t+1}) \tag{6}$$

In Equation (6), t and $z(t)$ represent the time and the corresponding mechanical parameters at time t, which are consistent with the previous definitions. The two unknown parameters a and b represent the fitting coefficients that need to be determined based on experimental data. The new model can be seen as a combination of hyperbolic function and arctangent function, and the accuracy of function fitting can be further improved by adding an item of arctangent function. The new model also inherits the advantage that its definition domain contains a point $t = 0$. At the same time, when t approaches infinity, $z(t)$ tends toward a certain value, which is in line with the fact that the mechanical properties of concrete in the later stage remain basically unchanged in practical engineering. This compensates for the defect of polynomial models that cannot converge when t approaches infinity, so this model can be used to predict the mechanical properties of concrete in the later stage. In addition, the prediction model only needs to fit two parameters, which is more convenient to use in practical engineering than the above three-parameter model. The physical meanings of parameters a and b are the weights of the hyperbolic function and the arctangent function in the sum function shown in Equation (6), respectively. From Equation (6), one obtains:

$$\lim_{t \to \infty} z(t) = a + b \cdot \frac{\pi}{4} \tag{7}$$

Therefore, the physical meaning of $a + b \cdot \frac{\pi}{4}$ is the stable mechanical parameter of concrete in the later stage, such as compressive strength or elastic modulus. Note that this new model only has two fitting parameters, a and b, that need to be determined via data fitting. When the new model is compared with the existing two-parameter models, it can be found that the new model is a combination of two types of curves (hyperbolic and arctangent curves), while the existing two-parameter models are all based on a single curve. The reason why the new model yields fewer errors compared to the existing models is that it introduces a new curve of the arctangent function, which is more in line with the growth law of mechanical properties of concrete at early age.

In summary, Table 1 presents the advantages/disadvantages of the above curve models for clarity. The improvement in fitting accuracy of the new model can be seen in the calculation results of multiple sets of experimental data in the next section.

Table 1. The advantages/disadvantages of various curve models.

Model	Number of Fitting Parameters	Definition Domain	Convergence of $z(t)$ When $t \to \infty$
Exponential model	2	Missing point with $t = 0$	Yes
Logarithmic model	2	Missing point with $t = 0$	No convergence
Hyperbolic model	2	Including point with $t = 0$	Yes
Polynomial model	3	Including point with $t = 0$	No convergence
Hybrid model	3	Including point with $t = 0$	Yes
New dual parameter model	2	Including point with $t = 0$	Yes

The next section describes the fitting parameter calculation method for the mathematical model described above. Firstly, a linear equation system of mechanical properties and age is established based on the measured data obtained from experiments, and then the corresponding fitting parameters are further calculated. Using n sets of experimental data, the linear system corresponding Equation (6) can be expressed as:

$$y = C \cdot x \tag{8}$$

$$y = \begin{Bmatrix} z(t_1) \\ \vdots \\ z(t_n) \end{Bmatrix}, \tag{9}$$

$$x = \begin{Bmatrix} a \\ b \end{Bmatrix} \tag{10}$$

$$C = \begin{bmatrix} \frac{t_1}{t_1+1} & \arctan(\frac{t_1}{t_1+1}) \\ \vdots & \vdots \\ \frac{t_n}{t_n+1} & \arctan(\frac{t_n}{t_n+1}) \end{bmatrix} \tag{11}$$

where $z(t_n)$ represents the mechanical parameter (elastic modulus or compressive strength) measured at time t_n. Clearly, the vector y and coefficient matrix C in Equation (8) can be obtained from the experimental data. Thus, the unknown fitting parameter vector x can be further calculated using the least square method. For this purpose, Equation (8) can be rewritten as

$$C^T y = (C^T C) \cdot x \tag{12}$$

From Equation (11), the solution of x is expressed as:

$$\hat{x} = (C^T C)^{-1} C^T y \tag{13}$$

The calculated $\hat{x}$ refers to the least squares estimate of x.

The fitted parameters obtained from the calculation can be used to obtain the specific curve equation of the prediction model, and the accuracy of the model prediction can be verified by comparing the difference between the predicted value and the experimental value. The difference between the predicted value and the experimental value is defined as:

$$\{\delta\} = |C \cdot \hat{x} - y| \tag{14}$$

where $\{\delta\} = (\delta_1, \cdots, \delta_n)^T$ is called as the residual vector. Specifically, δ_i is the residual error between the ith experimental data and the predicted value. Average and standard deviation of the residual errors can be calculated by:

$$\overline{\delta} = \frac{\delta_1 + \cdots + \delta_n}{n} \tag{15}$$

$$\sigma = \sqrt{\frac{\sum_{i=1}^{n} (\delta_i - \overline{\delta})^2}{n}} \tag{16}$$

where $\overline{\delta}$ is the average of the residual errors, σ is the corresponding standard deviation. The smaller the value of $\overline{\delta}$ and σ, the higher the fitting accuracy of the curve model.

3. Model Verification and Comparison via Concrete Experiment Data

The newly proposed evaluation model was validated using experimental data from references [28,30,34,36,39,41–45] (referred to as cases 1 to 11). The results were compared with the fitting results of existing logarithmic, hyperbolic and polynomial models to demonstrate the superiority of the model.

3.1. Evaluation of Compressive Strength

Case 1. In reference [36], 32 groups of concrete specimens with a water–cement ratio of 0.56 and 50 mm × 50 mm × 150 mm were cured under standard curing conditions. The components of concrete materials are listed in Table 2. The experimental data for the compressive strength are shown in Table 3. The logarithmic, hyperbolic, polynomial, and new models are used to fit the experimental data, and the fitting curves are shown in Figure 2. The fitting average error and standard deviation of these models are shown in Table 4. The specific fitting equations for the four models are as follows:

$$z(t) = 10.2190 + 21.3746\ln(t) \tag{17}$$

$$z(t) = \frac{t}{0.0822 + 0.0093 \cdot t} \tag{18}$$

$$z(t) = 17.6264 + 5.1571 \cdot t - 0.1102 \cdot t^2 \tag{19}$$

$$z(t) = 519.5090 \cdot (\frac{t}{t+1}) - 554.0748 \cdot \arctan(\frac{t}{t+1}) \tag{20}$$

Table 2. The components of concrete materials in case 1 (kg/m^3).

Ordinary Portland Cement	Standard Sand	Water	Water–Cement Ratio
460	800	180	0.56

Table 3. Compressive strength of concrete blocks in case 1 (MPa).

Age (Unit: day)	Compressive Strength	Age (Unit: day)	Compressive Strength
1	4.26	15	70.56
2	23.83	16	71.31
3	36.42	17	68.29
4	43.28	18	66.04
5	45.85	19	73.79
6	44.91	20	70.81
7	54.21	21	76.95
8	57.99	22	76.65
9	58.75	23	68.26
10	62.26	24	78.46
11	60.33	25	75.66
12	65.72	26	79.92
13	68.72	27	79.73
14	68.90	28	85.39

According to the average error and standard deviation of the four models in Table 4, it can be seen that the average error and standard deviation of the new model are significantly smaller than those of the hyperbolic, polynomial, and logarithmic models. Figure 2 shows that compared to the logarithmic, polynomial, and hyperbolic models, the prediction curve of the new model is closer to the experimental test data. The results indicate that the new model has better data fitting ability than the logarithmic, polynomial, and hyperbolic models and is more accurate and reliable in evaluating the compressive strength of concrete.

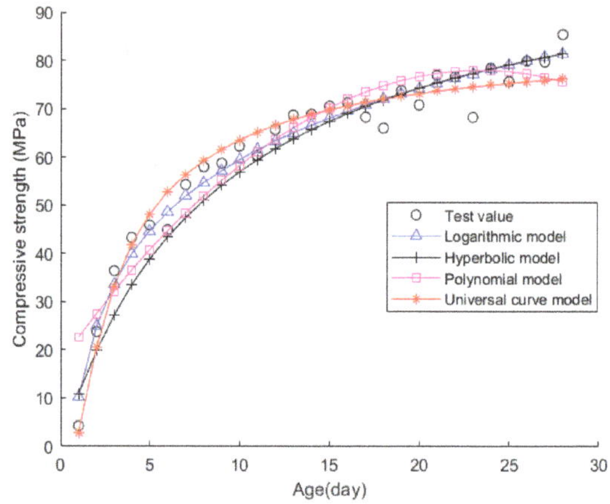

Figure 2. The fitting results of case 1.

Table 4. The fitting mean errors and standard deviations from the four models in case 1.

Index	Logarithmic Model	Hyperbolic Model	Polynomial Model	Universal Curve Model
Fitting mean error $\overline{\delta}$	1.4240	1.6451	4.3428	0.4712
Fitted standard deviation σ	0.2581	0.2895	0.8066	0.0669

Case 2. In reference [41], 14 groups of concrete specimens with 150 mm × 150 mm × 300 mm were manufactured and cured under standard and natural conditions. The components of concrete materials are listed in Table 5. The experimental data of compressive strength are shown in Table 6. Figures 3 and 4 illustrate the fitting curves for the four models. The fitting average errors and standard deviations for these models are shown in Tables 7 and 8. The specific fitting equations for the four models are as follows:

Table 5. The components of concrete materials in case 2 (kg/m^3).

Cement	Fine Aggregate	Coarse Aggregate	Water	Water Reducer	Admixture
357	712	1115	123	8.82	63

Table 6. Compressive strength of concrete blocks in case 2 (MPa).

Age (Unit: hour)	Natural Conservation Condition	Standard Maintenance Condition
12	4.9	2.7
16	14.5	10.7
20	19.0	16.0
24	22.9	19.1
28	24.9	22.4
32	27.2	25.3
36	28.7	26.1
40	29.6	27.6
44	30.3	30.0
48	30.7	30.4
52	31.2	30.6
56	32.9	31.6
60	33.7	33.2
64	34.0	32.3

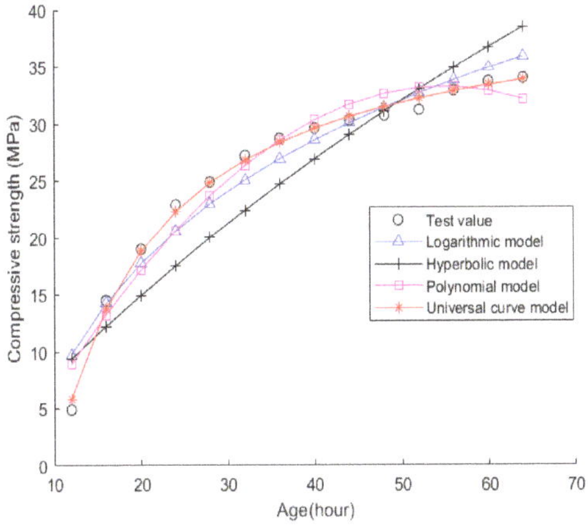

Figure 3. The fitting results for natural conservation conditions in case 2.

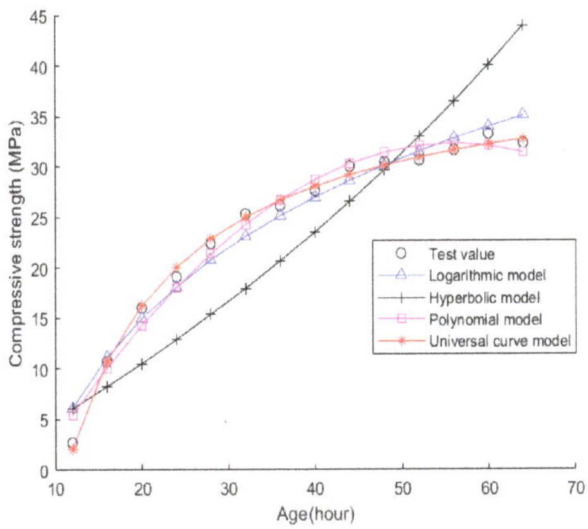

Figure 4. The fitting results for standard curing conditions in case 2.

Table 7. The fitting mean errors and standard deviations of the four models for natural conservation condition in case 2.

Index	Logarithmic Model	Hyperbolic Model	Polynomial Model	Universal Curve Model
Fitting mean error $\bar{\delta}$	1.0084	1.0436	0.8495	0.2083
Fitted standard deviation σ	0.2524	0.2214	0.2087	0.0497

Table 8. The fitting mean errors and standard deviations of the four models for standard maintenance condition in case 2.

Index	Logarithmic Model	Hyperbolic Model	Polynomial Model	Universal Curve Model
Fitting mean error $\bar{\delta}$	1.2761	1.5086	1.0147	0.2415
Fitted standard deviation σ	0.3257	0.2926	0.2571	0.0578

Natural conservation conditions:

$$z(t) = -29.0860 + 15.6257\ln(t) \quad (21)$$

$$z(t) = \frac{t}{1.1927 + 0.0073 \cdot t} \tag{22}$$

$$z(t) = -6.5801 + 1.4516 \cdot t - 0.0132 \cdot t^2 \tag{23}$$

$$z(t) = 1283.1235 \cdot (\frac{t}{t+1}) - 1581.0830 \cdot \arctan(\frac{t}{t+1}) \tag{24}$$

Standard maintenance conditions:

$$z(t) = -36.8725 + 17.2964 \ln(t) \tag{25}$$

$$z(t) = \frac{t}{2.1067 - 0.0101 \cdot t} \tag{26}$$

$$z(t) = -11.2280 + 1.5518 \cdot t - 0.0138 \cdot t^2 \tag{27}$$

$$z(t) = 1411.1477 \cdot (\frac{t}{t+1}) - 1744.6763 \cdot \arctan(\frac{t}{t+1}) \tag{28}$$

Tables 7 and 8 show that the new model outperforms the hyperbolic, logarithmic, and polynomial models regarding fitting mean error and standard deviation. Figures 3 and 4 visually demonstrate that the fitting curve of the new model aligns more closely with the experimental data. These findings indicate that the new model is a better fit for the test data than the logarithmic, hyperbolic, and polynomial models.

Case 3. In reference [28], 30 sets of concrete specimens were manufactured using C50 grade concrete and cured under standard curing conditions. The components of the concrete are listed in Table 9. The tested compressive strength is shown in Table 10. The four fitting curves are presented in Figure 5 using these experimental data. The fitting average errors and standard deviations are shown in Table 11. The specific fitting equations for the four models are as follows:

$$z(t) = 27.2818 + 8.9263 \ln(t) \tag{29}$$

$$z(t) = \frac{t}{0.0332 + 0.0165 \cdot t} \tag{30}$$

$$z(t) = 30.1059 + 2.0246 \cdot t - 0.0340 \cdot t^2 \tag{31}$$

$$z(t) = 122.6500 \cdot (\frac{t}{t+1}) - 82.8490 \cdot \arctan(\frac{t}{t+1}) \tag{32}$$

Table 9. The components of concrete materials in case 3 (kg/m^3).

Cement	Fine Aggregate	Coarse Aggregate	Water	Water Reducer	Admixture
400	382	1166	170	4.3	115

Table 10. Compressive strength of concrete specimens in case 3 (MPa).

Age (Unit: day)	Compressive Strength
1	24.0
2	32.1
3	39.6
5	45.7
7	46.8
11	47.6
14	49.9
28	56.8
40	58.4

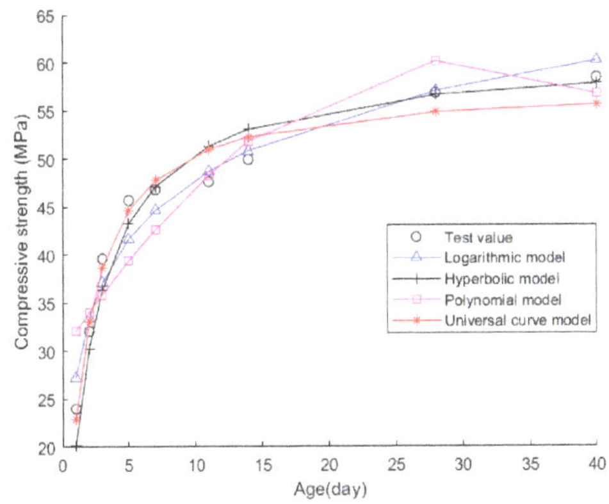

Figure 5. The fitting results of case 3.

Table 11. The fitting mean errors and standard deviations of case 3.

Index	Logarithmic Model	Hyperbolic Model	Polynomial Model	Universal Curve Model
Fitting mean error $\overline{\delta}$	0.1907	0.2214	0.3991	0.1231
Fitted standard deviation σ	0.0410	0.0492	0.0984	0.0163

Table 11 indicates that the new model has a smaller fitting mean error and standard deviation compared to the other models. The new model's average fitting error is approximately 35% lower than that of the logarithmic model, about 44% lower than that of the hyperbolic model, and about 35% lower than that of the polynomial model. The fitting standard error of the new model is roughly 40% for the logarithmic model, about 33% for the hyperbolic model, and about 16% for the polynomial model.

Case 4. In reference [42], 36 sets of 100 mm × 100 mm × 100 mm high-strength concrete specimens were manufactured and cured under standard curing conditions. The components of concrete materials are listed in Table 12. The experimental data of compressive strength are shown in Table 13, and the fitting curves are shown in Figure 6. The fitting average errors and standard deviations are shown in Table 14. The specific fitting equations for the four models are as follows:

$$z(t) = 71.7838 + 25.4840\ln(t) \tag{33}$$

$$z(t) = \frac{t}{0.0061 + 0.0075 \cdot t} \tag{34}$$

$$z(t) = 48.6435 + 21.4047 \cdot t - 1.4696 \cdot t^2 \tag{35}$$

$$z(t) = 81.4696 \cdot (\frac{t}{t+1}) + 68.7367 \cdot \arctan(\frac{t}{t+1}) \tag{36}$$

Table 12. The components of concrete materials in case 4 (kg/m^3).

Powder	Steel Fiber	Water
2130	160	196

Table 13. Compressive strength of concrete cubes in case 4 (Mpa).

Age (Unit: day)	Compressive Strength
0.5	43.303
1	71.012
2	99.915
3	108.909
4	114.274
5	114.300
6	118.566
7	119.790
8	121.774
9	122.116
10	122.919

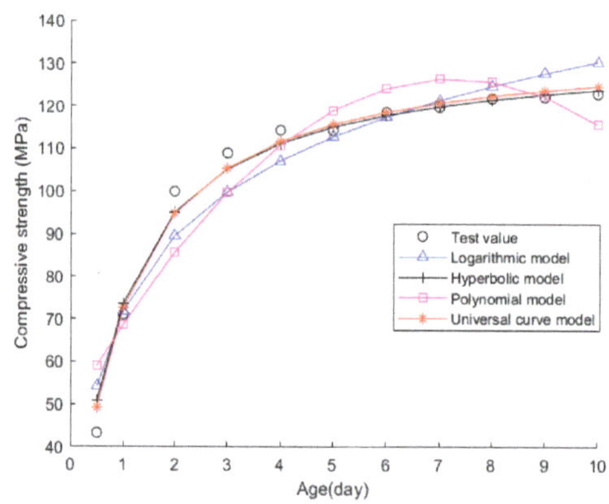

Figure 6. The fitting results of case 4.

Table 14. The fitting mean errors and standard deviations of case 4

Index	Logarithmic Model	Hyperbolic Model	Polynomial Model	Universal Curve Model
Fitting mean error $\overline{\delta}$	0.3023	0.1885	0.4164	0.1564
Fitted standard deviation σ	0.0703	0.0495	0.1001	0.0388

Table 14 shows that the fitting mean error and standard deviation of the new model are significantly smaller than those of the hyperbolic, logarithmic, and polynomial models. This again indicates that the new model can more accurately evaluate the compressive strength of concrete.

Case 5. In reference [30], some concrete specimens with 150 mm × 150 mm × 300 mm were manufactured, and the experimental data of compressive strength are shown in Table 15. The fitting curves for the four models are shown in Figure 7, and the fitting average errors and standard deviations are shown in Table 16. The specific fitting equations for the four models are as follows:

$$z(t) = 33.3919 + 9.2854\ln(t) \tag{37}$$

$$z(t) = \frac{t}{0.0229 + 0.0149 \cdot t} \tag{38}$$

$$z(t) = 37.3494 + 1.8490 \cdot t - 0.0265 \cdot t^2 \tag{39}$$

$$z(t) = 123.1825 \cdot (\frac{t}{t+1}) - 72.9417 \cdot \arctan(\frac{t}{t+1}) \quad (40)$$

Table 15. Compressive strength (MPa) of case 5.

Age (Unit: day)	Compressive Strength
1	25.63
2	40.71
3	48.50
5	52.30
7	53.20
10	55.00
20	62.30
28	64.00
50	65.00

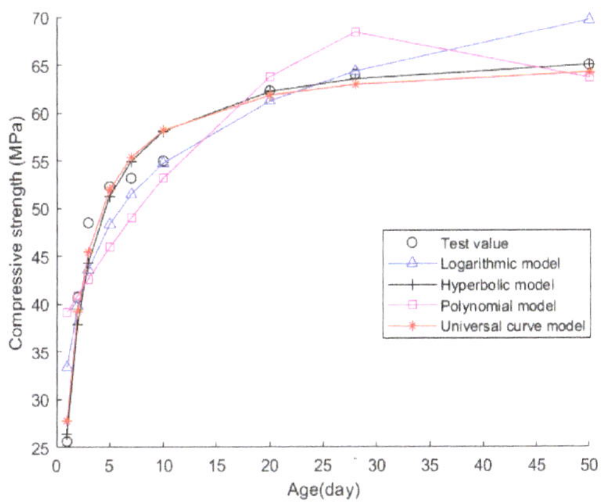

Figure 7. The fitting results of case 5.

Table 16. The fitting mean errors and standard deviations of case 5.

Index	Logarithmic Model	Hyperbolic Model	Polynomial Model	Universal Curve Model
Fitting mean error $\overline{\delta}$	0.3389	0.1337	0.5670	0.1332
Fitted standard deviation of error σ	0.0937	0.0309	0.1621	0.0276

From Table 16, it can be seen that the fitting accuracy of the new model is the highest. This indicates that the new model can more accurately evaluate the compressive strength of concrete.

3.2. Evaluation of Elastic Modulus

Case 6. In reference [30], Case 5 was followed as the experimental design scheme. Table 17 presents the experimental data of elastic modulus. The fitting curve, depicted in Figure 8, was obtained using four models. The average fitting error and standard deviation of the four models are stated in Table 18. The specific fitting equations for the four models are as follows:

$$z(t) = 29.5469 + 3.6042 \ln(t) \quad (41)$$

$$z(t) = \frac{t}{0.0196 + 0.0234 \cdot t} \quad (42)$$

$$z(t) = 31.1455 + 0.6973 \cdot t - 0.0098 \cdot t^2 \quad (43)$$

$$z(t) = -16.9105 \cdot (\frac{t}{t+1}) + 74.1154 \cdot \arctan(\frac{t}{t+1}) \tag{44}$$

Table 17. Elastic modulus (GPa) of case 6.

Age (Unit: day)	Elastic Modulus
1	26.90
2	32.10
3	35.30
5	37.10
7	37.32
10	37.54
20	40.39
28	41.65
50	42.10

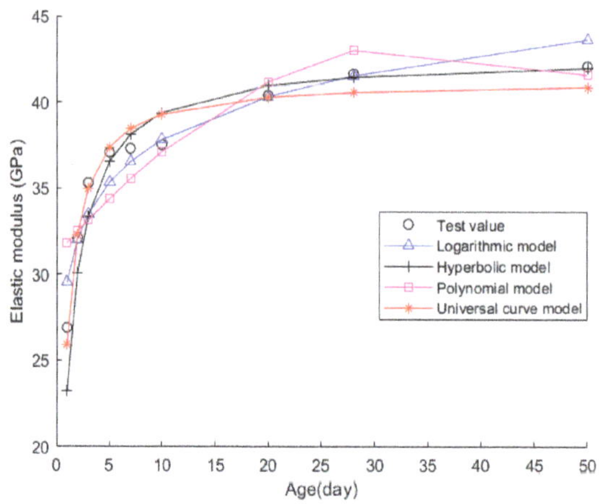

Figure 8. The fitting results of elastic modulus in case 6.

Table 18. The fitting mean errors and standard deviations for elastic modulus of case 6.

Index	Logarithmic Model	Hyperbolic Model	Polynomial Model	Universal Curve Model
Fitting mean error $\bar{\delta}$	0.1278	0.1701	0.2164	0.0787
Fitted standard deviation σ	0.0325	0.0425	0.0549	0.0158

According to the fitting average error and standard deviation of the four models in Table 18, it can be seen that the fitting average error and fitting standard error of the new model are smaller than those of the hyperbolic model, the logarithmic model, and the polynomial model. The average fitting error of the new model is about 61% of the logarithmic model, 46% of the hyperbolic model, and 36% of the polynomial model. Regarding fitting standard deviation, that of the new model is about 48% of that of the logarithmic model, about 37% of that of the hyperbolic model, and 28% of that of the polynomial model. The results indicate that the proposed new evaluation model is more accurate than the other three models in evaluating the elastic modulus of concrete.

Case 7. In reference [34], some concrete cube test blocks were manufactured with a design strength of 90 MPa. The mix proportions of concrete materials are listed in Table 19. The experimental data of elastic modulus are shown in Table 20, and the fitting curves are shown in Figure 9. The average fitting error and standard deviation of the models are shown in Table 21. The specific fitting equations for the four models are as follows:

$$z(t) = -9.6390 + 8.1827\ln(t) \tag{45}$$

$$z(t) = \frac{t}{0.8203 + 0.0278 \cdot t} \tag{46}$$

$$z(t) = 5.9162 + 0.4396 \cdot t - 0.0018 \cdot t^2 \tag{47}$$

$$z(t) = 1085.9326 \cdot (\frac{t}{t+1}) - 1341.0266 \cdot \arctan(\frac{t}{t+1}) \tag{48}$$

Table 19. The mix proportions of concrete materials in case 7.

Cement	Slag	Silica Fume	Water	Sand	Stone
0.5	0.4	0.1	0.23	1.2	1.8

Table 20. Elastic modulus of concrete cube in case 7 (GPa).

Age (Unit: hour)	Elastic Modulus
18	10.53
24	18.88
48	22.39
72	28.24
168	30.02

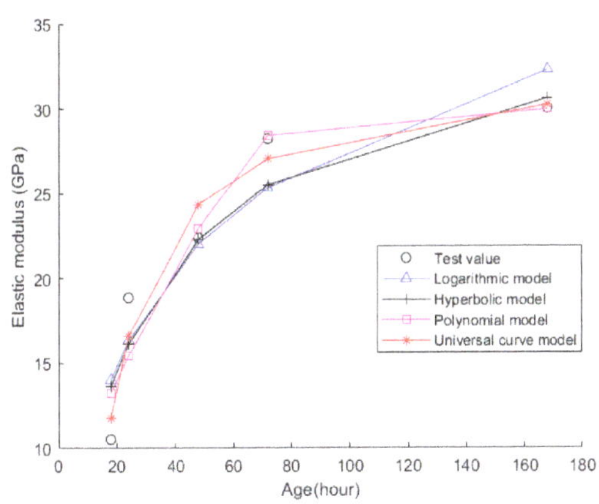

Figure 9. The fitting results of elastic modulus in case 7.

Table 21. The fitting mean errors and standard deviations of case 7.

Index	Logarithmic Model	Hyperbolic Model	Polynomial Model	Universal Curve Model
Fitting mean error $\bar{\delta}$	0.3788	0.3428	0.3174	0.1944
Fitted standard deviation σ	0.1194	0.1170	0.1182	0.0495

According to Table 21, the new model's fitting average error and standard deviation are smaller than those of the other models. It has been shown that the new model can more accurately evaluate the elastic modulus of concrete.

Case 8. The components of concrete materials are listed in Table 22. The experimental data of elastic modulus in reference [39] are shown in Table 23. The fitting curves of the four models are shown in Figure 10, and the corresponding fitting errors are shown in Table 24. The specific fitting equations for the four models are as follows:

$$z(t) = 8.6594 + 13.6630\ln(t) \tag{49}$$

$$z(t) = \frac{t}{0.0749 + 0.0175 \cdot t} \tag{50}$$

$$z(t) = 9.1545 + 4.8177 \cdot t - 0.1236 \cdot t^2 \tag{51}$$

$$z(t) = 344.8215 \cdot (\frac{t}{t+1}) - 367.0881 \cdot \arctan(\frac{t}{t+1}) \tag{52}$$

Table 22. The components of concrete materials in case 8 (kg/m^3).

Cement	Fine Aggregate (Machine Made Sand)	Coarse Aggregate	Water	Water Reducer	Admixture
465.9	478.8	1069.9	125.8	8.2	54.8

Table 23. Elastic modulus of concrete cube in case 8 (GPa).

Age (Unit: day)	Elastic Modulus
2	13.25
3	21.86
4	27.76
5	29.90
6	38.29
7	41.33
14	46.93
28	48.02

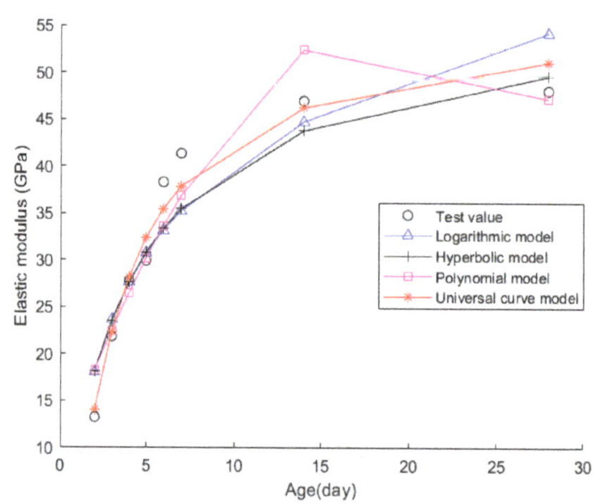

Figure 10. The fitting results of case 8.

Table 24. The fitting mean errors and standard deviations in case 8.

Index	Logarithmic Model	Hyperbolic Model	Polynomial Model	Universal Curve Model
Fitting mean error $\overline{\delta}$	0.4490	0.4344	0.4345	0.1667
Fitted standard deviation σ	0.1143	0.1177	0.1210	0.0293

Table 24 shows that the new model's average error and standard deviation are smaller than those of the hyperbolic, polynomial, and logarithmic models. This again indicates that the new model can better evaluate the elastic modulus of concrete than the existing models. Using Equation (7), the predicted long-term elastic modulus values in cases 6–8 are 41.30 MPa, 32.69 MPa, and 56.51 MPa, respectively. The coarse and fine aggregates in Cases 6 and 7 are both natural sand and gravel, while the fine aggregates in Case 8 are machine-made sand and the coarse aggregates are crushed stone. It can be observed that the type of aggregate has a significant impact on the elastic modulus of concrete, and the mechanism sand used in Case 8 may increase the elastic modulus. The research results of references [46,47] also show that the use of machine-made sand can improve the elastic modulus of concrete to a certain extent. Further research is needed in the future on the impact of aggregate types on elastic modulus of concrete.

3.3. High Strength Concrete and Lightweight Concrete

More types of concrete are further used to validate the new model to show its wider applicability and its practical implications. Cases 9 and 10 are high-strength concrete and Case 11 is lightweight concrete.

Case 9. In reference [43], a concrete mix design was developed by carefully combining different particle sizes of aggregates, setting the adjustment factor of cementitious materials, and determining parameters such as sand ratio, water–cement ratio, and dosage of admixtures. The relevant concrete material components for the three match ratios are listed in Table 25. The experimental data of compressive strength are shown in Table 26. Figures 11–13 present the fitting curves of the logarithmic, hyperbolic, polynomial, and new models for the three match ratios, respectively. In the Appendix A, Case 9 is also used as an example to elaborate in detail on the fitting process of the new model. The average fitting errors and standard deviations of these models are shown in Table 27. The specific fitting equations for the four models are as follows:

Table 25. The components of concrete materials in case 9 (kg/m^3).

Match Ratio	Adjustment Coefficient	Cement	Mineral Admixtures	Microsphere	Silica Fume	Spall	Artificial-Sand	Admixture	Water
1	1.0	182	26	26	17	1203	985	6.2	55
2	1.2	218	31	31	20	1169	956	7.4	66
3	1.5	273	39	39	26	1117	914	9.3	83

Table 26. Compressive strength of concrete blocks in case 9 (MPa).

Age (Unit: day)	Compressive Strength		
	Match Ratio 1	Match Ratio 2	Match Ratio 3
1	11.5	17.8	26.2
3	23.8	31.5	43.6
7	28.8	40.0	56.8
28	33.3	46.2	60.5
60	37.8	50.3	66.9
90	42.3	57.5	71.6

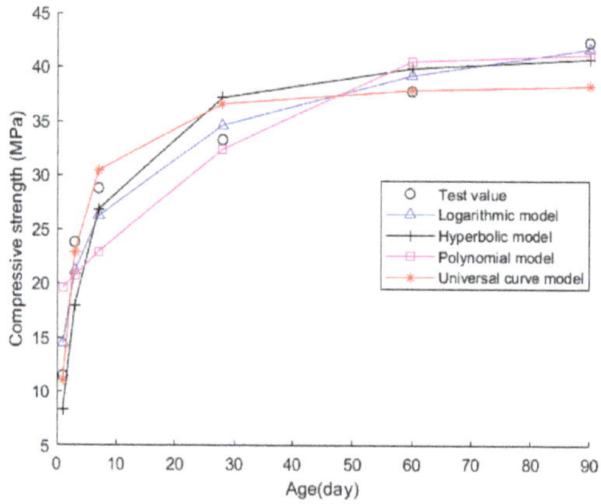

Figure 11. The fitting results for match ratio 1 in case 9.

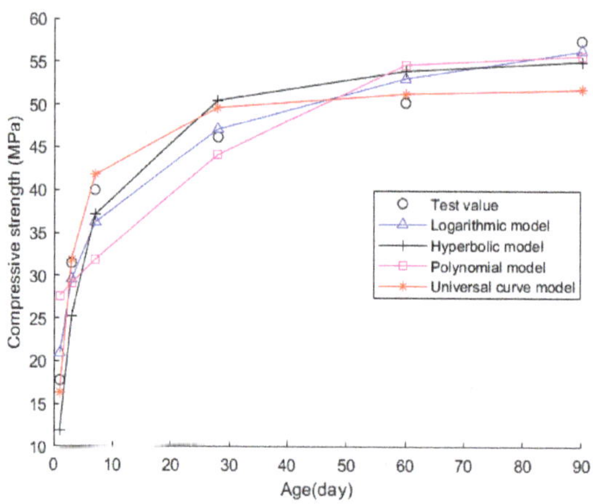

Figure 12. The fitting results for match ratio 2 in case 9.

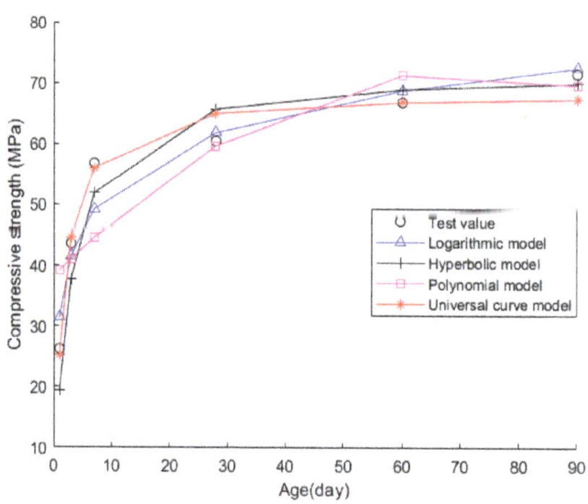

Figure 13. The fitting results for match ratio 3 in case 9.

Table 27. The fitting mean errors and standard deviations from the four models in case 9.

Match Ratio	Index	Logarithmic Model	Hyperbolic Model	Polynomial Model	Universal Curve Model
1	Fitting mean error $\overline{\delta}$	0.3013	0.3971	0.7499	0.1568
	Fitted standard deviation σ	0.0894	0.1024	0.2597	0.0380
2	Fitting mean error $\overline{\delta}$	0.2174	0.4132	0.6019	0.1561
	Fitted standard deviation σ	0.0579	0.1105	0.1989	0.0343
3	Fitting mean error $\overline{\delta}$	0.2500	0.3215	0.5478	0.1036
	Fitted standard deviation σ	0.0761	0.0879	0.1858	0.0281

For the first match ratio, one obtains

$$z(t) = 14.4787 + 6.0536\ln(t) \tag{53}$$

$$z(t) = \frac{t}{0.0964 + 0.0235 \cdot t} \tag{54}$$

$$z(t) = 19.0340 + 0.5839 \cdot t - 0.0038 \cdot t^2 \tag{55}$$

$$z(t) = 131.9067 \cdot (\frac{t}{t+1}) - 118.1620 \cdot \arctan(\frac{t}{t+1}) \tag{56}$$

For the second match ratio, one obtains

$$z(t) = 20.9103 + 7.8711\ln(t) \tag{57}$$

$$z(t) = \frac{t}{0.0667 + 0.0174 \cdot t} \tag{58}$$

$$z(t) = 26.8811 + 0.7510 \cdot t - 0.0048 \cdot t^2 \tag{59}$$

$$z(t) = 164.1922 \cdot (\frac{t}{t+1}) - 141.7374 \cdot \arctan(\frac{t}{t+1}) \tag{60}$$

For the third match ratio, one obtains

$$z(t) = 31.5073 + 9.1214\ln(t) \tag{61}$$

$$z(t) = \frac{t}{0.0379 + 0.0139 \cdot t} \tag{62}$$

$$z(t) = 38.2221 + 0.9551 \cdot t - 0.0067 \cdot t^2 \tag{63}$$

$$z(t) = 167.9351 \cdot (\frac{t}{t+1}) - 126.4422 \cdot \arctan(\frac{t}{t+1}) \tag{64}$$

To quantify the improved accuracy over existing models, Table 28 shows the relative ratios of fitting error and standard deviation for each model, when the error and standard deviation of the polynomial model are set to 1.

From Table 28, it can be seen that the relative values of the fitting error and standard deviation of the new model are the smallest, which are approximately 20% of the values of the polynomial model. These results indicate that the new model has significantly improved fitting accuracy compared to existing models. The practical implications of these results mainly lie in the precise correspondence between compressive strength and time can be obtained via the fitting equation of the new model. For example, when using concrete to manufacture large-volume box beams, the prestressing force of the steel bars should be applied as early as possible in order to prevent concrete cracks. In engineering practice, it is generally required that the compressive strength of concrete reach at least 25 MPa before prestressing can be applied, which requires accurate estimation of the corresponding time when the compressive strength reaches 25 MPa. This goal can be achieved via the specific fitting equation obtained from the new model. For concrete with the first match ratio, the time t corresponding to $z(t)$ = 25 MPa obtained from the fitting Equation (56) of the new model is t = 3.70 d. For concrete with the second match ratio, the time t corresponding to $z(t)$ = 25 MPa obtained from the fitting Equation (60) of the new model is t = 1.85 d. For concrete with the third match ratio, the time t corresponding to $z(t)$ = 25 MPa obtained from the fitting Equation (64) of the new model is t = 0.98 d. These results can be used to guide prestressed construction in bridge engineering.

Table 28. The relative ratios of fitting error and standard deviation in case 9 (based on polynomial model).

Match Ratio	Relative Ratios	Logarithmic Model	Hyperbolic Model	Polynomial Model	Universal Curve Model
1	Fitting mean error	0.4018	0.5295	1.0000	0.2091
	Fitted standard deviation	0.3442	0.3943	1.0000	0.1463
2	Fitting mean error	0.3612	0.6865	1.0000	0.2593
	Fitted standard deviation	0.2911	0.5556	1.0000	0.1724
3	Fitting mean error	0.4564	0.5869	1.0000	0.1891
	Fitted standard deviation	0.4096	0.4731	1.0000	0.1512

Case 10. In reference [44], two types of high-strength concrete test blocks were designed: natural aggregate concrete and self-compacting aggregate concrete. These test blocks contained three different design strengths. The 100 mm × 100 mm × 100 mm concrete specimens were manufactured and tested. The mix proportions of concrete materials are listed in Table 29. The experimental data of their compressive strength are shown in Table 30. Figures 14–19 show the fitting curves of the four models, and Table 31 presents the fitting average errors and standard deviations. The specific fitting equations for the four models are as follows:

Table 29. Mix proportions of concrete in case 10 (kg/m^3).

Number	Water	Cement	Fly Ash	Sand	Natural Aggregate	Water Reducer (%)	Water–cement Ratio (%)
NA-C-30	189.0	420	0	680.0	1110.0	0.07	45
NA-C-40	171.0	450	0	640.0	1138.0	0.33	38
NA-C-50	189.0	450	0	670.0	1090.0	0.10	42
NA-SCC-30	192.0	336	144	839.0	839.0	1.00	57
NA-SCC-40	176.8	364	156	826.6	826.0	1.20	49
NA-SCC-50	180.0	350	150	835.0	835.0	1.20	51

Table 30. Compressive strength of concrete blocks in case 10 (MPa).

Age (Unit: day)	Compressive Strength					
	NA-C-30	NA-C-40	NA-C-50	NA-SCC-30	NA-SCC-40	NA-SCC-50
7	21.67	29.43	29.75	19.37	24.87	30.73
28	44.17	50.71	53.40	35.33	49.40	53.20
56	45.70	52.75	55.47	40.53	55.60	55.75
90	47.22	53.86	54.72	40.57	57.91	57.13

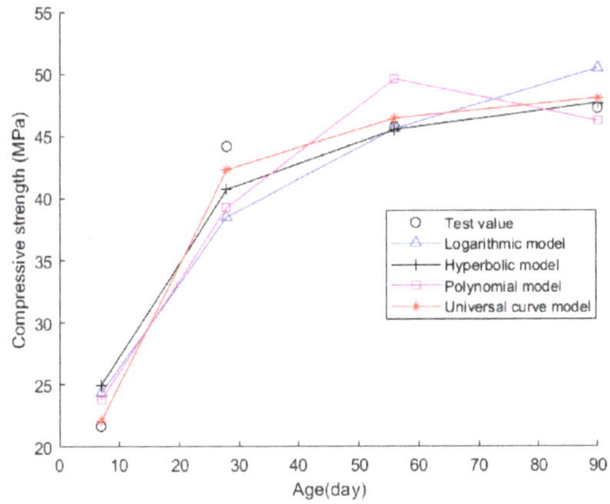

Figure 14. The fitting results of NA-C-30 in case 10.

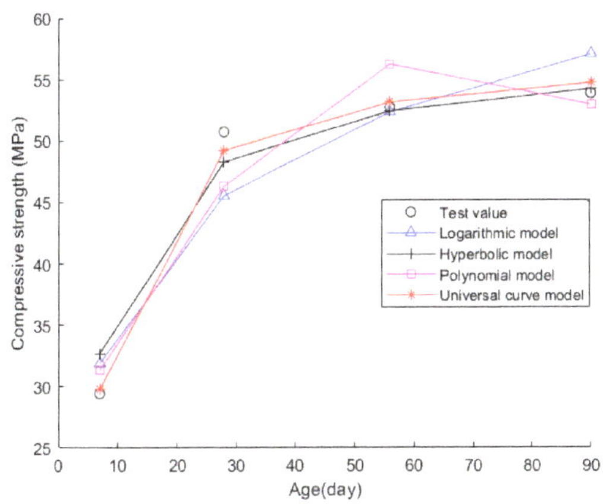

Figure 15. The fitting results of NA-C-40 in case 10.

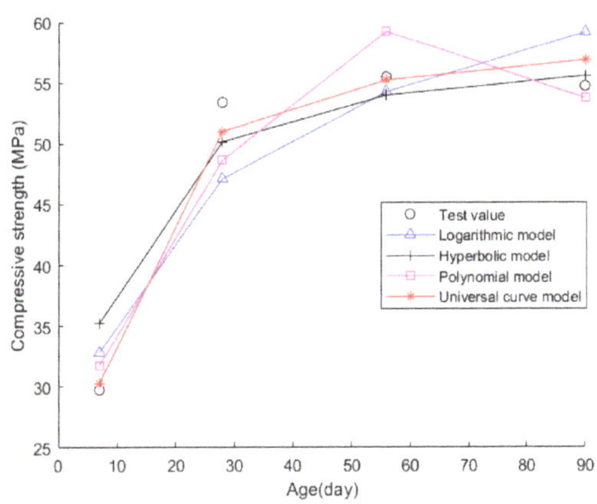

Figure 16. The fitting results of NA-C-50 in case 10.

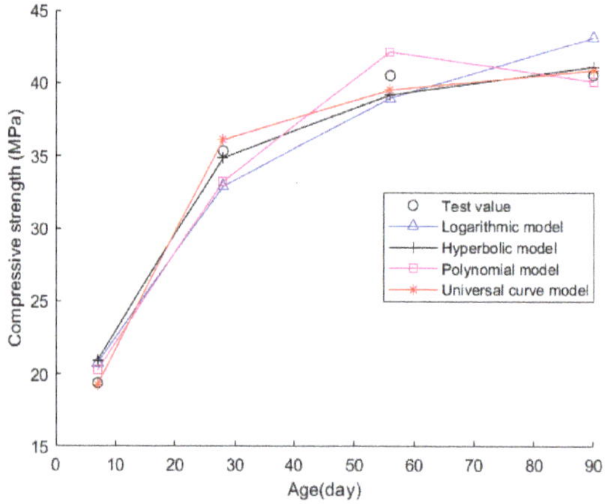

Figure 17. The fitting results of NA-SCC-30 in case 10.

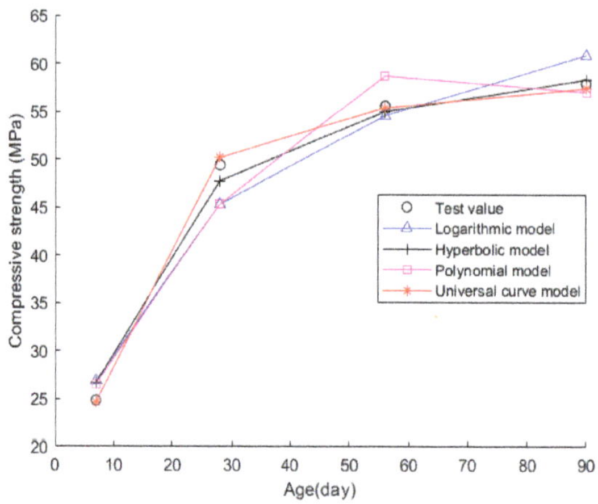

Figure 18. The fitting results of NA-SCC-40 in case 10.

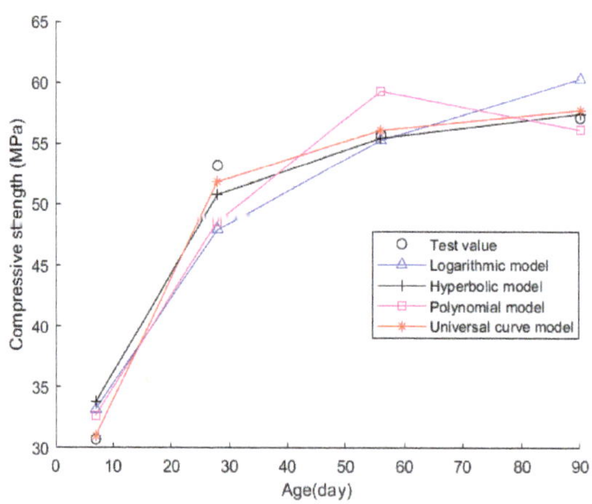

Figure 19. The fitting results of NA-SCC-50 in case 10.

Table 31. The fitting mean errors and standard deviations from the four models in case 10.

Number	Index	Logarithmic Model	Hyperbolic Model	Polynomial Model	Universal Curve Model
NA-C-30	Fitting mean error $\bar{\delta}$	0.1895	0.1700	0.1715	0.0517
	Fitted standard deviation σ	0.0583	0.0690	0.0395	0.0131
NA-C-40	Fitting mean error $\bar{\delta}$	0.1445	0.1201	0.1281	0.0362
	Fitted standard deviation σ	0.0409	0.0489	0.0296	0.0100
NA-C-50	Fitting mean error $\bar{\delta}$	0.1780	0.1970	0.1322	0.0627
	Fitted standard deviation σ	0.0423	0.0776	0.0301	0.0185
NA-SCC-30	Fitting mean error $\bar{\delta}$	0.1238	0.0899	0.0858	0.0343
	Fitted standard deviation σ	0.0153	0.0320	0.0205	0.0095
NA-SCC-40	Fitting mean error $\bar{\delta}$	0.1279	0.0811	0.1219	0.0195
	Fitted standard deviation σ	0.0303	0.0301	0.0291	0.0055
NA-SCC-50	Fitting mean error $\bar{\delta}$	0.1398	0.1091	0.1260	0.0295
	Fitted standard deviation σ	0.0394	0.0444	0.0293	0.0081

For natural aggregate concrete with a design strength of 30 MPa (NA-C-30), one obtains

$$z(t) = 4.3947 + 10.2281 \ln(t) \tag{65}$$

$$z(t) = \frac{t}{0.1451 + 0.0194 \cdot t} \tag{66}$$

$$z(t) = 17.1393 + 1.0002 \cdot t - 0.0075 \cdot t^2 \tag{67}$$

$$z(t) = 607.0926 \cdot (\frac{t}{t+1}) - 708.3099 \cdot \arctan(\frac{t}{t+1}) \tag{68}$$

For natural aggregate concrete with a design strength of 40 MPa (NA-C-40), one obtains

$$z(t) = 12.7301 + 9.8404 \ln(t) \tag{69}$$

$$z(t) = \frac{t}{0.0924 + 0.0174 \cdot t} \tag{70}$$

$$z(t) = 24.9121 + 0.9669 \cdot t - 0.0073 \cdot t^2 \tag{71}$$

$$z(t) = 564.4074 \cdot (\frac{t}{t+1}) - 645.6422 \cdot \arctan(\frac{t}{t+1}) \tag{72}$$

For natural aggregate concrete with a design strength of 50 MPa (NA-C-50), one obtains

$$z(t) = 12.8294 + 10.2890 \ln(t) \tag{73}$$

$$z(t) = \frac{t}{0.0787 + 0.0171 \cdot t} \tag{74}$$

$$z(t) = 24.4671 + 1.1061 \cdot t - 0.0087 \cdot t^2 \tag{75}$$

$$z(t) = 603.5390 \cdot (\frac{t}{t+1}) - 692.4944 \cdot \arctan(\frac{t}{t+1}) \tag{76}$$

For self-compacting aggregate concrete with a design strength of 30 MPa (NA-SCC-30), one obtains

$$z(t) = 3.7122 + 8.7625 \ln(t) \tag{77}$$

$$z(t) = \frac{t}{0.1778 + 0.0223 \cdot t} \tag{78}$$

$$z(t) = 14.7409 + 0.8316 \cdot t - 0.0061 \cdot t^2 \tag{79}$$

$$z(t) = 503.7077 \cdot (\frac{t}{t+1}) - 586.3501 \cdot \arctan(\frac{t}{t+1}) \tag{80}$$

For self-compacting aggregate concrete with a design strength of 40 MPa (NA-SCC-40), one obtains

$$z(t) = 0.9832 + 13.3191 \ln(t) \tag{81}$$

$$z(t) = \frac{t}{0.1544 + 0.0154 \cdot t} \tag{82}$$

$$z(t) = 18.6628 + 1.1919 \cdot t - 0.0085 \cdot t^2 \tag{83}$$

$$z(t) = 772.2900 \cdot (\frac{t}{t+1}) - 905.7117 \cdot \arctan(\frac{t}{t+1}) \tag{84}$$

For self-compacting aggregate concrete with a design strength of 50 MPa (NA-SCC-50), one obtains

$$z(t) = 12.5453 + 10.6228 \ln(t) \tag{85}$$

$$z(t) = \frac{t}{0.0926 + 0.0164 \cdot t} \tag{86}$$

$$z(t) = 25.8732 + 1.0270 \cdot t - 0.0077 \cdot t^2 \tag{87}$$

$$z(t) = 608.5387 \cdot (\frac{t}{t+1}) - 697.6181 \cdot \arctan(\frac{t}{t+1}) \tag{88}$$

To quantify improved accuracy over existing models, Table 32 shows the relative ratios of fitting error and standard deviation for each model, when the error and standard deviation of the hyperbolic model are set to 1.

Table 32. The relative ratios of fitting error and standard deviation in case 10 (based on hyperbolic model).

Number	Relative Ratios	Logarithmic Model	Hyperbolic Model	Polynomial Model	Universal Curve Model
NA-C-30	Fitting mean error	1.1147	1.0000	1.0088	0.3041
	Fitted standard deviation	0.8449	1.0000	0.5725	0.1899
NA-C-40	Fitting mean error	1.2032	1.0000	1.0666	0.3014
	Fitted standard deviation	0.8364	1.0000	0.6053	0.2045
NA-C-50	Fitting mean error	0.9036	1.0000	0.6711	0.3183
	Fitted standard deviation	0.5451	1.0000	0.3879	0.2384
NA-SCC-30	Fitting mean error	1.3771	1.0000	0.9544	0.3815
	Fitted standard deviation	0.4781	1.0000	0.6406	0.2969
NA-SCC-40	Fitting mean error	1.5771	1.0000	1.5031	0.2404
	Fitted standard deviation	1.0066	1.0000	0.9668	0.1827
NA-SCC-50	Fitting mean error	1.2814	1.0000	1.1549	0.2704
	Fitted standard deviation	0.8874	1.0000	0.6599	0.1824

From Table 32, one can observe that the relative values of the fitting error and standard deviation of the new model are the smallest, approximately 20%–30% of the values of the hyperbolic model. These results again indicate that the new model has significantly improved fitting accuracy compared to existing models. These results can also be used to accurately evaluate the corresponding time when the compressive strength of concrete will reach 25 MPa, in order to guide prestressed construction in practical engineering. Without loss of generality, for NA-C-50, the time t corresponding to $z(t)$ = 25 MPa obtained from the fitting Equation (76) of the new model is t = 5.63 d. For NA-SCC-50, the time t corresponding to $z(t)$ = 25 MPa obtained from the fitting Equation (88) of the new model is t = 5.50 d.

In order to explore the relationship between the fitting parameters (a and b) of the new model and the water–cement ratio (w/c) of concrete, Figure 20 shows the variation curves of parameters a, b, and $a + b \cdot \frac{\pi}{4}$ with the water–cement ratio. From Figure 20, it can be seen that the variation patterns of the absolute values of parameters a and b are basically consistent. When the water–cement ratio is about 50%, the long-term strength (i.e., $a + b \cdot \frac{\pi}{4}$) of concrete is relatively great, up to 60 MPa.

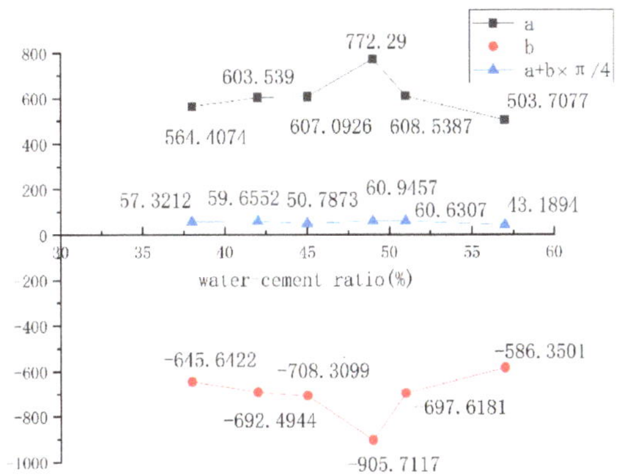

Figure 20. The variation curves of parameters a, b, and $a + b \cdot \frac{\pi}{4}$ with the water–cement ratio of case 10.

Case 11. In reference [45], some lightweight concrete specimens with dimensions of 100 mm × 100 mm × 100 mm were manufactured using CL30 grade concrete, cured under standard curing conditions. The composition of the concrete can be found in Table 33. The test results of the compressive strength for the three batches are shown in Table 34. Figures 21–23 show the fitted curves of the four models based on experimental data. The fitting average error and standard deviation are listed in Table 35. The fitting equations for the four models are as follows:

Table 33. Mix proportions of concrete in case 11 (kg/m^3).

Cement	Water	Sand	Coarse Aggregate
386	244	705	591

Table 34. Compressive strength of concrete blocks in case 11 (MPa).

Age (Unit: day)	Compressive Strength		
	Batch 1	Batch 2	Batch 3
3	22.1	18.8	16.8
7	32.0	28.5	27.7
14	35.2	35.7	34.0
21	38.3	36.8	38.0
28	39.7	37.3	38.7

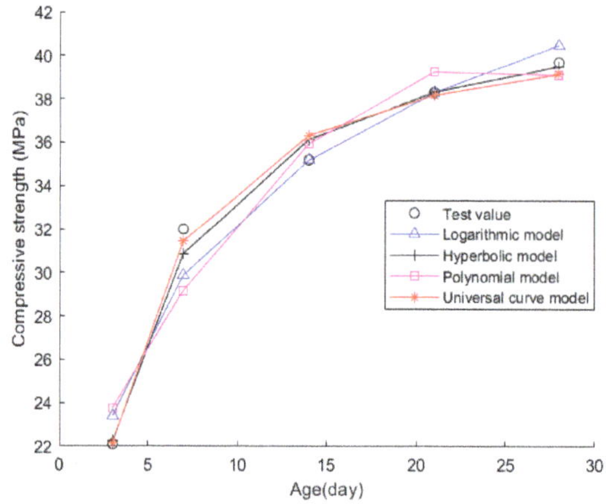

Figure 21. The fitting results of Batch 1 in case 11.

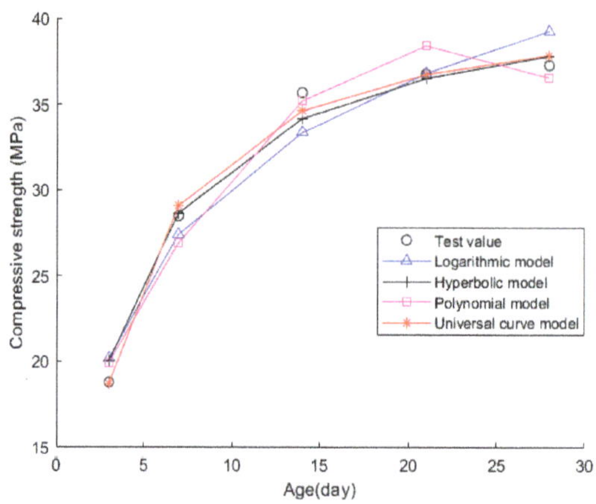

Figure 22. The fitting results of Batch 2 in case 11.

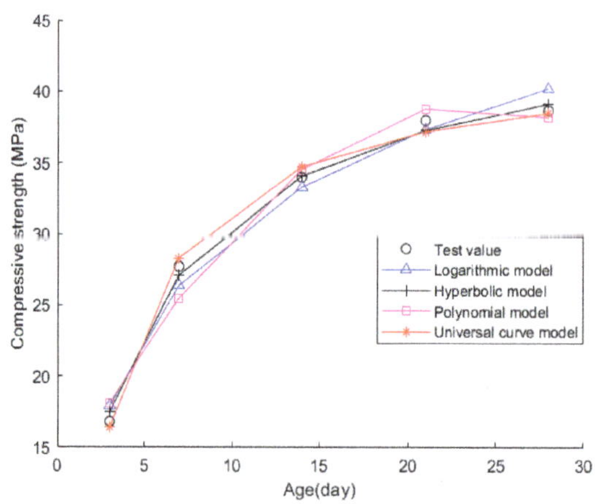

Figure 23. The fitting results of Batch 3 in case 11.

Table 35. The fitting mean errors and standard deviations of the four models in case 11.

Batch	Index	Logarithmic Model	Hyperbolic Model	Polynomial Model	Universal Curve Model
1	Fitting mean error $\overline{\delta}$ Fitted standard deviation σ	0.0908 0.0317	0.0448 0.0153	0.1215 0.0342	0.0390 0.0121
2	Fitting mean error $\overline{\delta}$ Fitted standard deviation σ	0.1182 0.0288	0.0798 0.0261	0.956 0.0210	0.0398 0.0116
3	Fitting mean error $\overline{\delta}$ Fitted standard deviation σ	0.0946 0.0204	0.0523 0.0148	0.1164 0.0347	0.0441 0.0076

For batch 1, one obtains

$$z(t) = 15.0016 + 7.6525\ln(t) \tag{89}$$

$$z(t) = \frac{t}{0.0660 + 0.0230 \cdot t} \tag{90}$$

$$z(t) = 18.9782 + 1.7046 \cdot t - 0.0352 \cdot t^2 \tag{91}$$

$$z(t) = 180.1339 \cdot (\frac{t}{t+1}) - 175.5051 \cdot \arctan(\frac{t}{t+1}) \tag{92}$$

For batch 2, one obtains

$$z(t) = 10.7864 + 8.5543\ln(t) \tag{93}$$

$$z(t) = \frac{t}{0.0789 + 0.0236 \cdot t} \tag{94}$$

$$z(t) = 13.5875 + 2.2685 \cdot t - 0.0517 \cdot t^2 \tag{95}$$

$$z(t) = 220.5330 \cdot (\frac{t}{t+1}) - 227.9646 \cdot \arctan(\frac{t}{t+1}) \tag{96}$$

For batch 3, one obtains

$$z(t) = 6.9371 + 9.9926\ln(t) \tag{97}$$

$$z(t) = \frac{t}{0.1061 + 0.0218 \cdot t} \tag{98}$$

$$z(t) = 11.5844 + 2.3295 \cdot t - 0.0492 \cdot t^2 \tag{99}$$

$$z(t) = 268.1057 \cdot (\frac{t}{t+1}) - 286.9670 \cdot \arctan(\frac{t}{t+1}) \tag{100}$$

To quantify the improved accuracy over existing models, Table 36 shows the relative ratios of fitting error and standard deviation for each model, when the error and standard deviation of the logarithmic model are set to 1.

From Table 36, one can observe that the relative values of the fitting error and standard deviation of the new model are the smallest, approximately 30%–40% of the values of the logarithmic model. These results indicate that the fitting equation obtained from the new model can more accurately evaluate the correspondence between compressive strength and time. The fitting equation of the new model can be used not only to guide prestressed construction, as in Cases 9 and 10, but also to guide the time for template removal. For example, if the template can be removed when the compressive strength reaches 15 MPa, the removal time for the first batch of concrete can be calculated from the fitting Equation (92) of the new model as t = 1.73 d. The removal time for the second batch of concrete can be calculated from the fitting Equation (96) as t = 2.31 d. The removal time for the third batch of concrete can be calculated from the fitting Equation (100) as t = 2.74 d.

Table 36. The relative ratios of fitting error and standard deviation in case 11 (based on logarithmic model).

Batch	Relative Ratios	Logarithmic Model	Hyperbolic Model	Polynomial Model	Universal Curve Model
1	Fitting mean error Fitted standard deviation	1.0000 1.0000	0.4934 0.4826	1.3381 1.0789	0.4295 0.3817
2	Fitting mean error Fitted standard deviation	1.0000 1.0000	0.6751 0.9063	8.0880 0.7292	0.3367 0.4028
3	Fitting mean error Fitted standard deviation	1.0000 1.0000	0.5529 0.7255	1.2304 1.7010	0.4662 0.3725

4. Conclusions

This article proposes a new curve model for evaluating the mechanical properties of concrete. This model can evaluate the mechanical properties of concrete at different ages by fitting a small amount of experimental data. The model was validated using multiple sets of experimental data. Based on the calculation results, the following main conclusions can be drawn:

- Logarithmic models and exponential models have similar advantages and disadvantages. Both models are incremental functions, which align with the law of mechanical properties of concrete increasing with time in engineering practice, and both lack a point of t = 0. The hyperbolic model can gradually approach a fixed value as time increases, which is consistent with the fact that the mechanical properties of concrete remain unchanged in the later stage. However, when dealing with some very discrete data, the fitting accuracy of the logarithmic, exponential, and hyperbolic models is not very good. The advantage of the polynomial model is that its domain of definition includes the point t = 0. However, its fitting function is a monotonically increasing function. It cannot be used for the prediction of the mechanical properties of concrete in the later stage.
- The new model proposed in this article can be seen as a combination of the hyperbolic and arctangent functions. This model makes up for the deficiencies of the existing models and includes the point t = 0. While it tends toward infinity, the function of this new model can approach a specific value, which is in line with the fact that the mechanical properties of concrete remain unchanged in the later stages. In addition, the new model only needs to fit two parameters, making it easier to use in practical engineering than the three-parameter model.
- Based on the case studies of multiple sets of experimental data, the results show that the new model proposed in this article can evaluate concrete's compressive strength and elastic modulus. This model can evaluate the mechanical properties of concrete under different curing environments and concrete strength levels and has a broader range of applications in practical engineering.

It should be noted that this article mainly discusses the law of changes in the mechanical properties of concrete over time, and other factors that affect the mechanical properties of concrete are not fully considered. The proposed model cannot simultaneously establish the relationship between multiple factors and the mechanical properties of concrete like machine learning. However, the proposed model only requires a small amount of experimental data to establish the mathematical equation for evaluation. In the case of insufficient experimental data, the proposed model can be combined with ANNs to simulate and generate more data for the training of ANNs, thereby improving the generalization ability of the ANN model. Further research is needed in the future to verify the fitting and evaluation capabilities of this new model under more complex conditions. This new model may also be applicable to other cementitious materials besides concrete. In subsequent studies, relevant experiments on other cementitious materials can be conducted to verify the feasibility of this dual parameter curve model.

Author Contributions: Conceptualization, Q.Y. and F.Q.; methodology, Q.Y. and X.W.; software, X.W. and X.P.; validation, Q.Y., X.W. and X.P.; investigation, Q.Y. and F.Q.; data curation, Q.Y., X.W. and X.P.; writing—original draft preparation, Q.Y., X.W., X.P. and F.Q.; writing—review and editing, Q.Y., X.W., X.P. and F.Q. All authors have read and agreed to the published version of the manuscript.

Funding: This work is supported by the Chongqing Transportation Science and Technology Project (Grant No. 2022-01), Zhejiang Public Welfare Technology Application Research Project (LGF22E080021), the Natural Science Foundation of China (52192663, 52008215), and the Major Special Science and Technology Project (2019B10076) of "Ningbo Science and Technology Innovation 2025".

Institutional Review Board Statement: Not applicable.

Informed Consent Statement: Not applicable.

Data Availability Statement: The data used to support the findings of this study are included in this study and also available from the corresponding author upon request.

Conflicts of Interest: The authors declare that there are no conflicts of interest. The funders had no role in the collection, analyses, or interpretation of data, or in the decision to publish the results.

Appendix A

Without losing generality, Case 9 can be taken as an example to elaborate in detail on the fitting process of the new model. The fitting process of other models is similar to that of this new model. As mentioned earlier, the experimental data in Case 9 have been provided in Table 26. Next, the new model is used to fit the compressive strength data of concrete specimens made according to the first mix ratio in Case 9 at different ages. Using the data from the first column of Table 26, the matrix C can be calculated using Equation (11) as

$$C = \begin{bmatrix} 0.5000 & 0.4636 \\ 0.7500 & 0.6435 \\ 0.8750 & 0.7188 \\ 0.9655 & 0.7679 \\ 0.9836 & 0.7771 \\ 0.9890 & 0.7799 \end{bmatrix} \tag{A1}$$

Using the data from the second column of Table 26, the vector y can be obtained using Equation (9) as

$$y = \begin{Bmatrix} 11.50 \\ 23.80 \\ 28.80 \\ 33.30 \\ 37.80 \\ 42.30 \end{Bmatrix} \tag{A2}$$

Substituting Equations (A1) and (A2) into (8), the linear system corresponding to the new model can be obtained as

$$\begin{Bmatrix} 11.50 \\ 23.80 \\ 28.80 \\ 33.30 \\ 37.80 \\ 42.30 \end{Bmatrix} = \begin{bmatrix} 0.5000 & 0.4636 \\ 0.7500 & 0.6435 \\ 0.8750 & 0.7188 \\ 0.9655 & 0.7679 \\ 0.9836 & 0.7771 \\ 0.9890 & 0.7799 \end{bmatrix} \cdot x \tag{A3}$$

Multiplying both sides of Equation (A3) by C^T, as shown in Equation (12), one obtains

$$\begin{bmatrix} 159.9672 \\ 129.2835 \end{bmatrix} = \begin{bmatrix} 4.4560 & 3.6205 \\ 3.6205 & 2.9475 \end{bmatrix} \cdot x \tag{A4}$$

From Equation (A4), the least squares solution of x can be obtained using Equation (13) as

$$\hat{x} = \begin{Bmatrix} 131.9067 \\ -118.1620 \end{Bmatrix} \tag{A5}$$

From Equation (A5), the two fitting parameters of the new model are obtained as $a = 131.9067$ and $b = -118.1620$. Therefore, the specific fitting equation for the new model corresponding to the first mix ratio of Case 9 is

$$z(t) = 131.9067 \cdot (\frac{t}{t+1}) - 118.1620 \cdot \arctan(\frac{t}{t+1}) \tag{A6}$$

The result shown in Equation (A6) is that shown in Equation (56).

References

1. Yang, Q.; Sun, Y.; Peng, X. Experimental Study on Mechanical Properties of Concrete at Super-Early Age. *Materials* **2022**, *15*, 7582. [CrossRef]
2. Yang, Q.; Peng, X.; Sun, Y. Shear Capacity Evaluation of the Recycled Concrete Beam. *Materials* **2022**, *15*, 3693. [CrossRef] [PubMed]
3. Yang, Q.W.; Ba, Z.; Zhao, Z.; Peng, X.; Sun, Y. Stability Evaluation Method of Hole Wall for Bored Pile under Blasting Impact. *Symmetry* **2022**, *14*, 79. [CrossRef]
4. Peng, X.; Yang, Q. Damage detection in beam-like structures using static shear energy redistribution. *Front. Struct. Civ. Eng.* **2023**, *16*, 1552–1564. [CrossRef]
5. Zhang, M.M.; He, Z.F. Analysis of the causes of cracking in the bottom plate of a prestressed continuous box girder. *Urban Roads Bridges Flood Control* **2023**, *10*, 110–114+119.
6. Xia, Y.J. Analysis of key construction technology points of high-rise building template engineering. *Chin. Constr.* **2023**, *11*, 135–137.
7. Jaf, D.K.I.; Abdulrahman, P.I.; Mohammed, A.S.; Kurda, R.; Qaidi, S.M.; Asteris, P.G. Machine learning techniques and multi-scale models to evaluate the impact of silicon dioxide (SiO_2) and calcium oxide (CaO) in fly ash on the compressive strength of green concrete. *Constr. Build. Mater.* **2023**, *400*, 132604.
8. Ahmed, H.U.; Mohammed, A.S.; Faraj, R.H.; Abdalla, A.A.; Qaidi, S.M.A.; Sor, N.H.; Mohammed, A.A. Innovative modeling techniques including MEP, ANN and FQ to forecast the compressive strength of geopolymer concrete modified with nanoparticles. *Neural Comput. Appl.* **2023**, *35*, 12453–12479. [CrossRef]
9. Trtnik, G.; Kavčič, F.; Turk, G. Prediction of concrete strength using ultrasonic pulse velocity and artificial neural networks. *Ultrasonics* **2009**, *49*, 53–60. [CrossRef]
10. Yousif, S.T.; Abdullah, S.M. Artificial neural network model for predicting compressive strength of concrete. *Tikrit J. Eng. Sci.* **2009**, *16*, 55–66. [CrossRef]
11. Duan, Z.H.; Kou, S.C.; Poon, C.S. Prediction of compressive strength of recycled aggregate concrete using artificial neural networks. *Constr. Build. Mater.* **2013**, *40*, 1200–1206. [CrossRef]
12. Nikoo, M.; Torabian Moghadam, F.; Sadowski, Ł. Prediction of concrete compressive strength by evolutionary artificial neural networks. *Adv. Mater. Sci. Eng.* **2015**, *2015*, 849126. [CrossRef]
13. Khashman, A.; Akpinar, P. Non-destructive prediction of concrete compressive strength using neural networks. *Procedia Comput. Sci.* **2017**, *108*, 2358–2362. [CrossRef]
14. Deng, F.; He, Y.; Zhou, S.; Yu, Y.; Cheng, H.; Wu, X. Compressive strength prediction of recycled concrete based on deep learning. *Constr. Build. Mater.* **2018**, *175*, 562–569. [CrossRef]
15. Feng, D.C.; Liu, Z.T.; Wang, X.D.; Chen, Y.; Chang, J.Q.; Wei, D.F.; Jiang, Z.M. Machine learning-based compressive strength prediction for concrete: An adaptive boosting approach. *Constr. Build. Mater.* **2020**, *230*, 117000. [CrossRef]
16. Muliauwan, H.N.; Prayogo, D.; Gaby, G.; Harsono, K. Prediction of concrete compressive strength using artificial intelligence methods. *J. Phys. Conf. Ser.* **2020**, *1625*, 012018. [CrossRef]
17. Asteris, P.G.; Mokos, V.G. Concrete compressive strength using artificial neural networks. *Neural Comput. Appl.* **2020**, *32*, 11807–11826. [CrossRef]
18. Bka, M.A.R.; Ngamkhanong, C.; Wu, Y.; Kaewunruen, S. Recycled aggregates concrete compressive strength prediction using artificial neural networks (ANNs). *Infrastructures* **2021**, *6*, 17.
19. Cui, X.; Wang, Q.; Zhang, R.; Dai, J.; Li, S. Machine learning prediction of concrete compressive strength with data enhancement. *J. Intell. Fuzzy Syst.* **2021**, *41*, 7219–7228. [CrossRef]
20. Kumar, A.; Arora, H.C.; Kapoor, N.R.; Mohammed, M.A.; Kumar, K.; Majumdar, A.; Thinnukool, O. Compressive strength prediction of lightweight concrete: Machine learning models. *Sustainability* **2022**, *14*, 2404. [CrossRef]

21. Pakzad, S.S.; Roshan, N.; Ghalehnovi, M. Comparison of various machine learning algorithms used for compressive strength prediction of steel fiber-reinforced concrete. *Sci. Rep.* **2023**, *13*, 3646. [CrossRef] [PubMed]
22. Cüneyt Aydin, A.; Tortum, A.; Yavuz, M. Prediction of concrete elastic modulus using adaptive neuro-fuzzy inference system. *Civ. Eng. Environ. Syst.* **2006**, *23*, 295–309. [CrossRef]
23. Demir, F. Prediction of elastic modulus of normal and high strength concrete by artificial neural networks. *Constr. Build. Mater.* **2008**, *22*, 1428–1435. [CrossRef]
24. Ahmadi-Nedushan, B. Prediction of elastic modulus of normal and high strength concrete using ANFIS and optimal nonlinear regression models. *Constr. Build. Mater.* **2012**, *36*, 665–673. [CrossRef]
25. Han, T.; Siddique, A.; Khayat, K.; Huang, J.; Kumar, A. An ensemble machine learning approach for prediction and optimization of modulus of elasticity of recycled aggregate concrete. *Constr. Build. Mater.* **2020**, *244*, 118271. [CrossRef]
26. Hasanzadeh, A.; Vatin, N.I.; Hematibahar, M.; Kharun, M.; Shooshpasha, I. Prediction of the mechanical properties of basalt fiber reinforced high-performance concrete using machine learning techniques. *Materials* **2022**, *15*, 7165. [CrossRef]
27. Yang, Y.; Chen, F.Y. Strength development and evaluation of high-strength concrete at early age. *China Concr. Cem. Prod.* **2004**, *137*, 1–4.
28. Zhao, S.B.; Zhao, Y.J.; Chen, H.J.; Liu, S.; Sun, Y.P. Application and study on time-dependent compressive strength and elastic modulus of C50 pumping concrete. *Concrete* **2015**, *303*, 98–101.
29. Yang, W.J.; Wang, Y. Diachronic variation model of compressive strength and elastic modulus of concrete at early age. *J. China Foreign Highw.* **2007**, *27*, 149–151.
30. Liu, G. Elastic modulus and strength test of C50 concrete. *Traffic World* **2020**, *20*, 37–38.
31. Zhao, Z. Proportion design and strength regularity of high strength concrete. *Concrete* **2011**, *264*, 101–103.
32. Carino, N.J.; Lew, H.S. The maturity method: From theory to application. In Proceedings of the Structures Congress and Exposition, Washington, DC, USA, 21–23 May 2001. 19p.
33. Viviani, M.; Glisic, B.; Smith, F.C. Three-day prediction of concrete compressive strength evolution. *ACI Mater. J.* **2005**, *102*, 231–236.
34. Jin, X.Y.; Shen, Y.; Li, Z.J. Experimental study on early aging characteristics of high-strength concrete. *China Concr. Cem. Prod.* **2003**, *5*, 5–7.
35. Ling, J.J. Compressive Strength and Nonlinear Prediction Model of High-grade Basalt Manufactured Sand Concrete. *J. Water Resour. Archit. Eng.* **2021**, *19*, 124–130.
36. Jia, M.; Han, Z.B. Analysis of Compressive Test of Silicate Concrete Based on Bridge Estimation. *J. North China Univ. Sci. Technol. Nat. Sci. Ed.* **2019**, *41*, 84–89.
37. Yeh, I.C. Generalization of strength versus water–cementitious ratio relationship to age. *Cem. Concr. Res.* **2006**, *36*, 1865–1873. [CrossRef]
38. Elaty, M. Compressive strength prediction of Portland cement concrete with age using a new model. *HBRC J.* **2014**, *10*, 145–155. [CrossRef]
39. Guo, Q.; Zhang, Z.; Pu, G.N.; Liu, B.; Sun, H.P. Experimental study on the time-age mechanism, compressive strength, and elastic modulus of concrete with sand and cement, and the constitutive model. *Highw. Eng.* **2023**, 1–10. Available online: https://kns.cnki.net/kcms2/detail/43.1481.U.20230529.1444.002.html (accessed on 31 October 2023).
40. Wang, W.; Yue, Q. The time variation law of concrete compressive strength: A review. *Appl. Sci.* **2023**, *13*, 4947. [CrossRef]
41. Zhao, M.Q.; Zhan, B.G.; Zhang, W.X.; Yang, Y.G.; Zhang, Z.Q.; Yang, Y.S. Application of maturity method to predict the strength of concrete at very early age. *J. Hefei Univ. Technol. Nat. Sci. Ed.* **2022**, *45*, 1528–1534.
42. Peng, X.; Yang, Q.; Cao, H.; Wang, H. Mathematical Model for Early-Aged UHPFRC Compressive Strength Changes. *Coatings* **2023**, *13*, 525. [CrossRef]
43. Liu, Y.C. Research on the preparation process and durability of ultra-early strength and high-strength concrete. *Concr. World* **2020**, *3*, 78–82.
44. Yao, D.L.; Yu, F.; Xie, G.F. Mechanical properties of self-compacting recycled concrete at different ages. *J. Shenyang Univ. Technol.* **2019**, *41*, 589–593.
45. Cheng, D.S.; Shang, J.L.; Sun, L.C.; Zong, W. Experimental research on early strength prediction of lightweight aggregate concrete. *Concrete* **2007**, *01*, 32–33+36.
46. Zhou, Y.J.; Chen, S.A.; Zhan, W.; Qin, M.Q.; Xu, W.B. Research on the mechanical and durability performance of fly ash concrete for box girder. *Constr. Technol.* **2015**, *44*, 16–18.
47. He, S.D.; Liu, L.X.; Wang, J.; Wang, L.L. Experimental study on the elastic modulus of mechanism sand concrete under different replacement rates. *J. Guangxi Univ. Nat. Sci. Ed.* **2010**, *35*, 698–701.

coatings

MDPI

Article

Study on Flexural Strength of Interface between Full Lightweight Ceramsite Concrete and Ordinary Concrete

Hongbing Zhu [1,2,3,*], Yu Xiao [1], Xiu Li [4], Ye Wang [1] and Siyu Wen [1]

[1] School of Urban Construction, Wuhan University of Science and Technology, Wuhan 430065, China; xiaoyu1572570451@163.com (Y.X.); iyewang@163.com (Y.W.); 18970520527@163.com (S.W.)
[2] High Performance Engineering Structure Research Institute, Wuhan University of Science and Technology, Wuhan 430065, China
[3] Hubei Provincial Engineering Research Center of Urban Regeneration, Wuhan University of Science and Technology, Wuhan 430065, China
[4] School of Transportation Engineering, Wuhan Technical College of Communications, Wuhan 430065, China; hubeilixiu@163.com
* Correspondence: zhuhongbing@wust.edu.cn

Abstract: The efficacy of full lightweight ceramsite concrete as a restorative material has been widely acknowledged, given its light weight, strength, and durability. However, the extent of its performance in repairing existing or old concrete remains uncertain. This study examined the reparation of flexural performance with full lightweight ceramsite concrete, using 14 different combinations of old and new concrete test blocks. The primary focus of the study was on investigating the flexural bond strength of the interface between the old and the new concrete. This included understanding the effects of the interfacial roughness, interfacial agent type, and concrete curing age of the concrete on the flexural strength. The test results showed that increasing the interface roughness from 0 mm to 5 mm resulted a restoration of the flexural strength of the sample by approximately 59%. Additionally, the flexural strength of the specimens was restored by 62%–78% of their original strength with the application of different types of interfacial agent. To rank the impact of these factors on the flexural strength, a univariate analysis of variance was conducted. This allowed us to establish a mathematical formula for calculating the flexural capacity of old and new concrete interfaces, taking the three aforementioned factors into account.

Keywords: full lightweight ceramsite concrete; interfacial roughness; interfacial agent; curing age; flexural strength

Citation: Zhu, H.; Xiao, Y.; Li, X.; Wang, Y.; Wen, S. Study on Flexural Strength of Interface between Full Lightweight Ceramsite Concrete and Ordinary Concrete. *Coatings* **2023**, *13*, 1383. https://doi.org/10.3390/coatings13081383

Academic Editor: Andrea Nobili

Received: 27 June 2023
Revised: 2 August 2023
Accepted: 3 August 2023
Published: 7 August 2023

1. Introduction

Numerous building structures annually encounter durability challenges due to adverse environmental conditions and external loads. For the safe operation of these structures, it is indispensable to repair and reinforce the whole or part of these structures. The patching and strengthening of concrete structures have thus emerged as crucial research topics in construction engineering [1,2]. Various bonding methods exist for concrete, most notably the sectional increase method [3] and the mechanical connection method [4]. The former, a widely adopted technique [5], is implemented by pouring new concrete onto existing concrete. However, this leads to the generation of new and old concrete interfaces, the most vulnerable parts of the repaired structure [6,7]. The mechanical properties of these interfaces critically influence the overall reliability of the structure, making their study an intriguing research subject.

In recent years, numerous experimental and theoretical studies have shed light on the factors affecting these interfaces' mechanical properties. Factors such as old concrete intensity [8], water content [9], new concrete intensity [10], the interval between new and old concrete pouring [11], interface roughness [12], surface-preparation methods [13], the use of interface agents [14], and implanted steel bars [15] have all been identified as key determinants. In addition, the size of the bonding surface also has a certain impact on the bond strength. Some studies have found that the incorporation of fibers in concrete can reduce the size effect due to the bridging effect of fibers [16]. Advanced studies proposed that new and old concrete interfaces can be considered as concrete material 100 microns wide, with their mechanical strength determined by their microstructure [17,18]. Detailed microscopic studies further illuminated these findings.

Despite these insights, most studies have focused on the tensile and shear strengths of interfaces, while the knowledge on their flexural strength remains limited [19]. Given the inevitable cold joint defects in concrete structures due to on-site construction complexities, the cold joint (the interface between new and old concrete) significantly weakens the component's flexural performance [20]. Meanwhile, under the action of bending load, the concrete bonding interface produces a complex stress state. This is expressed as partial tensile stress at the bonded surface, and the remainder is subjected to compressive stress. The areas with the greatest tensile stress may produce cracks [21]. Consequently, considering the interface's flexural strength is essential [22]. Flexural testing is considered to be a more convenient and practical bonding test by some scholars [23]. On one hand, based on the results obtained by Madani [24] and Farzad [9], flexural strength is lower than tensile strength and shear strength. It requires a greater level of attention. On the other hand, the JC/T 2381-2016 [25] specification recommends describing the bonding strength of an interface based on its flexural strength, as it better represents the interface's carrying capacity under bending and stretching. Hence, the study of the flexural strength of the interface offers immense research value.

Full lightweight ceramsite concrete is a widely used artificial lightweight aggregate concrete, renowned for its light weight, high strength [26], superior fire resistance [27,28], strong seismic performance [29], and excellent durability [30]. It also presents substantial economic benefits. Previous studies suggested that under a constant water–cement ratio, the early autogenous shrinkage of ceramsite concrete decreases as the pre-wetting humidity of ceramsite increases, due to its water absorption and desorption capacity [31]. However, the addition of ceramsite with high water absorption or an excessive amount of ceramsite may have a negative impact on the mechanical properties of ceramsite concrete, such as the tensile strength, compressive strength, and elastic modulus [32,33]. Additionally, it is considered a promising new type of cover material, particularly for the repair and reinforcement of old concrete [11,12,14]. Despite these benefits, research on full lightweight ceramsite concrete interfaced with ordinary concrete, particularly in terms of its flexural strength, has remained largely unexplored. Based primarily on the flexural strength, the effect of repairing ordinary lightweight ceramsite concrete is unknown. The influence of various factors on the flexural strength of this type of interface remains unclear, and an applicable calculation formula is yet to be established. Therefore, a comprehensive exploration of the flexural strength of this type of interface is warranted.

This paper aims to systematically investigate the flexural strength of the interface between full lightweight ceramsite concrete and ordinary concrete. To achieve this goal, the first involved measuring the flexural strength of this interface via a bending test. Subsequently, the influence law and the degree of various factors, including interfacial roughness, the type of interfacial agent, and the concrete curing age, on this strength are revealed. Finally, a calculation model for the flexural strength of the interface is established. The intention of this research is to provide a scientific reference for a concrete-repair project that utilizes full lightweight ceramsite concrete as the repair material.

2. Bending-Strength Test of Bonding Surface

2.1. Concrete Raw Materials and Mix Ratio

Ordinary Portland cement, P.O.42.5, was adopted. Its density was 3.15 g/cm^3 and its fineness was 6.5%. The mineral composition of cement clinker was C_3S, C_2S, C_3A, and C_3AF, and their contents were 45%, 25%, 12%, and 8%, respectively. The coarse aggregate used for preparing old concrete was ordinary crushed stone, and the coarse aggregate used for preparing new concrete was 900-grade crushed stone shale ceramsite. The ceramsite particles were pre-wetted and fully drained before use. Subsequently, a surface covering measure was used. River sand was used as fine aggregate in the preparation of old concrete. Among the fine aggregates used in construction of new concrete is ceramsite sand. The physical characteristics of aggregate are shown in Table 1. Before the experiment, the fine aggregate was screened through a 4.75-mm square-hole sieve, and its relevant performance parameters met the specification requirements. High-performance polycarboxylate water reducers were used in the admixture. There were four types of interface agent: cement paste, silica-fume cement paste, polymer, and epoxy resin. The type of cement used in cement slurry was the same as that of concrete. In addition, the main component of silicon fume was SiO_2, and its content was 96.74%. Its unit weight was 1600–1700 kg/m^3 and specific surface area was 20–28 g/m^2. The technical characteristics of polymer interface agent are shown in Table 2. The mixing-water type was tap water.

Table 1. Physical characteristics of aggregates.

Types of Aggregate	Particle Size (mm)	Apparent Density (kg/m^3)
Crushed stone	5~20	1517
Shale ceramsite	5~20	1760
Ordinary river sand	0~5	2641
Ceramsite sand	0~5	1479

Table 2. Technical characteristics of polymer-interface agents.

Type of Interface Agent	State	Color	Shear Bond Strength (MPa)	Tensile Bond Strength (MPa)	Solidification Time (d)
Polymer	Liquid	Milky white	1.5	1.0	1
Epoxy resin	Liquid	Milky white with a little yellow	1.6	0.8	2

The JGJ 55-2011 [34] and JGJ/T 12-2019 [35] provided specifications for the mix-ratio tests. Smooth preparation of C45 ordinary concrete and LC50 lightweight ceramsite concrete was achieved. The mix proportions are shown in Tables 3 and 4. Their compressive strength is shown in Table 5. After measurement and calculation, it was found that the density of full lightweight ceramsite concrete was 1741 kg/m^3, and its slump and gas content were 70 mm and 5.65%, respectively. The density of C45 ordinary concrete was 2417 kg/m^3, and its slump and gas contents were 75 mm and 6.71%, respectively.

Table 3. Mix proportion of C45 ordinary concrete (kg/m^3).

Gravel	Sand	Cement	Water	Water Reducing Agent
1050	565	605	230	0

Table 4. Mix proportion of LC50 full lightweight ceramsite concrete (kg/m^3).

Ceramsite	Ceramsite Sand	Cement	Water	Water Reducing Agent
620	600	550	154	5.1

2.2. *Experimental Design of Flexural Strength*

Using lightweight ceramsite concrete as a repair material, we studied the extent of the repair's recovery of flexural performance. Simultaneously, the effects of interfacial roughness, interfacial agent, and curing age of existing concrete on the flexural strength of the repaired interface were discussed. It was possible to produce 14 groups of specimens (C-3, J-1, and L-90 are the same group of test blocks) in this experiment (see Table 6). The dimension of each flexural specimen was 100 mm × 100 mm × 400 mm. First, the old concrete specimen (100 mm × 100 mm × 200 mm) was adopted from C45 ordinary concrete. In addition, the old concrete specimens were cured for 3 days, 7 days, 14 days, 28 days, and 90 days after demolding. In addition, their surfaces needed to be chiseled manually. Referring to the experimental design in the relevant research [14,36–38], the interface roughness was treated as 0 mm, 1 mm, 3 mm, and 5 mm, respectively. Subsequently, different interfacial agents were applied to the old concrete surface. Finally, one end of the 100 mm × 100 mm × 400 mm mold was placed. The LC50 full lightweight ceramsite concrete was poured into these molds for bonding to the C45 ordinary concrete specimen, as shown in Figure 1. After 28 days of standard curing (temperature 20 ± 2 °C, relative humidity above 95%), the new and old concrete specimens were successfully acquired.

Table 5. Compressive strength of C45 ordinary concrete and LC50 lightweight ceramsite concrete.

Curing Age (d)	C45 Ordinary Concrete				LC50 Lightweight Ceramsite Concrete			
	Sample 1	Sample 2	Sample 3	Average Value	Sample 1	Sample 2	Sample 3	Average Value
7	31.96	36.30	38.54	35.60	45.60	44.00	47.23	45.61
14	43.97	41.64	44.23	43.28	50.08	50.42	52.23	50.91
28	48.95	45.76	50.68	48.46	55.34	52.93	52.71	53.66

Figure 1. Bonding method of new and old concrete.

The bonding effect of new and old concrete was influenced by multiple factors. One of the most important signs of successful bonding was the excellent treatment of the bonding surface of new and old concrete. Therefore, the influence of roughness on the flexural performance of the interface between old and new concrete needs to be investigated. Using a sand-cone method, the roughness of the interface was measured, and the old concrete-base surface was treated manually with chisels. The roughness values of chiseling treatment were 0 mm, 1 mm, 3 mm, and 5 mm, respectively. The roughness of the bonding surface was calculated using Equation (1).

$$h = \frac{V}{S} \quad (1)$$

In the equation, h is the average depth of sand filling (mm), S is the surface area of the joint surface (mm^2), and V is the volume of sand filling (mm^3).

Table 6. Grouping and interfacial flexural strength of bonding specimens between full lightweight ceramsite concrete and existing concrete.

Group Number	Interfacial Agent Type	Interfacial Roughness (mm)	Existing Concrete Age (d)	Age of New Concrete (d)	Flexural Strength (MPa)				Objective
					Sample 1	Sample 2	Sample 3	Average Value	
Z-45	-	-	28	-	4.79	4.89	4.90	4.86	As a control group, it was used to evaluate the effect of the repair
Z-50			-	28	5.11	5.25	5.03	5.13	
C-0	Cement paste	0	90 + 28	28	2.89	2.85	2.84	2.86	Exploring the effect of interfacial roughness on the flexural strength of the interface
C-1		1	90 + 28	28	3.17	3.10	3.15	3.14	
C-3		3	90 + 28	28	3.33	3.28	3.14	3.25	
C-5		5	90 + 28	28	3.20	3.11	3.26	3.19	
J-0	-	3	90 + 28	28	3.06	2.98	3.02	3.02	Exploring the effect of interfacial agent types on the flexural strength of the interface
J-1	Cement paste	3	90 + 28	28	3.33	3.28	3.14	3.25	
J-2	Silica-fume cement slurry	3	90 + 28	28	3.35	3.41	3.38	3.38	
J-3	Polymer class	3	90 + 28	28	3.59	3.50	3.53	3.54	
J-4	Epoxy resin	3	90 + 28	28	3.79	3.75	3.80	3.78	
L-3	Cement paste	3	3 + 28	28	4.01	3.84	4.03	3.96	Exploring the influence of existing concrete's curing age on interfacial flexural strength
L-7		3	7 + 28	28	3.87	3.96	3.84	3.89	
L-14		3	14 + 28	28	3.72	3.52	3.74	3.66	
L-28		3	28 + 28	28	3.43	3.26	3.45	3.38	
L-90		3	90 + 28	28	3.33	3.28	3.14	3.25	

Note: Z-45 and Z-50 represent integral test specimens with strength grades of C45 and LC50, respectively. C-0, C-1, C-3, and C-5 represent test specimens with interfacial roughness of 0, 1, 3, and 5 mm, respectively. J-0, J-1, J-2, J-3, and J-4 represent test specimens without interfacial agent and coated with cement paste, silcrete slurry, polymer class, and epoxy resin interfacial agent, respectively. L-3, L-7, L-14, L-28 and L-90 represent the bonding test specimens of existing ordinary concrete with ages of 3, 7, 14, 28, and 90 days, respectively.

This experiment used a microcomputer-controlled electro-hydraulic servo universal testing machine to load. A four-point bending test was conducted to explore the influence of various factors on the flexural strength of prisms (100 mm × 100 mm × 400 mm). The loading speed of the testing machine was controlled at 0.05 MPa/s~0.08 MPa/s, and the specific loading mode is illustrated in Figure 2. The flexural strength of the interface between old and new concrete was calculated by using Equation (2).

$$f_f = \frac{Fl}{bh^2} \tag{2}$$

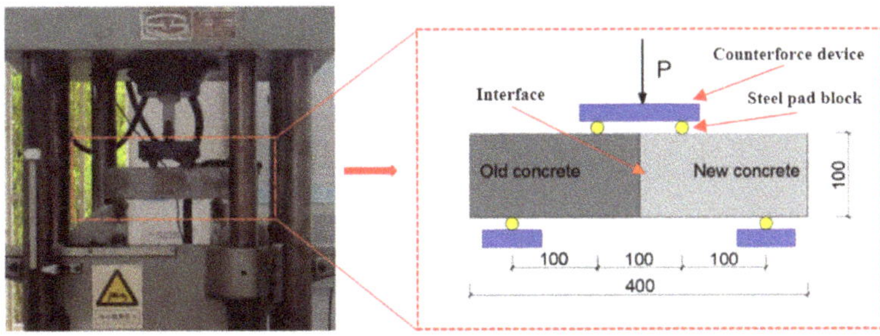

Figure 2. Bending test.

In the equation, f_f is the flexural strength of the concrete (MPa), F is the failure load of the specimen (N), l is the span between supports (mm), h is the cross-sectional height of the specimen (mm), and b is the cross-sectional width of the specimen (mm).

3. Results and Discussion

3.1. Test Phenomena and Strength

3.1.1. Experimental Phenomena of the Interface-Roughness Factor

According to Figure 3, the damage conditions of the interface at different roughness levels were different. Almost no sound was heard when the specimen with the interfacial roughness of 0 mm reached its ultimate load. Suddenly, the specimen broke at the point at which the new and old concrete interfaced. It can be observed from Figure 3a that the failure surface of the new and old concrete presented a smooth and complete state, and there was no aggregate attached to the fracture surface. When the specimen with the interfacial roughness of 1 mm reached the ultimate load, it made a slight cracking sound. Figure 3b shows the old concrete's fracture surface in black and white. The black part is mainly composed of ceramsite sand and shale ceramsite. This phenomenon showed that the interface between the new and old concrete had a bonding effect. The specimen with an interfacial roughness of 3 mm produced the sound of fine breakage during the loading process. It also produced the sound of clicking when the bearing capacity reached its limit. The specimen broke along the bonding surface of the new and old concrete. As shown in Figure 3c, more full lightweight ceramsite concrete was bonded on the failure surface of the old concrete. This shows that the new concrete and the old concrete combined very well. The failure state of the specimens with an interface roughness of 5 mm is displayed in Figure 3d. It can be seen that large areas of aggregate were exposed at the interface between old and new concrete.

3.1.2. Experimental Phenomena of Interfacial Agent Factors

The damage to the interface under different interface agents is indicated in Figure 4. The specimen without the interfacial agent did not produce a cracking sound accompanied by a uniform load increase. In contrast, the specimen coated with the interfacial agent generated a fine breaking sound. It was found that the specimen coated with the cement paste had more cement slurry on the fracture surface, as shown in Figure 4a,b. Figure 4c shows that the fracture surface had obvious black silcrete slurry. In addition, a distinct white gummy substance was observed on the failure surface, as depicted in Figure 4d,e. Simultaneously, some broken gravel could be seen on the broken torus in Figure 4e.

Figure 3. Failure states of specimens of different interfacial roughness levels: (**a**) failure state of a 0 mm-roughness specimen; (**b**) failure state of a 1 mm-roughness specimen; (**c**) failure state of a 3 mm-roughness specimen; (**d**) failure state of a 5 mm-roughness specimen.

Figure 4. Failure states of specimens with different types of interfacial agent: (**a**) failure state of specimen without interfacial agents; (**b**) failure state of specimen with cement-paste interfacial agent; (**c**) failure state of specimen with the silica-fume agent; (**d**) failure state of specimen with epoxy-resin interfacial agent; (**e**) failure state of specimen with a polymer interfacial agent.

3.1.3. Experimental Phenomena of Age Factors in Existing Concrete

Figure 5 illustrates the failure conditions of specimens with different ages of concrete. The bonds tested differed significantly between the curing times of 3 days, 7 days, 14 days, and 28 days, 90 days. The fracture surfaces of the specimens were black and white, as shown in Figure 5a–c. In addition, a large area of aggregate was exposed at the interface when specimens aged 3 days were broken. This phenomenon showed that the interface between the new and the old concrete had a certain flexural performance at this time. The failure conditions of the specimens at 28 days and 90 days were essentially the same. Simultaneously, some aggregate was exposed at the interface, as shown in Figure 5d,e.

(a) (b) (c) (d) (e)

Figure 5. Failure states of specimens at different ages of existing concrete: (**a**) failure state of specimen at 3 days; (**b**) failure state of specimen at 7 days; (**c**) failure state of specimen at 14 days; (**d**) failure state of specimen at 28 days; (**e**) failure state of specimen at 90 days.

3.1.4. Test Results

The flexural strength was used as an index to analyze the effects of the interfacial roughness, the interfacial agent, and the age of the existing concrete on the flexural performance. The specimens in Table 6 were grouped and their flexural strengths are shown.

3.2. The Effect of Repair

By calculating the data in Table 6, the flexural strength of the LC50 full lightweight ceramsite concrete was found to be equivalent to that of the C50 ordinary concrete. Its strength grade was one level higher than that of the original structure. Therefore, LC50 lightweight concrete can be used as a repair material for C50 ordinary concrete, according to GB 50367-2013 [39]. The interface between the old and the new concrete was the most vulnerable part of the structure, so the flexural strength at the interface could represent the flexural strength of the repaired structure. The group Z-45 was undamaged old concrete, with an average bending strength of 4.86 MPa. The groups of C-0~C-5 and J-0~J-4 were specimens repaired by interfacial-roughness and interface agents, respectively. Their flexural strengths ranged from 2.86 MPa to 3.96 MPa. It was seen from simple calculations that with the enhancement of the roughness, the flexural strength of the interface increased to varying degrees. It was possible to restore the flexural strength of the specimen to 59% of the undamaged specimen. The new and old concrete interfaces were coated with different interfacial agents. The flexural strength of the specimen was restored to 62%~78% of the undamaged strength.

3.3. *Influencing Factors and Rules of Flexural Strength*

3.3.1. Effect of Interfacial Roughness on Flexural Strength

It can be seen in Figure 6 that with the increase in the interfacial roughness, the flexural strength of the new and old concrete increased first and then decreased. The flexural strength of the interface reached its maximum when the interfacial roughness reached 3 mm. Moreover, a positive quadratic functional relationship between the two was observed through data fitting. The fitting relation is shown in Equation (3), and the R^2 is 0.93. It can be observed that, in Equation (3), when the roughness (average depth of the rough surface) is about 3 mm, the flexural strength reaches 3.31 MPa. Clearly, the interfacial roughness played an important role in the flexural strength of the interface. The reason for this phenomenon was that appropriate interfacial roughness can provide sufficient mechanical biting force to improve the flexural performance of the interface [40]. However, if the interfacial roughness is overtreated, microcracks occur in the near surface layer. In this way, the integrity of ab existing concrete structure is damaged and the interfacial bond strength decreases [41]. As a result, in relevant repair works, the roughness of the interface between new and old concrete should be controlled reasonably.

$$f_f = -0.05h^2 + 0.3h + 2.86 \tag{3}$$

where f_f represents the flexural strength of the interface (MPa) and h represents the interfacial roughness (mm).

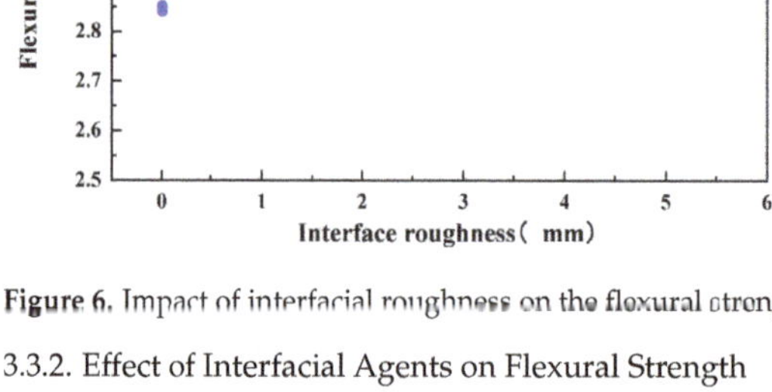

Figure 6. Impact of interfacial roughness on the flexural strength of interface.

3.3.2. Effect of Interfacial Agents on Flexural Strength

An investigation of the flexural strengths of specimens without and with various interfacial agents is presented in Figure 7. It can be observed that brushing the interfacial agent at the interface can effectively improve its flexural strength. The reason for this is that brushing the interfacial agent can improve the microstructure of the interfacial area, including the compactness, morphology, and distribution characteristics of hydration products [42]. Microlevel mechanical occlusion increases flexural strength at the interface by improving mechanical occlusion. In addition, the flexural strength of the interface was significantly improved after it was coated with an epoxy-resin interfacial agent. The use of the specimen restored its flexural strength to 77% of its pre-damage value. The influence sequence of the interfacial agents was polymer-class interfacial agent, silica-fume interfacial agent, and cement-paste interfacial agent. This is because polymer-based organic interfacial agents form films and cover aggregates and cement particles with their surface chemistry. Polymers, which form the film, prevent micro-cracks from developing and enhance their adhesion to concrete [5]. This showed that brushing the interfacial agent was an effective way to enhance the interface flexural strength.

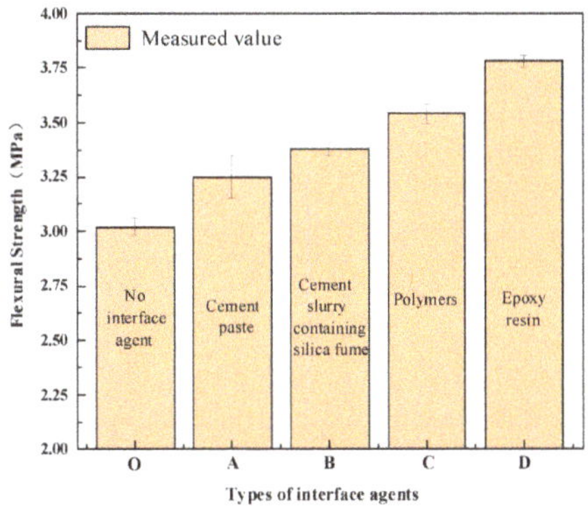

Figure 7. Impact of interfacial agents on the flexural strength of interface.

3.3.3. Effect of Curing Age on Flexural Strength of Existing Concrete

Figure 8 illustrates the effect of the differences in curing age of the existing concrete on the flexural strength of the interface between the new and the old concrete. It was found that the flexural strength of the existing concrete decreased rapidly when the curing age was less than 28 days. As the curing age continued to increase, the flexural strength decreased slowly with the growth of the curing age until it became stable. Due to the water membrane, the active ions in new concrete penetrate into the pores of existing concrete during the repair process. The existing concrete's pores are filled with reaction products when they react with cement. Eventually, a permeable layer is formed. However, with the increase in curing age, the existing concrete's porosity decreases gradually. Consequently, the permeable layer becomes thinner, reducing the adhesion between the new and the old concrete [43]. When the curing period exceeds 28 days, the concrete's strength stabilizes. It is rare for new concrete or old concrete to lose flexural strength with the aging process [44].

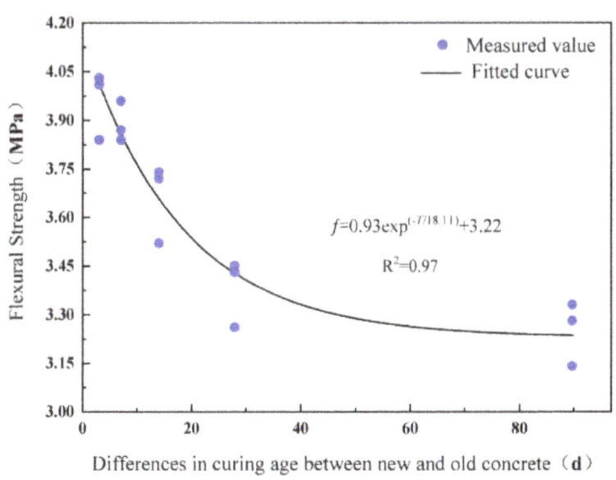

Figure 8. Impact of the differences in curing age of the existing concrete on the flexural strength of interface.

A positive negative exponential relationship was found between the two forms of concrete after fitting the data. The equation for this relationship is shown in Equation (4) and R^2 is 0.97. Therefore, the earlier the existing concrete was repaired in the project, the better the effect. The concrete's age was over 28 days, and the age of the repaired interface had little effect on its flexural strength.

$$f_f = 0.93exp^{(-T/18.11)} + 3.22 \tag{4}$$

where f_f is the flexural strength of the interface (MPa), and T represents the curing age of the existing concrete (d).

4. Analysis of Results

4.1. Degrees of Influence of Various Influencing Factors on New and Old Concrete

It can be concluded, based on the research presented above, that the flexural strength of the interfaces between the new and the old concrete was influenced by factors like the roughness, the type of interfacial agent, and the curing age. However, in actual repair projects, it might not be possible to consider these factors together for some reason. It may therefore be more helpful to guide related repair work if these factors can be quantified, as they affect the bending strength of the interface.

The difficulties above can be resolved very effectively through univariate variance analysis, based on previous findings [6,14,45]. With this method, F-values can be calculated for the interface between old and new concrete based on the impact of various factors on flexural strength. In Table 7, we present the calculation program and the results of the analysis of variance for the three influencing factors.

Table 7 clearly shows that the F values for the interfacial roughness, type of interfacial agent, and curing age of the new and old concrete were 49.37, 83.56, and 33.05, respectively. The data showed that the interfacial agent had the greatest impact on the flexural strength of the new and old concrete, followed by the interfacial roughness, and the existing concrete's curing age had the smallest impact. Therefore, it is recommended to use interfacial agents when repairing concrete in practical engineering projects.

4.2. Formula for Flexural Bearing Capacity of New and Old Concrete

Based on the analysis above, only a single factor was considered to influence the flexural strength of old and new concrete. In the actual repair work, the flexural strength of the interface between the new and the old concrete was often influenced by multiple factors. The comprehensive effects of the interfacial roughness, interfacial agents, and differences in curing ages were not considered in this experiment. In the process of repairing existing concrete, the flexural strength of new concrete can be determined through flexural tests. However, it is difficult to determine the flexural strength of old and new concrete interfaces, which requires the theory to be extended. The idea of establishing the equation while taking multiple factors into account is similar to those of Ding [46] and Huang [14]. Therefore, based on the flexural strength of the new concrete and considering three influencing factors, a formula for calculating the flexural strength of the interface between the new and the old concrete was derived.

$$f_f = K\alpha_1\alpha_2\alpha_3 f_c \tag{5}$$

In the equation, f_f is the flexural strength of the interface between the full lightweight ceramsite concrete and the ordinary concrete (MPa), K is the correction factor, α_1 is the influence coefficient of the interfacial roughness, α_2 is the influence coefficient of the interfacial agent, α_3 is the influence coefficient of the difference in curing age between the new and the old concrete, and f_c is the compressive strength (MPa) of full lightweight ceramsite concrete.

Table 7. Calculation formulas and results for each parameter.

Calculation Parameters	Calculation Formula	Interfacial Roughness	Interfacial Agent	Difference in Curing Age between New and Old Concrete
Within-group sum of squares	$SSE = \sum_{j=1}^{m}\left[\sum_{i=1}^{n_j}(x_{ij} - \bar{x}_j)^2\right]$	0.0348	0.03	0.1028
Sum of squares of deviations between groups	$SSA = \sum_{j=1}^{m} n_j(\bar{x}_j - \bar{x})^2$	0.6519	1.0029	1.3617
Within-group mean square deviation	$MSE = \frac{SSE}{n-m}$	0.0044	0.003	0.0103
Mean-square deviation between groups	$MSA = \frac{SSA}{m-1}$	0.2173	0.2507	0.3404
F value	$F = \frac{MSA}{MSE}$	49.37	83.56	33.05
$F_{0.025}(m-1, n-m)$	According to the F-distribution table	14.54	8.84	8.84
$F_{0.05}(m-1, n-m)$		8.85	5.96	5.96
$F_{0.1}(m-1, n-m)$		5.25	3.92	3.92
Level of influence	If $F > F_{0.025}$, this factor has an extremely significant impact on the flexural strength. If $F_{0.025} > F > F_{0.05}$, then this factor has a significant impact on the flexural strength. If $F_{0.05} > F > F_{0.1}$, then this factor has almost no effect on the flexural strength. If $F < F_{0.1}$, then this effect has no effect on the flexural strength.	Extremely significant impact	Extremely significant impact	Extremely significant impact

4.2.1. Influence Coefficient of Interface Roughness α_1

The 28-day-cured full lightweight ceramsite concrete had a compressive strength of 53.66 MPa, according to Table 5. It is possible to calculate the flexural strength of the interface between new and old concrete cured for 28 days with different interfacial-roughness values by using Equation (6).

$$\alpha_1 = \frac{-0.05h^2 + 0.3h + 2.86}{53.66} = -0.00093h^2 + 0.00559h + 0.0533 \tag{6}$$

where α_1 is the influence coefficient of interfacial roughness, and h is the interfacial roughness (mm).

4.2.2. Influence Coefficient of Interface Agent α_2

The values of the different interfacial agents in Table 6 were calculated by combining the data from groups J-0 to J-4. In the absence of interfacial agents, α_2 was taken as 0.0563. The α_2 was 0.0606, 0.0630, 0.0660, and 0.0704 when cement paste, silica fume, polymers, and epoxy resin were used as the interfacial agents.

4.2.3. Influence Coefficient of Difference in Curing Age between New and Old Concrete α_3

The same determination process for α_1, α_3 was obtained by a simple calculation, as shown in Equation (7).

$$\alpha_3 = \frac{0.93exp^{(-T/18.11)} + 3.22}{53.66} = 0.0173exp^{(-T/18.11)} + 0.0600 \tag{7}$$

where α_3 is the influence coefficient of the difference in curing age between new and old concrete, and T is the difference in curing age (d) between new and old concrete.

4.2.4. Determination of the Formula for Calculating the Flexural Strength of the Interface between New and Old Concrete

Firstly, Equations (6) and (7) were substituted into Equation (5) to obtain Equation (8). Based on this analysis, the K value was 266.959 and the R^2 was 0.99. A final calculation was performed using Equation (9) to determine the flexural strength at the interface between the new and the old concrete.

$$f_f = K(-0.00093h^2 + 0.00559h + 0.0533)\alpha_2(0.0173exp^{(-T/18.11)} + 0.0600)f_c \tag{8}$$

$$f_f = 266.959(-0.00093h^2 + 0.00559h + 0.0533)\alpha_2(0.0173exp^{(-T/18.11)} + 0.0600)f_c \tag{9}$$

where f_f is the flexural strength of the interface between the full lightweight ceramsite concrete and the ordinary concrete (MPa), h is the interfacial roughness (mm), and T is the difference in curing age between the new and the old concrete (d). When the interfacial agent is not used, the influence coefficient of the interfacial agent is taken as 0.0563. When using interfacial agents such as cement paste, silica-fume cement paste, polymer interfacial agent, and epoxy-resin interfacial agent, the influence coefficients of the interfacial agents are 0.0606, 0.0630, 0.0660, and 0.0704, respectively. The f_c is the compressive strength (MPa) of the full lightweight ceramsite concrete.

4.3. Calibration of Formulas

In order to calibrate the formula for calculating the flexural capacity of old and new concrete, as shown in Table 8, we collected four sets of data on the flexural properties of old and new concrete and substituted them into Equation (9). Finally, the calibrated data were compared with the original data. The data were taken from [47]. As shown in Figure 9, the change trends of the test results and calibration results were consistent. This indicated that the calibration results were accurate. However, there was a certain gap between the two flexural strength values. This may have been due to factors such as different aggregate particle sizes [48], different mix proportions [49] and different curing conditions for lightweight aggregate concrete [50]. In view of the results presented, the formula can be used in similar case studies.

Table 8. Calibration data for new and old concrete flexural capacity equations.

Group Number	Compressive Strength of Lightweight Aggregate Concrete (MPa)	Interfacial Roughness (mm)	Interfacial Agent	Curing Age (d)	Measured Value (MPa)	Calibration Value (MPa)
a	43.27	2	cement paste	28	1.03	2.71
b	53.70	1.5			1.07	2.96
c	53.70	2			1.52	3.36
d	63.50	2			2.01	3.97

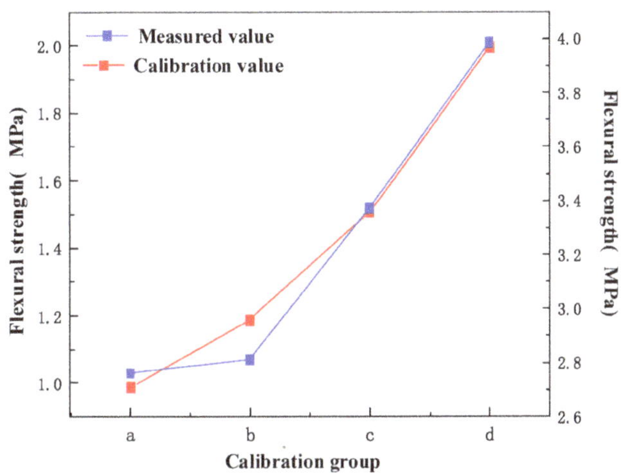

Figure 9. Calibration of flexural-strength-bearing-capacity equations for old and new concrete.

5. Conclusions

The following conclusions can be drawn from the above results and discussion.

(1) The repair of C45 ordinary concrete with LC50 lightweight ceramsite concrete can make the flexural strength of the repaired structure reach about 75% of that of an existing concrete structure. Moreover, the composite structure can fulfil relevant specifications after repair.

(2) As the interfacial roughness increased, the flexural strengths of the new and old concrete increased to varying degrees. When the roughness was 3 mm, the flexural strength reached its maximum value. However, if the treatment level of the interfacial roughness continued increase, this might have led to microcracks at the interface. This would have destroyed the integrity of the existing concrete and eventually led to a decrease in the flexural strength at the interface. Therefore, when the roughness of the new and old concrete interface was processed, the interfacial roughness should be reasonably controlled.

(3) By comparing the repair effects of different interfacial agents on the flexural strength of the interface between the new and the old concrete, it was found that the coating interfacial agent effectively improved the flexural performance of the new-and-old-concrete interface. In particular, the epoxy-resin interfacial agent had the greatest effect on the flexural strength of the interface, restoring the flexural strength of the specimen to 77% before damage occurred.

(4) The existing concrete's curing age and its interface's flexural strength exhibited a negative exponential relationship. When the curing age was less than 28 days, the flexural strength of the interface decreased rapidly. After 28 days of curing, the flexural strength decreased slowly with the increase in the age until it tended to become stable.
(5) Through the single-variable analysis of variance, it was found that the order of influence of three factors on the flexural strength of the interface was interfacial agent > interfacial roughness > existing concrete's curing age. Therefore, in the actual repair project, the use of interfacial agents to repair the interface should be given priority.
(6) The flexural strength of the interface between full lightweight ceramsite concrete and ordinary concrete was calculated using a formula. Three influencing factors were also taken into account in the formula. This study aimed to provide a scientific reference for the reinforcement of concrete projects with full lightweight ceramsite concrete.

Author Contributions: Conceptualization, H.Z. and Y.X.; methodology, H.Z.; software, X.L.; validation, Y.X. and S.W.; formal analysis, Y.X.; investigation, Y.X.; resources, Y.W. and S.W.; data curation, Y.X., Y.W. and S.W.; writing—original draft preparation, Y.X. and Y.W.; writing—review and editing, H.Z. and Y.X.; visualization, Y.W.; supervision, H.Z.; project administration, H.Z.; funding acquisition, H.Z. All authors have read and agreed to the published version of the manuscript.

Funding: This study was funded by the National Natural Science Foundation of China (grant number 52178182), Hubei Provincial Excellent Young and Middle-Aged Science and Technology Innovation Team Project of Colleges and Universities (grant number T2022002), The 14th Five Year Plan Hubei Provincial Advantaged Characteristic Disciplines (groups) Project of Wuhan University of Science and Technology (grant number 2023D0501).

Institutional Review Board Statement: Not applicable.

Informed Consent Statement: Not applicable.

Data Availability Statement: The data presented in this study are available on request from the corresponding author.

Conflicts of Interest: The authors declare no conflict of interest.

References

1. Allam, K.; Mosallam, A.S.; Salama, M.A. Experimental evaluation of seismic performance of interior RC beam-column joints strengthened with FRP composites. *Eng. Struct.* **2019**, *196*, 109308. [CrossRef]
2. Wu, C.; Li, V.C. CFRP-ECC hybrid for strengthening of the concrete structures. *Compos. Struct.* **2017**, *178*, 372–382. [CrossRef]
3. Cheong, H.K.; Macalevey, N. Experimental Behavior of Jacketed Reinforced Concrete Beams. *J. Struct. Eng.* **2000**, *126*, 692–699. [CrossRef]
4. Cao, X.; Yu, S.; Zheng, D.; Cui, H. Nail planting to enhance the interface bonding strength in 3D printed concrete. *Automat. Constr.* **2022**, *141*, 104392. [CrossRef]
5. He, Y.; Zhang, X.; Hooton, R.D.; Zhang, X. Effects of interface roughness and interface adhesion on new-to-old concrete bonding. *Constr. Build. Mater.* **2017**, *151*, 582–590. [CrossRef]
6. Huang, H.; Liu, B.; Xi, K.; Wu, T. Interfacial tensile bond behavior of permeable polymer mortar to concrete. *Constr. Build. Mater.* **2016**, *121*, 210–221. [CrossRef]
7. Liu, Y.; Wei, Y.; Ma, L.; Wang, L. Restrained shrinkage behavior of internally-cured UHPC using calcined bauxite aggregate in the ring test and UHPC-concrete composite slab. *Cem. Concr. Comp.* **2022**, *134*, 104805. [CrossRef]
8. Feng, S.; Xiao, H.; Li, H. Comparative studies of the effect of ultrahigh-performance concrete and normal concrete as repair materials on interfacial bond properties and microstructure. *Eng. Struct.* **2020**, *222*, 111122. [CrossRef]
9. Farzad, M.; Shafieifar, M.; Azizinamini, A. Experimental and numerical study on bond strength between conventional concrete and Ultra High-Performance Concrete (UHPC). *Eng. Struct.* **2019**, *186*, 297–305. [CrossRef]
10. Shaw, D.M.; Sneed, L.H. Interface shear transfer of lightweight-aggregate concretes cast at different times. *PCI J.* **2014**, *59*, 130–144. [CrossRef]
11. Wang, J.; Xiao, Z.; Zhu, C.; Feng, C.; Liu, J. Experiment on the bonding performance of the lightweight aggregate and normal weight concrete composite beams. *Case Stud. Constr. Mat.* **2021**, *15*, e00565. [CrossRef]
12. Zhu, H.; Fu, Z.; Wang, Y.; Zhang, N. Study on splitting tensile strength of interface between the full lightweight ceramsite concrete and ordinary concrete. *Case Stud. Constr. Mat.* **2023**, *18*, e01829. [CrossRef]
13. Afandi, M.; Yehia, S.; Landolsi, T.; Qaddoumi, N.; Elchalakani, M. Concrete-to-concrete bond Strength: A review. *Constr. Build. Mater.* **2023**, *363*, 129820. [CrossRef]

14. Huang, H.; Yuan, Y.; Zhang, W.; Gao, Z. Bond behavior between lightweight aggregate concrete and normal weight concrete based on splitting-tensile test. *Constr. Build. Mater.* **2019**, *209*, 306–314. [CrossRef]
15. Costa, H.; Carmo, R.N.F.; Júlio, E. Influence of normal stress and reinforcement ratio on the behavior of LWAC interfaces. *Constr. Build. Mater.* **2018**, *192*, 317–329. [CrossRef]
16. Cucuzza, R.; Aloisio, A.; Accornero, F.; Marinelli, A.; Bassoli, E.; Marano, G.C. Size-scale effects and modelling issues of fibre-reinforced concrete beams. *Constr. Build. Mater.* **2023**, *392*, 131727. [CrossRef]
17. Huang, H.; Yuan, Y.; Zhang, W.; Liu, B.; Viani, A.; Mácová, P. Microstructure investigation of the interface between lightweight concrete and normal-weight concrete. *Mater. Today Commun.* **2019**, *21*, 100640. [CrossRef]
18. Liu, S.; He, Z.; Hu, L. Interfacial microstructure between ultrahigh-performance concrete–normal concrete in fresh-on-fresh casting. *Constr. Build. Mater.* **2022**, *322*, 126476. [CrossRef]
19. Manawadu, A.; Qiao, P.; Wen, H. Characterization of Substrate-to-Overlay Interface Bond in Concrete Repairs: A Review. *Constr. Build. Mater.* **2023**, *373*, 130828. [CrossRef]
20. Deng, H.; Fu, H.; Li, H.; Shi, Y.; Huang, Z. Fracture performance and fracture characteristics of concrete members with cold joints: Numerical simulation. *Theor. Appl. Fract. Mec.* **2023**, *125*, 103868. [CrossRef]
21. Zhao, H.; Xie, Y.; Long, G.; Zhu, S.; Yang, K.; Tang, Z.; Umar, H.A.; Wu, Z. The effects of loading rates on the fracture characteristics at the bonding interface between steam concrete and self-compacting concrete using three-point bending test. *Theor. Appl. Fract. Mec.* **2023**, *125*, 103927. [CrossRef]
22. Zhao, Z.; Zhao, G.; Huang, C. Research on adhesive bending behavior of young on old concrete. *Chin. Civil Eng. J.* **2000**, *33*, 67–72. [CrossRef]
23. Li, Z.; Rangaraju, P.R. Effect of Surface Roughness on the Bond Between Ultrahigh-Performance and Precast Concrete in Bridge Deck Connections. *Transp. Res. Rec.* **2016**, *2577*, 88–96. [CrossRef]
24. Al-Madani, M.K.; Al-Osta, M.A.; Ahmad, S.; Khalid, H.R.; Al-Huri, M. Interfacial bond behavior between ultra high performance concrete and normal concrete substrates. *Constr. Build. Mater.* **2022**, *320*, 126229. [CrossRef]
25. *JC/T 2381-2016*; Repairing Mortar. Building Materials Industry Press: Beijing, China, 2016.
26. Wang, J.; Zhou, T.; Nie, S.; Zhang, X.; Shao, Y. Behavior of CFS shear walls infilled with lightweight ceramsite concrete under cyclic loading. *Structures* **2022**, *45*, 1833–1849. [CrossRef]
27. Xie, J.; Zhao, J.; Wang, J.; Huang, P.; Liu, J. Investigation of the high-temperature resistance of sludge ceramsite concrete with recycled fine aggregates and GGBS and its application in hollow blocks. *J. Build. Eng.* **2021**, *34*, 101954. [CrossRef]
28. Bao, J.; Zheng, R.; Sun, Y.; Zhang, P.; Cui, Y.; Xue, S.; Song, Q. A state-of-the-art review on high temperature resistance of lightweight aggregate high-strength concrete. *J. Build. Eng.* **2023**, *69*, 106267. [CrossRef]
29. Ma, S.; Gu, Y.; Bao, P. Seismic performance of ceramsite concrete T-shaped composite wallboard joints under cyclic loading. *Structures* **2021**, *33*, 4433–4445. [CrossRef]
30. Zhang, Y.; Hua, Y.; Zhu, X. Investigation of the durability of eco-friendly concrete material incorporating artificial lightweight fine aggregate and pozzolanic minerals under dual sulfate attack. *J. Clean. Prod.* **2022**, *331*, 130022. [CrossRef]
31. Ji, T.; Zheng, D.; Chen, X.; Lin, X.-J.; Wu, H.-C. Effect of prewetting degree of ceramsite on the early-age autogenous shrinkage of lightweight aggregate concrete. *Constr. Build. Mater.* **2015**, *98*, 102–111. [CrossRef]
32. Zhuang, Y.; Chen, C.; Ji, T. Effect of shale ceramsite type on the tensile creep of lightweight aggregate concrete. *Constr. Build. Mater.* **2013**, *46*, 13–18. [CrossRef]
33. Cui, T.; Ning, B.; Xia, X.; Zhang, C.; Jiang, Y. Shrinkage prediction model of high strength lightweight aggregate concrete based on relative humidity. *J. Build. Eng.* **2023**, *75*, 106932. [CrossRef]
34. *JGJ 55-2011*; Specification for Mix Proportion Design of Ordinary Concrete. China Construction Industry Press: Beijing, China, 2011.
35. *JGJ/T 12-2019*; Technical Standard for Application of Lightweight Aggregate Concrete. China Construction Industry Press: Beijing, China, 2019.
36. Shi, W.; Shafei, B. Bond characteristics between conventional concrete and six high-performance patching materials. *Constr. Build. Mater.* **2021**, *308*, 124898. [CrossRef]
37. Feng, S.; Xiao, H.; Ma, M.; Zhang, S. Experimental study on bonding behaviour of interface between UHPC and concrete substrate. *Constr. Build. Mater.* **2021**, *311*, 125360. [CrossRef]
38. Pereira Prado, L.; Carrazedo, R.; Khalil El Debs, M. Interface strength of High-Strength concrete to Ultra-High-Performance concrete. *Eng. Struct.* **2022**, *252*, 113591. [CrossRef]
39. *GB 50367-2013*; Code for Design of Strengthening Concrete Structure. China Construction Industry Press: Beijing, China, 2014.
40. Luo, Q.; Wang, W.; Wang, B.; Xu, S.; Sun, Z. Numerical study on interface optimization of new-to-old concrete with the slant grooves. *Structures* **2021**, *34*, 381–399. [CrossRef]
41. Courard, L.; Piotrowski, T.; Garbacz, A. Near-to-surface properties affecting bond strength in concrete repair. *Cem. Concr. Comp.* **2014**, *46*, 73–80. [CrossRef]
42. Luo, Q.; Qin, T.; Chen, Z.; Pang, B.; Qu, J.; Gao, Z. The influence of moisture and epoxy bonding agents on interfacial behavior between normal concrete substrate and ultrahigh performance concrete as a repair material: Experimental and molecular dynamics study. *Constr. Build. Mater.* **2023**, *372*, 130779. [CrossRef]

43. Xie, H.C.; Li, G.Y.; Xiong, G.J. Microstructure model of the interfacial zone between fresh and old concrete. *J. Wuhan Univ. Technol. Mater. Sci. Ed.* **2002**, *17*, 64–68. [CrossRef]
44. Hu, B.; Meng, T.; Li, Y.; Li, D.-Z.; Chen, L. Dynamic splitting tensile bond behavior of new-to-old concrete interfaces. *Constr. Build. Mater.* **2021**, *281*, 122570. [CrossRef]
45. Qiu, Y.; Sun, Q.; Feng, K.; Luo, A.; Wu, X.; Ba, M. Experimental study on the preparation of magnesium oxysulfate cement-based composites using waste locust powders to replace non-renewable magnesium oxide. *Constr. Build. Mater.* **2023**, *366*, 130129. [CrossRef]
46. Ding, F.; Wu, X.; Xiang, P.; Yu, Z. New damage ratio strength criterion for concrete and lightweight aggregate concrete. *ACI Struct. J.* **2021**, *118*, 165–178. [CrossRef]
47. Gao, Z. *Research on Bond Properties between Lightweight Aggregate Concrete and Ordinary Concrete Interface*; Chang'an University: Xi'an, China, 2019.
48. Sifan, M.; Nagaratnam, B.; Thamboo, J.; Poologanathan, K.; Corradi, M. Development and prospectives of lightweight high strength concrete using lightweight aggregates. *Constr. Build. Mater.* **2023**, *362*, 129628. [CrossRef]
49. Chi, J.M.; Huang, R.; Yang, C.C.; Chang, J.J. Effect of aggregate properties on the strength and stiffness of lightweight concrete. *Cem. Concr. Comp.* **2003**, *25*, 197–205. [CrossRef]
50. Karthik, J.; Surendra, H.J.; Prathibha, V.S.; Kumar, G.A. Experimental study on lightweight concrete using Leca, silica fume, and limestone as aggregates. *Mater. Today* **2022**, *66*, 2478–2482. [CrossRef]

coatings

MDPI

Article

Load Transfer Mechanism and Bond–Slip Behavior of Recycled Concrete-Encased Steel (RCES) Subjected to Cyclic Loading

Jun Gao [1], Guoliang Zhao [2] and Xin Zhang [1,*]

[1] School of Civil Engineering and Mechanics, Yanshan University, Qinhuangdao 066004, China
[2] Water Resources and Electric Engineering, Hebei University, Cangzhou 061001, China
* Correspondence: civil_xinzhang@163.com

Abstract: RCES composite structures, which combine the advantages of high bearing- and good anti-seismic capacity of ordinary steel reinforced concrete (SRC) with the characteristics of energy-saving and environmental protection of recycled concrete, are beneficial to promote the use of recycled concrete. However, the bond–slip behavior and load transfer mechanism are essential issues of RCES composite structures. This paper primarily focused on the load transfer mechanism and bond–slip behavior of RCES composite structures subjected to cyclic loading. A total of 14 specimens, which were designed with different replacement ratios of recycled concrete, cover thicknesses of H-shaped steel, transverse reinforcement ratios, and recycled concrete strengths, were tested to investigate the load transfer mechanism and interface damage. The results indicate that the whole loading procedure can be divided into four phases and four limit points. The bonding shear damage of a specimen develops rapidly, and most damage happened during 0~0.2 mm slip. The bonding stress values of chemical bonding stress τ_{ca}, friction resistance stress τ_f, and mechanical biting stress τ_m were calculated. Moreover, in order to reflect the influence of cyclic loading, a degradation factor ξ was proposed. The mean values of chemical bonding stress τ_{ca}: friction resistance stress τ_f: mechanical biting stress τ_m are 1:0.187:0.696, in which the chemical bonding stress is the largest, friction resistance stress is the second, and the mechanical biting stress is the least.

Keywords: recycled concrete-encased steel (RCES); bond–slip mechanism; full state analysis; interface damage mechanism; cyclic loading

Citation: Gao, J.; Zhao, G.; Zhang, X. Load Transfer Mechanism and Bond–Slip Behavior of Recycled Concrete-Encased Steel (RCES) Subjected to Cyclic Loading. *Coatings* **2022**, *12*, 1806. https://doi.org/10.3390/coatings12121806

Academic Editor: Enrico Quagliarini

Received: 5 October 2022
Accepted: 18 November 2022
Published: 23 November 2022

1. Introduction

In recent years, the construction industry has seen a rapid development. However, the rapid development of construction consumes a lot of non-renewable natural resources such as sand and stone. On the other hand, the demolition of many old buildings generates an astonishing amount of construction waste due to the rapid urbanization [1]. These wastes have a significant impact on the environment and mostly contribute to landfill saturation. In order to solve this problem, it is necessary to recycle the construction waste and reduce environmental degradation, as well as the negative effects due to construction processes [2]. Recycled concrete, which is made by using recycled coarse aggregates (RCA) to replace natural aggregate partially or completely, is believed to be a good way to solve this problem [2]. Recently, many countries have attached great importance to environmental protection and are paying great attention to research on recycled concrete. A larger number of research studies, performed worldwide, have focused on the mechanical properties or recycled concrete [3–11]. Most of the studies show that although some mechanical properties of RCA are inferior to those of natural concrete, it is appropriate for recycled concrete to be used in structural engineering [12].

In order to popularize the application of recycled concrete, studies concerning the beams [13–16], columns [17,18], beam-column joints [12,19], and frames [20,21] were investigated. The research results of most previous studies show that the anti-seismic behavior and failure mode of RCES structures are similar to those of ordinary reinforced concrete structures. However, the bearing capacity of RCES structures was reduced to some allowable extent. Nowadays, SRC structures have already been widely used in engineering construction due to their advantages, such as better seismic capacity and good ductility [22,23]. In this regard, RCES composite structures are proposed to broaden the application scope of recycled concrete [24]. It is believed that RCES composite structures, which combine the shape steel and recycled concrete, can achieve a better seismic performance, as the attractive features of SRC can make up for the shortcomings of recycled concrete. However, the bond–slip capacity is a key problem of RCES composite structures, which provide enough shear resistance at the interface and ensure that H-shaped steel and recycled concrete are working together [25,26]. Zhang et al. (2019) [27] investigated the flexural capacity of composite beams and found that about 27–35% deflection underestimation happened if the interlayer slip was neglected. Hotta et al. (1977) [28] studied the influence of the lateral restraint stress of concrete on the bond–slip behaviors between steel and concrete. Zheng et al. (2016) [29] studied the bond mechanisms and failure modes of SRRC composite structures. Bai et al. (2016) [30] studied the bond–slip capacity of RCES composite structures by using the orthogonal method and pull-out test. The results of the above investigation showed that the bond–slip failure process of RCES composite structures were similar to that of ordinary steel-reinforced concrete, both of which undergo four stress phases: no slip phase, slip phase, failure phase, and residual phase. The concrete cover thickness of H-shaped steel and recycled concrete strength have positive effects on the average bonding strength, while the replacement ratio of recycled coarse aggregates lead to an allowable decrease in average bonding strength. Chen et al. (2013) [31] designed 22 SRRC test specimens and carried out pull-out tests, in which they found that the ultimate bond stress and residual bond stress increased with the compressive recycled concrete strength. The bond stress of different parts of H-shaped steel is different; the web adhesion stress is the smallest, followed by the outer flange, and the inner flange adhesion stress is the largest. The bonding force at the interface of an SRRC specimen is exponentially distributed along the longitudinal direction of H-shaped steel. Chen et al. (2013) [32] tested 11 SRRC specimens with different replacement ratios of recycled coarse aggregates and found that the interface energy dissipation capacity of SRRC specimens is not significantly affected by the replacement ratio of recycled coarse aggregates when the replacement ratio is less than 70%. However, when the replacement ratio is greater than 70%, the energy dissipation coefficient increases with the replacement ratio. The shear stiffness of the interfacial bond decreases with the increase in the replacement ratio. Finally, a τ-s constitutive model was proposed. Most of the previous research, especially for the studies mentioned above, is focused on the investigation of bond strength and stress. There is little research on the degradation of bond capacity and no research on the full-stage analysis of load transfer mechanism at the interface between recycled concrete and steel. Studies generally indicate that the bond stress at the interface between concrete and H-shaped steel is composed of chemical bond stress, mechanical bite stress, and friction stress, but there are no quantitative calculation methods for the calculation of each part. Consequently, a series of RCES specimens were tested to investigate the load transfer mechanism, degradation of bond capacity, and to calculate the components of bond stress.

2. Test program

2.1. General

Materials used in this experiment include: (1) cement: ordinary Portland cement; (2) natural coarse aggregate (NCA): natural stones which were screened with continuous gradation from 0 to 25 mm; (3) recycled coarse aggregate (RCA): RCA comes from a recycled building material company; (4) fine coarse aggregate: the common river sand with a fineness modulus of 2.75; (5) water: city tap water; (6) water reducing agent: polycarboxylic type high performance water reducer; (7) channel steel: No. 10 U-shaped channel steel; (8) steel bar: Φ6 smooth bars were selected as the transverse reinforcement

(Φ means hot rolled plain steel bars with strength grade 300 MPa), ⏀12 ribbed steel bars were selected as the longitudinal reinforcement (⏀ means hot rolled ribbed steel bars with strength grade 335 MPa); and (9) steel plate: Q235 steel plate with thickness of 6 mm. The mechanical properties of shape steel and steel bars were tested according to the Chinese standard GB/T228, and the results are shown in Table 1.

In order to ensure the quality of recycled concrete and investigate the influence of coarse aggregate on the mechanical properties of recycled concrete, a series of material property tests have been performed on NCA and RCA. The test results show that for the NCA and RCA used in the experiment, bulk density (kg/m^3) is 1580:1330, apparent density (kg/m^3) is 2800:2570, 24(h) water absorption (%) is 0.1:4.8, crush index (%) is 5.0:13.0, porosity (%) is 44:48, and mud content (%) is 0.4:0.6. It should be noted that both the bulk and apparent density of NCA is bigger than that of RCA, which means the bulk of recycled aggregates have more voids. The 24(h)-water absorption of RCA is 48 times that of NCA. Moreover, the crush index of RCA is higher than NCA which means the strength of RCA is lower than NCA.

Table 1. Mechanical property of shape steel and steel bars.

Shape Steel and Steel Bars		Yield Strength f_y/MPa	Yield Strain $\varepsilon_y/10^{-6}$	Ultimate Strength f_u/MPa	Elastic Modulus E_s/MPa
Φ6 smooth bars		445	1878	490	2.12×10^5
⏀12 longitudinal bars		505	1927	715	2.09×10^5
Shape steel	Steel flange	359	1588	469	2.05×10^5
	Steel web	332	1450	429	2.13×10^5
	Steel plate	340	1201	445	2.15×10^5

2.2. Specimen Design

Fourteen RCES specimens were conducted, including 13 RCES cyclic loading specimens and 1 RCES pull-out loading contrast specimen. Six 150 mm × 150 mm × 150 mm concrete test cubs were cast and tested after 28 days of standard curing. Demolding of both RCES specimens and concrete cubs was carried out three days later after casting. Figure 1 shows the cross-section and elevation dimensions of typical RCES specimen.

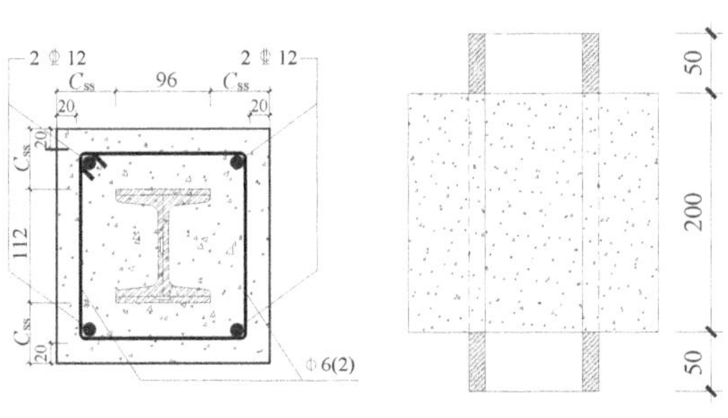

Figure 1. Design dimensions of typical RCES specimen, they should be listed as: (**a**) Cross-section of typical RCES specimen; (**b**) elevation of typical RCES specimen.

Table 2 shows the parameters investigated and details the mix properties of recycled concrete used in this paper. The effect of shrinkage was omitted due to the short-term test. The effect of steel surface treatment on adhesion was not studied in this paper and the steel used in this test was obtained directly from the factory. Four design parameters were mainly considered to investigate the bond capacity of RCES specimens as follows:

- Replacement ratio of recycled concrete (r): The chemical bonding force is an important part of the bond capacity at the interface. The magnitude of chemical bonding force depends on the amount and quality of cement colloid formed in the process of concrete mixing, curing, and finally coagulation. However, r has a great impact on the formation of cement colloid. Thus, specimens RCES-2, RCES-3, RCES-4, and RCES-5 with different r of 0%, 30%, 70%, and 100% were investigated. The other parameters remain constant.
- Thickness of recycled concrete cover (C_{ss}): The cover thickness provides much constraint to the interface and makes the slip difficult to happen. Specimens RCES-2, RCES-6, RCES-7, and RCES-8 with different concrete cover 50, 60, 70, and 80 mm were tested to investigate the relationship between bond capacity and restrain effect.
- Transvers reinforcement ratio (ρ_{sv}): The transvers reinforcement ratio also can provide constraint to the interface and improve the bond capacity to some extent. Specimens RCES-2, RCES-9, RCES-10, and RCES-11 were designed with different reinforcement ratios to study how the reinforcement improved the bond stress transferred at the interface.
- Compressive recycled concrete strength (f_c): The same as r, different concrete strengths require different mix proportions; different mix proportions influence the formation of cement colloid; which as a result affects the chemical bond stress. Therefore, some other factors, such as aggregate size, the type of cement, the additives and admixtures were omitted as the compressive strength was considered. In this test, specimens RCES-2, RCES-12, RCES-13, and RCES-14 with different compressive strengths of 31.90 MPa, 40.32 MPa, 45.44 MPa, and 50.05 MPa, respectively, were studied.

Table 2. Mix property and design details of test specimen.

Specimen	W/C	S_p (%)	Material Weight (kg·m^{-3})							r (%)	C_{ss} (mm)	ρ_{sv} (%)	f_c (MPa)	f_{cu} (MPa)
			Cement	W	RCA	NCA	S	Added Water	Water Reducer					
RCES-01	0.45	36	406	183	1158	0	651	55.6	4.07	100	60	0.77	C40	40.32
RCES-02	0.45	36	406	183	1158	0	651	55.6	4.07	100	60	0.77	C40	40.32
RCES-03	0.45	36	406	183	0	1158	651	–	4.07	0	60	0.77	C40	42.96
RCES-04	0.45	36	406	183	347.4	810.6	651	16.68	4.07	30	60	0.77	C40	41.44
RCES-05	0.45	36	406	183	810.6	347.4	651	38.92	4.07	70	60	0.77	C40	41.06
RCES-06	0.45	36	406	183	1158	0	651	55.6	4.07	100	50	0.77	C40	40.32
RCES-07	0.45	36	406	183	1158	0	651	55.6	4.07	100	70	0.77	C40	40.32
RCES-08	0.45	36	406	183	1158	0	651	55.6	4.07	100	80	0.77	C40	40.32
RCES-09	0.45	36	406	183	1158	0	651	55.6	4.07	100	60	0.62	C40	40.32
RCES-10	0.45	36	406	183	1158	0	651	55.6	4.07	100	60	1.23	C40	40.32
RCES-11	0.45	36	406	183	1158	0	651	55.6	4.07	100	60	1.54	C40	40.32
RCES-12	0.47	37	389	183	1151	0	676	55.3	3.90	100	60	0.77	C30	31.90
RCES-13	0.40	36	450	180	1132	0	637	54.4	4.50	100	60	0.77	C45	45.44
RCES-14	0.35	35	505	177	1116	0	601	53.6	5.06	100	60	0.77	C50	50.05

Note: W represent water; C represent Portland cement; S_p represent sand ratio; r represents replacement ratio of recycled concrete; C_{ss} represent cover thickness of H-shaped steel; ρ_{sv} represent transvers reinforcement ratio; f_c represent the design compressive strength of recycled concrete.

2.3. Loading Procedure and Instrumentation

This test was conducted at the key laboratory of structural engineering and earthquake resistance in Xi'an University of Architecture and Technology. The test setup and a typical RCES specimen were illustrated in Figure 2. Four linear variable displacement transducers (LVDTs) were placed vertically on the concrete splint at the end of recycled concrete (two at the top and two at the bottom) to measure the relative bond–slip along the vertical direction, as the slip was distributed unevenly along the longitudinal direction. In order to measure the inner slip of RCES specimens, two adhesive slip sensors were placed on the one-third place outside of the steel flange. The load machine, LVDTs, and adhesive slip sensors were calibrated before used. The loading device is mainly composed by eight parts: (1) fixture; (2) high-strength screw; (3) adhesive slip sensor; (4) splint; (5) high-strength bolt; (6) LVDT; (7) recycled concrete; and (8) H-shaped steel. The top of RCES specimen is defined as the loading end, while the bottom of the RCES specimen where H-shaped steel is fixed by the fixture and cannot move is defined as the fixed end, as presented in Figure 2. The H-shaped steel and recycled concrete were fixed by the bottom fixture and loading steel frame separately. As a result, the shear load at the interfacial transferred from concrete to steel through the bond effect.

The displacement control loading procedure was used in this test, as shown in Figure 3. The cyclic loading was loaded at a rate of 0.1 mm/s. Each level of displacement was loaded for three cycles to study the interface damage degradation. An amount of 1~2 kN was loaded at the beginning to keep all parts in close contact. Afterwards, the initial load was removed, and the loading was begun. The loading was terminated once 10 mm of relative slip was recorded at the fixed end.

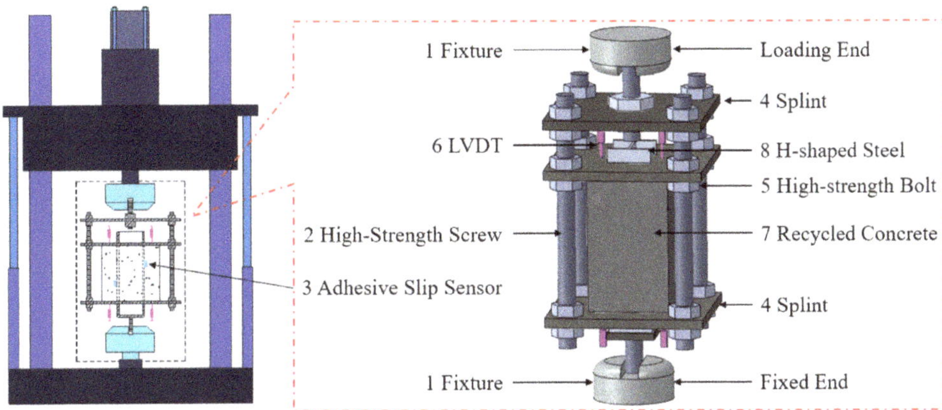

Figure 2. Test setup and loading specimen.

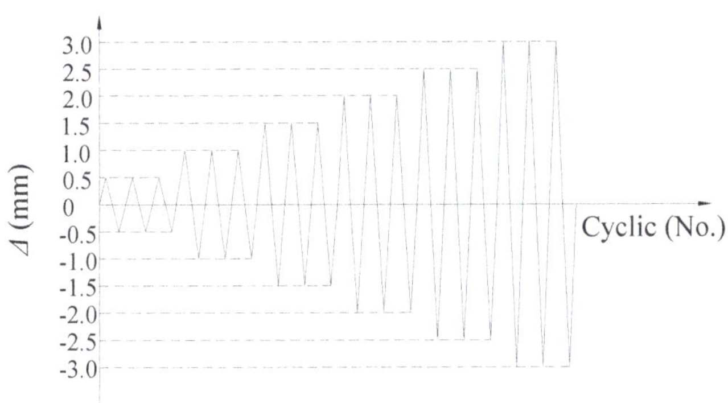

Figure 3. Loading procedure.

The distribution of strain gauges and adhesive slip sensors is presented in Figure 4. The measuring points are arranged by attaching strain gauges in the slot to prevent the strain gauges from being damaged during the loading process. Therefore, a composite H-shaped steel which is made up of two No. 10 channel steels and two 6 mm steel plates was used in this test.

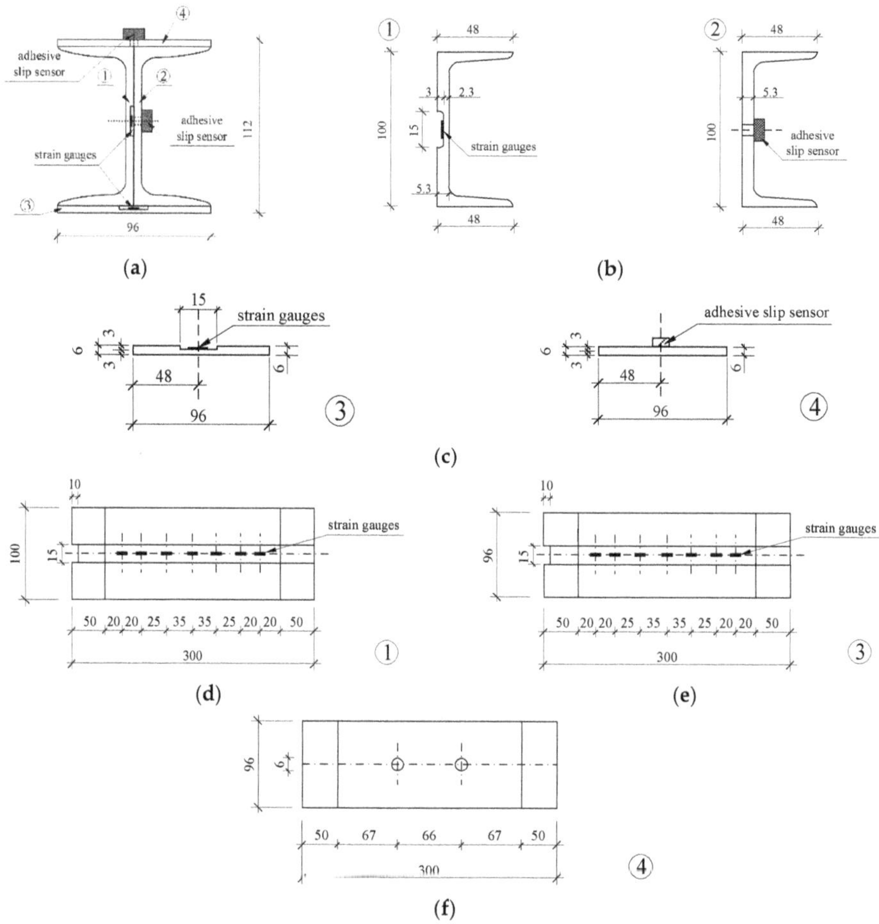

Figure 4. Measure arrangement of RCES specimen, they should be listed as: (**a**) Cross-section of H-shaped steel (mm); (**b**) Cross-section of channel steel (mm); (**c**) Cross-section of steel plate (mm); (**d**) Layout of strain gauges on web of channel steel (mm); (**e**) Layout of strain gauges on steel plate (mm); (**f**) Layout of adhesive sensors on steel plate (mm).

3. Results and Analysis

3.1. The Whole Loading Process

At the beginning of loading process, there was little damage at the bonding interface, the slip at both the top and bottom surface of specimen was very small, and the load increased dramatically. In this state, the transferred shear force at interface was mainly carried by chemical adhesive force and static friction resistance. However, the chemical adhesive force plays a major role. When the cyclic loading increased to 30~40% of ultimate load, the slip at the load end and fixed end gradually increased, specimen RCES-06, RCES-09, RCES-10, and RCES-12 showed obviously slip. As the cyclic load increased gradually, the other specimens begin to slip one after another. In contrast, RCES-03 and RCES-14 begin to slip later. This is because RCES-03 has a lower r and RCES-14 has a higher recycled concrete strength, both lower r and higher f_c affects the formulation of chemical adhesive stress and make it more difficult to slip at the interface. The slip increases with the increase in cyclic loading. At this stage, most H-shaped steel flanges showed concrete powder that was ground due to the sliding at the interface as presented in Figure 5b,d,e. The concrete covers of a few specimens were crushed and peeled off, as shown in Figure 5h. As the bond damage increased, the cyclic loading began slow down. The chemical bonding stress was completely damaged until the cyclic loading reached to the peak value. Therefore, the shear force at interface was carried by the frictional resistance and mechanical bite stress. Then, the cyclic load decreased rapidly and the slip increased correspondingly. In the end, the cyclic load remained the same, and the bond interface was completely destroyed. As the anchorage length of H-shaped steel is small, no bonding cracks occurred in this test. Most specimens generally failed due to the pull of the steel out of the concrete. Some typical failure modes which happened during the loading process are shown in Figure 5.

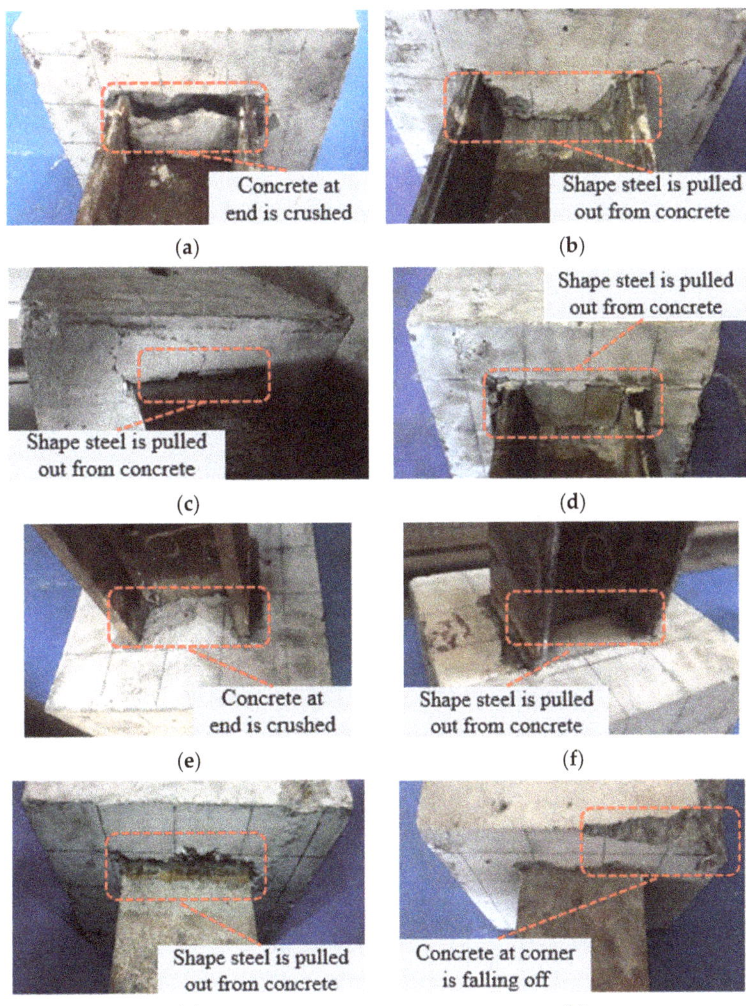

Figure 5. Typical failure modes of specimen, they should be listed as: (**a**) RCES-02; (**b**) RCES-03; (**c**) RCES-05; (**d**) RCES-08; (**e**) RCES-09; (**f**) RCES-10; (**g**) RCES-12; (**h**) RCES-13.

3.2. Analysis of Loading Procedure

The whole failure process of RCES specimens can be divided into two types by analyzing the full bond–slip skeleton curve, as shown in Figure 6. Four characteristic points were defined according to the variation of bonding force.

Micro slip point (A): The characteristic point A was defined using the geometric method as shown in Figure 7. Initially, a tangent line was made at point O which intersects the horizontal line BE at point E, then a vertical line was made through the E point that intersects the curve at point A, which is the micro slip point of the bond–slip curve. The bond–slip curve shows a linear relationship before point A [33]. The chemical bond stress reaches the ultimate value at point A.

Ultimate point (B): The peak point which the bond–slip curve reaches. As the cyclic loading increased gradually, the chemical adhesive stress was damaged completely. The transferred shear force at the interface is carried by the frictional resistance and mechanical bite stress.

Residual point (C): A line was drawn connecting point B and D. The point where the slope of tangent line in the bond–slip curve is equal to the slope of line BD is defined as the residual point C, as shown in Figure 7.

Failure point (D): The point reaches to 10 mm relative slip is defined as point D.

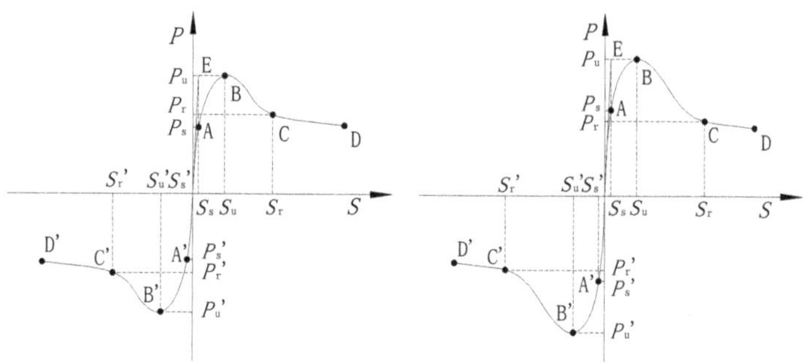

(a) (b)

Figure 6. Failure process of bond–slip skeleton curve, they should be listed as: (**a**) $P_s < P_r$; (**b**) $P_s > P_r$.

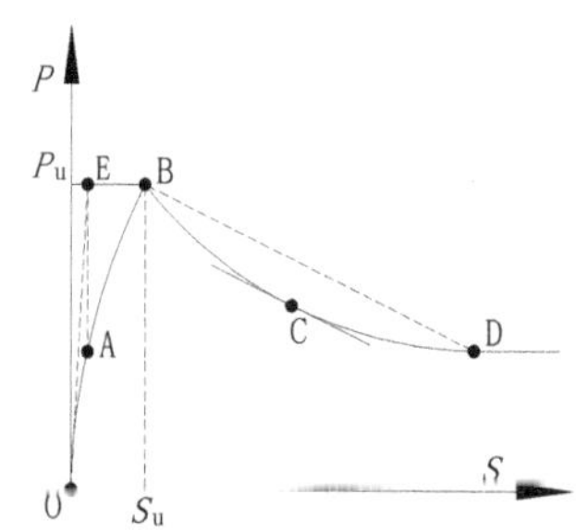

Figure 7. Method to defined characteristic point.

Correspondingly, there are four stages of the whole bond–slip curve: (1) micro slip phase (O–A); (2) slip development phase (A–B); (3) slip increases rapidly phase (B–C); and (4) residual phase (C–D).

(1) Micro slip phase (O–A): The cyclic load increases linearly with the slip from 0. As shown in Figure 8a, the bonding stress at the interface increases gradually with the cyclic load. However, the cyclic load applied in this phase is relatively small; the transferred shear force at the interface is mainly carried by the chemical bonding stress. As the cyclic load continues increasing, the bonding stress at a certain point of the interface increases to the limit value of chemical bonding stress. At this time, the specimen is in a critical slip state, as shown in Figure 8b. With the increase in cyclic load, the cement colloid was damaged and obvious slip happened at the interface.
(2) Slip development phase (A–B): In this phase, the bond–slip curve changes from a straight line to nonlinear line. Moreover, part of cement colloid was sheared and more damaged happened as shown in Figure 8c. The bonding slip zone diffused from fixed and loading end to the middle. The cement colloid damaged increased gradually so as to the slip zone as shown in Figure 8d. The chemical adhesive stress decreases gradually, while the mechanical bite stress and frictional resistance stress increase.
(3) Slip increases rapidly phase (B–C): As shown in Figure 8e, the cement colloid on the interface will be completely damaged after the cyclic load reached to P_u. With the increase in cyclic loading, the chemical adhesive stress was lost, and the axial load dropped steeply.
(4) Residual phase (C–D): In this phase, a concrete failure layer was formed as shown in Figure 8f. As the slip increased, the failure layer gradually grinds smooth. The mechanical bite stress gradually decreases and disappears, while the transferred shear force at the interface is carried by friction resistance stress.

Based on the analysis above, the bond stress, slippage and values of characteristic points were obtained and summarized in Table 3.

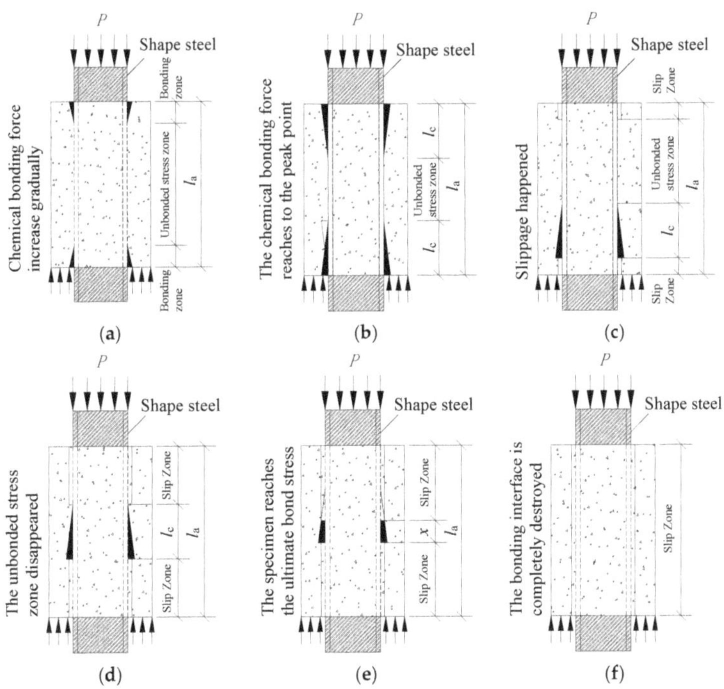

Figure 8. Failure process of specimen, they should be listed as: (**a**) Bonding stress increase; (**b**) slip critical state; (**c**) slippage happened; (**d**) the unbonded stress zone disappeared; (**e**) ultimate bonding state; (**f**) residual state.

Table 3. Values of characteristic bond strength.

Specimen	P_s (kN)	τ_s (MPa)	S_s (mm)	P_u (kN)	τ_u (MPa)	S_u (mm)	P_r (kN)	τ_r (MPa)	S_r (mm)
RCES-01	141.90	1.29	0.074	150.70	1.37	0.521	71.50	0.65	5.131
RCES-02	35.20	0.32	0.018	89.10	0.81	0.483	36.62	0.33	4.144
RCES-03	46.20	0.42	0.024	103.40	0.94	0.369	42.90	0.39	3.760
RCES-04	44.00	0.40	0.023	96.80	0.88	0.406	39.69	0.36	4.036
RCES-05	39.60	0.36	0.021	92.40	0.84	0.435	37.45	0.34	4.763
RCES-06	30.80	0.28	0.016	81.40	0.74	0.509	34.13	0.31	3.825
RCES-07	36.30	0.33	0.019	93.50	0.85	0.464	40.47	0.37	4.964
RCES-08	38.50	0.35	0.019	94.60	0.86	0.413	40.82	0.37	4.841
RCES-09	30.80	0.28	0.016	89.10	0.81	0.448	34.48	0.31	3.985
RCES-10	40.70	0.37	0.021	89.10	0.81	0.399	38.51	0.35	4.375
RCES-11	47.30	0.43	0.025	90.20	0.82	0.382	41.80	0.38	4.406
RCES-12	29.70	0.27	0.016	73.70	0.67	0.393	33.02	0.30	4.830
RCES-13	37.40	0.34	0.020	94.60	0.86	0.446	40.71	0.37	4.295
RCES-14	44.00	0.40	0.023	97.90	0.89	0.481	41.80	0.38	4.356

3.3. Interface Damage Analysis

In order to quantitatively describe and analyze the damage degree of bonding interface between concrete and steel, the secant stiffness K_i was used to define the damage factor D. The damage factor D can be calculated as Equation (1).

$$D = (K - K_i)/K \tag{1}$$

where: K is the initial bonding stiffness of specimen; D is the damage factor of specimen, $D = 0$ means there is no damage happened at the bonding interface; $D = 1$ indicates that the bonding interface is destroyed. The larger the value of D, the more serious the damage the bonding interface.

Figure 9 shows a relationship curve between damage factor and the slip value of a typical specimen. The whole bond–slip damage process can be divided into three phases according to the D–S curve. (1) Linear rapid damage phase (0~0.6): in this phase, the damage of specimen develops rapidly. D–S curve presenting a vertical line, D increases linearly with S. (2) Damage accumulation transition phase (0.6~0.9): damage factor D changes from linear to nonlinear and the development rate of damage slows down gradually. In this phase, the damage of the interface is serious, and the development of damage is basically complete. Usually, some bonding and splitting cracks will occur in this phase. (3) Damage stabilization phase (0.9~1.0): D–S curve shows a horizontal line parallel to x axis. Damage factor D does not change, and no new damage appeared.

Figure 10 shows the influence of different factors to D–S curves. (1) Replacement ratio of recycled concrete: r has a slight effect on the damage development of bonding interface as shown in Figure 10a. The D–S curves with different r will be different at the damage accumulation transition phase. This is because in this phase, most of the cement colloid on the bonding interface is broken, and at the same time, the recycled concrete appears to have bonding cracks, bonding splitting damage. (2) Cover thickness of H-shaped steel: at the initial damage stage, the D-S curves with different C_{ss} are basically the same. (3) Transvers reinforcement ratio: the transvers reinforcement ratio has little influence on the damage development of the bonding interface. At the initial loading stage, stirrups basically have little effect on the inhibition of damage. After the recycled concrete was cracked and damage, the stirrups provided lateral binding force to the recycled concrete and inhibited their mutual sliding. Stirrups enhanced the friction stress and mechanical bite stress, but had little influence on the damage development. (4) Strength of recycled concrete: the f_c has a great effect on the damage development of the bonding interface. Specimens with higher f_c had damage appear later. At the same damage degree, the specimen with higher strength had a small slip.

This is mainly because the higher the f_c is, the denser the concrete inside, the closer the bond between steel and concrete, and the interface is more difficult to damage.

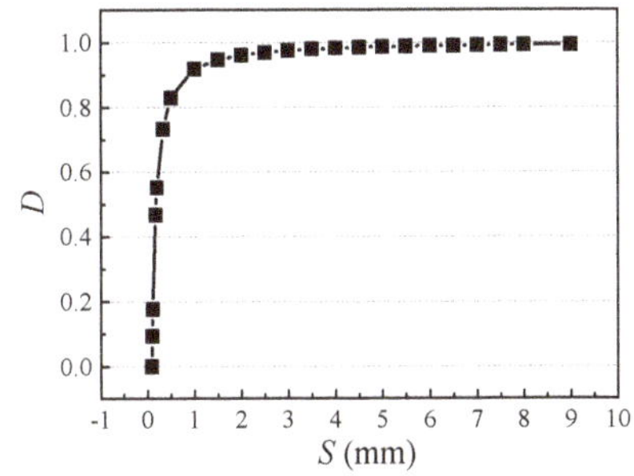

Figure 9. Damage factor vs. slip value curve.

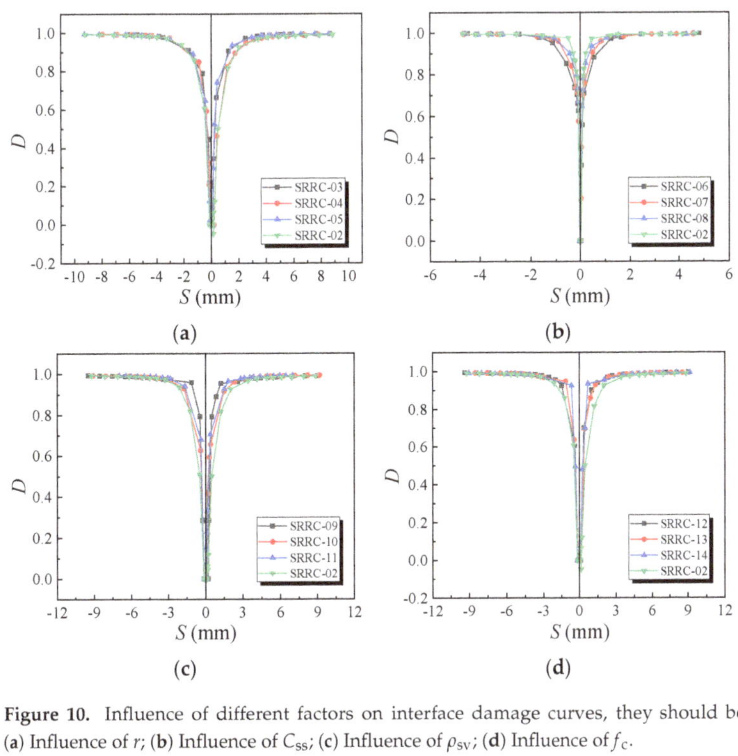

Figure 10. Influence of different factors on interface damage curves, they should be listed as: (**a**) Influence of r; (**b**) Influence of C_{ss}; (**c**) Influence of ρ_{sv}; (**d**) Influence of f_c.

4. Composition and Solution of Bonding Stress

4.1. Distribution of Bonding Stress

The bond between steel and recycled concrete is composed of chemical bonding stress τ_{ca}, friction resistance stress τ_f and mechanical biting stress τ_m. The chemical bonding stress comes from the chemical adsorption on the bonding interface of cement colloid in recycled concrete after water absorption and hardening, which mainly exists in the area where there is no relative sliding on the bonding interface. When the bonding interface is relatively sliding, the cement colloid is sheared, and the chemical bonding force is greatly reduced. The friction resistance stress is the friction resistance caused by the non-smooth surface after the sliding. The frictional resistance is mainly determined by the normal stress and friction coefficient on the bonding interface, and is mainly related to the transverse constraints such as the cover thickness of H-shaped steel, transverse reinforcement ratio, and the smoothness of the steel surface. The mechanical biting stress is mainly caused by the relative sliding of the bonding interface, and a layered concrete debris is generated after the cement colloid is sheared. The furrow effect occurs when relative slide happens. The mechanical biting stress depends mainly on the transverse restraint and the smoothness of the steel surface.

During the loading process, the bonding stress at the interface is not evenly distributed along the embedded length of the specimen. In order to analyze the stress transfer process of the bonding interface in detail, the embedded length l_a is divided into the slip zone and non-slip zone, as shown in Figure 11. The slip zone is the A region where the slip has already occurred, and the length is l_s. The non-slip zone is the region where no slip occurs, including the bonding diffusion zone B (the length of the bonding diffusion zone is defined as l_c) and the non-bonded stress zone C. Bonding diffusion zone B refers to a bonding diffusion length l_c which always exists in the non-slip zone during the process of shear transfer at the bonding interface. The bonding diffusion zone is located between the slip zone A and the no bonding stress zone C. Zone B moves from both ends of specimen to the middle with the repeated application of load until the bonding interface is completely destroyed.

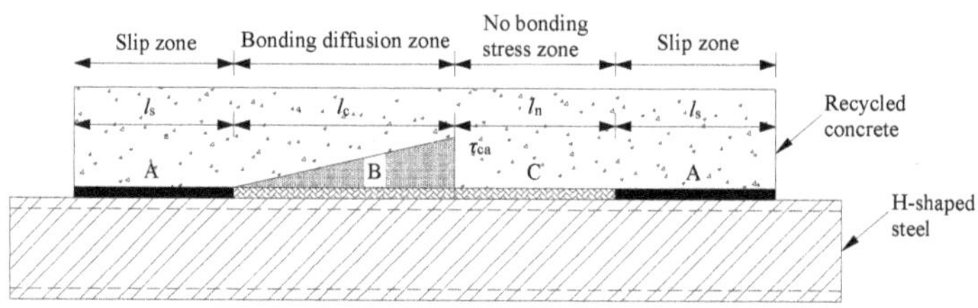

Figure 11. Division of the bonding interface.

4.2. Calculation of Bonding Stress

Cyclic loading has a great influence on the bonding stress at the interface. In the process of cyclic loading, continuous loading and unloading causes damage to the bonding interface, resulting in constant changes in the state of bonding interface, which affect the calculation of interface bonding stress. According to the test results of cyclic loading, the bonding strength of the specimen before reaching the peak value has a small degradation and basically presents a straight line, that is, the cyclic loading has a small effect on the bonding stress before peak load. Therefore, it can be approximated that cyclic loading has no effect on the chemical bonding stress τ_{ca}, friction resistance stress τ_f, and mechanical biting stress τ_m before peak loading. However, after peak load, especially in the region between the peak point and residual point, the bonding strength of the specimen degrades greatly, and the cyclic load has a great influence on the bonding stress of the interface. In this phase, the contact surface becomes smooth gradually due to repeated grinding, and the restraint effect of the lateral recycled concrete and transverse stirrups gradually weakens, which leads to the decrease in friction resistance stress τ_f and mechanical biting stress τ_m. Therefore, it can be considered that the mechanical biting stress and friction resistance stress of the bonding interface are weakened by cyclic loading, while the chemical bonding stress is not affected. In order to reflect the influence of cyclic loading on bonding stress, a degradation factor of bonding stress ξ was proposed. ξ can be calculated by Equation (2). Then the bonding stress composition of different characteristic loading points have the relationships shown in Equations (3) and (4).

$$\xi = \frac{\tau_u - \tau_r}{\tau_u} \tag{2}$$

$$\tau_{ca} = \tau_{ca}^{s} = \tau_{ca}^{u} \tag{3}$$

$$(\tau_f + \tau_m) = (\tau_f^s + \tau_m^s) = (\tau_f^u + \tau_m^u) = \xi(\tau_f^r + \tau_m^r) \tag{4}$$

where: τ_{ca}^{s} is the chemical bonding stress in micro slip phase; τ_{ca}^{u} is the chemical bonding stress at peak load; $(\tau_f^s + \tau_m^s)$ is the friction resistance stress and mechanical biting stress in micro slip phase; $(\tau_f^u + \tau_m^u)$ is the friction resistance stress and mechanical biting stress at peak load; and $(\tau_f^r + \tau_m^r)$ is the friction resistance stress and mechanical biting stress in residual phase.

Based on the analysis of Section 3.2, the load–slip skeleton curve can be divided into four different characteristic loading points and four loading phases. Correspondingly, there are five critical states in the whole loading process: (a) No damage stage; (b) A state; (c) B state; (d) C state; and (e) D state, as shown in Figure 12.

At the origin point O, the load was just started to load, and no shear force is transferred on the interface. There is no damage occurred in the contact surface and the bond stress distribution is shown in Figure 12a. with the axial load increase, the chemical bonding stress τ_{ca} and axial load balance each other appear in the bonding diffusion zone at the ends of specimen as shown in Figure 12b. With the application of cyclic loading, the bonding diffusion zone moves from both ends to the middle. The axial load increase linearly with the increase in slip zone. When the axial load increases to point A, the bonding diffusion zone has already moved to the middle of the specimen, and it is about to slide. The bonding stress distribution of the interface is shown in Figure 12c. At this critical point, the shear force transferred by the bonding interface is borne by the chemical bonding stress τ_{ca} in the bonding diffusion zone, friction resistance stress τ_f, and mechanical biting stress τ_m in slip zone. Based on the balance condition of force, the bonding stress at the interface can be calculated by Equation (5).

$$P_s = (\tau_f + \tau_m) \cdot (l_a - l_c) \cdot C_a + \int_0^{l_c} \frac{\tau_{ca}}{l_c} \tau C_a d\tau \tag{5}$$

where: l_a is the embedded length of H-shaped steel; l_c is the length of bonding diffusion zone; and C_a is the perimeter of steel cross-section.

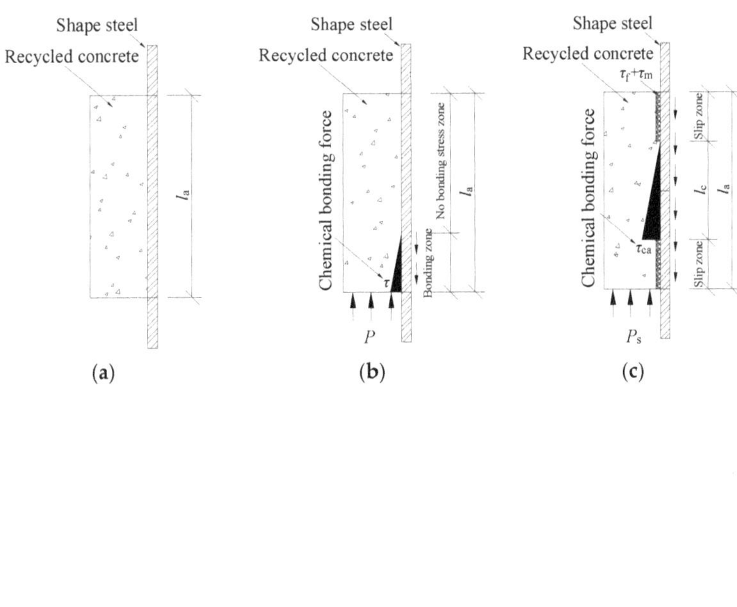

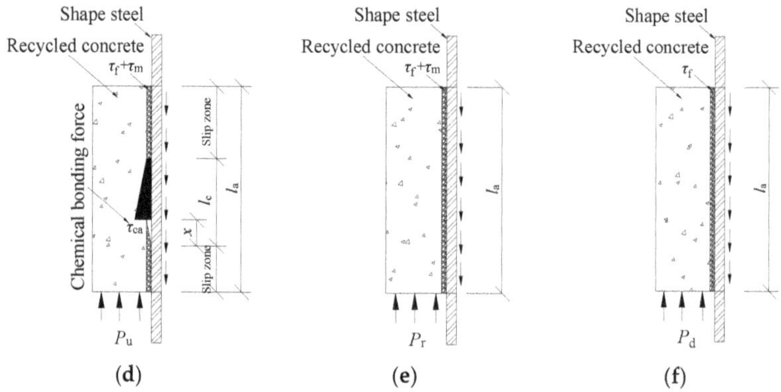

Figure 12. Diagram of specimen under different critical states, they should be listed as: (**a**) No damage stage; (**b**) Stress development state; (**c**) A state; (**d**) B state; (**e**) C state; (**f**) D state.

With the axial load increase, slip occurred in the bonding diffusion zone and the load–slip curve showed nonlinear growth. The increase rate of the axial load with the increase in slip zone slowed down. When the axial load increases to point B, part of the bonding diffusion zone has already slipped, the chemical bonding stress is reduced, while the friction resistance stress and mechanical biting stress are correspondingly increased. The area enclosed by the bonding stress curve and the embedded length of interface is the largest when the combined force of friction resistance stress and mechanical biting stress is equal to that of chemical bonding stress. At this time, the specimen reaches the peak value P_u. The relationship between peak load and bonding stress are shown in Equations (6)–(9).

$$P_u = (\tau_f + \tau_m) \cdot [l_a - (l_c - x)] \cdot C_a + \int_x^{l_c} \frac{\tau_{ca}}{l_c} \cdot \tau \cdot C_a d\tau \quad (6)$$

$$P_u = (\tau_f + \tau_m) \cdot (l_a - l_c + 0.5x) \cdot C_a + 0.5 \cdot \tau_{ca} \cdot l_c \cdot C_a \quad (7)$$

$$\frac{x}{l_c}\tau_{ca} = (\tau_f + \tau_m) \quad (8)$$

$$x = (\tau_f + \tau_m) \cdot l_c / \tau_{ca} \quad (9)$$

where: x is the sliding length of bonding diffusion zone in B critical state.

After B critical state, the bonding diffusion zone slips quickly, and the load drops rapidly. When the axial load increases to point C, all the bonding interfaces are destroyed, and the bonding diffusion zone disappears. The length of slip zone ls is equal to the embedded length of steel l_a, and the chemical bonding stress completely disappears. At the C state, the relationship between axial load and bonding stress are shown in Equation (10).

$$P_r = \xi_r(\tau_f + \tau_m) \cdot l_a \cdot C_a \quad (10)$$

where: ξ_r is the bond stress degradation factor corresponding to residual points.

The bond interface damage is basically complete after C critical state. When the axial load increases to the critical point D, the bonding interface is ground very smooth, and the mechanical biting stress generated by the furrow phenomenon is lost. The shear force transferred by the bonding interface is only borne by the friction resistance stress. Therefore, the relationship between the axial load and bonding stress is shown in Equation (11).

$$P_d = \xi_d \tau_f \cdot l_a \cdot C_a \quad (11)$$

where: ξ_d is the bond stress degradation factor corresponding to failure points.

The chemical bonding stress τ_{ca}, friction resistance stress τ_f, and mechanical biting stress τ_m can be obtained by substituting the characteristic bond strength into Equations (5)–(11), and the results are shown in Table 4.

Table 4. Composition and solution of bonding stress.

Specimen	l_c (mm)	l_c/l_a	x (mm)	τ_{ca} (MPa)	τ_m (MPa)	τ_f (MPa)	τ_m/τ_{ca}	τ_f/τ_{ca}
SRRC-02	190.72	95%	168.42	0.604	0.100	0.443	0.165	0.733
SRRC-03	179.92	90%	145.42	0.773	0.075	0.560	0.097	0.724
SRRC-04	182.99	92%	141.60	0.747	0.109	0.479	0.146	0.641
SRRC-05	191.76	96%	152.25	0.687	0.094	0.460	0.137	0.670
SRRC-06	181.92	91%	179.83	0.503	0.142	0.363	0.282	0.723
SRRC-07	181.88	91%	174.30	0.596	0.123	0.476	0.207	0.799
SRRC-08	174.36	87%	166.47	0.613	0.094	0.510	0.154	0.833
SRRC-09	223.92	112%	188.64	0.596	0.089	0.422	0.149	0.708
SRRC-10	162.31	81%	144.36	0.631	0.088	0.482	0.140	0.764
SRRC-11	126.47	64%	123.95	0.621	0.109	0.510	0.176	0.820
SRRC-12	179.40	90%	167.05	0.485	0.118	0.371	0.244	0.765
SRRC-13	176.20	88%	172.25	0.597	0.123	0.479	0.207	0.803
SRRC-14	168.22	84%	146.41	0.699	0.145	0.474	0.207	0.678
Average	178.01	89%	152.67	0.685	0.129	0.460	0.187	0.696

4.3. Parameter Analysis

4.3.1. Replacement Ratio of Recycled Concrete

Figure 13a shows the relationship between l_c/l_a (the length of bonding diffusion zone vs. the embedded length of steel) and the replacement ratio of recycled concrete under the same conditions of other analysis parameters. It can be seen from Figure 13a that the length of the bonding diffusion zone accounts for a large proportion of the embedded length of the specimen, reaching to more than 90%. This is partly due to the smaller embedded length designed in this test. l_c/l_a increases with the replacement ratio of the recycled concrete, but not obviously, which indicates that the length of bonding diffusion zone with different replacement ratios of recycled concrete are not the same. The length of the bonding diffusion zone of ordinary concrete is the smallest, which indicating that the cement colloid formed by ordinary concrete has a better adhesive effect. The cement colloid effect gradually decreased with the increase in the replacement ratio.

Figure 13b shows the influence of the replacement ratio on the chemical bonding stress and friction resistance stress. Both chemical bonding stress and friction resistance stress decrease with the increase in replacement ratio. However, the replacement ratio has little effect on friction resistance stress.

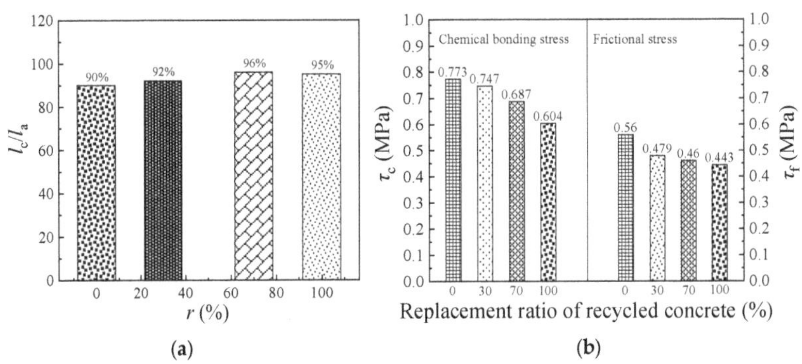

Figure 13. The effect of replacement ratio of recycled concrete on the bonding capacity, they should be listed as: (**a**) The effect of replacement ratio of recycled concrete on l_c/l_a; (**b**) effect on the chemical bonding stress and friction resistance stress.

4.3.2. Cover Thickness of Shape Steel

The influence of cover thickness of shape steel on the length of bonding diffusion zone vs. the embedded length of steel is shown in Figure 14a. It can be seen from Figure 14a that the l_c/l_a increases fist then decreases with the increase in cover thickness of the shaped steel. The maximum l_c/l_a value is 95% in SRRC-02, and the minimum l_c/l_a value is 87% in SRRC-08. Within a certain range, increasing the cover thickness of shape steel is beneficial to the bonding performance of the bonding interface.

Figure 14b shows the influence of the cover thickness of shape steel on the mechanical biting stress and friction resistance stress. The cover thickness of shape steel has a more significant influence on the friction resistance stress. The friction resistance stress gradually increases with the increase in the cover thickness of shape steel. The friction resistance stress corresponding to the four cover thicknesses (50, 60, 70 and 80 mm) are 0.363 MPa, 0.443 MPa, 0.476 MPa, and 0.510 MPa, respectively. That is, for every 10 mm increase in the cover thickness, the friction resistance stress will increase by 0.038 MPa on average. This is mainly because the cover thickness provides a strong lateral constraint for the bonding interface, increasing the normal pressure to the bonding interface, and as a result increasing the friction resistance.

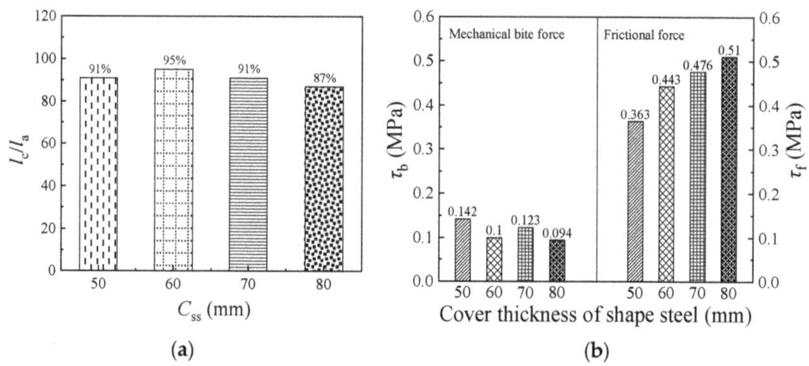

Figure 14. The effect of cover thickness of steel on the bonding capacity. (**a**) The effect of cover thickness of steel on l_c/l_a; (**b**) effect on the mechanical biting stress and friction resistance stress.

4.3.3. Reinforcement Ratio

Figure 15a shows the influence of reinforcement ratio on l_c/l_a. Compared with replacement ratio of recycled concrete and the cover thickness of shape steel, the reinforcement ratio has a significant influence on l_c/l_a. l_c/l_a decreases significantly with the increase in the reinforcement ratio, the l_c/l_a value of SRRC-09 is 112%, which is the biggest, the l_c/l_a value of SRRC-11 is 64%, which is smallest. It indicates that the reinforcement ratio has advantages in decreasing the bonding diffusion zone.

The influence of the reinforcement ratio on the chemical bonding stress τ_{ca}, friction resistance stress τ_f, and mechanical biting stress τ_m is shown in Figure 15b. Both chemical bonding stress and friction resistance stress increase with the increase in the reinforcement ratio. When the reinforcement ratio increases from 0.62% to 1.54%, the chemical bonding stress increases from 0.596 MPa to 0.621 MPa, with an increase of 0.025 MPa; while the friction resistance stress increases from 0.422 MPa to 0.510 MPa, with an increase of 0.088 MPa. Therefore, the reinforcement ratio has more influence on the friction resistance stress, but not a significant impact on the mechanical biting stress.

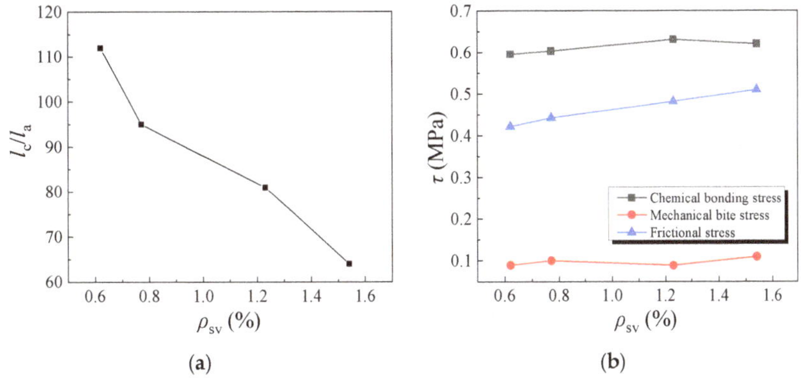

Figure 15. The effect of reinforcement ratio on the bonding capacity. (**a**) The effect of reinforcement ratio on l_c/l_a; (**b**) the effect of reinforcement ratio on the bonding stress.

4.3.4. Recycled Concrete Strength

Figure 16a shows the influence of recycled concrete strength on l_c/l_a. l_c/l_a decreases with the increase in recycled concrete strength to some extent, but not too much. Figure 16b shows the influence of recycled concrete strength on the chemical bonding stress and friction resistance stress. Both the chemical bonding stress and friction resistance stress increase with the increase in recycled concrete strength. As can be seen from Figure 16b, the recycled concrete strength has a relative great impact on the chemical bonding stress than friction resistance stress. The chemical bonding stress of SRRC-12 is 0.485, which is the smallest, while the chemical bonding stress of SRRC-14 is the biggest. The chemical bonding stress increases by 0.214 MPa when the concrete strength increased from C30 to C50. However, the influence of recycled concrete stress on frictional resistance is weak compared the impact on chemical bonding stress.

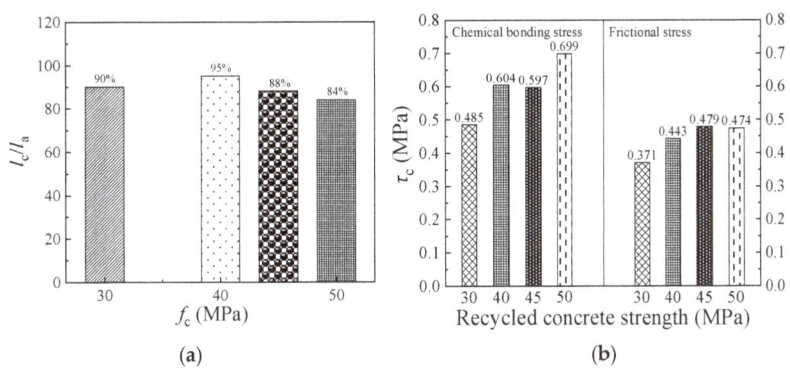

Figure 16. The effect of recycled concrete strength on the bonding capacity. (**a**) The effect of recycled concrete strength on l_c/l_a; (**b**) the effect of recycled concrete strength on chemical bonding stress and friction resistance stress.

5. Summary and Conclusions

This paper investigated the bond capacity and load transfer mechanisms between H-shaped steel and recycled concrete under cyclic loading. Fourteen specimens were tested and performed to carry out a full-stage analysis. A degradation factor of bonding stress ξ was proposed to investigate the bond degradation at the interface under cyclic loading. The following conclusions can be drawn based on the results and analysis of the test:

- Based on the effect of shear transfer on the bond capacity, the whole loading procedure can be divided into four phases with four limit points as follows: Micro slip phase (OA)→micro slip point (A)→slip development phase (AB)→ultimate point (B)→slip increasing rapidly phase (BC)→residual point (C)→residual phase (CD).
- The bond–slip damage process can be divided into three phases: (1) linear rapid damage phase (0~0.6); (2) damage accumulation transition phase (0.6~0.9); and (3) damage stabilization phase (0.9~1.0). The bonding damage develops rapidly, which reaches over 50% when the slip at interface is about 0.2 mm.
- The bonding stress between H-shaped steel and recycled concrete is composed of chemical bonding stress τ_{ca}, friction resistance stress τ_f, and mechanical biting stress τ_m. The mean values of the chemical bonding stress τ_{ca}: friction resistance stress τ_f: mechanical biting stress τ_m is 1: 0.187: 0.696. In the composition of bonding stress, the chemical bonding stress is the largest, friction resistance stress is the second-largest, and the mechanical biting stress is the least.
- The average ratio of the length of bonding diffusion zone vs. the embedded length of steel l_c/l_a is 89%. l_c/l_a increases with the increase in the replacement rate of recycled aggregate, and decreases with the increase in the thickness of the section steel protective layer, the reinforcement ratio and the strength of recycled concrete. The chemical bonding stress τ_{ca} and friction resistance stress τ_f decrease with the increase in the replacement ratio of recycled concrete, and increase with the increase in cover thickness of shape steel, reinforcement ratio, and recycled concrete strength. The effect of each design parameter on the mechanical biting stress is not obvious.

Author Contributions: Conceptualization, Methodology, Writing—Original draft, J.G.; Writing—Review & Editing, Resources, G.Z.; Supervision, Validation, X.Z. All authors have read and agreed to the published version of the manuscript.

Funding: The work presented herein was conducted as part of a comprehensive research of recycled concrete. A series of research studies about recycled concrete-encased steel frame were sponsored by National Natural Science Foundation of China (Grant No. 51178384 and Grant No. 51608435) and Scientific Research Project of Shanxi Province (Grant No. 2016JQ5113). The writers wish to thank to these sponsors.

Institutional Review Board Statement: Not applicable.

Informed Consent Statement: Not applicable.

Data Availability Statement: Not applicable.

Conflicts of Interest: The authors declare no conflict of interest

References

1. Ma, H.; Xue, J.Y.; Liu, Y.; Zhang, X. Cyclic loading tests and shear strength of steel reinforced recycled concrete short columns. *Eng. Struct.* **2015**, *92*, 55–68. [CrossRef]
2. Javier, E.L.; Sindy, S.P.; Belen, G.F.; Abella, F.M. Bond Behavior of Recycled Concrete: Analysis and Prediction of Bond Stress-Slip Curve. *J. Mater. Civ. Eng.* **2017**, *29*, 04017156. [CrossRef]
3. Casuccio, M.; Torrijos, M.C.; Giaccio, G.; Zerbino, R. Failure mechanism of recycled aggregate concrete. *Constr. Build. Mater.* **2008**, *22*, 1500–1506. [CrossRef]
4. Deepa, R.S.; Jithin, B. Strength and behaviour of recycled aggregate geopolymer concrete beams. *Adv. Concr. Constr.* **2017**, *5*, 145–154. [CrossRef]
5. Kim, S.W.; Yun, H.D. Influence of recycled coarse aggregates on the bond behavior of deformed bars in concrete. *Eng. Struct.* **2013**, *48*, 133–143. [CrossRef]
6. Prusty, R.; Mukharjee, B.B.; Barai, S.V. Nano-engineered concrete using recycled aggregates and nano-silica: Taguchi approach. Advances in concrete construction. *Adv. Concr. Constr.* **2015**, *3*, 253–268. [CrossRef]
7. Suman, S.; Rajasekaran, C. Mechanical properties of recycled aggregate concrete produced with Portland Pozzolana Cement. *Adv. Concr. Constr.* **2016**, *4*, 27–35. [CrossRef]
8. Tabsh, S.W.; Abdelfatah, A.S. Influence of recycled concrete aggregates on strength properties of concrete. *Constr. Build. Mater.* **2009**, *23*, 1163–1167. [CrossRef]

9. Thomas, C.; Setién, J.; Polanco, J.A.; Alaejos, P.; de Juan, A.S. Durability of recycled aggregate concrete. *Constr. Build. Mater.* **2013**, *40*, 1054–1065. [CrossRef]
10. Xu, J.; Wang, Y.; Ren, R.; Wu, Z.; Ozbakkaloglu, T. Performance evaluation of recycled aggregate concrete-filled steel tubes under different loading conditions: Database analysis and modelling. *J. Build. Eng.* **2020**, *30*, 101308. [CrossRef]
11. Corinaldesi, V. Mechanical and elastic behaviour of concretes made of recycled-concrete coarse aggregates. *Constr. Build. Mater.* **2010**, *24*, 1616–1620. [CrossRef]
12. Xue, J.; Zhai, L.; Bao, Y.; Ren, R.; Zhang, X. Seismic behavior of steel-reinforced recycled concrete inner-beam-column connection under low cyclic loads. *Adv. Struct. Eng.* **2017**, *21*, 631–642. [CrossRef]
13. Han, B.C.; Yun, H.D.; Chung, S.Y. Shear capacity of reinforced concrete beams made with recycled aggregate. *ACI Spec. Publ.* **2001**, *200*, 503–516.
14. Maruyama, I.; Sogo, M.; Sogabe, T. Flexural properties of reinforced recycled concrete beams. *Mater. Struct.* **2004**, *8*, 525–535.
15. Sato, R.; Maruyama, I.; Sogabe, T.; Sogo, M. Flexural behavior of reinforced recycled concrete beams. *J. Adv. Concr. Technol.* **2007**, *5*, 43–61. [CrossRef]
16. Ignjatović, I.S.; Marinković, S.B.; Mišković, Z.M.; Savić, A.R. Flexural behavior of reinforced recycled aggregate concrete beams under short-term loading. *Mater. Struct.* **2013**, *46*, 1045–1059. [CrossRef]
17. Breccolotti, M.; Materazzi, A.L. Structural reliability of eccentrically-loaded sections in RC columns made of recycled aggregate concrete. *Eng. Struct.* **2010**, *32*, 704–712. [CrossRef]
18. Choi, W.C.; Yun, H.D. Compressive behavior of reinforced concrete columns with recycled aggregate under uniaxial loading. *Eng. Struct.* **2012**, *41*, 285–293. [CrossRef]
19. Corinaldesi, V.; Letelier, V.; Moriconi, G. Behaviour of beam-column joints made of recycled-aggregate concrete under cyclic loading. *Constr. Build. Mater.* **2011**, *25*, 1877–1882. [CrossRef]
20. Xiao, J.Z.; Sun, Y.D.; Falkner, H. Seismic performance of frame structures with recycled aggregate concrete. *Eng. Struct.* **2006**, *28*, 1–8. [CrossRef]
21. Xue, J.; Zhang, X.; Ren, R. Pseudo-dynamic test investigation of recycled concrete-encased steel frame. *J. Build. Eng.* **2020**, *32*, 101834. [CrossRef]
22. Yang, Y.; Xue, Y.; Yu, Y.; Gao, F. Experimental study on seismic performance of partially precast steel reinforced concrete columns. *Eng. Struct.* **2018**, *175*, 63–75. [CrossRef]
23. Chen, C.; Wang, C.; Sun, H. Experimental study on seismic behavior of full encased steel-concrete composite columns. *J. Struct. Eng.* **2014**, *140*, 04014024. [CrossRef]
24. Zhang, X.; Xue, J.; Ren, R. Bond capacity between H-shaped steel and recycled concrete under cyclic loading. *Struct. Des. Tall Spec. Build.* **2021**, *30*, e1890. [CrossRef]
25. Jurkiewiez, B.; Meaud, C.; Michel, L. Nonlinear behaviour of steel concrete epoxy bonded composite beams. *J. Constr. Steel Res.* **2011**, *67*, 389–397. [CrossRef]
26. Tao, Z.; Song, T.Y.; Uy, B.; Han, L.H. Bond behavior in concrete-filled steel tubes. *J. Constr. Steel Res.* **2016**, *120*, 81–93. [CrossRef]
27. Zhang, Q.H.; Jia, D.; Bao, Y.; Dong, S. Flexural behavior of steel-concrete composite beams considering interlayer slip. *J. Struct. Eng.* **2019**, *145*, 04019084. [CrossRef]
28. Hotta, H.; Kihara, H.; Takiguchi, K. An empirical study on shear strength of steel encased reinforced concrete members concerning bond stress between steel and concrete. *J. Struct. Constr. Eng.* **1997**, *62*, 131–137. [CrossRef]
29. Zheng, H.H.; Chen, Z.P.; Su, Y.S. Research on mechanical behavior and bearing capacity calculation method of steel reinforced recycle aggregate concrete combined component with stud connecters. *Ind. Constr.* **2016**, *46*, 97–104. (In Chinese)
30. Bai, G.; Yin, Y.; Liu, C.; Han, Y. Experimental study on bond-slip behavior between section steel and recycled aggregate concrete in steel reinforced recycled concrete structures. *J. Build. Struct.* **2016**, *37*, 135–142. [CrossRef]
31. Chen, Z.; Zheng, H.; Xue, J.; Su, Y. Push-out test on interface bond slip behavior and analysis on bond strength between steel and recycled coarse aggregate concrete. *J. Build. Struct.* **2013**, *34*, 130–138.
32. Chen, Z.; Chen, Y.; Zheng, H.; Xue, J. Analysis of Influence Factors and Bond Strength of Steel-Recycled Aggregate Concrete Interface. *Ind. Build.* **2013**, *43*, 1–6. (In Chinese)
33. Zhou, Y.; Zheng, S.; Chen, L.; Long, L.; Wang, B. Experimental investigation into the seismic behavior of squat reinforced concrete walls subjected to acid rain erosion. *J. Build. Eng.* **2021**, *44*, 102899. [CrossRef]

coatings

MDPI

Article

Improvement of 3D Printing Cement-Based Material Process: Parameter Experiment and Analysis

Zihan Li [1], Huanbao Liu [1,2,*], Xiang Cheng [1,2], Ping Nie [1], Xianhai Yang [1], Guangming Zheng [1,2], Hongxing Su [1] and Wenyu Jin [1]

[1] College of Mechanical Engineering, Shandong University of Technology, Zibo 255000, China
[2] Shandong Provincial Key Laboratory of Precision Manufacturing and Non-Traditional Machining, Zibo 255000, China
* Correspondence: lhuanbao@sdut.edu.cn

Abstract: Three-dimensional printing concrete is a digital and automating construction technology, which is expected to solve a series of problems existing in the traditional construction industry, such as low automation, high labor intensity, low efficiency and high risk. However, there are still many technical and operational challenges. The purpose of this paper is to provide insights into the effects of process parameters on the geometry and stability of the printed layer. Firstly, a theoretical model is established to analyze the structure of the printed layer under different nozzle speeds, material flow rates and nozzle offset. Secondly, a slump test is carried out to select the optimal ratio suitable for 3D cement printers, and the specimen is printed under various conditions. Finally, based on the obtained parameters, multiple nozzles are used for printing, and a pressure value suitable for each nozzle in the nonlinear path is calculated. The experimental results show that theoretical model can sufficiently verify printing structure in different parameter intervals, and the process parameters (nozzle speed, material flow rate and nozzle offset) can be changed to achieve the best effect of cement-based material forming structure.

Keywords: 3D printing; cement-based mortar; theory model; process parameters; multi-nozzle structure

Citation: Li, Z.; Liu, H.; Cheng, X.; Nie, P.; Yang, X.; Zheng, G.; Su, H.; Jin, W. Improvement of 3D Printing Cement-Based Material Process: Parameter Experiment and Analysis. *Coatings* **2022**, *12*, 1973. https://doi.org/10.3390/coatings12121973

Academic Editor: Francesco Marra

Received: 16 November 2022
Accepted: 13 December 2022
Published: 16 December 2022

1. Introduction

The construction industry accounts for about 10% of total world output, but productivity in construction has barely increased since 1945 [1–3]. Compared with other industrial sectors (such as manufacturing, retail and agriculture), the productivity of the construction industry is extremely low: there are problems such as low technological content of production mode and lagging technological innovation. Meanwhile, problems such as the aging population and rising labor costs are gradually aggravated. Therefore, many researchers are urgently exploring solutions to change this situation.

With advent of Industry 4.0 [4], improving the degree of automation in the manufacturing stage of the construction process is expected to solve the problem of low productivity [5]. In this context of rapid development, construction 3D printing is an effective way to improve the current situation of the construction industry. Because the construction process has better degrees of freedom, lower labor costs and faster productivity, it has received wide attention [6–10]. At present, construction 3D printing has contour crafting [11,12], D-shape [13], concrete printing [14] and construction methods that are applicable to the field of architecture, and have also been successfully applied to artificial bridge construction, military shelters, rapid disaster relief shelters [15–18] and complex structures (an ultra-high-performance structure with complex geometric shapes designed by Gosselin et al. [19]). Different mechanical mechanisms and process parameters have certain influence on the shape, stability, material accumulation and fracture of the printed parts, that is, there are different forming effects, the geometric shape and local stability of the extrusion layer are key factors affecting the final deforming effect.

Many researchers have studied the influence of printing parameters on forming mechanism and obtained the corresponding optimal process parameters. Farahbakhsh, M et al. [20], by manipulating process parameters, the layer geometry was changed to simulate densification, and friction between successive layers was enhanced by periodically increasing the printed area or enhancing compaction during printing (Figure 1a,b). Carneau, P et al. [21] provide insights on the effect of printing parameters on the geometry and local stability of the extruded layer. The upper and lower limit requirements for controlling lamination are discussed from the influence of geometric and motion parameters, as well as from the study of the properties of fresh materials. (Figure 1c). Huang, X et al. [22] introduced the influence of printing parameters on molding quality and interlayer bond strength through single-factor experiment, and obtained corresponding printing parameters (Figure 1d). Chen, Y et al. [23] studied the influence of time interval and nozzle distance on the interlayer bond strength, and the results showed that extending the time interval could reduce bond strength, while only increasing the nozzle distance had limited influence on the bond strength (Figure 1e). Elistratkin, M et al. [24] studied the combined influence of factors such as layer thickness, nozzle speed and screw rotation speed on the extrusion and printing track parameters of building materials, and established the relationship between the above factors to improve the stability of the printing process and the final structural quality (Figure 1f). Based on the above discussion, the process parameters are the key technical indexes for building structure printing, which not only affect the interlayer bonding effect, but also have a certain impact on the appearance of the printed product. Therefore, it is necessary to carry out the research on the matching of the printing process parameters.

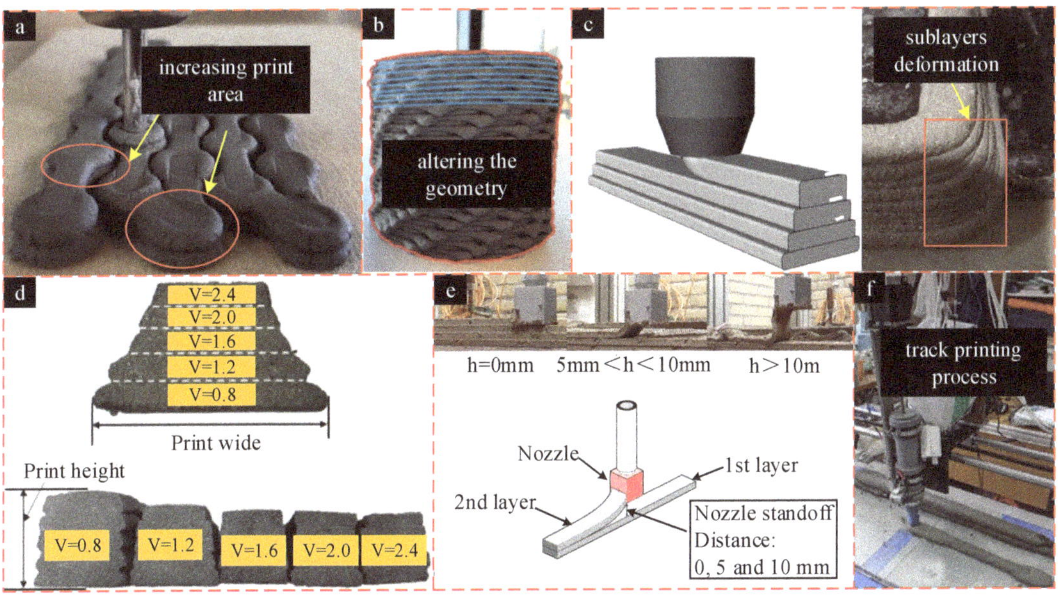

Figure 1. Effect of different printing parameters. (**a**) Increased surface contact between successive layers [20], (**b**) Altering the geometry [20], (**c**) Over pressing of the sublayers [21], (**d**) Influence of printing speed on concrete size [22], (**e**) Increasing the nozzle standoff distance could lead to the inaccurate layer deposition [23], (**f**) Track printing process [24].

In this paper, the influence of process parameters on material fluidity and molding effect was studied by single-factor experiment, taking tailing sand content, resting time, nozzle speed, pressure and nozzle offset as influencing factors, and slump, embryo strength, layer height and layer width as combining indicators. The self-developed 3D concrete printing equipment was utilized to analyze the matching relationship of print parameter, which can be adjusted to optimize printing parameter combination was obtained and provided reference for related research.

2. Materials and Methods

2.1. Materials

In order to determine the optimum sand content, a mixing ratio test was designed P.O 42.5 ordinary Portland cement was selected as the main cementitious, and the first grade fly ash, S95 grade slag powder and SF96 grade silicon powder were used as admixture to improve the micro pore structure, bonding performance, cement-based strength and pumping performance of mortar [25]. In addition to the above cementing materials, the chemical admixture used hydroxypropyl methyl cellulose ether and sodium gluconate retarder, which can greatly improve the plasticity and water retention of mortar and prevent drying and cracking. Tailing sand was selected for fine aggregate, and its physical properties were shown in Table 1. In order to ensure accuracy of water–binder ratio, tailing sand should be dried and sealed before use to ensure that the moisture content is 0. The water used for the experiment was tap water or medium water (medium water generally refers to reclaimed water, waste water or rainwater after proper treatment, to reach a certain quality index, can be used for beneficial water), and the printing experiment were carried out at room temperature. The process parameters were shown in Table 2.

Table 1. Physical properties of tailing sand.

Fine Aggregate	Fineness Modulus	Average Particle Size (mm)	Bulk Density (kg/m^3)	Mud Content (%)
Tailing sand	0.7	0.15	1400	0.4
Admixture	Fineness (45 micron standard sieve)	Average particle size (μm)	Density (g/cm^3)	Water content (%)
Fly ash	≤5%	5	2.6	≤1
Slag powder	≤3%	38	2.8	≤1
Silicon powder	≤3%	5	2.6	≤3

Table 2. Experimental material parameters.

Tailing Sand/%	Water/Cement Ratio	Cement/%	Mineral Admixture/%	Thickener/%	Retarder/%
0	0.39	80.2	19.8	0.05	0.3
10	0.39	80.2	19.8	0.05	0.3
20	0.39	80.2	19.8	0.05	0.3
30	0.39	80.2	19.8	0.05	0.3
40	0.39	80.2	19.8	0.05	0.3
50	0.39	80.2	19.8	0.05	0.3

The mixing process is as follows: Firstly, weigh out cement, admixtures, sand and admixtures. Secondly, first the cement and admixture mix and stir evenly, then pour the sand into the mix evenly, and then pour the admixture into the mixture, stir for 2–3 min after all mixing. Finally, gradually add water and stir.

2.2. Methods

2.2.1. 3D Cement Printing Equipment

A self-developed four-axis motion 3D cement printer was used in the experiment (Figure 2a,b). The X-axis, Y-axis and Z-axis motion mechanism was composed of linear modules. Stepping motors were used to control A axis rotation. Numerical control technology was used to control the stroke accurately. The printing system of the equipment was divided into two parts: one is the extrusion device, including air source (Figure 2c), ink cartridge (Figure 2d) and nozzle (Figure 2e). The other part is the Mach3 control system that controls the operation of equipment (Figure 2f). The specification of the 3D printing system was show in Table 3.

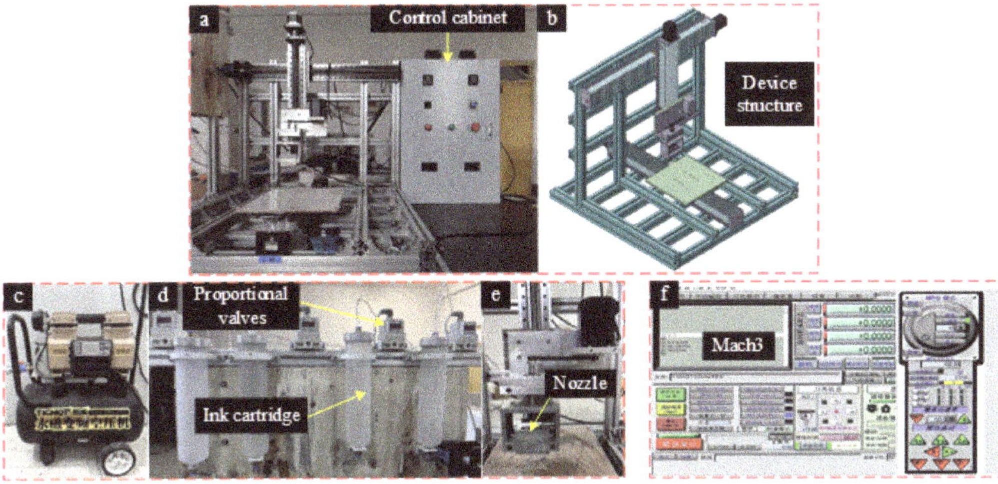

Figure 2. Three-dimensional cement printing equipment. (**a**) Three-dimensional concrete printing device, (**b**) Device structure, (**c**) Air source, (**d**) Ink cartridge, (**e**) Nozzle, (**f**) Control system.

Table 3. 3D printing system specifications.

Parameters	Value
Size specification (X × Y × Z)	800 × 800 × 600 mm^3
Walking accuracy	±1 mm
Print speed	0–3000 mm/min
Pressure range	0–1.0 MPa

2.2.2. Theory

Figure 3 summarizes the influence of cement-based material flow rate and nozzle speed on printing molding. The cement-based material can be squeezed through pressure control system, and the nozzle speed was controlled by stepper motor. With intermediate cement-based material flow-rate and nozzle speed, the width of the print layer will be similar to the nozzle diameter. In this case, the printing process can obtain more ideal specimen (Figure 3a). With a high flow-rate of cement-based material or low nozzle speed, the width of the print layer was often larger than the nozzle diameter (Figure 3b). A wider path will strengthen the bond between the layers and consequently achieve better mechanical properties, and on the other hand, the printing process can easily lead to material accumulation in narrow, non-linear paths. With a low flow-rate or high speed, the width of the print layer will be similar to or smaller than the nozzle diameter (Figure 3c). However, when the flow-rate was reduced (pressure value was less than 0.045 MPa), the cement-based material creates friction with the bottom surface, causing a dragging phenomenon and subsequent fracture (Figure 3d). The relationship between nozzle velocity and X, Y axis velocity is:

$$F = \sqrt{V_x^2 + V_y^2}\ (\mathrm{mm/s}) \quad (1)$$

$$\frac{F}{Vx} = \frac{S}{Lx} \quad (2)$$

$$\frac{F}{Vy} = \frac{S}{Ly} \quad (3)$$

where F is the theoretical speed of nozzle, in mm/s. S is the theoretical displacement of nozzle, in mm. V_x is the actual speed in the X-axis direction, in mm/s. L_x is the actual displacement in the X-axis direction, in mm. V_y is the actual speed in the Y-axis direction, in mm/s and L_y is the actual displacement in the Y-axis direction in mm.

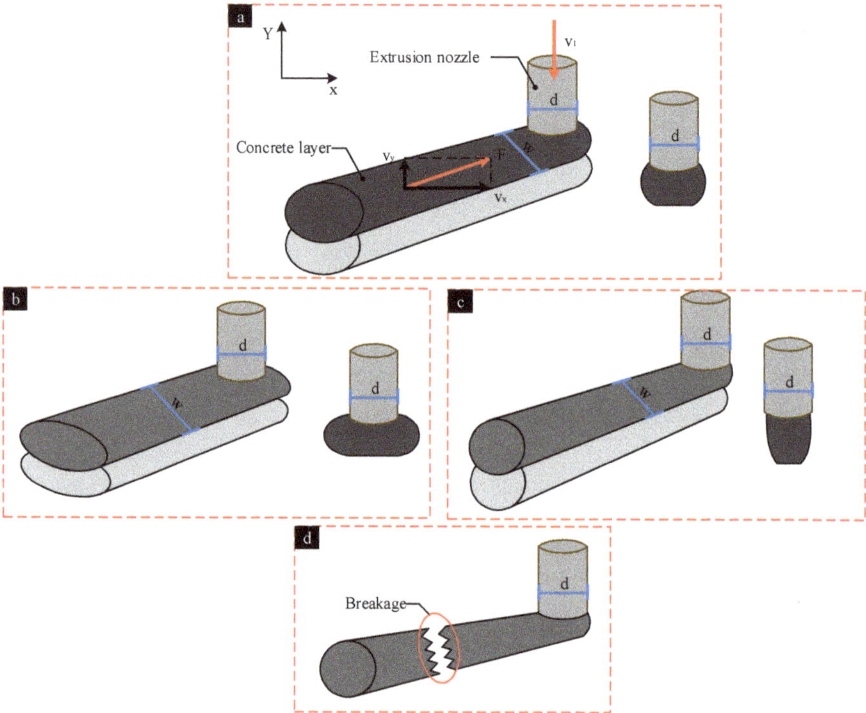

Figure 3. Influence of flow-rate of material and printing nozzle speed on molding. (**a**) Ideal state, (**b**) High flow-rate of material or low printing nozzle speed, (**c**) Low flow-rate of material or high printing nozzle speed, (**d**) Breakage state.

The material extrusion speed V_1 is:

$$t = \frac{S}{F}\,(s) \tag{4}$$

$$A = \pi r^2 b\,\left(mm^3\right) \tag{5}$$

$$V1 = \frac{A}{t}\,\left(mm^3/s\right) \tag{6}$$

where t is the printing time, in s. b is the height of material falling in the ink cartridge during the printing time, in mm. r is the radius of the ink cartridge, in mm and A is the volume of material extruded during the printing time in mm^3.

Figure 4 summarizes the influence of nozzle height on printing molding under optimal material flow-rate and nozzle speed. If the nozzle offset (H) was much larger than the nozzle diameter (d), the material gradually tilts forward, resulting in a layer height consistent with the nozzle offset, and the cross-section is trapezoidal (Figure 4a). In this case, the inter-layer contact area was smaller, and the inter-layer bond strength was weaker [21]. If H was equal to d, the surface will be slightly flattened by the nozzle, and a large inter-layer contact area can be obtained without destroying the shape (Figure 4b). If H was much smaller than d, the nozzle will push the cement-based material to both sides, forming thin layer (Figure 4c).

In this case, there may be some cracks and other defects in the spilled concrete on both sides. The relationship between nozzle offset and specimen width as

$$H = h\ (\text{mm}) \tag{7}$$

$$w = \frac{A}{SH}\ (\text{mm}) \tag{8}$$

where h is layer height, in mm. H is the nozzle offset and w is layer width in mm. When the volume of extruded material was constant, the layer height was inversely proportional to the layer width.

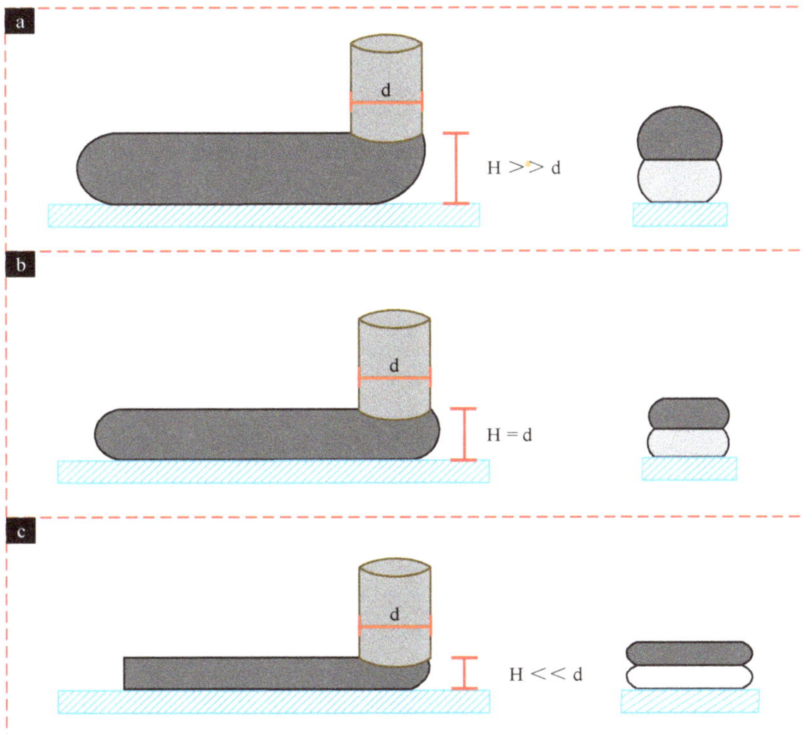

Figure 4. Influence of nozzle offset on molding. (**a**) The nozzle offset is much higher than the nozzle diameter, (**b**) The nozzle offset is equal to the nozzle diameter, (**c**) The nozzle offset is much smaller than the nozzle diameter.

2.2.3. Experimental Procedure

By controlling the content of tailing sand, the cement-based materials have well fluidity, extrudability and constructability. The tailing sand content was set to 0%–50% when the other ratios were unchanged (gradient value of 10%), as show in Table 2. The mixed cement-based material was loaded into a 100 × 100 mm cylinder (Figure 5a) and then tamped with a tamper stick. The cylinder was removed, and the specimen was gradually pressurized until the specimen severely deformed (Figure 5b). The pressure was then converted to strength (Pressure divided by contact area, F = mg, P = F/S). After the optimal content of tailing sand was selected, the slump and strength were tested after 0, 5, 10, 15 and 20 min, respectively, to select the optimal printing time.

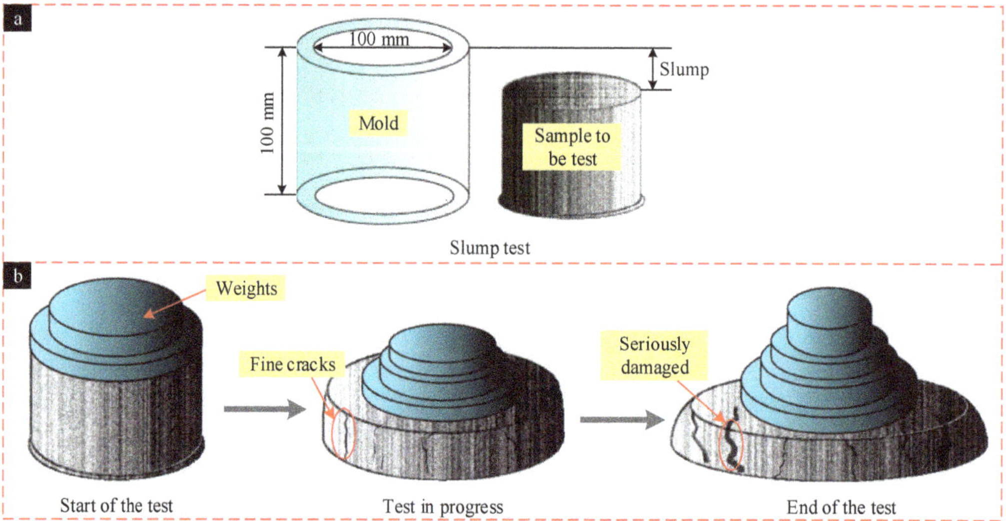

Figure 5. Slump and strength test theoretical model. (**a**) Slump test, (**b**) The test process of strength.

The theoretical model was verified under the support of optimal mix ratio. Experiments were designed to analyze the printing effect of cement-based materials under different parameters. Firstly, the influence of nozzle speed on molding effect was analyzed under the same pressure. The pressure was set to 0.045 MPa, and the nozzle speed was 1.667–13.333 mm/s (gradient value of 1.667 mm/s), as shown in Table 4. Secondly, the influence of pressure on the molding effect was analyzed at the same nozzle speed. The nozzle speed was set to 8.333 mm/s, and the pressure was set to 0.045–0.144 MPa (gradient value of 0.009 MPa), as shown in Table 5. Finally, the influence of different nozzle height on the molding effect was analyzed. The nozzle speed was set to 8.333 mm/s, the pressure was set to 0.117 MPa, the nozzle diameter was set to 10 mm and the nozzle height was set to 3, 5, 7.5, 10, 12 and 14 mm, as show in Table 6.

Table 4. Different nozzle speed under the same pressure.

Number	Nozzle Speed (mm/s)	Pressure Value (MPa)	Number	Nozzle Speed (mm/s)	Pressure Value (MPa)
1	1.667	0.045	5	8.333	0.045
2	3.333	0.045	6	10	0.045
3	5	0.045	7	11.667	0.045
4	6.667	0.045	8	13.333	0.045

Table 5. Different pressure at the same nozzle speed.

Number	Pressure Value (MPa)	Nozzle Speed (mm/s)	Number	Pressure Value (MPa)	Nozzle Speed (mm/s)
1	0.045	8.333	7	0.099	8.333
2	0.054	8.333	8	0.108	8.333
3	0.063	8.333	9	0.117	8.333
4	0.072	8.333	10	0.126	8.333
5	0.081	8.333	11	0.135	8.333
6	0.09	8.333	12	0.144	8.333

Table 6. Different nozzle offset at the same nozzle diameter.

Number	Nozzle Offset (mm)	Nozzle Diameter (mm)	Number	Nozzle Offset (mm)	Nozzle Diameter (mm)
1	3	10	4	10	10
2	5	10	5	12	10
3	7.5	10	6	14	10

3. Analysis and Discussion

3.1. Slump Analysis

Cement-based materials were usually extruded by nozzles, and fluidity was a key parameter to evaluate the extrudability of materials. Therefore, the good fluidity in the printing process can ensure that cement-based material was smoothly extruded continuously from the nozzle without break, and was easy to deposit and molding. Fluidity was usually expressed in terms of slump: the test process of slump and strength is shown in Figure 6. Slump refers to the height of automatic deposition of cement-based material (Figure 6a), and strength is converted by applied pressure (Figure 6b). The variation of strength and slump with tailing sand content was shown in Figure 7a. The slump of specimens decreased with the increase in tailing sand content, which indicated that constructability of cement-based materials can be effectively improved while the fluidity is reduced. At the same time, it can be seen from the figure that the strength increases almost linearly with the increase in tailing sand content. Combined with strength and slump analysis, the best printing results were obtained when the tailing sand content was 40%. The changes in slump and strength with time are shown in Figure 7b. With the increase in time, the higher strength of material, the better constructability of material. However, with the increase in time, the slump gradually decreased. When the time was longer than 15 min, the material was almost no slump, which will cause pipe blockage and was not conducive to printing. Therefore, 10 min after the stirring was the optimal printing time.

Figure 6. The test process. (**a**) Slump, (**b**) Strength.

3.2. Influence of Nozzle Speed and Pressure on Molding Effect

The extrudability of cement-based materials plays an important role in the experimental process, the pressure extrusion system is used to realize the on-demand extrusion of cement-based materials. Under high pressure conditions, the ink cartridges can burst due to excessive internal pressure and there are certain risks, cement-based materials were difficult to control, and material was difficult to extrude under low pressure. Therefore, in order to make the cement-based materials' extrusion stable and controllable, the pressure was set to 0.045 MPa. At different nozzle speeds, the specimen morphology has different widths and heights. The width and height of the print layer variation with the nozzle speed are shown in Figure 8a. With the increase in nozzle speed, the width and height showed a downward trend.

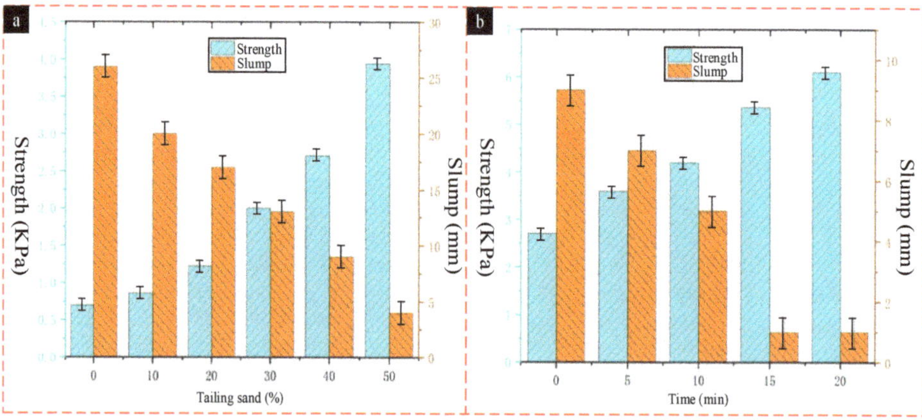

Figure 7. Variation of strength and slump (n = 8). (**a**) Variation of strength and slump with tailing sand content, (**b**) Variation of strength and slump with time.

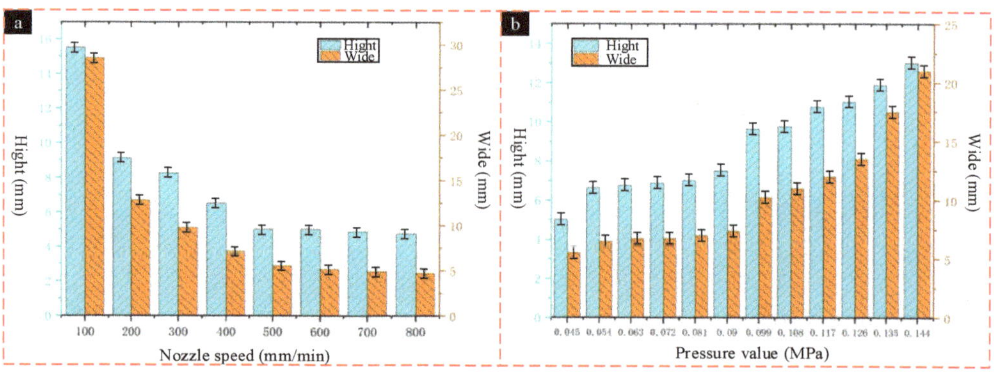

Figure 8. Cement-based material molding relationship. (**a**) Variation of layer height and wide with nozzle speed under the same pressure, (**b**) Variation of layer height and wide with pressure under the same nozzle speed.

In order to study the Influence of extrusion pressure on molding effect of cement-based materials and the optimal ratio of nozzle speed and extrusion speed, the nozzle speed was set to 8.333 mm/s (F500) according to the result in Figure 8a. Specimen had different morphology under different pressures. The variation of layer width and layer height with pressure are shown in Figure 8b. With the increase in pressure, the layer width and height showed an upward trend.

The printing effect of cement-based materials under different parameters is shown in Figure 9. When the nozzle speed was less than 3.333 mm/s (F200) and the pressure value was greater than 0.144 MPa, materials accumulation phenomenon will occur (Figure 9a). The surface of the specimen printed at low speed and high pressure was uneven, and materials accumulation can be obviously seen on the printing path (Figure 9d). However, an increase in layer width leads to increase in the contact area between layers, so the constructability was particularly outstanding in this case. On the contrary, when the nozzle speed was greater than 8.333 mm/s (F500) and the pressure value was less than 0.081 MPa, the shape of the printing specimen was not sufficient to support the printing of the next layer (Figure 9c). The specimens printed at high speed and low pressure have poor construction performance and were prone to collapse during printing (Figure 9f). When the nozzle speed was 8.333 mm/s and the pressure value was 0.117 MPa, the surface of the printing specimen was smooth and flat, and there was no material accumulation. At the same time, specimens has good constructability and ideal molding effect (Figure 9b,e).

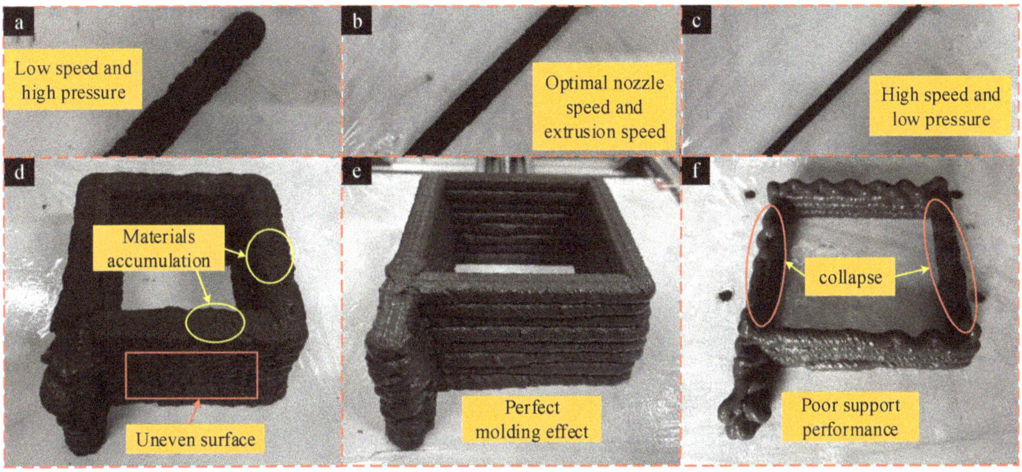

Figure 9. Printing effect of cement-based material under different parameters. (**a**) Shapes at low speed and high pressure, (**b**) Shape under the best matching parameters, (**c**) Shape at high speed and low pressure, (**d**) Printing molding effect under low speed and high pressure, (**e**) Printing molding effect under the best matching parameters, (**f**) Printing molding effect under high speed and low pressure.

3.3. *Influence of Nozzle Offset on Molding Effect*

Based on Section 3.2, the nozzle speed was set to 8.333 mm/s, the pressure was set to 0.117 MPa, and the nozzle diameter was set to 10 mm. Different nozzle offset lead to different deposition shapes. The changes in the layer width and layer height with the nozzle offset are shown in Figure 10. With the increase in the nozzle offset, the layer height showed an upward trend. When the nozzle offset was greater than or equal to the nozzle diameter, the layer height was consistent with the nozzle offset. According to Equation (8), the layer height was inversely proportional to the layer width, so the layer width showed downward trend.

The molding effect is shown in Figure 11. When the nozzle offset was less than 7.5 mm, compression condition of specimen was extremely serious, so the layer width was large (Figure 11a). When the nozzle offset was low for printing, the layer contact area was increased, and the constructability was improved. However, due to the severe extrusion of the printing layer, the materials will overflow on both sides of the printing path (Figure 11d). On the contrary, when the nozzle offset was greater than 10 mm, the layer height was larger and the layer width was smaller, which was not conducive to the support function of the lower layer to the upper layer, and higher the nozzle offset leads to inaccurate deposition, therefore, the constructability was poor (Figure 11c,f). In relative terms, the nozzle offset was set to be slightly smaller than the nozzle diameter. On the premise that the printing layer was not severely compressed, ensures good constructability, and at the same time, there was no material spillage during the printing process (Figure 11b,e).

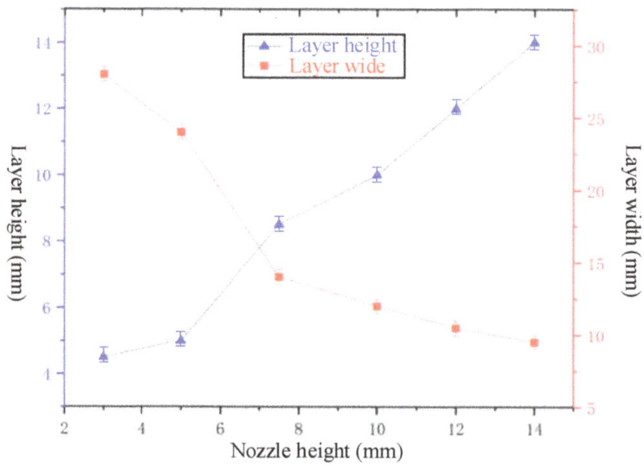

Figure 10. Effect of nozzle offset on layer height and layer width.

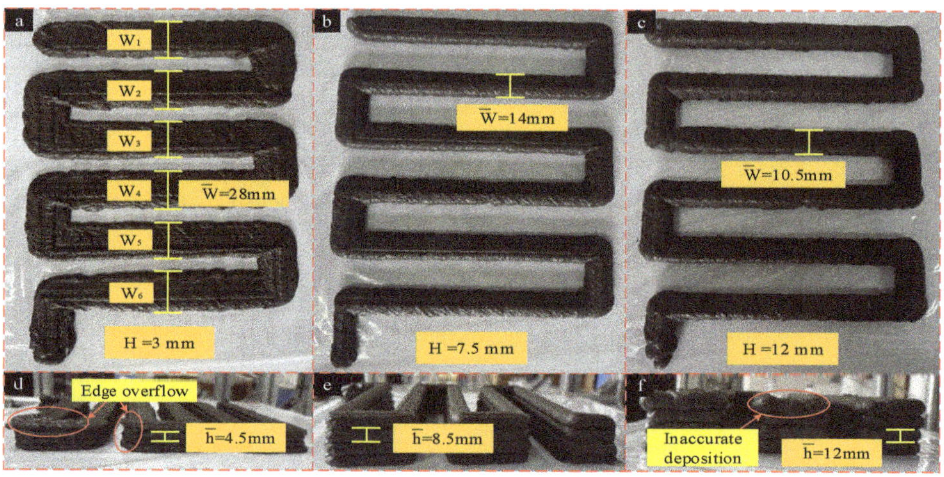

Figure 11. Printing effect of cement-based material under different nozzle offset. (**a**) Width of the print layer when the nozzle offset is 3 mm, (**b**) Width of the print layer when the nozzle offset is 7.5 mm, (**c**) Width of the print layer when the nozzle offset is 12 mm, (**d**) Height of the print layer when the nozzle offset is 3 mm, (**e**) Height of the print layer when the nozzle offset is 7.5 mm, (**f**) Height of the print layer when the nozzle offset is 12 mm.

3.4. Effect of Different Pressure on Multi-Nozzle Printing Molding

The nozzle device of the printer used in this paper can switch between single-nozzle extrusion and multi-nozzle extrusion by Mach3 control board. When multi-nozzle printing was carried out, different nozzles should be set different pressure values on the non-linear path to ensure that there was no materials accumulation. Based on the results of Sections 3.2 and 3.3, the nozzle speed of 8.333 mm/s, the pressure of 0.117 MPa, and the nozzle height of 7.5 mm were set as the basic parameters.

The molding effect of multi-nozzle printing as shown in Figure 12. When printing with double nozzles, the nozzle spacing can be adjusted, and the extrusion speed of each nozzle can be controlled by pressure, so as to meet the printing requirements of complex components. As shown in Figure 12a, the pressure value of inner ring was 0.099 MPa, and the pressure value of outer ring was 0.117 MPa. Therefore, the forming effect of the inner and outer rings was inconsistent. When the three nozzles are printed without spacing, the pressure value in the linear path was unified to 0.117 MPa. However, in the non-linear path, because the paths of the three nozzles are different, the pressure value of the nozzles should be adjusted to achieve the effect of no materials accumulation (Figure 12b). The relationship between material extrusion speed and arc length as

$$V_1 = \frac{lhw}{t}\ mm^3/s \tag{9}$$

where l is the arc length, in mm. layer height and layer width were fixed values, the volume of material required to calculate the nonlinear path, and the printing time was the same. Therefore, extrusion speed of different paths was obtained, and pressure value of each nozzle was determined according to the extrusion speed.

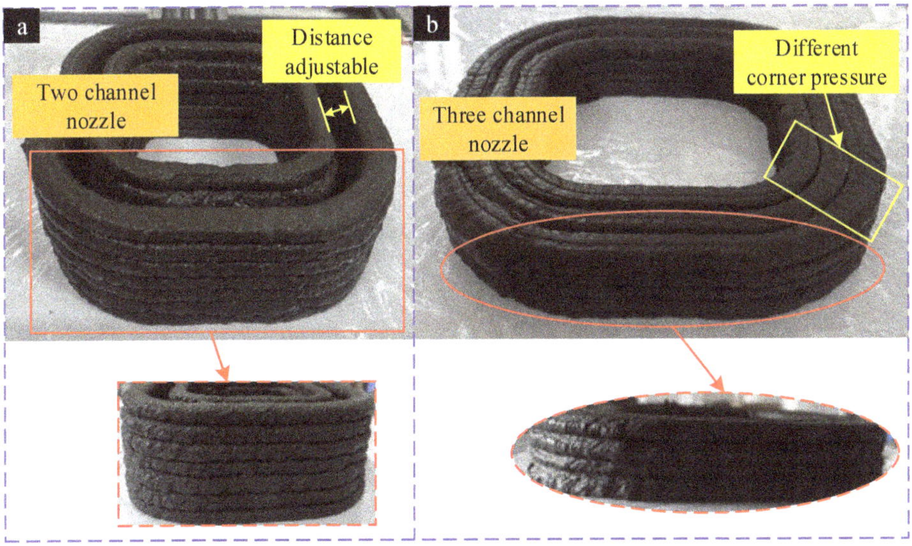

Figure 12. Multi-nozzle extrusion printing effect. (**a**) Double nozzle printing, (**b**) Three nozzle printing.

4. Conclusions

The 3D printing concrete process opens up a novel construction method, which will improve the complexity, diversity and geometric accuracy of products, and bring higher production efficiency to the construction industry. This paper aimed to investigate the influence of different printing parameters on the material printing molding through theoretical analysis and experimental methods. The conclusions are as follows:

(1) The success of the printed structure depends on the suitability of the properties of the cement-based material. Through strength and slump experiments, the optimal mix ratio suitable for 3D building printer was obtained, which can provide technical support for subsequent experiments.

(2) According to the influencing factors such as nozzle speed, material flow rate and nozzle offset, the theoretical models under different process parameters were analyzed, which can provide theoretical support for the subsequent parameter experiments.

(3) The experimental results show that the theoretical model can fully verify the print structure in different parameter intervals, the layer width and layer height can be controlled by changing the related parameters, which can effectively control the nozzle speed, material flow and nozzle offset to achieve the best effect of cement-based material forming structure.

Author Contributions: Z.L.: Data curation, Writing—Original draft preparation. H.L.: Conceptualization, Methodology, Writing—Reviewing and Editing. X.C.: Visualization, Software. P.N.: Writing—Reviewing and Editing. X.Y.: Investigation. G.Z.: Validation, Writing—Reviewing and Editing. H.S.: Formal analysis. W.J.: Data curation. All authors have read and agreed to the published version of the manuscript.

Funding: This research was funded by the Natural Foundation of Shandong Province (ZR2020QE179), National key research and development Program (2018YFB2001402-02) and the National Natural Science Foundation of China (No. 51505264).

Institutional Review Board Statement: Not applicable.

Informed Consent Statement: Not applicable.

Data Availability Statement: Not applicable.

Acknowledgments: The research undertaken was made possible by the equal scientific involvement of all the authors concerned.

Conflicts of Interest: The authors declare no conflict of interest.

References

1. Hossain, A.; Zhumabekova, A.; Paul, S.; Kim, J. A Review of 3D Printing in Construction and its Impact on the Labor Market. *Sustainability* **2020**, *12*, 8492. [CrossRef]
2. Crosthwaite, D. The global construction market: A cross-sectional analysis. *Constr. Manag. Econ.* **2000**, *18*, 619–627. [CrossRef]
3. Horta, I.; Camanho, A.; Johnes, J.; Johnes, G. Performance trends in the construction industry worldwide: An overview of the turn of the century. *J. Prod. Anal.* **2013**, *39*, 89–99. [CrossRef]
4. Hofmann, E.; Rüsch, M. Industry 4.0 and the current status as well as future prospects on logistics. *Comput. Ind.* **2017**, *89*, 23–34. [CrossRef]
5. Cai, S.; Ma, Z.; Skibniewski, M.J.; Bao, S. Construction automation and robotics for high-rise buildings over the past decades: A comprehensive review. *Adv. Eng. Inform.* **2019**, *42*, 100989. [CrossRef]
6. Bock, T. The future of construction automation: Technological disruption and the upcoming ubiquity of robotics. *Autom. Constr.* **2015**, *59*, 113–121. [CrossRef]
7. Rayna, T.; Striukova, L. From rapid prototyping to home fabrication: How 3D printing is changing business model innovation. *Technol. Forecast. Soc. Chang.* **2016**, *102*, 214–224. [CrossRef]
8. Jayathilakage, R.; Rajeev, P.; Sanjayan, J. Yield stress criteria to assess the buildability of 3D concrete printing. *Constr. Build. Mater.* **2020**, *240*, 117989. [CrossRef]
9. Adaloudis, M.; Roca, J.B. Sustainability tradeoffs in the adoption of 3D Concrete Printing in the construction industry. *J. Clean. Prod.* **2021**, *307*, 127201. [CrossRef]
10. Xiao, J.; Ji, G.; Zhang, Y.; Ma, G.; Mechtcherine, V.; Pan, J.; Wang, L.; Ding, T.; Duan, Z.; Du, S. Large-scale 3D printing concrete technology: Current status and future opportunities. *Cem. Concr. Compos.* **2021**, *122*, 104115. [CrossRef]
11. Zhang, J.; Khoshnevis, B. Optimal machine operation planning for construction by Contour Crafting. *Autom. Constr.* **2013**, *29*, 50–67. [CrossRef]
12. Khoshnevis, B. Automated construction by contour crafting—Related robotics and information technologies. *Autom. Constr.* **2004**, *13*, 5–19. [CrossRef]
13. Cesaretti, G.; Dini, E.; De Kestelier, X.; Colla, V.; Pambaguian, L. Building components for an outpost on the Lunar soil by means of a novel 3D printing technology. *Acta Astronaut.* **2014**, *93*, 430–450. [CrossRef]
14. Lim, S.; Buswell, R.; Le, T.; Austin, S.; Gibb, A.; Thorpe, T. Developments in construction-scale additive manufacturing processes. *Autom. Constr.* **2012**, *21*, 262–268. [CrossRef]
15. Delgado, J.M.D.; Oyedele, L.; Ajayi, A.; Akanbi, L.; Akinade, O.; Bilal, M.; Owolabi, H. Robotics and automated systems in construction: Understanding industry specific challenges for adoption. *J. Build. Eng.* **2019**, *26*, 100868. [CrossRef]
16. Wu, P.; Wang, J.; Wang, X. A critical review of the use of 3-D printing in the construction industry. *Autom. Constr.* **2016**, *68*, 21–31. [CrossRef]
17. Zhang, J.; Wang, J.; Dong, S.; Yu, X.; Han, B. A review of the current progress and application of 3D printed concrete. *Compos. Part A Appl. Sci. Manuf.* **2019**, *125*, 105533. [CrossRef]
18. Xu, J.; Ding, L.; Love, P.E. Digital reproduction of historical building ornamental components: From 3D scanning to 3D printing. *Autom. Constr.* **2017**, *76*, 85–96. [CrossRef]
19. Gosselin, C.; Duballet, R.; Roux, P.; Gaudillière, N.; Dirrenberger, J.; Morel, P. Large-scale 3D printing of ultra-high performance concrete—A new processing route for architects and builders. *Mater. Des.* **2016**, *100*, 102–109. [CrossRef]
20. Farahbakhsh, M.; Rybkowski, Z.K.; Zakira, U.; Kalantar, N.; Onifade, I. Impact of robotic 3D printing process parameters on interlayer bond strength. *Autom. Constr.* **2022**, *142*, 104478. [CrossRef]

21. Carneau, P.; Mesnil, R.; Baverel, O.; Roussel, N. Layer pressing in concrete extrusion-based 3D-printing: Experiments and analysis. *Cem. Concr. Res.* **2022**, *155*, 106741. [CrossRef]
22. Huang, X.; Yang, W.; Song, F.; Zou, J. Study on the mechanical properties of 3D printing concrete layers and the mechanism of influence of printing parameters. *Constr. Build. Mater.* **2022**, *335*, 127496. [CrossRef]
23. Chen, Y.; Jansen, K.; Zhang, H.; Rodriguez, C.R.; Gan, Y.; Çopuroğlu, O.; Schlangen, E. Effect of printing parameters on interlayer bond strength of 3D printed limestone-calcined clay-based cementitious materials: An experimental and numerical study. *Constr. Build. Mater.* **2020**, *262*, 120094. [CrossRef]
24. Elistratkin, M.; Alfimova, N.; Podgornyi, D.; Olisov, A.; Promakhov, V.; Kozhukhova, N. Influence of Equipment Operation Parameters on the Characteristics of a Track Produced with Construction 3D Printing. *Buildings* **2022**, *12*, 593. [CrossRef]
25. Jayswal, S.D.; Mungule, M. Performance assessment of Alccofine with silica fume, fly ash and slag for development of high strength mortar. *Front. Struct. Civ. Eng.* **2022**, *16*, 576–588. [CrossRef]

coatings

Article

Research on Dynamic Response and Construction Safety Countermeasures of an Adjacent Existing Line Foundation under the Influence of a New Railway Line

Shen Zuo [1], Tianyu Li [1], Jin Li [1,*], Peng Liu [2,*] and Xinzhuang Cui [3]

1 School of Civil Engineering, Shandong Jiaotong University, Jinan 250357, China; zuoshen2006@163.com (S.Z.); 20107020@stu.sdjtu.edu.cn (T.L.)
2 School of Civil Engineering, Central South University, Changsha 410075, China
3 School of Civil Engineering, Shandong University, Jinan 250061, China; cuixz@sdu.edu.cn
* Correspondence: lijinsdjtu@163.com (J.L.); liupeng868@csu.edu.cn or lop868@163.com (P.L.)

Abstract: The excavation of a new high-speed railway causes the side slope adjacent to the existing line foundation to become airborne, and the excessive dynamic deformation or cumulative deformation caused by the dynamic load of trains will affect the normal service of the subgrade, even leading to its instability. To date, there are no relevant experimental data regarding this, and there is also a lack of corresponding specifications. The only available numerical simulation research results need to be verified in practice. Therefore, this study relies on the Shanghai–Nanjing intercity high-speed railway construction project adjacent to the existing Beijing–Shanghai line to carry out a subgrade dynamic response test to ensure the safe operation of the existing line. The test obtained the vibration displacement, frequency, acceleration, and other parameters of the existing subgrade construction in three stages: subgrade excavation, pile formation, and subgrade filling. From the test results: During the test period, the vertical surface vibration displacement and vibration acceleration have a certain attenuation along the depth direction. In the stage of subgrade excavation, the vibration displacement and vibration acceleration generated are the largest. The vertical vibration displacement amplitude reaches 1.9 mm, and the horizontal vibration displacement amplitude reaches 0.15 mm. The vibration frequency of the roadbed under the action of the train load is concentrated in the range of 0–50 Hz, and the vibration energy at the peak value is relatively large, which reflects the load action frequency of the train, and the peak value is mainly concentrated in the range of 20–40 Hz. These results show that the maximum vibration response peak appears in the subgrade excavation stage, that is, the most dangerous stage of the existing subgrade. The vibration acceleration and vibration displacement of each dynamic response parameter are important in that they reflect the dynamic performance of the subgrade and establish the index control standard, which can be used as a control index for roadbed dynamic stability monitoring. The dynamic test of the subgrade state provides data support for the reasonable opening of the construction skylight and the protection of the excavation slope. Taking into account the impact of piling vibration, technical measures such as static pressure, jumping construction, and setting up stress relief holes are adopted. The test results and engineering measures ensure the safe operation of the existing subgrade, and have important theoretical significance for guiding the construction of the new subgrade adjacent to the existing line.

Keywords: intercity high-speed railway construction; adjacent to existing rail lines; dynamic test; pipe–pile construction

Citation: Zuo, S.; Li, T.; Li, J.; Liu, P.; Cui, X. Research on Dynamic Response and Construction Safety Countermeasures of an Adjacent Existing Line Foundation under the Influence of a New Railway Line. *Coatings* **2022**, *12*, 641. https://doi.org/10.3390/coatings12050641

Academic Editor: Andrea Nobili

Received: 1 April 2022
Accepted: 5 May 2022
Published: 7 May 2022

1. Introduction

The service life of large civil structures such as railways usually lasts for decades, or even hundreds of years. During the service period, structures are inevitably affected by external environmental erosion, material aging, etc., so that the engineering body becomes damaged to varying degrees. With the accumulation of these damages, the bearing capacity of engineering structures is reduced, and their safety and normal function are affected. In recent years, China has built a large number of high-speed railways, and it is inevitable that the new lines will run parallel to the existing lines in normal operation. The new Shanghai–Nanjing Passenger-Dedicated Line was built for public transportation between Nanjing and Shanghai. The total length of the line is 300.209 km, and the construction technical standard includes: a railway passenger dedicated line, a speed target value of 300 km/h; and a ballastless track. Some sections of the newly built railway are parallel to the existing railways in Beijing and Shanghai. The parallel section of the new and old lines is approximately 170 km long, the closest distance is only 4–5 m, and a drainage ditch is shared. During the construction of the Shanghai–Nanjing Line, a foundation pit was excavated close to the existing line causing the existing line foundation to be up in the air on one side, which significantly changed the stress field and displacement field of the existing line foundation. Therefore, compared with ordinary railways, the Shanghai–Nanjing section of the Beijing–Shanghai railway has more obvious characteristics of external interference. The existing Beijing–Shanghai line, which was built in the 1960s, is one of the busiest railway lines in the country, and has been in a saturated state for a long time. With the long-term effects of the construction of the new line and the long-term dynamic load of the trains, the service status of the existing roadbed needs to be carefully evaluated.

Research on the deformation of adjacent structures caused by construction has attracted the attention of engineers and technicians; for example, research on the stability and safety of railways under tunnels, which are now common [1]. Through the real-time monitoring of parameters, such as settlement deformation and horizontal displacement of adjacent structures, timely engineering measures can be carried out according to the monitoring to ensure the stability of structures during construction [2–4]. System safety assessment can be performed through efficient probabilistic stability analysis of engineering slopes along new railway lines [5]. The author carried out deformation monitoring of the existing line foundation, mainly based on the horizontal displacement and lateral stress changes, on the Shanghai–Nanjing Intercity Railway Project [6]. The above studies are all analyzed from a static point of view; however, research on the stability of the roadbed on one side facing up under the action of the train load belongs to the category of dynamic analysis. The dynamic response of the subgrade under the action of a dynamic load has become the basis for evaluating the subgrade condition [7,8], especially under conditions where the excavation on the side of the existing line base causes the stress field of the existing line to change. The dynamic deformation is an important evaluation index reflecting the state of the subgrade [9,10]. Ye et al. [11] combined field measurements and finite element calculations of dynamic response to analyze the amplitude characteristics of dynamic stress, dynamic deformation and vibration acceleration of ballastless track road foundations, and revealed the distribution law of stress and strain in the foundation bed under train load. Han [12] studied the dynamic characteristics of the roadbed of the Beijing-Shanghai high-speed railway, and concluded that the acceleration, dynamic stress, and displacement decrease nonlinearly with the increase of depth, and the dynamic stress in the roadbed increases with the increase of train speed. Bi [13], through the combination of three-dimensional finite difference software and theory, carried out the numerical simulation of the heavy-haul railway roadbed, analyzed the dynamic response law of the multi-layer roadbed system, and obtained the lateral distribution law of the dynamic stress on the roadbed surface. In the field of subgrade dynamic test research, in the early stages of the research on the dynamic characteristics of track subgrades, the test method was mostly used, and the dynamic response law of the subgrade was obtained by statistical analysis of the test data. Okumura analyzed the field test data of eight ordinary lines in Japan, and believed that the distance from the track, the structure of the track, the type of train, the speed of the train, the length of the train, and the natural vibration characteristics of the ground are the six main factors that affect the vibration level of the roadbed. Madshus proposed a semi-empirical model of subgrade low-frequency vibration, which includes five independent influencing factors: train type, speed, distance, track state, and the influence of structures. Roadbed dynamic tests have been conducted in China many times. Yang et al. [14] obtained the following from the actual measurement of subgrade dynamic stress and vibration: the irregularity of the line has a great influence on the subgrade dynamic stress, an increase in the driving speed can lead to an increase in the subgrade dynamic stress, and the subgrade dynamic stress and vibration acceleration show an exponential decay law in the depth direction. Mao [15] surveyed the ground vibration attenuation curves of 109 trains, and obtained empirical formulas for ground vertical and horizontal vibration attenuation consistent with the principles of vibration propagation energy diffusion and energy attenuation through composite regression. In research on the dynamic stability of roadbeds, Nie et al. [16] used the critical dynamic stress method (dynamic strength) and the dynamic shear strain method (dynamic deformation) to preliminarily evaluate the long-term dynamic stability of the red clay roadbed of the Wuhan–Guangzhou high-speed railway ballastless track. Wang [17] established a three-dimensional finite element numerical model of a ballastless track-roadbed system to simulate the running process of eight marshalled bullet trains. Combined with the measured data, the effects of track irregularity, train speed, axle load and depth on vertical dynamic stress were analyzed. Nie [18] developed the track-roadbed dynamic test system and applied it to the long-term dynamic performance tests of the dynamic response of many heavy-haul railway and high-speed railway subgrade projects.

In general, there are few studies on the dynamic performance of roadbeds. Restricted by objective conditions, most of the vehicle speeds in China's track subgrade dynamic response tests are not high, and cannot reflect well the subgrade dynamic characteristics of the trains at high speed. Due to the lack of high-speed test data, the dynamic response law of track subgrades cannot be summarized with high reliability. In addition, testing of the dynamic deformation of the subgrade requires the arrangement of measuring rods or embedded components in the subgrade drilling [19,20], which is costly and cumbersome, and is not suitable for promotion and application in the public works sector. Moreover, this type of subgrade dynamic response research is tested under the normal operating state of the subgrade body.

There are few relevant experimental studies on the dynamic testing of the existing line foundation under the influence of construction, and only numerical simulation methods can be used to analyze such problems. Bi et al. [21] used the finite element method to simulate a wheel-rail excitation load for the foundation pit construction of a nearby operating subway station in Shanghai, analyzed the dynamic response of the existing station structure during the excavation of the deep foundation pit, and used displacement and vibration acceleration to evaluate the subway's degree of impact. In addition, there are also research results on the use of numerical models to calculate the impact of foundation pit excavation on existing lines [22]; however, no experimental data were used to prove it.

In order to ensure the construction quality and construction period of the new line, and to ensure the safe operation of the existing line on the basis of static tests, this study carried out dynamic testing of the existing line foundation operation state close to the construction site. The test design takes the existing line operating trains as the vibration source, and the existing line base dynamic response as the goal to test the existing line base dynamic response; obtains the vibration displacement, frequency, acceleration and other subgrade dynamic performance indicators; and based on the test data, optimizes the construction technology and proposes construction safety measures adjacent to the existing subgrade. The test results and engineering measures provide a scientific basis for the safe operation and smooth construction of the existing line, and have important theoretical significance for guiding the construction of new subgrades adjacent to the existing line.

2. Project Overview

The nearest road section between the old and new lines was selected as the test section on the site to obtain data on the largest impact of new railway construction on the service status of the existing lines. The test section is located at DK95 + 256 of the Shanghai–Nanjing

Line, corresponding to K1274 + 128 of the existing Beijing–Shanghai Line, and is composed of silty clay with a thickness of 18–20 m underneath. The two lines are only 4–5 m away from each other and share a drainage ditch. The existing Beijing–Shanghai railway line was built in the 1970s using natural foundations. The post-construction settlement control of the newly built Shanghai–Nanjing Railway is strictly regulated, with the requirement being not greater than 15 mm. Therefore, the new pile-raft composite foundation adopted by the Ministry of Railways was introduced for soft soil foundation reinforcement. Taking into consideration that it was close to the existing line, the pile type adopted was that of static pressure construction, the pipe pile was 18 m long, the raft thickness was 0.5 m, and the gravel cushion was laid underneath. Temporary protection was provided before excavation and replacement, and a row of rail piles was installed outside the excavation range for protection. The distance between rail piles was 2.0 m, and the pile length was 5.0 m. The design section of the test section surface is shown in Figure 1.

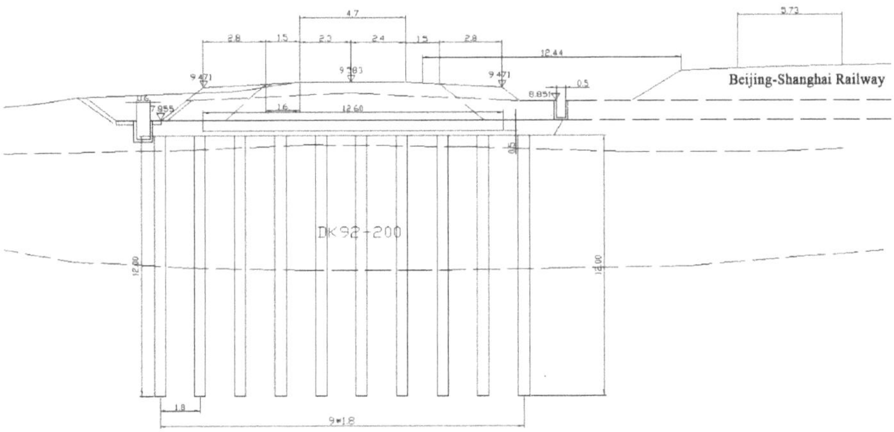

Figure 1. Schematic cross-section of two lines close to the test.

Subgrade dynamic analysis research is an important technical guarantee to reduce the damage of railway subgrades. The dynamic response of the subgrade includes the deformation, dynamic stress, acceleration, etc., of the subgrade. The magnitude and process of the dynamic response of the subgrade are directly related to the strength and fatigue characteristics of the subgrade, the cumulative deformation, and its dynamic stability [23,24]. Therefore, how to test and evaluate the stable state of the roadbed under the action of train power was the main problem to be solved.

3. On-Site Power Test

3.1. Test Content

The dynamic response of the subgrade under the action of a dynamic load (that is, a change in parameters such as vibration acceleration, vibration velocity, vibration flat rate, and vibration displacement of the subgrade) includes change over time, attenuation, and transmission and diffusion along the horizontal and vertical directions of space. The dynamic characteristics of the subgrade are exhibited by the changes in these dynamic parameters and directly affect the operation of the road. Field testing was undertaken to test the dynamic changes in these parameters over time and space under load. The dynamic test acquisition system is shown in Table 1.

Table 1. Dynamic testing of collecting systems.

Equipment Name	Quantity
Donghua DH-5935 dynamic signal test and analysis system	1
DH-5935 solid state acceleration shock pickup	4
imc16 channel dynamic signal tester	1
imc accelerometer	6
891-II type vibration sensor (horizontal, vertical)	6
RDP displacement sensor	5
Laptop	1
Desktop PC	1

The on-site in situ dynamic test mainly tested the following: (1) the distribution and changing law of soil dynamic parameters in the subgrade under a specific train operating speed; and (2) the distribution and variation law of dynamic displacement and overall displacement under different train operating speeds.

There was a large number of construction machines and personnel on the project site, and the test environment was very complex. During the test process, attention had to be paid to the safety of personnel and equipment, and various interferences in the process of power data acquisition had to be eliminated as much as possible. Combined with the actual situation on site, the overall test idea adopted an "embedded vibration measuring pile + sensor". Between the new line raft and the existing line base guardrail, three steel pipes with different depths were buried along the line, with depths of 1, 2, and 3 m, respectively, as permanent monitoring piles. In addition, at each construction stage, a section of temporary observation piles was buried.

The tested and operated trains were mainly the "Harmony" series of power-dispersed EMUs, with two 12-car trains as the standard. During the test, a particular person was responsible for measuring the speed of the vehicle and recording the vehicle information at the beginning/continuing/end of each construction stage. In addition, the test included parameter changes in the case of natural environment excitation, construction machinery vibration, etc.

3.2. Test Methods

When the existing train passed by, the test software was started to collect the vibration data. The channel adopted 1–4 piezoelectric sensor channels, which could obtain vibration acceleration, dynamic displacement, and vibration speed. The on-site power monitoring is shown in Figure 2.

(a) (b)

Figure 2. Field dynamic testing. (**a**) is data acquisition; (**b**) is data processing.

The data collection was designed according to the plan, and the sensor was fixed at the measuring point. In the cross-section direction, measuring points were arranged at different distances from the existing line to monitor the distribution law of the dynamic characteristics of the embankment soil in the cross-section direction. Acceleration sensors were arranged on the top of steel pipes with different depths to monitor the distribution of the dynamic characteristics of the embankment at different depths. Displacement sensors were arranged at different heights of steel pipes and slopes to monitor the distribution law of dynamic displacement when the train passed by. The layout of the measuring points is shown in Figure 3.

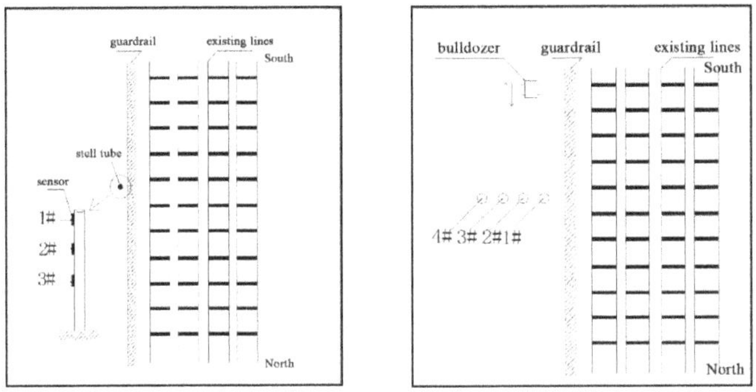

(a) (b)

Figure 3. Monitoring of measuring points. (**a**) is schematic of sensor placement; (**b**) is schematic diagram of steel pipe laying.

4. Analysis of Test Results

According to the collation and analysis of the test results, the data were analyzed in terms of vibration frequency, acceleration, velocity, and displacement. In the current vibration research, the vertical vibration generally dominated. Therefore, the vertical vibration was used as the evaluation index for the analysis in this paper.

4.1. Vibration Frequency

4.1.1. Subgrade Natural Frequency

The natural frequency of the structure is the basis of dynamic calculations. Before the dynamic calculation, frequency extraction and analysis must be carried out to obtain the mode shape and natural frequency of the structure. The natural vibration frequency of the existing line base depends on the subgrade filler, the subgrade soil properties, and the surrounding soil conditions. The dynamic test studies of high-speed railway foundations, such as Wuhan–Guangzhou and Qinshen, show that the [17,18] high-speed railway subgrade is an engineering structure with a low natural vibration frequency. The subgrade natural vibration is mainly a low-frequency natural vibration, ranging from 0 to 40 Hz, and there are five large peaks in between, namely 11.8, 20.1, 22.6, 27.9, and 30.9 Hz, which all represent the subgrade natural frequency. In a field resonance test of an embankment conducted in Japan on the Tokai Shinkansen, the resonance frequency of the embankment was measured at 15–20 Hz.

The train wheel group forms a load series through the subgrade section, so that the subgrade is subjected to repeated loads to generate forced vibrations, and the excitation frequency changes with the train speed. When the excitation frequency is equal to or close to the natural vibration frequency of the roadbed, resonance will occur and the vibration of the roadbed will be aggravated. Therefore, during the test, attention was paid to the coincidence of the natural frequency of the roadbed and the external excitation frequency.

4.1.2. Train Excitation Frequency

The dynamic loads imposed by trains on the roadbed are complex, and mainly include: the vibration of the sprung and unsprung mass of the vehicles, the impact of the trains on the track caused by the uneven surface of the wheel and the rail, and the periodic load caused by the movement of the wheel axle load, etc. These excitations can be regarded as the superposition of harmonic components of different frequencies, in which the shaft load movement mainly produces a low frequency (0–40 Hz). The unevenness of the wheel and rail mainly produces the middle frequency band (40–100 Hz). According to the measured data of the dynamic stress spectrum of the subgrade, the fundamental frequency (that is, the passing frequency of the vehicle) has the greatest impact on the subgrade. The self-power spectrum is obtained by filtering and eliminating the trend. The typical test spectrum is shown in Figure 4.

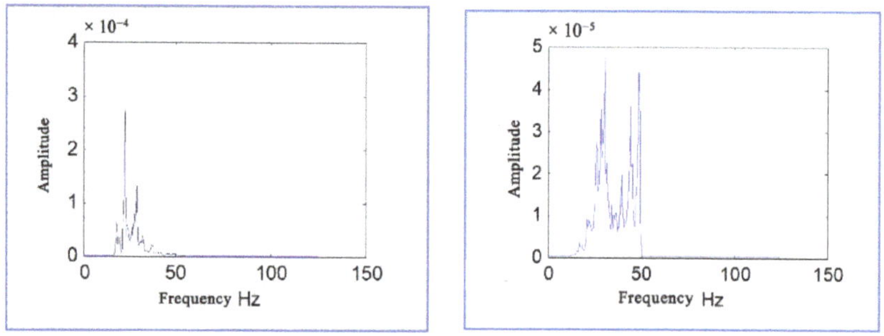

(a) (b)

Figure 4. Testing spectrum under train vibration load. (**a**) is Speed 60 km/h; (**b**) is Speed 120 km/h.

As shown in Figure 4, the vibration frequency of the subgrade under the train load is concentrated in the 0–50 Hz range, the vibration energy at the peak value is relatively large, reflecting the load action frequency of the train, and the peak value is mainly concentrated in the 20–40 Hz range. At different speeds, the soil response frequency has two peaks, verifying the double-peak phenomenon of the dynamic response of the high-speed railway subgrade mentioned in the literature [5]. The excitation frequencies all move to the high frequency direction with the increase in speed, and the energy corresponding to the peak value increases. The in situ monitoring went through the process of pouring the bearing plate, embankment filling, and preloading in the test section. The vibration frequency of the foundation soil caused by the train load was kept at 20–40 Hz. Among these, the excitation frequency at the foundation excavation stage is relatively low, maintained at approximately 20 Hz, which is close to the natural frequency of the roadbed obtained by the Japanese test in the existing research.

The impact or vibration frequency of the axle of high-speed trains on the roadbed is determined by the speed of the vehicle and the structure of the roadbed. The test vibration frequency is close to the natural vibration frequency of the roadbed. When the natural vibration frequency of the roadbed is consistent with the action frequency of the train, resonance will occur, which will intensify the vibration of the train and the track, increase the sinking deformation of the track, and affect the stability of high-speed driving. From the perspective of vibration frequency control, the safety monitoring of the dynamic performance of the existing line base should be further strengthened, and necessary engineering safety protection measures should be taken.

4.2. Vibration Displacement

Vibration displacement is generated briefly when a train passes by. Generally speaking, train speed, axle load, line smoothness, and the properties of roadbed materials are the main factors affecting vibration displacement. However, in the case of excavation at the immediate position, the stress state of the subgrade changes greatly, and attention should also be paid to the change in vibration displacement of the subgrade. The line/vehicle vertical coupled vibration analysis model is used by Southwest Jiaotong University to calculate the subgrade dynamics. It is believed that, under the action of the train load, the elastic deformation of the subgrade is the result of the interaction of all the parts of the wheel/rail/track bed/subgrade system. It is proposed that the dynamic elastic deformation value corresponding to subgrades with differing stiffness should be controlled within

3.5 mm. According to the compaction standards and wheel load conditions proposed in the "Interim Regulations on the Design of Bridges, Tunnels and Stations on the Beijing-Shanghai High-speed Railway Line", when the height of the embankment is 5–10 m, the elastic deformation of the subgrade surface is within the range of 1.32–2.25 mm.

As shown in Figures 5 and 6, during the construction excavation stage, the test vibration displacement was basically in the elastic stage, the vertical vibration amplitude reached 1.9 mm, and the horizontal vibration amplitude reached 0.15 mm. The maximum vibration displacement occurred in the surface soil of the slope during the excavation stage. After the embankment was filled in the later stage of the test, the vibration displacement of the foundation soil was basically in the 10–4 m range. Richart compiled data on the allowable vertical amplitude of vibration for a specific frequency, and found that when the operating frequency is 40 Hz (2400 cycles/min), the displacement amplitude limit is approximately 0.018 inches (0.457 mm). Based on this preliminary judgment, the dynamic characteristics of the foundation soil between the two lines tended to be stable.

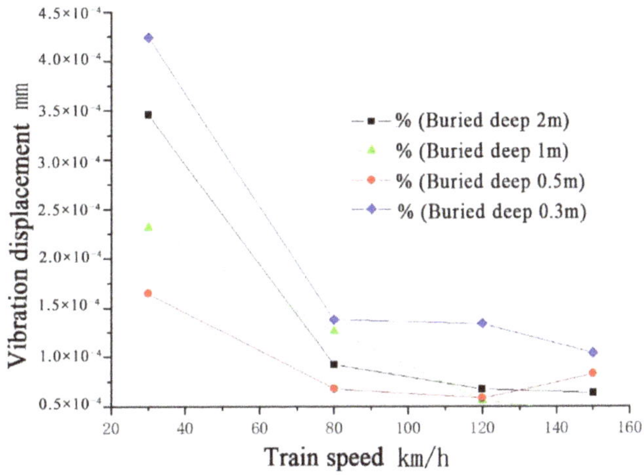

Figure 5. Foundation soil vibration displacement-train speed curve.

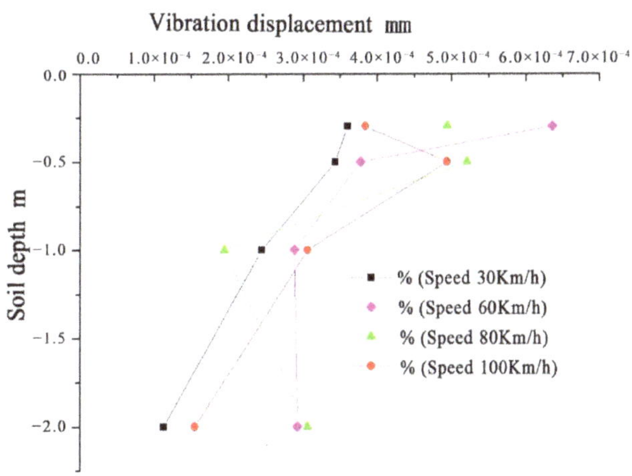

Figure 6. Foundation soil vibration displacement—depth distribution rule.

In addition, during the test period, the vertical vibration displacement of the ground had a certain attenuation along the depth direction. When the trains ran at low speed, the surface vibration mainly represented the local elastic deformation caused by the axle load, the surrounding surface fluctuation was very small, and the vibration was a quasi-static response. When the train speeds reached the foundation shear wave speed of 80 m/s, the vibration increased, the Mach effect was obvious, and the vibration influence range became larger.

The fast Fourier transform (FFT) was used to calculate the convolution integral, finding that when the excitation frequency is larger, the fluctuation of the surface displacement is larger. This can be explained as follows: after the vibration wave is transmitted to the underlying soil, it is reflected on the surface and interferes with the original wave. As the distance increases, the vertical amplitude will fluctuate to a certain extent, instead of being monotonically attenuated. This phenomenon increases with the excitation frequency, and the increase becomes more and more obvious.

As shown in Figure 7, taking the peak value of vibration displacement at the same measuring point in each construction stage, it is found that the vibration displacement generated is the largest in the stage of subgrade excavation, and the dynamic deformation gradually decreases with the completion of subgrade filling.

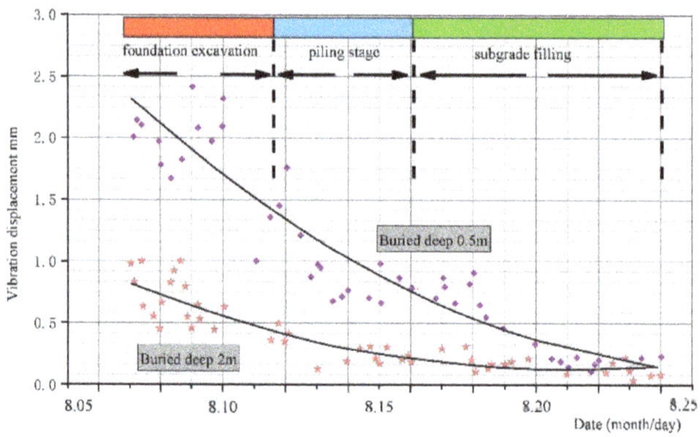

Figure 7. The vibration displacement peak trend of different construction stage vibrations.

4.3. Vibration Acceleration

The magnitude of the vibration acceleration of the subgrade is the main parameter for judging the effect of vibration on track damage [16], and the amplitude of the vibration acceleration is closely related to the dynamic characteristics of the subgrade and the frequency of load action. In Japan, vibration tests have been conducted on the subgrade surface of Shinkansen high-speed trains, and the maximum acceleration is 5–20 m/s^2. From the acceleration time–history curve, the peak periodicity of the acceleration waveform under the action of the dynamic load of the running train can be clearly distinguished.

The data show that there is a certain correlation between the peak vibration acceleration, the speed of the train, and the depth of the soil layer, showing a law similar to that of dynamic stress. When the train speed was 30 km/h, the vibration acceleration peak depth range corresponding to buried depths of 2, 1, 0.5, and 0.3 m was attenuated by 67%. At a burial depth of 0.5 m, the train speed increased from 30 to 120 km/h, and the ground response vibration acceleration peak increased by 44%. The vibration acceleration distribution and vibration acceleration-train speed change curves are shown in Table 2 and Figure 8.

Table 2. Vibration acceleration distribution.

Train Speed km/h	Buried Deep (m)	Content Test	Numeric Value m/s^2
30	2	Vibration acceleration	0.01029
	1		0.01736
	0.5		0.01874
	0.3		0.03202
120	0.5		0.03375

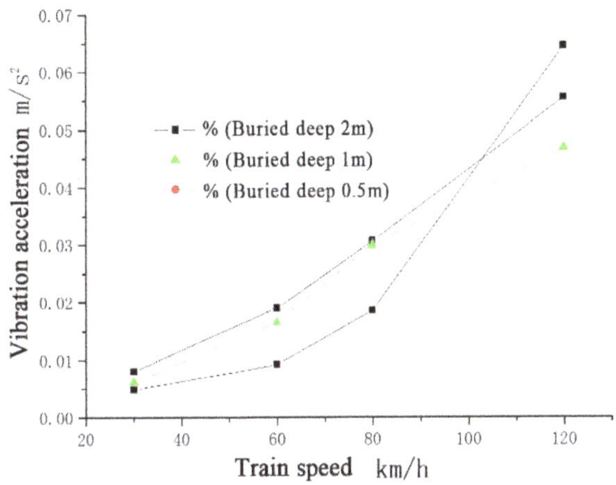

Figure 8. Vibration acceleration–train speed changing curves.

The foundation soil is non-uniform and non-isotropic, and is an inelastic medium. Vibration waves propagate in the soil layer, and different types of vibration waves will be reflected and refracted when they encounter the soil layer interface. In the process of wave propagation, even a very small plastic deformation will cause energy dissipation due to the internal friction of the soil. With an increase in the distance of the train excitation vibration source, the wave energy is dissipated, and the acceleration vibration amplitude is also obviously attenuated. The vibration acceleration-depth distribution is shown in Figure 9.

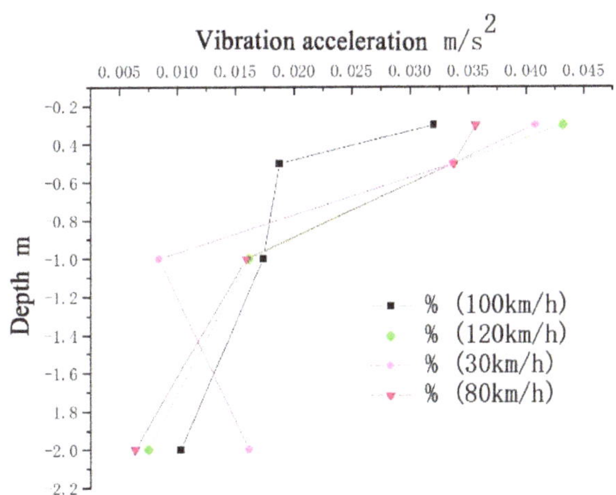

Figure 9. Vibration acceleration–depth distribution rule.

According to the regression analysis of the measured data, the vertical and horizontal vibration acceleration attenuation formulas of the foundation soil of the Shanghai–Nanjing and Beijing–Shanghai projects can be summarized as follows. The measured vibration acceleration curve is shown in Figure 10.

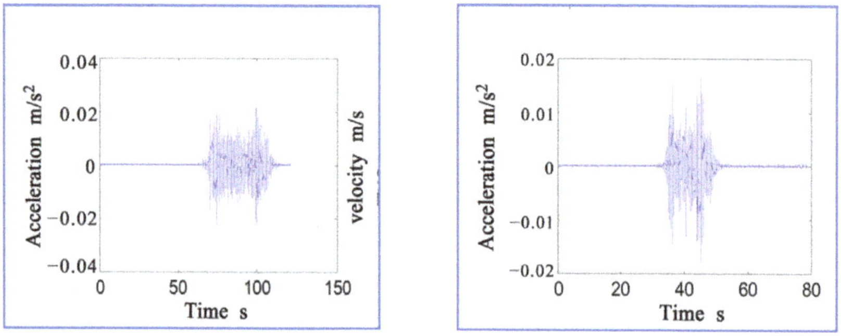

(**a**) (**b**)

Figure 10. Testing vibration acceleration curve. (**a**) is Speed 60 km/h; (**b**) is Speed 120 km/h.

Horizontal vibration acceleration distance attenuation formula:

$$a = 0.145 \times e^{-0.20s}$$

Vertical vibration acceleration attenuation formula:

$$a = 0.031 \times e^{-0.17s}$$

The variation law of the vibration velocity distribution can be obtained by the convolution and integration of the vibration acceleration, as shown in Figure 11.

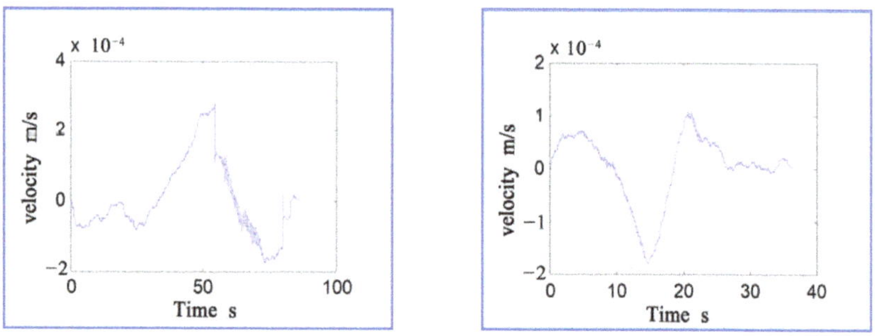

(**a**) (**b**)

Figure 11. Testing vibration velocity curve. (**a**) is Speed 60 km/h; (**b**) is Speed 120 km/h.

Many domestic and foreign scholars or codes have given structural vibration limits based on the peak ground vibration velocity (PPV). The blasting vibration effect is shown in Table 3.

German researchers obtained the relationship between vibration speed and the stability of railway subgrade at different depths using tests, as shown in Figure 12. The critical vibration speed of coarse-grained soil is approximately Vkrit = 30 Ip (mm/s), while for mixed-grained clay soil, it is Vkrit = 25 $Ip^{1.5}$ (mm/s).

Table 3. Blasting vibration effect.

Particle Velocity/mm·s^{-1}	Vibration Effect
<1	Hard to feel
1	Humans can feel weak vibrations
5	Makes people feel uncomfortable, vibrate
10	Disturbing, with obvious vibration
33	Gives a strong sense of vibration
50	Safe vibration limit of general civil residential buildings
100	Safe vibration limit of reinforced concrete structure and tunnel support structure
140	Cracks in rock medium and expansion of old cracks
190	General civil buildings are severely cracked and damaged
300	Vibration and shedding of rock in unsupported tunnel
600	New cracks in rock

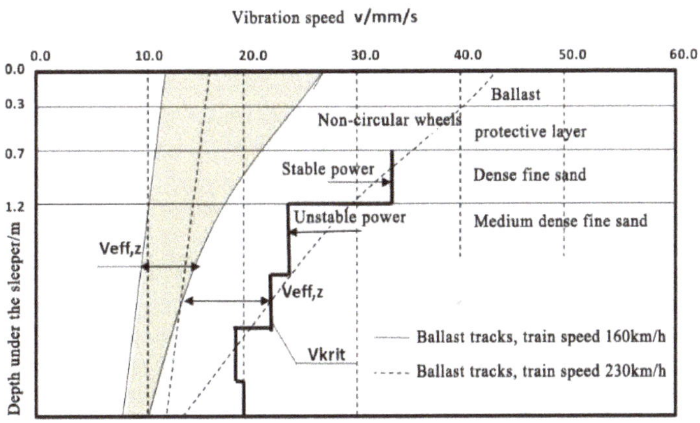

Figure 12. The dynamic stability and vibration velocity at different depths.

As peak velocity is less affected by frequency relative to peak acceleration and peak displacement, the dynamic test found that the attenuation of the vibration velocity in the horizontal amplitude is much faster than that in the vertical amplitude. The data show that the attenuation of soil vibration velocity along the depth direction is similar to the vibration acceleration, that the maximum vibration velocity of the surface is 5.04×10^{-3} m/s, and that the influence of vehicle speed on the vibration velocity is not obvious. Referring to the vibration velocity limit of the blasting vibration effect, there is an obvious vibration sense at the test site, but it is within the structural engineering safety limits.

Extracting the amplitudes of the dynamic response parameters in the three construction stages of roadbed excavation, piling, and roadbed filling, it can be seen that the vibration acceleration and vibration velocity show similar laws. That is, from the beginning of the excavation construction measurements, the vibration response tends to decrease, and the vibration response is the largest in the subgrade excavation stage. In the pile formation stage, due to the disturbance to the foundation soil, there is a slight rebound, and until the subgrade is filled, the vibration parameter gradually stabilizes. The trend of peak vibration acceleration in different construction stages is shown in Figure 13.

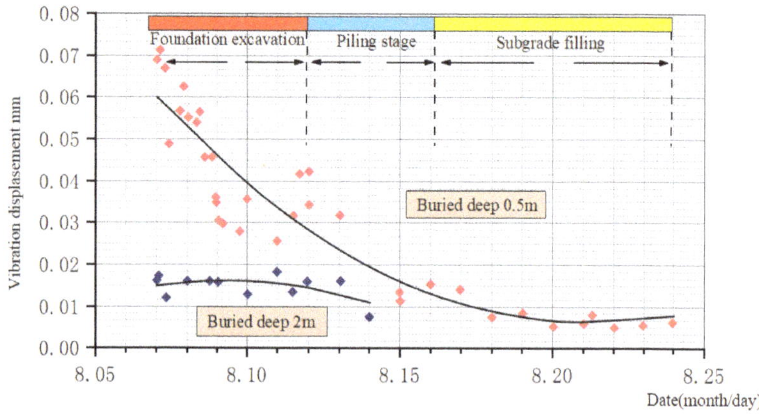

Figure 13. The vibration acceleration peak trend of different construction stage vibrations.

To sum up, the construction in the immediate position has a significant impact on the dynamic performance of the existing line foundation. Referring to the relevant regulations, the vibration displacement and vibration acceleration of the roadbed do not exceed the standard, but they specifically involve the dynamic stability of the existing line foundation. As there is no norm or experience in this area for reference, from the perspective of construction safety and existing line operation safety, close attention should be paid and construction safety measures should be adopted.

5. Test-Based Construction Safety Measures

5.1. Subgrade Excavation Stage

The stability of the roadbed was monitored through real-time dynamic tests during construction, and the test data were analyzed and fed back to the project site in real time. On one hand, protection measures for the subgrade slope were proposed, while at the same time, the road sections with abnormal stability and dangerous driving safety were found in the test to provide a basis for a reasonable opening time for the emergency construction skylight.

5.1.1. Excavation Roadbed Slope Protection

The test shows that the stability of the existing line foundation deteriorates after excavation. In addition, the exposed soil of the silty clay slope becomes looser under the influence of rain erosion and soaking, so the slope should be protected accordingly. On the project site, measures such as shotcrete plastering and hanging nets are mainly used to reinforce the excavation slope. Shotcrete plastering can effectively prevent the infiltration of rainwater and surface water, produce an embedded effect, improve cohesion, and improve the properties of the soil. Hanging nets can help strengthen the overall strength of the slope.

5.1.2. Reasonable Opening of Skylights and Determination of Construction Blocking Time

The "gap" required for maintenance and construction is reserved on the train operation map, and this gap is called the "skylight". The skylight is the comprehensive embodiment of the coordination between construction organization and transportation organization. In order to minimize the interference of the construction blockade on the existing line transportation, the arrangement of the blockade time is a core issue.

In the construction operation in the parallel section of the operation line, the normal operation of the existing line should be maintained as much as possible; however, from the perspective of safe driving, once the test finds that the stability of the subgrade is abnormal, emergency repairs should be carried out. The dynamic test provides a data basis for determining a reasonable opening time for the skylight, the timely forecasting of the road section with the subgrade stability exceeding the standard, and formulating the skylight opening time as soon as possible, so as to save valuable time for the transportation organization and the construction organization. The time actually occupied by the skylight consists of the following three parts: (1) the opening time of the fixed skylight; (2) the safety time reserved before and after the skylight; and (3) the additional time for the train to run after the skylight is opened.

5.2. Engineering Measures in the Piling Stage

In the Shanghai–Nanjing Intercity Railway DK92 + 700~DK93 + 515, DK95 + 404.3~DK95 + 619.54, and DK95 + 656.56~DK95 + 692, a total of 1065.62 m of subgrade foundation was reinforced with PC-A500 high-strength prestressed concrete pipe piles. The concrete strength grade was C60, the wall thickness of the pipe pile was 100 mm in a square layout, the pile spacing was 2.2~2.4 m, and the pile length was 15~37 m. The three sections of subgrade were all parallel to the existing Shanghai–Nanjing Railway, and the distance from the center of the outermost row of pipe piles to the center of the existing Shanghai–Nanjing Railway line was less than 20 m, with the nearest one being 6.8 m.

Engineering practice shows that prefabricated pipe piles are widely used because of their good bearing capacity, scale of production and construction, and easy detection of quality. The vibration, soil squeeze, and noise caused by the construction process have a negative impact on the surrounding environment; in particular, the stress waves generated by piling will cause a strong vibration of the surrounding soil, which will affect the normal use and safety of the surrounding buildings and underground facilities. A schematic diagram of the impact of piling is shown in Figure 14.

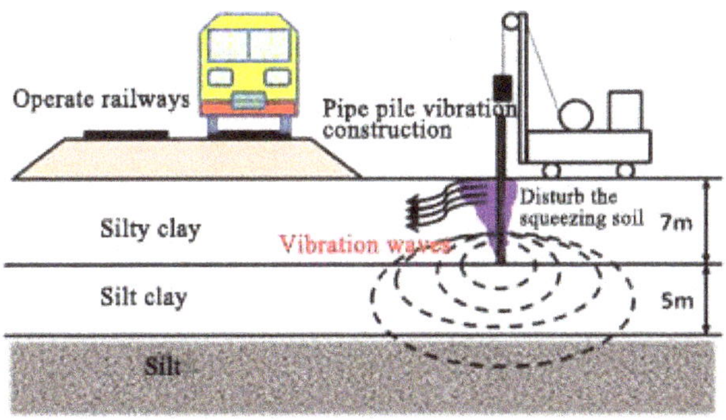

Figure 14. Piling influence diagram.

The vibration method of pile driving is similar to an earthquake or blasting. Energy is transmitted in the form of waves from the earthquake source through the rock and soil, thus causing the vibration of the foundation, existing bridges, and buildings. Although the duration of the single-click vibration is short, the amplitude and period of the impact energy load generated by the mechanical action are almost unchanged during the load action due to the small interval and the large number of strikes. According to the existing research results [25], the pile side soil can be divided into four areas according to the degree of disturbance from the pile driving process: Zone—the hard layer area along the edge of the pile body, which is formed under the action of the extrusion force along the pile, with most of the pore water pressure in the soil being squeezed out; Zone II—known as the remodeling of the compacted zone, where the soil mass is disturbed and reshaped under the action of the compressive force, and the compaction is improved, while the shallow soil above the ground uplifts; Zone III—the disturbance zone, where the thickness is approximately 3.5 times the diameter of the pile, the average density of the soil is reduced, the water content is large, and the pore water pressure is increased; and Zone IV—the less affected zone, where the additional stress and pore water pressure caused by the pile driving are relatively small, and the original structure of the soil remains basically unchanged.

In order to prevent the impact of piling construction on the vibration and extrusion of the existing line, it was proposed that the static pressure method be adopted to construct the prestressed pipe pile and that stress relief holes be set up on the side of the existing line with a spacing of 3 to 5 m and a diameter of 0.5 m, a depth not less than half the length of the pipe pile and not less than 10 m, and a general length of 10 to 19 m, which effectively minimizes the impact of the construction of adjacent prestressed pipe piles on the existing line. A schematic diagram of the optimization technology for pile construction is shown in Figure 15.

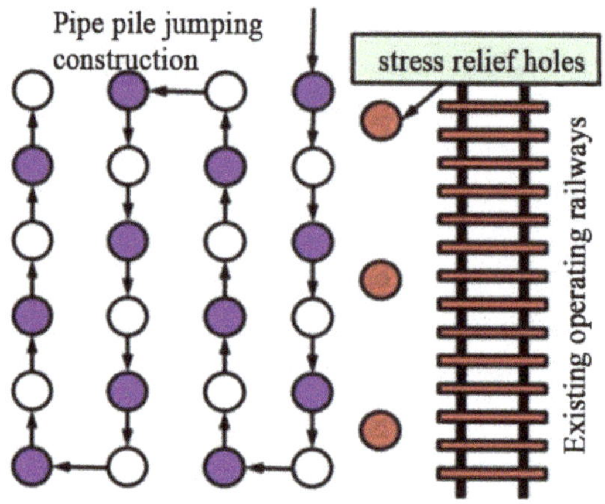

Figure 15. Pile construction optimization technology schemes.

The piling construction safety technology reduces the impact on the existing lines from three aspects:

(1) By setting up stress release holes on the side adjacent to the existing line, the soil-squeezing effect of the pipe pile construction can be effectively released, and the continuous transfer of soil to the existing line side can be blocked to ensure the safety of the existing line.
(2) Using static pressure construction instead of dynamic piling.
(3) The jumping sequence is adopted to form the pile, and the purdah is formed as soon as possible.

The specific construction technology is as follows. Before constructing the pipe piles, construct stress-relief holes first, using a gravel grouting pile machine to extract the soil, with a diameter of 50 cm, laying them at 1–2 m outside the pipe pile closest to the existing line, parallel to the existing line 3–5 m, with a hole depth of 50 cm. At half the length of the adjacent piles, and not less than 10 m, use a short-screw earth-boring rig to borrow soil to form a hole, filling it with gravel over time after the drilling is completed. After the construction of the stress relief holes is completed, the pipe pile is then constructed. Its setting enables a drainage channel to dissipate the excess pore water pressure generated by the construction of the pipe pile. When constructing the pipe piles, the same row of jumping is used; that is, the No. 1, 3, 5, 7, etc., piles are constructed first, and the No. 2, 4, 6, 8, etc., piles are constructed after 24 h of no change from roadbed observation and rail surface observation. The second row of piles and the third row of piles are constructed according to this method, so that the constructed pipe piles on the side of the existing line form a pile group enclosure, suppress the effect of soil squeeze on the side of the existing line, and provide sufficient dissipation time for the excess pore water pressure.

After the construction was completed, the displacement test showed that the construction of the adjacent pipe piles did not cause obvious extrusion deformation of the existing line. Due to the use of the static pressure construction method, there was no adverse vibration effect.

6. Conclusions

In this study, combined with a passenger-dedicated line construction project, experimental research on the dynamic stability of an existing line foundation under the influence of a new line construction was carried out, and the construction technology was optimized according to the experimental analysis. The following conclusions were obtained:

(1) A set of dynamic performance testing methods for railway subgrade in operation are proposed to obtain the dynamic performance indicators of the subgrade, such as vibration displacement, vibration frequency, and vibration acceleration of the existing subgrade under the influence of construction.

(2) During testing, the peak value of the vibration response parameter in the subgrade excavation stage was the largest, which was the most dangerous stage of the subgrade dynamic stability. Subgrade vibration acceleration as a common index of structural dynamics has similar standards to follow, which can directly reflect the dynamic performance of the test body, and vibration displacement can be used as a dynamic deformation condition to evaluate the dynamic stability of the subgrade. Therefore, it is suggested that these two parameters be used as the dynamic stability control index of the existing roadbed under the influence of excavation.

(3) During testing, the excitation frequency of the train to the subgrade is close to the natural frequency of the railway subgrade, and the possible resonance will lead to the aggravation of the subgrade vibration. This situation mainly occurs in the foundation excavation stage, and it was found that it is affected by the changes in the foundation stress field and displacement field. Based on this, it is suggested that the engineering construction department take preventative measures, such as strengthening the slope and speeding up the construction.

(4) Dynamic testing provides a forecast and an early warning for the stability of the subgrade slope, and supplies data support for the protection of the subgrade slope and the reasonable opening of skylights. The piling construction technology is optimized from the perspective of subgrade dynamic stability, and the adverse effects, such as the extrusion and vibration of the adjacent existing line foundation, are eliminated by setting up stress relief holes, static pressure pipe piles, and jumping construction.

The test results provide data support for the safety evaluation of the existing subgrade. The construction safety measures proposed based on the test results ensure the safe excavation of the adjacent existing subgrade, which has important theoretical significance for guiding the construction of new subgrades adjacent to the existing subgrades in the future.

Author Contributions: Conceptualization, S.Z. and J.L.; methodology, S.Z. and T.L.; software, P.L.; validation, P.L., J.L. and X.C.; formal analysis, T.L.; investigation, S.Z. and T.L.; resources, S.Z.; data curation, J.L. and P.L.; writing—original draft preparation, S.Z. and T.L.; writing—review and editing, J.L. and P.L.; visualization, X.C.; supervision, L.J and P.L.; project administration, T.L.; funding acquisition, S.Z. All authors have read and agreed to the published version of the manuscript.

Funding: This work was supported by the key R&D projects in Shandong Province under grant number 2019GSF111008 and the Shandong Provincial Department of Transportation Science and Technology Plan Project under Grant No. 2021B117, the National Natural Science Foundation of China (Grant Nos 51778630 and 52178182), China Railway Science and Technology Research and Development Plan Project (Grant Nos 2020-Major project-02, 2021-Major project-02, 2021-Key projects-11).

Institutional Review Board Statement: Not applicable.

Informed Consent Statement: Not applicable.

Data Availability Statement: The data that support the findings of this study are available from the corresponding author.

Conflicts of Interest: The authors declare no conflict of interest. The funders had no role in the design of the study; in the collection, analyses, or interpretation of data; in the writing of the manuscript, or in the decision to publish the results.

References

1. Huo, J.; Wang, B.; Zhou, S. Safety analysis of the foundation reinforcement scheme for subway shield tunnels passing through intercity railways. *China Railw. Sci.* **2011**, *32*, 71–77.
2. Xiong, Z. Horizontal displacement monitoring and reinforcement of a complex plane foundation pit supporting structure. *Geotech. Mech.* **2009**, *30*, 5.
3. Yang, L. Dynamic Prediction of Foundation Pit Support Displacement and Safety Monitoring. *Chin. J. Civ. Eng.* **1999**, *32*, 5.
4. Zhang, C.; Zhang, D.; Luo, J.; Wang, M.; Wu, J. Remote monitoring system in the construction of tunnels under subway stations under existing lines. *Geotech. Mech.* **2009**, *30*, 1861–1867.
5. Wang, T.; Luo, Q.; Li, Z.; Zhang, W.; Chen, W.; Wang, L. System safety assessment with efficient probabilistic stability analysis of engineered slopes along a new rail line. *Bull. Eng. Geol. Environ.* **2022**, *81*, 68. [CrossRef]
6. Xu, L. *Research on Construction Safety and Settlement Monitoring Technology of Shanghai-Nanjing Intercity Railway*; Central South University: Changsha, China, 2010.
7. Liang, B.; Sun, C. Analysis of double peak phenomenon in dynamic response of high-speed railway subgrade. *Chin. J. Civ. Eng.* **2006**, *39*, 117–126.
8. Zhan, Y.; Jiang, G. Field test research on dynamic response of subgrade with ballastless track pile-slab structure on Suiyu Line. *Rock Soil Mech.* **2009**, *30*, 832–839.
9. Liu, J.; Xiao, J. Experimental Study on the Stability of Railroad Silt Subgrade with Increasing Train Speed. *J. Geotech. Geoenvironm. Eng.* **2009**, *136*, 833–841. [CrossRef]
10. Wang, J.-J.; Chai, H.-J.; Hu, Y.-X.; Zhang, L.-J. Mechanism of dynamic instability for riverside subgrade. In Proceedings of the First International Conference on Transportation Engineering, Chengdu, China, 22–24 July 2007; American Society of Civil Engineerings: Chengdu, China, 2007; Volume 2, pp. 1933–1939.
11. Ye, Y.-S. Dynamic response characteristics of high-speed railway subgrade. *J. Railw. Constr.* **2015**, *55*, 7–12.
12. Han, Y.-K. Study on the dynamic response of high-speed railway subgrade. *J. Constr. Technol.* **2016**, *45*, 116–118+123.
13. Bi, Q.-C.; Wang, T.; Zhang, M.X. Numerical analysis of dynamic response of heavy-duty railway subgrade. *J. Hebei Inst. Civ. Eng. Archit.* **2018**, *36*, 6–10.
14. Ye, Y.; Yang, C. Dynamic stability analysis of soft soil subgrade under train load. In Proceedings of the 8th Soil Mechanics and Geotechnical Engineering Academic Conference of China Civil Engineering Society, Nanjing, China, 1 October 1999.
15. Mao, Y. Ground vibration characteristics and attenuation caused by transportation vehicles. *J. Build. Struct.* **1987**, *01*, 67–77.
16. Nie, Z.; Li, L.; Liu, B.; Ruan, B. Base vibration test analysis of Qinshen passenger dedicated line. *Chin. J. Rock Mech. Eng.* **2005**, *24*, 1067–1074.
17. Wang, Q.-Y. Load analysis of ballastless track subgrade packing power test of high-speed railway. *J. Railw. Stand. Des.* **2018**, *62*, 6–11. [CrossRef]
18. Nie, R.-S. Development and application of rail-subgrade dynamic test system. *J. Chin. Soc. Railw.* **2016**, *38*, 96–101.
19. Jiang, Q.; Tang, Z.; Tang, Z. Dynamic test analysis of subgrade under different vehicle speeds. *Subgrade Eng.* **2009**, *3*, 127–132.
20. Yang, X.; Gao, Y.; Liu, Z. On the detection of existing railway line bases. *Chin. J. Rock Mech. Eng.* **2003**, *22*, 2363–2367.
21. Bi, X.; Zhou, S. The influence of train vibration load on the deformation of the existing station adjacent to the deep foundation pit. *J. Tongji Univ.* **2004**, *32*, 1559–1604.
22. Hai, M.; Sun, Y.; Wang, B. Research on the impact of deep foundation pit excavation on adjacent subway stations. *J. East China Jiaotong Univ.* **2009**, *26*, 7–11.
23. Xiang, J.; He, D.; Zeng, J. Analysis of vibration response of different types of ballastless tracks under the action of high-speed trains. *Chin. J. Mech. Eng.* **2010**, *46*, 29–38. [CrossRef]
24. Qu, C.; Wang, Y.; Wei, J. Field test and analysis of subgrade vibration of Wuhan-Guangzhou high-speed railway. *Rock Soil Mech.* **2012**, *33*, 1451–1458.
25. Cao, Y.; Zhao, M. Monitoring and analysis of the impact of near-area blasting on bridge vibration. *Blasting* **2008**, *25*, 82–85.

coatings

Article

Experimental Study on Performance of Modified Cement-Based Building Materials under High-Water-Pressure Surrounding Rock Environment

Jun Yao [1,2], Di Feng [1,*], Zhikui Wang [3], Chengcheng Peng [1], Yonggang Zhang [4] and Lei Han [4]

1 College of Civil and Transportation Engineering, Hohai University, Nanjing 210098, China
2 College of Urban Construction and Transportation, Hefei University, Hefei 230601, China
3 China Railway Construction Underwater Shield Tunnel Engineering Laboratory, China Railway 14th Bureau Group Co., Ltd., Jinan 250101, China
4 Engineering Research Institute, China Construction Eighth Engineering Division Corp., Ltd., Shanghai 200122, China
* Correspondence: fengdi@hhu.edu.cn

Abstract: Traditional cement-based grouting materials have good reinforcement and anti-seepage effects on the surrounding rock under normal conditions, but the grouting effect is not ideal due to problems such as a long setting time, a low stone ratio, and poor crack resistance under high water pressure and in a dynamic water environment. In this study, we aimed to improve the physical properties, chemical properties, and microstructure of a cement-based slurry by forming a hydrogel through its chemical crosslinking with polyvinyl alcohol and boric acid as modifiers for the purpose of improving the permeability resistance of the surrounding rock grouting under high-water-pressure conditions, which can expand the function of traditional building materials. The grouting effect of the modified cementitious material on the surrounding rock was analyzed through indoor tests, the SEM testing of the performance of the modified slurry, the numerical calculation of the seepage field, and the application of the modified slurry in combination with the actual project to verify the water-plugging effect. The research findings demonstrate that (1) the additives boric acid and PVA can significantly speed up the slurry gel time, and the gel time can be controlled within 2–20 min to meet the specification requirements. (2) At a velocity of moving water > 1 m/s, the retention of the solidified modified slurry stone body reaches more than 80%. According to the SEM analysis, the structure of the solidified modified slurry stone body is dense and has good impermeability. (3) According to the numerical calculation analysis, the modified slurry can effectively change the seepage field of the surrounding rock and improve its seepage resistance. The water pressure outside the lining is reduced by 47%, 31%, and 22%, respectively, compared with no slurry, the pure cement slurry, and cement–water-glass grouting, and the indoor test and numerical simulation conclusions are consistent.

Keywords: cement-based material; polyvinyl alcohol; boric acid; impermeability

Citation: Yao, J.; Feng, D.; Wang, Z.; Peng, C.; Zhang, Y.; Han, L. Experimental Study on Performance of Modified Cement-Based Building Materials under High-Water-Pressure Surrounding Rock Environment. *Coatings* **2023**, *13*, 501. https://doi.org/10.3390/coatings13030501

Academic Editor: Valeria Vignali

Received: 3 February 2023
Revised: 18 February 2023
Accepted: 21 February 2023
Published: 24 February 2023

1. Introduction

With the rapid development of science and technology, building materials are increasingly widely used [1,2]. To increase permeability and protect the lining structure and the environment, the surrounding rock of the tunnel can be strengthened by grouting. Traditional grouting materials have good water-plugging and reinforcement effects in static environments, but the disadvantages of a slow setting time, dispersion resistance, and weak crack resistance of the traditional slurry will lead to a poor grouting effect in high-water-pressure and moving environments [3–6]. Therefore, an important research goal is to improve traditional cement-based grouting materials to meet the grouting requirements under high water pressure and in a dynamic water environment and to improve the application range of building materials.

The tunnel grouting material in the high-water-pressure surrounding rock environment needs to have good impermeability and injectability, and tunnel grouting materials have been studied by several researchers. Hua and Xie [7] conducted an indoor test to determine the fluidity, water barrier properties, and compressive strength of fly ash and a cement slurry and an engineering practice test to determine the grouting ratio. Zhang [8] analyzed the applicability and grouting effect of a single cement slurry, a double slurry composed of a cement slurry and water-glass slurry, and a fine cement slurry under different working conditions of karst tunnel curtain grouting in Gele Mountain, and the analysis showed that the double slurry grouting material composed of a cement slurry and water-glass slurry is more effective. Tian et al. [9] examined how fly ash, the water-reducing agent, and the water–cement ratio affected the grouting slurry's flowability, setting time, and compressive strength. Through optimization, a grouting slurry with a high degree of compatibility was created. Li et al. [10] developed a new grouting material with a good grouting capacity, good water retention, and high early strength. The results show that the elastic modulus, grouting volume, and pressure of the material all have an impact on ground settlement, according to the application results of a cross-river shield tunnel grouting project. Through indoor tests, Peng et al. [11] examined the effects of the water-glass admixture, boehmite content, and powder ratio on the slurry's performance. It was found that, compared with traditional materials, the double slurry has the advantages of a controlled setting time, high compressive strength, and environmental protection and is better used in river bottom tunnel grouting work. Zhang et al. [12] used bentonite and cement as base materials with a curing agent to obtain excellent flow, stability, and scouring resistance, which can be applied to meet engineering requirements in a moving water environment while being cost-effective and environmentally friendly. Zhang et al. [13] developed a grouting substance using cement, clay, meta-aluminate, and lignin. Indoor tests revealed that the substance has a quicker gel time, a lower exudation rate and volume shrinkage, a higher viscosity and anti-dispersion effect, and a better underwater grouting effect.

Through indoor tests, Zhu et al. [14] analyzed a slurry composed of cement, clay, and fly ash and tested its mechanical and rheological properties. The results show that the grouting material has good mechanical properties and impermeability. Liu et al. [15] prepared a kind of cement based on magnesium phosphate. The indoor test showed that GGBS significantly shortened the setting time, improved a certain amount of fluidity, produced a compact microstructure, and was environmentally friendly. Wan et al. [16] proposed water glass and bentonite as grouting materials and tested the mechanical properties and microscopic properties of the materials. The results indicate that the grouting materials have good viscosity. Wang et al. [17] studied the effect of PVA fibers on the mechanical properties and durability of rubber concrete through laboratory tests. The results show that PVA can significantly improve the impermeability and durability of rubber concrete. Zhang et al. [18] prepared a new type of double-fluid grouting material and analyzed the setting time, expansion rate, and mechanical strength characteristics of the grouting material through laboratory tests to meet the requirements of the grouting environment. Sun et al. [19] analyzed the influence of different proportions of organic materials such as fly ash, clay, waterborne polyurethane, and epoxy resin on the grouting performance of a cement slurry through a single-factor test, orthogonal test, and organic compound modification test. The results indicate that the addition ratio of clay and fly ash has a strong influence on the strength and stone ratio of the cement slurry.

Through indoor tests, Liu et al. [20] researched the effect of the PVA content on the performance of fly-ash-based cement grouting materials. It had a significant effect on the mechanical properties of the grouting material and could effectively improve the rock mass' hydraulic coupling characteristics. The impermeability of PVA to the grouting material was markedly improved, and it increased with increasing amounts of PVA content. Lu et al. [21] proposed porous sand, PVA, cement, and rubber grouting materials. The density, consistency, fluidity, exudation rate, consolidation shrinkage rate, setting time, and unconfined compressive strength of the slurry were tested in the laboratory. The results showed that the grouting material has good toughness and high strength.

In order to design grouting materials, many researchers have combined different grouting projects with their findings, obtained specific results based on the performance of cement bases, and added modified materials to the pure cement slurry to achieve the desired grouting anti-seepage and reinforcement effects under various working conditions.

However, there are still some issues within the current literature, such as the slurry's low stone solidification retention ratio, low compressive resistance, and insufficient crack resistance, which will result in a poor water-plugging effect when it is exposed to high water pressure or a moving water environment. In recent years, polyvinyl alcohol and boric acid have become increasingly applied modified materials, which are used to rapidly induce gelation via chemical crosslinking at certain concentrations. This mechanism improves the properties of cement-based materials so as to meet the grouting requirements under high water pressure and in moving water environments. In this study, an indoor test was performed to examine the effects of parameters such as the water–cement ratio, the concentration of polyvinyl alcohol, and the concentration of boric acid on the properties of the slurry setting time, stone solidification retention ratio, and compressibility in a moving water environment against the backdrop of the surrounding rock under high water pressure. Combined with scanning electron microscopy, the microstructural characteristics of the modified material's stone body were analyzed to verify whether the modified slurry achieved design expectations. Finally, the water-plugging effect of the cement–polyvinyl alcohol–boric acid slurry was examined by numerical calculations using a real-world engineering application as an example, and the consistency between the results of the indoor test and the numerical calculation was verified.

The main purpose of this work was to analyze the liquid properties of the grouting material, the properties of the stone body, and the water-plugging properties of the whole grouting ring after grouting through laboratory tests and numerical simulations. Further, tests were performed to verify whether the modified cement-based materials can meet the requirements of the surrounding rock grouting in a high-water-pressure environment from the characteristics of the liquid and solid material and the engineering application process in three stages. The results of this study can provide a comprehensive understanding of modified grouting materials in engineering practice.

2. Basic Performance Study of Single Cement Slurry

The primary component of composite grouting materials is the single cement slurry, which is enhanced by modified materials to meet the needs of various engineering environments, including requirements for grouting fluid impermeability and injectability [22]. As a result, the performance of the composite grouting material is directly influenced by the fundamental performance of the single cement slurry, and the following indoor test study was conducted to determine this fundamental performance.

2.1. Test Materials and Procedure

Since the water–cement ratio affects the primary characteristics of the common cement slurry, the pure cement slurry is primarily composed of silicate cement with different water–cement ratios (W/C). The pure cement slurry was prepared with five different water–cement ratios: 0.5:1 (50 g water, 100 g cement), 0.8:1 (80 g water, 100 g cement), 1:1 (100 g water, 100 g cement), 1.5:1 (150 g water, 100 g cement), and 2:1 (200 g water, 100 g cement). Each group of material tests was conducted three times, and the data in the figures in the manuscript are the average values of the three groups of tests.

Figure 1 shows the precise results from a rotary viscometer (Shanghai Jinglai Electronic Technology Development Co., Ltd., Shanghai, China) used to measure the viscosity of the cement slurry at various water–cement ratios. A viscometer (Shanghai Leiyun Test Instrument Manufacturing Co., Ltd., Shanghai, China) was used to determine the initial and final setting times of the cement slurry at various water–cement ratios, and the precise results are displayed in Figure 2. For standard curing, the cement slurry was formed into cubic specimens. Uniaxial compressive strength tests were then carried out at 3 d, 7 d, 14 d, and 28 d [23]. The results are shown in Figure 3. To calculate the stone solidification ratio, 100 mL of each type of slurry with one of the five water–cement ratios was placed in a measuring cylinder and left for 3 h. The liquid level was then read and recorded.

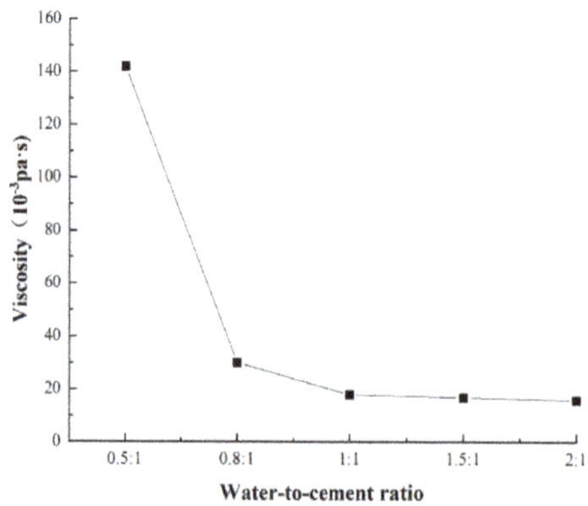

Figure 1. The influence of the water–cement ratio on the viscosity of the slurry.

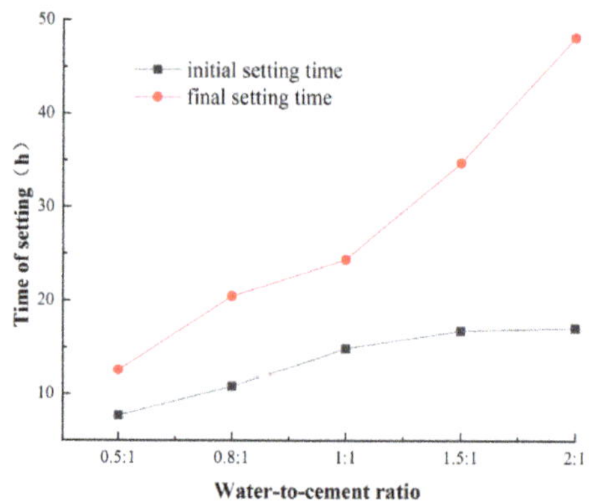

Figure 2. The influence of the water–cement ratio on the slurry setting time.

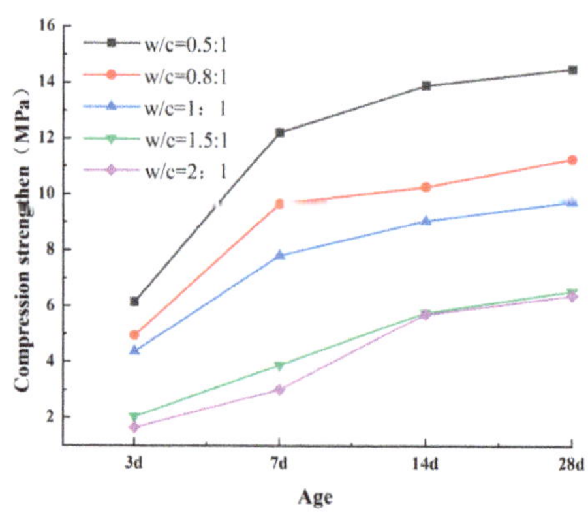

Figure 3. Compressive strength of the solidified slurry stone body.

Since the groundwater has a certain flow rate under the effect of high water pressure, the retention ratio of the stone body solidification of the slurry in a static water environment and a moving water environment after grouting must be taken into account in order to simulate the actual grouting environment. Cement slurries were prepared with water–cement ratios of 0.5:1, 1:1, and 2:1. After the initial setting of the material, it was placed on inorganic glass and in a homemade inorganic glass sink with a flow rate of 0.2, 0.4, 0.6, 0.8, 1.0, or 1.2 (m/s). The stone solidification retention ratio of the slurry was measured after 30 min in static water and moving water. The test primarily analyzed the water–cement ratio and water velocity on the slurry, combined with the anti-dispersion impact, and the test results are shown in Figures 4 and 5.

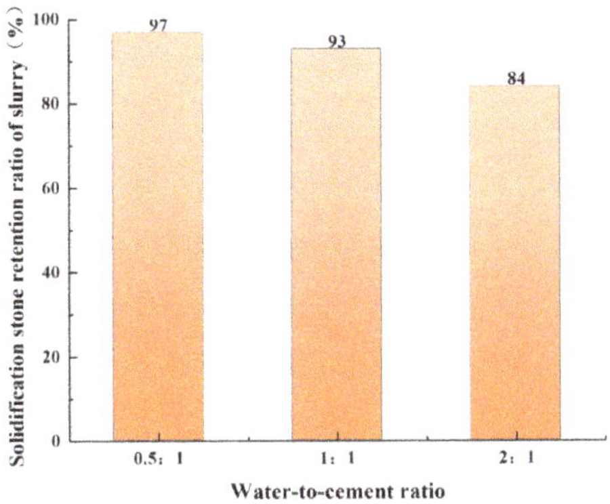

Figure 4. Stone solidification retention ratio of slurry in static environment.

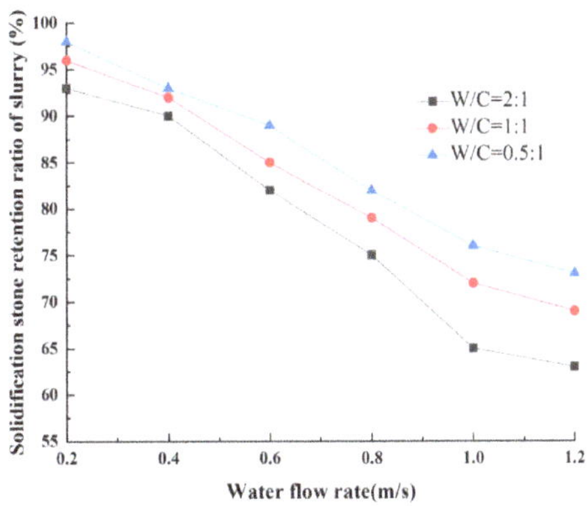

Figure 5. Stone solidification retention ratio of slurry in moving water environment.

2.2. Analysis of Experimental Results

According to Figure 1, as the water–cement ratio rises, the concentration of a single cement slurry decreases. The viscosity of the cement slurry decreases significantly from 140×10^{-3} to about 40×10^{-3}, or by about 70%, when the water–cement ratio is between 0.5 and 1.0. When the water–cement ratio is greater than 1, the viscosity of the slurry tends to reach a plateau and hardly change. The change is primarily caused by the relatively high water–cement ratio, which causes a complete reaction of cement hydration and prevents the viscosity from rising as the water ratio rises.

The water–cement ratio has a greater impact on the setting time of the pure cement paste, as shown in Figure 2. The initial setting time increases from 7 h to around 15 h as the water–cement ratio rises, and the final setting time rises as well. The initial to final setting time interval of the pure cement slurry varies with the water–cement ratio; for example, for a water–cement ratio of 0.5, the interval is approximately 5 h, whereas for a water–cement ratio of 2.0, the interval is approximately 30 h. This difference is primarily due to the higher cement content, which causes the slurry to set faster.

The solidification strength of the pure cement slurry is also influenced by the water–cement ratio, as shown in Figure 3, and the greater the compressive strength at the same age, the smaller the water–cement ratio. The strength growth is faster for the four water–cement ratios from 3 d to 7 d, and the solidification strength increases gradually after 7 d. The solidification strength of the cement slurry varies significantly depending on the water–cement ratio; for example, the solidification strength of a water–cement ratio of 0.5 is approximately 15 MPa, while that of a water–cement ratio of 2 is only approximately 6 MPa, a difference of approximately 2.5 times [24]. The solidification ratio of the pure cement slurry, which can reach about 99% with a high solidification ratio for all five water–cement ratios, is essentially unaffected by the water–cement ratio.

Figure 4 shows that the slurry with a water–cement ratio of 0.5 had the highest combined retention ratio of 97%, and the slurry with a water–cement ratio of 2 had the lowest combined retention ratio of 84% after being placed in still water for 30 min after the initial setting. This shows that the slurry with a high cement content had a better cementing effect in still water. Following the initial slurry setting with various water–cement ratios, Figure 5 illustrates the retention ratio of slurry stone solidification in moving water at various water velocities. As can be seen in the figure, the retention ratio of stone solidification decreases with the increasing moving water flow velocity and increases with the decreasing water–cement ratio at the same flow velocity. The stone solidification retention ratio for each water–cement ratio at a 1.2 m/s flow velocity is only 60%–70% of the stone solidification retention ratio at a 0.2 m/s flow velocity according to this law, which states that the slurry needs to have a certain carrier dispersion to ensure the grouting effect in the moving water environment. The grouting fluid should have better anti-dispersion properties under high-water-pressure conditions because the slurry is easily dispersed in the moving water environment compared to the static water environment.

3. Basic Performance Study of Cement–Polyvinyl Alcohol–Boric Acid Slurry

The pure cement slurry has good viscosity and a high ratio of solidified stone formation, but it also has a long setting time, numerous brittle material pores in the formed stone body, a low ratio of stone retention under high water pressure, and insufficient durability. Under high water pressure and with enhanced permeability, the pores of the solidified slurry stone body will further open up and form cracks, which will greatly reduce the water-plugging effect and prevent the expected grouting effect from occurring [25–27]. The use of the polyvinyl alcohol and cement slurry mixture, owing to the polyvinyl alcohol hydrogel's anti-cracking properties, enhances the performance of the grouting solution to achieve the expected effect of grouting. This improves the cement slurry setting time, improves the crack resistance of the grouted stone body, and enhances the durability of the grouting material [28,29].

3.1. Effect of Polyvinyl Alcohol Concentration on Gelation Time

Polyvinyl alcohol, or PVA, is a type of polymer with good water solubility that deforms well and has good toughness once it has gelled into a film. The PVA solution and cement liquid mix well, creating a gel film with good plasticity that also protects the cement from corrosion and increases the durability and crack resistance of the stone body [30–32].

Polyvinyl alcohol is insoluble in water at room temperature. Therefore, in this test, in order to dissolve it in water, it was heated and stirred with a magnetic stirrer. The temperature was set at 80 °C, and the solution was stirred for twenty hours. Different concentrations of the PVA solution and cement slurry volumes were mixed to configure PVA aqueous solutions with concentrations of 1%, 2%, 3%, 4%, and 5%, and the volume ratio of cement to PVA solution was 4:1. The inverted cup method was used to measure the water–cement ratio of the PVA solution with various concentrations and the time it took for the double-slurry gel to form. The specific results are shown in Table 1.

Table 1. Gel times of double slurries with different concentrations and water–cement ratios.

Concentration \ W/C	0.5:1	0.75:1	1:1	1.5:1
1%	189	654	1108	1598
2%	174	612	1097	1574
3%	154	563	1006	1528
4%	121	487	987	1503
5%	97	425	945	1458

Table 1 shows that the gel time of the cement–polyvinyl alcohol slurry ranges from 97 to 1458 s at various water–cement ratios and polyvinyl alcohol concentrations, all within a half-hour, which essentially satisfies the requirements for the gel time of the grouting material in grouting construction. The gel time decreased by up to 50% with an increase in the polyvinyl alcohol concentration at the same water–cement ratio, indicating that the polyvinyl alcohol concentration has a greater influence on the gel time. At the same polyvinyl alcohol concentration, the gel time increases with a higher water–cement ratio, in line with the law of change of the pure cement slurry.

3.2. Effect of Boric Acid on Cement–Polyvinyl Alcohol Slurry Properties

The cement–polyvinyl alcohol slurry was mixed with a boric acid solution to speed up the gelation process. As a type of chemical crosslinking agent for polyvinyl alcohol, boric acid is very sensitive to a PVA aqueous solution and can shorten the time it takes for the solution to gel and turn into a hydrogel. This meets the requirement for a slurry's controllable setting time under high water pressure.

In order to easily analyze the influence law of boric acid on the gel time of the cement–polyvinyl alcohol slurry, a 5% PVA solution concentration was used, with the ratio of the cement solution volume to the PVA solution volume = 4:1, and the boric acid–polyvinyl alcohol ratios were taken as 0.05, 0.1, 0.15, and 0.2. The gel times of different slurries were measured by the inverted cup method. At the same time, compressive tests were conducted on the solidified stone bodies at the age of 3 d for different slurries. The results obtained are shown in Figures 6 and 7.

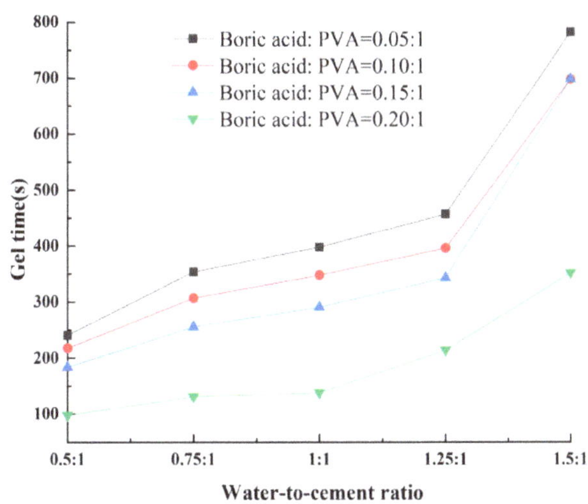

Figure 6. The influence of the water–cement ratio and boric acid on the slurry setting time.

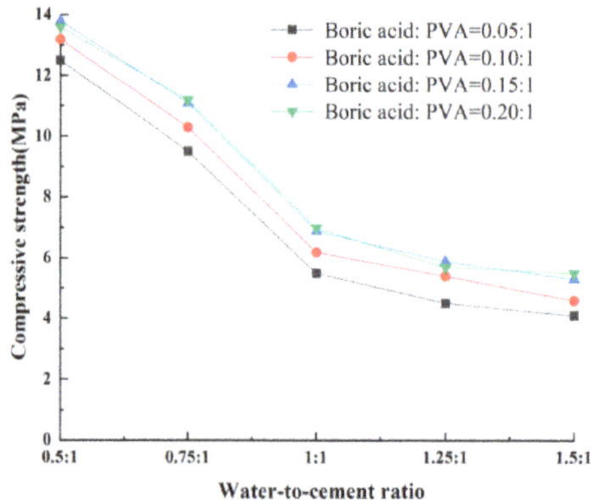

Figure 7. The influence of the water–cement ratio and boric acid on compressive strength.

As can be seen from Figure 6, when the ratio of boric acid to polyvinyl alcohol is less than 0.15, with an increase in the ratio, the time needed for the cement–polyvinyl alcohol slurry to gel is significantly shortened. PVA is very sensitive to boric acid, and a gel is formed between them through chemical crosslinking to accelerate the setting of the cement slurry. It can be seen from Figure 7 that both the water–cement ratio and boric acid ratio affect the compressive strength of the stone body. At the same water–cement ratio, the compressive strength increases with the increase in the boric acid ratio. When the boric acid ratio is higher than 0.15, the compressive strength increases slowly. BA ionizes $B(OH)_4$ in water, and $B(OH)_4$ crosslinks with the hydroxyl group on the PVA molecular chain to form a stable six-membered ring structure. Due to chemical crosslinking between boric acid and PVA, the compressive strength increases. The reaction process is shown in Figure 8.

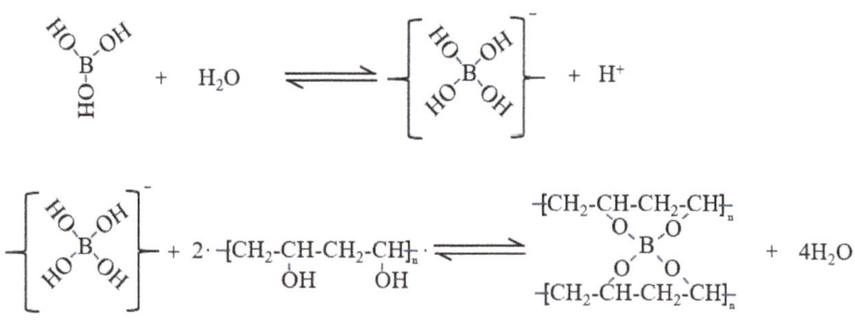

Figure 8. Reaction process of BA and PVA.

3.3. *Effect of Polyvinyl Alcohol Concentration on Stone Formation Ratio*

A cement solution with a water–cement ratio of 1:1 was mixed with 1%, 2%, 3%, 4%, and 5% polyvinyl alcohol solutions to form five cement–polyvinyl alcohol solutions with a volume ratio of cement solution to PVA solution of 4:1. In order to calculate the stone solidification ratio, 100 mL of each of the five water–cement ratio slurry mixtures was placed in a measuring cylinder. After three hours' rest, the liquid height was read and recorded. The overall stone solidification ratio is high, almost 100%. Figure 9 displays the results of measuring five solutions on a rotational viscometer to determine the viscosity of various slurries. From this figure, it is illustrated that the slurry viscosity generally increases with the concentration, but at concentrations greater than 3%, this growth is constrained. This is primarily due to the spatial grid structure that the solubilizer has created in the slurry, which no longer increases with the concentration.

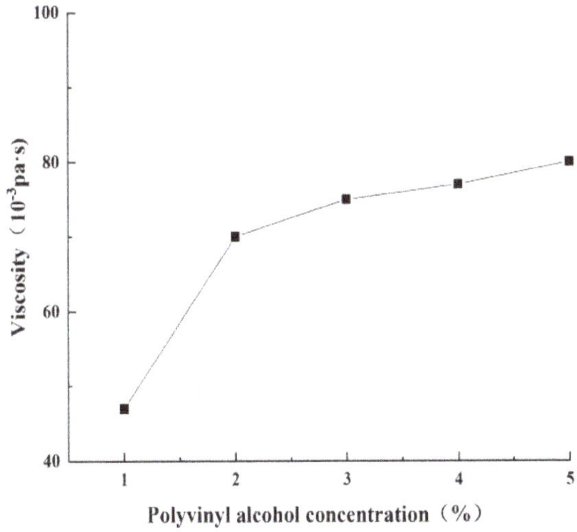

Figure 9. Effect of polyvinyl alcohol concentration on the slurry viscosity.

3.4. Cement–Polyvinyl Alcohol Slurry's Moving Water Retention Ratio

A cement solution with a water–cement ratio of 1:1 was mixed with polyvinyl alcohol solutions with concentrations of 1%, 2%, 3%, 4%, and 5%. The ratio of the cement liquid volume to polyvinyl alcohol solution volume was 4:1, creating five different cement–polyvinyl alcohol solutions. In order to measure the retention ratio of solidified slurry stones after 30 min and analyze the slurry's binding resistance to dispersion, the materials were first placed on inorganic glass and then placed in hydrostatic water and a homemade water tank with a specific flow rate. The results are shown in Figure 10.

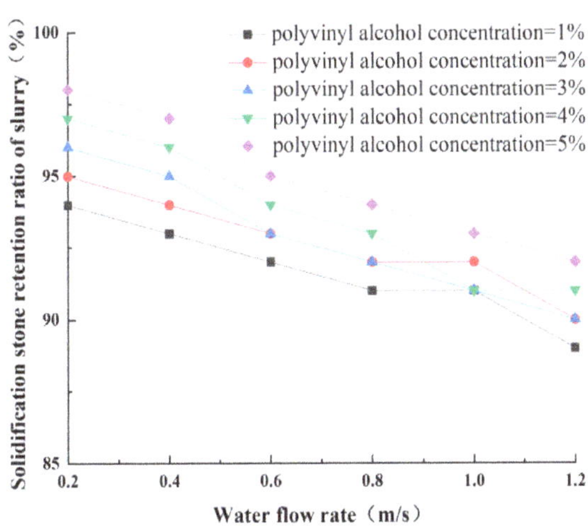

Figure 10. Effect of water flow rate on slurry solidification ratio.

Figure 10 demonstrates that the influence of different polyvinyl alcohol concentrations on the slurry's moving water retention rate is less, but the higher polyvinyl alcohol concentration results in a higher stone solidification retention ratio. In addition, the retention of the cement–polyvinyl alcohol slurry solidification ratio is higher in a moving water environment. This is primarily because the higher the polyvinyl alcohol concentration, the greater the gel formation and the higher viscosity of the slurry, which results in a higher retention ratio in the moving water environment.

3.5. Scanning Electron Microscope Analysis

Information on the solidified cement slurry and cement–polyvinyl alcohol slurry stones was obtained, as shown in Table 2, and a scanning electron microscopy test was carried out to observe the crack characteristics of both solidified slurry stones from a microscopic perspective. This was carried out to confirm that the crack resistance of cement–polyvinyl alcohol slurry nodules was improved [33]. A cold field-emission scanning electron microscope, model SU8010 from Hitachi, Japan, was used for the test. In order to conduct the test, samples from two solidified slurry stone bodies were first taken and pasted on conductive tape. The specimens were then placed in the apparatus for vacuum treatment and gold spraying. Finally, the specimens were scanned under a microscope to see the microscopic cracks.

Table 2. Preparation of slurry for electron microscope scanning.

No.	Slurry Type	Water–Cement Ratio	Additive Volume Ratio	External Additive Concentration	Age
A01	Cement slurry	1:1	/	/	3 d
A02	Cement–polyvinyl alcohol	1:1	4:1	5%	3 d

As seen in Figure 11a,b, the solidified cement slurry stones have more pores, which are randomly distributed throughout the stones. The porous and brittle nature of cement-based materials causes them to further develop when exposed to high water pressure, resulting in larger pores and the formation of fissures, which reduce the slurry stone body's bearing capacity and durability and interferes with the grouting effect.

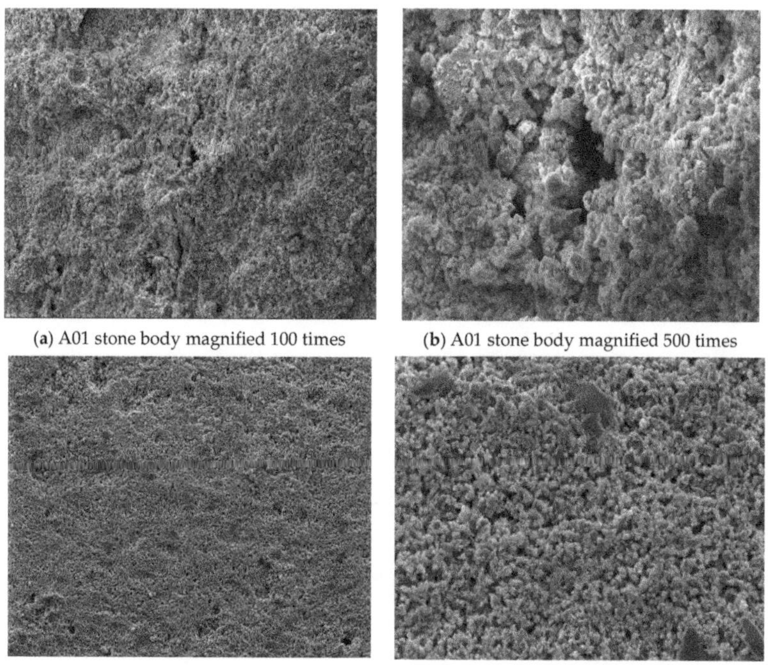

(**a**) A01 stone body magnified 100 times (**b**) A01 stone body magnified 500 times

(**c**) A02 stone body magnified 100 times (**d**) A02 stone body magnified 500 times

Figure 11. Scanning electron microscopy images of stone body.

The dense structure of the cement–polyvinyl alcohol slurry stone body in Figure 11c,d suggests that polyvinyl alcohol has some effect on the slurry stone body's ability to resist cracking. The solidified cement–polyvinyl alcohol slurry stone body has weak permeability and high compressive strength due to the microscopic characteristics of the slurry stone body combined with compressive strength characteristics, which meets the requirement for the water plugging of the surrounding rock grouting under high water pressure. Compared to the conventional cement slurry and cement–water-glass slurry, it is more appropriate for grouting requirements under high water pressure.

4. Project Application

The study area is located in +900 m of Qiandao Lake water diversion tunnel K22 in Hangzhou. This site has an upper rock layer of 100 m and IV~V surrounding rock, and its permeability coefficient indicates strong permeability. The groundwater level is at ground level, and the tunnel's central position head is 100 m. In order to prevent the excessive discharge of groundwater during tunnel construction, slurry plugging is applied to the surrounding rock [34–36]. The slurry ring thickness and slurry ring permeability coefficient are variables. The inner chamber of the cave is simplified as a circle, the equivalent inner diameter of the lining is 6.7 m, the equivalent outer diameter is 8.4 m, the lining permeability coefficient is 1.5×10^{-6} m/s, and the equivalent lining thickness is 0.85 m.

The tunnel excavation and grouting process was simulated using a two-dimensional seepage–stress coupling model created with the aid of the ABAQUS software. The model takes into account a hole influence range of 10 times its own diameter, and the model is designed to be 84 m wide and 130 m high. Horizontal displacement constraints are set on the model's left and right sides, while vertical displacement constraints are set at the bottom. Position heads are set at the left and right boundaries, while boundary heads are set at the top and bottom, and the top of the model is impermeable. The inner surface of the liner is a permeable surface, and the water pressure is set to 0. Four calculation conditions were established, as shown in Table 3, in order to analyze the impact of cement–polyvinyl alcohol–boric acid slurry injection, and Figure 11 depicts the seepage field's post-injection pore pressure cloud.

Table 3. Information of four working conditions and water pressure outside the liner.

Working Condition	Slurry Type	Slurry Ring Permeability Coefficient (m/s)	Water Pressure Outside the Liner (MPa)
Case 1	No grouting	1.0×10^{-5}	0.78
Case 2	Cement slurry	6.0×10^{-7} [37,38]	0.61
Case 3	Cement + water glass	1.8×10^{-7} [39,40]	0.54
Case 4	Cement + polyvinyl alcohol + boric acid	1.0×10^{-7} [41]	0.42

As seen in Figure 12a, when a tunnel is surrounded by rock without grout, the pore pressure cloud has the shape of an inverted funnel. The groundwater enters the tunnel after excavation, its level drops significantly, and the water pressure outside the lining is great. Excavation has a great influence on the seepage field. Figure 12d shows that after the injection of the cement–polyvinyl alcohol–boric acid slurry, the inverted funnel range of the pore pressure cloud decreases, and the drop in the groundwater level is smaller, which shows that tunnel excavation after grouting has little influence on the seepage field. This means that grouting has had a good effect on water plugging and achieved the expected plugging effect. When compared to Figure 12a, the pore pressure cloud diagram's inverted funnel shape is improved in Figure 12c,d, suggesting that grouting has a stronger anti-seepage effect. However, the anti-seepage effect varies because the grouting materials have different anti-seepage effects, and cement-based slurry stones are brittle materials that are vulnerable to cracking and have a minimal anti-seepage effect when subjected to external water pressure. Due to the function of the modifiers, the anti-seepage effect of the cement–polyvinyl alcohol slurry is strengthened.

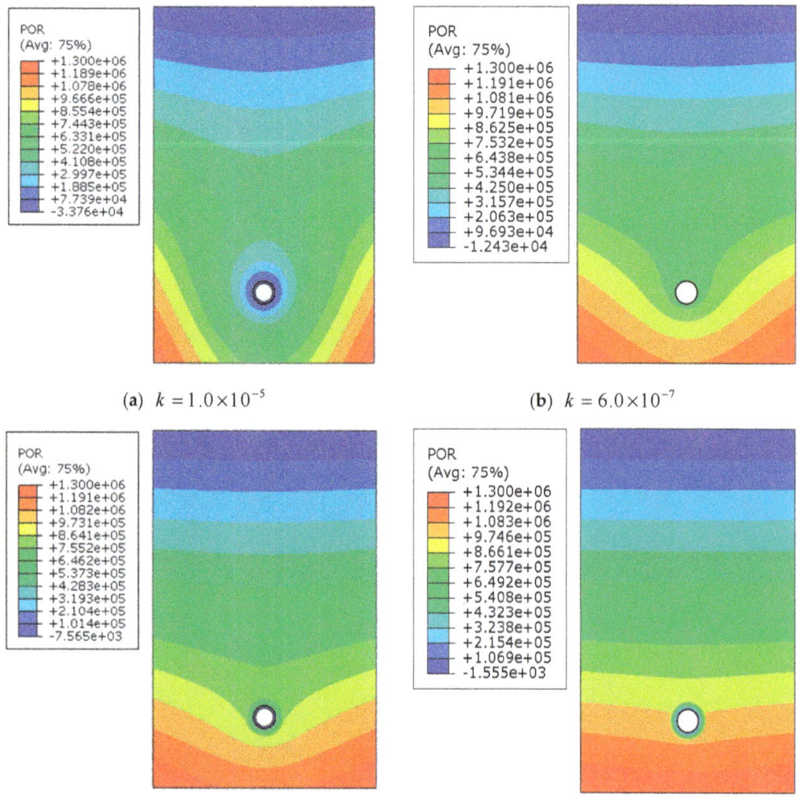

(**a**) $k = 1.0 \times 10^{-5}$ (**b**) $k = 6.0 \times 10^{-7}$

(**c**) $k = 1.8 \times 10^{-7}$ (**d**) $k = 1.0 \times 10^{-7}$

Figure 12. Pore pressure distribution clouds of the surrounding rock for the four working conditions.

According to Table 3, which displays the external water pressure of the lining under the four grouting conditions, the external water pressure of the lining starts to decrease once the permeability of the grouting ring is reduced, and it can withstand the external water pressure. In comparison to the first three working conditions, the external water pressure after the injection of cement–polyvinyl alcohol–boric acid grouting was reduced by 47%, 31%, and 22% after grouting with cement, polyvinyl alcohol, and boric acid, which perfectly demonstrates the improvement effect of the polyvinyl alcohol and boric acid modifiers. After grouting, a good water-plugging effect is achieved, which can reduce the water pressure outside the lining and reduce the discharge of groundwater.

5. Discussion

This study's findings indicate that when conventional grouting materials are used to seal off high water pressure around rocks, the grout's poor grouting performance and low retention ratio in a moving water environment endanger the tunnel's structural safety. As shown in Figures 6, 7, 10 and 11, the polyvinyl alcohol–boric acid–cement slurry controls the gel time through a modified concentration and volume ratio, a high retention ratio in moving water environments, and the good anti-seepage effect of the dense structure of the rock mass, and it had better grouting and water-plugging effects in an actual project of a water diversion tunnel, as depicted in Figure 12.

In this study, the laboratory test was mainly conducted to study the physical and mechanical properties of modified cement-based materials. To persuasively assess the grouting effect of the grouting fluid on the surrounding rock, differences in the strength and permeability of the surrounding rock before and after grouting were tested by on-site sampling, and the reinforcement and water-blocking effects of the modified grouting fluid were analyzed. In the numerical calculation, for simplicity, only the influence of the grouting ring's permeability coefficient on the seepage field was considered, and the external water pressure of the lining was calculated. In situ rock sampling was not carried out for the test, nor were numerical calculation results compared. These optimal steps should be carried out in the future.

In different engineering contexts, the requirements for the physical properties and mechanical properties of the grouting fluid vary greatly [42–44]. Therefore, the grouting solution should be designed with modified materials according to the project needs, and if possible, intelligent machine technology should be combined with grouting technology to improve the grouting effect. In the near future, grouting technology may have broader applications, such as the landslide management of slopes and other projects, giving full play to the role of grouting material reinforcement and seepage resistance [45].

6. Conclusions

In this study, traditional building materials were improved to meet the requirements of the surrounding rock grouting under high water pressure and in a dynamic water environment. Through indoor tests, SEM tests, and numerical simulation methods, the following main conclusions are obtained.

The physical and mechanical characteristics of a single cement slurry exhibit a certain degree of injectability, a high stone ratio, high compressive strength, and high viscosity. The water–cement ratio has a greater effect on a single slurry; the smaller the ratio, the quicker the slurry initially and finally sets, and the higher the annual retention ratio of stones in a moving water environment.

The performance of cement-based materials for grouting is enhanced by polyvinyl alcohol and boric acid. By chemically crosslinking polyvinyl alcohol and boric acid to create a polymer–cement glue, the gel time of the cement-based slurry is shortened, indicating that the gel time can be adjusted by modifying the polyvinyl alcohol concentration and boric acid content. The formation of the cement gel increases the density of the slurry stone structure, and SEM verification suggests that the stone body will be more effectively sealed in the surrounding rock and will perform better in increasing seepage resistance.

The seepage flow field's numerical calculation demonstrates that the cement-glue-modified material has a better water-plugging effect. The low permeability of the grouting ring protects the tunnel's idyllic surroundings by reducing groundwater seepage, lowering water pressure outside the lining, and preventing groundwater levels from dropping.

Author Contributions: Data curation, D.F.; numerical calculation, J.Y. and C.P.; experiment, J.Y., C.P. and Z.W.; project administration, Y.Z. and L.H.; writing—original draft, J.Y.; writing—review and editing, D.F. All authors have read and agreed to the published version of the manuscript.

Funding: Key Project of Natural Science Research Project of Higher Education Institutions in Anhui Province (Grant No. KJ2019A0822); National Natural Science Foundation of China (Grant No. 52179100).

Institutional Review Board Statement: Not applicable.

Informed Consent Statement: Not applicable.

Data Availability Statement: The data used to support the findings of this study are available from the corresponding author upon request.

Acknowledgments: The authors wish to thank JLL for advice on the experimental design.

Conflicts of Interest: No potential conflicts of interest were reported by the authors.

References

1. Assaad, J.J.; Gerges, N. Styrene-butadiene rubber modified cementitious grouts for embedding anchors in humid environments. *Tunn. Undergr. Space Technol.* **2018**, *84*, 317–325. [CrossRef]
2. Han, X.L.; Chen, X.M.; Jiao, H.Z.; Sun, G.D.; Dong, T.F. Research status and prospect of polymer-cement grouting materials. *Bull. Chin. Ceram. Soc.* **2018**, *37*, 2457–2462. [CrossRef]
3. Corson, L.; Reid, C.; Lunn, R.J.; El Mountassir, G.; Henderson, A.E.; Henderson, K.; Pagano, A.G.; Kremer, Y. Field validation of a detectable, magnetic, cementitious grout for rock fracture grouting. *Int. J. Rock Mech. Min. Sci.* **2021**, *145*, 104853. [CrossRef]
4. De la Varga, I.; Spragg, R.P.; Muñoz, J.F.; Helsel, M.A.; Graybeal, B.A. Cracking, bond, and durability performance of internally cured cementitious grouts for prefabricated bridge element connections. *Sustainability* **2018**, *10*, 3881. [CrossRef]
5. Li, X.; Cao, L.; Zhao, X. Assessment of potential impact of tunneling on the groundwater in Epi-Fissure-Karst-Zone and ecological environment. Environ. *Earth Sci.* **2011**, *66*, 967–976. [CrossRef]
6. Xie, X.-C.; Ceccato, F.; Zhou, M.-L.; Zhang, D.-M. Hydro-mechanical behaviour of soils during water-soil gushing in shield tunnels using MPM. *Comput. Geotech.* **2022**, *145*, 104688. [CrossRef]
7. Hua, X.Z.; Xie, G.X. Research on grouting material for bolt and grouting reinforcement of soft rock tunnel and its application. *Rock Soil Mech.* **2004**, *25*, 1462–1646.
8. Zhang, X.D. Study of Material for Curtain Grouting in Karst Tunnel. *Undergr. Space Eng.* **2005**, *3*, 432–434.
9. Tian, K.; Ding, Q.J.; Chen, Y.Q.; Luo, J. Optimal design of high volume fly ash simultaneous grouting material for shield tunnel. *Tunnel Constr.* **2007**, *27*, 26–29.
10. Xiao, L.; Zhang, Q.H.; Zhao, T.S.; Yang, J.L.; Yang, G.H. Study on mortar material of tail void grouting in slurry shield tunnel. *Chin. J. Undergr. Space Eng.* **2011**, *7*, 59–64. [CrossRef]
11. Liu, P.; Yan, C.; Sun, G.; Xi, L.; Liu, J. Experimental study on performance of grouting material for repairing surrounding rock in river intake tunne. *Geotech. Investig. Surv.* **2016**, *44*, 11–15+44.
12. Zhang, C.; Fu, J.; Yang, J.; Ou, X.; Ye, X.; Zhang, Y. Formulation and performance of grouting materials for underwater shield tunnel construction in karst ground. *Constr. Build. Mater.* **2018**, *187*, 327–338. [CrossRef]
13. Zhang, C.; Yang, J.; Fu, J.; Ou, X.; Xie, Y.; Liang, X. Performance evaluation of modified cement-sodium silicate grouting material for prereinforcing loose deposit tunnels. *J. Mater. Civ. Eng.* **2019**, *31*, 06019003. [CrossRef]
14. Zhang, W.; Zhu, X.; Xu, S.; Wang, Z.; Li, W. Experimental study on properties of a new type of grouting material for the reinforcement of fractured seam floor. *J. Mater. Res. Technol.* **2019**, *8*, 5271–5282. [CrossRef]
15. Liu, Y.; Chen, B. Research on the preparation and properties of a novel grouting material based on magnesium phosphate cement. *Constr. Build. Mater.* **2019**, *214*, 516–526. [CrossRef]
16. Wan, Y.; Zhu, Z.; Song, L.; Song, S.; Zhang, J.; Gu, X.; Xu, X. Study on temporary filling material of synchronous grouting in the middle of shield. *Constr. Build. Mater.* **2020**, *273*, 121681. [CrossRef]
17. Wang, J.; Dai, Q.; Si, R.; Guo, S. Investigation of properties and performances of Polyvinyl Alcohol (PVA) fiber-reinforced rubber concrete. *Constr. Build. Mater.* **2018**, *193*, 631–642. [CrossRef]
18. Zhang, J.; Guan, X.; Li, H.; Liu, X. Performance and hydration study of ultra-fine sulfoaluminate cement-based double liquid grouting material. *Constr. Build. Mater.* **2017**, *132*, 262–270. [CrossRef]
19. Sun, Y.; Zhang, P.; Yan, W.; Wu, J.; Yan, F. Grouting Material Development and Dynamic Grouting Test of Broken Rock Mass. *J. Mater. Civ. Eng.* **2022**, *34*, 04022072. [CrossRef]
20. Liu, B.; Yin, L.; Chen, J.; Li, M.; Tang, D.; Zhang, Y. Experimental study on the influence of PVA content on the performance of grouting material in deep stope. *Geofluids* **2021**, *2021*, 6684754. [CrossRef]
21. Lu, Y.; Li, C.; Zhang, X.; Huang, X.; Zhao, Z. A Workability Characterization of Innovative Rubber Concrete as a Grouting Material. *Materials* **2022**, *15*, 5319. [CrossRef] [PubMed]
22. Yang, Q.; Geng, P.; Wang, J.; Chen, P.; He, C. Research of asphalt–cement materials used for shield tunnel backfill grouting and effect on anti-seismic performance of tunnels. *Constr. Build. Mater.* **2021**, *318*, 125866. [CrossRef]
23. Le, H.; Sun, S. Effect of grouting materials and inclination angle of preexisting flaw on uniaxial compressive strength and failure mode of rock-like specimens. *Rock Soil Mech.* **2018**, *39*, 211–219. [CrossRef]
24. Ma, S.; Ma, D. Grouting Material for broken surrounding rock and its mechanical properties of grouting reinforcement. *Geotech. Geol. Eng.* **2021**, *39*, 3785–3793. [CrossRef]
25. Le, H.; Sun, S.; Wei, J. Influence of Types of Grouting Materials on Compressive Strength and Crack Behavior of Rocklike Specimens with Single Grout-Infilled Flaw under Axial Loads. *J. Mater. Civ. Eng.* **2019**, *31*, 06018022. [CrossRef]
26. Zhu, C.; Zhang, J.; Peng, J.; Cao, W.; Liu, J. Physical and mechanical properties of gypsum-based composites reinforced with PVA and PP fibers. *Constr. Build. Mater.* **2018**, *163*, 695–705. [CrossRef]
27. Zhang, B.; Gao, F.; Zhang, X.; Zhou, Y.; Hu, B.; Song, H. Modified cement-sodium silicate material and grouting technology for repairing underground concrete structure cracks. *Arab. J. Geosci.* **2019**, *12*, 680. [CrossRef]
28. Duque-Ossa, L.; Ruiz-Pulido, G.; Medina, D. Triborheological study under physiological conditions of PVA hydrogel/HA lubricant as synthetic system for soft tissue replacement. *Polymers* **2021**, *13*, 746. [CrossRef]
29. Yuan, Z.; Zhang, C.; Xia, C.; Wu, L.; Yu, Z.; Li, Z. Flexural properties of PVA fiber reinforced high ductility cementitious composites containing calcium carbonate whisker. *Constr. Build. Mater.* **2021**, *300*, 124329. [CrossRef]

30. Jiang, H.; Qiu, X. Performance assessment of a newly developed and highly stable grouting material for a completely weathered granite dam foundation. *Constr. Build. Mater.* **2021**, *299*, 123956. [CrossRef]
31. Pei, J.; Cai, J.; Zou, D.; Zhang, J.; Li, R.; Chen, X.; Jin, L. Design and performance validation of high-performance cement paste as a grouting material for semi-flexible pavement. *Constr. Build. Mater.* **2016**, *126*, 206–217. [CrossRef]
32. Zhang, P.; Su, Y.-L.; Fan, J.-J.; Feng, H.; Shao, J.; Guo, H.; Gao, D.-Y.; Sheikh, S.A. Experimental research on the mechanical behavior of grouted sleeves with fiber-reinforced grouting material under cyclic loading. *Structures* **2021**, *34*, 2189–2204. [CrossRef]
33. Xiao, C.; Zhang, Y. Structural investigation of poly(vinyl alcohol) fibers under axial compressional deformation. *Acta Polym. Sin.* **2002**, *18*, 791–794. [CrossRef]
34. Li, J.-F.; Liu, Y.-T.; Li, S.-J.; Song, Y. Experimental investigation of synchronous grouting material prepared with different mineral admixtures. *Materials* **2022**, *15*, 1260. [CrossRef]
35. Liu, D.; Xu, Q.; Tang, Y.; Jian, Y. Prediction of water inrush in long-lasting shutdown karst tunnels based on the HGWO-SVR model. *IEEE Access* **2020**, *9*, 6368–6378. [CrossRef]
36. Li, Z.; Li, S.; Liu, R.; Jiang, Y.; Zhang, Q.; Yang, L.; Sha, F.; Chen, B.; Wang, H. Model tests on grouting rein-forcement of water-rich broken rock mass. *Chin. J. Geotech. Eng.* **2016**, *37*, 1937–1946. [CrossRef]
37. He, Z.; Shen, A.; Guo, Y.; Lyu, Z.; Li, D.; Qin, X.; Zhao, M.; Wang, Z. Cement-based materials modified with superabsorbent polymers: A review. *Constr. Build. Mater.* **2019**, *225*, 569–590. [CrossRef]
38. Snoeck, D.; Steuperaert, S.; Van Tittelboom, K.; Dubruel, P.; De Belie, N. Visualization of water penetration in cementitious materials with superabsorbent polymers by means of neutron radiography. *Cem. Concr. Res.* **2012**, *42*, 1113–1121. [CrossRef]
39. Zhou, Y.; Wang, G.H.; Yuan, Y.F. Basic Properties and Engineering Application of Bentonite-Cement-Water Glass Grouting. *KSCE J. Civ. Eng.* **2020**, *24*, 2742–2750. [CrossRef]
40. Kazemian, S.; Prasad, A.; Huat, B.B.K.; Bazaz, J.B.; Aziz, F.N.A.A.; Ali, T.A.M. Influence of cement-sodium silicate grout admixed with calcium chloride and kaolinite on sapric peat. *J. Civ. Eng. Manag.* **2011**, *17*, 309–318. [CrossRef]
41. Zheng, K.; Xuan, D.; Li, J. Study on Fluid–Solid Characteristics of Grouting Filling Similar-Simulation Materials. *Minerals* **2022**, *12*, 502. [CrossRef]
42. Jiang, Y.; Xu, Z.; Geng, D.; Dong, J.; Liao, Y. Optimization of grouting material for shield tunnel antifloating in full-face rock stratum in nanchang metro construction in China. *Int. J. Géoméch.* **2022**, *22*, 04022009. [CrossRef]
43. Wang, S.; Wang, J.-F.; Yuan, C.-P.; Chen, L.-Y.; Xu, S.-T.; Guo, K.-B. Development of the nano-composite cement: Application in regulating grouting in complex ground conditions. *J. Mt. Sci.* **2018**, *15*, 1572–1584. [CrossRef]
44. Wang, C.; Zhang, D.; Chen, H.; Zhang, H.; Xiao, X.; Liu, Z. Preparation and properties of silicon-modified epoxy grouting material for repairing microcracks. *J. Mater. Civ. Eng.* **2022**, *34*, 04021479. [CrossRef]
45. Chang, Z.; Huang, F.; Huang, J.; Jiang, S.-H.; Zhou, C.; Zhu, L. Experimental study of the failure mode and mechanism of loess fill slopes induced by rainfall. *Eng. Geol.* **2021**, *280*, 105941. [CrossRef]

coatings

Communication

Blast Resistance of a Masonry Wall Coated with a Polyurea Elastomer

Xudong Zu, Taian Chen, Youer Cai, Zhengxiang Huang and Qiangqiang Xiao *

School of Mechanical Engineering, Nanjing University of Science and Technology, Nanjing 210094, China
* Correspondence: xiao_wawj@njust.edu.cn; Tel.: +86-13701459567

Abstract: The coating of polyurea elastomers on walls is a hotspot in the protection field. This work combines a numerical simulation with experimental validation to examine the blast resistance after coating a polyurea elastomer on a 370 mm wall under a contact explosion. Firstly, the failure limit, failure mode, and failure mechanism of the 370 mm unreinforced wall under different strength loads are studied. In the case of the contact explosion, the increase in size of the 370 mm wall blasting pit gradually stops after the dose is increased to more than 1 kg. Thereafter, the energy of the explosive load wis released by splashing wall fragments as well as by the deflection and movement of the wall. Spraying double-sided polyurea reinforcement on the wall can effectively improve the resistance to damages caused by exposure to explosive loads and can inhibit the damage to the surrounding personnel and equipment caused by flying structural debris. When the polyurea thickness on the front surface is 6 mm, the optimal thickness of the back surface should be 2 mm.

Keywords: masonry wall; polyurea elastomer; blast resistance; explosive loads

Citation: Zu, X.; Chen, T.; Cai, Y.; Huang, Z.; Xiao, Q. Blast Resistance of a Masonry Wall Coated with a Polyurea Elastomer. *Coatings* **2022**, *12*, 1744. https://doi.org/10.3390/coatings12111744

Academic Editor: Enrico Quagliarini

Received: 8 September 2022
Accepted: 17 October 2022
Published: 14 November 2022

1. Introduction

Terrorist acts on buildings have never disappeared in the modern world, where they routinely take place. Clay bricks are still a major component of modern buildings. Under detonation loading, the walls will suffer extensive damage. The lives of those residing in the buildings will be gravely affected when a large avalanche develops on the back of a wall. Therefore, it is vital to research ways to improve the anti-explosion performance of the walls without altering the building's fundamental design. According to Proter et al. [1], polyurea possesses excellent impact and rear resistance, as well as great application potential for anti-terrorism and explosive protection scenarios. Moreover, polyurea elastomers can strengthen a wall via spraying, and polyurea can cure and harden quickly, which is convenient during construction [2]. According to Bahei-El-Din et al. [3] and Tekalur et al. [4], spraying polyurea on a wall helps decrease the secondary damage produced by the wall's collapse by absorbing the shock energy. The impact of the polyurea and glass fiber-reinforced polymer (GFRP)-strengthened walls on the blast resistance was examined by Hrynyk and Myers [5]. The outcomes demonstrated that the wall reinforced with polyurea was the most efficient in lowering wall deflection and enhancing energy dissipation.

Researchers from around the world have studied wall blast resistance performance in an effort to reduce the damage caused by explosions to walls. Aghdamy et al. [6] studied the characteristics of the failure and collapse of concrete masonry walls reinforced by a nanoparticle polymer and aluminum foam under dynamic and impact conditions by combining experiments and simulations. Davidson et al. [2,7] analyzed the blast-resistant mechanism of a sprayed polyurea elastomer in strengthening masonry walls through a detonation test. Gattesco et al. [8] conducted experiments and numerical studies on the non-planar behavior of reinforced masonry walls. Fan et al. [9] simulated the response of and damage to brick-filled walls under explosive loads and reported multiple failure modes of masonry walls under different proportional distances. Zheng [10] studied the responses of non-porous masonry walls under explosive loads. Xu and Lu [11] simulated the explosive shock behavior of reinforced concrete walls and found that wall caving may be considerably reduced in concrete walls by strengthening the back surface. By simulating the enhancement effect of polymers on the masonry wall performance, Aghdamy et al. [5] drew the conclusion that polymers can significantly increase the resistance of masonry walls to explosive loads. Dinan et al. [12] determined the damage mechanism and conducted experimental research on coated polymer masonry walls. The results showed that the polymer coatings can significantly enhance the explosion-proof performance of the masonry walls. Zhu et al. [13] developed a detailed micro-numerical model that can predict the non-linear dynamic responses of both unreinforced clay brick masonry walls and polyurea-reinforced clay brick masonry walls under close-in blasting. Davidson et al. [2,7] carried out a series of field explosion experiments to investigate the blast resistance performance of a wall enhanced with polyurea. Iqbal et al. [14] conducted an experimental investigation on a coated polymer masonry wall to identify the damage mechanism. The outcomes demonstrated that the masonry wall's performance as an explosion-proof barrier was greatly improved by the coated polymer layer.

In our previous study [15], we found that the blast resistance can be strengthened by coating polyurea elastomer on a 240 mm wall through an experiment and numerical simulation, and showed that the reinforced wall will not collapse even if the wall is penetrated. In addition, the influence of the thickness of the coated polyurea on the blast resistance ability was analyzed, and the optimal structure was found. In this paper, the same research method was used to study a 370 mm wall. It was found that the coated polyurea elastomer protected the 370 mm wall, as the main load-bearing structure of the buildings, from penetrating damage and large-sizes cracks, and we also found the best structure. A 370 mm wall and 240 mm wall will be different in a building, and the destruction forms in the building will also be different. The purpose of the study was to prevent a 240 mm wall coated with a polyurea elastomer from collapsing during explosion and a 370 mm wall coated with a polyurea elastomer from penetrating damage. In summary, the combination of the research results from the two articles can provide a basis for how to enhance the responses of existing building targets to terrorist explosion damage.

2. Numerical Simulation

2.1. Blast-Resistant Performance of Clay Masonry Wall under Contact Explosion

In this section, the damage to the masonry wall under contact explosions with different quality charges is analyzed via numerical simulation.

2.1.1. Simulation Model

The masonry wall was described using a different model. The bricks and mortar were created from several material units and were viewed as separate parts in the solution. The simulation was conducted in accordance with the masonry regulations for masonry walls in the relevant standard GB 50003-2011, Code for the Design of Masonry Structures [16]. In the model, the masonry method is "one straight, one horizontal, and one cement mortar joint". The LS-DYNA finite element dynamic analysis software was used to simulate the damage effect of the entire TNT explosion on the wall. For the bricks and mortar we used the Lagrange algorithm, and for both explosives and air we used the ALE algorithm. The size of the single masonry wall was 1990 mm × 1260 mm × 370 mm. Given that the single wall was axisymmetric, a 1/2 model of the masonry wall was used for calculation. The 1/2 model of the masonry wall is shown in Figure 1.

In actual brick wall buildings, the foundation is generally excavated. The walls are built from about 0.5 m below the ground, and the masonry is firmly connected to the ground. Therefore, the part of the wall bottom close to the ground can be regarded as fully constrained during the numerical simulation modeling. There are tie bars on both sides or columns connected with the wall on the side, which can effectively limit the displacement and deformation of the wall. In order to make the wall flat and beautiful during construction, a thin layer of low-strength mortar (M5 and below) is often used to level the outside. In order to correspond to the actual situation, a full-constraint longitudinal strip is applied to the back burst surface of the 1/2 model to simulate the supporting effect of the adjacent walls on the masonry walls, so that the constraints on both sides of the wall are stronger than those on the upper and lower sides. The constraints in the X-axis direction are imposed on the symmetry plane, and non-reflective boundaries are imposed around the air domain (except for the symmetry plane). Due to the low strength and thin thickness of the plastering mortar outside the wall, its anti-explosion effect in the contact explosion is ignored.

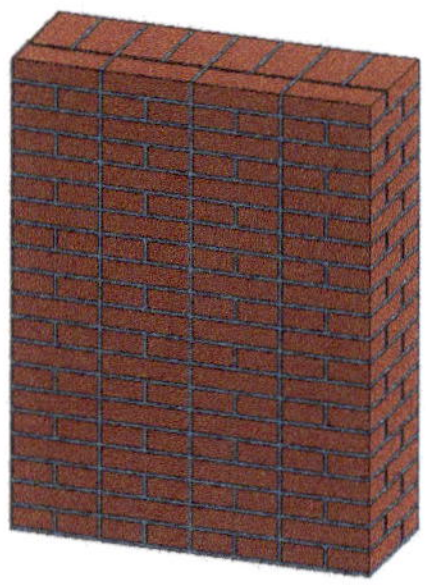

Figure 1. The 1/2 model and mesh generation of the masonry wall.

Since the duration of the explosion is very short, it is assumed that the clay brick block and mortar are well connected; that is, they are modeled in joint mode. In order to simulate the bonding, separation, and sliding between the bricks and mortar and to simulate the tensile failure and shear failure between the contact surfaces, TNTS is used to express the above relationship. The contact keyword is: * CONTACT_TIEBREAK_SURFACE_TO_SURFACE. The failure conditions of this keyword are:

$$\left(\frac{|\sigma_n|}{[\sigma_N]}\right)^2 + \left(\frac{|\sigma_S|}{[\sigma_S]}\right)^2 \geq 1 \tag{1}$$

where σ_n and σ_s are the calculated normal stress and calculated shear stress on the contact surface, respectively; $[\sigma_N]$ and $[\sigma_S]$ are the allowable normal stress and allowable shear stress on the contact surface, respectively, which can be taken as the elastic tensile strength and shear strength of the mortar material. As the tensile and shear strength between the masonry block and mortar of the brick masonry wall are relatively small, the allowable normal stress and allowable shear stress on the contact surface are 0.12 MPa and 0.14 MPa, respectively. After the failure of the contact surface, there will be dislocation and slippage between the block and mortar. At this time, the friction will prevent both from sliding. When the contact surface is dry, the friction coefficient μ is 0.7.

(1) Material parameters of Explosive and air.

The structural simulation model used the LS-DYNA ALE method to simulate the impact of the shock wave and the structure.

The JWL [17] equation of state for the explosives was used in the numerical simulation calculation;

$$P = A\left(1 - \frac{\omega}{R_1 V}\right)e^{-R_1 V} + B\left(1 - \frac{\omega}{R_2 V}\right)e^{-R_2 V} + \frac{\omega E}{V}, \tag{2}$$

where P is the pressure of the explosive detonation products; E is the internal energy per unit mass of the explosive; V is the relative volume; and A, B, R_1, R_2, and ω are the material parameters of the explosive. The material parameters and state equation parameters are provided in Tables 1 and 2.

Table 1. Material parameters of the explosives (g-cm-μs).

MID	RO	*D*	P_{CJ}	BETA	*K*	*G*	SIGY
—	1.63	0.693	0.27	0.00	0.00	0.00	0.00

Table 2. Parameters of the explosive (EOS_JWL) (g-cm-μs).

EOSID	*A*	*B*	R_1	R_2	OMEG	E_0	V_0
—	3.71	0.0743	4.15	0.95	0.3	0.07	1.00

(2) Material parameters of Masonry wall

The block bricks and mortar were described by the * MAT 096(MAT_-BRITTLE_DAMAGE) model in LS-DYNA in ANSYS 17.1. This model is primarily an anisotropic brittle damage model for concrete design and can be used for various brittle materials, including bricks and mortar. The restriction on the normal tractions is given by Equation (2) [17]:

$$\phi_t = (n \otimes n) : \sigma - f_n + (1 - \varepsilon) f_n (1 - exp[-H\alpha]) \leq 0 \tag{3}$$

where n is the smeared crack normal, ε is the small constant, f_n is the initial principal tensile strength of the material, H is the softening modulus, and α is an internal variable. Here, H is set automatically by the program; α measures the crack field intensity and is output in the equivalent plastic strain field.

The primary material parameters of block bricks and mortar were listed in Table 3.

Table 3. Material parameters of blocks and mortar.

Material	Density (kg/m^3)	Elastic Modulus/ MPa	Poisson's Ratio	Tensile Strength/ MPa	Shear Strength/ Mpa	Compressive Strength/ Mpa	Fracture Toughness (N/m)	Shear Transfer Coefficient
Block	1150	380	0.15	1.00	0.50	17.6	120	0.03
Mortar	2100	4644	0.25	1.76	0.90	9.0	140	0.03

2.1.2. Numerical Simulation Results

Figure 2 depicts the damage to the wall over time when the TNT was 1 kg. The left and middle figures show that the explosion load first caused a central blasting pit and longitudinal cracks on the blast front surface of the masonry wall and gradually formed divergent cracks along the mortar joint around the blasting pit at about 400 μs. The longitudinal cracks expanded and thickened. The figure on the right suggests that the blast back surface first formed a longitudinal crack, and the central mortar joint of the wall began to crack at 400 μs. As shown in Figure 2c–e, the cracks on the blast back surface show a divergent shape as a whole (except the rib plate).

Figure 3 shows the damage to the wall at t = 1000 μs under different explosive contact weights. The longitudinal cross-sectional view on the left shows that as the amount of charge gradually increased, the depth of the blasting pit increased and cracks were formed. The cracks in the vertical direction of the wall center, i.e., the direction of the axis of symmetry, were particularly deep. The vertical cross-sectional area S of the blasting pit of the brick wall under the contact explosions with different charges showed a gradual increase with increasing charge, and the volume of the blasting pit also showed an increasing trend. The destruction of the middle front surface showed that the strain gradually increased as the amount of charge increased. The area affected by the contact explosion also increased from spherical (Figure 3a) to petaloid (Figure 3b) and to square (Figure 3c,d). The deformation around the blasting pit diverged along the mortar joint in a radial shape (see Figure 3d–h). Sporadic avalanches appeared at the edges of the wall (see Figure 3a–h). The destruction of the back surface on the right side indicated that with the increasing charge, the deformation range of the back surface gradually increased, and the mortar joint was a weak link. Collapse and spalling also developed along the mortar joint but in the shape of a ring (except the rib plate). The central mortar joint was severely injured, forming a through crack.

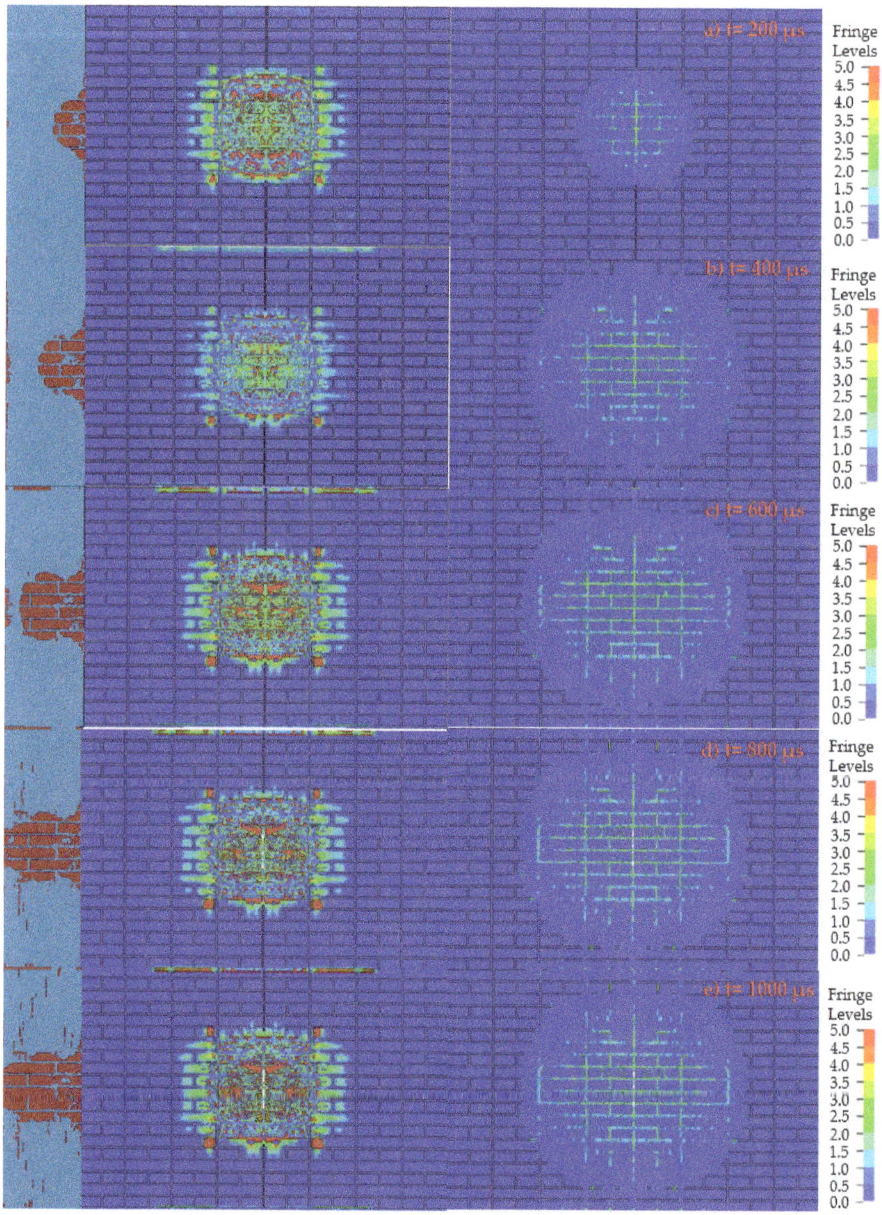

Figure 2. Damage to the 370 mm wall at different times. (**a**) 200 μs; (**b**) 400 μs; (**c**) 600 μs; (**d**) 800 μs; (**e**) 1000 μs.

Figure 4a–c illustrate the linear increases in the area, depth, and radius of the blast pit when the explosive weights increase from 0.25 kg to 1 kg. However, with the increase in mass of the charge, when the charge is greater than 1 kg, the area, depth, and radius of the explosive pit increase slowly and tend to be horizontal. This means that the charge mass of 1 kg is a critical state, and the damage form changes when it exceeds 1 kg.

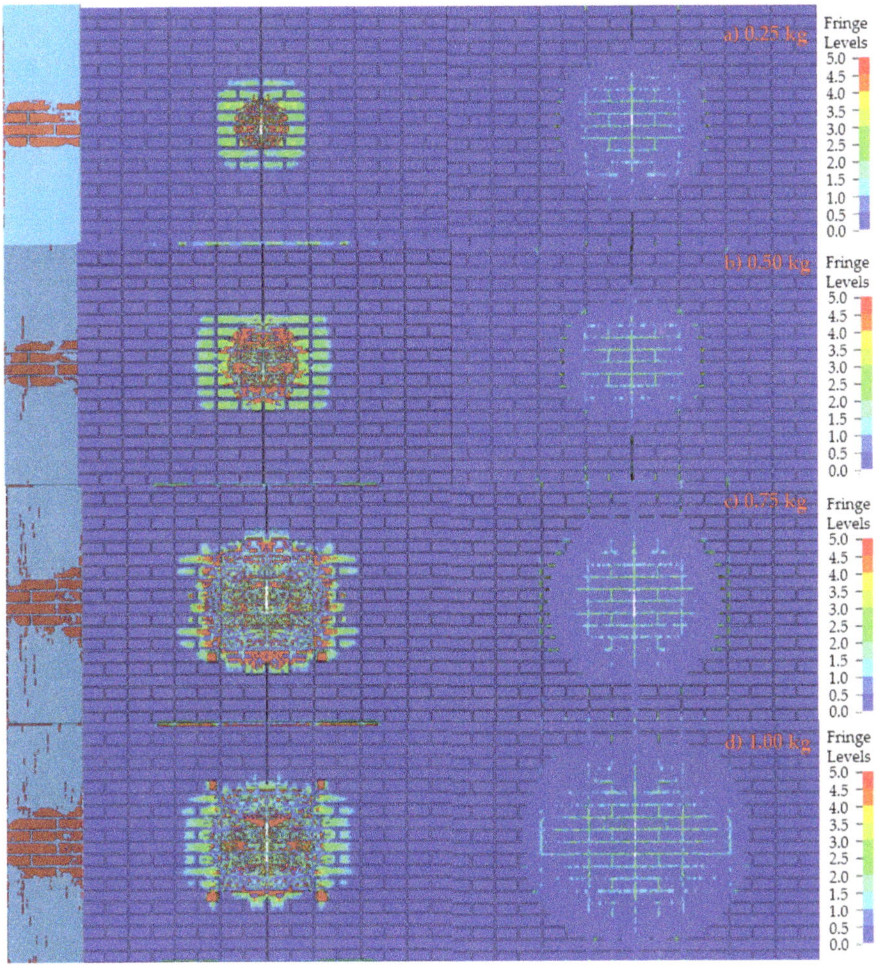

Figure 3. *Cont.*

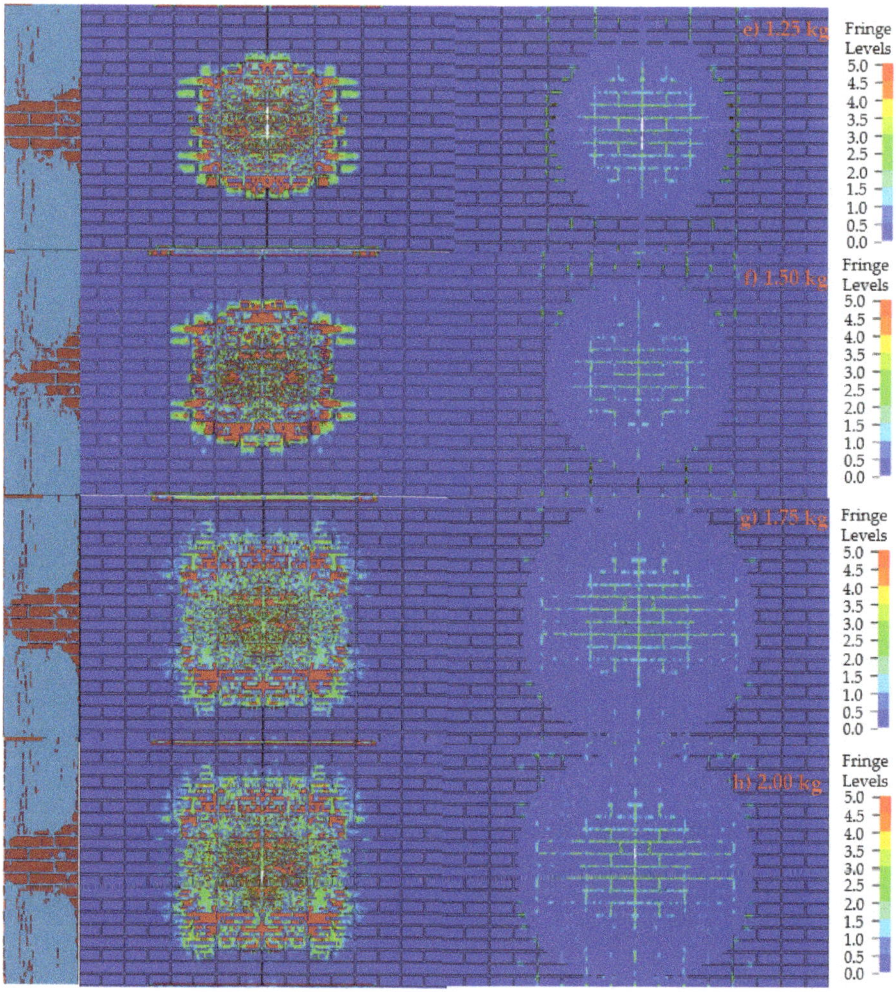

Figure 3. Damage to the 370 mm wall under different weights of TNT. (**a**) 0.25 kg; (**b**) 0.50 kg; (**c**) 0.75 kg; (**d**) 1.00 kg; (**e**) 1.25 kg; (**f**) 1.50 kg; (**g**) 1.75 kg; (**h**) 2.00 kg.

2.2. *Blast-Resistant Characteristics of Masonry Wall Sprayed with Polyurea Elastomer*

In this section, we detail the numerical simulations of the blast-resistant characteristics of masonry walls sprayed with different thicknesses of a polyurea elastomer.

2.2.1. Numerical Model

In the numerical simulation, the thickness of the polyurea reinforcement layer on the front surface remained unchanged at 6 mm, while the thickness of the back surface reinforcement layer was changed to study the influence of the thickness changes in the polyurea reinforcement layer on the failure of the masonry wall under contact explosion. The thickness of the polyurea reinforcement layer of the 370 mm wall model is shown in Table 4. The 1/2 model of the sprayed polyurea-reinforced 370 mm wall and the meshing of the large-scale surface of the polyurea coating are shown in Figure 5. The bonding force between the polyurea elastomer and the brick wall was obviously greater than the peel strength of the mortar, so the polyurea elastomer sprayed on the surface of the clay masonry wall and the wall were coupled by a common node to simulate their deformation coordination.

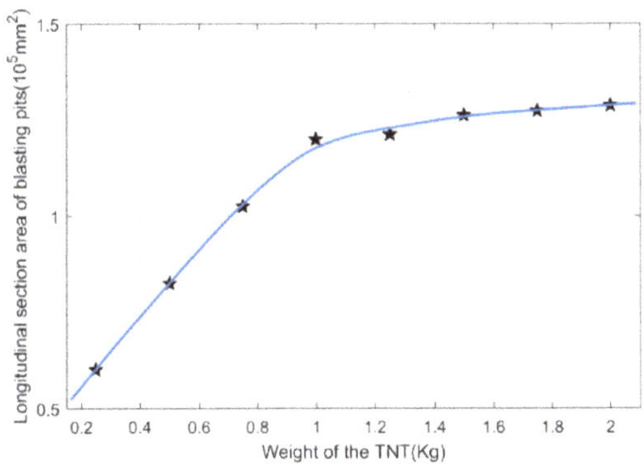

(**a**)

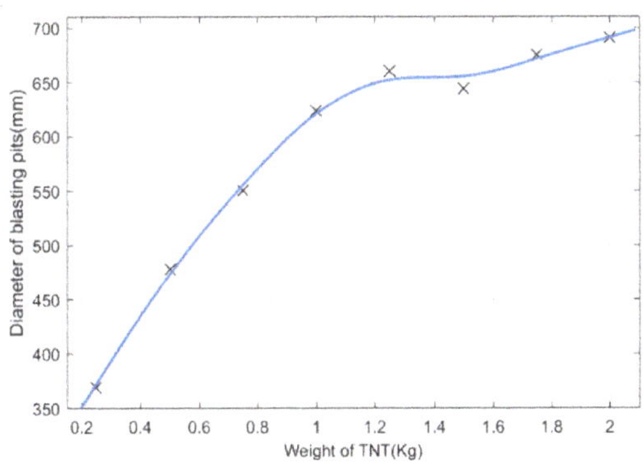

(**b**)

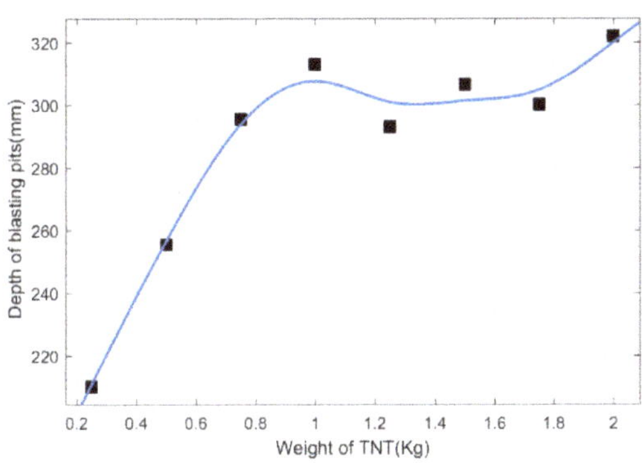

(**c**)

Figure 4. Statistics for the longitudinal section size of the 370 mm wall under different weights of TNT: (**a**) longitudinal section area of the blasting pits; (**b**) diameter of the blasting pits; (**c**) depth of the blasting pits.

Table 4. Thickness dimensions of the polyurea reinforcement layer of a 370 mm wall model.

Serial number	1	2	3	4
Polyurea thickness on back surface/mm	2	4	6	8

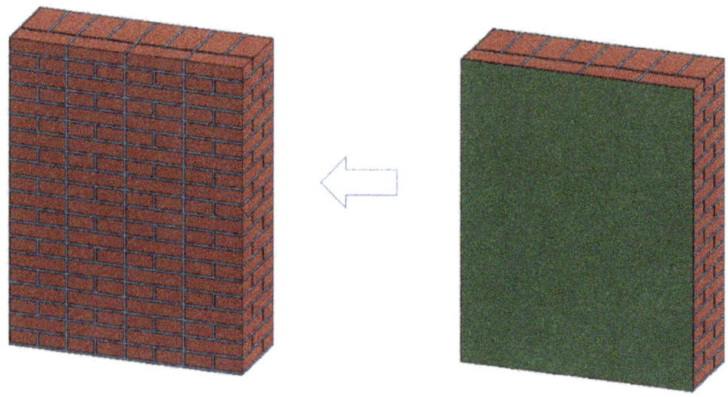

Figure 5. The 1/2 finite element model of the sprayed polyurea-strengthened 370 mm wall and the meshing of the polyurea layer.

Table 5 shows the mechanical properties of the polyurea elastomer [18].

Table 5. Parameters of the polyurea elastomer.

E/GPa	ρ/(g/cm^3)	v	σ_Y/GPa	E_Y/GPa	σ_0/GPa
230	1.02	0.4	1.38	3.50	13.8
Load curve 1			Load curve 2		
1.0×10^{-1}	2.3×10^{-4}		1.6×10^{-9}	5.0×10^{-5}	
3.0×10^{-1}	2.7×10^{-4}		1.6×10^{-7}	6.0×10^{-5}	
5.0×10^{-1}	3.5×10^{-4}		1.4×10^{-5}	8.0×10^{-5}	
8.0×10^{-1}	4.2×10^{-4}		8.0×10^{-5}	8.5×10^{-5}	
8.5×10^{-1}	4.9×10^{-4}		8.0×10^{-4}	1.1×10^{-4}	
9.0×10^{-1}	5.8×10^{-4}		2.25×10^{-3}	1.2×10^{-4}	
			16.5×10^{-3}	1.6×10^{-4}	
			9.0×10^{-3}	2.2×10^{-4}	

2.2.2. Numerical Simulation of the Polyurea-Reinforced 370 mm Wall

When the amount of TNT was 1 kg under working condition 2 (the thickness of the reinforcement layer on the front surface was 6 mm and the thickness of the reinforcement layer on the back surface was 4 mm), Figure 6 illustrates the 370 mm wall's damage over time. The figure on the left shows the front side of the reinforced wall before the explosion, which reflects the damage to the surface of the polyurea reinforcement layer before the explosion. The middle figure is only the stress cloud diagram of the wall, reflecting the damage of the inner wall of the masonry. The figure on the right shows the rear side of the reinforced wall after blasting, reflecting the damage to the polyurea reinforcement layer on the blasted surface.

From the front view on the left, after the contact explosion, the polyurea layer on the front surface stretched and dented quickly under the action of the explosion load and was quickly damaged to expose the masonry wall structure of the lower layer. The polyurea layer at the central symmetry plane showed wavy distortion (see Figure 6a). The breach of the polyurea layer on the blast front face expanded rapidly with time, and the polyurea layer around the breach was stacked and folded after extrusion and stretching (see Figure 6a–d). After 800 μs, when the central breach of the polyurea layer expanded to a certain extent, the size growth slowed down and some near-circular bulges appeared on the periphery of the stacked fold layer of the breach, leading to the uneven periphery of the folds (see Figure 6d,e).

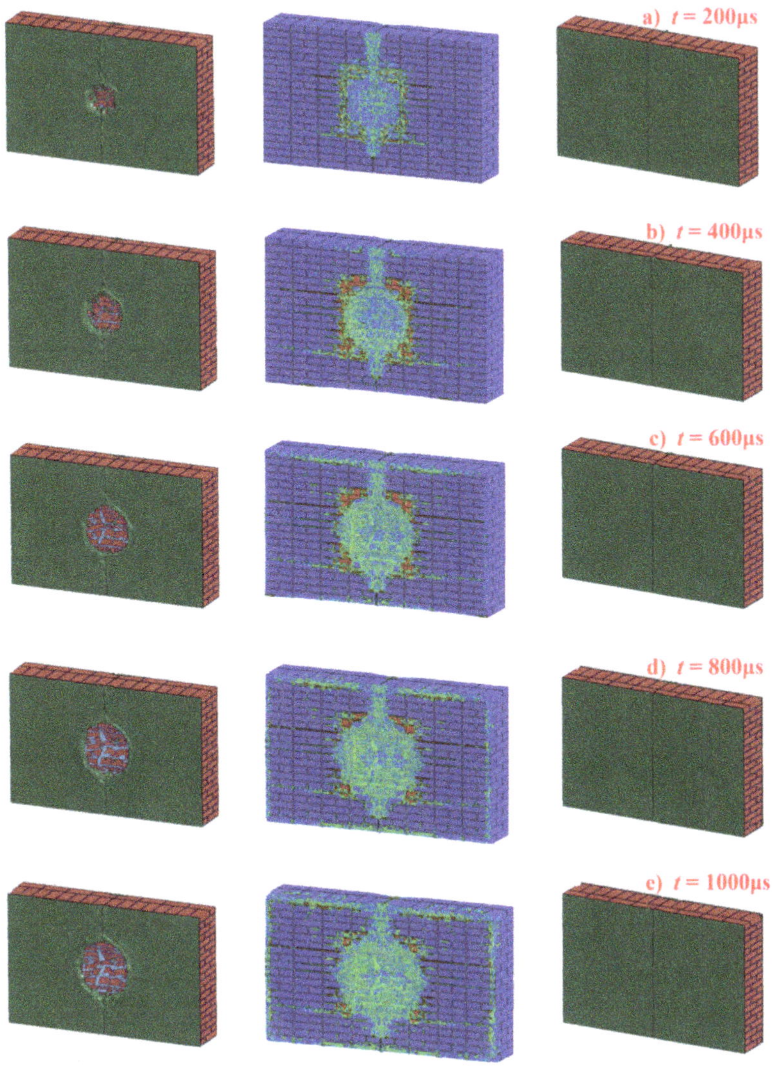

Figure 6. Destruction of the polyurea-reinforced 370 mm wall at different times. (**a**) 200 μs; (**b**) 400 μs; (**c**) 600 μs; (**d**) 800 μs; (**e**) 1000 μs.

Based on the stress diagram of the brick masonry in the middle, the explosive load first caused a rapidly expanding central blasting pit and large longitudinal cracks on the front surface of the masonry wall, with some divergent cracks formed along the mortar joint around the blasting pit (see Figure 6a). After 400 μs, the area and volume of the central blasting pit increased, radial cracks around it developed and were densely distributed, and the upper and lower sides of the wall were damaged (see Figure 6b,c). After 800 μs, the increase in size of the central blasting pit was no longer obvious, and a little peeling damage also appeared on the left and right sides of the wall (see Figure 6d,e). Based on a comparison of the figures on the left and in the middle, the area of the polyurea elastomer coating breach was always smaller than the area of the central blasting pit sandwiched by the masonry wall at any time.

The back surface on the right showed that the explosion load caused large longitudinal cracks in the center of the wall on the back surface, resulting in slight distortion in the top and middle of the back polyurea reinforcement layer (see Figure 6a,b). From 600 μs, the middle gray seam of the polyurea reinforcement layer on the back surface gradually showed lumpy protrusions under the action of impact waves, and the raised lumps merged with each other and expanded (see Figure 6c–e).

Figure 7 shows the damage to the polyurea reinforcement layer on both sides under contact explosion at t = 1000 μs. The back surface reinforcement layer has thickness k values of 2, 4, 6, and 8 mm, which correspond to the damage of working conditions 1–4, respectively, as shown by labels a through d. The figure on the left is the front surface of the reinforced wall, reflecting the damage of the polyurea reinforcement layer on the front surface; the figure on the right is the back side of the wall, reflecting the damage of the polyurea reinforcement layer on the back surface. Based on the front surface on the left under the same contact explosion load, the levels of deformation of the reinforcement layers from the explosions on the surfaces of the same thickness were very similar. The wavy distortions of the central symmetry plane were similar, and the sizes of the openings were almost the same. The polyurea layers around the break showed similar stacking, folding, and bulging after extrusion and stretching.

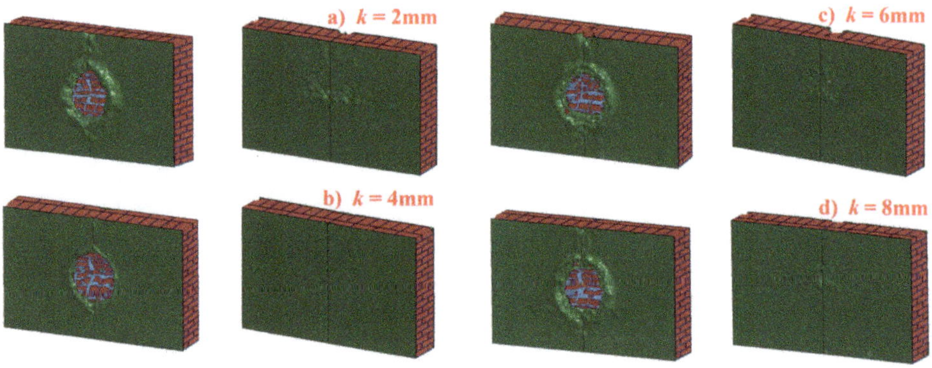

Figure 7. Damage of polyurea reinforcement layer under contact explosion at 1000 μs with different k values (thickness of the back face): (**a**) k = 2 mm; (**b**) k = 4 mm; (**c**) k = 6 mm; (**d**) k = 8 mm.

Based on the back surface on the right, when the polyurea reinforcement layer on the back surface was thin (i.e., in working conditions 1 and 2, as shown in Figure 7a,b), the distortion of the top and middle parts of the reinforcement layer was obvious, and the uplift of the reinforcement layer was small and dense, often distributed along the mortar joint. When the polyurea reinforcement layer was thick (i.e., in working conditions 3 and 4, as shown in Figure 7c,d), the distortion of the top and middle parts of the reinforcement layer was not obvious, and the uplifts of the reinforcement layer were larger but smaller in number, distributed in the corresponding central part of the blasting pit.

To more clearly see the damage to the reinforced masonry wall under contact explosions, Figure 8 was created without the polyurea reinforcement layers on each side. Only the damage to the bricks and mortar under the four working conditions in Table 4 is depicted in the image. The reinforced masonry wall's longitudinal section or symmetrical plane, blast front surface, and blast back surface are all destroyed on the figure's left, middle, and right sides, respectively.

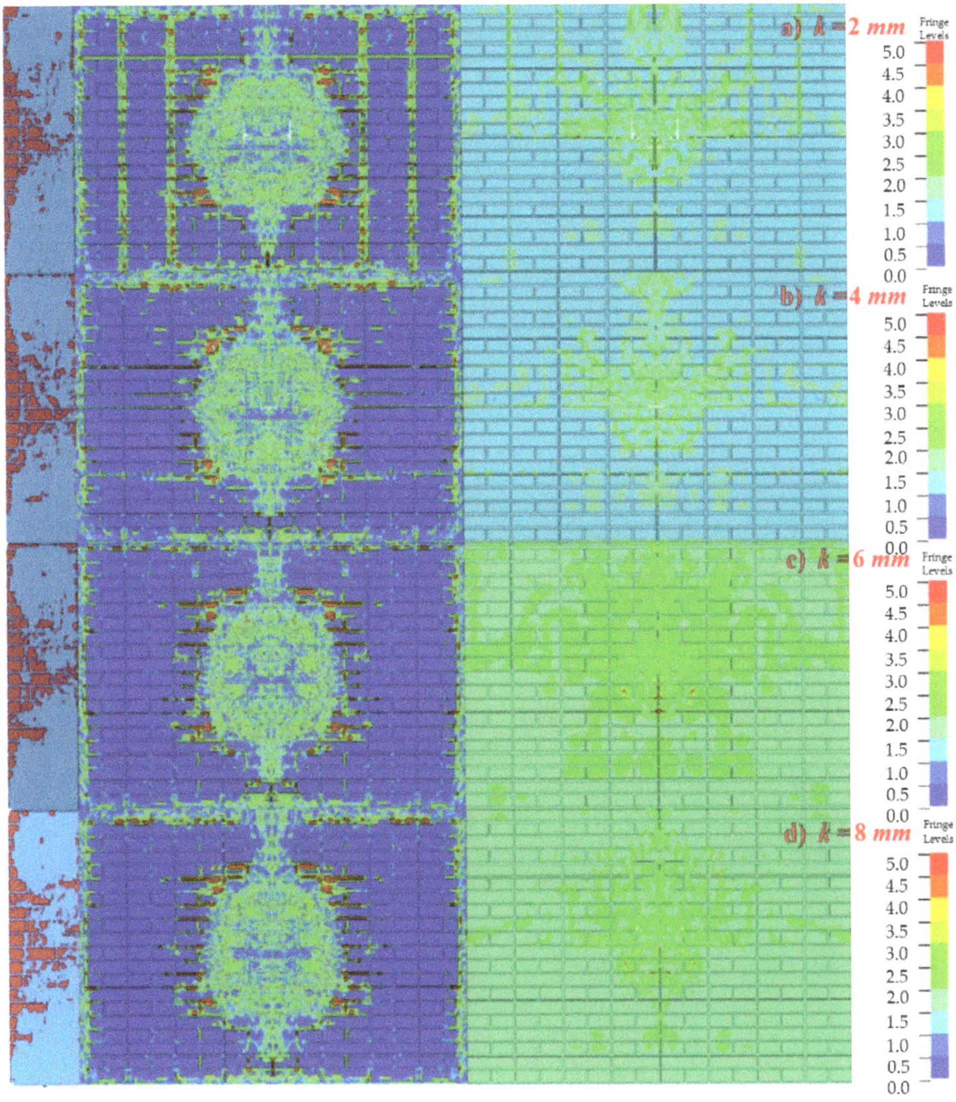

Figure 8. Damage to the reinforced 370 mm wall under contact explosion at 1000 μs with different k values (thickness of back face: (**a**) k = 2 mm; (**b**) k = 4 mm; (**c**) k = 6 mm; (**d**) k = 8 mm.

As shown in the longitudinal section on the left, when the thickness of the polyurea reinforcement layer on the front surface remained unchanged, the depth of the central blasting pit of the masonry wall did not change much with the increasing thickness of the back surface reinforcement layer, and the cracks were densely distributed inside the masonry wall. The damage to the front surface in the middle shows that the areas of the central blasting pit (circled in the figure) under the four working conditions were different. When the back surface was thin (i.e., in conditions 1 and 2, as shown in Figure 8a,b), the area of the blasting pit was larger. When the polyurea reinforcement layer was thicker (i.e., in conditions 3 and 4, as shown in Figure 8c,d), the area of the blasting pit was smaller; the spalling and cracking of the explosion-proof surface of the wall under the four working conditions were also different. Compared with conditions 3 and 4, the damage phenomena, such as the spalling and cracks on the blast front surface of the masonry wall concentrated in the mortar joint in conditions 1 and 2, were more obvious. The damage to the back surface on the right shows that under the four working conditions, the mortar joints were all weak links, and the collapse and spalling developed along the joints, with an overall radial trend (except for the rib plate). Compared with working conditions 3 and 4, where the polyurea reinforcement layer on the back surface was thicker, the strain on the back surface of the masonry wall in working conditions 1 and 2 was smaller.

By selecting the damaged areas along the bounding box, the size statistics of the same wall with different reinforcement methods under the same contact explosion load are shown in Table 6. As the thickness of the polyurea elastomer reinforcement layer on the blast back surface increased, the area of the central blasting pit decreased slightly; the values of working conditions 1 and 2 were close to those of working conditions 3 and 4. With the increasing thickness of the polyurea reinforcement layer on the blast back surface, the overall depth of the blasting pit did not change much, whereby the depths for working conditions 1, 2, and 4 were quite close, at about half of the thickness of the masonry wall.

Table 6. Size statistics for the destruction of the reinforced 370 mm wall under contact explosion.

Serial Number	Thickness of Front Surface/mm	Thickness of Back Surface/mm	Burst Area/mm^2	Burst Depth/mm
1	6	2	388,654	201
2	6	4	380,232	199
3	6	6	353,282	236
4	6	8	339,903	203

3. Experiment

3.1. Experimental Verification of 370 mm Wall under Contact Explosions

3.1.1. Test Plan and Arrangement

The test wall size was 2000 mm × 1200 mm × 370 mm, and a pair of ribs were symmetrically arranged on both sides of the wall to simulate the supporting effect of the surrounding walls on the actual building. The wall surface was smoothed with about 2 mm of mortar. The explosive used in the test was a TNT cylindrical compressed explosive with a charge density of 1.63 g/cm^3. Two charge specifications were used, i.e., 0.50 kg (test 1) and 1.00 kg (test 2). The basic dimensions were Φ100 mm × 39 mm and Φ100 mm × 78 mm. The wall was larger than the explosive to ensure that the explosive would not destroy the entire wall and to avoid the influence of the masonry wall and the size effect of the charge. Tests 1 and 2 correspond to the b and d working conditions in the numerical simulation, respectively. The test layout is shown in Figure 9.

Figure 9. Test layout.

3.1.2. Experimental Results and Analysis

The experimental results and measured parameters are shown in Figure 10. In Figure 10a, a blasting pit is shown in the center of the blast front surface of the masonry wall. Most of collapse damage occurred in the mortar position, resulting in a small amount of block fragments, and the fragments scattered outwards. The fragments fell below the blasting pit, and the fragments were all over the area in front of the test wall. Large cracks appeared on the left and right sides and the upper side of the blasting pit on the front surface of the wall in the horizontal direction. The cracks penetrated to the back surface of the masonry wall. Small spidery radioactive spalls appeared around the blasting pit on the front surface. In test 2, a large blasting pit appeared in the center of the front surface of the masonry wall (see Figure 10b–d). The collapse damage also mostly appeared in the mortar, resulting in large amounts of block fragments and flying debris. Large penetrating cracks appeared in the horizontal direction and upper side of the blasting pit on the front surface of the wall. The upper left part of the masonry wall collapsed backward along the large cracks, and the upper right part was staggered along the large cracks and deflected at a small angle of about 2.3°.

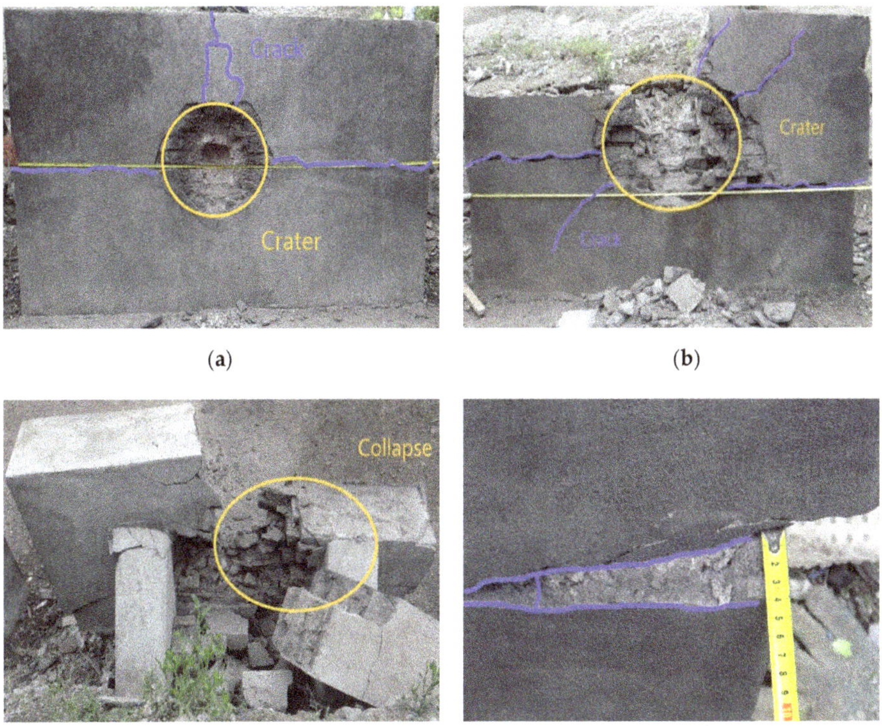

Figure 10. Damage of the wall in the test. (**a**) Test 1 blast front surface; (**b**) Test 2 blast front surface; (**c**) Test 2 blast back surface; (**d**) Test 2 Wall deflection.

3.2. Experimental Verification of Sprayed Polyurea to Reinforce 370 mm Wall under Contact Explosion

3.2.1. Test Plan and Arrangement

The masonry wall was the same as in the prior part, and a polyurea spray was utilized as the reinforcement. The TNT explosive with a density of 1.63g/cm^3, a mass of 1 kg, and dimensions of 100 mm × 78 mm was employed in the test. The reinforced 370 mm wall was sprayed with a 6-mm-thick polyurea reinforcement coating on the blast front surface and 2-mm- and 4-mm-thick polyurea reinforcement layers on the rear surface, respectively. Tests 1 and 2 correspond to working conditions 1 and 2 in the simulation, respectively, and the layout of the test site is shown in Figure 11.

Figure 11. Test site layout.

3.2.2. Experimental Results and Analysis

Figure 12 displays the damage to the reinforced wall during the test. The polyurea elastomer reinforcement layer coated on the surface was removed while maintaining the damaged state of the wall as much as possible. The right side of t figure demonstrates the damage to the wall when the reinforcing layer was removed. In tests 1 and 2, the center of the front surface in Figure 12a,b produced a hole with a small diameter, and on both sides of the hole there were several short tears. The area between the polyurea reinforcement layer and the brick wall "caught" some of the wall fragments, trash, and dust, which was visible through the hole in the reinforcement layer. Large blasting pits with shallow edges were formed in the center of the internal masonry wall, but the depth of the pits became larger in the part corresponding to explosive position. A large number of large radial cracks were distributed around the blast pit, with horizontal and vertical cracks penetrating the brick wall. In test 2, around the center blasting hole of the interior masonry wall, numerous little spidery fractures and spalls appeared along the mortar joints. Based on Figure 12c, the large cracks in the wall center of test 2 extended to the back surface, and the back surface of the internal masonry wall had cracks diverging from the center to the surroundings, where the horizontal and vertical cracks were the coarsest. The polyurea elastomer reinforcing layer on the blast back surface was not torn by massive fractures though. The reinforcing layer, which experienced a minor uplift under the displacement of the internal brick wall, blocked the huge fractures on top, as seen in the left half of Figure 12c.

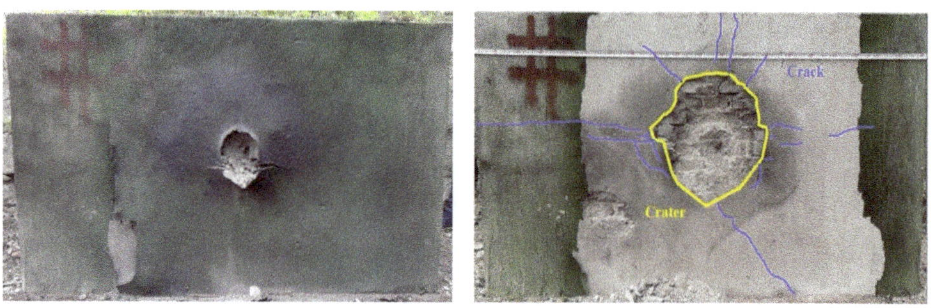

(a)

Figure 12. *Cont.*

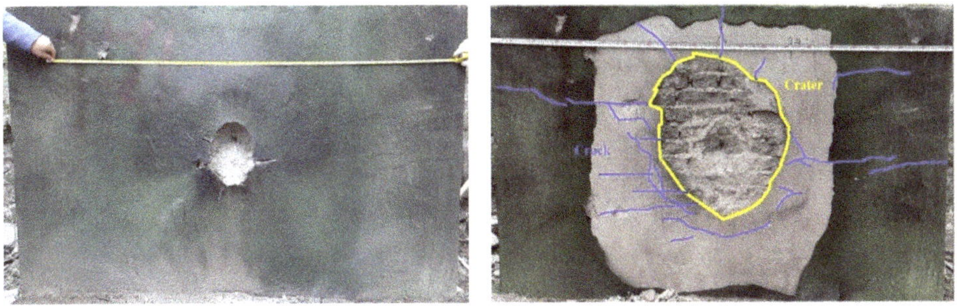

(b)

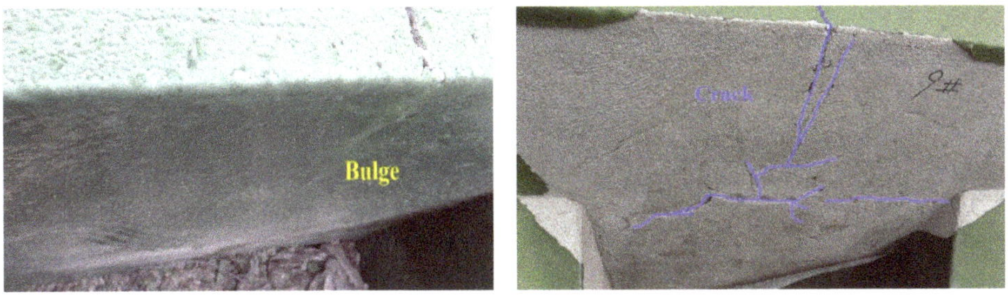

(c)

Figure 12. Damage to the reinforced 370 mm wall in the test. (**a**) Test 1 blast front surface; (**b**) Test 2 blast front surface; (**c**) Test 2 blast back surface.

4. Comparative Analysis

4.1. Comparative Analysis of Numerical Simulation and Experimental Results for an Unreinforced Wall

The numerical simulation showed that when the charge was less than 1 kg, blasting pits and longitudinal cracks predominated; when the charge was greater than 1 kg, the wall's front surface was covered in cracks and its back surface was covered in annular cracks along the mortar joint. In the test, the small charge in test 1 only formed blasting pits and a few cracks. However, not only did the increased charge in test 2 form larger blasting pits and thicker cracks, the complete penetration of the cracks also led to the brittle fracture of the wall. As such, dislocation, deflection, and collapse occurred under the action of the impact wave.

When the contact explosion occurred, the expansion wave and impact wave of the detonation products immediately reached the masonry wall, and the compressive stress wave caused serious damage to the front face of the wall. The central blasting pit and its surrounding cracks developed to different thicknesses along the mortar joint. The pressure surge propagated to the blast back surface to form a strong tensile wave, which caused the collapse and spalling of the back surface. The collapse and spalling caused by the strong tensile wave also developed along the mortar joint in a circular trend, forming penetrating cracks at the central mortar joint.

As shown in Figure 3, after the charge quantity exceeded 1 kg, the expansion of the blasting pit size gradually came to an end. According to the dislocation and collapse of the masonry wall in test 2, when the charge was large, the wall was penetrated and the energy of blast was converted into the displacement part of the wall. The dominant mode of energy release for the masonry wall changed as the charge increased; that is, when the charge was less than 1 kg, the damage to the masonry wall by explosives was mainly reflected in the cross-shaped cracks in the central blasting pits and in horizontal and vertical directions, and then when the charge exceeded 1 kg, the wall was penetrated and the damage to the masonry wall gradually spread, especially the damage to the surrounding mortar joints, until part of the mortar joints penetrated the wall and caused the wall's dislocation, deflection, and collapse.

4.2. Comparative Analysis of Numerical Simulation and Experimental Results for a Reinforced Wall

When the reinforced masonry wall was subjected to the contact explosion load, the numerical simulation depicted the process of large blasting pits forming, with small cracks forming at the edge of the pits and growing along the mortar joints, and coarse cracks gradually extending and even penetrating along the mortar joints in the horizontal and vertical directions. The general consistency of the damage state with the test results shows that the wall model with the polyurea reinforcement layer and the polyurea material model can accurately represent the dynamic response of a reinforced masonry wall under contact explosion load.

In the numerical simulation of the reinforced 370 mm wall, the depths of the blasting pits in the four working conditions were close. When the polyurea reinforcement layer on the back surface was thin, the deformation mainly appeared as the collapse of the masonry around the blasting pits and large-area cracks that developed along the mortar joints, except for in the central blasting pit. When the thickness of the polyurea reinforcement layer on the blast back surface was thick, the crack damages on the front surface of the wall converged.

Compared with the unreinforced 370 mm wall under a 1 kg TNT contact explosion (see Figure 3b–d), the damage to the masonry wall reinforced by the polyurea elastomer under the two working conditions was improved under the same conditions. After reinforcement, the cracks on the surface and other minor damages to the masonry wall increased. The blasting area of test 2 was significantly larger than that of the unreinforced wall, but the number of thick cracks penetrating the 370 mm wall was significantly reduced, with only the longitudinal cracks in the middle remaining. More importantly, the unreinforced masonry, which suffered severe penetrating damage under the contact explosion in the upper part of the wall, deflected and even collapsed, and large amounts of wall fragments, debris, and dust were scattered around the wall after splashing. After spraying the polyurea elastomer for reinforcement, the depth of the central blasting pit was suppressed effectively, the masonry wall remained complete under the contact explosion load, and the polyurea reinforcement layer on the blast back surface of the wall wrapped the debris generated by the wall damage effectively and prevented the splashing of the debris and other destructive elements.

Table 7 shows a simulation and test parameter comparison of the damages to the unreinforced 370 mm wall and the 370 mm wall reinforced by spraying the polyurea elastomer under contact explosion. The areas of the central blasting pit after reinforcement were all larger than those of the unreinforced 370 mm wall. With the increasing thickness of the polyurea reinforcement layer on the blast back surface, the area of the blasting pit slightly reduced; the blasting area in test 2 was close to the numerical simulation value (the error was about 10.3%), but the blasting area in test 1 was quite different from the numerical simulation value. The manually sprayed polyurea layer on the brick wall in the test was not uniform, leading to a certain degree of variation in the test results. Before the explosion test, the wall in the explosion test site of test 1 was improperly stored, and the polyurea reinforcement layer at the lower left corner was damaged and partially peeled off (see Figure 12a), causing a break in the lower left corner. Although the break was at the edge of the wall and far from the center of the contact explosive load, the break acted as a "vent." The right figure in Figure 12a shows that from the surroundings of the central blasting pit to the lower left corner, black detonation product residue was left on the masonry wall, which reduced the surface damage to the masonry wall in test 1 and decreased the area of the blasting pit.

Table 7. Comparison of the parameters for the destruction of the reinforced 370 mm wall under contact explosion.

Serial Number	Polyurea Thickness on Front Surface/mm	Polyurea Thickness on Back Surface/mm	Blasting Area/mm^2		Blasting Depth/mm	
			Numerical Simulation	Experimental Validation	Numerical Simulation	Experimental Validation
Unreinforced	0	0	292,247	—	315	Penetrated
Working condition 1/test 1	6	2	388,654	271,847	201	191
Working condition 2/test 2	6	4	380,232	344,577	199	196
Working condition 3	6	6	353,282	—	236	—
Working condition 4	6	8	339,903	—	203	—

Table 7 shows that the depth of the central blasting pit after reinforcement was less than the corresponding value of the unreinforced 370 mm wall. The simulation values for working conditions 1, 2, and 4 were close. Compared with the unreinforced 370 mm wall, the depth of blasting pit reduced significantly to about 36.4%. The numerical simulation values of blasting pit depths for working conditions 1 and 2 were in good agreement with the experimental values, with errors of 5.1% and 1.4%, respectively.

In summary, for working condition 1, the depth of the central blasting pit of the 370 mm wall was suppressed effectively, and the scattering of fragments, debris, and other destructive elements was prevented, thereby greatly improving the support and integrity of the wall. For working conditions 2 to 4, although a certain blast-resistant effect was achieved, the doubling of the thickness of polyurea layer on the back surface did not significantly increase the corresponding blast-resistant effect. Therefore, when the polyurea layer on the front explosion surface of the 370 mm wall was fixed at 6 mm, a coating of 2 mm of polyurea elastomer on the blast back surface can achieve a good reinforcement effect.

5. Conclusions

The following conclusions were reached after an analysis of the failure characteristics and damage processes of unreinforced and reinforced masonry walls under contact explosions:

(1) The destructional forms of the masonry wall under the contact explosions can be grouped into two kinds. When the explosive charge was below the critical value, the destruction of the masonry wall occurred near the explosive, forming a central blasting pit, a penetrating hole, and cross-cracks oriented in both horizontal and vertical directions. The other form of destruction slowly spread to the surrounding area, particularly the mortar joints; the spalling and collapse of the blast back surface also increased noticeably, meaning that eventually the wall would fall;

(2) The critical value of the TNT explosive for the 370 mm masonry wall under contact explosion was 1 kg. When the TNT explosive was over 1 kg, the destructional form of the masonry wall involved the splashing of wall fragments, deflection, and movement of the wall after the explosion;

(3) The polyurea elastomer sprayed on the front surface of the masonry wall can attenuate the shock waves generated by the TNT explosions effectively, and the polyurea elastomer sprayed on the back surface of the masonry wall can effectively restrain the wall breakage after the explosion. When the masonry wall was sprayed with the elastomer on the front and back surfaces, the wall's resistance to the damage caused by the explosive increased;

(4) When the thickness of the front polyurea layer was 6mm and the thickness of the back polyurea layer was 2 mm, the masonry wall had the best blast-resistant effect.

Author Contributions: Conceptualization, Q.X.; methodology, X.Z. and Z.H.; validation, Z.H.; test, Q.X., X.Z., and Z.H.; writing—original draft preparation, X.Z.; writing—review and editing, Y.C. and T.C. All authors have read and agreed to the published version of the manuscript.

Funding: This research was funded by the National Natural Science Foundation of China (grant No. 11872214).

Institutional Review Board Statement: Not applicable.

Informed Consent Statement: Not applicable.

Data Availability Statement: The data are contained within this article; more detailed data can be obtained from the corresponding author.

Conflicts of Interest: The authors declare no conflict of interest.

References

1. Porter, J.; Dinan, R.; Hammons, M.; Knox, K. Polymer coatings increase blast resistance of existing and temporary structures. *Amptiac Q.* **2002**, *6*, 47–52.
2. Davidson, J.; Porter, J.R.; Dinan, R.; Hammons, M.; Connell, J.D. Explosive testing of polymer retrofit masonry walls. *J Perform. Constr. Facil.* **2004**, *18*, 100–106. [CrossRef]
3. Bahei-El-Din, Y.A.; Dvorak, G.J.; Fredricksen, O.J. A blast-tolerant sandwich plate design with a polyurea interlayer. *Int. J. Solids Struct.* **2006**, *43*, 7644–7658. [CrossRef]
4. Tekalur, S.A.; Shukla, A.; Shivakumar, K. Blast resistance of polyurea based layered composite materials. *Compos. Struct.* **2008**, *84*, 271–281. [CrossRef]
5. Hrynyk, T.D.; Myers, J. Out-Of-Plane behavior of URM arching walls with modern blast retrofits: Experimental results and analytical model. *J. Struct Eng.* **2008**, *134*, 1589–1597. [CrossRef]
6. Aghdamy, S.; Wu, C.; Griffith, M. Simulation of Retrofitted Unreinforced Concrete Masonry Unit Walls under Blast Loading. *Int. J. Prot. Struct.* **2013**, *4*, 21–44. [CrossRef]
7. Davidson, J.S.; Fisher, J.W.; Hammons, M.I.; Porter, J.R.; Dinan, R.J. *Failure Mechanisms of Polymer-Reinforced Concrete Masonry Walls Subjected to Blast*; Air Force Research Laboratory Tyndall Air Force Base: Bay County, FL, USA, 2005.
8. Gattesco, N.; Boem, I. Out-of-plane behavior of reinforced masonry walls: Experimental and numerical study. *Compos. Part B Eng.* **2017**, *128*, 39–52. [CrossRef]
9. Fan, J.Y.; Fang, Q.; Zhang, Y.D. Numerical simulation on the anti-blast properties of masonry walls. *Prot. Eng.* **2011**, *33*, 35–40.
10. Guo, Y.R.; Zheng, H. Numerical Simulation of Polyurethane Strengthened Perforated Masonry Walls under Blast Loading. In *Advanced Materials Research*; Trans Tech Publications Ltd.: Wollerau, Switzerland, 2013.
11. Xu, K.; Lu, Y. Numerical simulation study of spallation in reinforced concrete plates subjected to blast loading. *Comput. Struct.* **2006**, *84*, 431–438. [CrossRef]
12. Dinan, R.; Fisher, J.; Hammons, M.I.; Porter, J.R. *Failure Mechanisms in Unreinforced Concrete Masonry Walls Retrofitted with Polymer Coatings*; Air Force Research Laboratory Tyndall Air Force Base: Bay County, FL, USA, 2003.
13. Zhu, H.; Wang, X.; Wang, Y.; Ji, C.; Wu, G.; Zhang, L.; Han, Z. Damage behavior and assessment of polyurea sprayed reinforced clay brick masonry walls subjected to close-in blast loads. *Int. J. Impact Eng.* **2022**, *167*, 104283. [CrossRef]
14. Iqbal, N.; Sharma, P.; Kumar, D.; Roy, P. Protective polyurea coatings for enhanced blast survivability of concrete. *Constr. Build. Mater.* **2018**, *175*, 682–690. [CrossRef]
15. Ji, L.; Wang, P.; Cai, Y.; Shang, W.; Zu, X. Blast Resistance of 240 mm Building Wall Coated with Polyurea Elastomer. *Materials* **2022**, *15*, 850. [CrossRef] [PubMed]
16. *GB 50003-2011*; Code for Design of Masonry Structures. General Administration of Quality Supervision, Inspection and Quarantine of the People's Republic of China: Beijing, China, 2011.
17. Hallquist, J.O. *LS-DYNA Keyword User's Manual*; Livemore Software Technology Corporation: Livermore, CA, USA, 2007.
18. Gan, Y.-D.; Song, L.; Yang, L.-M. Numerical Simulation for Anti-blast Performances of Steel Plate Coated with Elastomer. *Binggong Xuebao/Acta Armamentarii* **2009**, *30*, 15–18.

coatings

MDPI

Article

Effect of Planting Rebars on the Shear Strength of Interface between Full Lightweight Ceramsite Concrete and Ordinary Concrete

Hongbing Zhu [1,2,3,*], Yixue Duan [1], Xiu Li [4], Na Zhang [1] and Jingyi Chen [1,5]

1 School of Urban Construction, Wuhan University of Science and Technology, Wuhan 430065, China; dyx4885@163.com (Y.D.); zhangna212121jy@163.com (N.Z.); 13628659252@163.com (J.C.)
2 Institute of High Performance Engineering Structure, Wuhan University of Science and Technology, Wuhan 430065, China
3 Hubei Provincial Engineering Research Center of Urban Regeneration, Wuhan University of Science and Technology, Wuhan 430065, China
4 School of Transportation Engineering, Wuhan Technical College of Communications, Wuhan 430065, China; hubeilixiu@163.com
5 The National Prefabricated Construction Industry Base, Huasen Architecture and Engineering Design Consulting Company, Shenzhen 518000, China
* Correspondence: zhuhongbing@wust.edu.cn

Abstract: The use of full lightweight ceramsite concrete (FLWCC) for the repair of ordinary concrete (OC) structures has a good application prospect in the field of engineering structural strengthening. However, the interface here is less studied at present. For this purpose, 10 sets of FLWCC (new concrete)–OC (old concrete) specimens were produced for the shear test to test the bonding properties of the interface. The results showed that the damage form was changed from brittle damage to ductile damage after strengthening. It was proven that planting rebars in the interface could improve the shear performance. The interface shear strength increased with the number of rebars and it had a better effect after the number was more than 2. The strength was related to the rebar diameter and the maximum was obtained when the diameter was 8 mm. The most suitable spacing of the bars was 80 mm. The one-way analysis of variance (ANOVA) showed that the number of rebars had the greatest effect on shear strength followed by rebar diameter, while the effect of the spacing of the bars was less pronounced. Moreover, Fib's (2010) specification of the interface shear strength formula could be used for the calculation of FLWCC–OC.

Keywords: full lightweight ceramsite concrete; planting rebars; strengthening; new and old concrete; bonding interface; shear performance

Citation: Zhu, H.; Duan, Y.; Li, X.; Zhang, N.; Chen, J. Effect of Planting Rebars on the Shear Strength of Interface between Full Lightweight Ceramsite Concrete and Ordinary Concrete. *Coatings* **2023**, *13*, 1622. https://doi.org/10.3390/coatings13091622

Academic Editor: Paolo Castaldo

Received: 21 August 2023
Revised: 8 September 2023
Accepted: 12 September 2023
Published: 15 September 2023

1. Introduction

Concrete materials are widely used in construction and transportation. However, in the course of daily use, concrete buildings can fail to reach their expected lifespan because they suffer from problems such as creeping and freeze-thaw cycles [1]. Strengthening is usually used in engineering to extend its life. Therefore, the choice of repair materials has an important influence on the strength of the reinforced concrete structure [1]. With the innovation of construction materials, lightweight [2,3], energy-saving and environmentally friendly materials have been widely used in building structures [4]. FLWCC, for example, has been extensively researched and commonly used in practical engineering and repair fields [5–7]. FLWCC is a kind of lightweight aggregate concrete with shale ceramic granules and shale ceramic sand instead of crushed stone and river sand. Shale ceramic granules are made from shale as raw material and industrial waste is added during the roasting process. This reflects its environmental advantages. Ceramic granules have a large number of pores inside, which gives them the advantage of being lightweight and having thermal insulation. Therefore, when FLWCC is used in construction projects, it not only reduces the dead weight of the structure and saves the use of cement and steel but also reduces the burden of indoor HVAC (heating, ventilation and air conditioning). Because of the water retention properties of ceramic granules, when FLWCC is used to strengthen OC, the water inside the ceramic granules can further promote the hydration reaction after the hardening of the interface between the new and old concrete, thus making the interface stronger. At present, there is a lot of research on the composition of raw materials [8–10], force characteristics [11–13] and durability properties [14–16]. The application of FLWCC in the field of repair and strengthening can give full play to its potential to replace the shortcomings of traditional concrete, which is in line with the environmental protection concept of "carbon neutrality". And at the same time, it also brings a new direction for the topic of concrete structural strengthening.

The bonding interface is the weak part of the structure due to the discontinuity of aggregate particles in the transition region between the new and old concrete and the more complex load transfer [17]. The bonding strength of the interface is mainly influenced by the roughness of the interface, the Van der Waals force and the shrinkage and rebar difference between new and old concrete [18]. In recent years, scholars have conducted relevant studies mainly on roughness treatment [19], interface agent performance enhancement [20], high-tech repair material application [21] and rebar implantation [22]. Among them, the research direction of roughness, interface agent and high-tech repair materials is to improve the adhesion and friction of the bonding interface to improve the strength, which is to improve the bonding strength by mechanical occlusion force and its own strength and stiffness. Planting rebars in the interface can increase the ductility of the repaired structure relative to other methods, thus greatly improving the safety of the overall structure. Therefore, this method is widely used in strengthening works. In the process of rebar implantation, the influencing factors of bonding interface implantation are rebar diameter, rebar planting rates, spacing of bars, anchorage depth, properties and coating on rebars, etc. [23].

Even after controlling the influencing factors mentioned above, the effect of rebar implantation may show different results under different conditions. The five-zone three-layer model is a widely recognized microtheoretical basis for reinforced concrete [24]. Based on this microscopic model, many scholars have conducted research on the effect on bonding performance. The interfacial shear strength between the new and old concrete is the most important factor affecting the strengthening effect [25]. There are two dominant views on whether planting rebars strengthens the bonding ability: one is that scholars believe that rebars enhance the bonding effect. For example, Du [21] concluded from double-sided shear tests of OC strengthened with ultra-high performance RPC reinforcing mesh that the shear strength of the reinforced concrete was 1.37–3.11 times higher than before strengthening. Tian [26] carried out finite element simulations of the shear performance of bonding interfaces strengthened with rebars. They concluded that rebar implantation can inhibit the development of cracks. The rebar implantation makes the damage process of the new and old concrete structures ductile and it can increase the cracking load and damage load of the specimens to a certain extent. Yang [27] investigated the effects of different rebar planting rates, anchorage depths and different spacing of bars on the shear properties of the UHPC–stone interface. The results showed that the interfacial shear strength and ductility of the combined specimens increased step by step with the increase in the rebar planting rate when it was less than 0.269%.

Another view is that the presence of rebar breaks the continuity of the concrete near the bonding interface and affects the force transfer. Jiang [28] summarized from rebar pull-out tests that the average bonding strength of prefabricated connected members with notches is reduced by approximately 26%. The reduction in strength is explained by the discontinuity in the mechanical properties of the concrete around the rebars due to the presence of the bonding interface. Costa [29] evaluated the effect of normal stress and rebar planting rates on the shear resistance of the bonding interface of light aggregate concrete strengthened plain concrete. The tests illustrated that the bonding effect of the rebars comes into play only after the concrete near the bonding interface has cracked and bonding slip has occurred. The rebar planting rates do not contribute significantly to the maximum shear strength of the bonding interface. Goldyn [30] launched tests by strengthening plain concrete members with light aggregate concrete poured at different ages with bonding interfaces. Rebar implantation on the bonding interface causes a change in the damage mode of the member from brittle to ductile. The rebar planting rate was subjected to about 18%–23% of the stress at yield when the bonding interface was just cracked. This suggests that rebar can increase the post-peak residual stresses without significantly affecting the load-carrying capacity of the bonding interface.

The shear transfer mechanism of two layers of concrete in direct shear is complex, so it is necessary to evaluate the shear properties of the interface [31]. In response to the above research status, this paper investigated the shear performance of the bonding interface between FLWCC and OC. The rebar planting rate is varied by the number of rebars and diameter, so it is necessary to investigate which of the two ways enhances the interface shear best. Also, the spacing of the bars determines the rebar-concrete force transfer mode, so these three influencing factors are investigated in this experiment. The following work was performed using straight shear tests on the bonding interface. (1) The shear damage pattern and maximum shear strength under the influence of different factors were recorded and compared. (2) The most influential factors and factor levels were calculated and analyzed using one-way ANOVA combined with the analysis of the shear strength mechanism. (3) The predicted shear strength was predicted using the Fib (2010), the AASHTO (2010) and the ACI (2008) to derive the most appropriate expression for the specification expressing the present test situation. This paper expanded application areas for lightweight aggregate concrete. The properties of lightweight and thermal insulation give the ceramic granules a great research market with the potential to replace traditional concrete. This paper can provide macroscopic data reference for the study of the interface between old and new concrete and lay the foundation for the study of the microscopic interface of FLWCC–OC. Combined with the already available ceramsite concrete members' stress characteristics, durability, etc., it fills part of the research field of ceramsite.

2. Experimental Materials and Specimen Production

2.1. Experimental Materials

(1) Coarse aggregates

Ordinary crushed stone was used for OC in this study. The coarse aggregate of FLWCC used 900-grade crushed shale ceramsite pellets (as shown in Figure 1). Before concrete mixing, it was necessary to soak the ceramsite particles for 4 h and then dry for 12 h for prewetting treatment. The main physical properties of the coarse aggregates are shown in Table 1.

Table 1. Physical property parameters of coarse and fine aggregates.

Type of Aggregate	Particle Size (mm)	Volume Density (kg/m^3)	Apparent Density (kg/m^3)	Compressive Strength of Concrete Cylinder (MPa)	Water Absorption in 1 h (%)	Mud Content	Fineness Modulus
Ordinary crushed stone	5~20	814	2688	6.8	2.42	/	/
Ceramsite pellets	5~20	896	1553	9.7	6.3	/	/
Ordinary river sand	/	890	2641	/	/	≤2	3.5
Ceramsite sand	/	1570	1479	/	/	≤2	2.7

(a) (b)

Figure 1. Coarse aggregates. (**a**) Ordinary crushed stone. (**b**) Ceramsite pellets.

(2) Fine aggregate

The ceramsite granules were made of 800-grade shale ceramsite sand produced by Yichang Guangda Ceramsite Granule Products Limited company. Its relevant physical property parameters and the ordinary river sand are also shown in Table 1.

(3) Cement

All of the cement's physical property indexes meet the test specification requirements, as shown in Table 2.

Table 2. Physical property parameters of cement.

Type of Cement	Code Number	Density (g/cm^3)	Mineral Composition of Clinker				Fineness (%)
			C_3S	C_2S	C_3A	C_3AF	Margin of 80 μm Square Sieve
Ordinary silicate cement	P.O. 42.5	3.15	45	25	12	8	6.5

(4) Water-reducing agent

The water-reducing agent was HsC polycarboxylate-type superplasticizers, as shown in Table 3.

Table 3. Physical property parameters of water-reducing agent.

Appearance Color	PH Value	Specific Gravity	Solid Content (%)	Water Reducing Ratio (%)
Pale yellow	6~8	1.08 ± 0.02	40	25~35

(5) Rebar

The tested rebar type was HRB400 and the rebar diameters were 6 mm, 8 mm and 10 mm. The basic physical properties tested according to GB/T28.1-2010 [32] are shown in Table 4.

Table 4. Physical property parameters of rebar.

Category	Yield Strength (MPa)	Ultimate Strength (MPa)	Modulus of Elasticity (MPa)
HRB 400	435.5	671	2.03×10^5

2.2. Mix Proportion

The old concrete was configured with strength grade C45. According to GB 50367-2013 [33], all new concrete should be one level higher than the strength grade of the original structural concrete. In order to meet the practical requirements of the project, new concrete with a strength class of LC50 was used. The proportions of OC and FLWCC are shown in Table 5.

Table 5. Mixed proportion of FLWCC (Unit: kg/m^3).

Cement	Coarse Aggregates	Fine Aggregate	Water-Reducing Agent	Water
605	1050	565	/	230
550	620	600	5.1	154

2.3. Physical and Mechanical Properties

Some of the necessary indicators were measured according to GB/T50081-2019 [34]. The test results are shown in Table 6.

Table 6. Basic mechanical indices of concrete after 28 d maintenance.

Types of Concrete	Cube Compressive Strength (MPa)	Axial Compressive Strength (MPa)	Modulus of Elasticity (GPa)
LC50	53.66	46.53	3.067
C45	35.31	27.28	3.043

2.4. Specimen Design

The rebar connection of the interface is shown in Figure 2. The mode of arrangement was divided into the forms of 1, 2, 3 and 4 bars. The bonding performance under the action of number of rebars, rebar diameter (double rebars) and spacing of the bars (double rebars) were studied. The roughness of interface was processed using the manual chiseling method and the roughness was determined using the sand cone method. Three specimens were made in each group, which was for accuracy. The final results of the test should be taken as the average. Shear test groups are shown in Table 7.

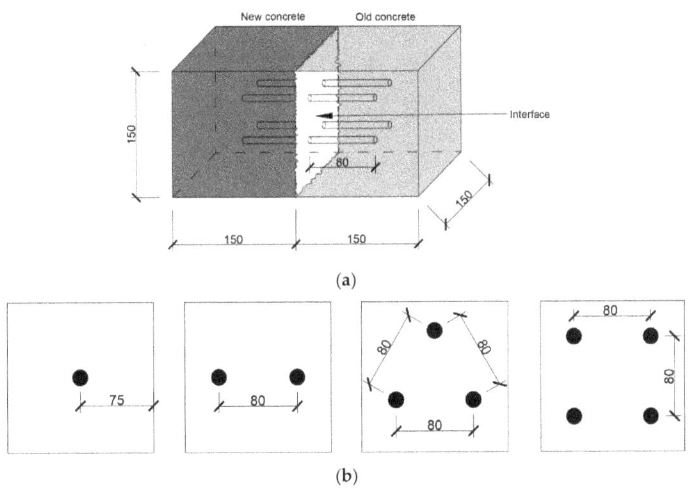

Figure 2. Design solutions for new and old concrete specimens (unit: mm). (**a**) Bonding specimens. (**b**) Rebar arrangement.

Table 7. Groups of shear tests.

Specimen Number	Number of Rebars	Rebar Planting Rate (%)	Spacing of Bars (mm)	Rebar Diameter (mm)	Anchorage Depth (mm)	Interface Roughness (mm)	Interface Agents	Number of Specimens
CZJ	/	/	/	/	/	/	/	3
SJ	/	/	/	/	/	2–3	cement paste	3
B1G1	1	0.223	C8	80	10 d	2–3	cement paste	3
B1G2	2	0.447	C8	80	10 d	2–3	cement paste	3
B1G3	3	0.670	C8	80	10 d	2–3	cement paste	3
B1G4	4	0.894	C8	80	10 d	2–3	cement paste	3
B2G1	2	0.251	C6	80	10 d	2–3	cement paste	3
B2G2	2	0.447	C8	80	10 d	2–3	cement paste	3
B2G3	2	0.698	C10	80	10 d	2–3	cement paste	3
B3G1	2	0.447	C8	60	10 d	2–3	cement paste	3
B3G2	2	0.447	C8	80	10 d	2–3	cement paste	3
B3G3	2	0.447	C8	100	10 d	2–3	cement paste	3

Note: The number of specimens was a shear test of each group of three specimens; "/" means that the group did not have this content. B1G1–B1G4 were the rebar planting rate groups. B2G1–B2G3 were the rebar diameter groups. B3G1–B3G3 were the spacing of bars groups. SJ was the cement net slurry treatment group. CZJ was the old concrete whole cast group. "d" was the diameter of the rebars.

2.5. Fabrication of Specimens

The poured interface of the old concrete was selected as the interface. After treatment with the manual chiseling method, rebars were then implanted on the interface. The method was scribing—drilling—clearing—filling with adhesives—implanting rebars—curing and maintenance. Before placing the new concrete, a cement paste was applied to the interface. The finished specimens are shown in Figure 3.

(a)

(b)

Figure 3. New and old concrete specimens. (**a**) After pouring. (**b**) After demolding.

2.6. Test Device and Loading Method

Four-point loading of distribution beams was used as the test loading method. And the main instruments of the test were a microcomputer-controlled electro-hydraulic servo universal testing machine, a loading sensing cell, etc.

The conceptual and field diagrams of the test setup are shown in Figure 4. It can be seen that the relationship between the distance d_2 between the far loading point and the bonding interface and the distance d_1 between the near loading point and the interface was $d_1 = \frac{2}{5}d_2$. The compact four-point shear loading was $p_1 = \frac{5}{2}p_2$ and the load at the interface could be calculated as $P_3 = \frac{3}{7}F$. That is to say, the shear force at the bonding interface was 3/7 times of the load loaded by the testing machine.

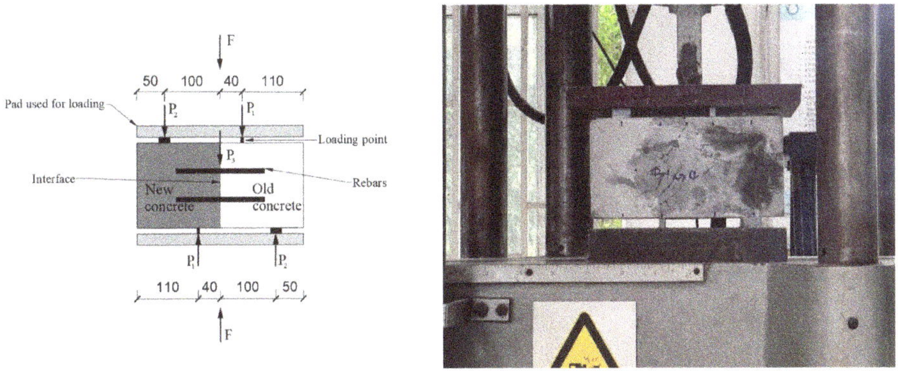

(**a**) (**b**)

Figure 4. Test configuration of concrete samples. (**a**) Test diagram (unit: mm), (**b**) Laboratory test setup.

3. Experimental Results

3.1. Destruction Form

The damage pattern of 12 groups of specimens in a direct shear test is shown in Figures 5–8. When the shear force of the interface without rebars group (SJ) reached its limit, the specimens cracked along the interface instantly. There was no obvious crack from a macroscopic point of view before fracture of the interface and its damage form was brittle damage. The damage pattern of the interface showed that the damage interface was in the same direction as the shear force action. The damage interface on the old concrete side was similar to the interface morphology after the old concrete was chiseled at the beginning. This fully proved that the interface was the weak part of shear damage resistance.

(**a**) (**b**)

Figure 5. Unstrengthened group shear damage. (**a**) CZJ group. (**b**) SJ group.

B1G1 B1G2 B1G3 B1G4

Figure 6. Shear damage with different number of rebars.

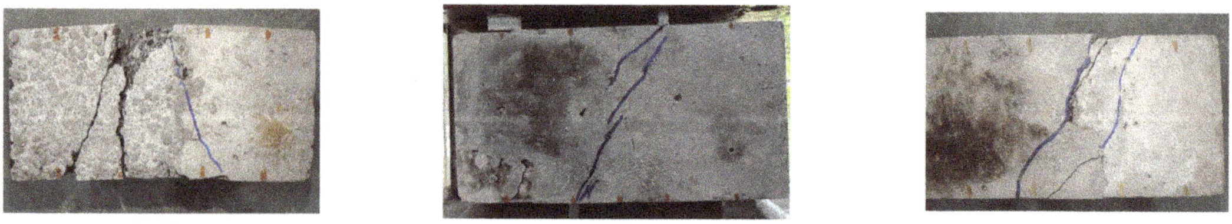

B2G1 B2G2 B2G3

Figure 7. Shear damage with different rebar diameters.

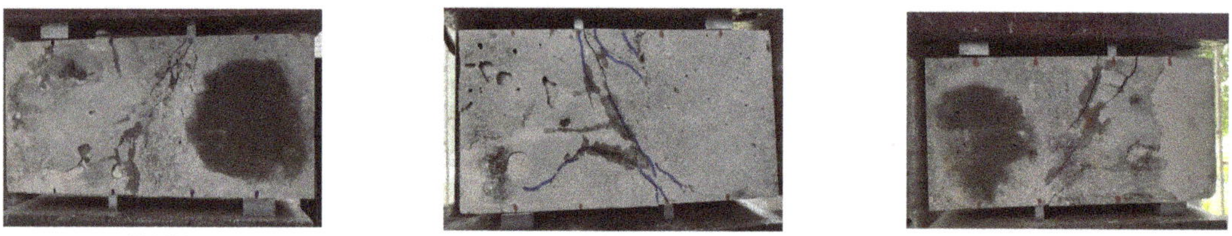

B3G1 B3G2 B3G3

Figure 8. Shear damage with different spacing of bars.

The damage of the bonding interface group (B1G1–B1G4, B2G1–B2G3, B3G1–B3G3) is shown in Figures 6–8. The whole damage process could be divided into four stages: At the beginning of loading, the specimens had no obvious changes. With the increase in load, small cracks appeared at the edge of interface. Upon continued loading, the cracks gradually penetrated, with their width widened, and a small amount of sheared aggregate was dislodged from the surface. When the ultimate load was finally reached, the rebars were not pulled out or cut off and the concrete was not partially pulled out.

Observation of the damage pattern showed that the damage pattern of the interface group was that cracks extended from the interface to the concrete on both sides caused by the shearing force. The destruction process of the specimens with different variables was approximately the same. The main cracks were the oblique cracks distributed between the shear supports, and some specimens also had a small number of micro cracks. This phenomenon was due to the axial pulling force of the rebars after cracks appeared in the specimens, which caused tensile stresses in the concrete around the rebars, therefore causing micro cracks.

What could also be seen was that aggregate shedding from the outer surface was a little more common in the groups with low rebar planting rates, such as B1G1, B1G2 and B2G1. It was clear to see that this phenomenon diminished as the rates were increased. This was due to the fact that the rebar bond to the concrete in the group with lower rebar planting rates had a smaller area of influence.

3.2. *Shear Strength Analysis*

The results of the shear strength tests at the interface are shown in Table 8. In summary, the shear strength was improved overall after 2 to 3 mm of manual chiseling method as well as rebar planting. The gain in shear strength rose from 0.48% to 50.52%. This indicated that the rebar planting played a good role in improving the shear resistance of the new and old concrete interface. A reasonable number of rebars, rebar diameter and spacing of the bars could significantly improve the shear strength so that the bonding performance could be enhanced.

Table 8. Results of shear test.

Group Number	Shear Load/kN	Shear Strength τ/MPa	Average Shear Strength/kN	$\frac{\tau}{\tau_{SJ}}$	Strength Gain Rate/%
CZJ	252.35 256.90 247.80	11.22 11.42 11.01	11.22	/	/
SJ	139.80 143.25 141.53	6.21 6.37 6.29	6.29	1	0
B1G1	142.30 143.60 141.00	6.32 6.38 6.27	6.32	1.005	0.5
B1G2	188.30 179.35 203.60	8.37 7.97 9.05	8.46	1.345	34.5
B1G3	181.30 201.45 225.90	8.06 8.95 10.04	9.02	1.434	43.4
B1G4	196.65 218.75 223.65	8.74 9.72 9.94	9.47	1.505	50.5
B2G1	142.23 141.01 143.45	6.32 6.27 6.38	6.32	1.005	0.5
B2G2	188.30 179.35 203.60	8.37 7.97 9.05	8.46	1.345	34.5
B2G3	147.40 168.50 187.40	6.55 7.49 8.33	7.46	1.185	18.5
B3G1	155.50 165.35 189.30	6.91 7.35 8.41	7.56	1.202	20..2
B3G2	188.30 179.35 203.60	8.37 7.97 9.05	8.46	1.345	34.5
B3G3	154.65 177.75 170.10	6.87 7.90 7.56	7.44	1.184	18.4

Note: τ_{SJ} was the shear strength of SJ group, $\frac{\tau}{\tau_{SJ}}$ was the ratio of shear strength of each group to SJ group.

1. Number of rebars

Figure 9 illustrates the influence pattern of the number of rebars on the shear strength. In conjunction with Table 8, the increase in shear strength was 0.5%, 34.5%, 43.4% and 50.5% when the number w increased from 0 to 4 (SJ, B1G1–B1G4). This indicated that the method of growing the rebar planting rate by increasing the number of rebars was effective in improving the shear strength of the interface. This finding was also in agreement with Yang [27]. This was because when the number of rebars increased, the tensile force generated by the rebars at the interface also increased. This led to an improvement in friction and pinning action at the interface.

Individually, B1G1 only increased by 0.5% compared to SJ, while B2G2 had a large increase of 34.5%. This implied that the degree of improvement in shear strength was not significant with a low rebar planting rate. When analyzed in conjunction with Yang's [27] study, this was because at a low rebar planting rate, the shear strength was mainly supported by the concrete and the rebars were unable to give full play to its ductile role. When the rate was high, the role of rebars in the middle and late stages of loading was more significant, thus the interface was strengthened by the rebars and the ductility was more obvious.

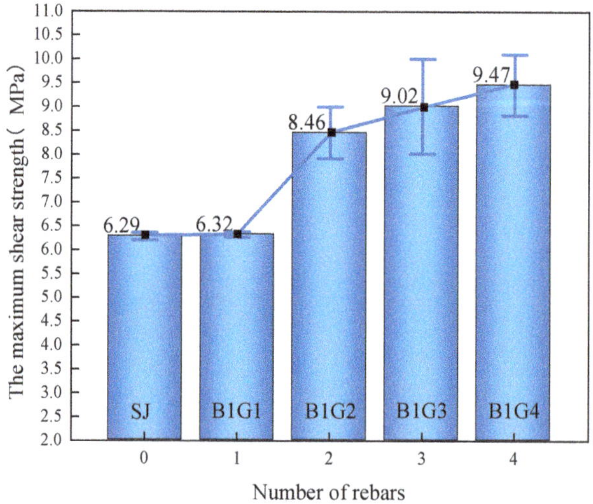

Figure 9. The maximum shear strength of the interface under the influence of the number of rebars.

2. Rebar diameter

As can be seen from Figure 10, the shear strength increased and then decreased with the growth of the rebar diameter. The shear strength reached its maximum at a diameter of 8 mm (B2G2), which was 34.5% higher than that of the SJ group. When the diameter was 10 mm (B2G3), the shear strength tended to decrease to 7.46 MPa, which was still larger than that of 6.32 MPa for 6 mm (B2G1). It could be seen that too large a rebar diameter did not improve the shear strength well. On the one hand, the increase in the diameter reduced the bonding performance with the concrete. On the other hand, it reduced the area of the concrete that was effectively loaded [35]. The selection of the appropriate rebar diameter in the actual project could better improve the interface bearing capacity.

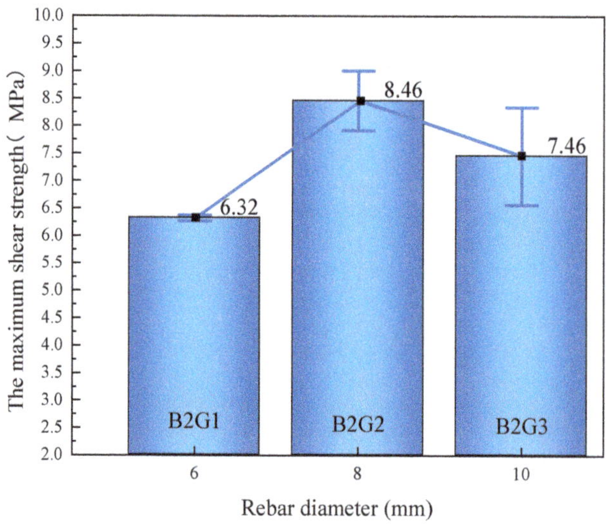

Figure 10. The maximum shear strength of the interface under the influence of the rebar diameter.

Combining B1G3 and B2G3, although the rebar planting rates were almost the same, the former had more than twice the rate of gain in shear strength than the latter. Figure 9 illustrates that the shear strength was positively correlated with the number of rebars within a certain range, while Figure 10 does not have this pattern. If the shear strength was to be increased by increasing the rebar planting rate, the number of rebars had a better gain than the rebar diameter. What was being suggested was that increasing the number of rebars in a strengthening project would be better than using larger diameters. Similarly, in practice, uniform sizes of smaller diameter rebar are also better constructed than different sizes of larger diameter.

3. Spacing of bars

Figure 11 depicts that the shear strength increased and then decreased with a rising spacing of the bars. This suggested that unsuitable spacing weakened the shear resistance of the rebars. The interface shear strength of the B3G2 group reached a maximum of 8.46 MPa. The shear strength of B3G1 was greater than B3G3. The reason for this phenomenon was the anchor group effect [36]. The internationally accepted method for calculating concrete post anchorage was the CCD method (concrete capacity design). It is shown in Figure 12, where h_{ef} is the anchorage depth. At a spacing of 60 mm (B3G1), the bearing areas of the two bars overlapped too much. As a result, the bars could not make good use of the concrete taper to transfer the load. When the spacing was 100 mm (B3G3), although the two bars could make better use of the concrete taper than with a spacing of 60 mm (B3G1), the edge distance of the rebar anchorage of 25 mm was too small for the cross-section of 150 mm × 150 mm. Hence, the cracks near the maximum stress of the bars developed toward the outer edge of the specimens, eventually leading to concrete splitting damage. Therefore, the spacing of 80 mm (B3G2) was the most appropriate in this test to achieve a balance in terms of edge distance and spacing compared to the other two groups. This conclusion was similar to that of Abbas [37]. The group anchor effect had to be taken into account in the actual project so that the connection between the rebar and concrete could be utilized as much as possible.

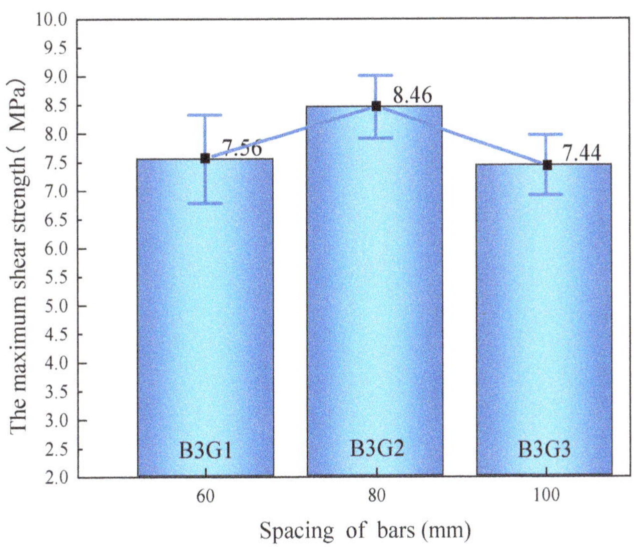

Figure 11. The maximum shear strength of the interface under the influence of the spacing of bars.

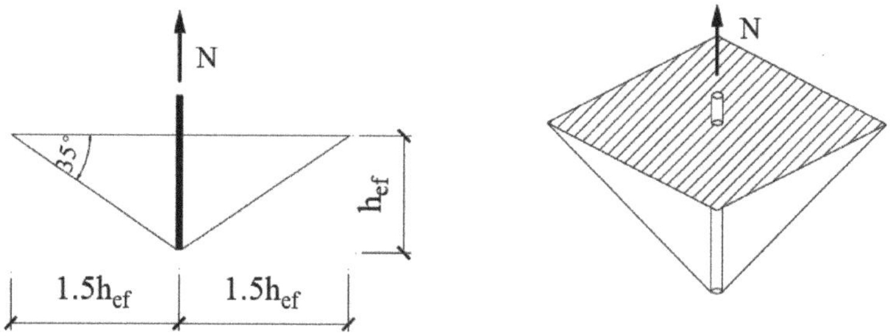

Figure 12. CCD algorithm calculation model.

3.3. One-Way ANOVA

The variation in the maximum shear strength reflected the effect of three influencing factors on the bonding performance. However, the strength of the influence of these factors was difficult to visualize. Therefore, in this section, one-way ANOVA was used to determine the strength and to further analyze which level of shear strength was greatest for a given factor.

The number of rebars (Factor *A*), rebar diameter (Factor *B*) and spacing of the bars (Factor C) were selected for significance analysis. The four levels of Factor *A*: A_1, A_2, A_3, and A_4 meaning the number of rebars 1 (B1G1), 2 (B1G2), 3 (B1G3) and 4 (B1G4), respectively. As above, the three levels of Factor *B*: B_1, B_2, B_3 meaning the diameter of rebar 6 mm (B2G1), 8 mm (B2G2) and 10 mm (B2G3). The three levels of Factor *C*: C_1, C_2, C_3 meaning the spacing of bars 60 mm (B3G1), 80 mm (B3G2) and 100 mm (B3G3). The formula and results are shown in Table 9.

Table 9. One-way analysis of variance table.

Source of Variance	Calculated Formula	Value Number of Rebars (Factor *A*)	Rebar Diameter (Factor *B*)	Spacing of Bars (Factor *C*)
Correction *C*	$\frac{T^2}{rl}$	830.17	494.77	550.53
Total sum of square S_T	$\sum\sum x_{ij}^2 - C$	20.81	9.07	4.21
Sum of effect squares S_A	$\frac{1}{l}\sum T_i^2 - C$	17.42	6.88	1.88
Error sum of squares S_e	$S_T - S_A$	3.39	2.19	2.34
Total degree of freedom df_T	$rl - 1$	11	8	8
Effect square and degree of freedom df_A	$r - 1$	3	2	2
Error square and degree of freedom df_e	$df_T - df_A$	8	6	6
Mean squared sum of effects $\overline{S}_A$	$\frac{S_A}{df_A}$	5.81	3.44	0.94
Mean square sum of errors $\overline{S}_e$	$\frac{S_e}{df_A}$	0.42	0.36	0.39
Statistic *F*	$\frac{\overline{S}_A}{\overline{S}_e}$	13.72 **	9.43 *	2.41
$F_{0.01}(df_A, df_e)$	Checklist	7.59	10.90	10.90
$F_{0.05}(df_A, df_e)$	Checklist	4.07	5.14	5.14

Note: "**" means that the effect of this level on the test results was highly significant; "*" means that this level had a significant effect on the results of the test.

In a further *F*-test, if $F < F_{0.05}$, it means that this level had a nonsignificant effect on the results of the test. If $F_{0.05} < F < F_{0.01}$, it means that this level had a significant effect on the results of the test (indicated by "*" in the following tables). If $F > F_{0.01}$, it means that the effect of this level on the test results was highly significant (indicated by "**" in the following tables).

The F-test can only test that the variation in the results stems mainly from the variation between levels. It cannot detect whether the difference between the level means is significant. To address this issue, multiple comparisons were further used to determine the significance of the difference between two horizontal means. As shown in Tables 10 and 11, Factor C was not subjected to multiple comparisons because its effect was not significant.

Table 10. Multiple comparison table of average shear strength under number of rebars.

Group Number	Number of Rebars	Level	Average $\overline{x}_i$	$\lvert\overline{x}_i - 9.47\rvert$	$\lvert\overline{x}_i - 9.02\rvert$	$\lvert\overline{x}_i - 8.46\rvert$
B1G1	1	A_1	6.32	3.14 *	2.69 *	2.14 **
B1G2	2	A_2	8.46	1.00	0.55	
B1G3	3	A_3	9.02	0.45		
B1G4	4	A_4	9.47			

Note: The minimum significant differences at the significant level of 0.05 and 0.01, respectively, were $LSD_{0.05} = t_{0.025}(df_e)S_{\overline{x}_i-\overline{x}_j} = 1.225$, $LSD_{0.005} = t_{0.005}(df_e)S_{\overline{x}_i-\overline{x}_j} = 1.7822$. "**" means that the effect of this level on the test results was highly significant; "*" means that this level had a significant effect on the results of the test.

Table 11. Multiple comparison table of average shear strength under rebar diameter.

Group Number	Rebar Diameter	Level	Average $\overline{x}_i$	$\lvert\overline{x}_i - 6.32\rvert$	$\lvert\overline{x}_i - 7.46\rvert$
B2G1	C8	B_2	8.46	2.14 **	1.00
B2G2	C10	B_3	7.46	1.14	
B2G3	C6	B_1	6.32		

Note: The minimum significant differences at the significant level of 0.05 and 0.01, respectively, were $LSD_{0.05} = t_{0.025}(df_e)S_{\overline{x}_i-\overline{x}_j} = 1.207$, $LSD_{0.005} = t_{0.005}(df_e)S_{\overline{x}_i-\overline{x}_j} = 1.828$. "**" means that the effect of this level on the test results was highly significant.

Table 9 demonstrates the formula and results of the calculations. The quantitative results of the completed analysis were used to accurately rank the strong and weak effects of the three factors and ten levels. These data tested the conclusions of the previous section.

4. Number of rebars

It is described in Table 9 that the F value of Factor A was 13.72, which was higher than $F_{0.01}$ and $F_{0.05}$. This indicated that the effect of the number of rebars on the improvement in the shear strength was highly significant. As can be seen from Table 10, among the four levels of Factor A, $|\overline{x}_1 - 9.47| = 6.44$ and $|\overline{x}_1 - 9.02| = 2.69$. So A_1 was significantly higher than A_2, A_3, and A_4, and more compared to the untreated (SJ). In addition, the enhancement effect of the number of rebars 4 (B1G4) was significantly higher than others.

5. Rebar diameter

The F value of Factor B was 9.43, which was less than $F_{0.01}$ and greater than $F_{0.05}$, indicating that the rebar diameter had a significant effect on the enhancement of the shear strength. Of the three levels of Factor B, $|\overline{x}_1 - 6.32| = 2.14$, $|\overline{x}_1 - 8.46| = 1.00$, so B_2 was significant over B_3 and the others were not.

6. Spacing of the bars

The F value of Factor C was 2.41, which was less than $F_{0.01}$ and $F_{0.05}$, suggesting that the spacing of the bars had no significant effect on the improvement in the shear strength. In general, when using the rebar implantation method to strengthen the OC, the spacing of the bars should satisfy the specifications and combine the actual conditions.

7. Summary

Combined with the above analysis, the interface shear performance was most influenced by the number of rebars. As two ways to increase the rebar planting rate, the former was a bit more significant. This was because a growth in the number of rebars not only quickly improved the planting rate, but also made the arrangement more stable. In a multiple rebar arrangement, the rebars were able to affect a larger area of concrete, resulting in a more uniform force. Smaller diameters but a greater number could be used while keeping the rebar planting rates constant.

The weakest influence was the spacing of the bars. The effects of the anchor groups on the rebar–concrete transfer mode should be considered during the structural design.

4. Shearing Mechanism and Comparison of Results

4.1. Analysis of the Shear Resistance Mechanism

Currently, the shear theories of shear resistance are based on the shear-friction theory proposed by Birkeland [38], and on this basis, Randle [39] established the formula for shear resistance. The currently accepted shear bearing capacity of the interface consists of three components: the force of cohesion $V_{u(coh)}$, the friction $V_{u(f)}$ and the pin force $V_{u(da)}$ [40].

$$v_u = \frac{V_{u(coh)} + V_{u(f)} + V_{u(da)}}{A_{cv}} \ (A_{cv} \text{ is the interface area})$$

The shear force transfer mechanism [21] is shown in Figure 13. When the interface undergoes relative slip u due to shear, the rebars are also driven and thus generate tension F_t and create an angle α with the rough interface. At this point, the component $F_t cos\alpha$ of the tensile force in the direction parallel to the interface is the pin force $V_{u(da)}$. Its component $F_t sin\alpha$ in the direction perpendicular to the interface is the pressurizing effect [39]. The concrete is produced under pressure because of the extrusion of the tensile rebar. The friction $V_{u(f)}$ generates on the rough bonding interface, which is $\mu F_t sin\alpha$, its magnitude depends on the roughness and pressure magnitude. Moreover, it decreases slightly with an increasing slip. In addition to the shear resistance, the interface itself has an adhesive cohesion $V_{u(coh)}$, consisting of Van der Waals forces, chemical action, interface tension and mechanical occlusion [41]. The above constituted the shear resistance of the interface.

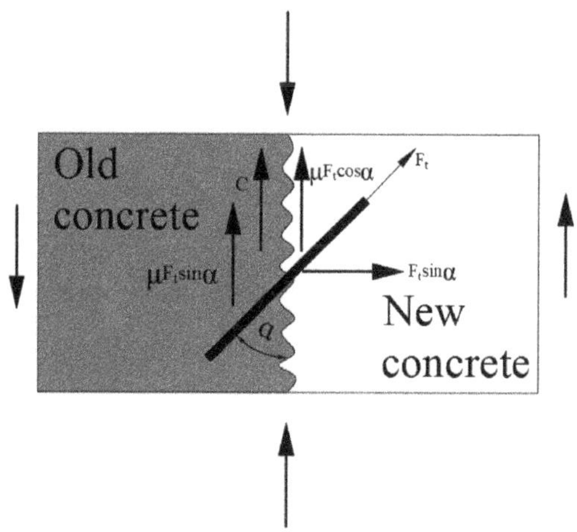

Figure 13. Shear force transmission diagram.

4.2. Analysis of the Application of Existing Shear Strength Calculation Formulae

Most of the existing well-established shear formulas are mostly based on the above theory, refined and extended in terms of quantification of roughness, correction of friction coefficient and different types of concrete [40]. Table 12 has shown some representative formulas for calculating the shear strength of the interface of new and old concrete at present.

Table 12. Typical formula for calculating the shear strength of the bonding interface of new and old concrete.

Specification	Calculation Formula	Annotation
Fib Model Code for Concrete Structures (2010) [42]	$\tau_u = \tau_a + \mu(\rho k_1 f_y + \sigma_n) + k_2\rho\sqrt{f_y f_c)}$	τ_a, c—Interface cohesion μ—Interface friction coefficient ρ—Bonding rate of interface σ_n—Normal stress caused by interface shear k_1—Bending coefficient k_2—Interaction coefficient between steel bar and concrete under dowel action f_y—Tensile strength of rebar f_c—Axial compression strength of concrete A_{sv}, A_{vf}—Total cross-sectional area of interfacial shear rebar A_{cv}—Interfacial area P_c—Positive interface pressure caused by external force
AASHTO LRFD Bridge Design Specifications (2010) [43]	$V_u = cA_{cv} + \mu(A_{sv}f_y + P_c)$	
ACI 318-08 (2008) [44]	$V_u = \mu A_{vf} f_y$	

Figure 14 expresses the ratio of theoretical value to test value. It can be seen that The Fib (2010) takes into account the coupling between friction and pin force. The shear strength value is considered to be equal to the algebraic sum of the force of cohesive, friction and pin force. It is worth mentioning that these three do not reach the maximum value at the same time during the shear resistance. The expression in the form of multiplication by a subfactor in front can better predict the shear strength in this test. For the FLWCC, the calculated value of Fib (2010) was conservative and the maximum error was controlled at about 25%, which was a better fit. The theoretical value of the AASHTO (2010) was lower and would be used to predict this test with some error. The reason was that although it used friction and pin force to resist external loads at the same time, the values of the bonding force and friction coefficient are taken roughly. The smallest theoretical value of the ACI (2008) was due to the fact that only the pinning action of the rebar is considered, ignoring the contribution of the concrete itself.

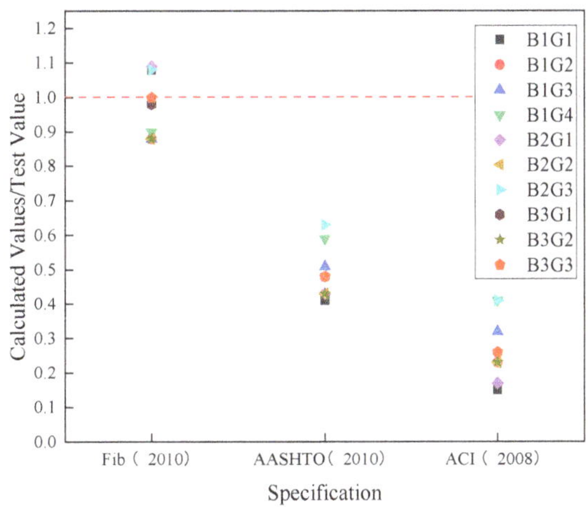

Figure 14. Calculated values/test values.

It can be seen that all three specifications take the total cross-sectional area of *Acv* as a major variable, which also corresponds to the results of the factor analysis in Section 3 of this paper. In other words, the shear strength is linearly related to the rebar planting rate and the rebar diameter. Definitely, there are Chinese specifications that codify the spacing of bars as an independent variable in the formula. Such as JTG/T J22-2008 [45]. But its calculated value differed greatly from the test value. The reason, as analyzed above, was that the effects of anchor groups affect mainly the way in which the forces are transferred between the rebar and the concrete. Thus the effect of this factor was difficult to quantify in different situations with different members, loading methods and different usage scenarios.

5. Conclusions

Based on the above results and discussions, the following conclusions can be drawn:

1. The form of shear damage at the unstrengthened FLWCC–OC interface was brittle damage. But the damage form becomes ductile after rebar implantation. This showed that rebar implantation could increase the ductility of the interface of FLWCC–OC.
2. The rebar planting rate had a significant effect on the interfacial shear strength. The strength reached a maximum of 9.47 MPa when the rate was 0.894%. Rebar was not effective at low rates. The interfacial shear strength was positively correlated with the number of rebars, which increased and then decreased with the increase in rebar diameter. The number of rebars enhanced the shear strength better than the rebar diameter for the same planting rate.
3. The shear strength was related to the spacing of the bars. In this test, the strength was maximum when the spacing was 80 mm. Too much or too little spacing affected the way the rebars transmitted force and thus the bonding action of the rebars.
4. The one-way ANOVA was used to analyze the effects of the number of rebars, rebar diameter and spacing of the bars on the interfacial shear strength. The results showed that the most obvious effect of the number of rebars was followed by rebar diameter, and the effect of the spacing of the bars was less significant.
5. The calculated strength of the Fib (2010) formula agreed well with the measured values in this text. It was recommended to refer to the Fib (2010) to calculate the interfacial shear strength of the rebar implantation when strengthening OC with FLWCC.

Author Contributions: Conceptualization, H.Z. and N.Z.; Methodology, Y.D. and H.Z.; Software, J.C.; Validation, N.Z. and Y.D.; Formal Analysis, Y.D.; Investigation, H.Z.; Resources, H.Z. and X.L.; Data Curation, Y.D. and N.Z.; Writing—Original Draft Preparation, Y.D.; Writing—Review and Editing, Y.D. and H.Z.; Visualization, N.Z. and J.C.; Supervision, H.Z.; Project Administration, H.Z.; Funding Acquisition, X.L. All authors have read and agreed to the published version of the manuscript.

Funding: This study was funded by the National Natural Science Foundation of China (grant number 52178182), Hubei Provincial Excellent Young and Middle aged Science and Technology Innovation Team Project of Colleges and Universities (grant number T2022002), "The 14th Five Year Plan" Hubei Provincial advantaged characteristic disciplines (groups) project of Wuhan University of Science and Technology (2023D0501, 2023D0503, 2023D0504).

Institutional Review Board Statement: Not applicable.

Informed Consent Statement: Not applicable.

Data Availability Statement: The data presented in this study are available on request from the corresponding author.

Conflicts of Interest: The authors declare no conflict of interest.

References

1. Jin, W.; Zhao, Y. *Durability of Concrete Structure.0s*; Science Press: Beijing, China, 2014; Volume 34, pp. 13–24. (In Chinese)
2. Ding, F.; Wu, X.; Xiang, P.; Yu, Z. New damage ratio strength criterion for concrete and lightweight aggregate concrete. *ACI Struct. J.* **2021**, *118*, 165–178.
3. Shariati, M.; Ramlisulong, N.H.; Sinaei, H.; Arabnejad Khanouki, M.M.; Shafigh, P. Behavior of channel shear connectors in normal and light weight aggregate concrete (experimental and analytical study). *Adv. Mat. Res.* **2011**, *168*, 2303–2307. [CrossRef]
4. Zheng, D.; Kou, J.; Wei, H.; Zhang, T.; Guo, H. Experimental study on flexural behavior of damaged concrete beams strengthened with high ductility concrete under repeated load. *Eng. Struct.* **2023**, *274*, 115203. [CrossRef]

5. Chen, M.; Li, Z.; Wu, J.; Wang, J. Shear behaviour and diagonal crack checking of shale ceramsite lightweight aggregate concrete beams with high-strength steel bars. *Constr. Build. Mater.* **2020**, *249*, 118730. [CrossRef]
6. Zhu, H.; Fu, Z.; Liu, P.; Li, Y.; Zhao, B. Shear behavior of stud-PBL composite shear connector for steel–ceramsite concrete composite structure. *Coatings* **2022**, *12*, 583. [CrossRef]
7. Zhu, H.; Fu, Z.; Wang, Y.; Zhang, N. Study on splitting tensile strength of interface between the full lightweight ceramsite concrete and ordinary concrete. *Case Stud. Constr. Mater.* **2023**, *18*, e01829. [CrossRef]
8. Guan, H.; Yu, J.; Umuhuza Kibugenza, A.; Sun, Q. Preparation of coal gangue ceramsite high-strength concrete and investigation of its physico-mechanical properties. *Sci. Rep.* **2022**, *12*, 16369. [CrossRef]
9. Wu, X.; Yang, J.; Lin, H.; Wang, Q.; Cao, Z. Mechanical Properties and Damage Mechanism of Shale Ceramsite Concrete after High-Temperature Treatment. *J. Mater. Civ. Eng.* **2022**, *34*, 04022123. [CrossRef]
10. Wang, S.; Yang, L.; Ma, X.; Mei, R.; Li, C.; Sun, Z. The formation of porous light ceramsite using Yellow River sediment and its application in concrete masonry production. *Case Stud. Constr. Mat.* **2022**, *17*, e1340. [CrossRef]
11. Guan, H.; Xia, Y.; Zhang, S.; Wu, J.; Wang, J.; Wang, B. A theoretical axial stress–strain model for gangue ceramsite concrete confined with FRP tube. *Constr. Build. Mater.* **2022**, *350*, 128864. [CrossRef]
12. Fang, Y.; Xu, Y.; Gu, R. Experiment and Analysis of Mechanical Properties of Lightweight Concrete Prefabricated Building Structure Beams. *Int. J. Concr. Struct. Mater.* **2022**, *16*, 5. [CrossRef]
13. Zhu, H.; Chen, H.; Fu, Z.; Guo, Z.; Liu, P.; Chen, Y. Experimental Study on Axial Compressive Load-Bearing Performance of RC Short Columns Strengthened by All-Light Ceramsite Concrete. *Appl. Sci.* **2022**, *12*, 11206. [CrossRef]
14. Wang, J.; Zhou, T.; Nie, S.; Zhang, X.; Shao, Y. Behavior of CFS shear walls infilled with lightweight ceramsite concrete under cyclic loading. *Structures* **2022**, *45*, 1833–1849. [CrossRef]
15. Xie, J.; Zhao, J.; Wang, J.; Huang, P.; Liu, J. Investigation of the high-temperature resistance of sludge ceramsite concrete with recycled fine aggregates and GGBS and its application in hollow blocks. *J. Build. Eng.* **2021**, *34*, 101954. [CrossRef]
16. Ma, S.; Gu, Y.; Bao, P. Seismic performance of ceramsite concrete T-shaped composite wallboard joints under cyclic loading. *Structures* **2021**, *33*, 4433–4445. [CrossRef]
17. Cheng, M.; Ma, J. Experimental Study on Shear Behavior of the Interface between New and Old Concrete with Reinforced KSCE. *J. Civ. Eng.* **2018**, *22*, 1882–1888.
18. Wang, X.; Dong, S.; Ashour, A.; Han, B. Bonding of nano inclusions reinforced concrete with old concrete: Strength, reinforcing mechanisms and prediction model. *Constr. Build. Mater.* **2021**, *283*, 122741. [CrossRef]
19. Hoła, J.; Sadowski, Ł.; Reiner, J.; Stach, S. Usefulness of 3D interface roughness parameters for nondestructive evaluation of pull-off adhesion of concrete layers. *Constr. Build. Mater.* **2015**, *84*, 111–120. [CrossRef]
20. Ouyang, J.; Guo, R.; Wang, X.; Fu, C.; Wan, F.; Pan, T. Effects of interface agent and cooling methods on the interfacial bonding performance of engineered cementitious composites (ECC) and existing concrete exposed to high temperature. *Constr. Build. Mater.* **2023**, *376*, 131054. [CrossRef]
21. Du, G.; Xu, B.; Bu, L. Experimental study and numerical simulation of the interfacial bonding performance of ordinary concrete reinforced with reactive powder concrete. *J. Build. Eng.* **2022**, *62*, 105352. [CrossRef]
22. Wang, Y.; Luo, W.; Zhou, X.; Shuai, F.; Ren, W.; Tan, J. Shear behaviour of a novel vertical grouting joint in precast concrete structure for wind turbines. *Eng. Fail. Anal.* **2023**, *143*, 106887. [CrossRef]
23. Reis, E.; de Azevedo, R.; Christoforo, A.; Poggiali, F.; Bezerra, A. Bonding of steel bars in concrete: A systematic review of the literature. *Structures* **2023**, *49*, 508–519. [CrossRef]
24. He, Y.; Zhang, X.; Hooton, R.; Zhang, X. Effects of interface roughness and interface adhesion on new-to-old concrete bonding. *Constr. Build. Mater.* **2017**, *151*, 582–590. [CrossRef]
25. Tang, H.; Yang, Y.; Li, H.; Xiao, L.; Ge, Y. Effects of chloride salt erosion and freeze–thaw cycle on interface shear behavior between ordinary concrete and self-compacting concrete. *Structures* **2023**, *56*, 104990. [CrossRef]
26. Tian, W.; Song, Z.; Xiao, C.; Zhu, R. Shear test and finite element simulation analysis of bonding interface of new and old concrete with planting bars. *Build. Struct.* **2019**, *49*, 127–133. (In Chinese)
27. Yang, J.; Chen, R.; Zhang, Z.; Zou, Y.; Zhou, J.; Wang, W. Effects of planting bar parameters on the shear resistance of UHPC-stone interface. *China Civ. Eng. J.* **2022**, *55*, 62–78. (In Chinese)
28. Jiang, Y.; Chen, H.; Nie, X.; Tao, M. Experimental study on bonding and anchorage of steel bars in precast concrete structures with new-to-old concrete interface. *Eng. Struct.* **2021**, *247*, 113086. [CrossRef]
29. Costa, H.; Carmo, R.N.F.; Júlio, E. Influence of normal stress and reinforcement ratio on the behavior of LWAC interfaces. *Constr. Build. Mater.* **2018**, *192*, 317–329. [CrossRef]
30. Goldyn, M.; Urban, T. Experimental Investigations on Interface between Ordinary and Lightweight Aggregate Concretes Cast at Different Times. *Materials* **2021**, *14*, 1664. [CrossRef]
31. Cattaneo, S.; Zorzato, G.; Bonati, A. Assessing method of shear strength between old to new concrete interface under cycling loading. *Constr. Build. Mater.* **2021**, *309*, 125160. [CrossRef]
32. *GB/T 228.1-2021*; Metallic Materials—Tensile Testing—Part 1: Method of Test at Room Temperature. State Administration for Market Reglation: Beijing, China, 2022; pp. 17–22. (In Chinese)
33. *GB 50367-2013*; Code for Design of Strengthening Concrete Structure. Ministry of Housing and Urban-Rural Development of the People's Republic of China: Beijing, China, 2014; pp. 7–11. (In Chinese)

34. *GBT50081-2019;* Standard for Test Methods of Concrete Physical and Mechanical Properties. Ministry of Housing and Urban-Rural Development of the People's Republic of China: Beijing, China, 2019; pp. 5–25. (In Chinese)
35. Wang, Q.; Chen, H.; Lu, C.; Wang, C.; Qian, K. Seismic performance of precast concrete column with different longitudinal reinforcement diameters under different lateral loading directions. *Eng. Struct.* **2022**, *266*, 114536. [CrossRef]
36. Jonak, J.; Siegmund, M.; Karpiński, R.; Wójcik, A. Three-Dimensional Finite Element Analysis of the Undercut Anchor Group Effect in Rock Cone Failure. *Materials* **2020**, *13*, 1332. [CrossRef] [PubMed]
37. Abbas, H.; Mohammed, A.A.; Hussein, E.; Tarek, A.; Aref, A.; Yousef, A. Effect of tie spacing and GFRP rebar diameter on compression behavior of concrete columns. *J. Build. Eng.* **2023**, *72*, 106740. [CrossRef]
38. Birkeland, P.W.; Birkeland, H.W. Connections in precast concrete construction. *ACI J. Proc.* **1966**, *63*, 345–368.
39. Randl, N. Investigations on Transfer of Forces between New and Old Concrete at Different Joint Roughness. Ph.D. Thesis, University of Innsbruck, Innsbruck, Austria, 1997. Available online: https://www.concrete.org/publications/internationalconcreteabstractsportal/m/details/id/7627 (accessed on 1 June 2023).
40. Santos, P.M.D.; Júlio, E.N.B.S. A state-of-the-art review on shear-friction. *Eng. Struct.* **2012**, *45*, 435–448. [CrossRef]
41. Huang, L.; Zhuo, W.; Gu, Y.; Shangguan, P.; Huang, X. Effect test of interface agent on bonding of new and old concrete. *J. Fuzhou Univ. (Nat. Sci. Ed.)* **2018**, *46*, 396–402. (In Chinese)
42. CEB-FIP. *Fib Model Code for Concrete Structures;* Ernst & Sohn: Berlin, Germany, 2010.
43. AASHTO. *AASHTO LRFD Bridge Design Specifications;* American Association of State Highway and Transportation Officials: Washington, DC, USA, 2010.
44. ACI Committee 318. *Building Code Requirement for Structural Concrete (ACI 318-08) and Commentary (ACI 318R-08);* American Concrete Institute: Detroit, MI, USA, 2008.
45. *JTG/T J22-2008;* Specifications for Strengthening Design of Highway Bridges. China Communications Press: Beijing, China, 2015. (In Chinese)

coatings

MDPI

Article

Nonlinear Numerical Analysis and Restoring force Model of Composite Joints with Steel Reinforced Recycled Concrete Columns and Steel Beams

Jing Dong *, Changling Shen, Hui Ma, Yunhe Liu and Xiaoran Cong

State Key Laboratory of Eco-Hydraulics in Northwest Arid Region, Xi'an University of Technology, Xi'an 710048, China

* Correspondence: skydongjing@xaut.edu.cn

Abstract: Cyclic loading tests were conducted for eight composite joints (five interior joints and three exterior joints) with steel reinforced recycled concrete (SRRC) columns and steel beams under cyclic loading. The axial compression ratio and recycled coarse aggregate (RCA) replacement percentage were considered as the main design parameters for the above composite joints. The seismic behavior of the composite frame joints with (SRRC) columns and steel beams were studied by numerical simulation and the constitutive models of recycled aggregate concrete (RAC), steel and reinforcement were given to the corresponding elements. The experimental results were used to validate the finite element (FE) model. The results indicated that the FE model can accurately predict the deformation and stress load-strain curves of the composite joints. Then, a FE parameter analysis was carried out on the seismic behavior of the composite joints. The RAC strength, steel strength and column steel web thickness were considered as the main parameters. Results pinpointed that the increase in RAC strength and steel web thickness of the columns significantly affected the horizontal bearing capacity of the composite joints. However, with the increase in RAC and steel strength, the deformation ability and ductility of composite joints were almost unaffected. The reasonable increase in column steel web thickness can improve the seismic ductility of composite joints, evidently. On this basis, considering the stress characteristics of joint specimens, a four broken line restoring force model for steel reinforced recycled concrete column steel beam composite frame joints was established. This model includes the skeleton curve model, the stiffness degradation pattern and hysteresis rules. The calculation curve was in good agreement with experimental curves, which can well reflect the hysteretic behavior of joints subjected to low cyclic repeated loading. The above results can be used as a reference for the seismic design of composite frame joints.

Keywords: steel reinforced recycled concrete columns; steel beams; joint; finite element analysis; restoring force model

Citation: Dong, J.; Shen, C.; Ma, H.; Liu, Y.; Cong, X. Nonlinear Numerical Analysis and Restoring force Model of Composite Joints with Steel Reinforced Recycled Concrete Columns and Steel Beams. *Coatings* **2022**, *12*, 1606. https://doi.org/10.3390/coatings12111606

Academic Editor: Andrea Nobili

Received: 23 September 2022
Accepted: 19 October 2022
Published: 22 October 2022

1. Introduction

With the acceleration of urbanization, a large amount of concrete demand leads to long-term over-exploitation of natural sand and stone materials, resulting in huge energy and natural resource consumption, and at the same time, the demolition of old buildings, new building construction and other human activities generated by the building waste accumulation. In addition, various natural disasters cause a large number of buildings to collapse and become damaged, and produce a large amount of construction waste, which seriously hinders the sustainable development strategy of the ecological environment, and bring serious problems to society, the economy and the ecological environment. Therefore, it is of great scientific significance and application value to develop a new type of structure system with good seismic performance and ecological environment friendliness. The emergence of recycled concrete can effectively solve the above problems [1]. It can reprocess waste concrete, recover its original mechanical properties to the greatest extent and form new building materials. This technology can not only reuse limited resources, but also solves some environmental problems. The promotion and application of recycled concrete technology can transform the traditional concrete production mode into the track of sustainable development.

The section steel recycled concrete column is a kind of combination form, a new type of composite member formed by configuring section steel, matching certain structural longitudinal bars and stirrups and, finally, by pouring recycled concrete. In steel structures, steel beams have good flexural performance. Compared with concrete beams, steel beams are lighter and can effectively reduce the section size of members, thus increasing the use space of the structure. In view of this, and in view of the characteristics of steel recycled concrete columns with good seismic performance and high bearing capacity, combined with the advantages of steel beams with good mechanical properties and simple construction, this paper puts forward a new structure system, the composite frame structure of steel recycled concrete columns and steel beams (SRRC). Numerous studies have been conducted on the performance of recycled aggregate concrete (RAC), including its preparation methods, physical and mechanical properties, durability, fatigue and bonding properties [2–4]. Ma et al. conducted experimental research and a theoretical analysis of RAC member sand structures [5–9]. Their results showed that the mechanical properties of RAC and its structures are inferior to those of ordinary concrete. This nature of RAC limits its engineering application, particularly in seismic structures, to a certain extent. Considering the good mechanical properties of steel and concrete composite structures, several scholars have attempted to apply RAC to composite structures to improve the mechanical properties of RAC structures and structural members. SRRC structures are among these steel and RAC composite structures [10–12]. In previous studies, SRRC columns that were tested under cyclic loading presented good seismic performance and high bearing capacity, and the results significantly promoted the application of RAC. As a new type of composite structure system, the restoring force model of the composite joints of SRRC was the basis for the analysis of the elastoplastic time history of the structure. The restoring force model should fully reflect the mechanical properties of the structure or component, such as strength decay, stiffness degradation, deformation ductility, energy dissipation capacity, damage evolution and so on, which was an important basis for the seismic performance analysis of the structure. The actual restoring force characteristic curve was very complex, so it was difficult to apply it directly to the analysis of the elastic–plastic seismic response of the structures. In order to facilitate the mathematical description and engineering application, the restoring force characteristic curve obtained by the test was abstracted and simplified in the form of mathematical modeling, and then the restoring force model of the structure or component was obtained. The research on the restoring force characteristics of SRRC was not mature. Only Xue et al. discussed the restoring force model of such joints under plane stress [13,14]. There is still a lack of research results on the restoring force model of SRRC.

Accordingly, the seismic performance of the composite joints has been further studied by the finite element method. The validity and accuracy of the refined numerical model were verified through experimental data. In addition, the influence of design parameters on the mechanical performance of composite joints was also investigated. On this basis, considering the failure process of the joints, a four-fold skeleton curve model of the composite joints was established by the regression fitting method. Moreover, through analyzing the stiffness degradation law and hysteretic rule in the whole loading and unloading process, a hysteretic curve restoring force model of the composite joint was proposed, which can provide reference for the engineering application of the joint.

2. Axial Compression Tests of the Composite Joints

2.1. Research Significance

The preparation and use of a large number of concrete materials lead to the long-term over-exploitation of natural sand and stone materials, resulting in huge energy and resource consumption. At the same time, the construction waste generated by the demolition of old buildings, new buildings and other human activities has piled up. In addition, earthquakes, tsunamis and floods and other natural disasters caused a large number of building collapses, a large amount of construction waste, etc., which to the sustainable development of our country has brought more and more serious social, economic, ecological and environmental problems. At present, construction waste is generally treated by stacking and burying, and its comprehensive utilization rate is too low. From the above, we can see that, on the one hand, it is the over-exploitation of natural non-renewable resources such as sand and stone. On the other hand, a large amount of construction waste in the natural and ecological environment is caused by serious pollution. Therefore, how to deal with and utilize these large amounts of construction wastes reasonably and effectively, so as to realize the resource-saving and environment-friendly society and the sustainable development of the economy, has become one of the problems that governments of all countries must face and solve. In view of the good seismic performance and high bearing capacity of the steel recycled concrete column, and the advantages of the steel beam, such as good mechanical performance, being light weight, of easy construction and being effective in the reduction of the section size of the members, a new structure, the composite structure of the steel recycled concrete column and steel beam, was proposed. Compared to traditional reinforced concrete structures and steel structures, the combination structure not only has high carrying capacity, good seismic performance, good durability and fire resistance, saves steel and other advantages, but also has the remarkable characteristics of renewable concrete environmental protection, and meets the requirements of our sustainable development; therefore, it has broad application prospects for development.

2.2. Design and Production of Specimens

The low cyclic loading test and shear strength of composite joints with SRRC columns and steel beams were studied in detail by the authors [15]. Eight joints were designed, including five interior joints and three exterior joints. The effects of the recycled coarse aggregate (RCA) replacement ratio and axial compression ratio on the seismic performance of the composite joints were mainly considered [16]. The design parameters of the composite joints are shown in Table 1. The details of the size and reinforcement of composite joints are presented in Figure 1. The composite joints production and completion process is shown in Figure 2. The mix of RAC materials is demonstrated in Table 2. The design parameters of each joint are shown in Table 3.

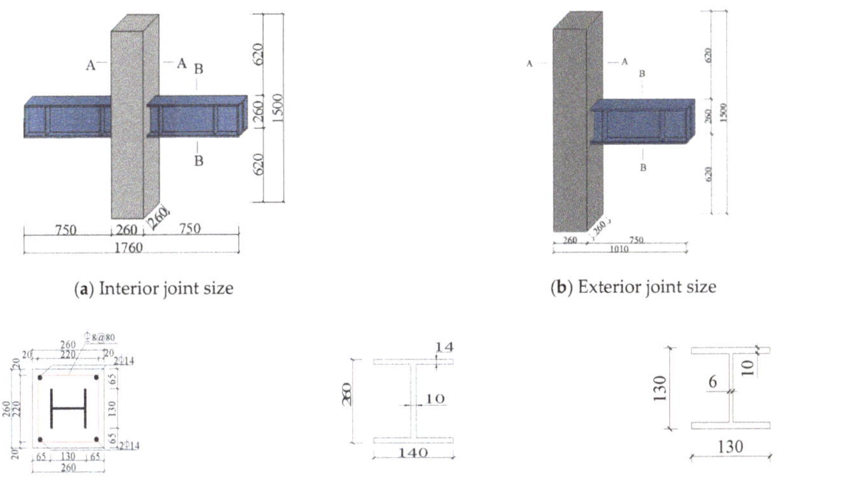

(**a**) Interior joint size (**b**) Exterior joint size

(**c**) A-A (**d**) B-B (**e**) Section of profile steel in column

Figure 1. Size and distributed steel of specimens (mm).

Table 1. Design parameters of specimens.

Design Parameter Level	Level 1	Level 2	Level 3
RCA replacement percentage r/%	0%	50%	100%
Axial compression ratio n	0.18	0.36	0.54

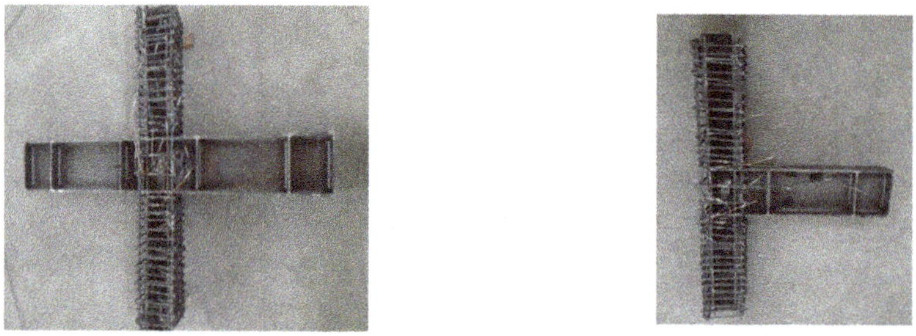

(**a**) Reinforced steel frame

(**b**) Support template of specimen

(**c**) Completed specimens

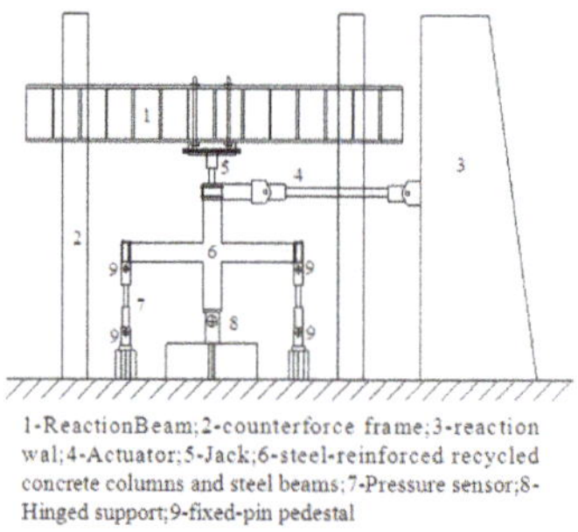

(**d**) Interior joint

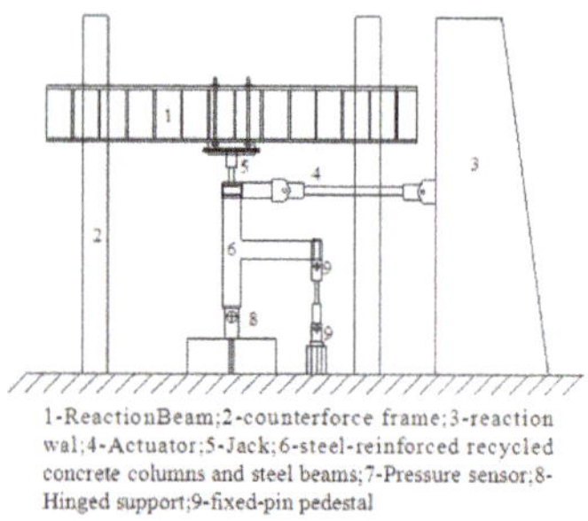

(**e**) Exterior joint

Figure 2. Schematic diagram of main manufacturing process and test device of test piece.

Table 2. Mix ratio of RAC materials.

RAC Strength Grade	RCA Replacement Percentage r/(%)	Mass of Each Component in Concrete per Cubic Meter (kg)					
		Water–Cement Ratio	Cement	Sand	NCA	RCA	Water
C40	0	0.43	464	585	1187	0	195
	50%	0.43	464	585	593.5	593.5	195
	100%	0.43	464	585	0	1187	195

Table 3. Design parameters of joint specimens.

Specimen Number	RAC Strength Grade	RCA Replacement Percentage r (%)	Axial Compression Ratio n	Profile Steel Ratio ρ_a (%)	Joint Forms
CFJ1	C40	0	0.36	4.8	Interior joint
CFJ2	C40	50	0.36	4.8	Interior joint
CFJ3	C40	100	0.36	4.8	Interior joint
CFJ4	C40	100	0.18	4.8	Interior joint
CFJ5	C40	100	0.54	4.8	Interior joint
CFJ6	C40	100	0.18	4.8	Exterior joint
CFJ7	C40	100	0.36	4.8	Exterior joint
CFJ8	C40	100	0.54	4.8	Exterior joint

Note: $r = m_{RAC}/m$, m_{RAC} is the weight of RAC; m is the total weight of the aggregates; $\rho_a = A_a/A$, A_a is the area of the profile steel in the column; A is the total area of the cross section of the column.

2.3. Failure Modes of Specimens

From the experimental results, the typical shear failure of each composite joint occurs under low cyclic load. There were obvious shear deformation and "X" shaped cross-diagonal cracks that appeared in the joint core region. From the influence of the two designed parameters on the cracking load of joints, the RCA replacement rate has little effect, while the axial compression ratio has a great influence on the cracking of joint core region. The cracking load of joints with different RCA replacement rates was approximately 70 kN, and there was no significant difference in the crack development process and distribution characteristics. The cracking load of joints with a large axial compression ratio was significantly greater than that of joints with a small axial compression ratio. The failure characteristics of CFJ1–CFJ8 are recorded in Figure 3.

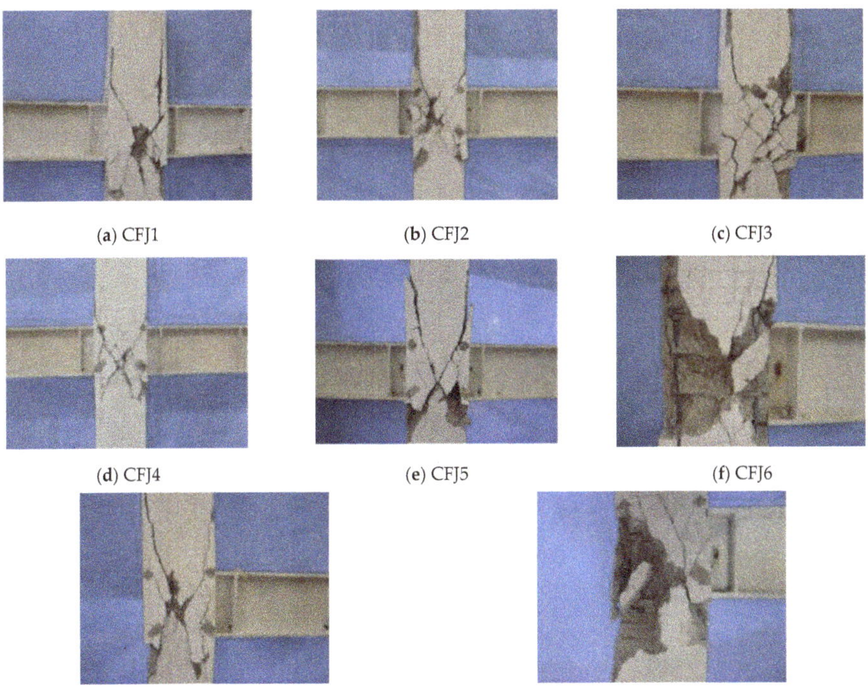

(**a**) CFJ1 (**b**) CFJ2 (**c**) CFJ3
(**d**) CFJ4 (**e**) CFJ5 (**f**) CFJ6
(**g**) CFJ7 (**h**) CFJ8

Figure 3. Failure modes of joints.

2.4. Comparison with Previous Studies

In order to verify the accuracy of the test and the correctness of the results, the following conclusions are drawn by comparing the experimental results and analysis in the previous literature:

(1) The failure modes of the joints in the reinforced steel recycled concrete frame are similar to those of the ordinary reinforced steel concrete frame, both of which are shear failure in the core area of the joints, and the shear deformation in the core area is significant [17].
(2) The failure modes of the specimens under repeated cyclic loads were the same, and obvious "X"-shaped cross-oblique cracks were formed along the diagonal. The failure modes are shear baroclinic failure, flexural shear failure and bending failure. However, due to the different structures of the specimens, the results can only serve as a reference [18].
(3) The hysteretic curves of the joints of the recycled steel concrete frame are roughly spindle-shaped, which are similar to the hysteretic curves obtained in the references, and the mechanical properties are in line with the mechanical properties of the steel frame and reinforced concrete frame [19].

3. Model Descriptions of the Composite Joints

3.1. Constitutive Models of the Materials in the Joints

(1) Constitutive relationship of steel

ABAQUS software 6.16 (Dassault Systèmes, Vélizy-Villacoublay, France) enables a variety of constitutive models for steel materials. At present, the widely used constitutive models include elastic–plastic, whole curve, bilinear and three-fold line models. The steel used in the test was Q235 low carbon steel, which belongs to the isotropic elastic–plastic material with an obvious yield point. The stress-strain curve generally adopts a simplified secondary plastic flow model. This model encapsulated that the longitudinal stress of the steel directly enters the strengthening stage after reaching the yield stress, while the strength of the plastic flow stage was unchanged. The secondary plastic flow model can not only consider the strengthening effect of steel, but also reflects the real material characteristics. The stress-strain curve of steel is shown in Figure 4. The mathematical expression of the stress-strain is determined by Equation (1):

$$\sigma_i = \begin{cases} E_s\varepsilon_i & \varepsilon_i \le \varepsilon_y \\ f_s & \varepsilon_y < \varepsilon_i \le \varepsilon_{st} \\ f_s + \zeta E_s(\varepsilon_i - \varepsilon_{st}) & \varepsilon_{st} < \varepsilon_i \le \varepsilon_u \\ f_u & \varepsilon_i > \varepsilon_u \end{cases} \quad (1)$$

where σ_i is equivalent stress of steel, f_s is the yield strength of steel, f_u is the ultimate strength of steel, E_s is elastic modulus of steel, E_t is the tangent modulus of steel, ζ is the strengthening coefficient ζ = 1/216, ε_i is equivalent strain of steel, ε_y is the strain of steel at yield, ε_u is the strain at the limit of steel, ε_{st} = 12ε_y, ε_u = 120ε_y.

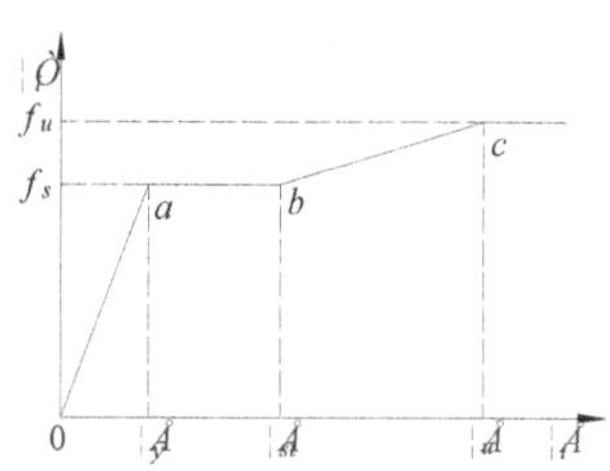

Figure 4. Stress-strain curve of steel.

The Poisson's ratio of steel is assumed to be:

$$v_s = \begin{cases} 0.285 & \varepsilon_i \leq 0.8\varepsilon_y \\ 1.075(\sigma_i/f_s - 0.8) + 0.285 & 0.8\varepsilon_y < \varepsilon_i \leq \varepsilon_y \\ 0.5 & \varepsilon_i > \varepsilon_y \end{cases} \tag{2}$$

The above steels all adopt Miss yield criterion, which is defined as a relatively general yield criterion. In a 3D space, the Miss yield surface is a cylindrical surface with $\sigma_1 = \sigma_2 = \sigma_3$ as the axis. On a 2D plane, the yield surface is an ellipse. Any stress state inside the yield surface is elastic, and any stress state outside the yield surface will cause yield. Miss yield criterion is suitable for materials with good toughness.

The spatial representation of Miss yield criterion is shown in Figure 5. The mathematical expression of equivalent stress is:

$$\sigma_s = \sqrt{\frac{1}{2}\left[(\sigma_1 - \sigma_2)^2 + (\sigma_2 - \sigma_3)^2 + (\sigma_1 - \sigma_3)^2\right]} \tag{3}$$

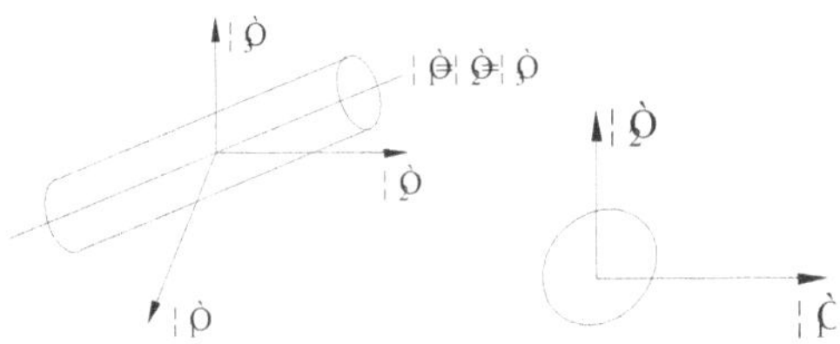

Figure 5. Schematic diagram of Miss yield criterion.

In the formula: $\sigma_1, \sigma_2, \sigma_3$ are three principal stresses.

(2) Hysteretic curve model of reinforcement

Existing studies have stated that the direct manifestation of the bond-slip between concrete and reinforcement in structural hysteretic curve analysis is the emergence of the "kneading" phenomenon. As a result, how to accurately describe the kneading phenomenon is one of the difficulties in numerical analysis. After comparison with the VECCHIO model of reference [20], the M-P model of reference and hysteretic curve model of the reference based on CLOUGH frame [21,22], Professor Zihu Fang of Shenzhen University established a hysteretic model of reinforcement that considered the bond-slip effect [23,24]. In this paper, the steel model was selected to simulate the hysteretic behavior of composite joints. The model is described in Figure 6.

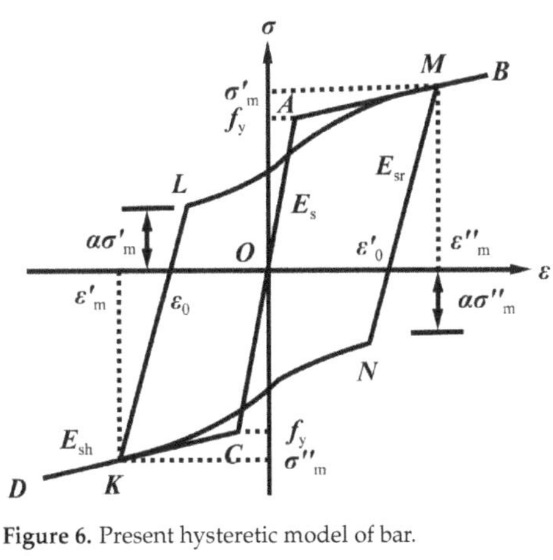

Figure 6. Present hysteretic model of bar.

(3) RAC material

The RAC was mainly distributed in the column, while the SRRC column was mainly composed of RAC, reinforcement and section steel. Therefore, the RAC part will be subject to different forms of constraints, which can be roughly divided into the section steel constraint part and reinforcement constraint part. The strength of the two constraints on RAC was different, so the reinforcement constraint area was called the weak constraint area, and the section steel constraint area was called the strong constraint area [25]. Figure 7 shows the distribution details of the SRRC column section constraint areas. For the convenience of the model, referring to the simplified method of Ellobody [26], the arc region in Figure 7 is simpiflied as a straight line. Based on this linear division method, the section constraint area of the SRRC column was divided as shown in Figure 8. The stress-strain curves of different concrete areas are presented in Figure 9.

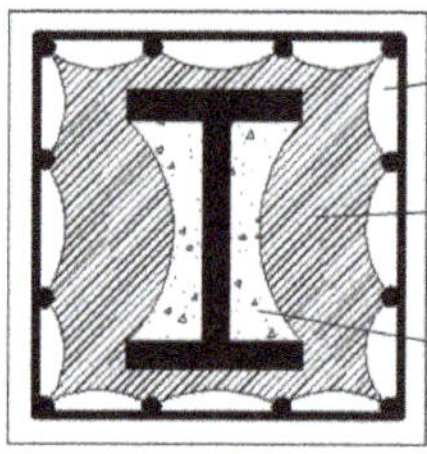

Figure 7. Section divisions of composite column.

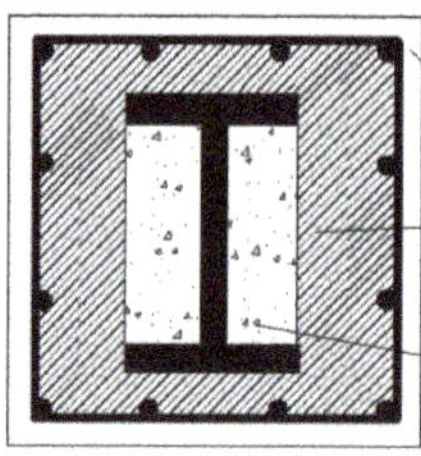

Figure 8. Regions simplified for composite sections.

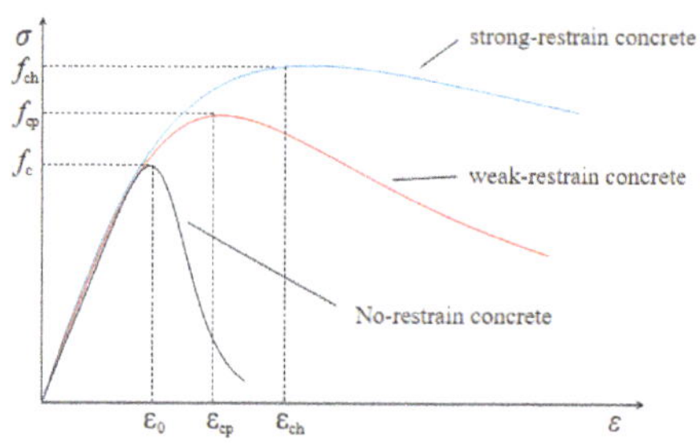

Figure 9. Stress-strain curves for concrete.

f_c is the axial compressive strength of RAC and ε_0 is the peak strain of RAC; f_{cp} is the peak stress of concrete in the weak constraint area and ε_{cp} is the peak strain of concrete in the weak constraint area; f_{ch} is the peak stress of strong constraint concrete and ε_{ch} is the peak strain of strong constraint concrete.

3.2. *Definition of Concrete Damage*

The plastic damage model is provided in the simulation of the concrete reciprocating load. Under the reciprocated load, the damage index of the concrete model gained a significant impact on the simulation results [27]. In order to improve the accuracy of ABAQUS in a nonlinear calculation, domestic and foreign scholars have included the damage of concrete in the FE simulation [28–30]. At present, there are many damage models for concrete. The damage factor was calculated by the Mander method. The formula was stated as follows:

$$d_c = 1 - \frac{E_{sec2}}{E_c} \tag{4}$$

Determined peak strain ε_{cc} and reverse unloading point (ε_{un}, σ_{un}) of concrete are the key to solving the damage factor by the Mander method.

The yield function and plastic potential function of materials are the critical factors affecting the performance of RAC. This paper uses meridian plane (p, q) and partial plane (q, θ) to represent it. The yield condition criterion is scheduled as follows:

$$\sigma_1 \leq 0 : f = q - 3\alpha p + \gamma\sigma_1 - (1-\alpha)\sigma_{ci} = 0 \tag{5}$$

$$\sigma_1 > 0 : f = q - 3\alpha p + \beta\sigma_1 - (1-\alpha)\sigma_{ci} = 0 \tag{6}$$

Figures 10 and 11 show the plane stress yield surface and the stress yield surface on the deviatoric plane of the concrete plastic damage model.

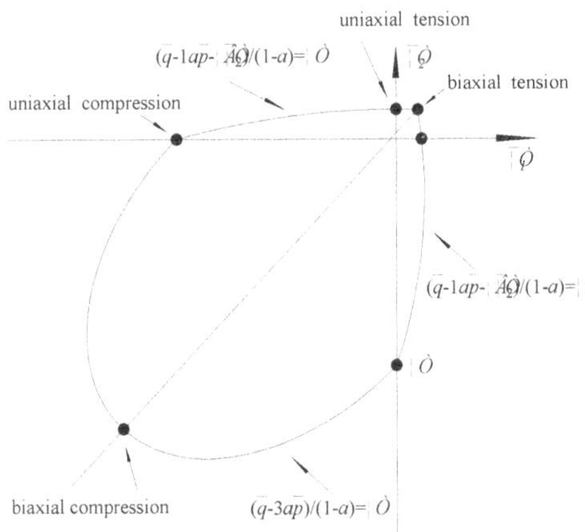

Figure 10. Yield surface in plane stress state.

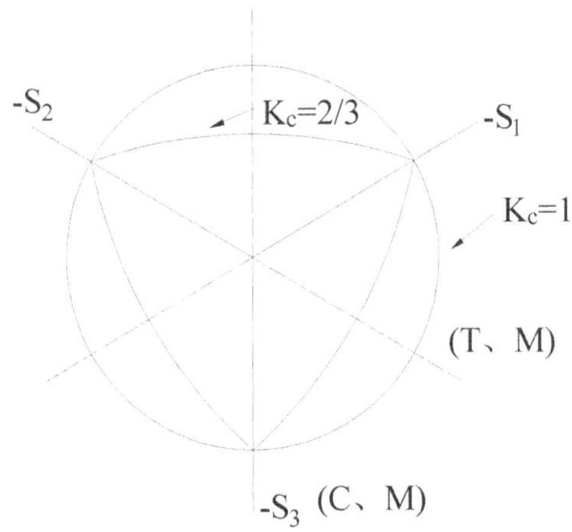

Figure 11. Yield surface in deviatoric stress state.

3.3. Interaction between the Materials

The steel reinforced part was embedded into the concrete through the embedded function in ABAQUS. There were no anti-sliding structural measures such as studs between the steel and RAC. Consequently, it was necessary to define the interaction between steel reinforced and RAC. Kinematic contact was selected for the normal direction contact of the two contact surfaces, which was also called "hard" contact. In other words, in addition to mutual contact, there was no interaction force between the surfaces as separation after contact was allowed. The Coulomb friction model was adopted for the bond-slip in the tangential direction, while the friction coefficient was set to 0.25 according to the research results [31,32].

3.4. Description of Mesh and Elements

The type of C3D8R element was adopted in the steel beam, built-in column steel and RAC, which was an eight-node hexahedron reduction integral solid element. Element type T3D2 was applied for the longitudinal reinforcement and stirrup in the composite column and the double-node three-dimensional truss element. The mesh size was 50 mm × 50 mm. The mixed model of these two element types not only improves the calculation accuracy, but also ensures the calculation speed to some extent. Figure 12 describes the FE model of the composite joints element mesh.

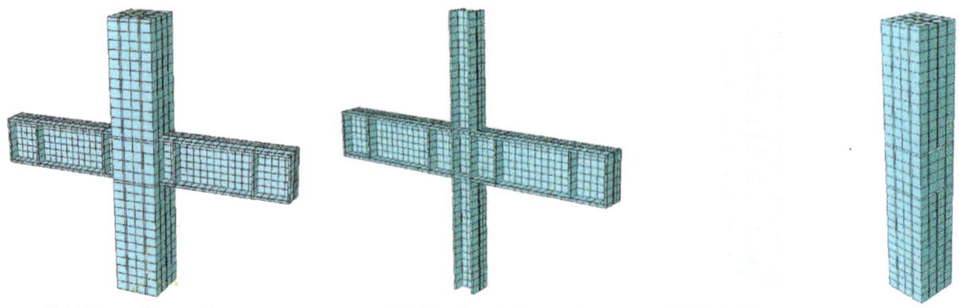

(**a**) Overall specimen (**b**) Steel reinforced (**c**) Reinforcement (**d**) RAC

Figure 12. Mesh division of composite joints.

3.5. Axial Loading Method and Boundary Conditions

The overall FE model was modeled according to the size of the test. The boundary conditions of the model were consistent with the load conditions of the actual test. The horizontal reciprocated load and axial force were applied on the top of the column. The beam end was connected by connected rods, and the bottom of the column was hinged. The hybrid loaded method of force and displacement was adopted in the test. However, considering that it was easier to ensure calculation convergence by using displacement control in FE simulation, displacement control was used to replace the mixed control in FE simulation. Finally, the Newton iteration method was conducted to calculate the FE model.

4. Verification of FE Model on the Composite Joints

4.1. Comparative Analysis of Experimental Results and Simulation Results

Through the modeling calculation of CFJ1–CFJ8 composite joints, the load-displacement hysteretic curves and skeleton curve of each composite joint were obtained. The comparison between the test results and the FE simulation results is shown in Figures 13 and 14. For the skeleton curve, it is the track of the maximum peak value of horizontal force reached by each cyclic loading that reflects the different stages of force and deformation of components and their characteristics (strength, stiffness, ductility, energy dissipation and collapse resistance, etc.), and it is also an important basis for determining the characteristic points in the restoring force model.

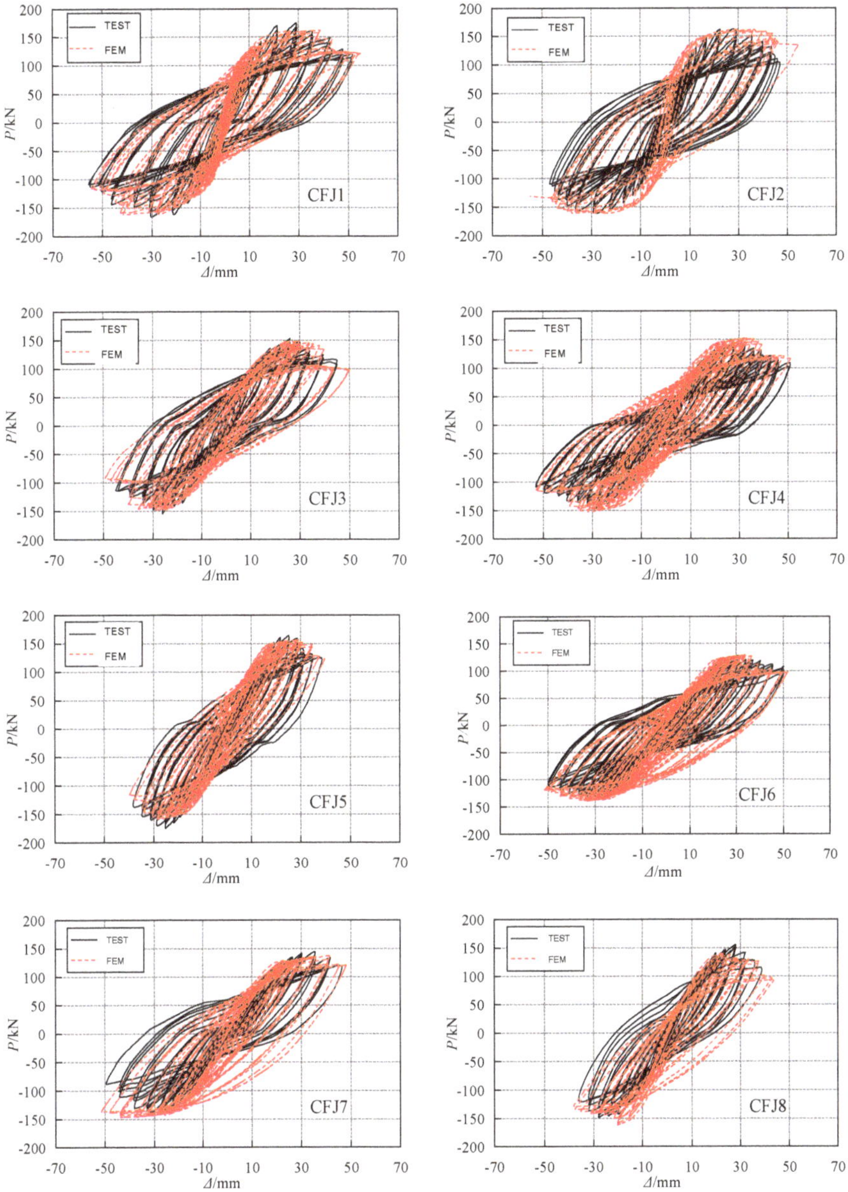

Figure 13. Comparison of test and simulation results of hysteretic curves of composite joints.

The hysteretic curves in Figure 13 and the skeleton curves in Figure 14 pinpoint that the simulation results of the hysteretic characteristics of the composite joints are basically in good agreement with the experimental results. However, compared with the experimental curves, the pinch effect of the simulated hysteretic curve of the exterior joints (CFJ6–CFJ8) is far from expectation. The unloading stiffness of the simulated curve is relatively larger. The main reason may be that the exterior joints are asymmetric structures. In addition, the cumulative damage of positive loading is greater than that of negative loading in the simulation. When the cumulative damage of positive loading reaches a certain degree, the damage degree of continuous loading to the positive is greater than that of negative loading. As a result, the unloading stiffness of the simulated curve is larger. This phenomenon is also reflected in the skeleton curves of the exterior joints. The stiffness of the simulated curves under negative loading is slightly larger than that of the experimental curves. In general, the numerical simulation is in good agreement with the experimental results, which verifies the availability and correctness of the numerical simulation method, and can be used as a reliable numerical method for further analysis in the following paragraphs.

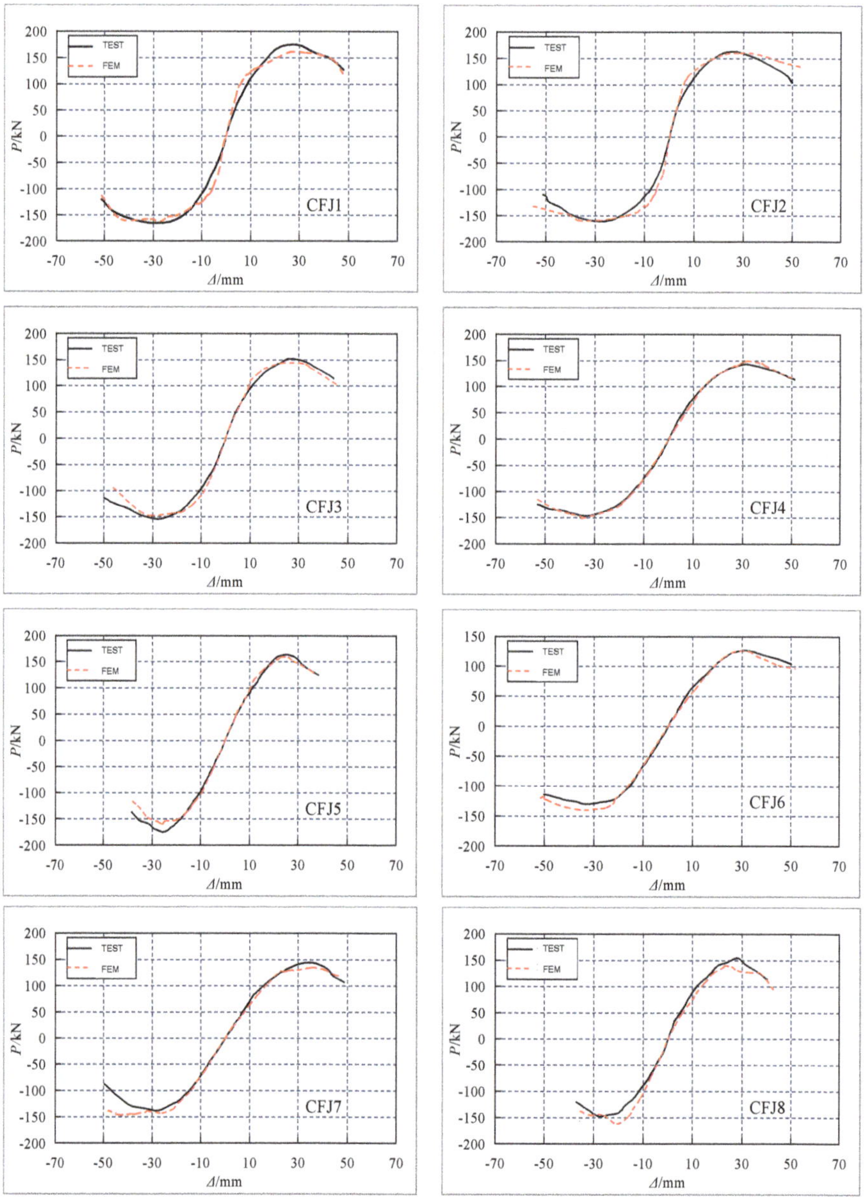

Figure 14. Comparison of test and simulation results of skeleton curves of composite joints.

4.2. Comparison of the Characteristic Values of the Composite Joints

Table 4 presents a comparative analysis on the load and corresponding displacement values of the composite joints obtained by numerical simulation and test at each characteristic point. It indicates that the load error of each characteristic point is basically within 6%, indicating that the bearing capacity consistency is fine, but some displacement differences are relatively large. The reasons may be as follows. First, there was a gap between each test equipment in the real test, resulting in a certain assembly error. Second, the bond-slip between steel bar, section steel and RAC in FE simulation was analyzed under the ideal condition, but in fact, the bond-slip of each part was quite complicated during the test process. Therefore, there is a certain difference in the bond-slip between the FE simulation and the real test. On the whole, there are some differences between the numerical analysis and the test results, but they are also within the acceptable range. It shows that the FE simulation results possess good accuracy.

Table 4. Comparison of test and simulation results of bearing capacities and displacements.

Specimen Number		Yield Load P_y/kN	Yield Displacement Δ_y/mm	Peak Load P_m/kN	Peak Displacement Δ_m/mm	Ultimate Load P_u/kN	Ultimate Displacement Δ_u/mm
CFJ1	TEST	131.15	13.37	170.34	27.04	144.77	44.66
	FEM	128.21	11.54	160.86	26.53	141.98	44.89
	δ/%	−2.24%	−13.69%	−5.57%	−1.89%	−1.93%	0.52%
CFJ2	TEST	121.65	12.79	162.14	28.67	137.42	41.32
	FEM	137.51	12.98	159.74	28.94	144.22	43.64
	δ/%	13.04%	1.49%	−1.48%	0.94%	4.95%	5.61%
CFJ3	TEST	113.67	12.94	153.19	27.28	130.21	40.51
	FEM	103.27	12.89	145.19	31.87	136.06	39.68
	δ/%	−9.15%	−0.39%	−5.22%	16.83%	4.49%	−2.05%
CFJ4	TEST	91.83	12.98	144.68	32.42	123.14	49.88
	FEM	94.82	12.15	150.64	35.08	116.86	50.51
	δ/%	3.26%	−6.39%	4.12%	8.20%	−5.10%	1.26%
CFJ5	TEST	108.76	12.25	168.93	25.05	143.59	34.94
	FEM	115.38	13.74	158.64	28.63	140.53	37.52
	δ/%	6.09%	12.16%	−6.09%	14.29%	−2.13%	7.38%
CFJ6	TEST	88.76	14.08	128.68	31.83	109.38	50.16
	FEM	90.82	14.32	135.56	33.61	120.85	49.82
	δ/%	2.32%	1.70%	5.35%	5.59%	10.49%	−0.68%
CFJ7	TEST	95.56	14.21	141.03	31.78	119.88	43.33
	FEM	90.66	13.15	140.16	35.71	115.26	44.17
	δ/%	−5.13%	−7.46%	−0.62%	12.37%	−3.85%	1.94%
CFJ8	TEST	108.06	13.39	150.85	27.45	128.22	35.01
	FEM	110.86	11.53	150.36	22.02	127.68	37.59
	δ/%	2.59%	−13.89%	−0.32%	−19.78%	−0.42%	7.37%

Note: Difference δ = (simulation results—test results) × 100%/test results.

4.3. Deformation and Stress Characteristics

The deformation characteristics and failure modes of all composite joints obtained by FE analysis are basically consistent with the test results. Therefore, in order to avoid repeated statements, the results of interior joint CFJ5 and exterior joint CFJ7 are representative and typical, which are exhibited in Figures 15 and 16.

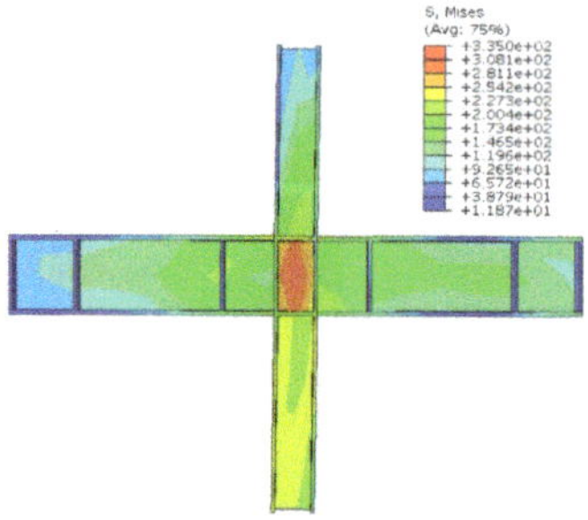

(**a**) Test damage (**b**) Steel reinforced (Unit: MPa)

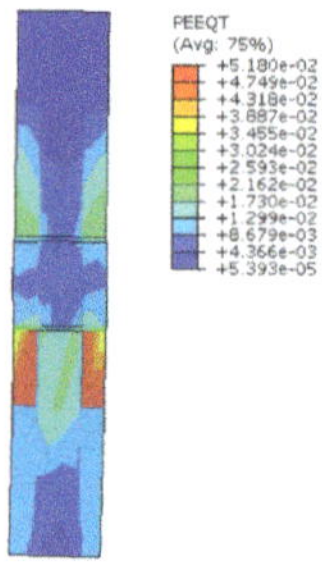

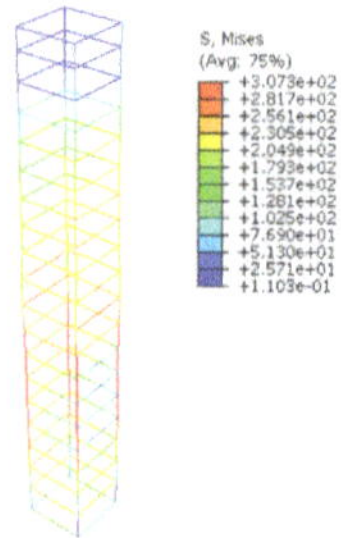

(**c**) RAC column (**d**) Reinforcement (Unit: MPa)

Figure 15. Comparison of test and simulation results of CFJ5.

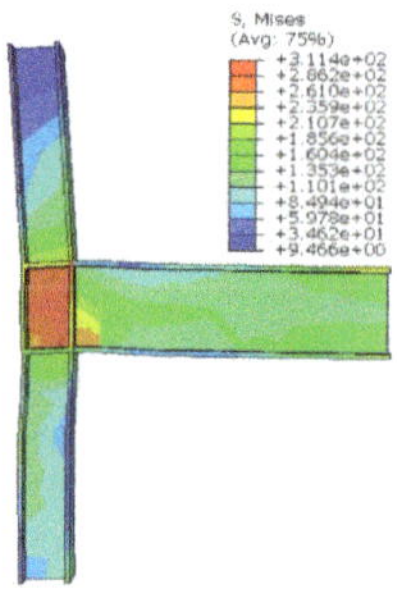

(**a**) Test damage (**b**) Steel reinforced (Unit: MPa)

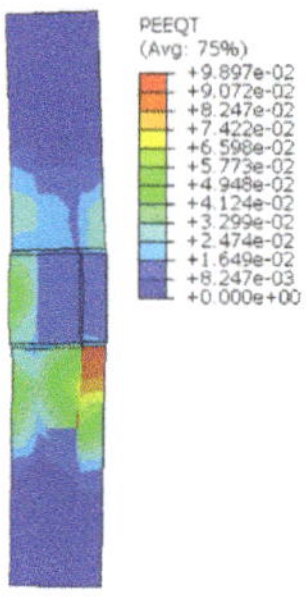

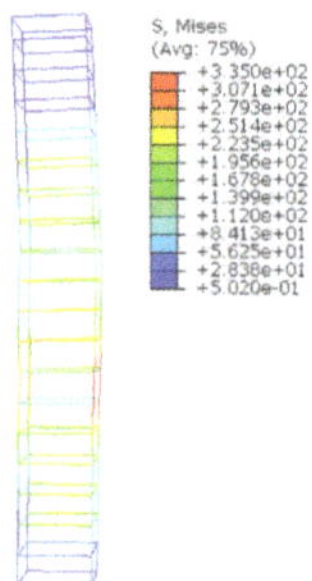

(**c**) RAC column (**d**) Reinforcement (Unit: MPa)

Figure 16. Comparison of test and simulation results of CFJ7.

In ABAQUS, the damage of concrete is mainly checked by two indicators: compression equivalent plastic strain (PEEQ) and tension equivalent plastic strain (PEEQT). Therefore,

PEEQT was used to reflect the failure of RAC. Figures 15 and 16 inform that when the joint was under ultimate load, the RAC in the core area of the joint formed a banded compression zone, which roughly formed oblique cracks along the diagonal direction. The compression mechanism of the oblique compression rod of the composite joint was consistent with the "X" shape diagonal cracks produced in the core area of the joint in the test. Figures 15 and 16 demonstrate the Mises stress diagrams of the steel reinforced and reinforcement cage of the composite joint model under ultimate load. It can be concluded that the yield of the steel web in the core area of the joint was gradually expanded from the middle local area to the outside. When the steel web was yielded in the core area of the joint, there was no plastic hinge at the end of the steel beam and column. It showed that the damage area mainly occurred in the core area of the joint, which was consistent with the experimental failure process. In summary, the FE model established can accurately reflect the stress of the composite joint, which can be applied to subsequent parameter analyses.

5. Parametric Analysis of FE on the Composite Joints

Due to the limitations of test conditions, financial and human resources, it is not practical and economical to carry out experimental research on each parameter affecting seismic performance. Moreover, the influence of other parameters on the seismic performance of the composite joints needs to be investigated further. Therefore, the influence of other parameters on the seismic performance of the composite joints was analyzed by ABAQUS 6.16 software in this paper.

From the load displacement skeleton curve in the above model verification, it can be concluded that the skeleton curve of the joint can reflect the full stress process of the composite joint in detail. Therefore, considering the calculation efficiency in FE parameter analysis, the pushover analysis was carried out with the interior joint as the basic specimen. The RAC strength, yield strength of section steel and web thickness of column steel web are the main analytical parameters.

5.1. RAC Strength

When including the influence of RAC strength factors, only changing this single parameter in numerical analysis leads to other parameters remaining unchanged. The RAC strength f_{cu} was 35 MPa, 45 MPa, 55 MPa and 65 MPa. The numerical simulation results are shown in Figure 17. Table 5 lists the influence of RAC strength on each characteristic point of the composite joints.

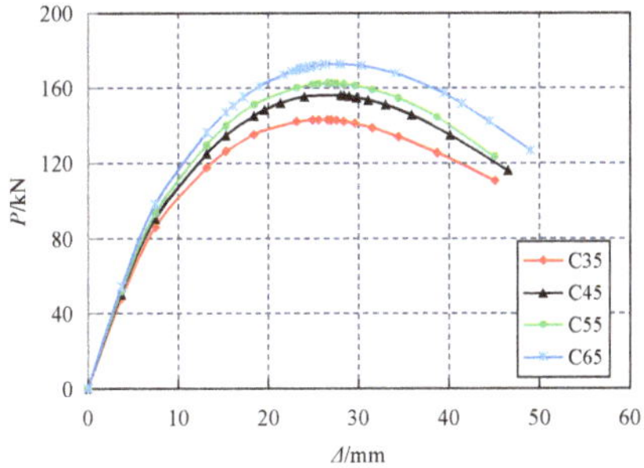

Figure 17. Influence of RAC strength on composite joints.

It can be estimated from Figure 17 and Table 5 that the strength of RAC has little effect on the initial stiffness of the composite joint. The horizontal bearing capacity of the composite joints increased with the increase the strength of RAC, while the maximum increase was 20.87% compared with the peak bearing capacity of the composite joints with strength grade C35, which indicated that the increase of the strength of RAC can significantly improve the strength of the composite joint. The displacement corresponding to the peak load was similar, indicating that the strength of RAC has little effect on the deformation capacity of the composite joints. At the end of loading, the horizontal load also gradually increased with the increase of RAC strength, and the skeleton curve had a significant decline. The ultimate load of each specimen decreased by 22.75%, 25.60%, 24.01% and 26.64% relative to the peak load.

Table 5. Influence of RAC strength on characteristic points of composite joints.

RAC Strength	Yield Load P_y/kN	Yield Displacement Δ_y/mm	Peak Load P_m/kN	Peak Displacement Δ_m/mm	Ultimate Load P_u/kN	Ultimate Displacement Δ_u/mm
C35	117.39	13.12	142.85	25.59	110.35	45.08
C45	124.98	14.13	156.16	28.04	116.18	46.53
C55	129.55	13.98	162.36	26.59	123.38	45.08
C65	136.08	14.87	172.67	27.99	126.67	49.06

5.2. Yield Strength of Steel

Steel accounts for a large proportion of composite joints. The influence of steel strength on joints cannot be ignored. Therefore, Q235, Q345, Q390 and Q420 steels commonly applied in practical engineering were selected as the influence parameters of numerical simulation. The numerical simulation results were recorded in Figure 18, and the influence of yield strength of different types of steel on each characteristic point of composite joint is listed in Table 6.

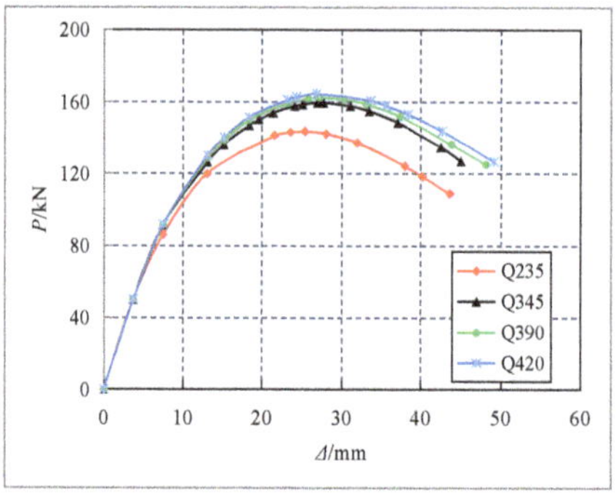

Figure 18. Influence of yield strength of profile steel on composite joints.

Table 6. Influence of yield strength of profile steel on characteristic points of composite joints.

Yield Strength of Steel	Yield Load P_y/kN	Yield Displacement Δ_y/mm	Peak Load P_m/kN	Peak Displacement Δ_m/mm	Ultimate Load P_u/kN	Ultimate Displacement Δ_u/mm
Q235	120.04	13.12	143.71	25.32	109.06	43.63
Q345	130.34	14.51	159.57	27.68	127.03	44.99
Q390	130.74	13.98	162.41	27.12	125.21	48.11
Q420	132.66	14.33	164.64	26.76	126.87	49.09

It can be seen from Figure 18 and Table 6 that the horizontal bearing capacity of the composite joints increases slightly with the increase of steel yield strength. Compared with the Q235 steel composite joint, the peak bearing capacity of the Q345, Q390 and Q420 steel composite joints was increased by 11.04%, 13.02% and 14.57%, respectively. However, there was little difference among them. The displacements of the joints with different strengths were similar, indicating that the yield strength of steel has little influence on the deformation capacity of the composite joints. The skeleton curves of the strength joints of all types of steel were significantly reduced. The ultimate load of the combined joints is 24.11%, 20.39%, 22.91%

and 22.94% lower than the peak load. It can be seen that the deterioration of the composite joints was basically gentle, demonstrating that the yield strength of steel had little effect on the ductility of the joints, and showed good ductility in general.

5.3. Thickness of Column Steel Web

In order to study the influence of different column steel web thicknesses on the performance of composite joints, this paper took d = 4~10 mm for the parameter influence analysis, and d was taken as 4 mm, 6 mm, 8 mm and 10 mm. The numerical simulation results are shown in Figure 19. Table 7 lists the influence of column steel web thickness on each characteristic point of the composite joints.

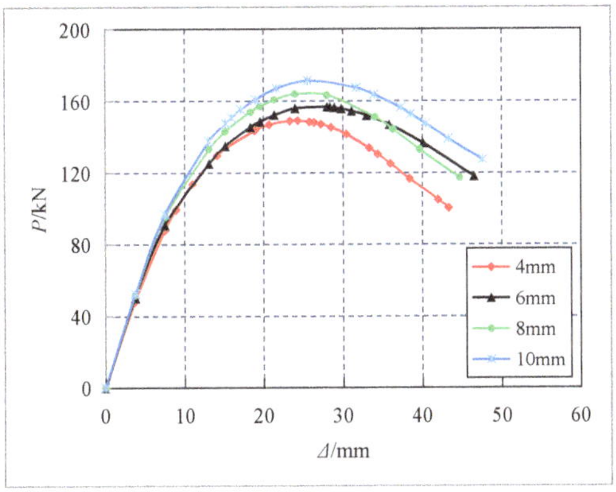

Figure 19. Influence of profile web thickness on composite joints.

Table 7. Influence of profile web thickness on characteristic points of composite joints.

Thickness d/mm	Yield Load P_y/kN	Yield Displacement Δ_y/mm	Peak Load P_m/kN	Peak Displacement Δ_m/mm	Ultimate Load P_u/kN	Ultimate Displacement Δ_u/mm
4	117.39	13.12	149.14	24.37	100.33	43.34
6	124.98	14.13	156.79	28.04	118.04	46.53
8	129.55	13.98	164.11	24.03	117.39	44.74
10	136.08	14.87	171.41	25.66	127.45	47.59

It can be estimated from Figure 19 and Table 7 that the column steel web has little effect on the stiffness of the initial elastic section of this type of composite joints, but has great influence on the peak bearing capacity and the degradation of the later bearing capacity. Compared with the peak bearing capacity of the composite joints with 4 mm web thickness, the peak bearing capacity of the composite joints with 6 mm, 8 mm and 10 mm web thicknesses increased by 5.13%, 10.04% and 14.93%, respectively, demonstrating that the thickness of column steel web had a significant effect on improving the strength of the composite joints. At the end of loading, the ultimate horizontal bearing capacity of the column increases with the increase of the steel web thickness, but the decline section of the skeleton curve was still obvious and the decline trend was similar. In summary, the increase of column steel web thickness can improve the horizontal bearing capacity of composite joints, while reasonable web thickness can also improve the seismic ductility of composite joints.

6. Restoring Force Model of Composite Joints

6.1. Establishing Method

The principal stress process and characteristics of composite joints were analyzed by low-cycle cyclic loading tests on the inner and outer joints of composite joints under different RAC replacement rates and axial compression ratios. Combined with the test hysteretic curve and skeleton curve, the restoring force model of interior joints and exterior joints in the composite joints was proposed by using the fitting method of test data, which provided reference for the elastic–plastic seismic response analysis of the composite joints and frame.

The following assumptions are mainly the simplification and summary of the principal stress process and characteristics of composite joints, the experimental hysteretic curves and skeleton curves. They should be made first before established the restoring force model of composite joints with SRRC columns and steel beams.

(1) Assume that the elastic limit point of the specimen coincides with the cracking point of the composite joints;
(2) Before the cracking point, the unloading stiffness and loading stiffness of the composite joints were the same, and after the elastic section, the stiffness decreased with the increase of displacement;
(3) The skeleton curve adopted elastic stage, crack working stage, yield strengthening stage and failure stage four-fold line model.

6.2. Establishment of Skeleton Curve Model

The skeleton curve obtained by the actual test was a continuous smooth curve without an obvious yield point. When established the restoring force model of skeleton curve, it should be simplified to a quadrangle model with stiffness degradation. The details are stated as follows:

The characteristic points of skeleton curve were cracking point A (P_c, Δ_c); yield load point B (P_y, Δ_y); peak load point C (P_{max}, Δ_{max}); and limit load point D (P_u, Δ_u), where yield load point B was calculated by the "general yield moment method", and limit load point D was 85% of the peak load. Due to the differences of skeleton curves of various composite joints, dimensionless processing was carried out for skeleton curves of various composite joints first, as shown in Figure 20.

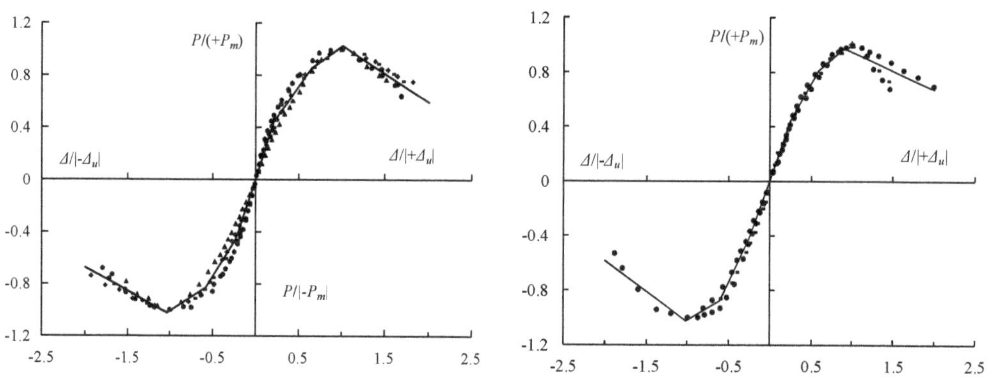

(a) Interior joint (b) Exterior joint

Figure 20. Dimensionless skeleton curve.

It can be estimated from Figure 20 that the data points of the skeleton curve of interior joints and exterior joints after dimensionless processing have good regularity, as shown in the real line in the graph. The linear regression method was applied to fit the data points of each stage of the skeleton curve. The four-line skeleton curve model was shown in Figure 21. OA was the elastic section from the positive loading to the cracking point, while AB was the working section with cracks under positive loading, and BC was the yield strengthening stage under positive loading, while CD was the failure stage under positive loading. OA' was the elastic section from reverse loading to cracking point, AB was the working section with cracks under reverse loading, $B'C'$ was the yield strengthening stage under reverse loading and $C'D'$ was the failure stage under reverse loading. The fitting process of skeleton curves of interior joints and exterior joints in each stage is exhibited in Figure 22. The linear equations and stiffness in each stage of the skeleton curve model are shown in Tables 8 and 9.

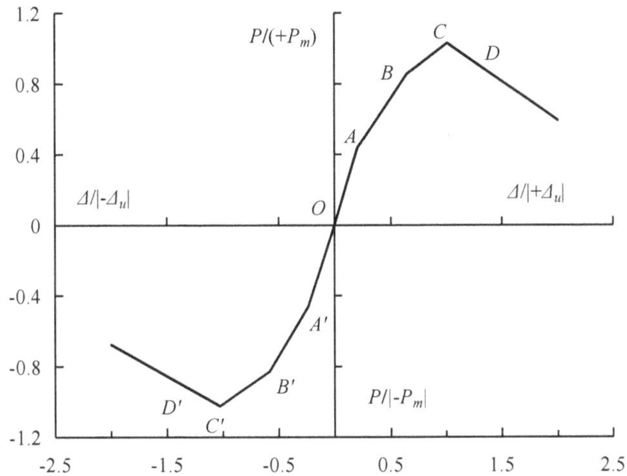

Figure 21. Skeleton curve model with four-line.

Table 8. Linear equation and stiffness of each stage in skeleton curve model of interior joint.

Line Segment	Regression Equation	Angle to Displacement Axis	Rigidity
OA	$P/(+P_m) = 2.1151\Delta/(+\Delta_m)$	65°	2.1151
AB	$P/(+P_m) = 0.9534\Delta/(+\Delta_m) + 0.2407$	44°	0.9534
BC	$P/(+P_m) = 0.465\Delta/(+\Delta_m) + 0.555$	25°	0.465
CD	$P/(+P_m) = -0.4409\Delta/(+\Delta_m) + 1.4745$	−24°	−0.4409
OA'	$P/\lvert -P_m \rvert = 1.9697\Delta/\lvert -\Delta_m \rvert$	63°	1.9697
$A'B'$	$P/\lvert -P_m \rvert = 1.0582\Delta/\lvert -\Delta_m \rvert -0.2126$	47°	1.0582
$B'C'$	$P/\lvert -P_m \rvert = 0.4369\Delta/\lvert -\Delta_m \rvert -0.5749$	24°	0.4369
$C'D'$	$P/\lvert -P_m \rvert = -0.3568\Delta/\lvert -\Delta_m \rvert -1.3909$	−20°	−0.3568

Table 9. Linear equation and stiffness of each stage in skeleton curve model of exterior joint.

Line Segment	Regression Equation	Angle to Displacement Axis	Rigidity
OA	$P/P_m = 1.7705\Delta/\Delta_m$	60	1.7705
AB	$P/P_m = 1.1003\Delta/\Delta_m + 0.1707$	48	1.1003
BC	$P/P_m = 0.5188\Delta/\Delta_m + 0.5066$	27	0.5188
CD	$P/(+P_m) = -0.4409\Delta/(+\Delta_m) + 1.4745$	−15	−0.2675
OA'	$P/\lvert -P_m \rvert = 1.9697\Delta/\lvert -\Delta_m \rvert$	59	1.6875
$A'B'$	$P/\lvert -P_m \rvert = 1.0582\Delta/\lvert -\Delta_m \rvert -0.2126$	51	1.2427
$B'C'$	$P/\lvert -P_m \rvert = 0.4369\Delta/\lvert -\Delta_m \rvert -0.5749$	20	0.3605
$C'D'$	$P/\lvert -P_m \rvert = -0.3568\Delta/\lvert -\Delta_m \rvert -1.3909$	−24	−0.4395

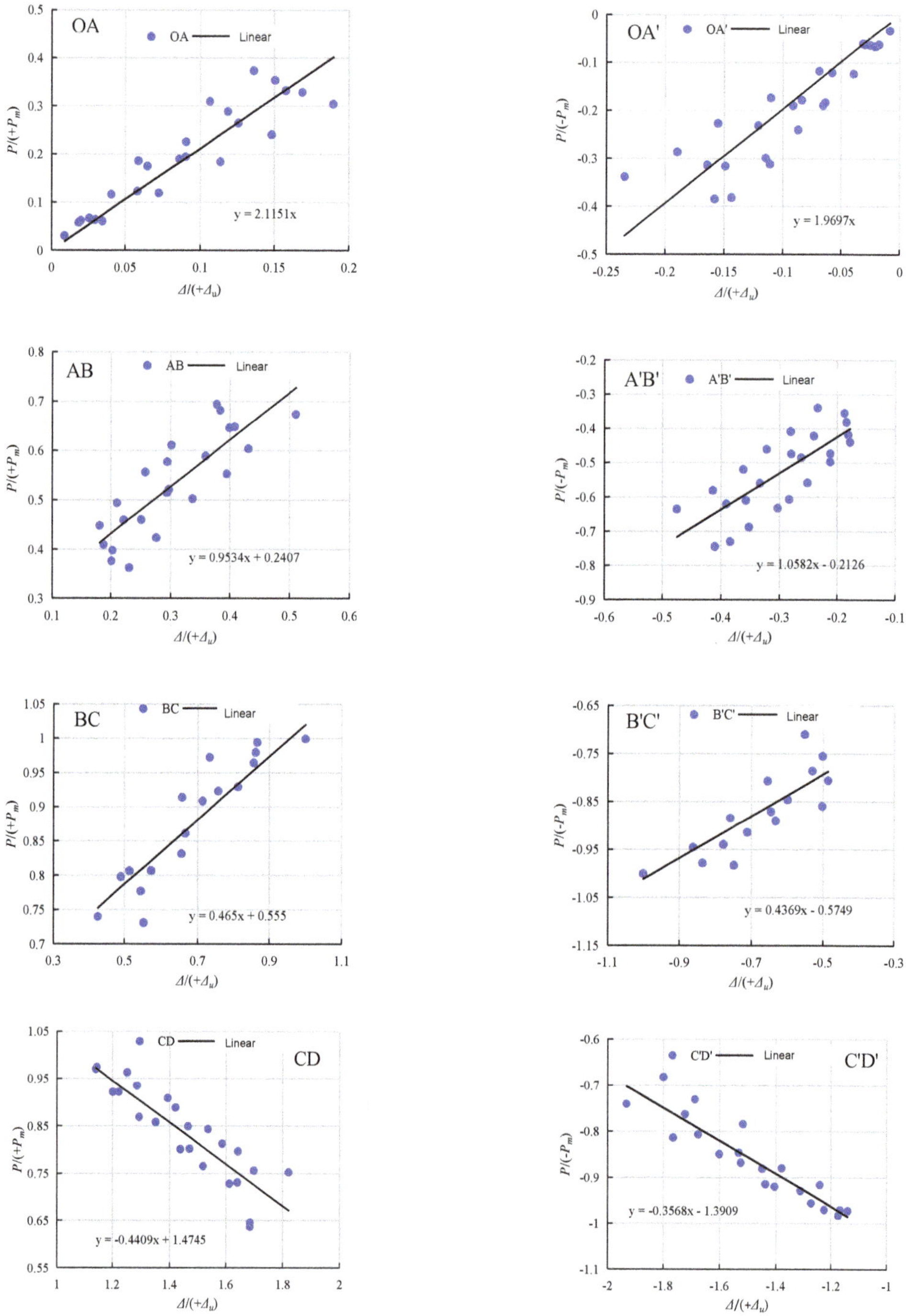

(**a**) Each stage fitting process of interior joint skeleton curve

Figure 22. *Cont.*

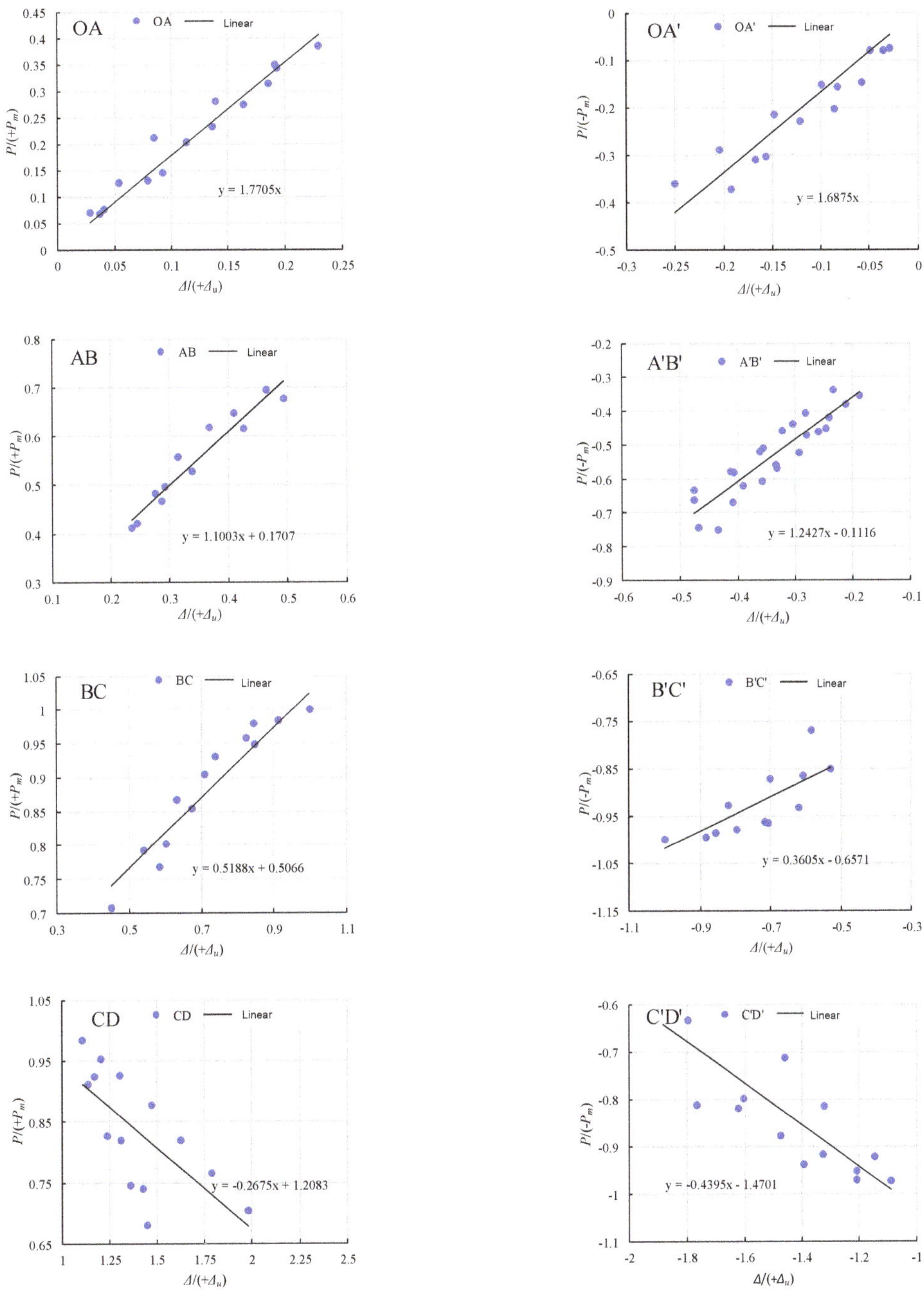

(**b**) Each stage fitting process of exterior joint skeleton curve

Figure 22. Fitting process of skeleton curve.

6.3. Stiffness Degradation Rule

Through the low-cyclic loading test of composite joints, it was found that the stiffness of the composite joints decreases gradually with the increase of load and displacement loading stages. The skeleton curve restoring force model simplifies the whole loading process of the composite joints into four stages, and nonlinear fitting was carried out for the loading stiffness and unloading of the composite joints at each stage. The variation of the forward and reverse unloading stiffness of the composite joints was described in detail below. K_{12}, K_{34} and K_{56} represent the unloading stiffness under forward loading, while $K_{1'2'}$, $K_{3'4'}$ and $K_{5'6'}$ represent the unloading stiffness under reverse loading, as presented in Figure 23.

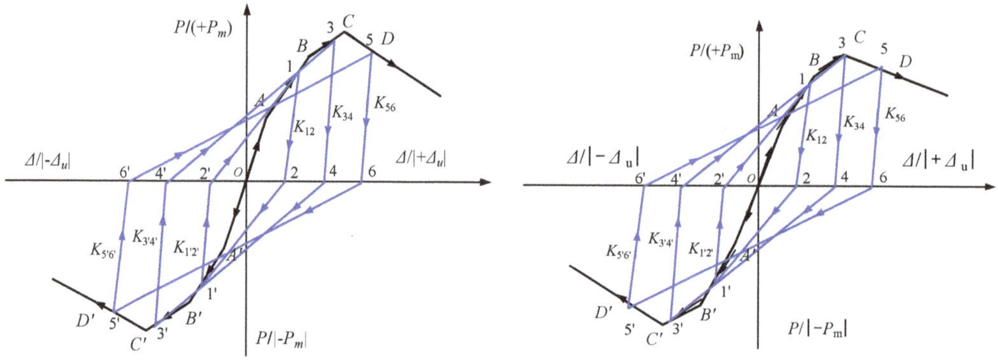

Figure 23. Stiffness degradation rule of specimen.

The unloading stiffness K_{12}, AB section was the working stage of positive loading with cracks. When unloading before positive loading to A, the unloading route was along OA, and the unloading stiffness was the initial stiffness K_0 of the composite joints. After the unloading of section AB, the unloading route was carried out along 12, and the unloading stiffness was K_{12}. The slope K_{12} of forward unloading route 12 was obtained by linear fitting and all measured data between unloading point 1 and point 2 when unloading load was 0. Then, the unloading stiffness of the AB section of each composite joint was normalized, while the nonlinear regression fitting was carried out by power function. The nonlinear relationship curve between K_{12}/K_0 and $\Delta_1/(+\Delta_m)$ was obtained, where K_0 was the initial stiffness of the positive loading elastic section, or the slope of the straight line obtained by linear fitting of the measured data of the positive loading elastic section of the skeleton curve. Δ_1 was the displacement corresponding to positive loading point 1. The above method was adopted to nondimensionalize the unloading stiffness of each test. Moreover, the power function was used for nonlinear regression fitting, as shown in Figure 24.

Based on the experimental research, combined with the obtained hysteretic curve and skeleton curve, the restoring force model of composite joints proposed was shown in Figure 25. The model considered the four stress stages of the composite joints during the whole loading process in the form of four broken lines, and fitted the stiffness degradation law of each stage, which can truly reflect the stiffness degradation law of the composite joints in each stress stage and was suitable for the elastic–plastic analysis of composite joints and other structures.

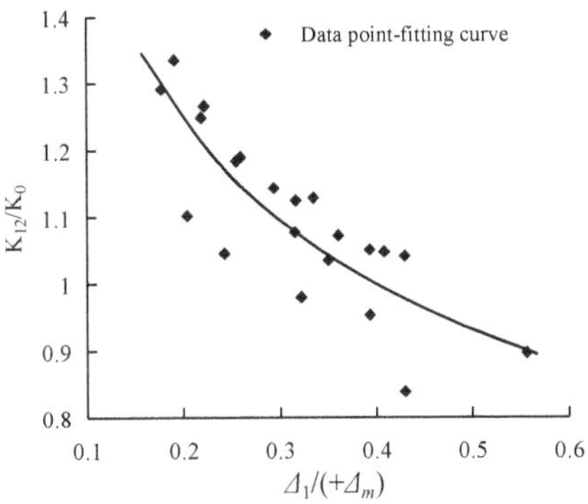

$K_{12}/K_0 = 0.7465[\Delta_1/(+\Delta_m)]^{-0.319}$

(**a**) Interior joint

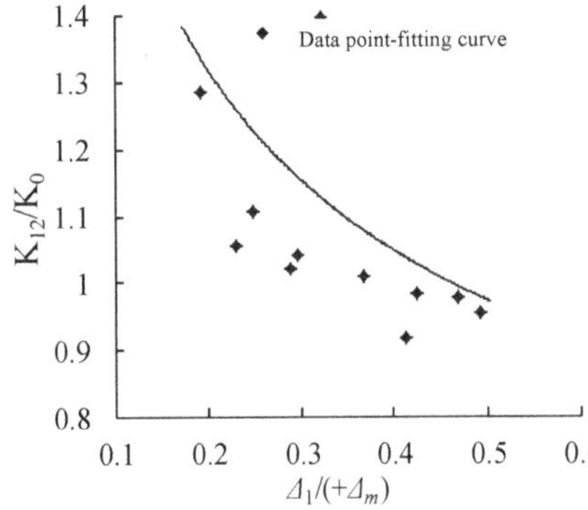

$K_{12}/K_0 = 0.7743[\Delta_1/(+\Delta_m)]^{-0.331}$

(**b**) Exterior joint

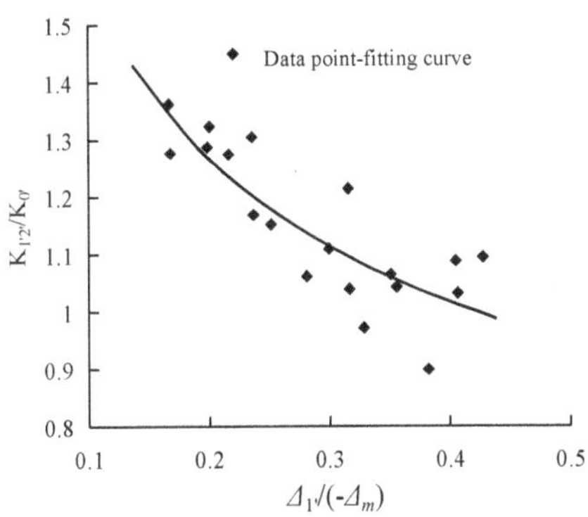

$K_{1'2'}/K_{0'} = 0.76[\Delta_{1'}/(-\Delta_m)]^{-0.318}$

(**c**) Interior joint

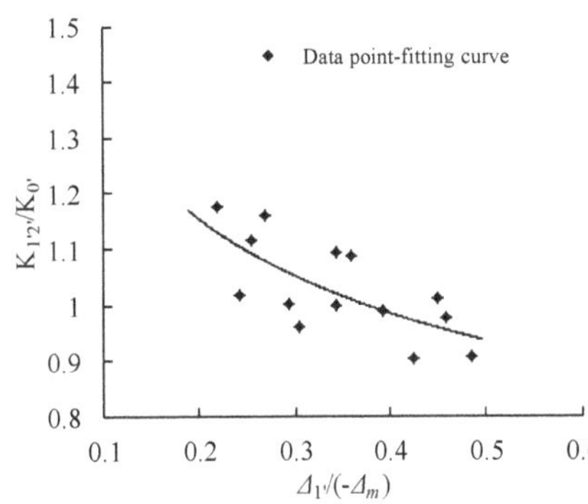

$K_{1'2'}/K_{0'} = 0.7996[\Delta_{1'}/(-\Delta_m)]^{-0.228}$

(**d**) Exterior joint

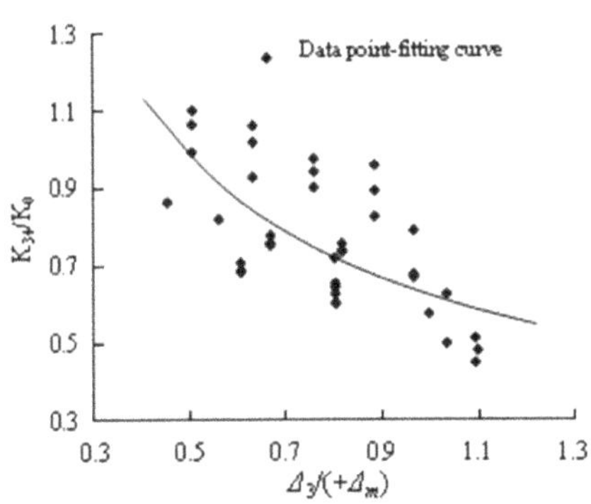

$K_{34}/K_0 = 0.6228[\Delta_3/(+\Delta_m)]^{-0.659}$

(**e**) Interior joint

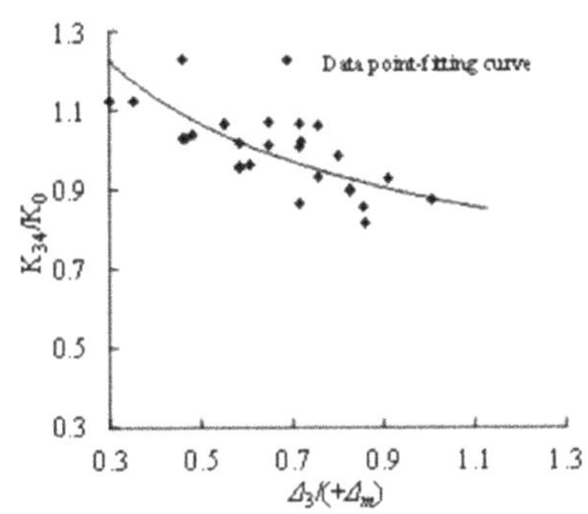

$K_{34}/K_0 = 0.8796[\Delta_3/(+\Delta_m)]^{-0.276}$

(**f**) Exterior joint

Figure 24. *Cont.*

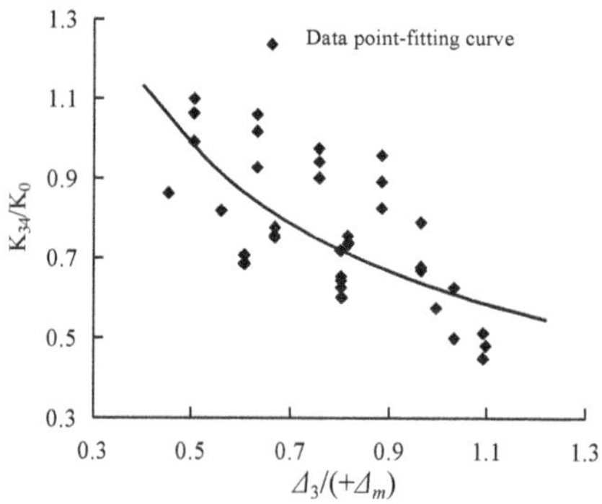

$K_{3'4'}/K_{0'} = 0.7679[\Delta_{3'}/(-\Delta_m)]^{-0.723}$

(**g**) Interior joint

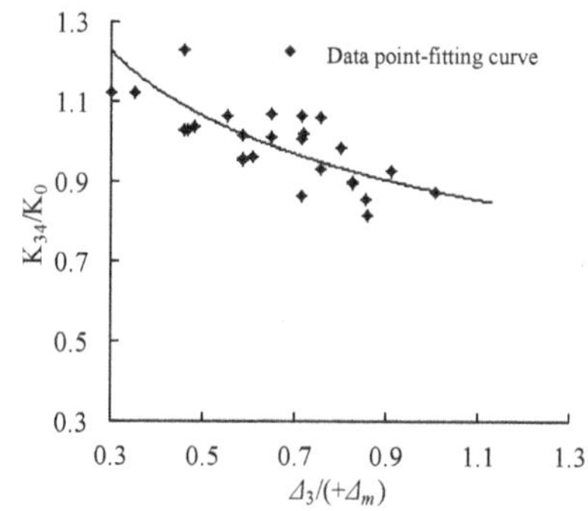

$K_{3'4'}/K_{0'} = 0.8407[\Delta_{3'}/(-\Delta_m)]^{-0.016}$

(**h**) Exterior joint

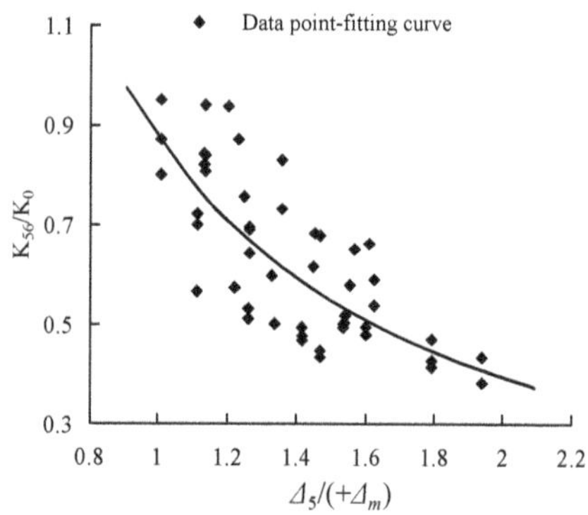

$K_{56}/K_0 = 0.8747[\Delta_5/(+\Delta_m)]^{-1.148}$

(**i**) Interior joint

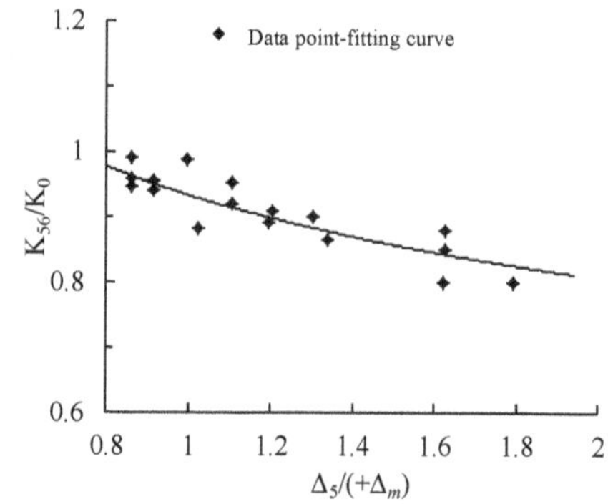

$K_{56}/K_0 = 0.9284[\Delta_5/(+\Delta_m)]^{-0.211}$

(**j**) Exterior joint

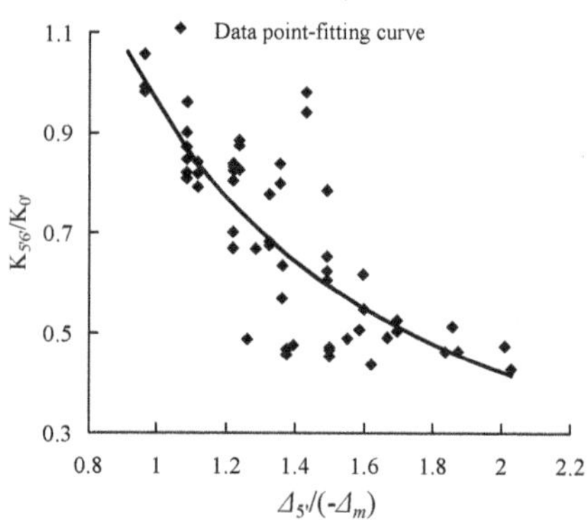

$K_{5'6'}/K_{0'} = 0.9567[\Delta_{5'}/(-\Delta_m)]^{-1.178}$

(**k**) Interior joint

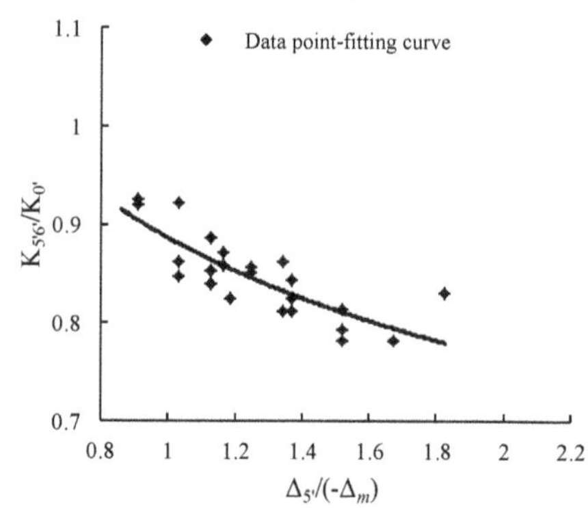

$K_{5'6'}/K_{0'} = 0.8846[\Delta_{5'}/(-\Delta_m)]^{-0.248}$

(**l**) Exterior joint

Figure 24. Stiffness degradation rule curve of unloading stiffness.

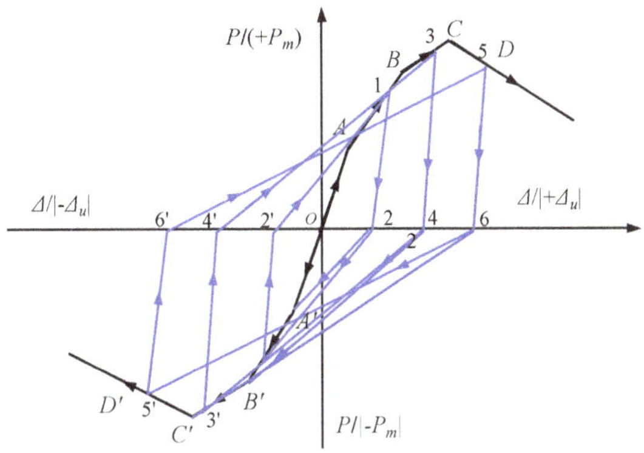

Figure 25. Four-line restoring force model of stiffness degradation.

The restoring force model was described as follows:

(1) The restoring force model was obtained by combining the main feature points (cracking point, yield point, peak point and limit point) of the test skeleton curve and by dimensionless processing and fitting of the test data. The coordinates were dimensionless coordinates.

(2) The skeleton curve of the restoring force model was divided into four stages. In the figure, OA and OA′ were elastic sections under positive and negative loading, respectively, AB and A′B′ were working sections with cracks under positive and negative loading, respectively and BC and B′C′ were yield strengthening sections under positive and negative loading, respectively, while CD and C′D′ were failure sections under positive and negative loading, respectively.

(3) When the specimen is loaded and unloaded in the elastic segment (OA and OA′), the unloading stiffness was the elastic stiffness of the specimen. When the specimen was loaded to the elastic segment (AD and A′D′), the unloading stiffness gradually degrades.

Therefore, the restoring force model established in this paper mainly has the following characteristics. In the quasi-static test, the loading and unloading stiffness of the composite joints before reaching the cracking load were both initial stiffness. After reaching the cracking load, the stiffness degradation phenomenon of the composite joints is considered. After cracking, the stiffness of forward and reverse loading gradually degrades with the increase of displacement, and the degradation rate accelerates with the loading process.

6.4. Validation of Restoring Force Model

According to the regression equations of the skeleton curve model at each stage shown in Table 8, the established skeleton curve model was compared with the experimental skeleton curve, as exhibited in Figure 26.

The calculated hysteretic curve of each composite joint can be acquired by the four-fold restoring force model proposed in Figure 20. The calculated hysteretic curve was compared with the experimental hysteretic curve, such as Figure 27.

From Figures 26 and 27, it can be seen that the skeleton curve and hysteretic curve calculated by both interior joints and exterior joints were in good agreement with the overall trend of the test skeleton curve and hysteretic curve. The results demonstrated that the four-fold restoring force model proposed can well reflect the hysteretic characteristics of composite joints with SRRC columns and steel beams under low cyclic loading.

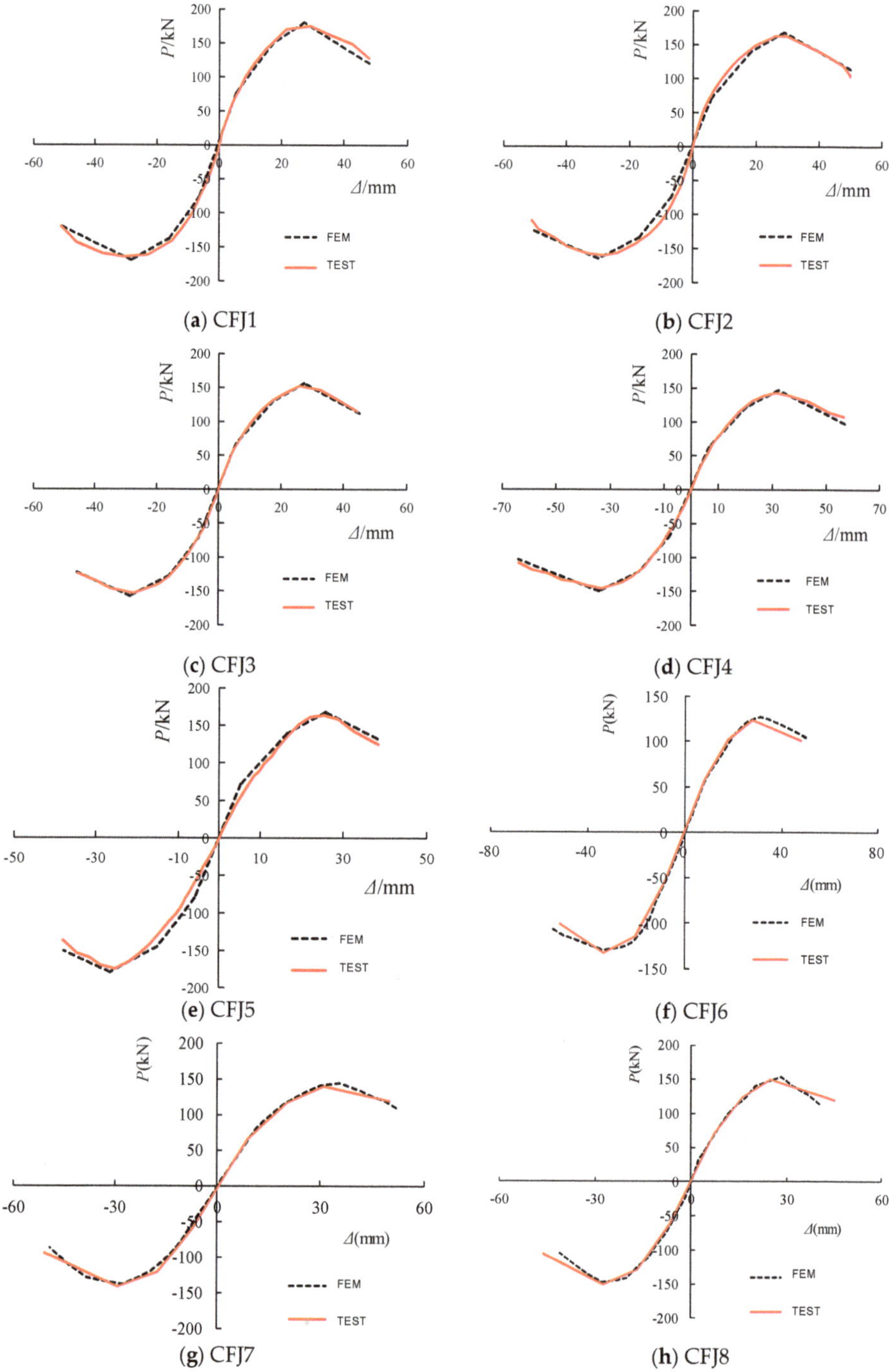

Figure 26. Comparison of skeleton curve between calculation results and test results.

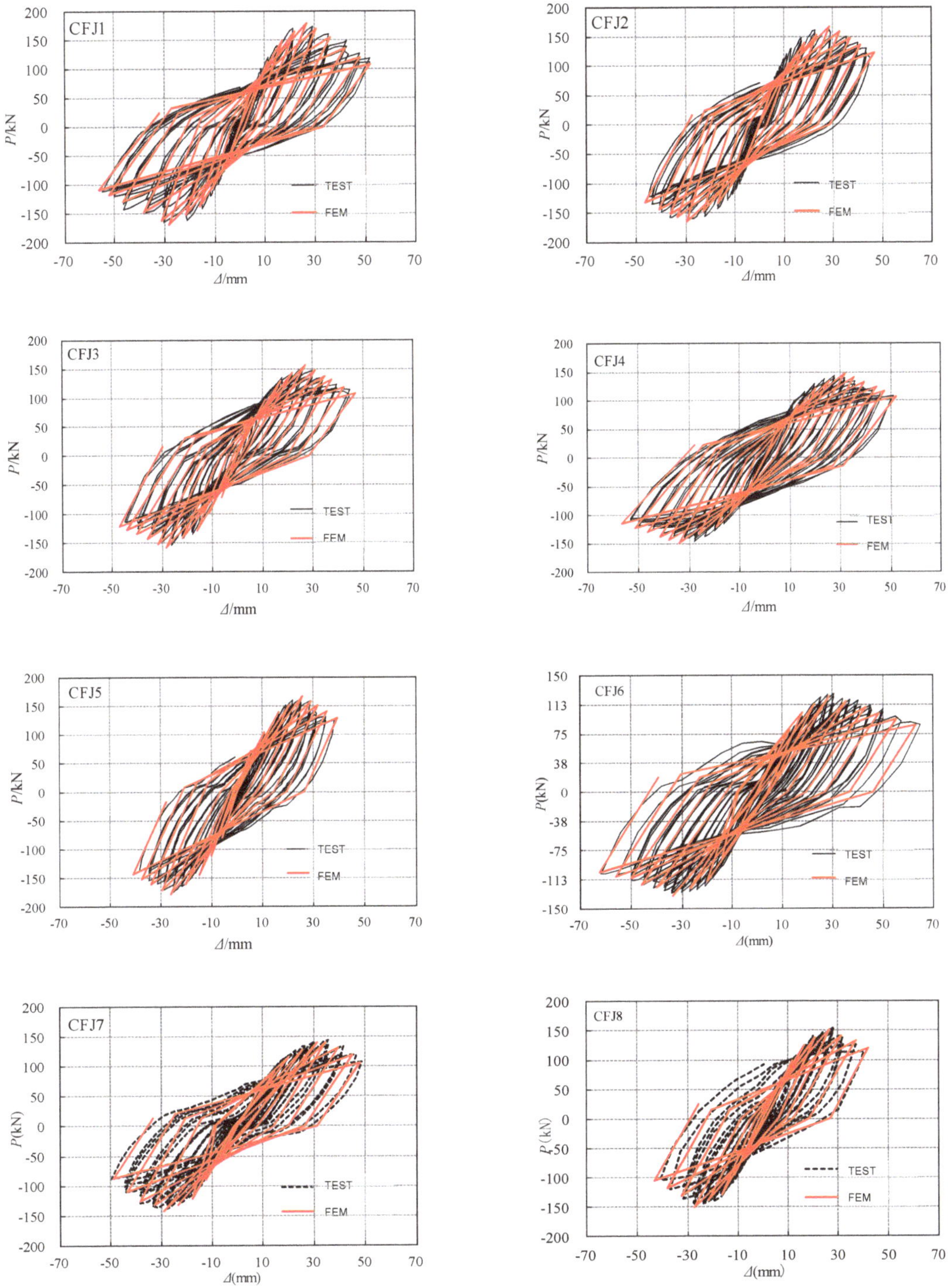

Figure 27. Comparison of hysteretic curve between calculation results and test results.

7. Results of the Study

7.1. Conclusions

Based on the experimental study, the failure mechanism of composite joints with SRRC columns and steel beams is studied, the mechanical model is established and the calculation method of bearing capacity is verified. The refined numerical model of composite joints with SRRC columns and steel beams was established, and the design parameters were extended and analyzed. The main conclusions are as follows:

(1) The finite element calculation results are in good agreement with the experimental results, which shows that the simulation method in this paper has good rationality and applicability. The modeling method can provide a reference for the numerical analysis of these kind of joints.
(2) With the increase of RAC strength, the horizontal bearing capacity of the composite joints increases significantly, but the stiffness of the elastic section increases slightly, and the RAC strength has little effect on the deformation capacity of composite joints.
(3) The increase of steel yield strength has no obvious effect on the horizontal bearing capacity and ductility of the composite joints.
(4) The increase of the thickness of column steel web can improve the horizontal bearing capacity of composite joints, and reasonable thickness can improve the seismic ductility of the composite joints.
(5) By fitting the experimental data of eight composite joints, a hysteretic curve restoring force model of composite joints under repeated load was proposed. The hysteretic characteristics of the composite joints under low cyclic loading can be well reflected by restoring the force model curve.

7.2. Recommendations

On the one hand, the model has good seismic performance. On the other hand, it can solve the problem of waste concrete disposal. It meets the requirements of national green environmental protection and sustainable development and has broad development prospects. In this paper, the composite frame joints of reclaimed steel column and steel beam are studied by combining experiment and theory. Although some achievements have been made, there are still many problems to be further studied.

(1) The influence of the bond slip between the steel and concrete, the friction coefficient value and how to extend the restoring force model proposed for other typologies of composite joints should be investigated systematically and deeply in future work.
(2) The proposed nodal restoring force model in the combined framework is mainly proposed based on the test results. Although the model curve is in good agreement with the test curve, the number of specimens is small and has not been verified by numerical simulation, which needs to be further studied and modified.
(3) The composite joints of the reclaimed steel reinforced concrete column and steel beam studied in this paper are planar joints with a strong axis direction, and the subsequent mechanical properties of the composite frame joints in the form of spatial joints need to be studied.
(4) At present, there is a certain basis for the research on the mechanical properties of recycled concrete and related components, but the research content on the durability of recycled concrete is scant, and should be further studied in the future.

Author Contributions: Conceptualization, J.D.; Data curation, C.S. and X.C.; Formal analysis, H.M.; Methodology, J.D. and Y.L.; Project administration, Y.L.; Writing—original draft, J.D. and C.S. All authors have read and agreed to the published version of the manuscript.

Funding: The research was financially supported by the National Natural Science Foundation of China P.R. (No. 52109167), the Scientific Research Program funded by the Education Department of Shaanxi Provincial Government (No. 21JK0803), Program 2022TD-01 for Shaanxi Provincial Innovative Research Team and the Doctoral Research start-up funding project supported by Xi'an University of Technology (No. 451120004), all of whom are gratefully acknowledged.

Institutional Review Board Statement: Not applicable.

Informed Consent Statement: Not applicable.

Data Availability Statement: Not applicable.

Conflicts of Interest: The authors declare no conflict of interest.

Abbreviations

SRRC	Steel Reinforced Recycled Concrete
RAC	Recycled Aggregate Concrete
FE	Finite Element
RCA	Recycled Coarse Aggregate
List of Symbols	
r	RCA replacement percentage
n	Axial compression ratio
ρ_a	Profile steel ratio
σ_i	Equivalent stress of steel
f_s	Yield strength of steel
f_y	Yield strength of steel
E_s	Elastic modulus of steel
E_t	Tangent modulus of steel
ζ	Profile steel index
ε_i	Equivalent strain of steel
ε_y	Strain of steel at yield
ε_u	Strain at the limit of steel
σ_s	Stress of steel
ε_0	Peak strain
ε_{cc}	Peak strain
f_{cp}	Peak stress of concrete in the weak constraint area
ε_{cp}	Peak strain of concrete in the weak constraint area
f_{ch}	Peak stress of strong constraint concrete
ε_{ch}	Peak strain of strong constraint concrete
$\sigma_1, \sigma_2, \sigma_3$	Three principal stresses
d_c	Damage factor

References

1. Xue, J.; Zhai, L.; Bao, Y.; Ren, R.; Zhang, X. Cyclic loading tests and shear strength of steel reinforced recycled concrete beam–column inner joints. *Struct. Des. Tall Spec. Build.* **2018**, *27*, e1454. [CrossRef]
2. Pan, Z.; Zheng, W.; Xiao, J.; Chen, Z.; Chen, Y.; Xu, J. Shear behavior of steel reinforced recycled aggregate concrete beams after exposure to elevated temperatures. *J. Build. Eng.* **2022**, *48*, 103953. [CrossRef]
3. Dong, H.; Li, Y.; Cao, W.; Guo, Y. Seismic behavior of full-scale steel reinforced recycled concrete columns under high axial compression ratio. *Structures* **2021**, *29*, 1882–1897. [CrossRef]
4. Dong, J.; Ma, H.; Zou, C.; Liu, Y.; Huang, C. Finite element analysis and axial bearing capacity of steel reinforced recycled concrete filled square steel tube columns. *Struct. Eng. Mech.* **2019**, *72*, 43–60.
5. Ma, H.; Jia, C.; Xi, J.; Dong, J.; Zhang, X.; Zha, Y. Cyclic loading test and nonlinear analysis on composite frame consisting of steel reinforced recycled concrete columns and steel beams. *Eng. Struct.* **2021**, *241*, 112480. [CrossRef]
6. Liu, Z.; Xue, J.; Qi, L.; Gao, L. Experimental study on seismic performance of steel–reinforced recycled concrete frame with infill wall. *Struct. Des. Tall Spec. Build.* **2020**, *29*, e1744. [CrossRef]
7. Ke, X.; Yang, C.; Su, Y.; Jun, L. Experimental study on eccentric compressive performance of embedded steel tube confined reinforced recycle concrete composite columns. *J. Basic Sci. Eng.* **2020**, *28*, 354–365.
8. Nassiraei, H.; Rezadoost, P. Static capacity of tubular X-joints reinforced with fiber reinforced polymer subjected to compressive load. *Eng. Struct.* **2021**, *236*, 112041. [CrossRef]

9. Xue, J.; Li, H.; Ren, R.; Zhang, X. The bond behavior of steel reinforced recycled concrete under cyclic reversed load. *Eng. Struct.* **2021**, *248*, 113238. [CrossRef]
10. Toshikazu, T.; Sozen, M.A.; Nielsen, N.N. Reinforced Concrete Response to Simulated Earthquakes. *J. Struct. Div.* **1970**, *96*, 2557–2572.
11. Nassiraei, H.; Rezadoost, P. Stress concentration factors in tubular T/Y-joints strengthened with FRP subjected to compressive load in offshore structures. *Int. J. Fatigue* **2020**, *140*, 105719. [CrossRef]
12. Valente, M.; Castiglioni, C.A.; Kanyilmaz, A. Welded fuses for dissipative beam-to-column connections of composite steel frames: Numerical analyses. *J. Constr. Steel Res.* **2017**, *128*, 498–511. [CrossRef]
13. Xue, J.; Yang, Y.; Qi, L. Restoring force model of steel-reinforced recycled concrete frames infilled with recycled concrete blocks. *J. Build. Eng.* **2022**, *60*, 105173. [CrossRef]
14. Bai, G.; Han, Y.; Liu, C. Resilience model test of recycled concrete beam-column joints. *Ind. Build.* **2016**, *46*, 42–46 + 62. (In Chinese)
15. Ma, H.; Dong, J.; Liu, Y.; Yang, D. Cyclic loading tests and shear strength of composite joints with steel-reinforced recycled concrete columns and steel beams. *Eng. Struct.* **2019**, *199*, 109605. [CrossRef]
16. Vymazal, T.; Hrabová, K.; Teplý, B.; Kocáb, D. Sustainability quantification of concrete with recycled aggregate. *AIP Conf. Proc.* **2020**, *2210*, 020018.
17. Ma, H.; Gan, X.; Xi, J.; Dong, J.; Zhang, P.; Zhao, Y. Pseudo Static Test research on Seismic Behavior of Steel reinforced recycled Concrete Columns-steel Beams Combination Fram. *J. Basic Sci. Eng.* **2022**, *30*, 603–617.
18. Xue, J.; Ma, H.; Liu, Y. Experimental study on seismic performance of steel reinforced recycled concrete columns under low-cyclic reversed loading. *China Civ. Eng. J.* **2014**, *47*, 36–46. (In Chinese)
19. Xue, J.; Bao, Y.; Ren, R.; Wang, G. Experimental study on seismic performance of steel reinforced recycled concrete inner-frame joints under low-cyclic reversed loading. *China Civ. Eng. J.* **2014**, *47*, 1–8. (In Chinese)
20. Vecchio, F.J. Towards Cyclic Load Modeling of Reinforced Concrete. *Struct. J.* **1999**, *96*, 193–202.
21. Filippou, F.C.; Popov, E.P.; Bertero, V.V. *Effects of Bond Deterioration on Hysteretic Behavior of Reinforced Concrete Joints*; University of California, Berkeley: Berkeley, CA, USA, 1983.
22. Wang, C.; Xu, H.; Zhang, J. An improved displacement based seismic design method of prestressed fabricated concrete frame. In Proceedings of the 2020 5th International Conference on Minerals Source, Geotechnology and Civil Engineering (MSGCE 2020), Hong Kong, China, 17–19 April 2020; pp. 360–374.
23. Fang, Z.; Zhen, Y.; Li, X. Reinforcement hysteresis model for reinforced concrete structures. *J. Wuhan Univ. (Eng. Ed.)* **2018**, *51*, 613–619. (In Chinese)
24. Fang, Z.; Zhou, H.; Lai, S.; Xie, Q. ABAQUS bond-slip element of reinforced concrete under cyclic loading. *J. Wuhan Univ. (Eng. Ed.)* **2014**, *47*, 527–531. (In Chinese)
25. Mander, J.B.; Priestley, M.J.N.; Park, R. Theoretical Stress-Strain Model for Confined Concrete. *J. Struct. Eng.* **1988**, *114*, 1804–1826. [CrossRef]
26. Mirza, S.A.; Skrabek, B.W. Statistical Analysis of Slender Composite Beam-Column Strength. *J. Struct. Eng.* **1992**, *118*, 1312–1332. [CrossRef]
27. Thanh, C.; HoangLe, M.; Thanh, S. A nonlinear concrete damaged plasticity model for simulation reinforced concrete structures using ABAQUS. *Fract. Struct. Integr. Int. J. Ital. Group Fract.* **2022**, *16*, 232–242. [CrossRef]
28. Zhang, J.; Wang, Q.; Hu, S.; Wang, C. Parameter verification of ABAQUS concrete damage plasticity model. *Build. Struct.* **2008**, *8*, 127–130. (In Chinese)
29. Jean, L. A Continuous Damage Mechanics Model for Ductile Fracture. *J. Eng. Mater. Technol.* **1985**, *107*, 83–89.
30. Huang, W.; Yang, T.; Chen, Y.; Ge, P.; Zhou, L. Study on Damage Constitutive Model of Recycled Aggregate Concrete in Code for Design of Concrete Structures and Development in ABAQUS. *J. Phys. Conf. Ser.* **2021**, *1759*, 012002.
31. Jia, Z.-L.; Wang, W.-D.; Shi, Y.-L.; Xian, W. Performance of steel-reinforced concrete-filled square steel tubular members under sustained axial compression loading. *Eng. Struct.* **2022**, *264*, 114464. [CrossRef]
32. Balaty, P.; Gjelsyilk, A. Coefficient of riction for steel on concrete at high normal stress. *J. Mater. Civ. Eng.* **1990**, *2*, 46–49. [CrossRef]

coatings

MDPI

Article

Cyclic Performance of Prefabricated Shear Wall Connected to Columns by Rectangular Concrete-Filled Steel Tube Keys

Zhijun Zhou [1,*], Ming Li [1,*], Qian Wu [1,*], Shuang Yuan [2] and Li Zhang [3]

[1] School of Civil Engineering, Shenyang Jianzhu University, Shenyang 110168, China
[2] School of Construction Engineering, Tianjin University, Tianjin 300072, China
[3] School of Civil Engineering, Shenyang Urban Construction University, Shenyang 110000, China
* Correspondence: me572742890@163.com (Z.Z.); cemli@sjzu.edu.cn (M.L.); dingxianxing@126.com (Q.W.); Tel.: +86-024-24691800 (Q.W.)

Abstract: A prefabricated frame–shear wall structure is a major structure in an assembled building. To find a method of connecting a shear wall and columns that can both reduce the amount of wet work required and maintain adequate stiffness, we propose connecting the shear wall to the vertical frame using rectangular concrete-filled steel tube keys (RCFSTKs). Static tests of a cast-in-place frame–shear wall structure without keys and a prefabricated structure with RCFSTKs were conducted to compare their seismic performance. The feasibility of the new method was analyzed. Finite element models were then set up to determine if plain concrete blocks between RCFSTKs could be removed and to identify the influence of different parameters on the cyclic performance of the assembled structure. The results show that the use of RCFSTK is practical. Compared with a dimensionally similar cast-in-place shear wall–column construction, a prefabricated shear wall connected to columns by RCFSTKs has a fuller hysteresis curve, better ductility, slightly higher energy dissipation, and slightly slower degradation of stiffness and bearing capacity. The removal of inter-key concrete blocks significantly reduces bearing capacity and initial stiffness. The cyclic performance of the assembled structure is primarily influenced by the number of RCFSTKs, the thickness of the steel tube key wall, and the axial compression ratio, with less effect from key unit height, steel strength, and concrete grade in RCFSTKs.

Keywords: RCFSTKs; frame column; shear wall; fabricated concrete structure; cyclic performance

Citation: Zhou, Z.; Li, M.; Wu, Q.; Yuan, S.; Zhang, L. Cyclic Performance of Prefabricated Shear Wall Connected to Columns by Rectangular Concrete-Filled Steel Tube Keys. *Coatings* **2022**, *12*, 1584. https://doi.org/10.3390/coatings12101584

Academic Editor: Paolo Castaldo

Received: 14 September 2022
Accepted: 16 October 2022
Published: 19 October 2022

1. Introduction

Assembled concrete structures have the advantages of fast construction, high production efficiency and good engineering quality. They have been widely used in various buildings. A frame–shear wall structure is composed of frames and shear walls. Due to the combination effect [1], it has the advantages of flexible building layout, large spacing and good lateral resistance. An assembled frame–shear wall structure combines the advantages of the assembled concrete structure and the frame–shear wall structure, and it has been proved to have good cyclic performance [2,3]. This promotes their application in areas requiring seismic fortification.

Prefabricated shear walls, prefabricated frame beams and columns are the important components of a prefabricated frame–shear wall structure. How to connect them is the key technical problem of this structure. Two widely used assembly methods for such a structure exist in China. One is to prefabricate a shear wall, frame columns, and frame beams as a complete unit for subsequent assembly; this method is suitable when the wall, columns, and beams are small. Researchers [1–5] conducted a series of studies on this kind of structure and found that its cyclic performance was similar to that of a cast-in-place one. The second method is to prefabricate a wall, columns, and beams as independent units for subsequent assembly; this method is suitable when these components are larger. The feasibility of this method has been demonstrated by Zhao et al. [6]. There is less research into assembled frame–shear wall structures from other countries where the assembly method mainly used is to attach shear walls to the outside of the frame plane [7,8].

We study the second case. For the second method, it is necessary to study how to ensure a rigid connection between the shear wall and columns. Tong et al. [9] investigated a single-story structure of walls and columns connected with straight bars, U-shaped bars, straight U-shaped bars, and link bars using compression–shear tests and static tests. Gao et al. [10–12] examined the feasibility of connecting the shear wall and columns with U-bar in the graduate student apartment building of Shenyang Jianzhu University. Anion et al. [13] proposed setting keys on the surface of the wall and columns. The preceding studies all require pouring concrete or grout in or around the connection interface, which necessitates a relatively large amount of wet work. To reduce the amount of wet work and increase industrialization, Chinese researchers proposed using pure steel members to connect walls and columns. Gao et al. [14] investigated the mechanical properties of this type of structure using numerical models and found that its stiffness was very low and did not meet the level required to resist deformation. So, it becomes a topic of concern to find a method of connecting a shear wall and columns that can both reduce the amount of wet work required and maintain adequate stiffness.

Because the method using pure steel members to connect walls and columns can not meet the requirement of stiffness, we propose a method that uses rectangular concrete-filled steel tube keys (RCFSTKs). This paper investigates the cyclic performance of a prefabricated shear wall and column unit connected by RCFSTKs to determine the feasibility of using RCFSTKs in connecting prefabricated components in order to develop a wall and column connection method that facilitates more rapid assembly.

2. Structure of Wall and Columns Connected by RCFSTKs

Frame–shear wall structure showing the wall and columns connected by RCFSTKs is as Figure 1. The part embedded in the wall–column interface consists of a steel shear plate, a bottom steel plate, and an anchor steel plate welded together (Figure 1a). The RCFSTK is formed by filling the steel tube with concrete (Figure 1b). The plain concrete block inserted between the keys is shown in Figure 1c. If there are no RCFSTKs or plain concrete blocks, there will be gaps between the surface of the column and shear wall, which will reduce the bearing capacity and initial stiffness. The embedded parts are fixed at the sides of the wall and the columns. During assembly, the RCFSTKs are first welded between the steel shear plates, and then the plain concrete blocks are placed between the RCFSTKs, as shown in Figure 1; the joint gap is then filled with grout. This method can provide enough stiffness because the gap of column and shear wall are all filled with steel, concrete or grout. In addition, it can also reduce the wet work because the plain concrete blocks are prefabricated and only a small amount of wet gout is used. The internal force between the shear wall and the column includes shear, axial tension and compression force along the RCFSTKs and moment. The tension force is transferred by the steel of RCFSTKs, the moment is transferred by the RCFSTKs, and the other forces are transferred by the RCFSTKs, plain concrete blocks and grout.

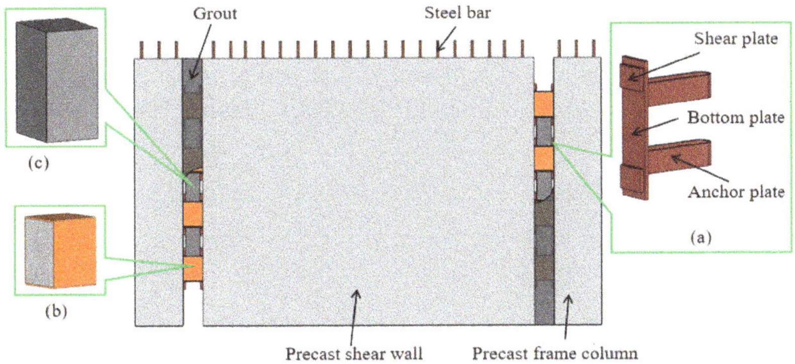

Figure 1. Frame–shear wall structure showing the wall and columns connected by RCFSTKs. (**a**) Embedded part; (**b**) RCFSTKs; (**c**) Plain concrete block.

3. Test Program

3.1. Test Specimen Design

A testing program was designed for specimens of two different single-story shear wall structures, XJ and ZP. Wall XJ was a single-story cast-in-place shear wall, while wall ZP was a prefabricated shear wall attached to the supporting columns with RCFSTKs. The dimensions are shown in Table 1.

Table 1. Design parameters of specimens.

Wall ID	Specimen Type	Wall and Column Connection Method	Wall Section $b \times h$ (mm)	Column, Loading Beam, and Ground Beam Sections $b \times h$ (mm)
XJ	cast-in-place	cast-in-place	1080 × 200	400 × 400
ZP	prefabricated	RCFSTKs	800 × 200	400 × 400

The dimensions and reinforcement of the two specimens are shown in Figures 2 and 3. The concrete grade was C40; the steel plates were 4 mm thick Q235; the stirrups were HPB300; the tension bars were HRB400.

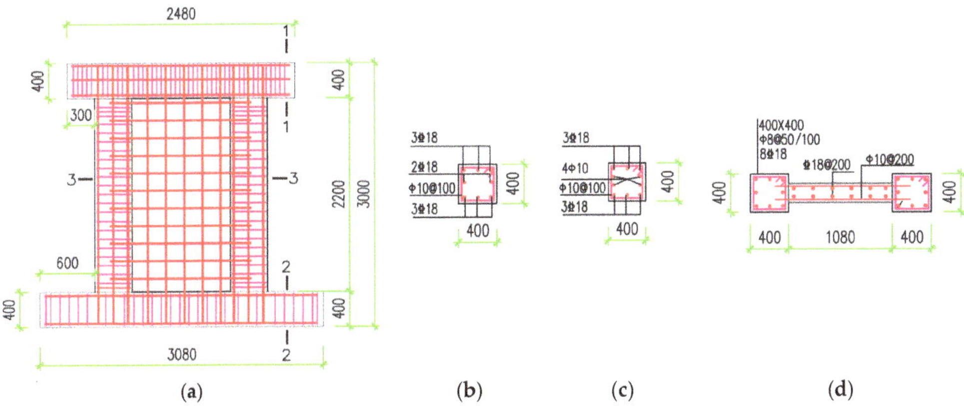

Figure 2. Geometric dimensions and reinforcement of specimen XJ. (**a**) Geometric dimensions and reinforcement of specimen XJ; (**b**) 1–1; (**c**) 2–2; (**d**) 3–3.

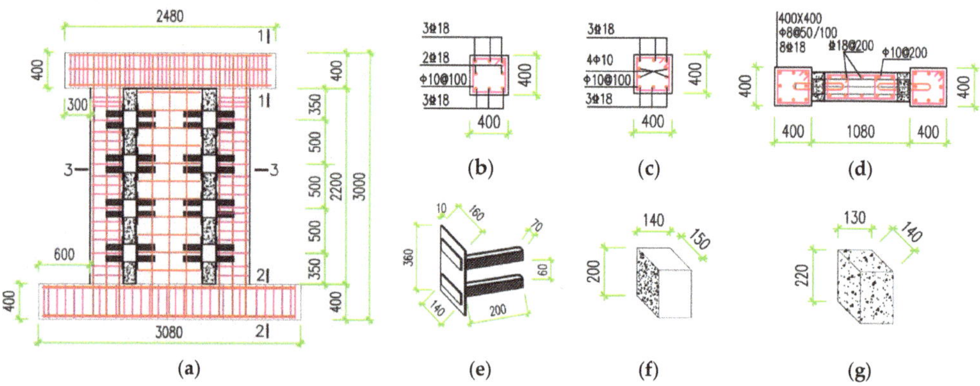

Figure 3. Geometric dimensions and reinforcement of specimen ZP. (a) Geometric dimensions and reinforcement of specimen ZP; (b) 1–1; (c) 2–2; (d) 3–3; (e) Embedded part; (f) RCFSTK; (g) Plain concrete block.

3.2. Mechanical Properties

The testing results of the mechanical properties are shown in Table 2.

Table 2. Mechanical properties of concrete, steel bars, and steel plate.

Materials	Modulus of Elasticity E (10^5 N/mm^2)	Yield Strain ε_y ($\mu\varepsilon$)	Yield Strength f_y (MPa)	Tensile Strength f_u (MPa)	Cube Compressive Strength of Concrete f_{ck} (MPa)
Concrete	0.34	-	-	-	42
Steel plate	2.1	1304	274	352	-
Steel bar (A8)	1.4	3100	434	584	-
Steel bar (C10)	2.2	2400	528	604	-
Steel bar (C18)	1.7	2335	397	563	-

3.3. Loading Program

The test loading device is shown in Figure 4. The specimen was fixed to the test bench by lateral struts, ground beams, ground anchor bolts, and compression beams. The vertical load was applied through the reaction beam, jack, pressure sensor, steel distribution beam, and loading beam, and a sliding plate support was set between the pressure sensor and the steel distribution beam. The horizontal load was applied through the loading frame, an MTS system horizontal actuator, loading head, tension bar, and loading plate.

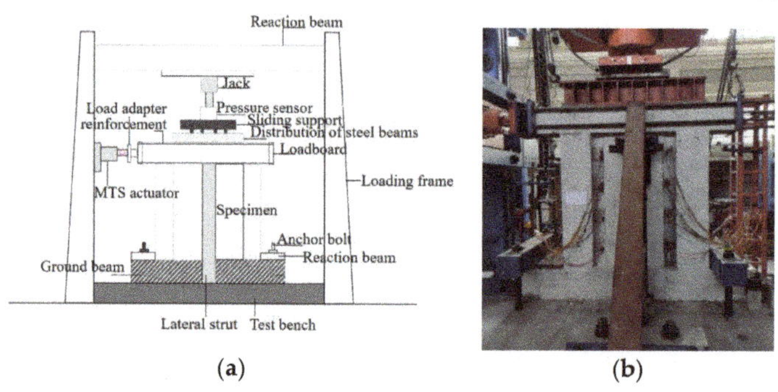

Figure 4. Test loading device. (a) Loading device; (b) Live loading photo.

3.4. Loading System

The test loading system is shown in Figure 5. Before the test, a pre-load was applied to check the equipment for connection reliability and data measurement accuracy. In test loading, the predetermined vertical load was applied and remained constant, and then the horizontal load was applied. The controlled loads were 1000 kN vertical and 200 kN horizontal. The horizontal load was cycled once per stage, and the vertical load was calculated using an axial compression ratio of 0.4, totaling 4000 kN. Until the specimen yielded (at the horizontal load F_y), the loading was changed to cyclic displacement loading, with the grade difference of yield displacement Δ and cycling twice per stage. The test ended when the bearing capacity dropped to 85% of the peak load [15,16]. The push direction was positive, while the pull direction was negative.

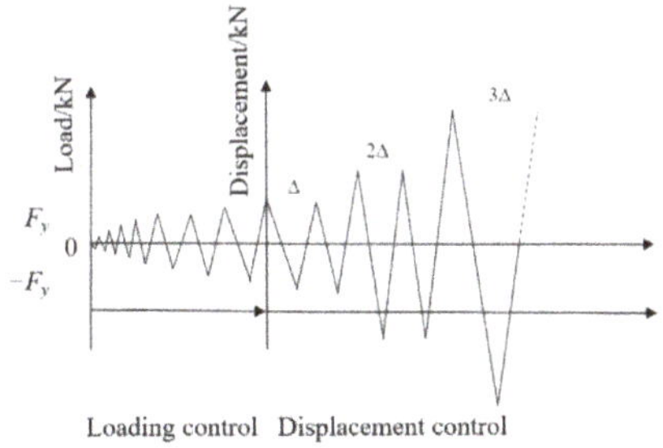

Figure 5. Horizontal loading system.

3.5. Measurement Program

Figure 6 shows the arrangement of sensors for testing specimen ZP. Six displacement gauges measured the horizontal displacement of the loading beam (D1), the horizontal displacement of the shear wall (D2–D4), and the vertical and horizontal displacements of the ground beam (D5–D6). Eight reinforcement strain gauges were arranged to measure the strain of longitudinal bars at the outer corners of the column footings (S1–S2, S1′–S2′) and at the outer footings of the walls (S3–S4, S3′–S4′). Four strain gauges were arranged at the upper and lower ends of the steel plates of RCFSTK A. Four strain gauges (A1–A4) measured the strain generated by bending stress, and one strain flower (A5) was placed at the center of the side steel plate to measure the strain generated by shear stress. The other RCFSTKs (B–H) were arranged with the same strain gauge and flower patterns as RCFSTK A.

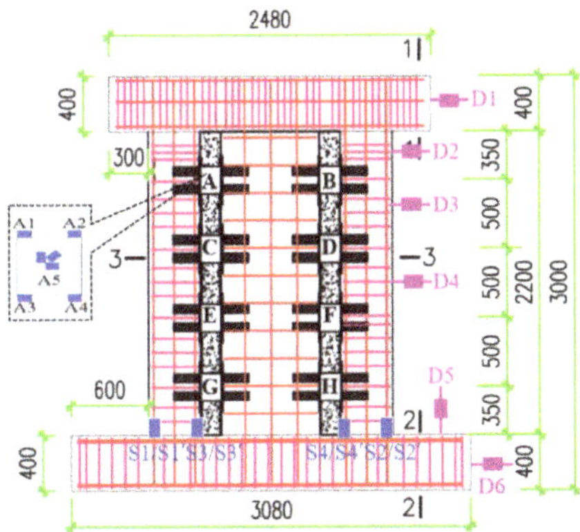

Figure 6. Arrangement of specimen ZP displacement meters and strain gauge; the displacement meters and steel strain gauge arrangement of specimen XJ were the same as for specimen ZP.

4. Specimen Damage and Comparative Analysis

4.1. Specimen Damage

4.1.1. Specimen XJ

There was little sign of damage as loading commenced. When loaded to −750 kN, a slight diagonal crack appeared in the upper left corner of the shear wall. When loaded to 1200 kN, a diagonal crack appeared in the lower right corner of the shear wall. When loaded to −1200 kN, a vertical crack appeared in the middle left side of the shear wall. When loaded to 1600 kN, the specimen yielded, and the loading became displacement loading. When loaded to −1Δ, a longer diagonal crack appeared in the right side of the shear wall oblique cracks. When loaded to +2Δ, oblique cracks appeared at the foot of the right frame column, and as displacement increased, oblique cracks developed and extended to the left and right sides of the shear wall. When loaded to −2Δ, additional horizontal bending cracks appeared in the frame column, and the foot of the left column, the right side of the wall, and the right column were obviously damaged. When loaded to +3Δ, the foot of the column and the foot of the wall were seriously damaged, and the bearing capacity was <85% of the maximum. When the displacement returned to zero, the test ended. The final damage of the specimen XJ is shown in Figure 7.

Figure 7. Final damage of specimen XJ.

4.1.2. Specimen ZP

There was also little sign of damage as loading started. When loaded to 600 kN, vertical cracks appeared in the center of the shear wall. When loaded to 1000 kN, vertical cracks appeared in the horizontal direction of the shear wall at about a quarter of the wall length. When loaded to 1200 kN, diagonal cracks appeared at the upper part of the foot of the frame column. When loaded to 1600 kN, the specimen yielded, and the loading became displacement loading. When loaded to −1Δ, horizontal cracks appeared in the lower left part of the shear wall, and diagonal cracks appeared at the lower end of the vertical cracks and reinforced concrete blocks in the middle of the shear wall. When loaded to +2Δ, more diagonal cracks appeared in the shear wall, and diagonal cracks appeared in the upper part of the right frame column. When loaded to −2Δ, more uniform damage occurred in both the shear wall and frame column. When loaded to +3Δ, cracks developed and extended rapidly in the specimen, and the damage was obvious. The bearing capacity was <85% of the maximum value. When the displacement returned to zero, the test was stopped. The final damage of the specimen ZP is shown in Figure 8.

Figure 8. Final damage of specimen ZP.

4.2. Comparative Analysis of Damage

The damage to specimens XJ and ZP were compared. The locations of the first crack in the frame column were different. The first crack in XJ appeared at the foot of the frame column, while the first crack in ZP appeared at the upper part of the column foot and was more evenly distributed vertically along the column. The vertical distributions of cracks in the two types of shear walls were different: the cracks in ZP were more evenly distributed along the height of the wall, while the cracks in XJ were distributed mainly in the lower part of the wall. The reason for these differences is that when an RCFSTK transfers the internal forces between the frame column and the shear wall, it produces a concentrated force perpendicular to the frame column. This changes the location where the maximum bending moments of the frame columns and the shear wall appear so that both the columns and the wall of ZP are subjected to more uniform bending moments in the vertical direction than XJ. Therefore, the fabricated frame–shear wall structure using RCFSTKs to connect the wall and columns, ZP, withstands the applied force better than XJ.

The sequence of the occurrence of vertical and diagonal cracks in the shear wall differed between specimens XJ and ZP. In XJ, diagonal cracks appeared first and then vertical cracks; in ZP, that order was reversed. The corresponding horizontal loads were different for the appearance of vertical cracks, with the loads being 1200 kN for XJ and 600 kN for ZP. The load was greater for XJ than for ZP because the elastic modulus of the grout was less than that of the concrete, and there might also be gaps in the joint parts introduced during assembly, which together led to the uncoordinated vertical deformation of the assembly.

5. Analysis of Test Results

5.1. Load–Displacement Hysteresis Curve

The load–displacement hysteresis curves of specimens XJ and ZP are shown in Figure 9.

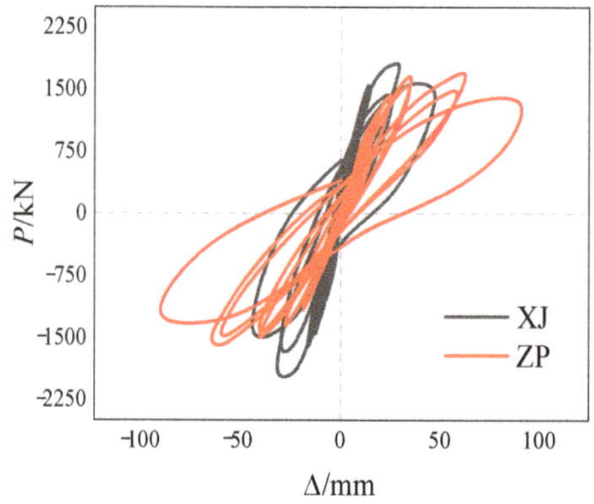

Figure 9. Load–displacement hysteresis curves of specimens XJ and ZP.

5.2. Skeleton Curve

Figure 10 shows the load–displacement skeleton curves of specimens XJ and ZP. Table 3 lists the corresponding characteristic index values. Characteristic index values were defined according to Reference [17]: peak load is the maximum load on the specimen; damage load is 85% of the peak load; yield load is calculated by the energy equivalence method; the ductility coefficient is the ratio of damage displacement to yield displacement; initial stiffness is the ratio of 20% of yield load to the corresponding displacement.

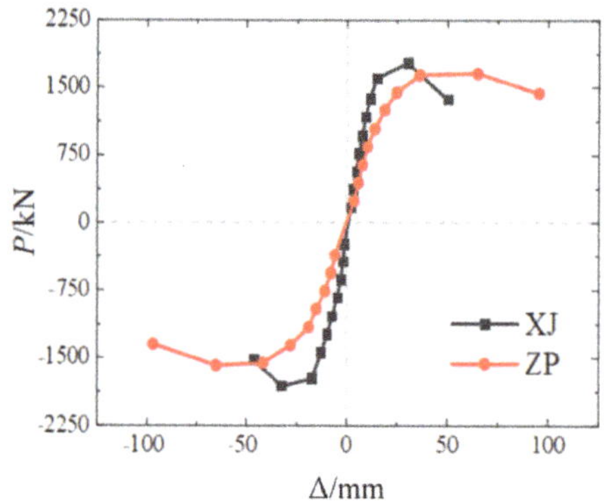

Figure 10. Load–displacement skeleton curves of specimens XJ and ZP.

Table 3. Characteristic index values in the load–displacement skeleton curves of specimens XJ and ZP.

Specimen ID	Yield Load (kN)	Ultimate Load (kN)	Failure Load (kN)	Initial Stiffness (kN/mm)	Ductility Ratio
XJ	1583	1769	1503	131	2.7
ZP	1443	1653	1405	101	3.3

Figure 10 and Table 3 illustrate that: the load–displacement skeleton curves of each specimen have distinct elastic, elastoplastic, and decreasing sections. The skeleton curve of ZP was below that of XJ in both elastic and inelastic rising sections; yield load, ultimate load, failure load, and initial stiffness were 9%, 6%, 6%, and 23% less, respectively. However, the inelastic decreasing section of ZP was significantly above XJ and decreased slowly; the ductility ratio was 22% greater than that of XJ. This was due to the elastic modulus of the grout being less than that of the concrete. Voids might also have been introduced into the joint parts during assembly, which resulted in a lower initial stiffness of the fabricated structure and less bearing capacity than that of the cast-in-place structure, but the deformation capacity and ductility of the fabricated structure increased.

In summary, the initial stiffness of the fabricated frame shear wall using RCFSTKs to connect the wall and columns (ZP) was less than that of the cast-in-place structure (XJ). However, the difference in initial stiffness was not great, and decreasing the stiffness or adjusting factors that affect stiffness enable the design requirements to be met. The difference between the bearing capacity of ZP and that of XJ was small, while the ductility of ZP was significantly better than that of XJ.

5.3. Stiffness Degradation

The stiffness degradation curves of specimens XJ and ZP are shown in Figure 11. The secant stiffness K (referred to simply as stiffness) indicates the stiffness degradation of the specimens; it is calculated by Equation (1) [18,19]:

$$K = \frac{|P_i| + |-P_i|}{|\Delta_i| + |-\Delta_i|} \tag{1}$$

where P_i and $-P_i$ are the forward and reverse ultimate cyclic loads in cycle i, and Δ_i and $-\Delta_i$ are the displacements corresponding to these loads.

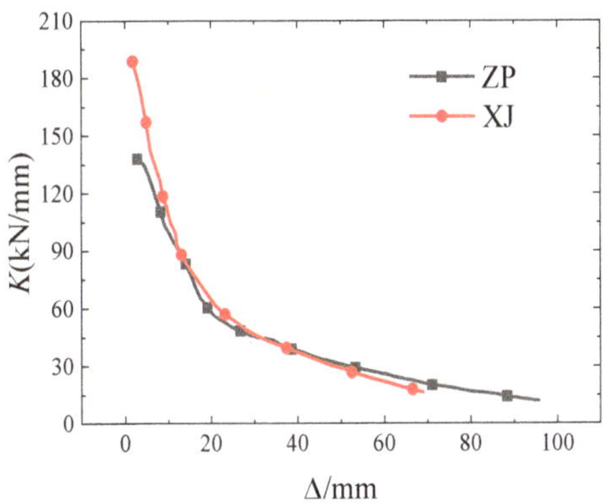

Figure 11. Stiffness degradation curves of specimens XJ and ZP.

Figure 11 shows that at the beginning of loading, the slopes of the stiffness degradation curves of specimens XJ and ZP were steep, indicating that the stiffness of each specimen decreased rapidly. When the displacement was <5 mm, the stiffness of ZP decreased less rapidly than that of XJ. Halfway through loading, the slopes of the stiffness degradation curves of each specimen became less steep, indicating that the stiffness of each specimen decreased slowly; differences between XJ and ZP in stiffness were small. In the later stage of loading, the slopes of the stiffness degradation curves of each specimen further decreased, implying that stiffness degradation changed slowly for both specimens and tended to be the same.

In conclusion, the rate of stiffness degradation of ZP, with the wall attached to the frame columns using RCFSTKs, was slower than that of XJ at the early stage of loading, and the difference in stiffness degradation between XJ and ZP decreased as the load increased and finally converged. The reason for the difference in stiffness degradation at the early stage of loading is similar to that given in the Section Skeleton Curve; the eventual convergence of stiffness degradation is due to the pores in the joint part of ZP being gradually compacted under the horizontal reciprocating load. The difference in elastic modulus between grout and concrete has little effect on the stiffness of the overall structure in the inelastic phase.

5.4. Bearing Capacity Degradation

The bearing capacity degradation coefficients of specimens XJ and ZP are shown in Table 4. The bearing capacity degradation coefficient indicates the degree of continuous degradation in structural bearing capacity caused by plastic deformation of the structure during equal amplitude cyclic loading. It is defined as the ratio of the second to the first cyclic ultimate load [20,21].

Table 4. Bearing capacity degradation coefficients of specimens XJ and ZP.

Specimen ID	1Δ	−1Δ	2Δ	−2Δ
XJ	0.94	0.94	0.89	0.86
ZP	0.95	0.92	0.96	0.96

Note: Since the damage of the third cycle was severe and the bearing capacity had decreased to <85% of the ultimate load, the third cycle was only cycled once.

It can be seen from Table 4 that in cycles +1Δ and −1Δ, the bearing capacity degradation coefficients of specimens ZP and XJ were similar. The reason is that in the first cycle, specimen deformation was primarily elastic deformation. In cycles +2Δ and −2Δ, the bearing capacity degradation coefficient of ZP was greater than that of XJ. The reason is that there were gaps between the RCFSTKs, the plain concrete blocks, and the wall and columns. The bearing capacity degradation of ZP is therefore greater than that of XJ.

5.5. Energy Dissipation

The energy dissipation coefficients E of specimens XJ and ZP are shown in Table 5. E are calculated for the cycle of the first line in the table. They measured the energy dissipation capacity of the specimens [22,23]. E is the ratio of the total energy ($S_{(ABC + CDA)}$) to the elastic energy ($S_{(OBE + ODF)}$) of a hysteresis loop. A greater value of E indicates a greater energy dissipation capacity of the structure. The equation for E is shown as Equation (2), and a graph is shown in Figure 12.

$$E = \frac{S_{(ABC + CDA)}}{S_{(OBE + ODF)}} \tag{2}$$

Table 5. Energy dissipation coefficients of specimens XJ and ZP.

Specimen ID	1Δ	2Δ	3Δ
XJ	0.58	0.69	1.23
ZP	0.58	0.75	1.28

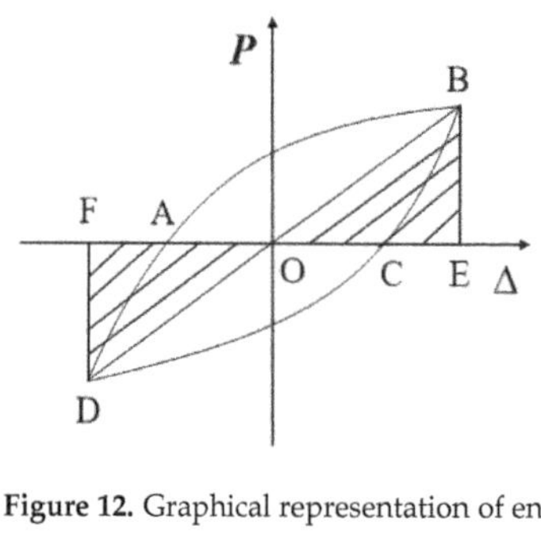

Figure 12. Graphical representation of energy dissipation coefficient calculation.

Table 5 shows that the energy dissipation capacity of specimen ZP was greater than that of XJ, but the overall difference was not significant with the maximum difference <10%. This is because the RCFSTK bonds in ZP affect structural energy dissipation, but the proportion of RCFSTK bonds in the overall structure is small, so the increase in energy dissipation capacity is insignificant.

5.6. Steel Strain and Steel Tube Stress

To analyze stress conditions in the frame columns and shear walls during the loading of specimens XJ and ZP, the load–strain curves of the reinforcements of each specimen at S1–S2 (at the foot of the column, shown in Figure 6) and S3–S4 (at the foot of the wall) were obtained (Figure 13).

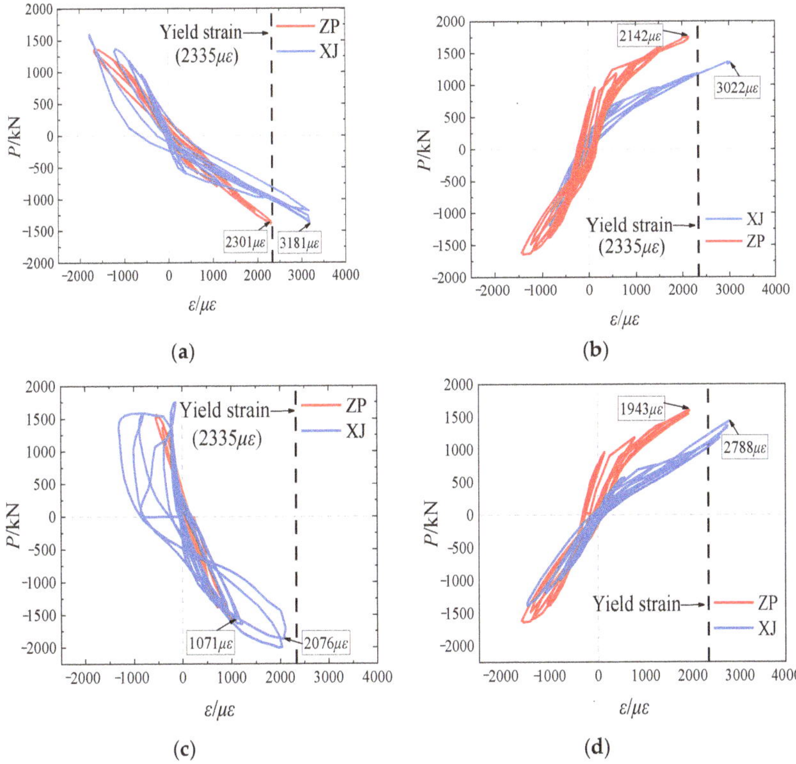

Figure 13. Typical load–strain curves for the steel bars of each specimen. (**a**) Strain of steel bar at S1; (**b**) Strain of steel bar at S2; (**c**) Strain of steel bar at S3; (**d**) Strain of steel bar at S4.

Figure 13 shows that the reinforcement in both the frame column and the shear wall of specimen XJ yielded or nearly yielded during loading, whereas the reinforcement at the same locations in ZP did not. This indicates that the overall force of the prefabricated frame–shear wall structure using RCFSTKs is uniform, and the force dissipation is better than that of the cast-in-place structure.

To analyze the stress conditions in the interface of the wall and the column for specimen ZP, the load–strain curves of the steel plates in RCFSTKs A1–D1 (upper and lower steel plate ends, shown in Figure 6) and the load–stress curves of the steel plates in E5–H5 (side steel plate centers) were plotted (the stresses were calculated from the strains measured by the strain gauge flowers) (Figures 14 and 15).

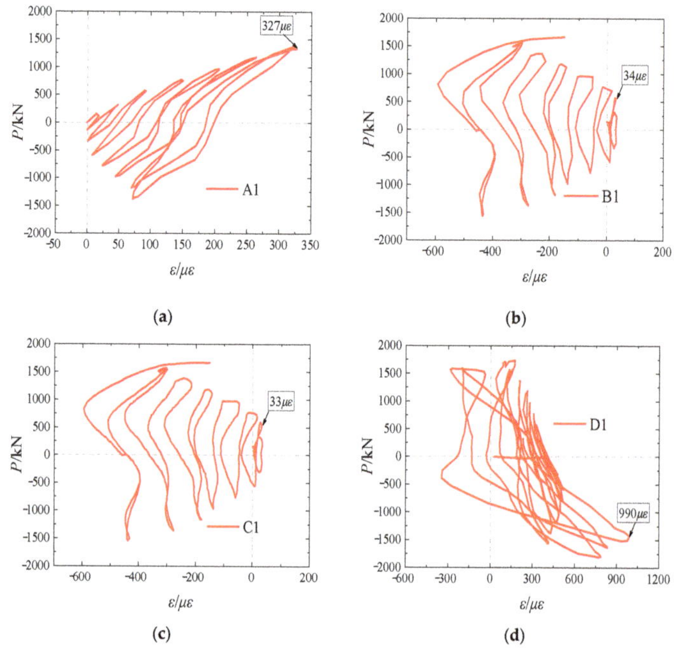

Figure 14. Load–strain curves of the steel plates at the top and bottom of the RCFSTKs. (**a**) Steel plate strain at RCFSTK A1; (**b**) Steel plate strain at RCFSTK B1; (**c**) Steel plate strain at RCFSTK C1; (**d**) Steel plate strain at RCFSTK D1.

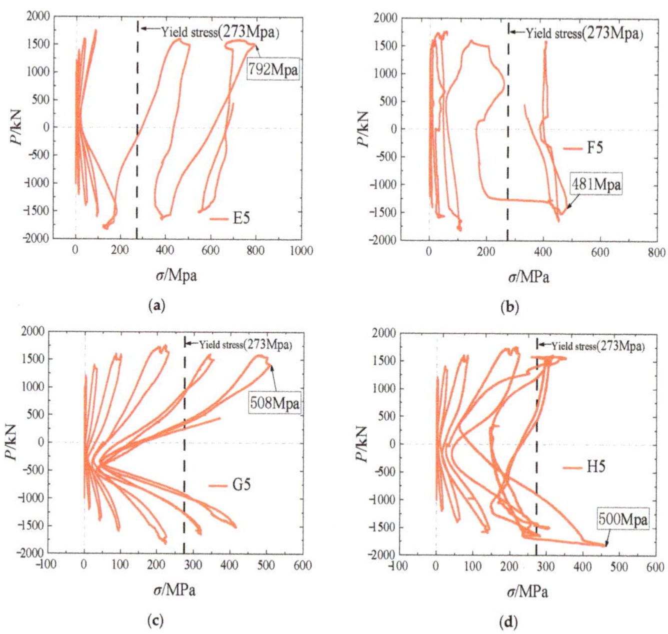

Figure 15. Load–stress curves of the steel plates at the side center of the RCFSTKs. (**a**) Steel plate stress at RCFSTK E5; (**b**) Steel plate stress at RCFSTK F5; (**c**) Steel plate stress at RCFSTK G5; (**d**) Steel plate stress at RCFSTK H5.

Figure 14 shows that the maximum strain values at the top and bottom ends of the steel plates of the RCFSTK were much lower than the yield strain of the steel plates, so the plates did not yield. Figure 15 shows that the maximum stress values at the center of the side steel plates of RCFSTK were greater than the yield stress of the steel plates, so the side steel plates yielded. This means that RCFSTK failure is predominantly shear deformation and is capable of coordinating the shear deformation of the shear wall and columns.

6. Finite Element Parametric Analysis

6.1. Finite Element Modeling Methods and Verification

We set up several finite element models of the test specimen walls using Abaqus, as described in References [24–27]. The model boundary conditions were consistent with the test arrangement. The attributes of C3D8R were assigned to the concrete solid units, the linear steel units were T3D2, and the steel plate shell units were S4R. Concrete was used in the plastic damage model, steel in the bilinear model, and steel plate in the trilinear model [28]. The material behavior is shown as follows: Figure 16a is the concrete compression behavior; Figure 16b is the concrete tension behavior; Figure 16c is the reinforcement; Figure 16d is the steel plate. The material parameters in the nonlinear principal structure model were defined according to the measured material properties shown in Table 2. The steel bar, anchoring steel plate, and bottom steel plate were embedded in the concrete of the wall and columns using Embedded Region. The end of the steel tube and the bottom steel plate of the embedded elements were connected by binding (Tie). Friction contact with a friction coefficient of 0.6 was set for the interfaces between the steel tube and the concrete, between the concrete in the steel tube and the embedded elements, and between the plain concrete block and the steel tube.

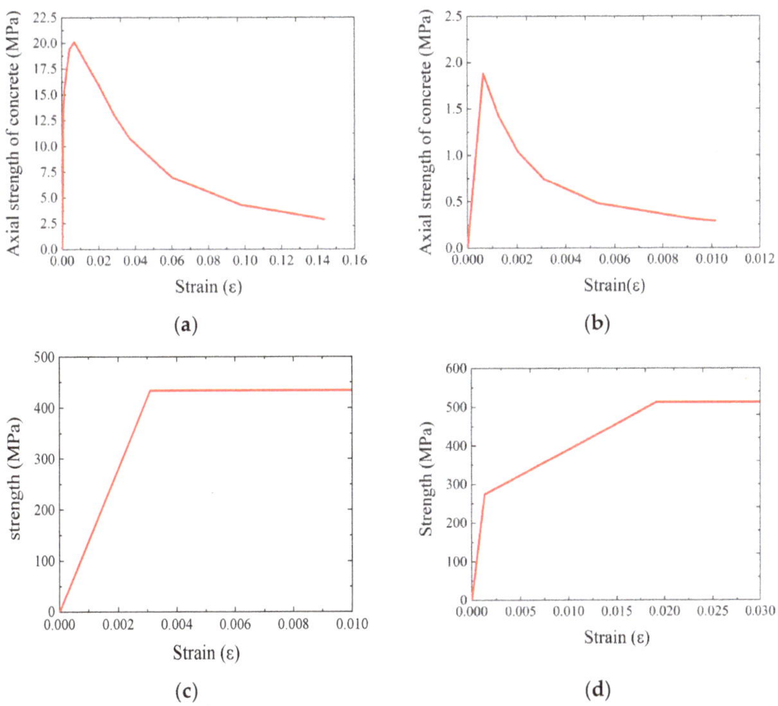

Figure 16. Constitutive Relations. (**a**) Concrete compression behavior. (**b**) Concrete tensile behavior. (**c**) Reinforcement behavior. (**d**) Steel plate behavior.

The finite element simulation and physical test load–displacement hysteresis curves and skeleton curves are shown in Figures 17 and 18, respectively, for comparison. The characteristic index values of the test and simulated load–displacement skeleton curves are shown in Table 6. The error of each index was generally $\leq$10%, and the maximum was <14.6%. These results show that the model is accurate and can be used for subsequent studies.

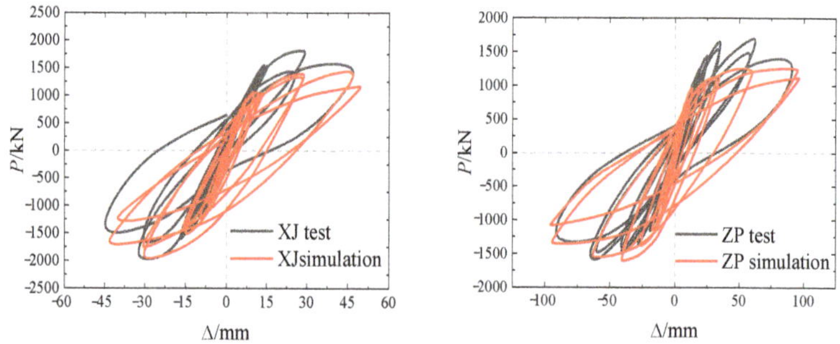

(a) (b)

Figure 17. Physical test and simulated hysteresis curves for comparison. (**a**) XJ; (**b**) ZP.

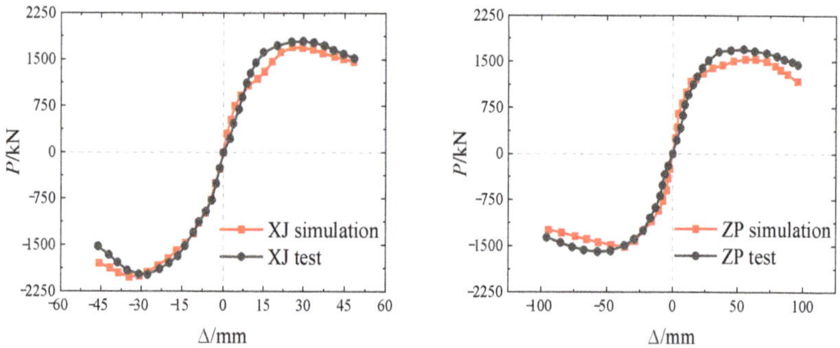

(a) (b)

Figure 18. Physical test and simulated load–displacement curves for comparison. (**a**) XJ; (**b**) ZP.

Table 6. Comparison results of physical tests and simulation.

Wall ID	Result	Load (kN) Yield	Load (kN) Ultimate	Load (kN) Failure	Initial Stiffness (kN/mm)	Ductility Ratio	Energy Dissipation Coefficient
XJ	test	1583	1769	1503	131	2.7	1.23
	simulation	1353	1658	1409	143	2.9	1.31
	error	14.5%	6.3%	6.3%	9.2%	7.4%	6.5%
ZP	test	1443	1653	1405	101	3.3	1.28
	simulation	1298	1519	1291	112	3.7	1.34
	error	10.0%	8.1%	8.1%	10.9%	12.1%	4.7%

6.2. Necessity of Both RCFSTKs and Plain Concrete Blocks

Although using RCFSTKs and plain concrete blocks to connect the wall and columns can reduce the wet work, making the connection is not straightforward. To determine whether the RCFSTKs and plain concrete blocks must be installed together, as in the physical test, or whether, to simplify construction, the RCFSTKs could be installed without concrete blocks or concrete blocks could be installed without RCFSTKs, we set up an Abaqus model of ZP and three models with equivalent dimensions to simulate the effects of modifying the shear wall–column interface on skeleton curves and characteristic indexes. Details of the models are shown in Table 7.

Table 7. Model characteristics.

Model	Model Type	Model Characteristics
ZP	Assembly	Set plain concrete blocks between RCFSTKs
JJWK	Assembly	No plain concrete blocks between RCFSTKs
QZYK	Assembly	Gaps between the wall and columns
QZWK	Assembly	No gaps between the wall and columns

The load–displacement skeleton curves of the four models are shown in Figure 19, and the characteristic index values of each model are shown in Table 8.

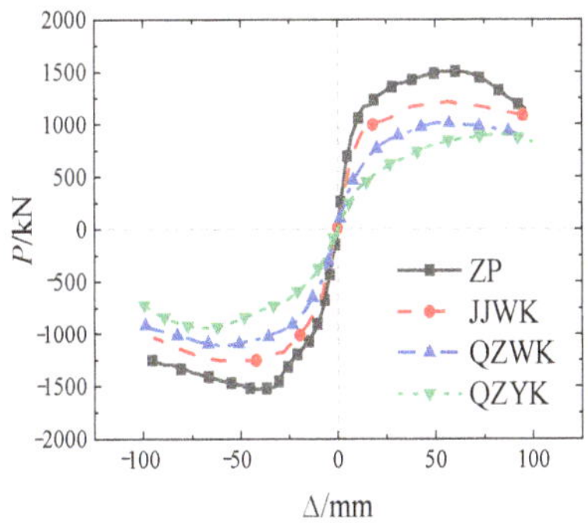

Figure 19. Load–displacement skeleton curves of each model for comparison.

Table 8. Characteristic indexes given by load–displacement skeleton curves for each model.

Model	Ultimate Load (kN)	Failure Load (kN)	Yield Load (kN)	Initial Stiffness (kN/mm)
ZP	1519	1291	1298	138
JJWK	1193	1014	951	102
QZYK	884	752	683	43
QZWK	1028	874	833	66

Figure 19 and Table 8 show that, in absolute terms, the model load–displacement skeleton curve of JJWK (without plain concrete blocks between RCFSTKs) was significantly lower than that of ZP. The yield load, ultimate load, failure load, and initial stiffness of JJWK were 26.7%, 21.5%, 21.5%, and 26.1% less than those of ZP. The load–displacement skeleton curve of QZWK (without RCFSTKs and without gaps between the wall and columns) was lower than that of ZP, and the yield load, ultimate load, failure load, and initial stiffness were all significantly less than those of ZP. The QZWK curve was higher than that of specimen QZYK (with gaps between the wall and column) but lower than the ZP curve, and its yield load, ultimate load, failure load, and initial stiffness were 35.8%, 32.3%, 32.3%, and 52.2% less than those of curve ZP. These results indicate that it is essential that the wall–column connection includes both RCFSTKs and plain concrete blocks; if not, the bearing capacity and initial stiffness of the wall and column connection will be significantly insufficient.

The corresponding plastic strain clouds of each model were obtained for the breaking points of the models (Figure 20) to determine the causes of the preceding results.

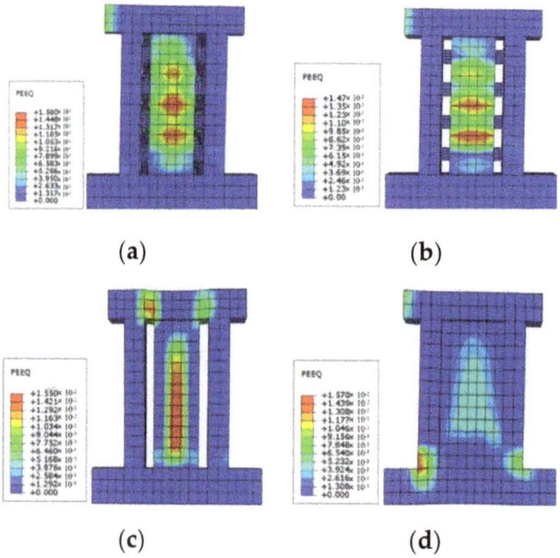

Figure 20. Concrete plastic strain cloud charts for each model. (**a**) Model ZP; (**b**) Model JJWK; (**c**) Model QZYK; (**d**) Model QZWK.

Figure 20 shows that the concrete damage mechanisms of models JJWK and ZP were similar. Both show the damage to the shear wall was distributed between the RCFSTKs, which indicates that the RCFSTKs coordinate the deformation between walls and columns. Because there were no plain concrete blocks between RCFSTKs in JJWK, the initial stiffness and bearing capacity were less than those of ZP. The damaged area of QZWK was concentrated at the foot of the column and showed prominent bending damage; the shear wall was less affected by the forces, so its load–displacement skeleton curve was higher than that of QZYK. In QZYK, there was a gap between the wall and the column, so the loaded beam and shear wall both show the damage caused by the action of axial force (i.e., the damaged area was distributed vertically), and its load–displacement skeleton curve was the lowest.

6.3. *Analysis of Influencing Factors*

We set up 20 models to study the effects of various factors on the cyclic performance of the shear wall using RCFSTKs. Parameters that were varied in the models were the number of RCFSTKs n, RCFSTK unit height h, RCFSTK steel tube wall thickness t, the axial compression ratio of ZP ζ, the reinforcing steel strength Q, and the concrete grade G. The values of the characteristic indexes of the load–displacement skeleton curves for each model are shown in Table 9. Only one parameter was varied in each model and the default parameter values were the settings for model ZP.

Table 9. Values of characteristic indexes of load–displacement skeleton curves for each model.

Parameter Value	Yield Load (kN)	Ultimate Load (kN)	Failure Load (kN)	Initial Stiffness (kN/mm)	Ductility Ratio	Energy Dissipation Coefficient
$n = 2$	1224	1486	1263	95	2.7	0.39
$n = 4$	1242	1508	1282	158	2.8	0.47
$n = 6$	1413	1885	1603	160	3.0	0.58
$n = 8$	1421	1889	1606	160	3.2	0.67
$h = 15$	1356	1874	1593	160	3.0	0.65
$h = 20$	1421	1889	1606	160	3.2	0.67
$h = 25$	1533	2010	1709	162	2.8	0.62
$t = 4$	1280	1519	1291	112	3.7	1.34
$t = 8$	1421	1889	1606	160	3.2	0.67
$t = 12$	1228	1755	1492	162	3.8	0.84
$\zeta = 0.2$	1209	1698	1443	160	3.6	0.68
$\zeta = 0.4$	1421	1889	1606	160	3.2	0.67
$\zeta = 0.5$	1507	1984	1686	160	2.8	0.54
Q-Q235	1421	1889	1606	160	3.2	0.67
Q-Q390	1568	1998	1698	163	2.9	0.63
Q-Q420	1593	2048	1741	167	2.8	0.61
G-C25	1562	2050	1743	142	2.8	0.61
G-C30	1553	1909	1623	153	2.9	0.63
G-C40	1421	1889	1606	160	3.2	0.67
G-C50	1411	1879	1597	164	3.4	0.73

6.3.1. Influence of the Number of RCFSTKs

Figure 21 shows the load–displacement skeleton curves for each model with vertical and cyclic loading when the number of RCFSTKs (n) was 2, 4, 6, or 8.

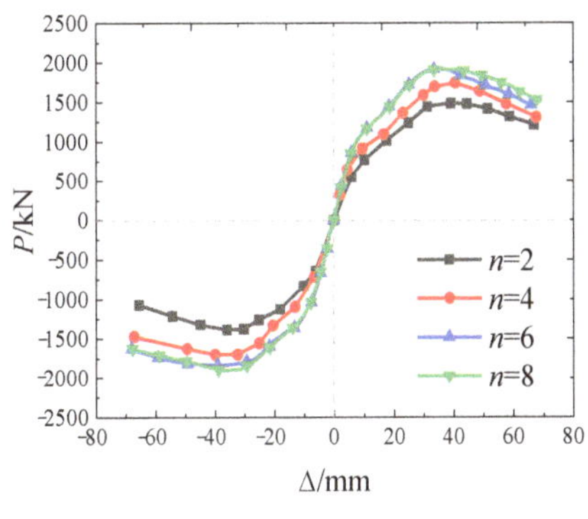

Figure 21. Load–displacement skeleton curves for models with different numbers of RCFSTKs.

Figure 21 and Table 9 show that bearing capacity increased as the number of RCFSTKs increased. When the number of RCFSTKs increased from 2 to 6, the values of all characteristic indexes of the specimen increased; the yield load, ultimate load, and failure load of the models increased by 15.4%, 26.9%, and 26.9%; the initial stiffness increased by 68.4%; and the ductility ratio and energy dissipation coefficient increased by 11.1% and 48.7%. When the number of RCFSTKs increased from 6 to 8, the energy dissipation coefficients increased by 15.5%, and the other characteristic index values increased by <5%. This is explained as follows. When the number of RCFSTKs is small, they cannot coordinate the deformation between walls and columns. The values of each characteristic index therefore increase as the number of RCFSTKs increases. When the number of RCFSTKs is sufficient to coordinate the deformation between walls and columns, further increasing the number becomes equivalent to increasing the steel content inside the structure that resists structural stress, and only the energy dissipation capacity and ductility of the structure increase.

In summary, an increase in the number of RCFSTKs significantly influenced the cyclic performance of the structure. An increase within a particular range will enhance various cyclic performance indexes of the structure. However, beyond this range, continuing to increase the number of RCFSTKs will only increase the ductility and energy dissipation capacity of the structure.

6.3.2. Influence of RCFSTK Unit Height

Figure 22 shows the load–displacement skeleton curves for the models in which the unit height of the RCFSTKs was varied (h = 15, 20, and 25 cm) under vertical load and cyclic loads with other parameters were unchanged.

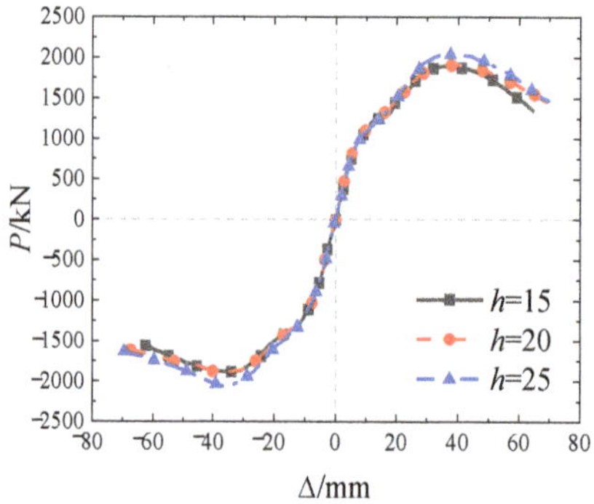

Figure 22. Load–displacement skeleton curves for models with different RCFSTK unit height.

Figure 22 and Table 9 show that there were no significant differences in the load–displacement skeleton curves of the models. When the unit height of the RCFSTK was increased from 15 to 25 cm, the yield load, ultimate load, and failure load of the specimen increased by 13.1%, 7.3%, and 7.3%, respectively. Since the RCFSTK deformation was primarily shear deformation, increasing its unit height improved its resistance to shear deformation, so the overall structural load carrying capacity was enhanced. The initial stiffness, ductility ratio, and energy dissipation coefficient of the models all increased and then decreased as the unit height increased. However, the overall change was small. When the RCFSTK unit height increased from 15 to 20 cm, the ductility ratio increased by 6.7%, but the other index values increased by <3%. When the RCFSTK unit height increased from 20 to 25 cm, the ductility ratio decreased by 12.5%, and the energy dissipation coefficient decreased by 7.3%; all other index values decreased by <4%. The main reason for this is that an increase in RCFSTK unit height is effectively an increase in steel content inside the structure, so the index values increase. However, when the unit height exceeds a particular value, deformation and damage of the steel plate are reduced. The damage is transferred to the concrete, so the index values decrease. In summary, the change in RCFSTK unit height has little effect on the cyclic performance of the structure in general; as the section height increases, the bearing capacity of the structure improves, and the ductility ratio and energy dissipation capacity first increase and then decrease; the initial stiffness is almost unchanged.

6.3.3. Influence of Wall Thickness of RCFSTKs

Figure 23 shows the load–displacement skeleton curves of the models in which the RCFSTK steel tube wall thickness t was 4, 8, and 12 mm under vertical load and cyclic loads with all other parameters unchanged.

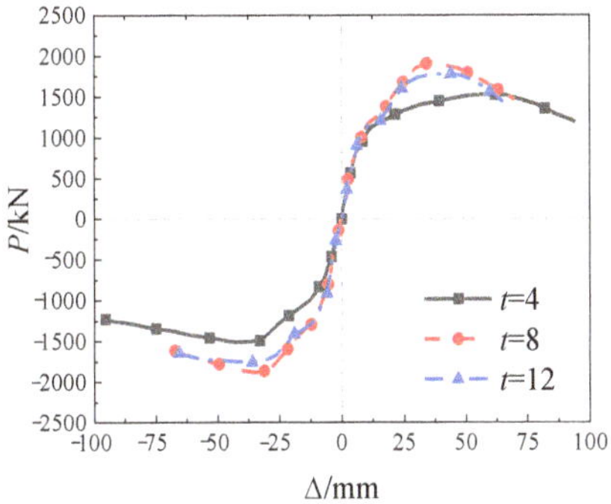

Figure 23. Load–displacement skeleton curves for models with different RCFSTK steel tube wall thickness.

Figure 23 and Table 9 show that the wall thickness of the steel tube has an obvious impact on the load–displacement skeleton curve. The yield load, ultimate load, and failure load of the models first increased and then decreased, and when the wall thickness of the steel tube increased from 4 to 8 and then to 12 mm, the values of the indicators first increased respectively by 11.0%, 24.4%, and 24.4% and then decreased by 13.6%, 7.1%, and 7.1%. The initial stiffness of the models consistently increased; when the wall thickness of the steel tube increased from 4 to 12 mm, initial stiffness increased by 15.2%. The ductility ratios of the models and the energy dissipation coefficients first decreased and then increased. When the wall thickness of the steel tube increased from 4 to 8 and then to 12 mm, these indicators first decreased respectively by 15.8% and 50% and then increased by 18.8% and 25.4%. The principal reason is that the RCFSTKs undergo shear deformation, and increasing the wall thickness of the steel plate improves the capability of the RCFSTK to resist shear deformation. However, as the wall thickness increases, the damage is transferred to the concrete, so the index value has a turning point.

In summary, these three models show that the change in steel tube wall thickness has a pronounced effect on the seismic performance of the structure. As the steel tube wall thickness increases, the bearing capacity of the structure first increases and then decreases, while the ductility and energy dissipation capacity first decrease and then increase, and the initial stiffness always increases.

6.3.4. Influence of Wall Thickness of RCFSTKs

Figure 24 shows the load–displacement skeleton curves of each model under vertical and cyclic loads with axial compression ratios (ζ) 0.2, 0.4, and 0.5, and other parameters unchanged.

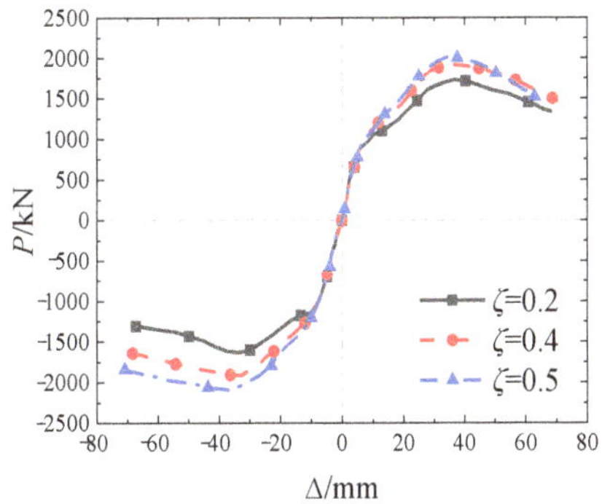

Figure 24. Load–displacement skeleton curves of models with different axial pressure ratios.

Figure 24 and Table 9 show that the load–displacement skeleton curves of the models differed significantly with different axial pressure ratios. When the axial pressure ratio increased from 0.2 to 0.5, the yield load, ultimate load, and failure load of the models increased by 24.6%, 16.8%, and 16.8%, respectively, while the ductility ratio and energy dissipation factor decreased by 22.2% and 20.6%, respectively, with no change in initial stiffness. The reasons for these changes are similar to those for the changes in the cast-in-place structure caused by different axial pressure ratios. Change in axial pressure ratio has a pronounced effect on the cyclic performance of the structure. As the axial pressure ratio increases, the bearing capacity of the structure increases, while the ductility and energy dissipation capacity decrease, and the initial stiffness remains unchanged.

6.3.5. Influence of RCFSTK Steel Strength

Figure 25 shows the load–displacement skeleton curves of each model with RCFSTK steel strengths (Q) of Q235, Q390, and Q420 under vertical load and cyclic load and all other parameters remained unchanged.

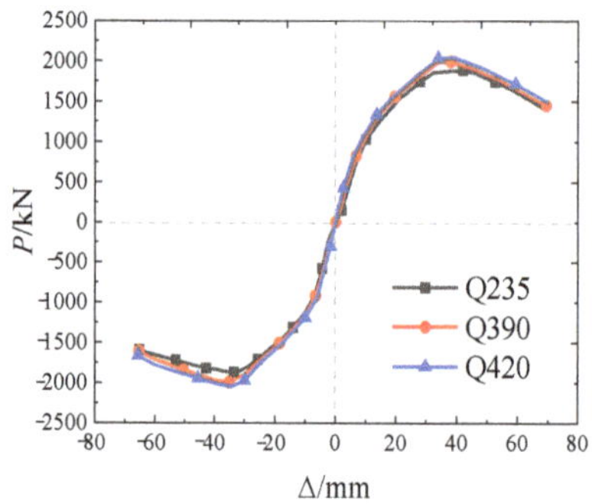

Figure 25. Load–displacement skeleton curve of models with different RCFSTK steel strengths.

Figure 25 and Table 9 show that the load–displacement skeleton curves of the specimens differed slightly. When the RCFSTK steel strength increased from Q235 to Q420, the yield load, ultimate load, failure load, and initial stiffness of the models increased by 12.1%, 8.4%, 8.4%, and 4.1%, respectively. The displacement ductility ratio and energy dissipation coefficients decreased by 12.5% and 9.0%. The principal reason for these changes is that the RCFSTKs are mainly subjected to shear deformation and increasing steel strength can increase RCFSTK resistance to shear stress and thus decrease deformation. Overall, the change in the steel strength of the RCFSTKs has little effect on the cyclic performance of the structure. As steel strength increases, bearing capacity and initial stiffness of the structure increase, while the ductility ratio and energy dissipation capacity decrease.

6.3.6. Influence of RCFSTK Concrete Grade

Figure 26 shows the load–displacement skeleton curves of each model with concrete grades in RCFSTK (G) of C25, C30, C40, and C50 under vertical and cyclic loads and other parameters remained unaltered.

Figure 26 and Table 9 show that the load–displacement skeleton curves of the models differed slightly. When the strength of the concrete inside the steel tube increased from C25 to C50, the yield load, ultimate load, failure load, and initial stiffness of the specimen increased by 9.6%, 8.4%, 8.4%, and 12.3%, respectively, and the displacement ductility ratio and energy dissipation coefficient decreased by 21.4% and 19.7%, respectively. The reason for these changes is similar to the reason for changes in RCFSTK steel strength. The change in RCFSTK concrete grade generally had little effect on the cyclic performance of the models. As the concrete grade of the shear wall and columns increases, the bearing capacity and the initial stiffness of the structure increase, while the ductility ratio and energy dissipation capacity decrease.

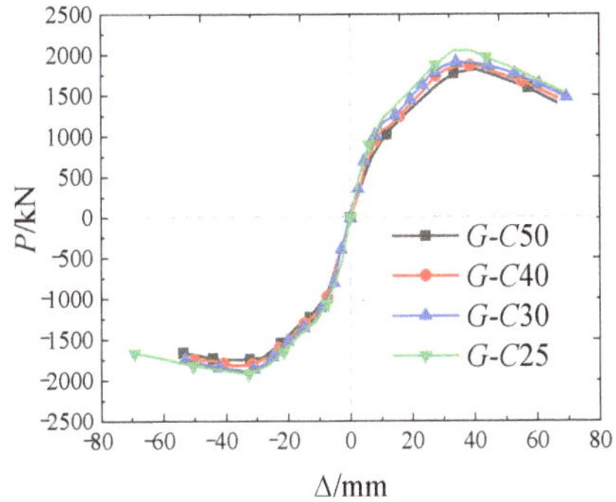

Figure 26. Load–displacement skeleton curves of models with different RCFSTK concrete grades.

7. Conclusions

To find a method of connecting a shear wall and columns that can both reduce the amount of wet work required and maintain adequate stiffness, we propose connecting the shear wall to the vertical frame using rectangular concrete-filled steel tube keys (RCFSTKs). Static tests and finite element models were used to study their cyclic performance. The feasibility of the new method was analyzed. In addition, the influence of different parameters on the cyclic performance of the assembled structure was identified. The following conclusions can be obtained.

(1) In terms of transferring force, RCFSTK produces a concentrated force perpendicular to the frame column when transferring the internal force between the column and the wall. This force changes the locations of the maximum bending moments of the column and the wall, making the bending moments more uniform in the vertical direction and improving the mechanical properties of the wall and the column.

(2) The RCFSTKs yielded in the center of their lateral steel plates, while the ends of the steel plates on the upper and lower surfaces did not yield. This indicates that the RCFSTKs are predominantly subjected to shear deformation and are important in coordinating shear deformation of the walls and columns.

(3) The prefabricated frame–shear wall structure using RCFSTKs, compared with the cast-in-place structure, has a fuller hysteresis curve, greater total energy consumption, better ductility, slightly higher energy consumption capacity, and slightly lower stiffness degradation and bearing capacity. The difference in bearing capacities was slight, so practical use of the proposed prefabricated structure is feasible.

(4) Removing the RCFSTKs or the plain concrete blocks between them, or both, will significantly decrease the bearing capacity and initial stiffness of the prefabricated column–wall unit by 21.5%–52.2%. Thus, both RCFSTs and the concrete blocks between them are necessary. How to simplify the construction merits further investigation.

(5) The cyclic performance of the prefabricated structure is influenced largely by the number of RCFSTKs, steel tube wall thickness, and axial compression ratio while to a lesser extent by the key height, steel strength, and concrete grade of the RCFSTKs.

Author Contributions: Conceptualization, M.L. and Z.Z.; methodology, Z.Z.; software, Q.W.; validation and formal analysis, M.L. and Z.Z.; data curation, S.Y. and L.Z.; writing—original draft preparation, Z.Z.; writing—review and editing, Q.W. All authors have read and agreed to the published version of the manuscript.

Funding: This research was funded by the National Natural Science Foundations of China (Grant No. 51978422), the General Project of Liaoning Provincial Department of Education (Grant No. LJKZ0561).

Institutional Review Board Statement: Not applicable.

Informed Consent Statement: Not applicable.

Data Availability Statement: Not applicable.

Conflicts of Interest: The author declares no potential conflicts of interest regarding the publication of this article.

References

1. Ma, J.W. Study on Seismic Behavior of the New Full Precast Concrete Shear-Wall Frame Strucures. Master's Thesis, Southeast University, Nanjing, China, 2018.
2. Ma, J.W.; Mo, C.; Jiang, S.T.; Yin, W.Y.; Jin, R.C.; Zheng, Q. Study on Construction Method and Experimental Verification of Fully-precast RC Frame-shear Wall Structures. *Constr. Technol.* **2017**, *46*, 30–34, 52. [CrossRef]
3. Ma, J.-W.; Pan, J.-L.; Yin, W.-Y.; Liu, S.-C.; Mo, C. Experimental study on seismic performance of full precast shear wall-frame structures with reinforcement spliced by grout-filled sleeves. *Eng. Mech.* **2017**, *34*, 178–187. [CrossRef]
4. Mo, C. Study on Seismic Performance of Fully Precast Frame-Shear Wall Structure. Master's Thesis, Southeast University, Nanjing, China, 2017.
5. Ma, J.; Pan, J.; Yin, W.; Liu, S.; Jiang, S. Experimental study on seismic behavior of wholly precast RC frame-shear wall structure. *J. Build. Struct.* **2017**, *38*, 12–22. [CrossRef]
6. Zhao, W.; Tong, J.; Yuan, S.; Guo, Y. Research on the connection method of column-wall member in precast frame-shear wall structure. *J. Shenyang Jianzhu Univ.* **2015**, *31*, 408–417.
7. Bournas, D.; Negro, P.; Molina, F.; Viaccoz, B.; Magonette, G. *Pseudodynamic Testing of the SAFECAST 3-Storey Precast Concrete Building*; JRC Technical Report EUR; Institute for the Protection and Security of the Citizen, Joint Research Centre: Ispra, Italy, 2012.
8. Buddika, H.S.; Wijeyewickrema, A.C. Post-tensioned hybrid precast wall-frame buildings for seismic regions. *J. Struct. Eng.* **2014**, *142*, 04016021. [CrossRef]
9. Tong, J.X. Research of the Connection Methods between Frame Beam Column and Shear Wall Based on the Shear Test. Master's Thesis, Shenyang Jianzhu University, Shenyang, China, 2015.
10. Gao, Q. Structure Forms and Assembly Technologies on Precast Frame-Shear Wall Structure. Master's Thesis, Shenyang Jianzhu University, Shenyang, China, 2014.
11. Gao, Q.; Zhao, W.J. Experimental study on performance of grouted sleeve connection with surfacing forming under high stress repeated tension-compression load. *J. Build. Struct.* **2022**, *43*, 220–227. [CrossRef]
12. Gao, Q.; Zhao, W.J. Experimental study on performance of grouted sleeve connection with surfacing forming under uniaxial tensile load. *J. Build. Struct.* **2022**, *43*, 208–219. [CrossRef]
13. Ioani, A.M.; Tripa, E. Structural behavior of an innovative all-precast concrete dual system for residential buildings. *PCI J.* **2012**, *57*, 110–123. [CrossRef]
14. Gao, J. Study on Performance of Concrete Filled Steel Tubular Composite Keyway Connection in Fabricated Frame-Shear Structures. Master's Thesis, Shenyang Jianzhu University, Shenyang, China, 2019.
15. Zhao, Y.; Wang, E.C.; Kong, D.Z.; Kong, D.P.; Li, Y.C. Experimental Research on Seismic Performance of New Low-rise Precast Concrete Shear Walls. *China Concr. Cem. Prod.* **2021**, *2*, 69–73. [CrossRef]
16. Chen, Q.J.; Liang, J.J.; Cai, J.; Li, Z.; He, A.; Zuo, Z.L. Investigation on seismic behavior and influencing parameters of steel-jacket strengthened columns with recycled aggregate concrete. *Eng. Mech.* **2019**, *36*, 206–217, 256. [CrossRef]
17. *JGJ 101-1996*; Specificating of Testing Methods for Earthquake Resistant Building. Architecture & Building Press: Beijing, China, 1996.
18. Sun, X. Experimental research on seismic behavior of single span concrete-filled steel tubular frame with reinforced concrete beams. *J. Build. Struct.* **2009**, *30*, 142–146, 156. [CrossRef]
19. Zheng, S.-S.; Liu, W.; Qin, Q.; Zhang, C.-C.; Dong, L.-G.; Li, Q.-Q. Experimental research on seismic behavior of rc frame beams under different loading protocols. *Eng. Mech.* **2018**, *35*, 109–117. [CrossRef]
20. Li, Q.; Li, Q. Seismic performance study of new shaped plate applied in self-centering steel frame. *J. Suzhou Univ. Sci. Technol.* **2017**, *30*, 1–6. [CrossRef]
21. Rong, X.; Chen, L.; Zhang, J.-X. Experimental Study on Seismic Performance of Bottom-Flange-Bolted and Top-Flange-Welded Prefabricated CFST Column to Steel Beam Joints. *J. Northeastern Univ.* **2020**, *41*, 1348–1355. [CrossRef]
22. Weichang, L.; Mingwei, L.; Zhijian, Y.; Hongping, Y. Research on mechanical properties of square steel tube column-I-beam joints connected by T-section steel under reciprocating load. *Ind. Arch.* **2018**, *48*, 135–141. [CrossRef]
23. Lu, Y.; Liu, L.; Cui, Q.; Cao, C.; Yang, Y. Research on the influence of shear reinforcement on the seismic performance of pre-compressed prefabricated joints. *Ind. Arch.* **2022**, *52*, 105–111. [CrossRef]
24. Wei, H.; Li, Q. Experimental study on seismic behavior of prefabricated RC shear walls with horizontal joints welded by steel plates. *J. Build. Struct.* **2020**, *41*, 77–87. [CrossRef]
25. Wang, W.; Chen, B.; Ou, Y.; Jiang, X. Experimental study on seismic performance of steel truss and concrete composite shear wall. *J. Build. Struct.* **2021**, *42* (Suppl. S1), 330–340. [CrossRef]
26. Saleh, H.M.; Salman, A.A.; Faheim, A.A.; El-Sayed, A.M. Sustainable composite of improved lightweight concrete from cement kiln dust with grated poly(styrene). *J. Clean. Prod.* **2020**, *277*, 123491. [CrossRef]

27. Saleh, H.M.; Eskander, S.B. Innovative cement-based materials for environmental protection and restoration. In *New Materials in Civil Engineering*; Butterworth-Heinemann: Giza, Egypt, 2020; pp. 613–641.
28. Zong, Z.H.; Ge, J.P.; Yang, Q.Y. Nonlinear finite element analysis of the concrete filled square steel tubular column to steel beam connections under low-cycle reversed loading. *J. Build. Struct.* **2006**, *27*, 75–81. [CrossRef]

MDPI
St. Alban-Anlage 66
4052 Basel
Switzerland
www.mdpi.com

Coatings Editorial Office
E-mail: coatings@mdpi.com
www.mdpi.com/journal/coatings

www.ingramcontent.com/pod-product-compliance
Lightning Source LLC
Chambersburg PA
CBHW041934260926
51850CB00018B/167
9783725802777